WASTES: SOLUTIONS, TREATMENTS AND OPPORTUNITIES III

Wastes: Solutions, Treatments and Opportunities
Conference Selected Papers

ISSN 2640-9623
eISSN 2640-964X

Volume 3

Aims & Scope:
Wastes: Solutions, Treatments and Opportunities is an international conference that takes place every two years, organized by CVR – Centre for Waste Valorization since 2011. The Wastes Conferences aim at bringing together academia and industry experts from the Waste Management and Recycling sectors, from around the world, offering state of the art knowledge and sharing experiences with all in attendance.

The Wastes Selected Papers Series presents the latest theoretical research and innovative technologies and techniques in the waste field, as well as tools and strategies to improve corporate environmental performance.

SELECTED PAPERS FROM THE 5TH INTERNATIONAL CONFERENCE WASTES: SOLUTIONS, TREATMENTS AND OPPORTUNITIES, COSTA DA CAPARICA, LISBON, PORTUGAL, 4–6 SEPTEMBER 2019

Wastes: Solutions, Treatments and Opportunities III

Editors

Cândida Vilarinho & Fernando Castro
University of Minho, Guimarães, Portugal

Margarida Gonçalves & Ana Luísa Fernando
Faculty of Sciences and Technology, NOVA University of Lisbon, Caparica, Portugal

CRC Press
Taylor & Francis Group
Boca Raton London New York Leiden

CRC Press is an imprint of the
Taylor & Francis Group, an **informa** business
A BALKEMA BOOK

CRC Press/Balkema is an imprint of the Taylor & Francis Group, an informa business

Typeset by MPS Limited, Chennai, India

Published by: CRC Press/Balkema
Schipholweg 107C, 2316 XC Leiden, The Netherlands
e-mail: Pub.NL@taylorandfrancis.com
www.crcpress.com – www.taylorandfrancis.com

ISBN: 978-0-367-25777-4 (Hbk)
ISBN: 978-0-429-28979-8 (eBook)
DOI: https://doi.org/10.1201/9780429289798

Wastes: Solutions, Treatments and Opportunities III – Vilarinho et al. (Eds)
© 2020 Taylor & Francis Group, London, ISBN 978-0-367-25777-4

Table of contents

Wastes: Solutions, Treatments and Opportunities III – Vilarinho et al. (Eds)
© 2020 Taylor & Francis Group, London, ISBN 978-0-367-25777-4

Preface

Wastes: Solutions, Treatments and Opportunities III contains selected papers presented at the 5th edition of the International Conference Wastes: Solutions, Treatments and Opportunities, that took place on 4–6 September 2019, in Costa da Caparica, Portugal.

The 2019 edition was jointly organized by CVR – Centre for Waste Valorisation and the Faculty of Sciences and Technology, NOVA University of Lisbon.

The Wastes conference, which takes place biennially, is a prime forum for sharing innovation, technological development and sustainable solutions for the waste management and recycling sectors around the world, counting on the participation of experts from academia and industry.

The papers included in this book cover a wide range of topics, including: Wastes as construction materials; Wastes as fuels; Waste treatment technologies; MSW management; Recycling of wastes and materials recovery; Environmental, economic and social aspects in waste management; Life cycle assessment; Circular economy and wastes refineries; Logistics, policies, regulatory constraints and markets in waste management.

All the articles were individually reviewed by members of the Scientific Committee of the Conference, who also contributed to the editing of the 94 papers presented. The editors wish to thank all reviewers, namely:

Alberto Coz
Ana Andres Payan
Ana Cristina Pontes de Barros Rodrigues
Ana Júlia Cavaleiro
Ana Silveira
Ana Vera Machado
Anabela Leitão
André Mota
André Ribeiro
Annalisa Tassoni
António Alberto S. Correia
António Brito
António José Roque
António Roca
Benilde Mendes
Carla Martins
Carlos Afonso
Carlos Nogueira
Castorina Silva Vieira
Célia Ferreira
Conceição Paiva
Cristina Queda
Eduardo Ferreira
Felipe Macias
Feliz Mil-Homens
Fernanda Margarido
Graça Martinho
Helder Gomes
Hugo Silva
Ignacio Álvarez Lanzarote
Ignacio Pérez
Isabel Ferreira
Isabel Marques
Javier Escudero Sanz
Javier Viguri
Joana Carvalho
Joana Dias
João Almeida
João Labrincha
Joel Oliveira
Jorge Araújo
José A. C. Silva
José Alcides Peres
José Barroso de Aguiar
José Carlos Teixeira
José Ricardo Carneiro
José Teixeira
Ligia Pinto
Madalena Alves
Magdalena Borzecka
Manuel Afonso Magalhães da Fonseca Almeida
Margarida M. J. Quina
Maria Alcina Pereira
Maria de Lurdes Lopes

Miguel Brito
Nídia Caetano
Nuno Cristelo
Nuno Lapa
Paulo Brito
Raul Fangueiro
Rosa Quinta-Ferreira
Silvia Nebra
Susete Martins-Dias
Teresa Carvalho
Tiago Miranda
Vasco Fitas

With this book we expect to contribute to the dissemination of state of the art knowledge on waste treatment, recycling and valorization, focused on real application studies that contribute to the development of sustainable pathways for a circular economy.

The Editors,

Cândida Vilarinho
Fernando Castro
Margarida Gonçalves
Ana Luísa Fernando

Wastes: Solutions, Treatments and Opportunities III – Vilarinho et al. (Eds)
© 2020 Taylor & Francis Group, London, ISBN 978-0-367-25777-4

Adhesion analysis of waste cork dust as filler for bituminous mixtures

A.R. Pasandín, J.J. Galán-Díaz & I. Pérez
Department of Civil Engineering, E. T. S. I. Caminos, Canales y Puertos, Universidade da Coruña (UDC), La Coruña, Spain

ABSTRACT: Suberin is a biopolymer that can be found in high amounts in natural cork. As is well known, polymers can be used as bitumen modifiers to improve the properties of bituminous mixtures. This paper describes laboratory adhesion analysis that was conducted to analyse the use of waste cork dust as a filler in hot-mix asphalt (HMA). The performance of the waste cork dust as a filler was compared with that of a typically used natural filler and with Portland cement. These three fillers were characterized by their grain size distributions, morphological studies, and X-ray fluorescence spectroscopy experiments. The adhesion between the aggregates and the binder was analysed using two adhesion tests: a "boiling water test" and using the "rolling bottle method". Both tests led to the same result: the waste cork dust was a better filler than the natural filler.

1 INTRODUCTION

Cork is the name given to the bark of the cork oak (Quercus Suber L.). Portugal has 34% of the world's cork oaks, followed by Spain with 27% (APCOR n.d.).

Cork is a 100% natural and sustainable material because once the bark is removed from the cork oak, it is regenerated again, and a new bark can be extracted every nine years (Jardin et al. 2015). Using different products (e.g., cork stoppers for wine), 280,000 tonnes of cork are consumed annually (Cordeiro et al. 1998). Of this, approximately 20–30% is waste, which is mainly in the form of cork dust. This fraction currently lacks industrial interest; thus, it is usually employed for combustion to produce energy. Thus, there is a certain granulometric fraction of waste cork with the potential for reuse.

The cork contains 45% of the biopolymer, suberin (APCOR n.d.). The effects derived from the use of polymers as bitumen-modifying agents are well known: a decrease in thermal susceptibility with a consequent larger resistance to permanent deformation as well as fatigue and temperature cracking (Sun & Lu 2006). Considering the beneficial action of polymers in asphalt and that there is a large amount of waste cork dust (rich in suberin biopolymer), there is a great potential to use this waste in bituminous mixtures, not only as a bitumen modifier but also as an aggregate and mineral filler.

To date, only few researchers have investigated the use of cork in bituminous mixtures. There has been evidence of several investigations in which cork is used to modify bitumen. For example, although it does not fit into the field of road paving, a patent (US 2099131 A) is reported, in which a mastic of bitumen and granulated cork is used to make the top of tiles (Miller 1937).

Regarding the use of cork as an aggregate, the work by Pereira et al. (2013) stands out. These researchers replaced 5% by volume of natural aggregate with cork particles from 1 mm to 4 mm to manufacture hot-mix asphalt (HMA) type AC14bin. The mixtures that were obtained showed an excellent resistance to permanent deformation, adequate water sensitivity, lower stiffness, and fatigue life that is lower than that of the control mixture.

However, to date, no study has been published in which waste cork dust has been used as a filler for the manufacture of HMA.

Table 1. Main properties of the hornfels.

Property	Standard	Results
ρa (g/cm^3)	EN-1097-6 (AENOR 2014)	2.65
WA_{24} (%)	EN 1097-6 (AENOR 2014)	0.7
FI (%)	EN 933-3 (AENOR 2012a)	13
LA (%)	EN 1097-2 (AENOR 2010a)	20

Table 2. Main properties of the bitumen.

Property	Standard	Results
Penetration grade (0.1 mm)	EN-1426 (AENOR 2015a)	59
Ring and ball softening point (°C)	EN 1427 (AENOR 2015b)	50.7

2 AIMS AND SCOPE

The objective of this investigation was to analyse the feasibility of using waste cork dust from the manufacturing of wine stoppers as a filler for HMA. The motivation of this investigation was twofold. First, we aimed to solve the environmental problems arising from the energy recovery of these types of wastes. Second, cork is rich in suberin, a vegetable biopolymer that can improve the performance of bituminous mixtures. Hence, we analysed the possible binder-aggregate adhesion improvements that the use of waste cork dust may entail. To comply with this objective, the characteristics and affinities of the three fillers were compared: waste cork dust, natural filler, and Portland cement.

3 MATERIALS AND METHODS

3.1 *Materials*

Next, we describe the materials used in this investigation.

3.1.1 *Aggregates*

A typical siliceous commercial quarry aggregate was used in this study, which was a crushed hornfel. The hornfel was supplied by a local contractor. The main properties of the hornfel were evaluated. Table 1 shows the bulk specific gravity (ρa), the water absorption (W_{24}), the sand equivalent (SE), the Los Angeles abrasion coefficient (LA), and the flakiness index (FI).

3.1.2 *Filler*

As mentioned above, three types of filler were used in this preliminary investigation: waste cork dust, natural filler, and Portland cement type CEM I 52.5 R.

3.1.3 *Bitumen*

A typically used penetration-grade bitumen B50/70 was selected for this investigation. Table 2 shows the basic properties of this bitumen.

3.2 *Methods*

Next, the methods used in this investigation are described.

3.2.1 *Filler characterization*

Some filler properties may affect the binder–aggregate adhesion. In this regard, the following properties were analysed:

- The grain size distributions of the three filler grains were analysed following the EN-933-10 (AENOR 2010b). Air jet sieving was used to perform this test.
- A qualitative evaluation of the geometric characteristics of the filler particles was conducted. This morphological study was conducted by means of scanning electron microscopy (SEM).
- The elemental composition of the three filler particles was determined via X-ray fluorescence spectroscopy (XRF) (Bruker S4 Pioneer fluorescence spectrometer).

3.2.2 *Binder–aggregate adhesion*

To analyse the affinity between the aggregate and the binder, two types of tests were carried out. First, a boiling water test was performed following the ASTM D3625 standard (ASTM 2005) Second, a test of the rolling bottle was performed according to the procedure described in the EN 12697-11 standard (AENOR 2012b).

In the boiling water test, a sample of loose bituminous mixture made with an aggregate fraction of 8/11.2 mm was left for 10 min in boiling water, and we analysed the amount of bitumen that detached from the surface of the aggregate. The visual observation was made just when the mixture separated from the water and after 24 h. This second observation was made because during the entire time period, the loose mixture became dry and the surface of the aggregate that remained coated could be observed with higher accuracy.

In the rolling bottle test, a sample of loose bituminous mixture, which was also made with the 8/11.2 mm aggregate, was introduced into a bottle with distilled water. The bottle rotated at a constant speed for several hours. The amount of bitumen that was released from the surface of the aggregate was analysed after 24 h of rotation.

In both tests, the greater the proportion of aggregate surface that remained coated with bitumen, the better adhesion between the aggregate and the binder and, therefore, the better expected performance the bituminous mixture will have against water.

In both tests, to analyse the influence of the natural filler, cork dust, and cement, a 4% filler (dry aggregate) was used.

In addition, for the boiling water test, Kennedy et al. (1984) indicated that loose asphalt mixtures that retain less than 70% of bitumen on the aggregate surface are moisture susceptible. Kim & Coree (2005) indicated that this percentage must be 95%. In the case of the rolling water test, there are no requirements on the percentage of aggregate surface that should be coated with bitumen after the rotation periods. Therefore, this test is useful only for comparison.

4 RESULTS AND DISCUSSION

4.1 *Filler characterization*

Figure 1 shows the grain size distribution of the three tested fillers according to the air jet sieving results. This figure also includes the upper and lower limit specified by the Spanish General Technical Specifications for Road and Bridges (PG-3) (MFOM 2008), in which the grain size of the filler must be inscribed to be suitable for use in bituminous mixtures.

As shown in Figure 1, the three fillers (the waste cork dust, the natural filler, and the cement) are suitable for the manufacture of bituminous mixtures according to the PG-3. Nevertheless, Figure 1 also shows that the natural filler is finer grade than the cement and that the cement is finer grade than the waste cork dust. The filler activity is related to fineness because a finer filler typically has a higher activity. Thus, we expected a lower activity in the case of the waste cork dust. This could negatively affect the physical affinity between the aggregate and the binder.

Figure 2 show the SEM images of the three fillers. As shown in the figure, the waste cork dust (Fig. 2a) appears to be less caked with finer scales than the natural filler (Fig. 2b) and the

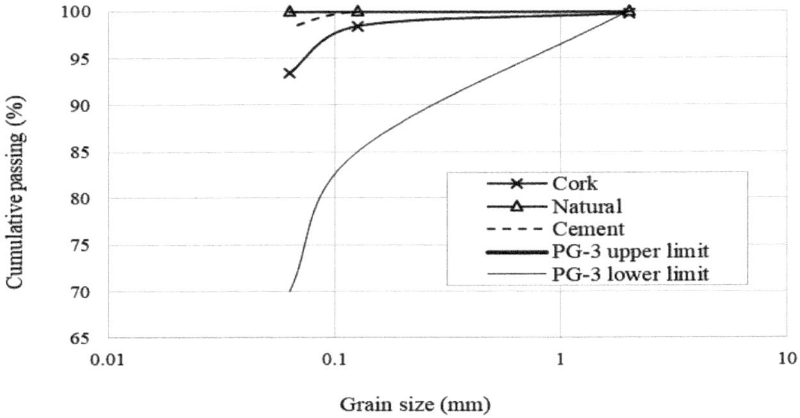

Figure 1. Grain size distribution of the three fillers.

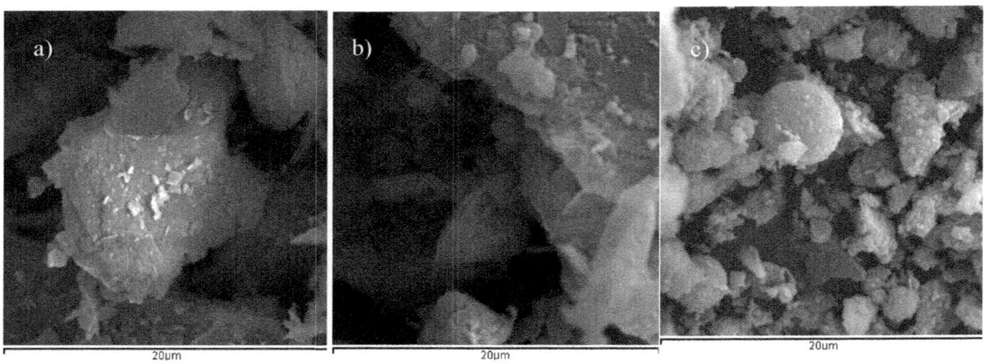

Figure 2. SEM images of a) the waste cork dust, b) natural aggregate and c) Portland cement.

Portland cement (Fig. 2c). This property could positively affect the affinity between the binder and the aggregate because it is easier to form heterogeneous mixtures if the filler does not tend to agglomerate.

A high SiO_2 percentage is usually associated with poor adhesion (Bagampadde 2004). In this regard, the XRF test showed that the natural filler has 57.4% of SiO_2, the waste cork dust has 56.3% of SiO_2, and the Portland cement has 17.7% of SiO_2. Thus, it is expected that the binder aggregate adhesion is better when the Portland cement is the filler, followed by the waste cork dust and then the natural filler.

A high CaO percentage is associated with adequate adhesion (Bagampadde 2004). The XRF test indicated that the natural aggregate has 0.39% of CaO, the waste cork dust has 0.45% of CaO, and the Portland cement has 58.0% of CaO. Again, the CaO percentages confirm the above expected adhesion results.

4.2 *Binder–aggregate adhesion*

Figure 3 shows the results of the boiling water test for the three tested fillers: waste cork dust, natural filler, and Portland cement. As mentioned above, two visual observations were made: when the loose mixture left the water and 24 h later. As mentioned above, the second observation was the one that was considered the most appropriate because the sample was dry and it was easier to appreciate

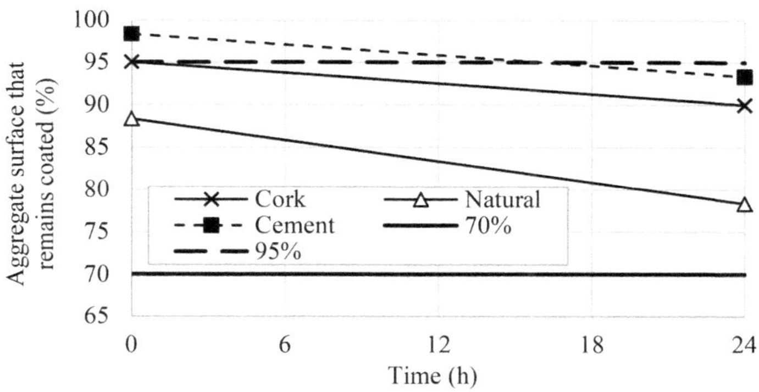

Figure 3. Boiling water test results.

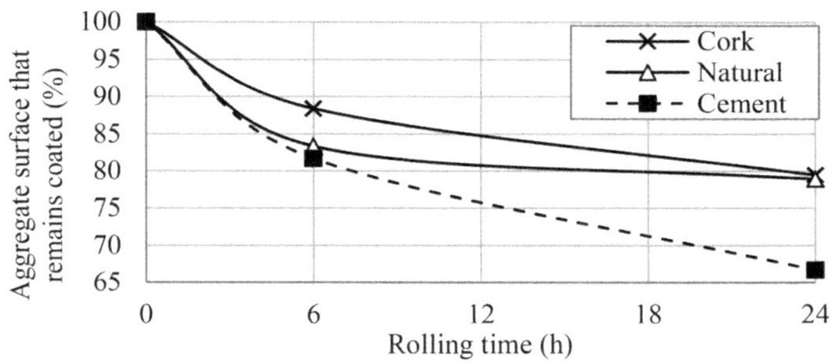

Figure 4. Rolling bottle method results.

the degree of coating of the aggregate by the bitumen. According to this last observation, Figure 3 shows that the better coating results were obtained when Portland cement was used as the filler (93%), followed by the waste cork dust (90%) and the natural filler (78%). The results obtained when the waste cork was used as the filler were very similar to those obtained with the Portland cement.

Both fillers showed results higher than the 70% required by some authors, but none of them showed results higher than the 95% required by other authors. Nevertheless, results for the use of Portand cement and waste cork dust as fillers were very close to this percentage.

Figure 4 shows the results obtained for the rolling bottle test for the three tested fillers. As shown, the use of waste cork dust as the filler resulted in higher coating results (79.4%) after 24 h of rotation. The natural filler was second (78.9%), and the Portland cement as the filler was third (66.7%).

Clearly, the different mechanisms that prevailed in each of the two tests produced some differences in the results. The boiling water test results are adequate to represent the effect that high temperatures can have on bituminous mixtures in warm climates. High temperatures can reduce the viscosity of the binder and affect its adhesion. The rolling bottle method results are adequate to represent the effect of mechanical damage on the adhesion in cold climates. In the case of the use of waste cork dust as a filler, both tests yielded very satisfactory results in terms of aggregate–binder adhesion. Thus, this filler could be used in both warm and cold climates.

5 CONCLUSIONS

The use of waste cork dust as a filler for the manufacture of bituminous mixtures is suitable in terms of grain size distribution and in terms of aggregate–binder adhesion. This is the case in both cold and warm climates. Nevertheless, further investigation is needed.

ACKNOWLEDGEMENTS

The authors would like to thank the Servizos de Apoio á Investigación (SAI) (*Research support services*) of the Universidade da Coruña (UDC) for performing the SEM and XRF tests. The authors would also like to thank Probigasa for supplying the natural aggregates and Repsol for supplying the bitumen.

REFERENCES

AENOR. Asociación Española de Normalización y Certificación. UNE-EN 933-3 Tests for geometrical properties of aggregates. Determination of particle shape. Flakiness index. Madrid, Spain, 2012a (in Spanish).

AENOR. Asociación Española de Normalización y Certificación. UNE-EN 1097-2 Tests for mechanical and physical properties of aggregates. Methods for the determination of resistance to fragmentation. Madrid, Spain, 2010a (in Spanish).

AENOR. Asociación Española de Normalización y Certificación. UNE-EN 1097-6 Tests for mechanical and physical properties of aggregates. Determination of particle density and water absorption. Madrid, Spain, 2014 (in Spanish).

AENOR. Asociación Española de Normalización y Certificación. UNE-EN 1426 Bitumen and bituminous binders. Determination of needle penetration. Madrid, Spain, 2015a (in Spanish).

AENOR. Asociación Española de Normalización y Certificación. UNE-EN 1427 Bitumen and bituminous binders. Determination of the softening point. Ring and Ball method. Madrid, Spain, 2015b (in Spanish).

AENOR. Asociación Española de Normalización y Certificación. UNE-EN 12697-11 Bituminous mixtures. Test methods for hot mix asphalt. Determination of the affinity between aggregate and bitumen. Madrid, Spain, 2012b (in Spanish).

AENOR. Asociación Española de Normalización y Certificación. UNE-EN 933-10 Tests for geometrical properties of aggregates – Part 10: Assessment of fines – Grading of filler aggregates (air jet sieving). Madrid, Spain, 2010b (in Spanish).

APCOR. (n.d.). http://www.apcor.pt/ accessed 10 december 2018.

ASTM. D 3625-96. Standard Practice for Effect of Water on Bituminous-Coated Aggregate Using Boiling Water, 2005.

Bagampadde, U. "On investigation of stripping in bituminous mixtures". Tesis doctoral. Karlstad University, 2004.

Cordeiro, N., Belgacem, M.N., Silvestre, A.J.D., Neto, C.P., Gandini, A. 1998. Cork suberin as a new source of chemicals: 1. Isolation and chemical characterization of its composition.International Journal of Biological Macromolecules, 22(2), 71–80.

Jardin, R.T., Fernandes, F.A.O., Pereira, A.B., de Sousa, R.A. 2015. Static and dynamic mechanical response of different cork agglomerates. Materials & Design, 68, 121–126. https://doi.org/10.1016/j.matdes.2014.12.016.

Kennedy, T.W., Roberts, F.L., Anagnos, J.N. 1984. Texas boiling test for evaluating moisture susceptibility of asphalt mixtures (No. FHWA-TX-85-63+ 253-5). Center for Transportation Research, Bureau of Engineering Research, University of Texas at Austin.

Kim, S., Coree, B.J. 2005. Evaluation of hot mix asphalt moisture sensitivity using the Nottingham asphalt test equipment (No. IHRB Project TR-483). Iowa State University. Center for Transportation Research and Education.

MFOM, Ministry of Public Works. Article 542 (Asphalt Concrete) of the General Technical Specifications for Road and Bridge Works (PG3) from the Spanish Ministry of Public Works. Madrid, Spain, 2008 (in Spanish).

Miller, S. P. 1937. U.S. Patent No. 2,099,131. Washington, DC: U.S. Patent and Trademark Office.

Pereira, S.M., Oliveira, J.R., Freitas, E.F., Machado, P. (2013). Mechanical performance of asphalt mixtures produced with cork or rubber granulates as aggregate partial substitutes. Construction and Building Materials, 41, 209–215. https://doi.org/10.1016/j.conbuildmat.2012.12.005.

Sun, D., Lu, W. 2006. Phase morphology of polymer modified road asphalt. Petroleum science and technology, 24(7), 839–849. 10.1081/LFT-200043780.

Wastes: Solutions, Treatments and Opportunities III – Vilarinho et al. (Eds)
© 2020 Taylor & Francis Group, London, ISBN 978-0-367-25777-4

Nutrient extraction alternatives from mixed municipal waste compost

M. Fernández-Delgado, E. del Amo, M. Coca, M.T. García-Cubero & S. Lucas
Department of Chemical and Environmental Engineering, Sustainable Processes Institute, Valladolid University, Valladolid, Spain

ABSTRACT: Conventional and microwave extraction, using water as solvent, were tested for the recovery of nutrients from mixed municipal waste compost (MMWC). The effect of compost to water ratio (10, 20, 40 %w/V), extraction time (6, 24, 48 h) and temperature (30, 45 and 60°C) has been studied for conventional extraction and particle size, temperature (60, 80 and 100°C) and solid/liquid ratio (10, 20, 40 %w/V) for microwave extraction. Maximum TOC and TN concentrations of 16.8 g/L TOC and 2.3 g/L TN were obtained for optimal conventional extraction (40% w/V, 45°C and 24 h). Regarding microwave extraction, the optimum was reached operating for 2.5 min with 40% w/V, 80°C and milled MMWC (dp <0.5 mm) obtaining TOC and TN concentrations of 11.9 g/L and 2.5 g/L, respectively. The microwave extraction provides a very attractive and green alternative for the nutrients extraction of MMWC with an important reduction of the extracting time.

1 INTRODUCTION

The generation of bio-waste in Europe has increased due to the growth of the population and the industry. Only in Europe more than 88 million tons of municipal waste is generated (COM/2010/235/FINAL). Every day in the Europe and Central Asia area approximately 255,000 tons/day of municipal solid waste (MSW) are generated, and it is expected that by 2025 this generation of MSW will increase by 40% (Meena et al. 2019). This indicates the need to manage the bio-waste so that it does not end up in landfill or incineration. The European Waste Framework Directive (Directive 2018/851) encourages the Member States to collect and recycle bio-waste separately in order to reduce greenhouse gas emissions. The collection of bio-waste separately means that the organic matter can be composted with the purpose of obtaining a high-quality compost that can be used in agricultural soils (Lin et al. 2018). However, the use of the organic fraction of MSW obtained from triage on mechanical-biological treatment plants treating mixed household waste produced compost of low quality (Farrell and Jones 2009).

In this way, the valorisation of the organic matter of MSW is still an essential societal challenge. Moreover, so, it is necessary to develop new technologies for the valorisation of mixed municipal waste compost (MMWC).

The nutrients composition interval of MMWC are 5–35 g total nitrogen/kg dry matter (DM), 1–10 g total phosphorous /kg DM, 2–40 g potassium /kg DM, 2–100 g calcium /kg DM, 0.5–10 g sulphur /kg DM, 0.5–5 g magnesium/kg DM and 2-20 g sodium/kg DM (Hargreaves et al. 2008). These ranges make MMWC suited for agricultural use as organic fertiliser because MMWC has enough nutrients (nitrogen, phosphorous and potassium) for the development and growth of plants.

Nitrogen (N), phosphorus (P) and potassium (K) are essential for intensive agriculture. Unfortunately, the reserves of P and K are running out, which implies an increase in the prices of fertilisers. Synthetic N-fertilisers are the primary source of nitrogen for soils, and it is estimated that the Haber-Bosch process is responsible for 2% of the world's energy consumption (Smil 2001). It is possible that the P mineral reserves are exhausted in 60–100 years, which would cause a significant impact on agriculture since 90% of P extracted from the mines is used for fertilisers (Childers et

Table 1. The composition of MMWC.

Components	Units	Valour
Humidity	%w/w TM*	23,5 ± 0,1
Organic Matter	%w/w TM	41,6 ± 0,4
Total Organic Carbon (TOC)	%w/w DM**	28,7 ± 0,2
Total Nitrogen (NT)	%w/w DM	2,1 ± 0,3
Phosphorus as P_2O_5	%w/w DM	1,2 ± 0,1
Potassium as K_2O	%w/w DM	1,9 ± 0,2

*TM (Total Matter) **DM (Dry Matter)

al. 2011). On the other hand, the K worldwide demand surpasses 40 million tons for agricultural use, which corresponds to 65% of the current K production (FAO 2017).

Synthetic fertilisers are being overcome by organic fertilisers, which contain all the nutrients needed for crops without the need to add any chemical compound. Using organic matter improves the soil characteristics and reduces the negative impacts derived from the massive overuse of chemical fertilisers (Pergola et al. 2018). Organic fertilisers are obtained after the composting of agricultural residues (e.g. wheat straw), sewage sludge, organic residues of animals origin (e.g. pig manure) or the organic fraction of the MSW (Masters 1991; Proietti et al. 2016).

The aim of this paper is the comparison of the efficiency of conventional extraction and microwave extraction for the recovery of organic carbon (TOC) and nutrients (NPK) from MMWC. The purpose is to maximise the concentration of nutrients to obtain organic liquid fertilisers that can be used in agriculture in compliance with legal requirements. For this, the critical variables of each type of extraction have been studied. For conventional extraction, the solid/liquid ratio, the operating time and the extraction temperature have been the operating parameters studied. For the microwave extraction, the variables analysed were the solid/liquid ratio, the extraction temperature and the particle size of the solid (original MMWC, dp <20 mm and milled MMWC, dp <0.5 mm).

2 MATERIALS AND METHODS

2.1 *Raw material*

The MMWC used was kindly donated by Resíduos do Nordeste (Mirandela, Portugal). It was freeze stored at −18°C until use and dried in an oven at 70°C before the experiments. The main characteristics of the compost are shown in Table 1.

The solids samples dried and sieved to <20 mm were analysed as follows: Humidity and organic matter by gravimetry. Total nitrogen and total carbon by elemental analysis by LECO CHN-2000 analyser and total phosphorus by wet digestion with H_2SO_4 and HNO_3 followed by phosphorus measurement in extracts by spectrophotometry (HITACHI U 2000) according to the molybdenum blue method (APHA, 2012). Potassium was analysed by ICP Optical Emission Spectrometry (ICP-MS with Agilent HP 7500c Octopolar Reaction System) and ICP Mass Spectrometry (Atomic emission spectrophotometer ICP-OES Radial Simultaneous Agilent 725-ES) after microwave digestion (Milestone Ultrawave).

2.2 *Conventional extraction*

Conventional extraction with water was conducted using 250 mL sealed flasks in which the appropriate amount of compost and 100 mL of water were blended to achieve the required solid-liquid ratios (10, 20 and 40 %w/V). The extraction was carried out in an orbital shaker (Comecta Optic Ivymen system) at fixed temperature (30, 45 and 60°C), 200 rpm of agitation and extraction times ranging from 6 h to 48 h. After centrifugation for 15 min, the corresponding supernatants were carefully recovery and analysed. Experiments were performed in duplicate.

2.3 *Microwave extraction*

The microwave extraction of MMWC with water at 10%, 20% and 40% (w/V) was conducted at fixed temperature (60, 80 and 100°C) and with original MMWC (<20 mm) and milled (<0.5 mm). The extraction was carried out in a Multiwave PRO SOLV reactor 50 Hz with Rotor type 16HF100 (Anton Paar GmbH, Austria, Europe). Pressure vessels, fitted with magnetic stirrers and whose capacity volume was of 100 mL, were made of ceramic and PTFE-TFM. An IR sensor was used to register the temperature of all vessels continuously. MMWC and water were mixed at a corresponding solid/liquid ratio (appropriate amount of MMWC and 50 mL of water) in each of the pressure vessels of the multiwave reactor. The reactor was heated at a rate of 7°C/min and the reaction time (2.5 min) was considered when the microwave had reached the desired temperature.

2.4 *Analytical methods*

The liquid samples were analysed as follows: the total organic carbon (TOC) and total nitrogen (TN) was determined by TOC-V 5000 analyser (TOC-VCSH Shimadzu). The total Kjeldahl nitrogen (TKN) was analysed by acid digestion with H_2SO_4 and distillation with BUCHI KJELFLEX K-360 distillatory (APHA 2012). The phosphorus and potassium were analysed with the methods previously described in section 2.1.

2.5 *Data analysis*

ANOVA tests were carried out with *Statgraphics Centurion XVIII* version. ANOVA test was used to conclude the statistical differences at a confidence level of 95% ($p < 0.05$). Tukey's multiple range tests have allowed identifying means that are significantly different from each other.

3 RESULTS AND DISCUSSION

3.1 *Conventional extraction*

In this work, the effect of compost to water ratio (10, 20, 40 %w/V), extraction time (6, 24, 48h) and temperature (30, 45 and 60°C) on extraction yield of TOC and TN has been studied (see Table 2).

Higher amounts of solvent increase the extraction recoveries of TOC and TN from 7.9 to 12.5 % for TOC and from 21.3 to 30.7% for TN when S:L ratio increases from 10 to 40% w/V. The maximum TOC and TN concentrations have been reached with 40% w/V (9.1 g/L for TOC and 1.1 g/L for TN). The operation with high solid:solvent ratios contributes to reducing energy costs associated with subsequent concentration and purification steps.

Longer extraction times contribute to slightly improve on extraction recoveries being this effect more noticeable for the TN (19% for 6 h to 21.4% for 48 h). However, extending the extraction time from 24 to 48 h increases the TOC concentration only from 9.1 g/L to 9.2 g/L remaining constant the TN concentration (1.1 g/L). From an economical and technical point of view, the optimal extraction time has been fixed in 24 h because longer times only cause small improvement in extraction recoveries and concentrations values.

Higher extraction temperatures enhanced TOC and TN extraction recoveries. The extraction yield increases from 7.9 to 14.8% for TOC and from 21.3 to 27.6% for TN when temperature goes up from 30 to 60°C. However, no significant differences have been found when the temperature increases from 45°C to 60°C remaining approximately constant the TCO and TN concentrations and recoveries.

3.2 *Microwave extraction*

The effect of particle size of compost (original MMWC and milled MMWC), extraction temperature (60, 80 and 100°C) and solid/liquid ratio (10, 20, 40 %w/V) on concentrations and recoveries of

Table 2. Concentrations and recoveries for TOC and TOC in the conventional extraction. (A) Effect of S/L ratio, (B) Effect of extraction time and (C) Effect of extraction temperature.

			Concentrations (g/L)		Recovery (%)	
S/L (%w/V)	t (h)	T (°C)	TOC	TN	TOC	TN
S/L RATIO (%w/V)						
10	24	30	3.60 ± 0.04	0.64 ± 0.01	12.54 ± 0.31	30.67 ± 0.52
20			7.10 ± 0.05	1.03 ± 0.05	12.37 ± 0.37	25.06 ± 2.59
40			9.09 ± 0.02	1.11 ± 0.01	7.91 ± 0.20	21.29 ± 0.11
Extraction time (h)						
40	6	30	9.02 ± 0.01	1.00 ± 0.01	7.86 ± 0.65	18.94 ± 0.04
	24		9.09 ± 0.02	1.11 ± 0.01	7.91 ± 0.20	21.29 ± 0.11
	48		9.24 ± 0.05	1.13 ± 0.01	7.94 ± 0.29	21.41 ± 0.10
Extraction temperature (°C)						
40	24	30	9.09 ± 0.02	1.11 ± 0.01	7.91 ± 0.20	21.29 ± 0.11
		45	16.80 ± 0.07	2.31 ± 0.12	14.59 ± 0.07	27.46 ± 0.51
		60	16.97 ± 0.08	2.32 ± 0.23	14.77 ± 0.09	27.55 ± 0.28

Table 3. Concentrations and recovery for TOC and TN in the microwave extraction. (A) Effect of particle size. (B) Effect of extraction temperature and (C) Effect of S/L ratio.

			Concentrations (g/L)		Recovery (%)	
Particle size*	S/L (%w/V)	T (°C)	TOC	TN	TOC	TN
Particle size						
Original MMWC	10	60	3.09 ± 0.04	0.69 ± 0.01	10.77 ± 0.02	32.86 ± 0.41
Milled MMWC			3.48 ± 0.05	0.86 ± 0.01	12.12 ± 0.03	40.79 ± 0.51
Extraction temperature (°C)						
Milled	10	60	3.48 ± 0.05	0.86 ± 0.01	12.12 ± 0.03	40.79 ± 0.51
		80	4.20 ± 0.05	0.94 ± 0.01	14.63 ± 0.03	44.92 ± 0.56
		100	4.64 ± 0.06	0.88 ± 0.01	16.16 ± 0.04	41.66 ± 0.52
S/L RATIO (%w/V)						
Milled	10	80	4.20 ± 0.05	0.94 ± 0.01	14.63 ± 0.03	44.92 ± 0.56
	20		7.76 ± 0.10	1.23 ± 0.02	13.51 ± 0.04	29.26 ± 0.37
	40		11.89 ± 0.06	2.49 ± 0.15	10.35 ± 0.14	29.66 ± 0.77

*Original MMWC (dp <20 mm) and Milled MMWC (dp <0.5 mm)

TOC and TN have been studied (Table 3). The microwave extraction was carried out at fixed extraction time of 2.5 min.

Higher extraction recoveries for TOC and TN were achieved operating with smaller particle size due to the enhance of extraction mass transfer rates. The maximum concentrations of TOC and TN were found with milled MMWC (3.5 g/L for TOC and 0.9 g/L for TN).

Higher extraction temperatures enhanced TOC and TN extraction recoveries being this effect more remarkable for TOC. Taking into account the energy costs of the extraction process, a temperature of 80°C is enough to achieve high concentrations of TOC (4.2 g/L) and TN (0.94 g/L) in the extracts.

Extracts with higher concentrations of TOC and TN were found operating with a lower proportion of solvent as was expected. The optimal S/L ratio was 40% w/V due to it allows obtaining more concentrated extracts in TOC and TN (11.9 g/L and 2.5 g/L, respectively) while minimising the costs associated with the subsequent concentration process.

4 CONCLUSIONS

The optimum operating conditions for the conventional extraction of nutrients with water from MMWC were S:L ratio = 40% w/V, T = 45°C and t = 24 h. These conditions allowed to obtain concentrations of 16.8 g/L TOC and 2.3 g/L TN. For microwave extraction with water for 2.5 min the most favourable conditions were S:L ratio = 40% w/V, T = 80°C and milled MMWC (dp <0.5 mm) obtaining TOC and TN concentrations of 11.9 g/L and 2.5 g/L, respectively. The microwave extraction provides a very attractive and green alternative to conventional techniques for the nutrients extraction of MMWC with an important reduction of the extracting time.

ACKNOWLEDGEMENTS

The work was funded by the programme of cooperation INTERREG V-A Spain-Portugal (POCTEP) 2014-2020 and European Regional Development Fund (ERDF) (Project 0119_VALORCOMP_2_P).

REFERENCES

APHA. 2012. *Standard Methods for the Examination of Water and Wastewater*. 21th ed. Washington, DC, USA.: American Public Health Association.

Childers, Daniel L., Jessica Corman, Mark Edwards, and James J. Elser. 2011. "Sustainability Challenges of Phosphorus and Food: Solutions from Closing the Human Phosphorus Cycle." *BioScience* 61 (2): 117–24. https://doi.org/10.1525/bio.2011.61.2.6.

COM/2010/235/FINAL. 2010. "Communication from the Commission to the Council and the European Parliament on Future Steps in Bio-Waste Management in the European Union." Brussels. https://eur-lex.europa.eu/legal-content/EN/TXT/PDF/?uri=CELEX:52010DC0235&from=EN.

Directive 2018/851. 2018. "DIRECTIVE (EU) 2018/851 OF THE EUROPEAN PARLIAMENT AND OF THE COUNCIL of 30 May 2018 Amending Directive 2008/98/EC on Waste (Text with EEA Relevance)." 2018. https://eur-lex.europa.eu/legal-content/EN/TXT/PDF/?uri=CELEX:32018L0851&from=ES.

FAO. 2017. "World Fertilizer Trends and Outlook to 2020." Rome. 2017. http://www.fao.org/3/a-i6895e.pdf.

Farrell, M., and D. L. Jones. 2009. "Critical Evaluation of Municipal Solid Waste Composting and Potential Compost Markets." *Bioresource Technology* 100 (19): 4301–10. https://doi.org/10.1016/j.biortech.2009.04.029.

Hargreaves, J.C., M.S. Adl, and P.R. Warman. 2008. "A Review of the Use of Composted Municipal Solid Waste in Agriculture." *Agriculture, Ecosystems & Environment* 123 (1–3): 1–14. https://doi.org/10.1016/J.AGEE.2007.07.004.

Lin, Long, Fuqing Xu, Xumeng Ge, and Yebo Li. 2018. "Improving the Sustainability of Organic Waste Management Practices in the Food-Energy-Water Nexus: A Comparative Review of Anaerobic Digestion and Composting." *Renewable and Sustainable Energy Reviews* 89 (February): 151–67. https://doi.org/10.1016/j.rser.2018.03.025.

Masters, Gilbert M. 1991. *Introduction to Environmental Engineering and Science*. Prentice Hall. https://books.google.es/books?id=VhhSAAAAMAAJ&redir_esc=y.

Meena, M. D., R. K. Yadav, B. Narjary, Gajender Yadav, H. S. Jat, P. Sheoran, M. K. Meena, et al. 2019. "Municipal Solid Waste (MSW): Strategies to Improve Salt Affected Soil Sustainability: A Review." *Waste Management* 84: 38–53. https://doi.org/10.1016/j.wasman.2018.11.020.

Pergola, Maria, Alessandro Persiani, Assunta Maria Palese, Vincenzo Di Meo, Vittoria Pastore, Carmine D'Adamo, and Giuseppe Celano. 2018. "Composting: The Way for a Sustainable Agriculture." *Applied Soil Ecology* 123 (February): 744–50. https://doi.org/10.1016/J.APSOIL.2017.10.016.

Proietti, Primo, Roberto Calisti, Giovanni Gigliotti, Luigi Nasini, Luca Regni, and Andrea Marchini. 2016. "Composting Optimization: Integrating Cost Analysis with the Physical-Chemical Properties of Materials to Be Composted." *Journal of Cleaner Production* 137 (November): 1086–99. https://doi.org/10.1016/J.JCLEPRO.2016.07.158.

Smil, Vaclav. 2001. *Enriching the Earth: Fritz Haber, Carl Bosch, and the Transformation of World Food Production*. MIT Press.

Wastes: Solutions, Treatments and Opportunities III – Vilarinho et al. (Eds)
© 2020 Taylor & Francis Group, London, ISBN 978-0-367-25777-4

Effect of chemical additives on the regeneration of waste lubricant oil

C.T. Pinheiro, M.J. Quina & L.M. Gando-Ferreira
CIEPQPF, Department of Chemical Engineering, University of Coimbra, Portugal

C.M. Cardoso
SOGILUB – Sociedade de Gestão Integrada de Óleos Lubrificantes Usados, Lda., Lisboa, Portugal

ABSTRACT: Coagulation phenomena can occur in certain types of waste lubricant oils (WLO) during regeneration processes involving alkaline treatments, causing plant shutdowns. In this context, the main objectives of this work were to study the effect of additives on the coagulation of WLO and to evaluate the possibility of removing the additives causing coagulation by solvent extraction. Calcium sulfonate, a detergent additive commonly used in engine oils was considered. It was observed that the presence of this additive at certain concentrations leads to the coagulation of WLO. Extraction tests were performed using 1-butanol. Results showed that it is possible to reverse the phenomena by removing the additives. However, when coagulation is due to the intrinsic nature of the base oil, as is the case of ester-based lubricants, it is not possible avoid coagulation. Therefore, in some case, a solvent extraction treatment can increase the regeneration potential of WLO.

1 INTRODUCTION

Waste lubricant oil (WLO) is the most significant liquid hazardous waste stream in Europe. The most economic and environmental sound way to manage WLO is regeneration in which high quality base oils can be recovered a limitless number of times. In the EU, about 13% of the base oil consumed is obtained from regenerated WLO, which helps to reduce the dependence on natural resources, limit waste and harmful emissions, and promote economic growth (GEIR 2016).

The current objectives towards a circular economy represent a challenge to continuously maximize the regeneration rate of WLO. The quality of this waste is a limiting factor regarding its potential for regeneration (Pinheiro et al. 2017a). The technical performance of the process is seriously affected by the feedstock's chemical composition, especially in technologies using alkaline treatments (Kupareva et al. 2013). The addition of a strong alkali agent, such as KOH, is a common treatment adopted to enhance the removal of contaminants and reduce plant fouling. However, under certain conditions, a gel can be produced, reducing the lubricant's fluidity and causing blocking problems in the plant (Audibert 2006). The occurrence of this phenomenon, called coagulation, is highly detrimental for the regeneration process (Pinheiro et al. 2017b). Consequently, WLO with these characteristics cannot be subject to regeneration and must be managed by alternative ways according to the waste treatment hierarchy, such as energy recovery.

Previous studies have reported that WLO mainly composed by synthetic ester type base oils are responsible for the coagulation phenomena. Certain additives have also been pointed out as possible causes of coagulation (Pinheiro et al. 2017b). However, the role of additives in coagulation has never been addressed in detail. Additives are compounds added to the base oil to improve their natural properties, promote desirable properties, inhibit adverse effects and increase oil stability (Rudnick 2009). In some lubricants, chemical additives represent only 1% of the total mass of oil (e.g., compressor oils or hydraulic oil), while others (e.g., gearing or metalworking) may contain up to 30% (Braun 2007). The additives can form soaps after the interaction with KOH which impart a

Table 1. Classification of WLO according to the result of the coagulation test.

Coagulation class	Observation
A	Oil matrix remains unaffected.
B	recipitate formation.
C	Gel-type oil formation.

thickening effect in the oil. Thickening agents are generally a metallic soap formed by saponification of a fatty acid, but can also be a non-soap product such as a polymer (Martín-Alfonso et al. 2009). Therefore, additives such as zinc dithiophosphates, nitrogen containing compounds, such as succinimides, calcium sulfonates, amine antioxidants or polymethacrylates might contribute to the coagulation. Consequently, knowing the nature of additive interaction and their impacts on the regeneration process can improve the process. It is important to identify these components and find ways to remove them from the WLO before the regeneration process based on alkaline treatments. Solvent extraction is a technology that can be used to remove additives (degraded or not), carbonaceous particles or polymeric compounds, in order to recover the base oil (Reis & Jerónimo 1988).

In this context, this work aims to evaluate the effect of one of the main additives used in engine oils on the coagulation phenomena of WLO and to assess the possibility of removing additives by solvent extraction.

2 MATERIALS AND METHODS

2.1 *Materials*

The lubricant additive studied was a common detergent used in engine oils – calcium sulfonate, which was kindly supplied by Galp, a Portuguese lubricant manufacturer. To evaluate the effect of calcium sulfonate on coagulation, two WLO samples were tested – one collected directly from a garage producer and one pretreated sample, in which water and sediments were previously removed. The samples were selected ensuring that the coagulation test result was initially class A (see Section 2.2). Furthermore, solvent extraction tests were carried out in 12 WLO samples whose initial coagulation result was a class C (see Section 2.2). After collection, all WLO samples were stored at room temperature in the dark to maintain their integrity until analysis. 1-butanol (Carlo Erba, 99%) was selected as extraction solvent due to its high capacity for additive removal and since it is the greenest solvent for the regeneration of WLO (Pinheiro et al. 2018b).

2.2 *Coagulation test*

The regeneration industry has been struggling with the fact that during alkaline treatments, some WLO forms a gel type oil that can cause plant plugging. The coagulation test is an empirical method often used to predict potential adverse saponification reaction of WLO during the alkaline treatment. Potassium hydroxide is added in excess to ensure the detection of these reactions and avoid blocking problems in the regeneration plant. The test involves the addition of a KOH aqueous solution (50 wt. %) to the WLO in a ratio of 1/10 (w/w) and then heated up to 180°C under magnetic stirring. The mixture is then cooled down to room temperature. The outcome of the test is one of the classes presented in Table 1. WLO of coagulation classes B and C hamper the regeneration process.

Table 2. Results of the coagulation test after the addition of calcium sulfonate.

Calcium sulfonate (wt. %)	Coagulation class	
	Producer WLO	Pre-treated WLO
0	A	A
1	A	A
2	C	C
4	B	B
6	B	B
8	B	B

2.3 *Solvent extraction procedure*

The extraction procedure was adapted from the centrifugal tube technique described by Reis & Jerónimo (1988). A polar solvent, 1-butanol, was used as the extracting agent, to which 3 g/L of KOH was added to promote a higher flocculation capacity (Pinheiro et al. 2018a). Centrifugal tubes were filled with a mixture of 20 g of waste oil sample and solvent in a solvent/oil ratio of 3:1. The samples were stirred by magnetic stirring at 500 rpm and during 20 min to ensure adequate mixing. The tubes were centrifuged at 4000 rpm for 20 min. After centrifugation, the sludge phase (additive, impurities and carbonaceous particles) was separated from the mixture solvent/oil. The solvent was recovered by vacuum distillation in a rotary evaporator and the extracted base oil was further collected and used for the coagulation test.

2.4 *FTIR*

Fourier Transform Infrared (FTIR) spectroscopy spectra were recorded in transmittance mode using potassium bromide (KBr) pellets, prepared using a pneumatic press. One drop of lubricant was placed on the top of one pellet, which was placed into an appropriate sample holder and immediately analyzed in a PerkinElmer FTIR. 64 scans/sample with a resolution of 4 cm^{-1} were carried out in the 4000–500 cm^{-1} range. Every analysis was performed using a different pellet and before each measurement, a background spectrum was obtained using one clean pellet.

3 RESULTS AND DISCUSSION

3.1 *Effect of calcium sulfonate on the coagulation phenomena*

Detergents such as calcium sulfonate are metal salts of organic acids which typically are grouped into a reverse micellar structure. This structure avoids the formation of deposits on metal surfaces, by suspending the polar materials in the oil. In addition, some detergents may have the ability to neutralize combustion or oxidation acid products (Ahmed & Nassar 2011).

Increasing amounts of calcium sulfonate (1–8 wt.%), according to the typical range applied in engine oils (Rizvi 2009) were added to class A samples. Subsequently, the coagulation test was performed in each mixture and the results are listed in Table 2.

The results showed that both in the producer and in the pre-treated oils, the addition of 1% calcium sulfonate did not cause significant changes in the matrix behavior. Figure 1 illustrates the results obtained for the pre-treated WLO sample. On the other hand, increasing the concentration of additive to 2% caused its transformation to a class C oil, where a uniform gel-type oil is obtained. In fact, the presence of detergent additives in lubricant oils may intensify the thickening effect of soaps resulting from saponification reactions (Lashkhi et al. 1992).

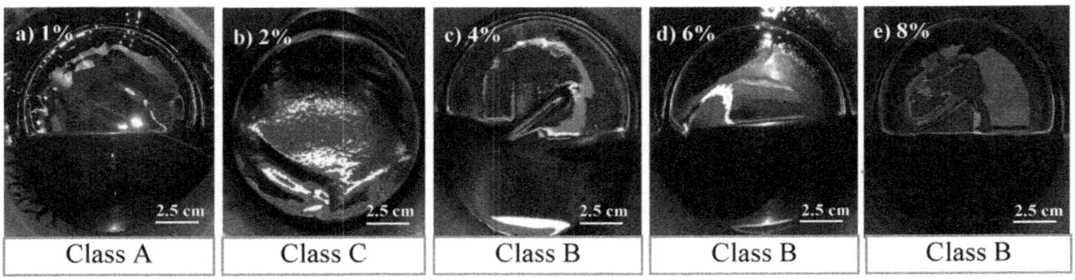

Figure 1. Coagulation test results of the pre-treated oil after the addition of a) 1%, b) 2%, c) 4%, d) 6% and e) 8% of calcium sulfonate.

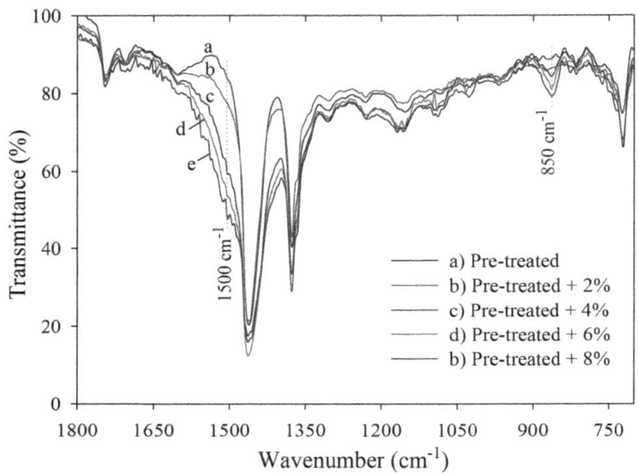

Figure 2. IR spectra of the pretreated oil and its mixture with different concentrations of calcium sulfonate.

However, the addition of increasing amounts of calcium sulfonate (Figure 1c – e) resulted in the formation of a precipitate phase (class B). The saturation of the mixture with an excess of the additive may have occurred, where colloidal aggregate particles that are too large to be held in suspension in the fluid may have formed (Rudnick 2009). These conditions lead to the precipitation of the gel formed. Therefore, it can be concluded that the presence of additives such as calcium sulfonate can effectively promote the coagulation phenomena. Nevertheless, the concentration of additive present is a determining factor.

FTIR spectra (between 1800–700 cm^{-1}) of the original pre-treated sample and of the mixtures after addition of additive (before the coagulation test was performed) are depicted in Figure 2.

The presence of calcium sulfonate in lubricant oils is typically detected at about 850 and 1500 cm^{-1}, as indicated in Figure 2. Increasing peak height in these ranges is observed with increasing additive concentration. However, by comparing the pre-treated original sample (class A) and after the addition of 2% calcium sulfonate (class C), the peak heights are practically indistinguishable. Thus, the detection of coagulation phenomena due to the presence of additives is extremely difficult by simple observation of infrared spectra.

3.2 *Removal of additives by solvent extraction*

The basic principle of the solvent extraction process consists of the different solubility of waste oil constituents in the solvent. The solvent has a greater affinity for the oil fraction, the aggregation

Table 3. Results of the coagulation test for different oils before and after solvent treatment.

	Coagulation class	
Sample	Sample as received	After extraction
S1	C	C
S2	C	A
S3	C	B
S4	C	C
S5	C	C
S6	C	C
S7	C	B
S8	C	C
S9	C	C
S10	C	B
S11	C	B
S12	C	B

and flocculation of insoluble particles occur, and their separation is possible. The heavier phase contains non-polar additives/polymers (Awaja & Pavel 2006). The coagulation test results obtained of the WLO samples as received and after an extraction stage are presented in Table 3.

The results show that for samples S1, S4–S6, S8 and S9 it was not possible to change the coagulation result after the extraction procedure. On the other hand, the behavior of samples S3, S7 and S10–S12 was altered from a class C to a class B oil. In this case, the precipitate formed was similar to an insoluble soap. Figure 3a illustrates the FTIR spectra of oils S1, S4, S7, and S10. Although the behavior of the first two samples is different from the last ones, the spectra are very similar. In general, all the samples show multiple bands in the 2980–2830 cm^{-1} region associated with $-CH_2$ e CH_3 stretching vibrations and peaks at 720, 1380 and 1460 cm^{-1} corresponding to –CH and =CH groups (Zi?ba-Palus et al. 2001). These bonds are characteristic of aliphatic components from hydrocarbons. Additionally, all samples show intense peaks at 1750 cm^{-1} associated with carbonyl groups (C=O) and in the range from 1250–1100 cm^{-1}, which indicates the presence of synthetic ester type oils. Therefore, all these samples should be constituted by synthetic ester-based oils. However, the ones that change from a class C to a class B oil should be of semisynthetic type, constituted by a mineral fraction, that remains unaffected after the alkaline treatment, and a synthetic ester-based fraction which undergo saponification reactions. The formed soap becomes insoluble, tending to precipitate out of the oil. Therefore, it is possible to conclude that ester-like or phosphate ester-type lubricants will still result in coagulation (class C) or precipitation (class B), regardless of extracting additives that may also contribute to this phenomenon.

On the other hand, it was possible to change the coagulation result from class C to class A after the extraction of samples S2. The FTIR spectra before and after extraction, as well as the resulting sludge, are shown in Figure 3b. Comparing to the previous spectra, the peaks associated with aliphatic components are also present, but the intense carbonyl peak and bands in the range from 1250–1100 cm^{-1}, indicating that this waste oil might have a mineral origin. The spectra show that peaks typically attributed to the presence of additives in lubricating oils are removed or reduced after the extraction. Examples of additives include zinc dialkyldithiophosphate – ZDDP (990 cm^{-1}), nitrogen compounds such as succinimides (1230 e 1630 cm^{-1}), calcium sulfonates (850 cm^{-1}) or polymethacrylates (990, 1710 cm^{-1}). Figure 3b shows that these functional groups are removed from the WLO and transferred to the sludge phase. Therefore, the solvent extraction tests allow concluding that the presence of specific additives can be responsible for the coagulation of WLO.

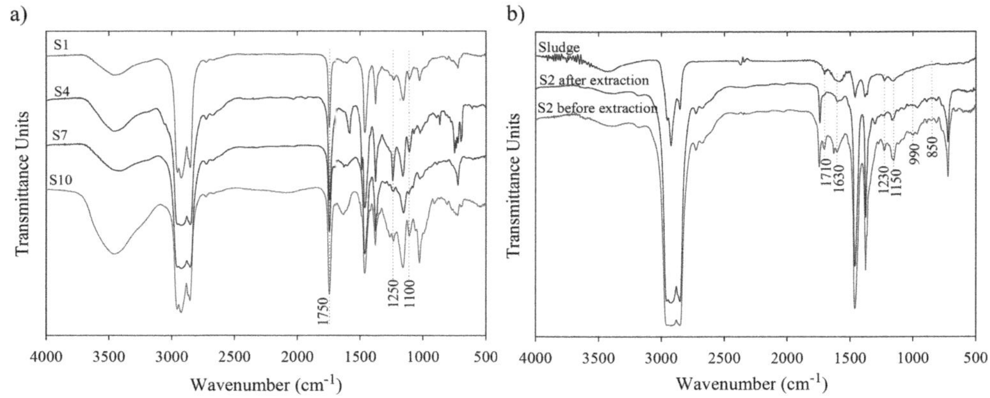

Figure 3. FTIR spectra of samples a) S1, S4, S7 and S10 before extraction. b) S2 before and after extraction, and of the obtained sludge.

4 CONCLUSIONS

The studies developed with calcium sulfonate, a detergent commonly used in engine lubricants, allowed to confirm that the presence of this additive in waste oils can lead to coagulation during alkaline treatments. The concentration of additive is a determining factor for the occurrence of this phenomena. Extraction tests with 1-butanol have shown that when the coagulation is due to the presence of additives, it is possible to remove them and obtain a waste oil that can be regenerated. On the other hand, in cases where the coagulation is associated with the base oil type, such as esters and phosphate esters, the coagulation or precipitation of soap will still occur after extraction. The decision whether it is suitable to make treatment by extraction to avoid coagulation after the addition of the alkaline agent can be supported by FTIR.

ACKNOWLEDGEMENT

The authors gratefully acknowledge the financial support of SOGILUB – Sociedade de Gestão Integrada de Óleos Lubrificantes Usados, Lda.

REFERENCES

Ahmed, N.S. & Nassar, A.M. 2011. Lubricating oil additives. In Kuo, C.-H. (ed.), *Tribology – Lubricants and lubrication*: 249–626. Shanghai: InTech.

Audibert, F. 2006. *Waste engine oils. Rerefining and energy recovery*. Amsterdam: Elsevier Science & Technology Books.

Awaja, F. & Pavel, D. 2006. *Design aspects of used lubricating oil re-refining*. Amsterdam: Elsevier.

Braun, J. 2007. Additives. In Mang, T. & Dresel, W. (eds), *Lubricants and lubrication*: 88–118. Weinheim: WILEY-VCH Verlag.

GEIR 2016. Waste Framework Directive revision: European waste oil re-refining industry position.

Kupareva, A., Mäki-Arvela, P. & Murzin, D.Y. 2013. Technology for rerefining used lube oils applied in Europe: A review. *Journal of Chemical Technology and Biotechnology* 88(10): 1780–1793.

Lashkhi, V.L., Fuks, I.G. & Shor, G.I. 1991. Colloid chemistry of lubricating oils (in the conditions of use). *Chemistry and Technology of Fuels and Oils* 27(6): 311–319.

Martín-Alfonso, J.E., Moreno, G., Valencia, C., Sánchez, M.C., Franco, J.M. & Gallegos, C. 2009. Influence of soap/polymer concentration ratio on the rheological properties of lithium lubricating greases modified with virgin LDPE. *Journal of Industrial and Engineering Chemistry* 15(5): 687–693.

Pinheiro, C.T., Ascensão, V.R., Cardoso, C.M., Quina, M.J. & Gando-Ferreira, L.M. 2017a. An overview of waste lubricant oil management system: Physicochemical characterization contribution for its improvement. *Journal of Cleaner Production* 150: 301–308.

Pinheiro, C.T., Ascensão, V.R., Reis, M.S., Quina, M.J. & Gando-Ferreira, L.M. 2017b. A data-driven approach for the study of coagulation phenomena in waste lubricant oils and its relevance in alkaline regeneration treatments. *Science of the Total Environment* 599–600: 2054–2064.

Pinheiro, C.T., Pais, R.F., Quina, M.J. & Gando-Ferreira, L.M. 2018a. Regeneration of waste lubricant oil with distinct properties by extraction-flocculation using green solvents. *Journal of Cleaner Production* 200: 578–587.

Pinheiro, C.T., Quina, M.J. & Gando-Ferreira, L.M. 2018b. New methodology of solvent selection for the regeneration of waste lubricant oil using greenness criteria. *ACS Sustainable Chemistry & Engineering* 6(5): 6820-6828.

Reis, M.A. & Jerónimo, M.S. 1988. Waste lubricating oil rerefining by extraction-flocculation. 1. A scientific basis to design efficient solvents. *Industrial & Engineering Chemistry Research* 27(7): 1222–1228.

Rizvi S.Q.A. 2009. Detergents. In Rudnick, L.R. (ed.), *Lubricant additives: Chemistry and applications*:123–142. Boca Raton: CRC Press.

Rudnick, L.R. 2009. *Lubricant additives: Chemistry and applications.* Boca Raton: CRC Press.

Zięba-Palus, J., Kościelniak, P. & £ącki, M. 2001. Differentiation of used motor oils on the basis of their IR spectra with application of cluster analysis. *Journal of Molecular Structure* 596(1–3): 221–228.

Wastes: Solutions, Treatments and Opportunities III – Vilarinho et al. (Eds)
© 2020 Taylor & Francis Group, London, ISBN 978-0-367-25777-4

Pilot test involving pulp and paper industry wastes in road pavements

H. Paiva & F. Simões
Department of Civil Engineering/CICECO, University of Aveiro, Aveiro, Portugal

M. Morais & V.M. Ferreira
Department of Civil Engineering/RISCO, University of Aveiro, Aveiro, Portugal

ABSTRACT: Waste recycling is part of the circular economy concept that goes from the redesign of processes, products, business models with resource efficiency. Bituminous mixtures were developed with paper and pulp industry wastes (dregs and grits). These mixtures will be used as a surface course in a pilot large-scale demonstrator of an on-going European Circular Economy project (Paperchain). To scale up from laboratory to the industrial level, mix designs were designed, tested, technical and legal constraints were assessed in order to carry on the pilot test. This paper discusses relevant results so that the pilot test can be carried out with success in an environmentally safe way. Results show mix designs that have the possibility to be scaled up with success, using dregs and grits as alternative raw materials of natural filler and fine aggregates present in traditional bituminous mixtures for road construction.

1 INTRODUCTION

Circular economy aims to develop new, economically viable and environmentally efficient products and services based on optimal perpetual cycles. It deals with minimizing resource extraction, maximizing reuse, increasing efficiency and developing new business models.

The Pulp and Paper industry (PPI) sector is resource intensive and produces every year 11 million tonnes of waste in Europe. PPI wastes can become a valuable alternative raw materials for other natural resource intensive industries such as construction or others. Green liquor dregs and grits are wastes coming from pulp and paper production usually disposed in landfills. Although these wastes are non-hazardous according to its composition, it is important to find solutions for its valorization or recycling. For instance, (Novais et al. 2018) showed that it is possible to incorporate successfully dregs as an aggregate in geopolymer formulations. Dregs are non-reactive or present poor reactivity (Novais et al. 2018, Saeli et al. 2018). (Siqueira & Holanda 2018, Siqueira & Holanda 2013) have also used dregs and grits as raw materials to replace traditional limestone material in ceramic wall tiles and also used grits as a raw material for replacing Portland cement in cement bricks.

Dregs and grits have been added in sealing layers to cover mine tailings to improve the layers quality (Makitalo et al. 2015, Jia Yu et al. 2014) and also used in cement clinker production (Castro et al. 2009) as well as in bituminous mixtures (Passadín et al. 2016). In this last application, (Pasandín et al. 2016) found that the introduction of dregs as filler in bituminous mixtures promoted a small degradation in the mixtures final properties. However, good results were obtained by incorporating dregs as filler and grits as filler or fine aggregate in bituminous mixtures (Modolo et al. 2010).

This paper is based on work developed within the European project entitled Paperchain (H2020 project n° 730305), a large-scale demonstrator project in the area of circular economy. Globally, this project aims to validate and demonstrate solutions, namely its scalability from the laboratory to the industrial dimension, using pulp and paper industry wastes valorization in construction materials sectors. In this paper, the waste valorization process focuses in one of those sectors, the bituminous mixtures for road pavements.

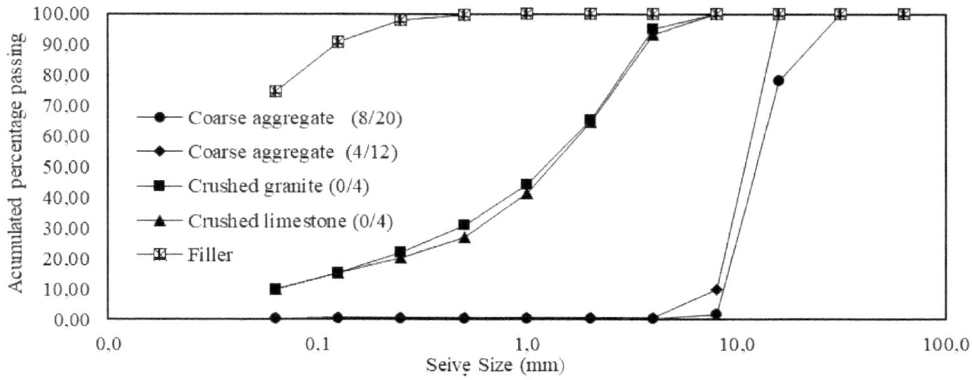

Figure 1. Characterization of current natural raw materials.

Table 1. Chemical composition for dregs and grits (ICP-OES).

Parameter (unit)	Dregs	Grits
Cl (mg/l)	$6,0*10^3$	$1,2*10^3$
Ca (mg/kg)	$1,3*10^5$	$4.9*10^5$
Na (mg/kg)	$2,4*10^5$	$2,1*10^4$
Mg (mg/kg)	$6,5*10^4$	$2,6*10^3$
K (mg/kg)	$4,3*10^3$	$7*10^2$
P (mg/kg)	83,3	$3,3*10^3$
Fe (mg/kg)	$4,1*10^3$	$1,8*10^3$
Si (mg/kg)	$5,4*10^3$	$3,4*10^3$
Al (mg/kg)	$4,7*10^3$	$4,2*10^3$
Cu (mg/kg)	154,4	<LOQ
Cd (mg/kg)	<LOQ	<LOQ
Ni (mg/kg)	92,9	6,4
Zn (mg/kg)	206,5	5,1
Cr (mg/kg)	42,3	5,7

2 EXPERIMENTAL WORK

2.1 *Materials characterization*

In this preliminary study, a company that will produce the bituminous mixtures with the pulp and paper industry wastes supplied their standard bituminous mixture. The standard bituminous mixture (for an AC14 surface) constitution involves a bitumen (35/50) as a binder, a crushed limestone and a crushed granite aggregate up to 4 mm, two coarse granite aggregate (4/12 and 8/20 mm) and a filler which is recovered from the process. The materials and bituminous mixture are according to the requirements of the specifications of *Infrastructures of Portugal* guidebook (EP 2014). The aggregates particle size distribution is presented in Figure 1.

The pulp and paper industry wastes in this study were dregs and grits. Dregs and grits were first dried at 100° C in a laboratory oven. Dregs and grits have around 50% and 20% of moisture, respectively, measured according to EN 12880:2000. The 2014/955 Dec 18th decision classifies these wastes with LER code 030302 (non-hazardous). Wastes were chemically characterized, using Inductively Coupled Plasma Optical Emission Spectrometry (ICP-OES) (Table 1).

After drying, both wastes still need to be deagglomerated. For the grits the grinding was manual and mechanical, in the laboratory for the experimental tests. Particle size distribution was

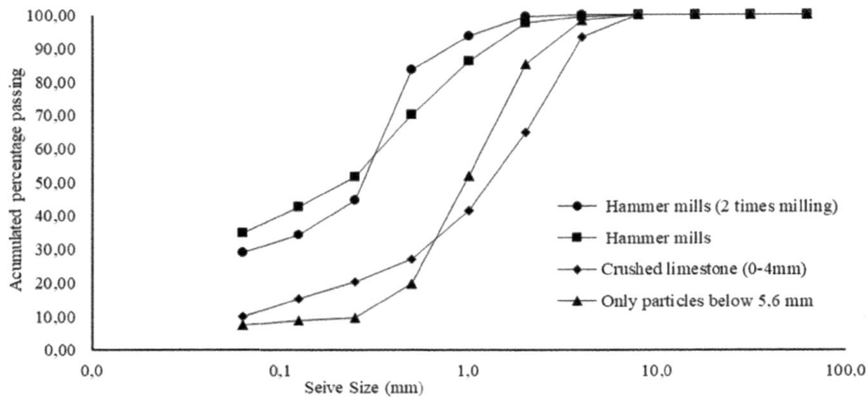

Figure 2. Particle size distribution of grits and crushed limestone.

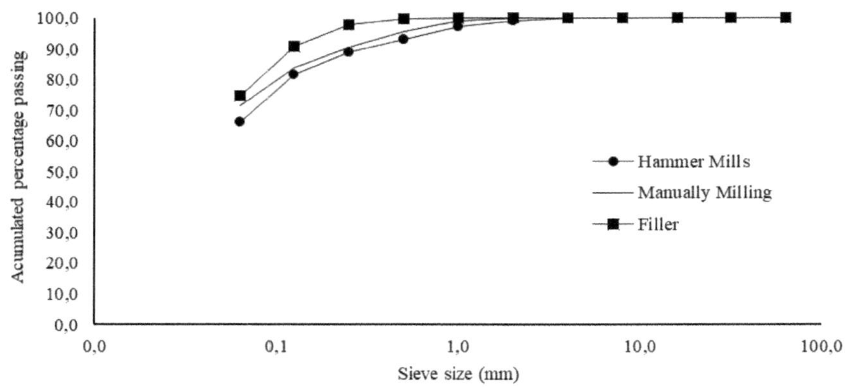

Figure 3. Particle size distribution of dregs and natural used filler.

determined according to EN 933-1:2012. Figure 2 presents the different particle size distributions resulting for the grits, where one is similar to crushed limestone particle size distribution. Grits chemical characterization showed that this waste consists of near 95% of calcium carbonate, similar to the crushed limestone composition.

Figure 3 show that dregs particle size distribution is similar to the filler particle size distribution. According to *Infrastructures of Portugal* guidebook (EP, 2014), a filler must be 70% below 63 μm, which means that grits could also be used as filler replacement if properly ground.

Other properties, as blue methylene and density were determined for dregs and grits. Blue methylene test determines the amount of clay present in the material and has to be lower than 10g/kg. Blue methylene was 3,8 g/kg for dregs and 0,05 g/kg for grits. Particle density is 2,70 g/cm^3 for grits and 2,89 g/cm^3 for dregs, which is close to the limestone or granite values, usually used as aggregates or fillers.

The drying and grinding of these materials at laboratory scale seems easy but, when transfered to industrial scale, some technical and legal constraints might appear. This may demand, for instance, that a specific company is brought to deal with drying, grinding and transport operations for dregs and grits. This company needs to have a specific legal license or permit for these wastes management.

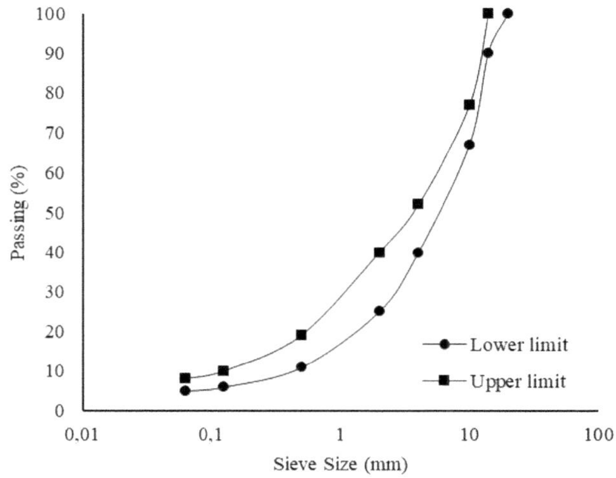

Figure 4. Particle size range/ spindle requirements for an AC 14 surface.

Table 2. Requirements of bituminous mixture for an AC14 surface course.

Properties	Limits	Units
Number of blows on each sample side	75	–
Stability	7,5 to 15	kN
Flow	2 to 4	mm
Marshall quotient	$\geq$3	kN/mm
Conserved Strength	$\geq$80	%
Air voids content	3 to 5	%
Voids content in the mineral aggregate	$\geq$14	%
Minimum Binder Content	4	%

Concerning dregs, they undergo a filter-press operation in the PPI factory in order to eliminate part of their absorbed water. The filter pressed dregs plates (600×600 mm^2) are very compact and hardened after drying.

2.2 *Bituminous mixtures characterization and tested mix designs*

An AC 14 surface mixture, for a road surface course, has requeriments to fulffil, namely, particle size range of agregates mixture and physical propeties of agregates mixture with binder. Figure 4 shows the lower and upper limit of particle size range of agregates mixture for an AC 14 surface bituminous mixture.

The optimum bitumen content (OBC) was determined for the standard mix design (mixtures having 4 to 6% were made and optimum content corresponds to the ones with higher bulk density, stability and a porosity limit between 3 and 5%). All other mix designs were produced with this optimum bitumen content. All mix designs satisfy the particle size range for an AC14 surface (Figure 4). Table 2 presents the bituminous mixture requirements. The Marshall test was made according to the EN12697-34:2012 standard. Table 3 shows the diferent mix designs made for this study (dried and grinded in laboratory).

Table 3. Different bituminous mixtures studied.

Mix designs	Coarse aggregate (%) 8/20	4/12	Fine aggregate (%) 0/4 (granite)	0/4 (limestone)	Filler (%)	Grits (%)	Dregs (%)
Standard (Std)	19	34	22	23	2	0	0
Grits replacing 5% crushed limestone (G5CL)	19	34	22	18	2	5	0
Grits replacing 10% crushed limestone (G10CL)	19	34	22	13	2	10	0
Grits and Granite replacing all crushed limestone (G9CL)	19	34	48	0	0	9	0
Grits replacing filler (G2F)	19	34	22	23	0	2	0
Dregs replacing filler (D2F)	19	34	22	23	0	0	2

Table 4. Results for studied mix designs with grits or dregs.

Mix design	Std	G5CL #	G5CL ##	G9CL ###	G10CL ##	D2F *	D2F **	G2F	Limits
Bitumen (%)	5	5	5	5	5	5	5	5	≥4
Maximum density (kg/m^3)	2440	2443	2441	2454	2451	2442	2440	2445	–
Bulk density (kg/m^3)	2347	2347	2361	2325	2352	2305	2331	2325	–
Air voids content (%)	3,8	3,9	3,3	5,3	4,1	5,6	4,5	4,9	3 to 5
Voids content in the mineral aggregate (VMA) (%)	15,2	15,3	14,8	16,5	15,5	16,8	15,8	16,2	≥14
Bitumen saturation (%)	74,9	74,5	77,7	68,3	73,5	66,7	71,8	69,7	–
Flow (mm)	4,0	4,0	4,1	3,0	3,3	3,7	4,0	4,0	2 to 4
Stability (kN)	12,0	14,9	15,1	14,8	16,3	11,8	12,0	15,5	7 to 15
Marshall quotient (Stability/Flow) (kN/mm)	3,0	3,7	3,7	4,9	5,0	3,2	3,0	3,9	≥3

*(# hammer mill (2 times); ### Only particles below 5,6 mm; ##hammer mill; * hammer mill; **manual milling)*

3 RESULTS AND DISCUSSION

It is possible to observe in Table 4 the results for the different compositions prepared for this scale up test. Besides the standard mix design (Std) with no waste, which acts as a reference, grits replaced 5, 9 and 10% of the crushed limestone (fine aggregate). Dregs and grits were also used to replace all the filler (2%). Different millings were performed to adjust the particle size of the wastes to incorporate on the bituminous mixtures. Manual or hammer mills and sieving were used to achieve proper particle size.

Observing the results in Table 4, only the mix designs G5CL##, G9CL### and D2F* didn't comply completely with all the demanded requirements (expressed in the Table 4 last column as Limits). Regarding mix design G5CL## presents a slightly higher flow than the limit. This behaviour can be justified by the bitumen saturation and air voids content (porosity) values. This composition has the highest value of bitumen saturation and the lowest of air voids content. Probably 5% of bitumen is too much for this material mixture. By decreasing the bitumen content, porosity increases and flow decreases.

Concerning G9CL### mix design, the air voids content (porosity) is higher than 5% (limit). Although there is no traditional filler and crushed limestone in this composition (granite replaced crushed limestone since its particle size distribution is very similar, 0 to 4 mm), grits addition with particle size distribution from 0 to 2 mm promotes a mixture with higher amount of fine material, replacing the filler effect. Fine particles require more bitumen as can be observed in Table 4. This composition shows the lower value of bitumen saturation and the higher value of porosity. In this case, it needs a higher percentage of bitumen. Finally, the D2F* mix design shows a porosity (air voids) higher than the limit. Compared with highly disaggregated D2F** mix design, in D2F* there are small agglomerates of dregs particles with air inside them increasing overall porosity. All the other compositions in Table 4 could be used as an AC14 surface course in asphalt road.

Although these mix designs may be used in bituminous mixtures with acceptable results there are still issues to be addressed in the scale up process from lab to industrial application. Indeed this project, as a large demonstrator, will have to scale up two selected mix designs with grits and dregs to follow up together with the standard mix design for comparision. The setup of this experiment requires also that a licensed waste manager is introduced in the process between the PPI waste location and the mix design and final application. This waste manager has to control the drying and grinding operations when and if needed. If necessary the waste producer (PPI) might have to laminate the filter-pressed dregs to reduce its size in order to help the drying and grinding operations. On the other hand, grits are an easier material to manage. Nevertheless, tests at industrial scale of drying and grinding have to be done to determine final particle distribution and its incorporation rate in the final bituminous mixture.

Furthermore, prior to applying the asphalt mixtures in the pilot road pavement section, performance tests will be carried out, such as water sensitivity, rutting resistance, stiffness modulus, and fatigue cracking resistance. In situ operations related to waste transfer, treatment and incorporation in selected mix designs as well as other operation variables are going to be discussed before the pilot test planned for the last quarter of 2019.

4 CONCLUSIONS

In order to prepare the scale up process from laboratory to industrial level, mix designs incorporating dregs and grits from the pulp and paper industry were made and characterized. Determination of optimum bitumen content for each composition is extremely important since it affects various properties such as porosity and deformation. Dregs can be used as filler if they are properly deagglomerated. On the other hand, grits can be applied to replace either the filler or as a partial replacement of the fine aggregate, depending on its particle size distribution.

The execution of the pilot test at industrial scale will serve as a large scale demonstrator in the Paperchain H2020 project will also require dealing with larger amounts of wastes in an environmentally safe way. Hence, a proper licensed waste manager will be introduced in the process to transport, dry and grind the wastes as required to incorporate in the mix designs to be tested and compared with a traditional standard one as a reference. These mix designs will be deposited as an asphalt surface course in a 250 meters road section that will be monitored and serve as a demonstrator in the following project years.

ACKNOWLEDGEMENTS

The authors wish to acknowledge the support of the European Union Horizon2020 Research and Innovation programme for the funding of the Paperchain project under grant agreement n° 730305. The authors also wish to acknowledge the contribution of their partners in the Paperchain project, namely, The Navigator Company and RAIZ (S. Pereira, P. Branco and P. Pinto) the infrastructures company Megavia SA (O. Cafofo, M. Vala and T. Marcelino) and also to their former collaborator for previous results (R. Modolo).

REFERENCES

Castro, F., Vilarinho, C., Trancoso, D., Ferreira, P., Nunes, F. & Miragaia, A. 2009. Utilization of pulp and paper industry wastes as raw materials in cement clinker production. *International. Journal of Materials Engineering Innovation* 1(1): 74–90.

Estradas de Portugal SA. 2014. Caderno de Encargos Tipo Obra: 14.03 – Pavimentação Características dos materiais. vol. V.

Jia Y., Maurice, C. & Ohlander, B. 2014. Effect of the alkaline industrial residues fly ash, green liquor dregs, and lime mud on mine tailings oxidation when used as covering material. *Environmental Earth Sciences* 72: 319–334.

Novais, R.M., Carvalheiras, J., Senff, L. & Labrincha, J.A. 2018. Upcycling unexplored dregs and biomass fly ash from the paper and pulp industry in the production of eco-friendly geopolymer mortars: A preliminary assessment. *Construction and Building Materials* 184: 464–472.

Saeli, M., Novais, R.M., Seabra, M.P. & Labrincha, J.A. 2018. Green geopolymeric concrete using grits for applications in construction. *Materials Letters* 233: 94–97.

Makitalo, M., Macsik, J., Maurice, C. & Ohlander, B. 2015. Improving Properties of Sealing Layers Made of Till by Adding Green Liquor Dregs to Reduce Oxidation of Sulfidic Mine Waste. *Geotechnical Geological Engineering* 33: 1047–1054.

Modolo, R., Benta, A., Ferreira, V.M. & Machado, L.M. 2010. Pulp and paper plant wastes valorization in bituminous mixes. *Waste Management* 30: 685–696.

Pasandín, A.R., Perez, I., Ramírez, A. & Cano, M. M. 2016. Moisture damage resistance of hot-mix asphalt made with paper industry wastes as filler. *Journal of Cleaner Production* 112: 853–862.

Siqueira, F.B. & Holanda, J.N.F. 2018. Application of grits waste as a renewable carbonate material in manufacturing wall tiles. *Ceramics International* 44: 19576–19582.

Siqueira F.B. & Holanda J.N.F. 2013. Reuse of grits waste for the production of soil cement bricks. *Journal of Environmental Management* 131: 1–6.

Wastes: Solutions, Treatments and Opportunities III – Vilarinho et al. (Eds)
© 2020 Taylor & Francis Group, London, ISBN 978-0-367-25777-4

Different extraction methods to produce antioxidants from agro-industrial wastes

G. Squillaci, F. Veraldi, F. La Cara & A. Morana
Research Institute on Terrestrial Ecosystems, National Research Council of Italy, Naples, Italy

ABSTRACT: This study investigated different extraction techniques for production of antioxidants from chestnut shells (CS), a solid waste produced by the chestnut peeling process. Conventional and pressurized liquid extractions, ultrasound and microwave-assisted extractions were tested in presence of water as solvent, and the effect of the extraction time and CS concentrations were examined. The most efficient method was pressurized liquid extraction. The highest yields of antioxidant species, expressed in mg/mL of extract, were obtained for 10% CS after 60 min of incubation: 0.836 ± 0.018 for total phenols, 0.297 ± 0.003 for *ortho*-diphenols, 0.173 ± 0.002 for flavonoids and 0.545 ± 0.018 for tannins. Measurement of the radical scavenging activity indicated that all the extracts exhibited an antioxidant power stronger than that showed by commonly used antioxidants. HPLC analysis revealed only quantitative differences among the extracts, and gallic acid was the main simple phenol detected independently of the extraction method used. Its amount ranged from 159.66 ± 0.31 to 284.14 ± 0.72 μg/mL extract.

1 INTRODUCTION

The increasing interest in searching for new natural sources of antioxidant compounds is justified by the need to find suitable alternatives to synthetic antioxidants, whose safety is currently discussed (Thorat et al. 2013). In this context, the large amount of wastes produced by the anthropic activities represents a cheaply available and underexploited resource of bioactive compounds. Agro-industrial wastes create problems of disposal but, at the same time, many of them are precious sources of antioxidant compounds that can be recovered and used in several fields, such as food, pharmaceutical and cosmetic industries. Consequently, their exploitation will allow to reach a double beneficial effect: the production of blends of powerful and natural antioxidants, and the material valorization of wastes that represent a source of environmental pollution.

Chestnut tree (*Castanea* sp.) is recognized as one of the most remarkable trees in the world due to its economic importance. In 2017 Italy, the main chestnut producer in the European Union with 38% of the total European production yielded about 52,356 tonnes (FAOSTAT 2019). The chestnut processing chain produces, through the peeling step, large quantities of outer and inner shells that represent approximately 10% of the chestnut whole weight. More in detail, Italy produces about 5200 tonnes of shell/year that are usually burned in the factories, in order to overcome disposal problems. As an alternative use, this residue can represent a valuable material for production of bioactive extracts since it contains phenolic compounds with antioxidant, anti-cancer and anti-inflammatory properties (Squillaci et al. 2018).

The search for eco-friendly methodologies of extraction that avoid or limit the use of organic solvents to safeguard health and environment is very active. Moreover, methods such as Microwave-Assisted Extraction (MAE) and Ultrasound-Assisted Extraction (UAE), that reduce the extraction time and the phytochemical degradation that can occur during long extraction processes, are valid alternatives used for the extraction of natural compounds (Muñiz-Márquez et al. 2013, Valdés et al. 2015). As it is well ascertained that extraction operative parameters (temperature, extraction time, solid/liquid ratio, type and particle size of the residues) can also affect the extraction yields, it is extremely important to establish the optimal operational conditions.

The aim of the present work was to compare the efficiency of different extraction procedures in recovering antioxidant compounds from chestnut shells (CS). In this work, Conventional Liquid Extraction (CLE), Pressurized Liquid Extraction (PLE), MAE, and UAE were tested as extraction methods and the influence of process variables (extraction time, temperature and CS concentration) was evaluated. The effectiveness of extractions was assessed by the estimation of the total phenols, *ortho*-diphenols, flavonoids and tannins in the extracts. The antioxidant power was evaluated by measuring the radical scavenging capacity, and a molecular analysis by HPLC was carried out in order to identify the compounds contained in the extracts.

2 MATERIALS AND METHODS

2.1 *Chemicals*

Folin-Ciocalteu reagent and Na_2CO_3 for phenolic compounds determination, HCl, $NaNO_2$ and Na_2MoO_4 for *ortho*-diphenols determination, $AlCl_3 \cdot 6H_2O$ for flavonoids and cinchonine hemisulfate for tannins determination, 2,2′-diphenyl-1-picrylhydrazyl free radical (DPPH·), catechin, caffeic acid, ascorbic acid (AA), quercetin (Q), and butylhydroxytoluene (BHT) were purchased from Sigma-Aldrich Co. (Milano, Italy). Methanol, acetic acid and acetonitrile for HPLC analyses were from Carlo Erba Reagents (Milano, Italy).

2.2 *Extraction from chestnut shells*

Burned chestnut shells (CS) from "Brulage" peeling process were kindly provided by a food factory located in Montoro Inferiore (Avellino, Italy). CS were dried in oven at 55°C until reaching constant weight, and powdered using a food homogenizer (type 8557-54, Tefal, France). Using water as solvent, the following extractions procedures were carried out: 1) boiling water (CLE) under continuous and vigorous stirring; 2) autoclave (121°C) (PLE); 3) Microwave-Assisted Extraction (MAE) (frequency 2450 MHz, power 1000 W, input 1080 W – output 700 W); 4) Ultrasound-Assisted Extraction (UAE) in ultrasonic bath at 40, 50 and 60°C (frequency 28-34 kHz, power 80-180 W, 230 V) (Ultrasonic Falc, Treviglio, Bergamo). All the extractions were carried out for 20, 40, and 60 min, using 2.5, 5, and 10% (w/v) CS. After extraction, the obtained suspension was cooled on ice, centrifuged at 3,220 *g* for 1 h at 4°C (Eppendorf 5810R), and the supernatant was recovered. The solid residue was rinsed with a volume of water equivalent to that lost during the extraction, and the resulting suspension was centrifuged as described above. Both supernatants were combined in order to restore the original water volume.

2.3 *Characterization of CS extracts and antioxidant activity*

Total phenolic (TPC), *ortho*-diphenolic (*o*DPC), flavonoid (FC), and tannin (TC) contents were determined as described by Squillaci et al. (2018). Analysis of the phenolic compounds was performed by HPLC (250 × 4.6 mm, 5.0 μm, RP Luna C18 (2) column, Phenomenex Inc. Castelmaggiore, Italy), as reported in Squillaci et al. (2018). The following standards were used for identification and quantification of the compounds: gallic acid, protocatechuic acid, chlorogenic acid, epicatechin, syringic acid, *p*-coumaric acid, ellagic acid, ferulic acid, scopoletin and sinapic acid (Sigma-Aldrich Co. (Milano, Italy).

Antioxidant activity was estimated by measuring the radical scavenging activity (RSA) through the DPPH assay, as reported in detail by Squillaci et al. (2018). BHT, Q, and AA were used as antioxidant reference compounds. All tests here described were performed in triplicate and results were expressed as mean ± Standard Deviation (SD).

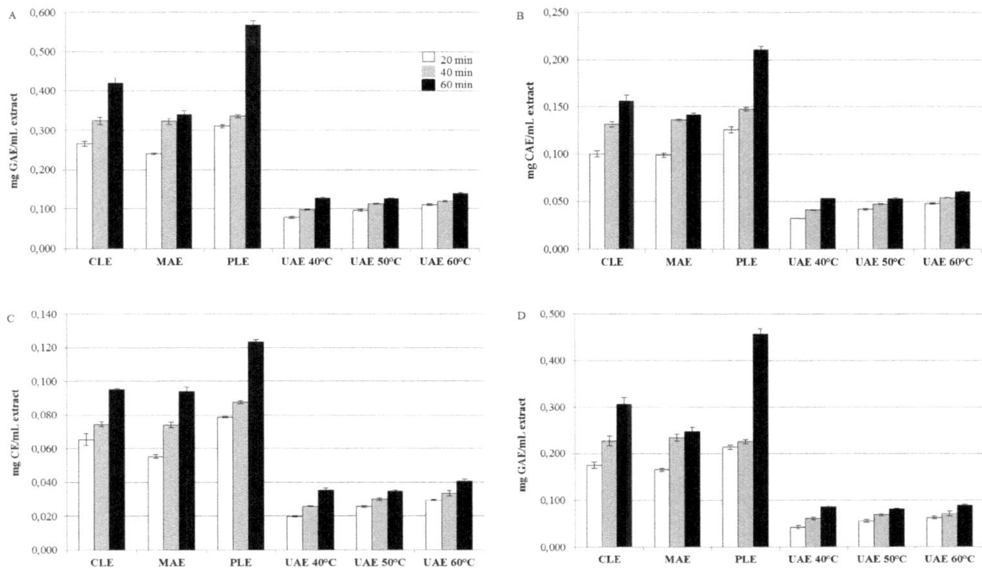

Figure 1. Total phenols (A), *ortho*-diphenols (B), flavonoids (C) and tannins (D) in extracts obtained from 2.5% (w/v) CS by different methods: CLE, Conventional Liquid Extraction; MAE, Microwave-Assisted Extraction; PLE, Pressurized Liquid Extraction; UAE, Ultrasound-Assisted Extraction. GAE, Gallic Acid Equivalents; CAE, Caffeic Acid Equivalents; CE, Catechin Equivalents. Values are mean ± SD ($n = 3$).

3 RESULTS AND DISCUSSION

3.1 *Characterization of CS extracts*

Aqueous extracts were obtained from burned CS using multiple extraction procedures at different operational conditions. The efficiency of the extraction methods was evaluated by measuring the amount of TPC, *o*DPC, FC and TC in the extracts, and the results are shown in Figures 1–3 according to the percentage concentration of CS used. The TPC ranged from 0.078± 0.003 (UAE 40°C, 2.5% CS, 20 min, Fig. 1A) to 0.836 ± 0.018 mg GAE/mL extract (PLE, 10% CS, 60 min, Fig. 3A) and tannins were the main class of phenolic compounds present in the extracts, in agreement with previous results showing that CS are rich in tannins (Aires et al. 2016). TC was comprised between 0.043 ± 0.003 (UAE 40°C, 2.5% CS, 20 min, Fig. 1D) and 0.545 ± 0.018 mg GAE/mL extract (PLE, 10% CS, 60 min, Fig. 3D), while the lowest and the highest tannin percentage values, with respect to the TPC, were 50.32% and 80.32% in extracts from UAE 60°C, 5% CS, 20 min and PLE, 2.5% CS, 60 min, respectively. The *orhto*-diphenols are not usually estimated in chestnut extracts, but their concentration was measured because such type of molecules have strong antioxidant power due to the presence of hydroxyl groups in *ortho* position that increase their antioxidant activity through the stabilization of the phenoxyl radical (Natella et al. 1999). The values were included between 0.032 (UAE 40°C, 2.5% CS, 20 min, Fig. 1B) and 0.297 ± 0.003 mg CAE/mL extract (PLE, 10% CS, 60 min, Fig. 3B). FC ranged from 0.020±0.001 (UAE 40°C, 2.5% CS, 20 min, Fig. 1C) to 0.173 ± 0.002 mg CE/mL extract (PLE, 10% CS, 60 min, Fig. 3C).

All the extractions were performed with the same solvent to ensure that any differences found among the tested methods were independent of the solvent used. Water was used as solvent for the following reasons: it is non-toxic and has low cost and low environmental impact. Furthermore, its safety makes it suitable for use in the health care sector (Petigny et al. 2015).

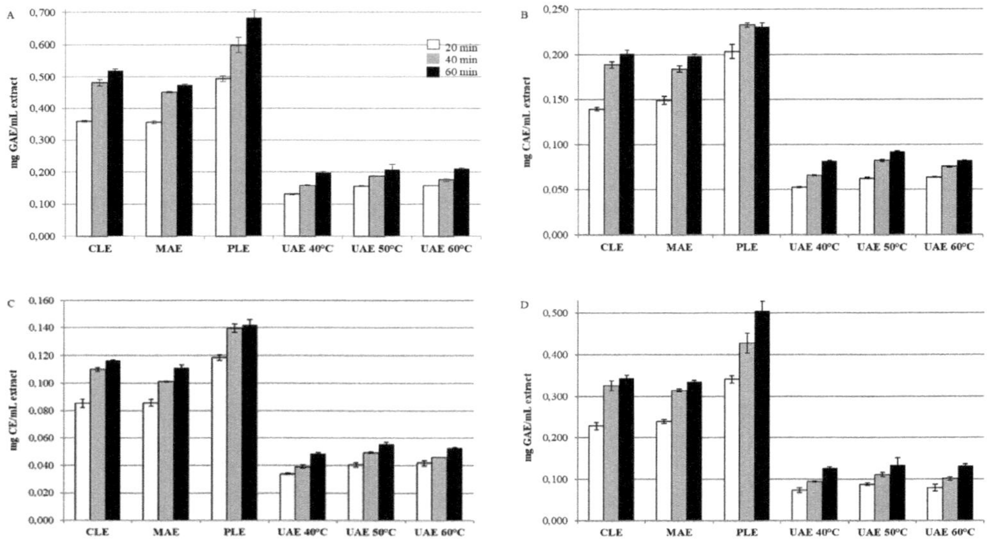

Figure 2. Total phenols (A), *ortho*-diphenols (B), flavonoids (C) and tannins (D) in extracts obtained from 5% (w/v) CS by different methods. For abbreviations, see the caption of Figure 1. Values are mean ± SD ($n = 3$).

As general result, TPC, *o*DPC, FC, and TC followed the trend PLE>CLE≥MAE>UAE 60°C>UAE 50°C>UAE 40°C for extractions in 2.5 and 5% CS, and PLE>MAE>CLE>UAE 60°C>UAE 50°C>UAE 40°C for extraction in 10% CS.

Since MAE is regarded as a technology with numerous advantages (ease of handle, fast heating of the solvents and short residence time), it was decided to further investigate the efficiency of this technique in the recovery of the antioxidant compounds. The extraction yield obtained by MAE was comparable to that of CLE, with a slight improvement at 10% CS.

Temperature is a significant parameter for obtaining high quantities of phenolic compounds since high extraction temperatures are usually related to high extraction yields. Our results are in agreement with those reported by Baiano et al. (2014) because the extraction yield of the different classes of phenolic compounds also increased with increasing temperature. Under our experimental conditions, PLE was the best extraction method in terms of phenolic extraction yield. The positive effect of the temperature was clearly observed for the UAE method as well; in fact, the phenolic yield of the extracts increased when UAE operating temperature was increased from 40 to 60°C. The extraction time also plays an important role within the same extraction procedure, because the phenolic content was enhanced when residence time increased from 20 to 60 minutes.

Generally, the choice of the best extraction method is not easy since many factors contribute to the result. As an example, UAE from *Salvia officinalis* L. leaves gave a phenolic yield higher than CLE in water and 30% ethanol, while the result was opposite in acetone (Dent et al. 2015).

The identification of the molecules contained in the extracts was performed by RP-HPLC through comparison of their retention times with corresponding standards. Samples from treatment of 10% CS for 60 min were chosen for the analysis. As shown in Table 1, no qualitative differences were detected in the extracts. Gallic acid was the main simple phenol with amounts varying from 159.66 ± 0.31 (UAE 40°C) to 284.14 ± 0.72 μg/mL extract (PLE). Protocatechuic acid was the second most abundant compound in the extracts; the highest and the lowest amounts were detected in the extracts from PLE (39.95 ± 0.13 μg/mL extract) and UAE 50°C (23.16±0.11 μg/mL extract), respectively. Minor phenolic compounds were also detected. Among them, epicatechin and chlorogenic acid were the most representatives. Ellagic acid was detected in the extracts prepared by ultrasound method in lower amount compared to CLE, PLE, and MAE.

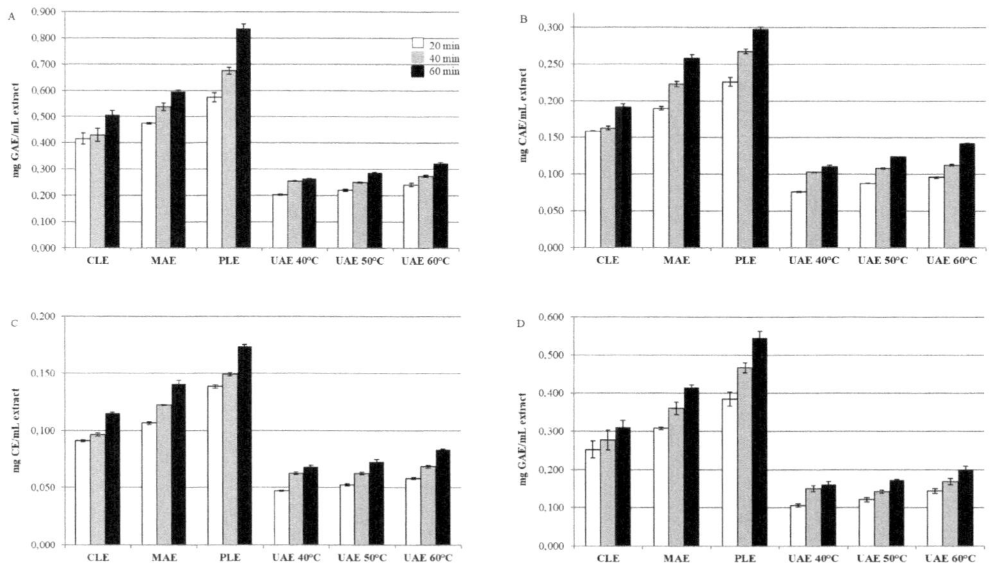

Figure 3. Total phenols (A), *ortho*-diphenols (B), flavonoids (C) and tannins (D) in extracts obtained from 10% (w/v) CS by different methods. For abbreviations, see the caption of Figure 1. Values are mean ± SD ($n = 3$).

Table 1. Phenolic compounds identified in CS extracts by RP-HPLC*.

	PLE (μg/mL)	CLE (μg/mL)	MAE (μg/mL)	UAE 40°C (μg/mL)	UAE 50°C (μg/mL)	UAE 60°C (μg/mL)
GA	284.14 ± 0.72	226.12 ± 0.30	224.58 ± 0.45	159.66 ± 0.31	166.54 ± 0.28	166.60 ± 0.34
PCA	39.95 ± 0.13	30.34 ± 0.15	32.10 ± 0.02	23.39 ± 0.19	23.16 ± 0.11	23.94 ± 0.13
CGA	2.15 ± 0.01	1.82 ± 0.01	2.26 ± 0.05	1.40 ± 0.01	0.98 ± 0.07	0.88 ± 0.06
EP	2.55 ± 0.02	2.86 ± 0.02	2.44 ± 0.07	2.58 ± 0.03	2.60 ± 0.10	0.88 ± 0.04
SyA	0.44 ± 0.02	0.52 ± 0.01	0.40 ± 0.02	0.20 ± 0.01	0.16 ± 0.01	0.16 ± 0.02
EA	3.26 ± 0.02	1.8 ± 0.02	2.08 ± 0.06	0.71 ± 0.02	0.77 ± 0.01	0.65 ± 0.03
p-CA	0.95 ± 0.04	0.77 ± 0.01	0.82 ± 0.02	0.53 ± 0.01	0.60 ± 0.02	0.58 ± 0.01
SA	0.50 ± 0.02	0.41 ± 0.02	0.52 ± 0.01	0.40 ± 0.03	0.36 ± 0.01	0.53 ± 0.02
FA	0.16 ± 0.01	0.08 ± 0.01	0.17 ± 0.01	0.22 ± 0.02	0.14 ± 0.01	0.34 ± 0.01
S	0.32 ± 0.02	0.26 ± 0.01	0.36 ± 0.01	0.32 ± 0.02	0.26 ± 0.02	0.46 ± 0.02

*Extracts from 10% CS and 60 min. GA, gallic acid; PCA, protocatechuic acid; CGA, chlorogenic acid; EP, epicatechin; SyA, syringic acid; EA, ellagic acid; *p*-CA, *para*-coumaric acid; SA, sinapic acid; FA, ferulic acid; S, scopoletin.

All the identified compounds are known to possess antioxidant power, but their occurrence and amount were greatly variable among the chestnut shell extracts described up to now. As already stated, this may depend on several factors, but gallic acid was usually present and was among the most abundant phenolic compounds identified in aqueous extracts (Aires et al. 2016).

3.2 *Antioxidant activity*

The antioxidant power of the chestnut extracts was investigated as a function of their radical scavenging activity (RSA) through the evaluation of the decreased absorbance of the free radical DPPH·. As the antioxidant power is mainly ascribed to the presence of the hydroxyl groups on the

Table 2. CS extracts selected for measurement of the antioxidant activity.

Extraction method	CS (% w/v)	Extraction time (min)	TPC (mg/mL)	*o*DPC/TPC (%)
CLE	5	60	0.519 ± 0.004	38.73
PLE	10	60	0.836 ± 0.018	35.53
MAE	10	40	0.536 ± 0.015	41.60
UAE 40°C	10	60	0.262 ± 0.002	41.98
UAE 50°C	10	60	0.286 ± 0.001	43.36
UAE 60°C	10	60	0.322 ± 0.004	43.79

TPC, total phenolic content; *o*DPC, *ortho*-diphenolic content. For abbreviations of the extraction methods, see the caption of Figure 1. Values are mean ± SD ($n = 3$).

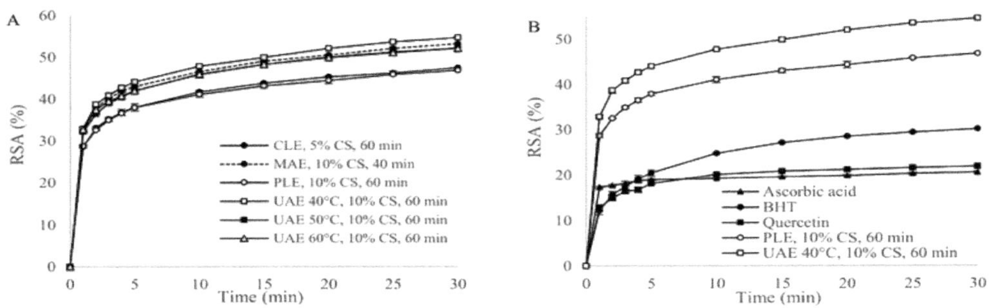

Figure 4. Free radical scavenging activity of CS extracts (A) and comparison with antioxidant reference compounds (B). DPPH assay was performed with 2 μg GAE of extracts or pure compounds. Values are mean ±SD ($n = 3$).

phenolic molecules (Jung et al. 2017), were selected for this analysis the extracts with the highest total phenolic content obtained with each extraction procedure (Table 2).

As shown in Figure 4A, the kinetic profile of RSA was similar in all the extracts. The scavenging activity was fast in the first minutes, slowing down as the time of the assay increased. After 1 minute of assay, the RSA (%) was comprised between 28.5% (CLE) and 32.9% (UAE 40°C), reaching values ranging from 47.3% (CLE) to 54.5% (UAE 40°C) after 30 minutes. It is noteworthy that two extracts containing a high quantity of phenolic compounds (CLE and PLE) exhibited the lowest antioxidant activity after 30 minutes (47.3% and 46.7%, respectively). This result, initially unexpected, may be related with their low *o*DPC/TPC percentage ratio (Table 2), thus confirming that the *ortho*-diphenolic fraction plays a key role in the attribution of the antioxidant power of a blend of phytochemicals.

In Figure 4A are clearly observable two groups, each of them formed by extracts with similar *o*DPC/TPC percentage ratio and radical scavenging activity. The group with higher antioxidant power corresponds to the group with higher *o*DPC/TPC ratio.

The extracts prepared by UAE exhibited a better antioxidant power in comparison to the extracts prepared at higher temperature. If the use of high extraction temperatures can favour the release of phenolic molecules, it can also modify the chemical structure of thermolabile phenols, reducing their chemical properties (Dai & Mumper, 2010). This could explain why methods like CLE and PLE, that provided a high TPC, showed lower antioxidant activity. Furthermore, it could be possible that MAE did not alter the chemical features of the natural compounds as the extract exhibited a RSA close to those of ultrasound extractions.

The antioxidant activity of the extracts was compared with those of known antioxidants, chosen as reference compounds. In Figure 4B, the extracts with the highest and the lowest RSA were compared with the following antioxidants used in several applications: Q, AA, and BHT. It is noteworthy that even the extract showing the lowest RSA among all those tested (PLE), had an antioxidant power superior to that of the reference compounds. Because of the same amount of GAE (2 μg) was used for both extracts and standards, the stronger antioxidant power of the extracts is to ascribe to the presence of different compounds acting in a synergic mode.

This finding supports the actual tendency of consumers that are even more interested in natural preparation for achieving health benefits.

4 CONCLUSIONS

The investigations on different extraction methods demonstrated that processes performed at high temperature provided extracts with highest phenolic, flavonoid, *ortho*-diphenolic and tannin contents (CLE, MAE and PLE). On this base, it was possible to conclude that temperature is an important operational parameter to take into account. The increase of the extraction time also gave an improvement of the extraction yields. At the same time, the analysis of the antioxidant power showed that the extracts prepared at lower temperature were endowed with a stronger radical scavenging activity. These findings indicate that the most suitable extraction technique to apply depends on the objective requested. As an example, if the goal is to obtain tannins for adhesives or tanning purposes, an extraction technique, which offers high yields, is preferable. Alternatively, if a blend of antioxidants is the target, techniques operating at lower temperature, such as UAE, should be chosen.

REFERENCES

Aires, A., Carvalho, R. & Saavedra, M.J. 2016. Valorization of solid wastes from chestnut industry processing: extraction and optimization of polyphenols, tannins and ellagitannins and its potential for adhesives, cosmetic and pharmaceutical industry. *Waste Management* 48: 457–464.

Baiano, A., Bevilacqua, L., Terracone, C., Conto, F. & Del Nobile, M.A. 2014. Single and interactive effects of process variables on microwave-assisted and conventional extractions of antioxidants from vegetable solid wastes. *Journal of Food Engineering* 120: 135–145.

Dai, J. & Mumper, R.J. 2010. Plant phenolics: extraction, analysis and their antioxidant and anticancer properties. *Molecules* 15(10): 7313–7352.

Dent, M., Dragović-Uzelac, V., Elez Garofulić, I., Bosiljkov, T., Ježek, D. & Brnčić, M. 2015. Comparison of conventional and ultrasound-assisted extraction techniques on mass fraction of phenolic compounds from sage (*Salvia officinalis* L.)". *Chemical and Biochemical Engineering Quarterly* 29(3): 475–484.

FAOSTAT, Food and Agriculture Organization of the United States: http://www.fao.org/faostat/en/#data accessed on 25 February 2019.

Jung, K., Everson, R.J., Joshi, B., Bulsara, P.A., Upasani, R. & Clarke, M.J. 2017. Structure-function relationship of phenolic antioxidants in topical skin health products. *International Journal of Cosmetic Science* 39(2): 217–223.

Muñiz-Márquez, D.B., Martínez-Ávila, G.C., Wong-Paz, J.E., Belmares-Cerda, R., Rodríguez-Herrera, R. & Aguilar, C.N. 2013. Ultrasound-assisted extraction of phenolic compounds from Laurus nobilis L. and their antioxidant activity. *Ultrasonics Sonochemistry* 20(5): 1149–1154.

Natella, F., Nardini, M., Di Felice, M. & Scaccini, C. 1999. Benzoic and cinnamic acid derivatives as antioxidants: structure-activity relation. *Journal of Agricultural and Food Chemistry.* 47(4): 1453–1459.

Petigny, L., Özel, M.Z., Périno, S., Wajsman, J. & Chemat, F. 2015. Water as green solvent for extraction of natural products. In F. Chemat & J. Strube (eds.), Green Extraction of Natural Products: 237-263. Wiley-VCH Verlag GmbH & Co.

Squillaci, G., Apone, F., Sena, L.M., Carola, A., Tito, A., Bimonte, M., De Lucia, A., Colucci, G., La Cara, F. & Morana, A. 2018. Chestnut (*Castanea sativa* Mill.) industrial wastes as a valued bioresource for the production of active ingredients. *Process Biochemistry* 64: 228-236.

Thorat, I.D., Jagtap, D.D., Mohapatra, D., Joshi, D.C., Sutar R.F. & Kapdi, S.S. 2013. Antioxidants, their properties, uses in food products and their legal implications. *International Journal of Food Studies* 2(1): 81-104.

Valdés, A., Vidal, L., Beltrán, A., Canals, A. & Garrigós, M.C. 2015. Microwave-Assisted Extraction of Phenolic Compounds from Almond Skin Byproducts (*Prunus amygdalus*): A Multivariate Analysis Approach. *Journal of Agricultural and Food Chemistry.* 63(22): 5395–5402.

Wastes: Solutions, Treatments and Opportunities III – Vilarinho et al. (Eds)
© 2020 Taylor & Francis Group, London, ISBN 978-0-367-25777-4

Effect of temperature and time on the phenolic extraction from grape canes

G. Squillaci, L.A. Giorio, N.A. Cacciola, F. La Cara & A. Morana
Research Institute on Terrestrial Ecosystems, National Research Council of Italy, Naples, Italy

ABSTRACT: Grape canes from *Vitis vinifera* L. cultivars of Campania region (Italy): "Aglianico", "Fiano", and "Greco" were considered in order to valorise them through the extraction of phenolic compounds provided with antioxidant activity. Since many operational parameters influence the extraction process, here different extraction conditions were compared. Grape canes (5% w/v) were extracted in water at different temperatures (25, 50, 75 and 100° C) and extraction times (10, 20, 40 and 60 min). The highest yields, in terms of mg/g dry extract, were obtained by "Greco" at 100°C and 20 min: 110.51±5.46 for total phenols, 63.29±1.35 for *ortho*-diphenols, 77.31 ± 3.59 for tannins, and by "Greco" at 50°C and 40 min for flavonoids: 29.37 ± 0.73. The extracts from "Greco" at different temperatures showed antioxidant activity comparable to that of butylhydroxytoluene and quercetin chosen as reference compounds. HPLC analysis revealed that gallic acid and catechin were the most abundant phenolic compounds detected in the extracts.

1 INTRODUCTION

In the last years, there was a growing interest toward the utilization of agro-industrial wastes for the production of high added value products. Usually, solid residues are burned or used for composting even if they still contain valuable molecules that can be utilizable for many applications (Morana et al. 2017; Caputo et al. 2018).

Although the early research on active molecules was mainly directed towards pure compounds, there is an alternative tendency focusing on mixtures of natural molecules. Moreover, the increasing demand for biologically active molecules has encouraged the research of new sources of natural compounds. In this perspective, the large amount of agro-industrial residues generated every year represents a valuable resource for the production of bioactive compounds.

Grape (*Vitis vinifera* L.) belongs to the family of *Vitaceae*, and owing to the benefits on human health and the economic importance, it is one of the fruit crops broadly grown in many areas of the world. In 2017, about 23,735,725 tonnes of grape were produced in the European Union, with Italy being the leader producer (7,169,745 tonnes) (FAOSTAT 2019). In Italy, the grape cultivation represents one of the main agricultural activities, with 762,072 hectares used for this purpose (ISTAT 2019), and the production chain generates large quantities of by-products whose management creates disposal problems. However, these residues (skin, cane, stalk, pulp and seed) are rich in valued bioactive compounds, as phenolic acids, flavanols, flavonols, anthocyanins, and stilbenoids (Obreque-Slier et al. 2010). In the Campania region, 29,330 hectares are dedicated to the grape cultivation, and "Aglianico", "Fiano" and "Greco", typical cultivars of this territory, produce high quality red (Aglianico) and white (Fiano and Greco) wines. Grape canes represent the main solid waste from vineyard, and it is estimated that about 1.5–2.5 tonnes/hectare are annually produced during the vine pruning process. According to this, about 43,995–73,325 tonnes/year of grape canes are produced in Campania that could be exploited as a cheap source for the extraction of bioactive molecules. These compounds, thanks to their strong antioxidant power, are of great interest because can be employed in food, cosmetic and pharmaceutical industries. Furthermore, they possess anti-microbial, anti-inflammatory and anti-cancer activities (Jang et al. 1997, Teixeira et al. 2014).

The aim of this study was to valorise the grape canes from "Aglianico", "Fiano" and "Greco" cultivars through the extraction of phenolic compounds by a low cost and eco-friendly method. Because the operative conditions of an extraction process (temperature, solvent, liquid/solid ratio, and extraction time), and the particle size can influence the result, we investigated the effect of the temperature and the extraction time on the phenolic compounds yield. The efficiency of the procedure was evaluated by estimating the yields of the total phenols, *ortho*-diphenols, flavonoids and tannins. The antioxidant power was evaluated by the radical scavenging activity assay and a molecular analysis by HPLC was performed in order to identify the compounds present in the extracts.

2 MATERIALS AND METHODS

2.1 *Chemicals*

Folin-Ciocalteu reagent and Na_2CO_3 for phenolic compounds quantification, HCl, $NaNO_2$ and Na_2MoO_4 for *ortho*-diphenols assay, $AlCl_3 \cdot 6H_2O$ for flavonoids and cinchonine hemisulfate for tannins evaluation, 2,2′-diphenyl-1-picrylhydrazyl free radical (DPPH·), catechin, caffeic acid, quercetin (Q), butylhydroxytoluene (BHT), HPLC standards, and NaOH were purchased from Sigma-Aldrich Co. (Milano, Italy). Methanol and solvents for HPLC analyses were from Carlo Erba Reagents (Milano, Italy).

2.2 *Extraction of phenolic compounds*

A local company kindly provided the grape canes. They were collected in February 2018 and came from the following grape cultivars: "Aglianico", "Fiano", and "Greco". The material was cut into pieces of 0.5-1.0 cm, and dried in oven at 55 °C until reaching constant weight. Then, the samples were milled with a MF10 IKA mill (Werke GmbH & Co. KG), and sieved to screen particles until 500μm in size. The extraction was performed as follows: 5% (w/v) grape canes were suspended in 50 mL of deionized water and incubated under continuous stirring at 50°C for 10, 20, 40 and 60 min. Alternatively, the suspension was incubated at 25, 50, 75 and 100°C for 20 min. After cooling on ice and centrifuging at 18,000 rpm for 1 h at 4°C (Sorvall RC6 plus), the supernatant (extract) was lyophilized in an Edwards Modulyo freeze-dryer (Edwards, Cinisello Balsamo, Milano, Italy) and the dry extract (DE) was stored at 4°C.

2.3 *Characterization of grape cane extracts and antioxidant activity*

For spectrophotometric assays and HPLC analysis, 10 mg of DE were dissolved in 1 mL of PBS. Total phenolic (TPC), *ortho*-diphenolic (*o*DPC), flavonoid (FC), and tannin (TC) contents were determined as described by Squillaci et al. (2018) and expressed as mg GAE/g DE. HPLC analysis of the phenolic compounds was performed using a reversed-phase Luna C18 (2) column (250 × 4.6 mm, 5.0 μm, Phenomenex Inc. Castelmaggiore, Italy) as reported in Squillaci et al. (2018). Results were expressed as mg of phenolic compound/g DE.

Antioxidant activity was estimated by measuring the radical scavenging activity (RSA) of 5 μg of extract or standard through the DPPH assay, as reported in detail by Squillaci et al. (2018). BHT and Q were used as antioxidant reference compounds.

2.4 *Statistical analysis*

All tests were performed in triplicate and expressed as mean ± Standard Deviation (SD) calculated by Microsoft Excel 2013. Statistical analysis was carried out by GraphPad Prism (version 5).

Significant differences were determined by two-way analysis of variance (ANOVA) with Bonferroni post-tests. Mean values were considered not significantly different at $p \geq 0.05$.

3 RESULTS AND DISCUSSION

3.1 *Characterization of grape cane extracts*

In the present work, aqueous extracts from grape canes of different cultivars typical of the Campania territory were obtained by different temperatures and extraction times. The efficiency of the extraction methods was evaluated by measuring the amount of TPC, *o*DPC, FC and TC in the extracts, and the results are shown in Figure 1 (variable: temperature) and Figure 2 (variable: time). The amount of the phenolic classes analysed was different not only in relation to the different conditions used, but also in function of the cultivars investigated. The extracts from "Fiano" and "Greco" contained the lowest and the highest amount of compounds, respectively at any condition of temperature and time.

The TPC ranged from 45.59 ± 1.83 (Fiano 25°C, Fig. 1A) to 110.51±5.46 mg GAE/g DE (Greco 100°C, Fig. 1A), the FC varied from 7.07±0.46 (Fiano 25°C, Fig. 1C) to 29.37 ± 0.73 mg GAE/g DE (Greco 40 min, Fig. 2C), whereas the TC was comprised between 20.62 ± 1.36 (Fiano 25°C, Fig. 1D) and 77.31 ± 3.59 mg GAE/g DE (Greco 100°C, Fig. 1D). It is not frequent to determine the amount of *ortho*-diphenols in natural extracts but, as they represent a class of molecules provided with high antioxidant power, we decided to estimate their quantity as well. The *o*DPC showed values comprised between 18.63 ± 0.50 (Fiano 25°C, Fig. 1B) and 63.29 ± 1.35 mg GAE/g DE (Greco 100°C, Fig. 1B). As displayed in Figure 1A, extraction temperatures of 25 to 50 °C did not affect the total phenol content significantly ($p \geq 0.05$). A temperature increase to 75 and 100°C led to a significant increase, particularly in the cultivar "Greco" with a TPC, measured at 100°C, 1.56-folds higher than 25°C. A similar result was observed for the TC (Fig. 1D): this fraction was significantly affected by the extraction temperature from 25 to 100°C in cultivars "Aglianico"

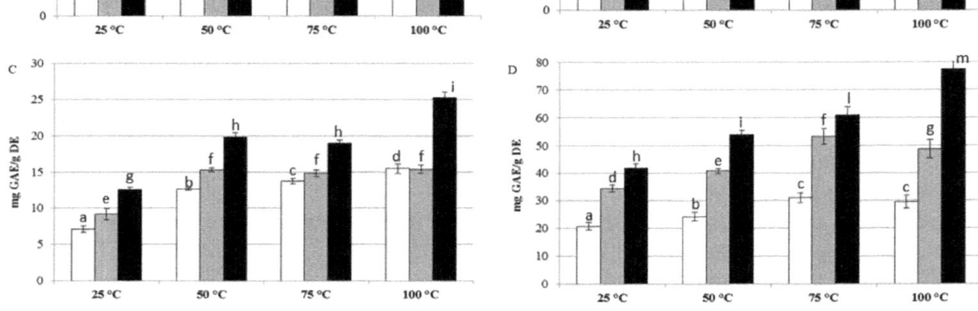

Figure 1. Total phenols (A), *ortho*-diphenols (B), flavonoids (C) and tannins (D) in grape cane extracts from *Vitis vinifera* cultivars "Aglianico, "Fiano" and "Greco" (5% w/v) obtained at 20 min and different temperatures. GAE, Gallic Acid Equivalents. Values are mean ± SD ($n = 3$). Bars with the same letters are not significantly different at $p \geq 0.05$(A: a-b, c-e, f-h; B: a-c, d-e, f-i; C: a-d, e-f, g-i; D: a-c, d-g, h-m).

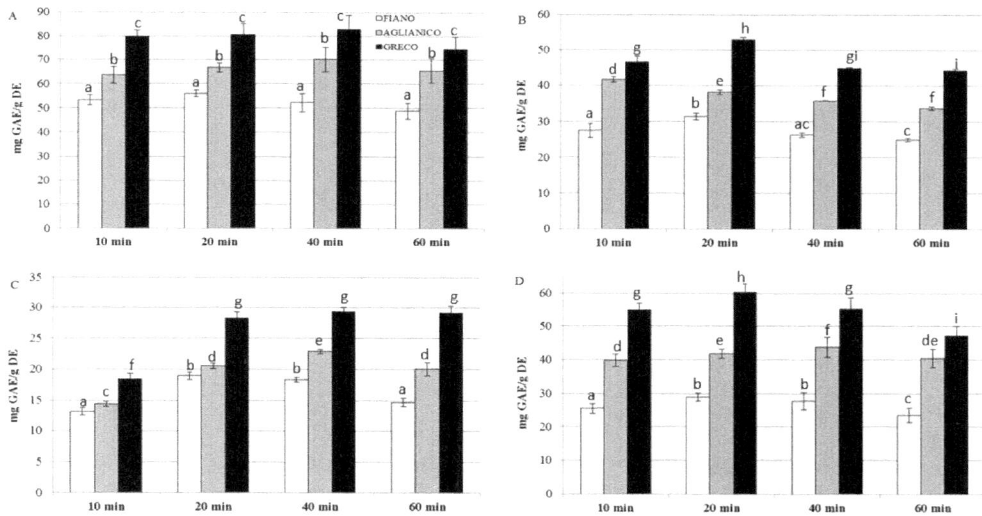

Figure 2. Total phenols (A), *ortho*-diphenols (B), flavonoids (C) and tannins (D) in grape cane extracts from *Vitis vinifera* cultivars "Aglianico, "Fiano" and "Greco" (5% w/v) obtained at 50°C and different extraction times. GAE, Gallic Acid Equivalents. Values are mean ± SD ($n = 3$). Bars with the same letters are not significantly different at $p \geq 0.05$(A: a, b,c; B: a-c, d-f, g-i; C: a-b, c-e, f-g; D: a-c, d-f, g-i).

and "Greco" ($p<0.001$). The tannin content increased 1.84-folds from 25 to 100°C in the "Greco" extract, while the increase in "Aglianico" extract was 1.42-folds.

It is widespread opinion that phenols degradation occurs when the temperature increases. This is not always true as many factors can concur to this effect. Our results show that all the phenolic classes examined rise with the temperature, in agreement with findings of other authors. Marete et al. (2009) registered a phenolic increase in a watery extract from *Tanacetum parthenium* from 20 to 100°C. The phenolic content almost doubled from 17 to 31 mg GAE/g dry matter.

Variation of the extraction time was considered in the present work, as it is as an important aspect for saving energy and lowering process costs. In our tests, the extraction time was a variable not significantly affecting the total phenol content ($p \geq 0.05$) (Fig. 2A). Significant increase of *o*DPC was measured at 10 and 20 min. "Fiano" and "Greco" extracts reached the highest yield after 20 min (31.34 ± 0.92 and 53.06 ± 0.54 mg GAE/g DE, respectively) ($p<0.001$), while "Aglianico" after 10 min (41.63±0.75 mg GAE/g DE) (Fig. 2B). Even for FC and TC, the trend of molecules release was similar, attesting that a prolonged extraction time did not allow an improvement of the yields, but led to their decrease at 60 min. Our findings could be explained by Fick's second law of diffusion: the equilibrium between solute concentration in solid and liquid phases is reached after some time, and any further increase of time has not effects. Mokrani & Madani (2016) described a not significant increase of TPC in extracts from peach fruit from 30 to 180 min, and a decrease from 180 to 270 min.

The identification of the phenolic molecules in the extracts was performed by RP-HPLC. Gallic acid, protocatechuic acid and catechin were the majority phenols measured in all extracts. "Fiano" grape cane extracts contained the lowest amount of gallic acid both at different temperatures (Table 1) and extraction times (Table 2) ranging from 1.53 to 2.16 mg/g DE, while this phenol varied from 2.24 to 3.30 mg/g DE in the other extracts.

Minor phenols as ellagic acid,*p*-coumaric acid and ferulic acid were more represented in the extracts prepared at 50°C and at different times (Table 2). In extracts prepared at different temperatures, they were absent at 75°C, whereas ferulic and ellagic acids were not observed at 100°C as

Table 1. Phenolic compounds identified in grape cane extracts prepared at different temperatures.

	GA	PCA	C	EA	*p*-CA (mg/g DE)	FA	R
25°C							
A	2.72 ± 0.01	0.16 ± 0.01	1.57 ± 0.01	nd	0.19 ± 0.00	0.10 ± 0.00	0.26 ± 0.01
F	1.63 ± 0.02	0.26 ± 0.00	1.01 ± 0.02	nd	0.09 ± 0.01	0.02 ± 0.00	0.06 ± 0.01
G	2.60 ± 0.02	0.16 ± 0.00	0.70 ± 0.02	nd	0.12 ± 0.00	nd	0.19 ± 0.01
50°C							
A	2.85 ± 0.03	0.62 ± 0.01	0.86 ± 0.00	0.01 ± 0.00	0.09 ± 0.01	0.07 ± 0.02	0.27 ± 0.00
F	1.74 ± 0.00	0.59 ± 0.02	1.83 ± 0.05	0.01 ± 0.00	0.04 ± 0.01	0.09 ± 0.02	0.10 ± 0.02
G	2.67 ± 0.00	0.90 ± 0.04	3.85 ± 0.20	0.05 ± 0.00	0.06 ± 0.00	0.05 ± 0.00	0.20 ± 0.01
75°C							
A	2.78 ± 0.04	1.17 ± 0.02	0.51 ± 0.01	nd	nd	nd	0.50 ± 0.01
F	1.76 ± 0.03	0.80 ± 0.01	0.40 ± 0.01	nd	nd	nd	nd
G	2.63 ± 0.05	0.95 ± 0.03	nd	nd	nd	nd	0.26 ± 0.01
100°C							
A	3.30 ± 0.07	0.72 ± 0.03	0.75 ± 0.05	0.02 ± 0.00	0.05 ± 0.02	nd	0.60 ± 0.06
F	2.16 ± 0.04	0.70 ± 0.01	0.56 ± 0.01	nd	0.06 ± 0.00	nd	0.19 ± 0.00
G	3.01 ± 0.03	1.08 ± 0.02	1.26 ± 0.03	nd	0.07 ± 0.00	nd	1.50 ± 0.03

A, gallic acid; PCA, protocatechuic acid; C, catechin; EA, ellagic acid; *p*-CA, *para*-coumaric acid; FA, ferulic acid; R, resveratrol. A, Aglianico; F, Fiano; G, Greco; nd, not detected.

Table 2. Phenolic compounds identified in grape cane extracts prepared at different extraction times.

	GA	PCA	C	EA	*p*-CA (mg/g DE)	FA	R
10 min							
A	2.73 ± 0.14	0.53 ± 0.01	2.49 ± 0.12	0.06 ± 0.00	0.08 ± 0.01	0.13 ± 0.01	0.44 ± 0.14
F	1.55 ± 0.06	0.54 ± 0.03	1.77 ± 0.02	0.02 ± 0.00	0.04 ± 0.00	nd	0.08 ± 0.00
G	2.43 ± 0.17	0.91 ± 0.05	1.74 ± 0.10	nd	0.05 ± 0.01	nd	0.22 ± 0.02
20 min							
A	2.85 ± 0.03	0.62 ± 0.01	0.86 ± 0.00	0.01 ± 0.00	0.09 ± 0.01	0.07 ± 0.02	0.27 ± 0.00
F	1.74 ± 0.00	0.59 ± 0.02	1.83 ± 0.05	0.01 ± 0.00	0.04 ± 0.01	0.09 ± 0.02	0.10 ± 0.02
G	2.67 ± 0.00	0.90 ± 0.04	3.85 ± 0.20	0.05 ± 0.00	0.06 ± 0.00	0.05 ± 0.00	0.20 ± 0.01
40 min							
A	2.70 ± 0.22	0.66 ± 0.05	0.99 ± 0.18	0.02 ± 0.00	0.12 ± 0.01	0.15 ± 0.04	0.28 ± 0.00
F	1.53 ± 0.19	0.55 ± 0.00	0.27 ± 0.04	0.01 ± 0.00	0.05 ± 0.00	0.06 ± 0.00	0.06 ± 0.00
G	2.24 ± 0.00	0.82 ± 0.00	nd	0.05 ± 0.00	0.07 ± 0.00	0.06 ± 0.01	0.16 ± 0.00
60 min							
A	3.28 ± 0.04	0.58 ± 0.01	2.17 ± 0.04	0.01 ± 0.00	0.09 ± 0.00	0.04 ± 0.01	0.32 ± 0.00
F	1.71 ± 0.21	0.62 ± 0.00	1.25 ± 0.08	0.01 ± 0.00	0.03 ± 0.00	0.11 ± 0.01	0.06 ± 0.01
G	3.01 ± 0.02	1.04 ± 0.04	4.19 ± 0.08	0.05 ± 0.00	0.07 ± 0.00	0.10 ± 0.00	0.18 ± 0.00

GA, gallic acid; PCA, protocatechuic acid; C, catechin; EA, ellagic acid; *p*-CA, *para*-coumaric acid; FA, ferulic acid; R, resveratrol. A, Aglianico; F, Fiano; G, Greco; nd, not detected.

well with the exception of 0.02 mg/g DE of ellagic acid measured in "Aglianico" (Table 1). Resveratrol was detected in variable amounts in the extracts: from 0.06 to 1.50 mg/g DE. This molecule, belonging to the group of stilbenes, has gained significant attention because of its health promoting benefits (Springer & Moco 2019). The highest quantity was found in "Greco" grape cane extract at 100°C. The occurrence and amount of the identified compounds are variable among the extracts from grape canes described up to now, so that a comparison is very difficult. Zhang et al. (2011) identified gallic acid, resveratrol, catechin and protocatechuic acid in methanol acidified grape cane extracts from Chines grape cultivars.

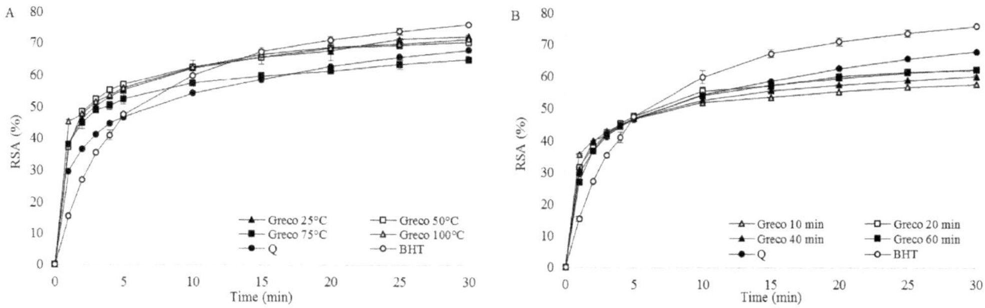

Figure 3. Free radical scavenging activity of "Greco" grape canes extracted at different temperatures for 20 min (A) and different extraction times at 50°C (B), and comparison with antioxidant reference compounds. Values are mean ± SD ($n = 3$).

3.2 *Antioxidant activity*

"Greco" extracts were chosen to evaluate the antioxidant power as a function of their radical scavenging activity (RSA). The kinetic profile of RSA (%) of the extracts prepared at different temperatures is shown in Figure 3A. They exhibited strong antioxidant power as the activity raised rapidly at the beginning of the assay, and after 5 min the RSA (%) was more than 50% for all extracts. After this time it was higher than the RSA (%) of the antioxidant standards BHT and Q (55.26% 25°C; 57.06% 50°C; 52.45% 75°C, 55.97% 100°C; 47.54% BHT, and 46.73% Q) and after 30 min, only the extract prepared at 75°C showed an antioxidant power lower than the standards, whereas the remaining extracts exhibited values comprised between the RSA of BHT and Q. The kinetic profile of RSA (%) of the extracts obtained at different extraction times is shown in Figure 3B. They exhibited an antioxidant activity lower than standards, but however significant as the RSA (%) ranged from 57.67% (10 min) to 62.31% (60 min) after 30 min of assay.

4 CONCLUSIONS

The studies on the extractions from cultivars "Aglianico", "Fiano" and "Greco" grape canes at different temperatures and extraction times indicated that the temperature increase enhanced the phenolic yields, while a prolonged time produced negative effects, as the amount of phenols in the extracts decreased. Among the cultivars investigated, "Greco" provided the highest amount of total phenols, *ortho*-diphenols, flavonoids and tannins. The analysis of the antioxidant power showed that the extracts from "Greco" were endowed with a significant radical scavenging activity, making them interesting candidates for several industrial applications. Studies on the antimicrobial and anticancer properties of the extracts are currently in progress.

REFERENCES

Caputo, L., Quintieri, L., Cavalluzzi M.M., Lentini, G. & Habtemariam, S. 2018. Antimicrobial and Antibiofilm Activities of Citrus Water-Extracts Obtained by Microwave-Assisted and Conventional Methods. *Biomedicines* 6: 70.

FAOSTAT, Food and Agriculture Organization of the United States: http://www.fao.org/faostat/en/#data accessed on 27 February 2019.

Jang, M., Cai, L., Udeani, G.O., Slowing, K.V., Thomas, C.F., Beecher, C.W.W., Fong, H.H.S., Farnsworth, N.R., Kinghorn, A.D., Mehta, R.G., Moon, R.C. & Pezzuto J.M. 1997. Cancer Chemopreventive Activity of Resveratrol, a Natural Product Derived from Grapes. *Science* 275(5297): 218–220.

Marete, J., Jacquier, C. & O'Riordan, D. 2009. Effects of extraction temperature on the phenolic and parthenolide contents, and colour of aqueous feverfew (*Tanacetum parthenium*) extracts. *Food Chemistry* 117(2): 226–231.
Mokrani, A. & Madani, K. 2016. Effect of solvent, time and temperature on the extraction of phenolic compounds and antioxidant capacity of peach (*Prunus persica* L.) fruit. *Separation and Purification Technology* 162: 68–76.
Morana, A., Squillaci, G., Paixão, S.M., Alves, L., La Cara, F. & Moura, P. 2017. Development of an Energy Biorefinery Model for Chestnut (*Castanea sativa* Mill.) Shells. *Energies* 10: 1504.
Obreque-Slier, E., Peña-Neira, A., López-Solís, R., Zamora-Marín, F., Ricardo-da Silva, J.M. & Laureano O. 2010. Comparative Study of the Phenolic Composition of Seeds and Skins from Carménère and Cabernet Sauvignon Grape Varieties (*Vitis vinifera* L.) during Ripening. *Journal of Agricultural and Food Chemistry* 58(6): 3591–3599.
Springer, M. & Moco, S. 2019. Resveratrol and Its Human Metabolites-Effects on Metabolic Health and Obesity. *Nutrients* 1(1)pii: E143.
Squillaci, G., Apone, F., Sena, L.M., Carola, A., Tito, A., Bimonte, M., De Lucia, A., Colucci, G., La Cara, F. & Morana, A. 2018. Chestnut (*Castanea sativa* Mill.) industrial wastes as a valued bioresource for the production of active ingredients. *Process Biochemistry* 64: 228-236.
Teixeira, A., Baenas, N., Dominguez-Perles, R., Barros, A., Rosa, E., Moreno, D.A. & Garcia-Viguera, C. 2014. Natural Bioactive Compounds from Winery By-Products as Health Promoters: A Review. *International Journal of Molecular Sciences* 15(9): 15638–15678.
Zhang, A., Fang, Y., Wang, H., Li, H. & Zhang, Z. 2011. Free-Radical Scavenging Properties and Reducing Power of Grape Cane Extracts from 11 Selected Grape Cultivars Widely Grown in China. *Molecules* 16: 10104–10122.

Wastes: Solutions, Treatments and Opportunities III – Vilarinho et al. (Eds)
© 2020 Taylor & Francis Group, London, ISBN 978-0-367-25777-4

CHP from dual-fuel engine using biogas of anaerobic digestion of dairy manure

N. Akkouche & F. Nepveu
DRT/CEA Tech Region/DGDO (1-7R30-0-R), CEA Grenoble, Nantes, France

K. Loubar, M.E.A. Kadi & M. Tazerout
Laboratoire GEPEA, UMR 6144 CNRS, IMT Atlantique, Nantes, France

ABSTRACT: Biogas is produced using solid-state anaerobic digestion of dairy manure. When the moisture content varies from 70% to 83%, the methane yield potential rises from 63 to 90 Nm^3/ton-VS, and the retention time decreases from 67 to 43 days. The optimal production rate of methane is 2.01 Nm^3/ton-VS/day. Increasing number of the digesters, the fluctuation range of methane production rate goes from ±1.92 to ±0.02 Nm^3/ton-VS/days when number of digester goes from one to nine. CHP generation has been studied using dual-fuel engine. By varying the biogas-air dosing intake, the production of 3.15 kW of break power and 4.94 kW of thermal power was obtained. The optimal operating condition involves consumption of 3.36 10^{-4} kg/s of biogas and 1.45 10^{-4} kg/s of diesel. This biogas flow rate can be achieved through the digestion of the effluents of a farm that can produce 197 kg of dried manure per day.

1 INTRODUCTION

If the use of gaseous fuel is widespread in two-fuel spark ignition (SI) engines, the high CO2 content of the biogas produced by anaerobic digestion, especially at the beginning of the digestion reaction, disadvantages it as a fuel in the internal combustion engines (Makareviciene et al. 2013).

Putting existing diesel engines into dual-fuel mode, using biogas as primary fuel, has both environmental and economic advantages (Rosha, Dhir, et Mohapatra 2018). The operating improvement of the compression ignition (CI) engines in dual-fuel operation mode has been studied by several authors (Abd-Alla et al. 2000).

Biogas, usually produced by fermentation or anaerobic digestion (AD) of organic matter, is a cleaner and potentially renewable fuel. It consists mainly of methane (CH4) and carbon dioxide (CO2), small traces of carbon monoxide (CO), hydrogen (H2), oxygen (O2) and hydrogen sulphide (H2S) (Jingura et Matengaifa 2009).

In dual-fuel operation mode and under higher engine loads (above 80%), the brake specific energy consumption (BSFC), which is in the order of 300 g/kWh is slightly lower than that of conventional diesel mode, whereas the brake thermal efficiency (BTE) in dual-fuel mode (about 30%), is considerably lower than that of diesel mode (about 34%). Increasing the CH4 content of the biogas, which raises the heat release rate, leads to a significant increase in the BTE (Kim et al. 2018).

The lower carbon content of CH4 compared to petroleum-based diesel reduces the exhaust gases emissions (Henham et Makkar 1998). Several researchers have confirmed that the relative homogeneous charge and the lower cylinder temperature in dual-fuel mode have the advantage of significant reduction of NOx and smoke emissions. As regard the HC and CO emissions, they are higher when the biogas substitution is high, especially if its percentage of CO2 is considerable (Kalsi et Subramanian 2017).

Table 1. Gompertz and experimental parameters of SS-AS.

Parameters	MC70	MC76	MC83
$P[NL/kg_{VS}]$	63	72	90
$R_m[NL/(kg_{VS} \cdot day)]$	2	4	4
$\lambda[day]$	13.9	8.9	7.6
$T95[days]$	66.7	38.2	43.2
Effective period [days]	52.8	29.3	35.6

In this study, the biogas used is that produced from anaerobic digestion (AD) of manure. AD is now considered an attractive and innovative technology for odor reduction, manure stabilization with mass and volume reduction, green electricity generation and reduction of greenhouse gas methane emissions (Khalid et al. 2011). Where the volatile solids content of the raw material is high, solid-state anaerobic digestion (SS-AD) is used. This system is used to treat mixture, consisting of dairy manure and pine tree sawdust bedding material (Kim et al. 2018).

In order to simulate the performance of the Micro-CHP unit, the experimental results of anaerobic digestion, where the effect of the feedstock moisture content is studied, are recovered from the literature (Kim et al. 2018), while the dual-fuel engine tests have was carried out on a single-cylinder Lister Petter engine, whose maximum break power at 1500 rpm is 4.5 kW.

2 MATERIALS AND METHODS

2.1 *Feedstock*

The experimental results of the anaerobic digestion were inspired by the literature, thanks to the experimental work of Kim et al (Kim et al. 2018), carried out on mixtures of sawdust and dairy manure. In a lab-scale anaerobic reactor with a volume of 1.8 L (1.5 L working volume), three manure feedstock's were studied at a digestion temperature of 37°C. The effect of moisture content was investigated by adding water to the raw material to prepare three types of manures with different moisture contents (70%, 76% and 83%), called M70, M76 and M83, respectively.

2.2 *Biogas production*

The experimental results from the digesters were checked to model the instantaneous production of biogas. The cumulative biogas production in SS-AD can be described by means the modified Gompertz model (Kim et al. 2018) as follows:

$$M(t) = P \cdot exp\left[-exp\left(\frac{R_m \cdot exp(1)}{P}(\lambda - t) + 1\right)\right] \tag{1}$$

Where $M[NL/kg_{VS}]$ is the instantaneous cumulative methane production;
$P[NL/kg_{VS}]$ is the methane yield potential;
$R_m[NL/(kg_{VS} \cdot day)]$ is the maximum methane production rate;
$\lambda[day]$ is the duration of the lag phase;
$t[day]$ is the time at which the cumulative methane production is calculated.

The estimated parameters are listed in table 1.

The time needed to produce 95% of the methane potential (T95) is an important AD performance indicator. The effective periods for methane production (calculated by subtracting the lag phase from the T95 value) were 53, 29, and 36 days at 70, 76, and 83% MC, respectively.

Regarding the production of CO2, it will be inspired by another source where the instantaneous composition of biogas is given ("Anaerobic co-digestion of cattle manure and meadow grass Effect

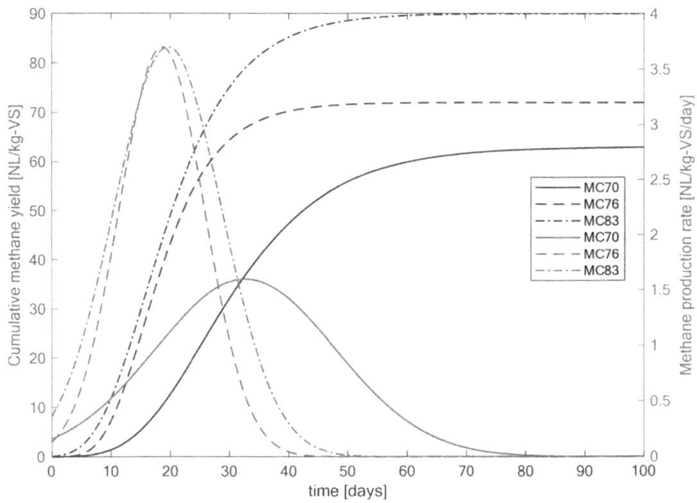

Figure 1. Effect of initial feedstock moisture content on cumulative methane yields from dairy manure.

of serial con?gurations of continuous stirred tank reactors (CSTRs).pdf", s. d.). The given composition of biogas, will be used and will be considered independent of the moisture content of the studied feedstock during its SS-AD.

The cumulative methane yield as well as its instantaneous volume percentage, reported to the input charge in volatile solids (VS), are represented in figure 1. It will be used in the Micro-CHP unit as a flow of biogas injected into the cylinder of the dual-fuel engine.

2.3 *Dual-fuel engine tests*

The engine tests were carried out in a single cylinder Lister-Petter TS1 diesel engine, operating in dual fuel mode. The experimental scheme as well as the acquisition and control systems were already presented by several researchers (Aklouche et al. 2017).

The optimum operating mode of engine refers to operation under engine loads ranging from 60 to 80% of the rated power output (4.5 kW). Figure 2 shows the heat release in the cylinder, when engine load is 70 % and biogas air dosing values go up to its maximum value. Indeed, the maximum dosing value goes from 17% to 8% when the methane biogas content increases from 40 to 60%.

The intake biogas-air dosing (*D*) is determined as a ratio of the biogas flow rate ($\dot{m}_{bg}$) to the air flow rate (($\dot{m}_a$). It is given by the following formula:

$$D = 100 * \frac{\dot{m}_{bg}}{\dot{m}_a} \tag{2}$$

In order to express the performance of the Micro-CHP unit, it is essential to take into account three main quantities:

- the brake thermal efficiency (BTE), which reflects the mechanical efficiency of the unit;
- the exhaust gas efficiency (EGE), which reflects the first part thermal efficiency of the unit;
- the cooling heat efficiency (CHE), which reflects the heat transmitted to the cooling fluid through the walls of the cylinder.

Figure 3 shows the evolution of the engine brake thermal efficiency as well as the exhaust gas thermal efficiency versus intake biogas-air dosing.

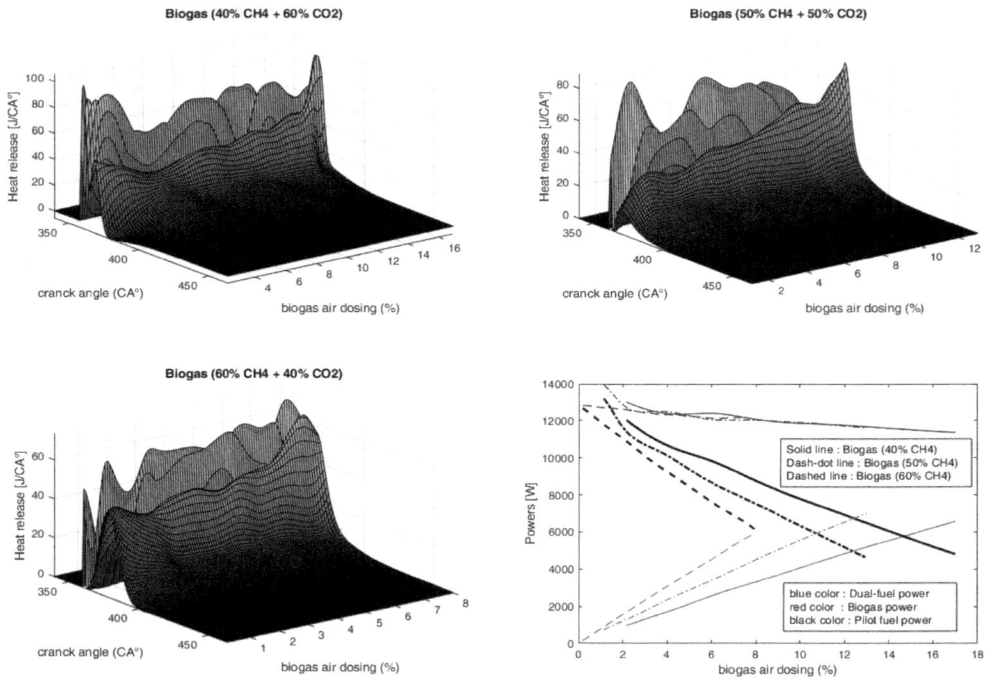

Figure 2. Heat release and primary power versus intake dosing and biogas methane content.

The BTE is given by the following formula:

$$BTE = \frac{L*4500}{\dot{m}_{bg}*LHV_{bg}+\dot{m}_f*LHV_f} = \frac{L*4500}{\dot{m}_{df}*LHV_{df}} \tag{3}$$

Where

$L = 0.7$ is the engine load;

$(\dot{m}_{bg}, \dot{m}_f, \dot{m}_{df})$ are mass flow rates of biogas, diesel and dual-fuel respectively. $(LHV_{bg}, LHV_f, LHV_{df})$ are their lower mass heating.

The exhaust gas thermal efficiency is given by the following formula:

$$EGE = \frac{(\dot{m}_{bg}+\dot{m}_f+\dot{m}_a)*Cp_{eg}*T_{eg}}{\dot{m}_{bg}*LHV_{bg}+\dot{m}_f*LHV_f} = \frac{\dot{m}_{eg}*Cp_{eg}*T_{eg}}{\dot{m}_{df}*LHV_{df}} \tag{4}$$

Where: $(\dot{m}_a, \dot{m}_{eg})$ are the mass flow rates of air and exhaust gas respectively.

(T_{eg}, Cp_{eg}) are the temperature and specific heat capacities of the exhaust gas.

3 MICRO-CHP PERFORMANCE

The optimization of the anaerobic digestion of the effluents is often ensured by means of the setting up of several anaerobic digestion chambers (caissons), operating in parallel mode. This ensures relatively constant production of biogas. In fact, in order to minimize the quantitative and qualitative fluctuation of the biogas production, the outlets of the anaerobic digestion chambers are connected using collector.

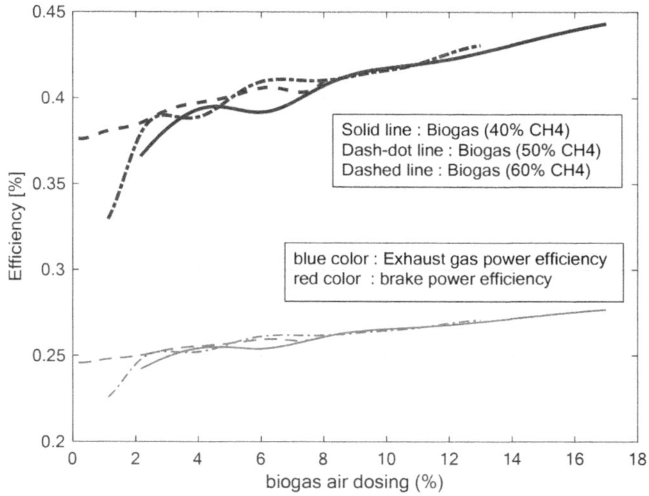

Figure 3. Engine brake and exhaust gas thermal efficiency versus intake dosing and biogas methane content.

In this study, we only use the results of the anaerobic digestion of MC83, which presents a maximum production of methane (Figure 1). Figure 4 shows methane rate production from the outlet of the collector, from several caissons.

The results show that increasing number of boxes greatly reduces the fluctuation of the output of the collector. Indeed, when using a single reactor caisson, the cumulative output varies between 0 and 4 Nm3/ton-VS/days. The cumulative production, given on average plus fluctuation (difference), is 2.01 ± 1.92 Nm3/ton-VS/days. This fluctuation is ± 0.09, ± 0.03 and ± 0.02 Nm3/ton-VS/days when using three, six and nine reactor caissons respectively. In the case where the fluctuation is relatively low, the outlet of the caissons collector can be directly connected to the engine intake. Figures 5 and 6 summarize both the mass flow rate introduced into the engine as a function of the intake biogas air dosing and the mass flow rate of the biogas leaving the collector of the nine SS-AD reactor caissons.

A biogas air dosing intake of 8% (figure 6), which has relatively high mechanical and thermal efficiencies (figure 3), implies a biogas flow of $3.36\ 10^{-4}$ kg/s. i.e. 70 wt% of the dual fuel flow rate injected into the engine.

As regard the biogas production from the nine reactor caissons (Figure 5), it is estimated to $4.684\ 10^{-5} \pm 0.04\ 10^{-5}$ kg/s/ten-VS. In order to ensure the necessary flow to supply the engine ($3.36\ 10^{-4}$ kg/s), the production of the farm must be equal to 7.17 tons of volatile solid per AD period (43.2 days). This comes back to a daily production of 166 kg-VS/day. It also represents production of 197.15 kg of dried manure per day. So, with a loading period of 4.8 days, each caisson must contain 946.32 kg of dried manure, and kept closed and connected to the collector for 43.2 days.

4 CONCLUSION

In this study, this study studies the use of biogas from SS-AD as a primary fuel in a dual fuel engine. The digester feedstock (influent) is a bedded pack dairy manure at three different moisture levels normally found on farms. After 85 days of digestion, digesters containing dairy manure at 83% MC outperformed digesters containing dairy manure with 70% and 76% MCs in terms of VS reduction and biogas production. However, the hydraulic retention time (HRT) for anaerobic digesters, which serves as the needed time to produce 95% of the methane potential (T95), is reduced up to 43 days when 83% MC is used. In the case of the use of a single digestion reactor caisson, the flow rate of

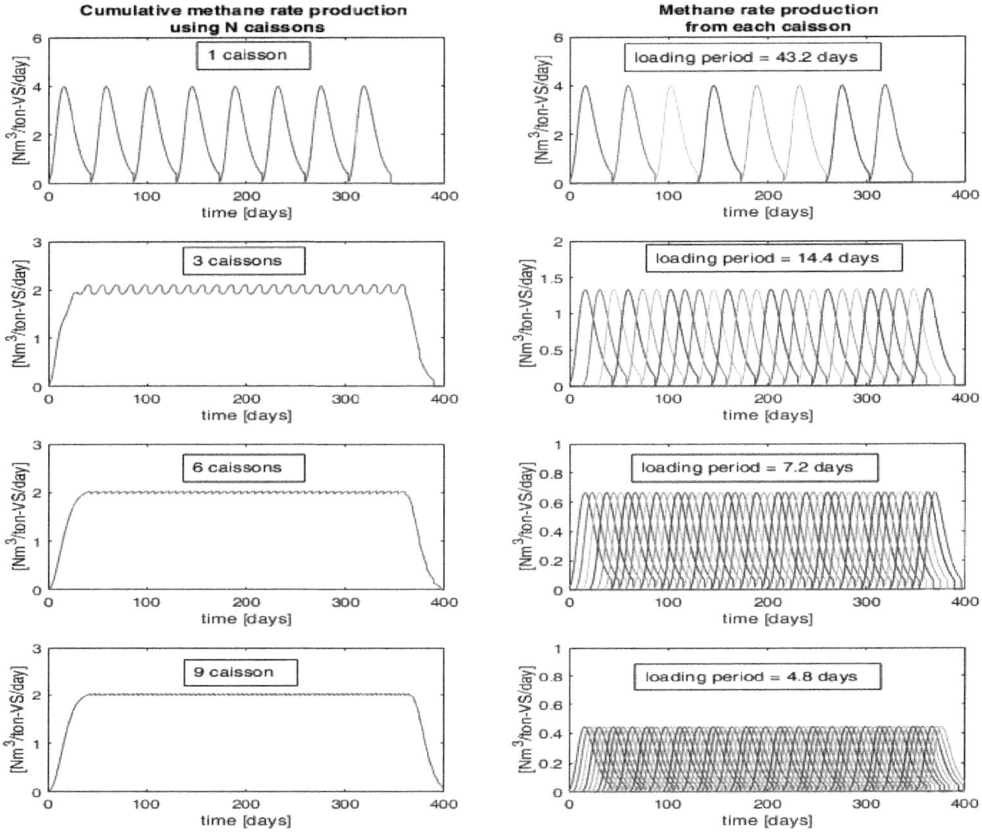

Figure 4. Methane rate production according to the number of installed caissons.

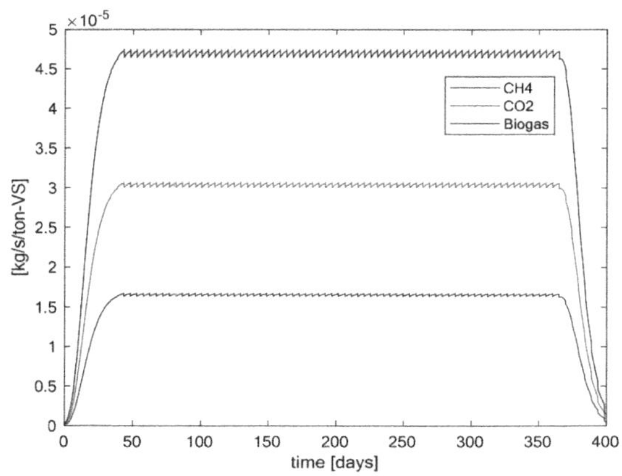

Figure 5. Biogas flow rate from the nine SS-AD reactors.

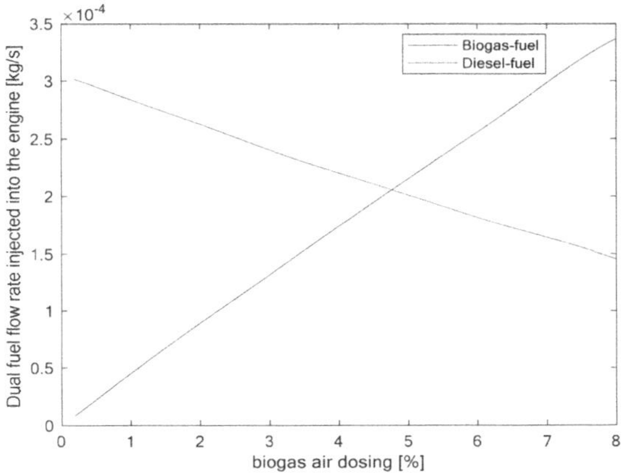

Figure 6. Dual fuel flow rate injected into the engine.

methane leaving the reactor is 2.01 ± 1.92 Nm3/ton-VS/day. It varies between 0 and 4 Nm3/ton-VS/day. This fluctuation (± 1.92) is reduced to ± 0.02 Nm3/ton-VS/days when the caisson reactor number is increased to 9 caissons.

Using this biogas, consisting of 60% methane and 40% carbon dioxide, the production of 3.15 kW of break power using the dual-fuel engine (single-cylinder Lister Peter engine), operating under optimal operating conditions, necessities to ensure a biogas air dosage intake of 8%. This operating point involves a consumption of 3.36 10-4 kg/s of biogas (primary fuel) and 1.45 10–4 kg/s of diesel (pilot fuel). This consumption of dual fuel, which represents a consumption power of 12.04 kW, also ensures an exhaust thermal power output of 4.94 kW. This implied an EGE of 41.04% and a BTE of 26.16%.

This biogas flow (3.36 10-4 kg/s) can be produced using SS-AD of the effluents of a farm, which can produce 197.15 kg of dried manure per day. This is done by means of 9 caissons reactors, for which each of them ensures digestion of manure generated during 4.2 days (i.e. 946.32 kg of dried manure in each reactor caisson).

REFERENCES

Abd-Alla, G H, H A Soliman, O A Badr, et M F Abd-Rabbo 2000. "Effect of Pilot Fuel Quantity on the Performance of a Dual Fuel Engine ". *Energy Conversion &Management* 41:559–572.

Aklouche, F.Z., K. Loubar, A. Bentebbiche, S. Awad, et M. Tazerout 2017. "Experimental Investigation of the Equivalence Ratio Influence on Combustion, Performance and Exhaust Emissions of a Dual Fuel Diesel Engine Operating on Synthetic Biogas Fuel ". *Energy Conversion and Management* 152: 291–99.

Henham, A, et M K Makkar 1998. "Combustion of Simulated Biogas in a Dual-Fuel Diesel Engine ". *Energy Conversion & Management* 39 (16-18): 2001–9.

Jingura, Raphael M., et Rutendo Matengaifa 2009. "Optimization of Biogas Production by Anaerobic Digestion for Sustainable Energy Development in Zimbabwe". *Renewable and Sustainable Energy Reviews* 13 (5): 1116–20.

Kalsi, Sunmeet Singh, et K.A. Subramanian 2017. "Effect of Simulated Biogas on Performance, Combustion and Emissions Characteristics of a Bio-Diesel Fueled Diesel Engine". *Renewable Energy* 106 : 78-90.

Khalid, Azeem, Muhammad Arshad, Muzammil Anjum, Tariq Mahmood, et Lorna Dawson 2011. "The anaerobic digestion of solid organic waste". *Waste Management* 31 (8): 1737–44.

Kim, Eunjong, Seunghun Lee, Hyeonsoo Jo, Jihyeon Jeong, Walter Mulbry, Shafiqur Rhaman, et Heekwon Ahn 2018. "Solid-State Anaerobic Digestion of Dairy Manure from a Sawdust-Bedded Pack Barn: Moisture Responses". *Energies* 11 (3): 484.

Makareviciene, Violeta, Egle Sendzikiene, Saugirdas Pukalskas, Alfredas Rimkus, et Ricardas Vegneris 2013. "Performance and Emission Characteristics of Biogas Used in Diesel Engine Operation ". *Energy Conversion and Management* 75: 224–233.
Rosha, Pali, Amit Dhir, et Saroj Kumar Mohapatra 2018. "Influence of Gaseous Fuel Induction on the Various Engine Characteristics of a Dual Fuel Compression Ignition Engine: A Review". *Renewable and Sustainable Energy Reviews* 82: 3333–49.

Wastes: Solutions, Treatments and Opportunities III – Vilarinho et al. (Eds)

Capture of CO_2 in activated carbon synthesized from municipal solid waste compost

M. Karimi, L.F.A.S. Zafanelli, J.P.P. Almeida & J.A.C. Silva
Instituto Politécnico de Bragança, Bragança, Portugal

A.E. Rodrigues
Universidade do Porto, Porto, Portugal

G.R. Ströher
Universidade Tecnológica Federal do Paraná – Apucarana, Paraná, Brasil

ABSTRACT: In this study, municipal solid waste composts obtained from mechanical biological treatment has been considered as a source of adsorbents for CO_2 capture. Three samples derived from the maturated compost in the municipal solid wastes were modified to produce activated carbon. The first sample was treated with sulfuric acid, the second one was thermally treated at 800°C and the last one was modified chemically and thermally with sulfuric acid and at 800°C. Then, the CO_2 uptake capacity of prepared samples was measured through breakthrough adsorption experiments at the post combustion operational conditions to collect isotherm data. Also a fixed bed adsorption mathematical model was developed by applying mass and energy balances. Results showed the municipal solid wastes have an excellent capacity to be considered as source of adsorbent for CO_2 capture also the mathematical model is able to predict breakthrough data.

1 INTRODUCTION

1.1 *CO_2 capture*

Global warming has been one of the major concerns of mankind in the recent decades, which it needs significant attempts to reduce the greenhouse gases (GHGs) emissions (Karimi et al., 2014). Among all GHGs, CO_2 has the main role, which has contributed to several adverse effects on the ecosystem and environment, and if the current dangerous level of the GHGs is not controlled, it can face the life on this planet with several serious challenges. According to reports, the coal and natural gas fired power plants released 11.1 Gt of carbon dioxide, nearly 30% of the total global emissions in 2012 (Stocker et al., 2013; IEA). In this way, the combustion of coal, also oil and natural gas industries including naphtha refineries (Iranshahi et al., 2014; Karimi et al., 2018a), and petro-chemical complexes (Karimi et al., 2016), are the main industrial sources of CO_2 emissions. As consequence of these industrial activities, the CO_2 percentage has exceeded 50 ppm in the atmosphere from the maximum allowable level in the pre-industrial period until now (280 ppm to 400 ppm) (Karimi et al., 2018b). Thus, strict policies, better strategies and more attention for capturing and sequestering CO_2 are required.

1.2 *Municipal Solid Wastes Management*

Municipal solid waste (MSW) is a term usually referred to the unwanted or useless solid materials originated from the combined residential, industrial and commercial activities in the urban areas. The characteristics and qualities of the MSWs depend on several parameters, including: the income, lifestyle and living standards of region's inhabitants also, climate and natural resources of region. For example, the generated solid wastes in the humid, tropical, and semitropical areas are rich with plant debris, while in the areas with seasonal climate changes it has more ash. Thus, the classification of MSWs can be based on the origin, content or hazardous potentials, but the category which considers the content (organic or inorganic materials) is the most popular one (Hoornweg and Bhada-Tata, 2012). There are some main treatment techniques for solid wastes such as employing extremely high temperatures, dumping on the land, also applying the biological processes to treat the wastes and producing the compost, which is one of the most popular strategies (Karak et al., 2012). The results can be contributed to reduce the GHGs emission, conserve the natural resources, energy saving, environment protection and finally, a healthier life.

2 MATERIALS & METHODS

2.1 *Synthesis of Activated Carbon from compost*

The employed compost was obtained in mechanical biological treatment plants for municipal solid waste, supplied by the company "Resíduos do Nordeste, EIM". In order to homogenize and remove the soluble compounds and suspended solids, the compost was first mixed with water and washed. Then, the first sample was prepared by carbonization at 800 ^{o}C (C-800). The second sample was synthesized by treating the washed compost with sulfuric acid. In addition, the other samples were prepared by treating with sulfuric acid before the carbonization at 800°C.

2.2 *Experimental breakthrough measurements*

The measurement of uptake capacities of the prepared samples for CO_2 capture was performed through the breakthrough experiments in the fixed bed adsorption unit at LSRE-LCM. To this goal, an adsorption column was packed with prepared samples, then, the adsorption process takes place by passing the gas mixture (career gas and CO_2) on the fixed bed, which has been putted in the oven and it has a constant partial pressure and temperature. Then, the mass flow rate was continuously measured at the output of the packed bed by TCD, until getting the saturation condition at output of adsorption column. More details about this unit can be found in author's previous studies (Karimi et al., 2018b).

2.3 *Objective*

In this work, based on the scopes of Carbon Capture and Storage (CCS) technique and municipal solid waste management a novel strategy has been proposed. In this way, the obtained compost in the mechanical biological treatment from municipal solid wastes has been considered as a source of adsorbents for CO_2 capture and its performance for CO_2 capture evaluated through breakthrough adsorption experiments.

3 MODELLING OF BREAKTHROUGH DATA

Breakthrough data for CO_2 adsorption, were modeled with mass and energy balances based on the following assumptions:

- The ideal gas law was assumed for the gas phase.
- The pressure drop was considered negligible.
- The adsorption equilibrium was described by Langmuir's isotherm;
- The external resistance and macropore diffusion can be combined in a global resistance to a lumped model for the adsorbent particle (Linear Driving Forge model – LDF).

Accordingly, the mass and energy balances results in the of partial differential equations (PDEs):

Global mass balance:

$$\frac{\partial F}{\partial z} + \varepsilon_b \frac{\partial C}{\partial t} + (1 - \varepsilon_b) \sum_{i=1}^{n} \frac{\partial \bar{q}_i}{\partial t} = 0 \tag{1}$$

Component mass balance (Adsorbate specie):

$$-\varepsilon_b D_{ax} \frac{\partial}{\partial z}\left(C \frac{\partial y_i}{\partial z}\right) + \frac{\partial (F y_i)}{\partial z} + \varepsilon_b \frac{\partial (C y_i)}{\partial t} + \rho_p (1 - \varepsilon_b) \frac{\partial \bar{q}_i}{\partial t} = 0 \tag{2}$$

Mass transfer (LDF Model):

$$\frac{\partial \bar{q}_i}{\partial t} = k_{LDF}(q^* - \bar{q}_i) \tag{3}$$

Energy balance:

$$-K_{ax} \frac{\partial^2 T}{\partial z^2} + F c_{pg} \frac{\partial T}{\partial z} + \varepsilon_b C c_{pg} \frac{\partial T}{\partial t} +$$

$$+(1 - \varepsilon_b) a_p h_p (T - T_s) + a_c h_w (T - T_w) = 0 \tag{4}$$

Energy balance in the solid phase:

$$c_{ps} \frac{\partial T_s}{\partial t} = a_p h_p (T - T_s) + \sum_{i=1}^{n} (-\Delta H_i) \frac{\partial \bar{q}_i}{\partial t} \tag{5}$$

The initial and boundary conditions are summarized below:
Initial condition:

$$t = 0: \forall z; \;\; F = F_f; \;\; C = C_f; \;\; \bar{q}_i = 0; y_i = 0; \;\; T = T_s = T_f;$$

Boundary condition:

$$z = 0, t > 0: \qquad F = F_f, \;\; F y_{if} = F y_i - \varepsilon_b D_{ax} C \frac{\partial y_i}{\partial z},$$

$$F c_{pg} T_f = F c_{pg} T - K_{ax} \frac{\partial T}{\partial z} \;\; z = L, t > 0: \qquad \frac{\partial y_i}{\partial z} = 0 \;\; \frac{\partial T}{\partial z} = 0$$

Here, L, ε_b and $\overline{q_i}$ are the length of adsorption column, void fraction and the average adsorbed concentration of absorbable species i in the solid phase, respectively. Also, y_i, D_{ax} and q^*represent the molar gas fraction of solute i, the coefficient of axial dispersion and the adsorbed phase concentration of species i in equilibrium with gas phase. In addition,k_{LDF} is called the Linear Driving Force (LDF). To solve the considered model, the set of partial differential equations (PDEs) are converted into a system of ordinary and algebraic differential equations (DAEs) via method of lines (MOL) and solved by numerical integrator ode15s method, available in the MATLAB library. The ode15s integrator was considered as a variable order method and robust strategy to solve stiff equations (Shampine & Reichelt, 1997). The MOL method is based on a library of differentiation routines provided by Schiesser & Griffiths, 2009 This model is also important for the design of cyclic processes for CO_2 capture (e.g Pressure Swing Adsorption). In addition, this model can be coupled with proper isotherm (Langmuir isotherm) that describe the measured CO_2 adsorption experimental data

Table 1. Elemental analysis of the prepared samples.

Sample	C (%)	H (%)	S (%)	N (%)	Remaining (%)	Ashes (%)
C-800	17.5	0.4	0.4	0.0	81.6	80.6
C-S	20.1	2.3	0.6	1.7	70.4	34.3
C-S-800	20.5	1.9	0.4	1.4	75.8	65.9

Table 2. Textural properties of materials determined from BET and t-plot methods.

Sample	Burn-off (%)	S_{BET} $(m^2 \cdot g^{-1})$	S_{ext}	V_{mic} S_{mic}	$(mm^3 \cdot g^{-1})$	V_{mic}/V_{Total} (%)	W_{mic} (nm)
C-800	39.9	77	52	25	12	14.0	1.9
C-S	59.6	11	11	0	0	0.0	–
C-S-800	76.3	279	56	223	92	53.4	1.6

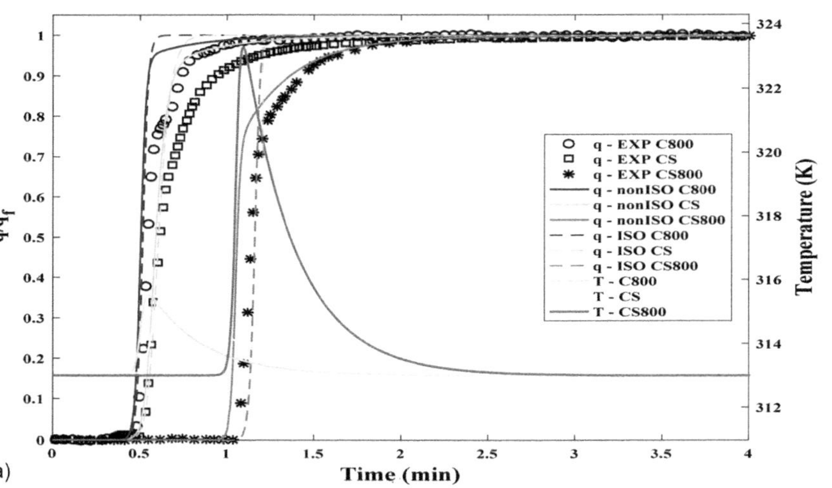

(a)

Figure 1. Breakthrough adsorption RUNs in samples C-800, C-S and C-S-800 at 40 C and total pressure: (a) 1 bar, (b) 3 bar and (c)5 bar. Points are experimental data and lines model predictions.

4 RESULTS & DISCUSSION

The textural properties of the materials were determined from N_2 adsorption–desorption isotherms at 77 K, obtained in a Quantachrome NOVA 4200e adsorption analyzer. The specific surface area was calculated using the S_{BET} method. The external surface area (Sext) and the micropore volume (V_{mic}) were obtained by the t-method (thickness was calculated by employing ASTM standard D-6556-01). The total pore volume (V_{Total}) was calculated at $p/p_0 = 0.98$. The microporous surface area (S_{mic}) was determined as the subtraction of Sext from SBET and the average pore width (W_{mic}) by approximation ($W_{mic} = 4\, V_{mic}/S_{mic}$). The results of elemental analysis and textural properties of the prepared samples are presented in Table 1 and Table 2, respectively.

The breakthrough adsorption runs on the synthesized samples were performed at 40°C and different pressures (in the range of 1-5 bar) to collect the isotherm data. Figure 1 (a-e) shows

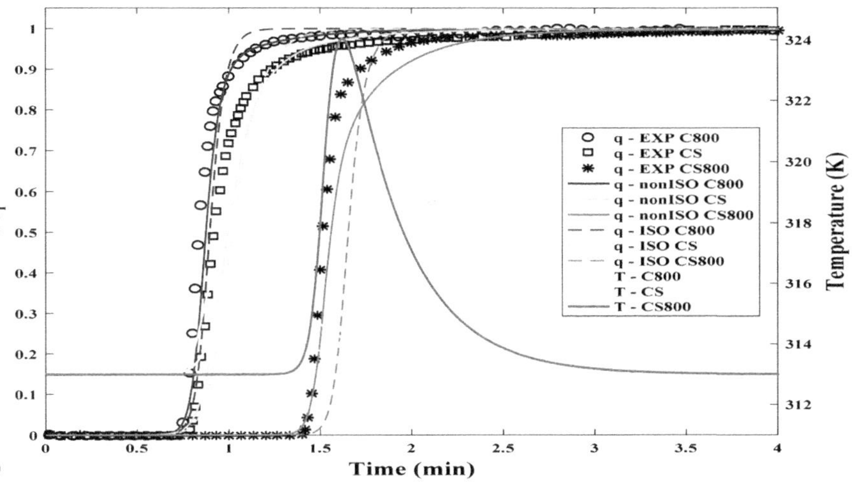

(b)

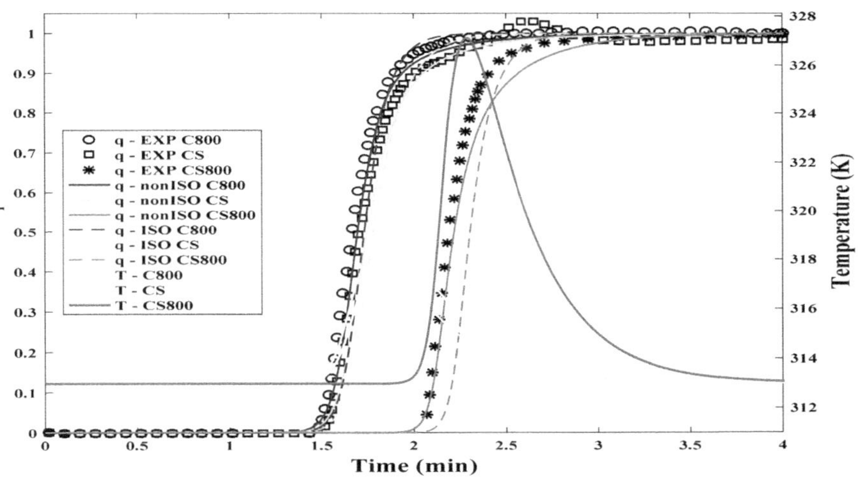

(c)

Figure 1. *(Continued)*

the examples of breakthrough experiments (points) with the mathematical modelling (lines). It is clear that the developed model is capable to describe the experimental data being a valuable tool to design cyclic processes for CO_2 capture. In addition, maximum loadings (around 2.5 mmol/g) can be considered reasonable value for CO_2 capture (since municipal waste has been considered) when compared to literature. Accordingly, results are very promising in view of developing a cyclic process for CO_2 capture using municipal solid wastes as source of activated carbon. It also clear from the Figures that sample C-S-800, is the best sample for CO_2 capture since the breakthrough time is much higher than the other ones. In addition, as can be observed, the enhancement of adsorption pressure has a positive effect on the adsorption capacity, which by increasing the pressure, the loading of different prepared samples has increased.

5 CONCLUSION

In this study, the potential of municipal solid wastes as a source of adsorbents for CO_2 capture were investigated at the post-combustion operational conditions. The composts from municipal solid wastes are converted to activated carbons and then breakthrough experiments were performed to collect isotherm data and simultaneously, the mathematical model was developed to fit the experimental results in view of the design of cyclic processes for CO_2 capture. The equilibrium adsorption capacity of the considered samples revealed that the adsorption capacity of the sample treated with sulfuric acid and thermally calcinated is the best one with CO_2 loadings comparable with commercial carbon materials. Finally, it can be concluded that the obtained results through this work can be significant for CO_2 capture and reducing the global warming, also high added value to the waste by using these materials as a source of Activated Carbons.

ACKNOWLEDGMENT

This work is a result of the projects "VALORCOMP" (0119_VALORCOMP_2_P), funded by FEDER through Programme INTERREG V A Spain - Portugal (POCTEP) 2014–2020, and "AIProcMat@N2020 - Advanced Industrial Processes and Materials for a Sustainable Northern Region of Portugal 2020", with the reference NORTE-01-0145-FEDER-000006, supported by NORTE 2020, under the Portugal 2020 Partnership Agreement, through FEDER; and POCI-01-0145-FEDER-006984 – Associate Laboratory LSRE-LCM funded by FEDER through COMPETE2020 – POCI – and by national funds through FCT.

REFERENCES

Hoornweg, D. Bhada-Tata, P. 2012. What a waste: a global review of solid waste management, urban development & local government unit, World Bank, Washington DC, 20433, USA.

Iranshahi, D. Karimi, M. Amiri, S. Jafari, M. Rafiei, R. Rahimpour, M.R. 2014. Modeling of naphtha reforming unit applying detailed description of kinetic in continuous catalytic regeneration process. Chem. Eng. Res. Des. 92, 1704–1727.

Karak, T. Bhagat, R.M. Bhattacharyya, P. 2012. Municipal solid waste generation, composition, and management: the world scenario. Crit. Rev. Environ. Sci. Technol. 42, 1509–1630.

Karimi, M. Rahimpour, M.R. Iranshahi, D. 2018a. Enhanced BTX production in refineries with sulfur dioxide oxidation by thermal integrated model, Chem. Eng. Technol. 41, 1746-1758.

Karimi, M. Rahimpour, M.R. Rafiei, R. Jafari, M. Iranshahi, D. Shariati, A. 2014. Reducing environmental problems and increasing saving energy by proposing new configuration for moving bed thermally coupled reactors. J. Nat. Gas Sci. Eng. 17, 136-150.

Karimi, M. Rahimpour, M.R. Rafiei, R. Shariati, A. Iranshahi, D. 2016. Improving thermal eficiency and increasing production rate in the double moving beds thermally coupled reactors by using differential evolution (DE) technique. Appl. Therm. Eng. 94, 543–558.

Karimi, M. Silva, J.A.C. Gonçalves, C.N.d.P. Diaz de Tuesta J.L. Rodrigues, A.E. Gomes, H.T. 2018b. CO2 capture in chemically and thermally modified activated carbons using breakthrough measurements: experimental and modeling study. Ind. Eng. Chem. Res. 57, 11154–11166.

Schiesser, W. E. Griffiths, G.W. 2009. A Compendium of Partial Differential Equation Models: Method of Lines Analysis with Matlab Installation of Matlab files Running an Application. https://doi.org/10.1017/CBO9780511576270.

Shampine, L.F. Reichelt, M.W. 1997. The MATLAB ODE Suite. SIAM J. Sci. Comput., 18(1), 1–22.

Stocker, T.F. Qin, D. Plattner, G.K. Tignor, M. Allen, S.K. Boschung, J. Nauels, A. Xia, Y. Bex, V. Midgley, P.M. 2013. In Climate Change 2013: The Physical Science Basis. Contribution of Working Group I to the Fifth Assessment Report of the Intergovernmental Panel on Climate Change, Cambridge University Press, New York, US.

Wastes: Solutions, Treatments and Opportunities III – Vilarinho et al. (Eds)
© 2020 Taylor & Francis Group, London, ISBN 978-0-367-25777-4

Biogas: Olive mill wastewater as complementary substrate of piggery effluent

A. Neves, L. Ramalho, L.B. Roseiro, A. Eusébio & I.P. Marques
LNEG-Unit of Bioenergy, Lisboa, Portugal

ABSTRACT: The anaerobic digestion of a peculiar piggery effluent (PE), with a high organic content (93 g/L), was carry out using olive mill wastewater (OMW) as complementary substrate. From the different tested conditions – [100%PE], [70%PE+30%OMW], [50%PE+50%OMW], [20%PE+80%OMW] – units containing only PE and the lowest proportion of OMW in the mixture (30% OMW), provided the highest biogas volume of about 780 mL (70% CH_4). Comparatively, identical quantities of each substrate ([50%PE+50%OMW]) generates some gas (320 mL, 60% CH_4), understood as the result of an adaptation process by the microbial consortium, while the [20%PE+80%OMW] condition provided even less gas volume (120 mL, 6% CH_4), probably due to the antimicrobial capacity of the phenolic compounds in OMW, confirming the negative influence of using so high OMW proportion.

1 INTRODUCTION

Portugal owns large breeders (at least 400 pigs and 100 sows) which manage more than two thirds of the pigs fattened before slaughtering, being in a similar situation to countries such as the Czech Republic, Estonia, Ireland, Greece and Cyprus (EUROSTAT 2014). The production of pigs in Portugal was about 2,165,000 heads in 2017, counting for 1.2% of the Europe production (FAOSTAT 2019). This activity produces high volumes of piggery effluent, which is an unbalanced and, consequently, a potential inhibiting substrate to the anaerobic digestion process, mainly due to high ammonia concentrations. A huge research was done in the scope of inhibiting recalcitrant effluents, involving several arrangements of various pre-treatments processes, which result in very expensive procedures not always achieving good results. The concept of "effluents complementarity" was study and successfully applied to organic effluents by anaerobic digestion, as the case of the OMW treatment. The OMW was digested without chemical correction or pre-treatment by means of a feeding strategy involving an OMW fraction of 83% (v/v) in admixture with complementary substrate (Marques 2001; Gonçalves et al. 2012). Sampaio et al. (2012) confirmed that the drawbacks of the raw OMW characteristics could be easily overcome by applying the same feeding approach through heterotrophic microalgae (*Chlorella protothecoides*) as OMW complementary effluent. Later on, Assemany et al. (2018) reported the beneficial effect of substrates complementarity on anaerobic processes since the replacement of algal biomass by OMW (10%, v/v) provided a three-fold increase in methane production compared to algal digestion alone. According to Sampaio et al. (2011), the use of complementarity concept allows to obtain a stable anaerobic digestion able to degrade OMW in its original composition (100%, v/v), making the process simpler, more flexible and cheaper. The objective of this work was to optimize operational conditions of the anaerobic digestion of a peculiar high organic content piggery effluent by means of the effluent complementarity concept, using OMW as complementary substrate.

Table 1. Scope of the anaerobic digestion experiment.

Substrates composition (%, v/v)		
PE	OMW	Designation
100	0	[100%PE]
70	30	[70%PE + 30%OMW]
50	50	[50%PE + 50%OMW]
20	80	[20%PE + 80%OMW]

Table 2. Effluents characteristics: COD and solids.

	COD (g/L)	Total solids (g/L)	Volatile solids (g/L)	Total nitrogen (mg/L)	Ammonium nitrogen (mg/L)
PE	93 ± 5	47.4 ± 0.79	31.9 ± 0.6	4900 ± 277	3206 ± 20
OMW	106 ± 1	31.8 ± 0.04	26.1 ± 0.2	213 ± 16	1.4 ± 1.6

Table 3. Effluents characteristics: pH values and volatile fatty acids.

	pH	Acetic acid (mg/L)	Propionic acid (mg/L acetic)	Isobutyric acid (mg/L acetic)	Butyric acid (mg/L acetic)	TOTAL VFA (mg/L acetic)
PE	7.27	1373.0	561.8	2157.7	1545.7	5638.2
OMW	5.09	213.0	64.0	110.4	159.5	546.9

2 MATERIALS AND METHODS

2.1 *Substrates*

The substrates assayed were: piggery effluent (PE) collected in *Valorgado* Company (Salvaterra de Magos, Portugal), using the liquid fraction after main solids removal by a solid-liquid separator, and olive mill wastewater (OMW) collected from an olive mill (Rio Maior, Portugal) using three-phase continuous extraction process.

2.2 *Anaerobic digestion experimental set-up*

Anaerobic digestion assays were carried out in batch mode, under 37 ± 1°C, by using glass reactors with 165 mL total volume. Tested conditions were run in triplicate as designated in Table 1.

2.3 *Analytical and chromatograph methods*

Performance of the process was monitored by analytical characterizations of all samples and by the volume and quality of the obtained biogas. Total and volatile solids (TS, VS), chemical oxygen demand (COD), total nitrogen (Kjeldahl, TN), ammonium (NH_4^+-N), and pH, were assayed according to Standard Methods (APHA 2012). Biogas production was monitored by means of a pressure transducer, while the gas composition and volatile fatty acids (VFA) were analysed by chromatographic techniques (Varian 430-GC, TDC; HP-5890, FID, ASTM Standard Method 2000). The characterization of substrates is shown in Tables 2–4.

Table 4. Effluents characteristics: phenols and antioxidant activity.

	Total phenols (g GAE/L)	Antioxidant activity (mmol TEAC)	DPPH Inhibition (%)
PE	0.89 ± 0.001	1.11 ± 0.012	80.9 ± 0.70
OMW	3.12 ± 0.017	0.80 ± 0.069	60.9 ± 2.97

2.4 *Quantification of total phenolics content*

Total phenolics content was determined by the Folin-Ciocalteu colorimetric method (Singleton & Rossi 1965), according to an improved procedure described by Hagerman et al. (2000). Total phenols were expressed as g GAE (gallic acid equivalents)/L by comparison to the gallic acid standard curve. Results were obtained in triplicate.

2.5 *DPPH radical scavenging assay for Antioxidant activity evaluation*

Radical scavenging activity against stable DPPH radical (2,2-diphenyl-2-picrylhydrazyl hydrate) was determined spectrophotometrically. When DPPH reacts with an antioxidant compound, which can donate hydrogen, it is reduced. The changes in color (from deep-violet to light-yellow) were detected at 515 nm on a UV/visible light spectrophotometer. Radical scavenging activity of samples was measured by a modified method (Brand-Williams et al. 1995). The decreasing of the DPPH solution absorbance indicated an increase of the DPPH radical-scavenging activity. The experiment was carried out in triplicate. Radical scavenging activity (%) was calculated by the equation 1.

$$\%\text{DPPH inhibition} = [(Abs_b - Abs_f)/\, Abs_b] \times 100 \quad (1)$$

where Abs_b is the absorption of blank ($t = 0$ min) and Abs_f is the absorption of tested solution ($t = 30$ min).

The antioxidant activity was expressed in mmol TEAC (Trolox Equivalent Antioxidant Capacity), by preparing a standard curve using Trolox as an antioxidant standard.

3 RESULTS AND DISCUSSION

3.1 *Effluents*

This particular piggery effluent used is an unusual very concentrated substrate, whose organic matter content resembles OMW concentrations. So, the presence of such high organic concentrations (93–106 g/L COD: Table 2) indicates that there is a great potential for biogas/methane production. On the other hand, this effluent presents complementary characteristics in terms of composition that can be used advantageously to balance the suitable conditions of the anaerobic digestion process. Effectively, the inhibitory capacity of OMW, due to the total phenols concentration ($\approx$3 g/L), associated with the acid pH (5.0) (Tables 3 and 4), can be minimized by piggery effluent addition. Furthermore, OMW nitrogen absence (evaluated through total and ammonia nitrogen contents) may also be compensated by the presence of the high nitrogen content of piggery effluent (4.9 g/L total, 3.2 g/L ammonium nitrogen: Table 2).

3.2 *Biogas/methane production*

Biogas production was registered in all tested mixtures without any "lag" phase (Figure 1).

The biogas production started immediately and a similar accumulated volume ($\approx$120 mL) was observed in all units, elapsed 13 days. From then on, and over the remaining experimental time,

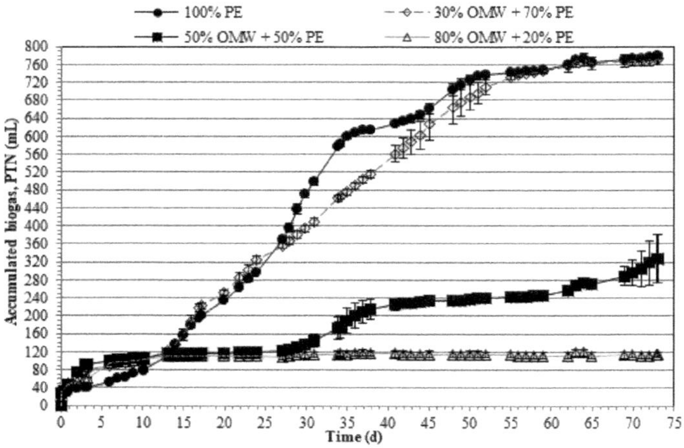

Figure 1. Biogas production.

Table 5. Biogas composition.

Substrates	Methane (% v/v)	Carbon dioxide (% v/v)
[100%PE]	71.00 ± 0.26	29.00 ± 0.26
[70%PE + 30%OMW]	70.27 ± 0.57	29.73 ± 0.57
[50%PE + 50%OMW]	60.30 ± 6.68	39.70 ± 6.68
[20%PE + 80%OMW]	5.57 ± 8.86	94.43 ± 8.86

units containing [100%PE] and [70%PE + 30%OMW] showed a similar behaviour, providing the highest accumulated biogas amount obtained (≈780 mL). For the other two trials, only by the 30th experimental day, the assay [50%PE + 50%OMW] was able to generate some gas but it did not exceed the average value of 320 mL. This behaviour can be understood as the result of an earlier process (before the day 30th) of microorganism adaptation to the effluents mixture, which allowed further increases in gas production. Comparatively, the gas production absence from the assay involving a volume participation of 80% OMW, suggests that under the tested operating conditions, around 50% v/v is the limiting OMW quantity in the blend. Concerning biogas composition, the units [100%PE] and [70%PE + 30%OMW] showed the highest methane concentrations (around 70% CH_4: Table 5), being in conformity with the previous observations and indicating the presence of a healthy methanogenic archaea population in both assays.

The value of 60% methane in biogas from [50%PE + 50%OMW] mixture may result from the already referred adaption process of the remaining consortium that had maintained the capacity of converting the substrate and produce biogas/methane. As expected, a very poor biogas was obtained in [20%PE + 80%OMW], confirming the negative influence on anaerobic consortium if so high OMW proportions are included in the influent.

3.3 *Removal capacity*

The highest amount of organic material removal was observed, as expected, in the units [100%PE] and [70%PE+30%OMW], with [100%PE] being less efficient than [70%PE + 30%OMW]. Removal values of 63% against 75% of initial COD (93 and 81 g/L) were respectively reached (Table 6). Solids concentrations of 40–47 g/L TS and 28–32 g/L VS ([100%PE] and [70%PE + 30%OMW] units (Table 6) were removed in proportion of 17–21% (TS) and 30–36% (VS).

Table 6. Chemical oxygen demand and solids.

Substrates	COD initial (g/L)	COD removal (%)	Total Solids initial (g/L)	Total Solids removal (%)	Volatile Solids initial (g/L)	Volatile Solids removal (%)
[100%PE]	93 ± 5	62.5	47.4 ± 0.79	16.5	31.9 ± 0.56	29.5
[70%PE + 30%OMW]	81 ± 3	74.6	39.6 ± 0.04	21.0	27.8 ± 0.03	35.6
[50%PE + 50%OMW]	77 ± 3	47.6	38.0 ± 0.05	16.1	27.7 ± 0.00	22.0
[20%PE + 80%OMW]	73 ± 1	28.9	32.8 ± 0.60	20.7	25.4 ± 0.61	25.2

Table 7. Volatile fatty acids

Substrates	Acetic acid (mg/L) initial	Acetic acid (mg/L) final	Total VFA (mg/L acetic) (%) initial	Total VFA final	Total VFA removal
[100%PE]	1373	688	5638.2	831.4	85.3
[70%PE + 30%OMW]	2314	1190	3962.0	1761.5	55.5
[50%PE + 50%OMW]	2175	2061	3683.9	6675.5	−81.2
[20%PE + 80%OMW]	1801	757	2792.2	1328.5	52.4

Table 8. Nitrogen.

Substrates	pH initial	pH final	Total nitrogen (g/L) initial	Total nitrogen (g/L) final	Ammonia nitrogen (g/L) initial	Ammonia nitrogen (g/L) final
[100%PE]	7.27	8.05	4.90 ± 0.280	n/d	3.21 ± 0.019	3.86 ± 0.158
[70%PE + 30%OMW]	6.90	7.94	1.83 ± 0.024	1.84 ± 0.075	1.35 ± 0.113	2.55 ± 0.297
[50%PE + 50%OMW]	6.66	7.25	1.35 ± 0.004	1.32 ± 0.024	0.83 ± 0.029	0.95 ± 0.012
[20%PE + 80%OMW]	6.23	5.73	0.58 ± 0.012	0.58 ± 0.039	0.34 ± 0.014	0.29 ± 0.004

VFA contained in digesting substrate of [100%PE] and [70%PE+30%OMW] were more efficiently converted than the other and removal values of 85 and 56% were registered, respectively (Table 7). As known, most of the methane formed in the anaerobic digestion process results from the acetic acid. In the two present cases, the contribution of acetic acid in total of VFA at baseline was 24 and 58%, meaning that the remaining VFA correspond to higher-chain acids than acetic and of more difficult degradation, such propionic and butyric (data not shown), justifying the slower initial gas production of these units (Figure 1). This situation presumes the existence of a buffer capacity in the medium capable of maintaining the process balance, for which the high amounts of nitrogen in effluents had certainly a positive effect on the digestion process. Initial nitrogen amount (total and ammonia nitrogen; Table 8) decreased with PE decreasing volume, as expected, due to the presence of OMW, characterized by low nitrogen levels.

Initial concentrations of total nitrogen were more or less maintained after anaerobic digestion while those relative to ammonia nitrogen were increased due to protein material degradation during the process. The only exception concerns the digestions with the highest proportions of OMW (80%), which always had the worst results and where the decrease of ammonia nitrogen content

Table 9. Total phenols and antioxidant activity.

	Total phenols (g GAE/L)		Antioxidant activity (mmol TEAC)		DPPH Inhibition (%)	
Substrates	initial	final	initial	final	initial	final
[100%PE]	0.89 ± 0.001	0.86 ± 0.001	1.11 ± 0.01	0.64 ± 0.10	80.9 ± 0.70	30.3 ± 8.32
[70%PE+30%OMW]	1.69 ± 0.001	1.10 ± 0.000	1.07 ± 0.02	0.39 ± 0.08	78.6 ± 0.29	9.7 ± 6.58
[50%PE+50%OMW]	2.18 ± 0.000	1.52 ± 0.003	0.99 ± 0.02	1.34 ± 0.03	73.7 ± 1.37	69.1 ± 1.66
[20%PE+80%OMW]	2.69 ± 0.002	2.50 ± 0.001	0.91 ± 0.01	1.35 ± 0.02	68.1 ± 0.21	69.5 ± 1.50

confirms the deficient anaerobic digestion functioning, indicating an insufficient activity to perform the degradation/conversion of the organic matter. Accordingly, this mixture emphasized its acidic characteristics during digestion, in which the pH evolved from 6.2 to 5.7 (Table 8), testifying the presence of an imbalanced process in the [20%PE+80%OMW] unit, because of the inhibitory conditions establishment. In the other assays, the anaerobic process led to increased pH values, in a range close to neutrality (6.7–7.3 to 7.3–8.1).

The phenolics content (Total phenols, TP; Table 9) rise as the PE amount decreases in the initial substrates, indicating that OMW is mainly responsible for this increase and, subsequently, the highest TP content was found in admixtures containing OMW. The opposite is observed in relation to the antioxidant activity (Table 9). Doesn't seem to be significant difference for the initial values of the antioxidant activity (ranging 0.9–1.1 mmol TEAC and 68-81 % DPPH radical inhibition). One should bear in mind that different type of phenolics compounds is involved in both PE and OMW, and also, total phenolic content doesn't necessarily correlates with antioxidant activity. Nevertheless, after digestion, it was verified that the lowest antioxidant values were observed in units with highest PE amount ([100%PE] and [70%PE + 30%OMW]).

TP concentrations of 1.7 and 2.2 g/L were removed in proportions of 30 and 35% for [70%PE + 30%OMW] and [50%PE + 50%OMW], respectively, while [20%PE + 80%OMW] unit presented a poorer removal amount (7%) after digestion. No significant change was observed for TP concentration in piggery digestion. However, the antioxidant activity decreased to nearly half of the initial value, revealing that most likely, other compounds than phenolics in PE, which are degraded during digestion, may be responsible for this activity. Therefore, these decrease in antioxidant activity for the anaerobic effluent of [100%PE] and [70%PE + 30%OMW], indicates the loss of compounds of interest, however, the beneficial characteristics associated with these parameters were still evident in the digested material. In the two other assays ([50%PE + 50%OMW] and [20%PE + 80%OMW]), there was a maintenance or even an increase of the initial values for the antioxidant activity, revealing the importance of the OMW composition for this parameter. From the results obtained, one can infer that it is possible to valorise energetically this peculiar PE, alone or in admixture with a small amount of OMW, through the biogas production, and take additional advantage of the remaining compounds of interest present in the digested material.

4 CONCLUSIONS

Piggery effluent can be degraded under anaerobic conditions alone or in admixture with 30% OMW (v/v). The highest gas volumes (≈780 mL biogas, 70% CH_4) were obtained in both experiments. OMW is a substrate with high inhibiting capacity against the anaerobic digestion probably due to the known antimicrobial capacity of phenolic compounds in it, and only a small fraction such as 30% (v/v) can be added to PE without damaging the process.

The low biogas production obtained after using a proportion of 50% (v/v) OMW in the blend confirms the inhibitory characteristic of this substrate. However, it was observed that microbiota has the capacity to adapt to substrate disadvantageous characteristics. Then under the same operational condition, a possible extension of experimental time could provide better results. From the available data, there is no relevant advantage in associating OMW with this peculiar PE.

ACKNOWLEDGEMENTS

This work was financed by national funds through the *FCT – Fundação para a Ciência e a Tecnologia, I.P.* under the project ERANETLAC/0001/2014, GREENBIOREFINERY – Processing of brewery wastes with microalgae for producing valuable compounds. The authors would like to thank *Valorgado* and olive oil mill company of Rio Maior (Portugal) for effluents, and Natércia Santos for laboratory assistance.

REFERENCES

Assemany P., Marques I.P., Calijuri M.L., Silva T.L., Reis A. 2018. Energetic Valorization of Algal Biomass in a Hybrid Anaerobic Reactor. *J. Environ. Management* 209: 308–315.

APHA. 2012. Standard Methods for examination of water and wastewater, Washington DC.

ASTM D1946-90. 2000. Standard Practice for Analysis of Reformed Gas by Gas Chromatography. ASTM International, West Conshohocken, PA.

Brand-Williams W., Cuvelier M.E., Berset C. 1995. Use of a Free Radical Method to Evaluate Antioxidant Activity Lebensm.-Wiss. u.-Technol. 28: 25–30.

Eurostat. 2014. Pig farming sector – statistical portrait 2014. Access in 27 february 2019, available: http://ec.europa.eu/eurostat/statisticsexplained/index.php/Pig_farming_sector -statistical_portrait_2014

FAOSTAT. 2019. Access in 27 february 2019, available: http://www.fao.org/faostat/en/#data/QA.

Gonçalves M.R., Freitas P., Marques I.P. 2012. Bioenergy Recovery from Olive Mill Wastewater in a Hybrid Reactor. Biomass and Bioenergy 39: 253–260.

Hagerman A., Harvey-Muller I., Makkar H.P.S. 2000. Quantification of tannins in tree-foliage – a laboratory manual for the FAO/IAEA co-ordinated research project. In: Joint FAO/IAEA Working Document, 4–6, IAEA, Vienna.

Marques I.P. 2001. Anaerobic digestion treatment of olive mill wastewater for effluent re-use in irrigation. Desalination 137: 233–239.

Sampaio M., Marques I.P. 2012. Direct Anaerobic Digestion of Olive Oil Mills Wastes using Microalgae as a Complementary Substrate. Proc. of Symposium "Olive Oil Mills Wastes and Environmental Protection" 16–18 October 2012, Chania, Crete.

Sampaio M., Gonçalves M.R., Marques I.P. 2011. Anaerobic digestion challenge of raw olive mill wastewater. Bioresour. Technol. 102(23): 10810–10818.

Singleton V.L. and Rossi J.A. 1965. Colorimetry of total phenolics with phosphomolybdicphosphotungstic acid reagents. Am. J. Enol. Vitic. 16: 144–158.

Wastes: Solutions, Treatments and Opportunities III – Vilarinho et al. (Eds)
© 2020 Taylor & Francis Group, London, ISBN 978-0-367-25777-4

Hybrid anaerobic reactor: Brewery wastewater and piggery effluent valorisation

A. Neves, L.B. Roseiro, L. Ramalho, A. Eusébio & I.P. Marques
LNEG-Unit of Bioenergy, Lisboa, Portugal

ABSTRACT: A hybrid anaerobic reactor (HAR) operated to digest brewery wastewater, complemented with piggery effluent (60% and 40% v/v, respectively), under three hydraulic retention times: HRT: 5.7, 3.0 and 1.0 d. Along the first phase, the biogas of 0.9 $LL^{-1}d^{-1}$ evolved to 1.2 $LL^{-1}d^{-1}$ with a methane content of 77–78%. The HRT reduction to 3 days promoted the production until 2.1 $LL^{-1}d^{-1}$, with methane proportions of 79.5%. The operation with an even lower HRT (1 d) allowed to obtain a higher biogas (2.9 $LL^{-1}d^{-1}$, 79.5% CH_4). HAR was successfully applied to the mixture digestion, even with a HRT as low as 1 day. Phenolic compounds with antioxidant capacity in effluent adds value, besides the supply of biogas/methane.

1 INTRODUCTION

The production of beer has an important impact on the economic sector in Portugal. In 2017, 744 million liters of beer were produced in Portugal (Eurostat 2018). However, the brewery industry generates large volumes of highly polluted water, about 3–10 L of waste effluent per L of beer produced are generated depending on the production and specific water usage (Simate et al. 2011). The high pollution potential of these effluents is due to their organic load (sugars, soluble starch, ethanol and volatile fatty acids), suspended solids content and the presence of phosphorus and nitrogen (Raposo et al. 2010). The present work aims to valorise the wastewaters from the breweries through anaerobic digestion process using a hybrid anaerobic reactor, with the addition of another waste flow used as a complementary substrate.

2 MATERIALS AND METHODS

2.1 *Substrates*

Brewery wastewater (BWW) was collected from the Sociedade Central de Cervejas e Bebidas (SCC) brewery (Vialonga, Portugal). Piggery effluent (PE) was collected in Valorgado Company (Salvaterra de Magos, Portugal). Both effluents were previously subjected to the removal of most of the solid fraction.

2.2 *Experimental set-up*

Anaerobic digestion was carried out in the hybrid anaerobic reactor (HAR, Figure 1), designed in LNEG and tested with different organic materials. It is equipped with a packed bed selected in previous studies (Marques 2001), which was placed in the upper section of the column and occupying only 1/3 of reactor height. No device separator of solid/liquid/gas was installed and no substrate recycling was provided. Hybrid with a total volume of 2 L (1.7 L working volume), was functioning in up-flow mode, under semi-continuous conditions, fed by means of a peristaltic pump. The operational temperature was maintained at 37°C using a water jacket.

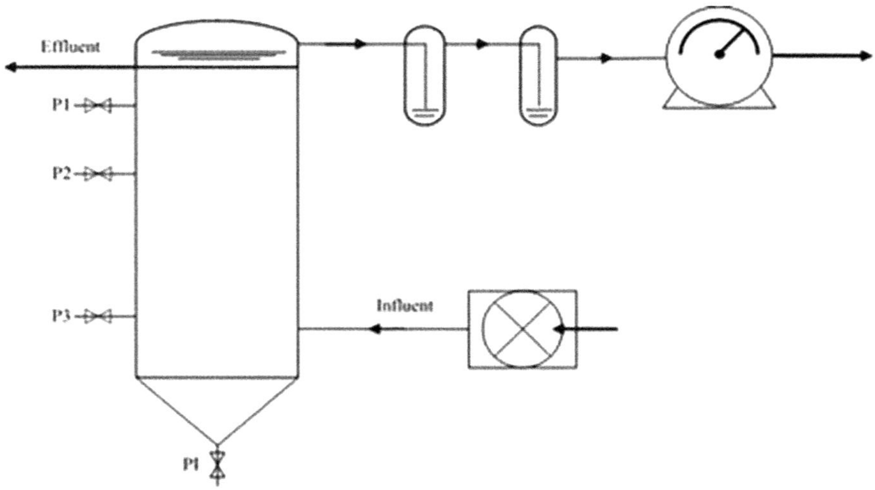

Figure 1. Hybrid anaerobic reactor.

Table 1. Scope of the anaerobic digestion experiment.

Assay	Time (d)	HRT (d)	Substrate composition (%, v/v)	
			BWW	PE
I	0-19	5.7	60	40
II	20-37	3.0	60	40
III	41-47	1.0	60	40

After a start-up phase (data not shown), it operated with different hydraulic retention times (5.7–1.0 days) to digest BWW and PE admixture in proportions of 60% and 40% v/v, respectively (Table 1). A wet gas meter was used to measure the biogas production volume, which is expressed to standard conditions for temperature and pressure (0°C, 1 bar).

2.3 *Analytical and chromatograph methods*

The performance of the anaerobic process was monitored by the influents and effluents analytical characterizations and by the volume and quality of the biogas obtained. Total and volatile solids (TS, VS), chemical oxygen demand (COD), total nitrogen (Kjeldahl, TN), ammonium (NH_4^+-N), and pH, were assayed according to Standard Methods (APHA 2012). Biogas production was monitored daily, by means of a pressure transducer, while the gas composition and volatile fatty acids (VFA) were analysed weekly by chromatographic techniques (Varian 430-GC, TDC; HP-5890, FID), according to ASTM Standard Method (D1946–90 2000). The characterization of substrates is shown in Table 2.

2.4 *Quantification of total phenolics content*

Total phenolics were determined using the Folin-Ciocalteu colorimetric method (Singleton & Rossi 1965), according to an improved procedure described by Hagerman et al. (2000). Briefly, the sample (or water for blank) (0.1 mL) was made up to 0.5 mL with distilled water and mixed with 1/1 (v/v)

Table 2. Effluents characteristics.

Parameters/Effluents	BWW	PE
pH	5.10	7.27
COD (g/L)	7.37 ± 0.00	93.22 ± 5.01
Total Solids (g/L)	3.6 ± 0.09	47.4 ± 0.79
Volatile Solids (g/L)	1.3 ± 0.09	31.9 ± 0.56
Total nitrogen (g/L)	0.025 ± 0.004	4.9 ± 0.277
Ammonium nitrogen (g/L)	0.007 ± 0.002	3.2 ± 0.019
Total phenolics (g/L)	0.004 ± 0.001	0.89 ± 0.001
Antioxidant activity (mmol TEAC)	0.04 ± 0.07	1.11 ± 0.01
Antioxidant activity DPPH inhibition (%)	6.7 ± 4.93	80.9 ± 0.70
Volatile fatty acids-VFA:		
Acetic acid (mg/L)	2272.00	1373.00
Propionic acid (mg/L acetic)	551.21	561.75
Isobutyric acid (mg/Lacetic)	66.79	2157.74
Butyric acid (mg/L acetic)	237.17	1545.72
Total VFA (mg/L acetic)	3127.18	5638.21

diluted Folin-Ciocalteu reagent (0.25 mL) and 20% $Na_2CO_3.10H_2O$ (1.25 mL). Absorbance was measured at 725 nm after 40 min incubation at room temperature. Total phenolics were expressed as g GAE (gallic acid equivalents)/L by comparison to a gallic acid standard curve. Results were obtained in triplicate.

2.5 *DPPH radical scavenging assay for Antioxidant activity evaluation*

Radical scavenging activity against stable DPPH radical (2,2-diphenyl-2-picrylhydrazyl hydrate) was determined spectrophotometrically according to a modified method (Brand-Williams et al. 1995). Briefly, a solution of DPPH in methanol (60 μM) was prepared daily, before UV measurements. This solution (1950 μL) was mixed with the sample solution (50 μL) (or methanol as blank) and vortexed. The samples were kept in the dark for 30 min at room temperature and then the decrease in absorption was measured. The decreasing of the DPPH solution absorbance indicated an increase of the DPPH radical scavenging activity. The experiment was carried out in triplicate. Radical scavenging activity (%) was calculated by the equation 1.

$$\%\text{DPPH inhibition} = [(Abs_b - Abs_f)/\ Abs_b] \times 100 \quad (1)$$

where, Abs_b is the absorption of blank (t = 0 min) and Abs_f is the absorption of tested solution (t = 30 min).

The antioxidant activity was expressed in mmol TEAC (Trolox Equivalent Antioxidant Capacity), by preparing a standard curve using Trolox as an antioxidant standard.

3 RESULTS AND DISCUSSION

3.1 *Effluents*

Brewery wastewater (BW) is a dilute substrate with very low organic matter (7 g/L COD), showing small nitrogen amounts and an acid pH (Table 2).

These characteristics are justified since the brewing unit effluent was previously submitted to a pre-treatment before collection as substrate for the present work, which probably led to the removal of a large fraction of the material contained in the raw sewage. In the presence of such substrate, it was necessary to find a solution that would allow creating an adequate balance to the

Table 3. Biogas composition.

Assay	HRT (d)	Organic load (g L^{-1} d^{-1})	Biogas production (L L^{-1} d^{-1})	Methane (% v/v)
I	5.7	5.28	1.16 ± 0.09	77.80 ± 0.00
II	3.0	9.97	2.16 ± 0.06	79.45 ± 0.05
III	1.0	33.60	2.91 ± 0.06	79.50 ± 0.00

establishment of the anaerobic digestion process. Mixing in another effluent had the advantage of the joint digestion, making the process economically more sustainable. However, just a simple addition does not solve the drawback of the imbalance composition, thus, an effluent with opposite characteristics had to be found, and PE is apparently advantageous to serve this purpose.

PE used to digest with BWW was markedly the opposite and could be applied as BWW complementary substrate. PE is a very concentrate substrate, having a high content in COD (93 ± 5 g/L) and solids. This effluent is characterized by a high nitrogen content (4.9 g L^{-1} TN, 3.2 g L^{-1} NH_4^+: Table 2), presents a pH near the neutrality range and also shows some phenolics content (TP) and antioxidant activity corresponding to about 80% inhibition of DPPH radical. Both effluents had a diverse VFA composition, with different acid species, as shown in Table 2. Comparatively, in terms of volatile composition, BWW is more stable and attractive for anaerobic digestion than PE, since the acetic acid is present in greater amounts than other acids, and almost doubles the concentration found in PE (2.27 versus 1.37 g L^{-1}).

3.2 *Biogas/methane production*

A biogas production of 0.9 L L^{-1} d^{-1} was obtained after 11 days of hybrid operation, under a HRT of 5.7 days. The biogas obtained was mostly composed of methane, and concentrations of 77-78% where consistently registered (Assay I, Table 3).

Maintaining the same conditions, it was observed that extending the experimental time until approximately 18 days, allowed an increase in biogas production to 1.2 L L^{-1} d^{-1} (77-78% CH_4). Reducing HRT to 3 days and operating with a loading rate of about 10 g L^{-1} d^{-1}, an increase in biogas volume was again attained, in which, mean values of 2.1 L L^{-1} d^{-1} obtained after one month, increased slightly to 2.2 L L^{-1} d^{-1} after a further week (35-37 days). The biogas quality was also improved, reaching values of about 80% (Assay II, Table 3). The highest gas volumes - 2.9 L L^{-1} d^{-1} of biogas and 2.3 L L^{-1} d^{-1} of methane - were obtained by applying a retention of 1 day (Assay III, Table 3).

Biogas production was increased as the residence time decreased, which means that even the highest loads of the last experimental period (HRT = 1 day) were comfortably digested. The quality of the biogas, which has been maintained or even slightly increased to concentrations of about 80%, indicates that the consortium preserves the equilibrium of interdependence and that the methanogenic *archaea* are in good conditions and capable of operating under different amounts of organic matter, keeping the working stability.

3.3 *Removal capacity*

The addition of the piggery effluent allowed the hybrid digester to operate with feed concentrations of 30–34 g/L (Table 4), which were not converted in the same extent for each of the experimental phases. The COD removal of 52% (Assay I) decreased in the following two assays to values of 12–26%, contradicting the progressive increase in gas production. Unexpected results were also obtained concerning the solids (Table 4), where the influent concentrations were removed/converted in such differentiated amounts. This can be justified by the decrease in the HRT giving an increase

Table 4. Chemical oxygen demand and solids.

	COD		Total Solids		Volatile Solids	
Assay	initial (g/L)	removal (%)	initial (g/L)	removal (%)	initial (g/L)	removal (%)
I	29.9 ± 0.94	51.84	22.5 ± 1.26	7.6	14.8 ± 0.97	21.6
II	29.9 ± 0.94	11.71	22.5 ± 1.26	54.2	14.8 ± 0.97	67.6
III	33.6 ± 2.60	25.60	22.5 ± 0.20	9.8	14.1 ± 0.23	13.5

Table 5. Volatile fatty acids.

	Acetic acid (mg/L)		Total VFA (mg/L acetic) (%)		
Assay	initial	final	initial	final	removal
I	2627.00	237.00	4235.60	534.95	87.37
II	2627.00	152.70	4235.60	195.49	95.38
III	1852.30	586.50	3078.80	1095.64	64.41

Table 6. Nitrogen.

	pH		Total nitrogen (g/L)		Ammonium nitrogen (g/L)	
Assay	initial	final	initial	final	initial	final
I	7.00	8.13	1.047 ± 0.008	1.086 ± 0.024	0.748 ± 0.008	0.753 ± 0.004
II	7.00	7.91	1.047 ± 0.008	0.823 ± 0.016	0.748 ± 0.008	0.703 ± 0.004
III	7.38	7.80	n/d	n/d	0.787 ± 0.004	0.802 ± 0.018

n/d – not determined

in the flow velocity and, consequently, some particles have been dragged from inside the reactor to the effluent and counted as non-degraded material.

The highest VFA concentrations (2.6 g/L, Table 5) available during the first two trials (Assays I and II), were more efficiently degraded than the lower amount (ca. 1.9 g/L) present in the last experimental phase.

VFA removals of 87–95% and 64% were registered, respectively. It is interesting to note that, in all situations, the influent acetic acid concentration corresponds to more than half of the acids total content present in the feed. As it is known, acetic acid is the most important in methane formation and, consequently, the good methane production obtained in any of the experimental phases is supported by the presence of the acetic acid in prevailing amounts, in the influent, and by the respective good removal efficiency. Accordingly, an identical good removal was also observed concerning other volatile acids (propionic, isobutyric and butyric), that need an efficient working capacity by the hydrogenotrophic population to convert them. No acidic pH values were recorded, conversely, a slight increase was found from 7.0–7.4 to 7.8–8.1 after anaerobic digestion (Table 6).

The initial total nitrogen content usually decreases during the anaerobic process due to degradation of organic matter, which, consequently, will generally result in the ammonium concentration increase in the effluent. From the data obtained, the organic matter degradation process did not have great impact on the nitrogen concentrations change. Amounts of 1.05 g/L TN and 0.75–0.79

Table 7. Total phenols and antioxidant activity.

Assay	Total phenols (g GAE/L) (%) initial	final	removal	Antioxidant activity (mmol TEAC) initial	final	DPPH inhibition (%) initial	final
I	0.37	0.23	37.8	0.85 ± 0.044	0.23 ± 0.081	60.5 ± 2.96	0.0 ± 6.57
II	0.37	0.23	37.8	0.85 ± 0.044	0.41 ± 0.067	60.5 ± 2.96	12.6 ± 4.05
III	0.36	0.33	8.3	1.11 ± 0.053	0.81 ± 0.081	55.0 ± 3.24	37.0 ± 4.71

g/L NH_4^+ in the influent evolved to effluent values of 0.82–1.09 g/L TN and 0.70–0.80 g/L NH_4^+ (Table 6).

Total phenol content was determined initially in order to evaluate the potential toxicity of these compounds towards biogas production, which are normally a limiting factor. On the other hand, the recognised antioxidant capacity of these compounds might add some value to the residual substrate after biogas production. Total phenol removal was observed in all experimental phases and it was accompanied by a decrease in the antioxidant capacity of the substrate during the process (Table 7). This means that the presence of phenols in the feed did not cause any inhibition in the process, therefore, no additional costs were necessary to remove them from the substrate.

However, this phenols reduction had as a consequence the loss of properties of this type of compounds, with respect to their potential commercial interest, namely, as antioxidants. Nevertheless, despite the decrease in the antioxidant activity, the effluent obtained after anaerobic digestion did not completely loose its properties as some antioxidant activity remained in the digested flow, which might be considered for further valorisation.

4 CONCLUSIONS

A progressive increase in biogas and methane productions were observed by decreasing the residence time but maintaining the same ratio of the two substrates. Biogas volumes of 1.2, 2.1 and 2.9 L L^{-1} d^{-1} and proportions of methane from 77% to 80% were registered when operating the hybrid anaerobic digester under HRT of 5.7, 3.0 and 1.0 days, respectively. The phenolic load of the piggery effluent didn't affect biogas/methane production, and their presence in the effluent thereof can add some value to the process. Thus, wastewaters from the breweries were successfully anaerobically digested with piggery effluent by using this hybrid anaerobic reactor.

ACKNOWLEDGEMENTS

This work was financed by national funds through the FCT – Fundação para a Ciência e a Tecnologia, I.P. under the project ERANETLAC/0001/2014, GREENBIOREFINERY - Processing of brewery wastes with microalgae for producing valuable Compounds. The authors would like to thank Sociedade Central Cervejas e Bebidas (SCC) brewery, Portugal, and olive mill of Rio Maior, Portugal, for effluents, and Natércia Santos for laboratory assistance.

REFERENCES

APHA-American Public Health Association. 2012. Standard Methods for examination of water and wastewater, Washington DC.

ASTM D1946-90. 2000. Standard Practice for Analysis of Reformed Gas by Gas Chromatography. ASTM International, West Conshohocken, PA.
Brand-Williams W, Cuvelier ME, Berset C. 1995. Use of a free radical method to evaluate antioxidant activity. *Lebenson Wiss Technol*. 28: 25–30.
EUROSTAT. 2018. Happy Beer day! Access in 27 feb 2019, available: https://ec.europa.eu/eurostat/en/web/products-eurostat-news/-/EDN-20180803-1.
Hagerman A., Harvey-Muller I., Makkar H.P.S. 2000. Quantification of tannins in tree-foliage – a laboratory manual for the FAO/IAEA co-ordinated research project. In: *Joint FAO/IAEA Working Document*, 4–6, IAEA, Vienna.
Marques I.P. 2001. Anaerobic digestion treatment of olive mill wastewater for effluent re-use in irrigation. *Desalination* 137: 233-239.
Raposo M.F.J., Oliveira S.E., Castro P.M., Bandarra N.M., Morais R.M. 2010. On the Utilization of Microalgae for Brewery Effluent Treatment and Possible Applications of the Produced Biomass. *J. Inst. Brew.* 116: 285–292.
Simate G.S., Cluett J., Iyuke S.E., Musapatika E.T., Ndlovu S., Walubita L.F., Alvarez A.E. 2011. The treatment of brewery wastewater for reuse: State of the art. *Desalination* 273: 235–247.
Singleton V.L. and Rossi J.A. 1965. Colorimetry of total phenolics with phosphomolybdicphosphotustic acid reagents. *Am. J. Enol. Vitic.* 16: 144-158.

Wastes: Solutions, Treatments and Opportunities III – Vilarinho et al. (Eds)
© 2020 Taylor & Francis Group, London, ISBN 978-0-367-25777-4

Winery wastes: A potential source of natural dyes for textiles

B. Lagoa, L. Campos & T. Silva
Instituto Politécnico de Coimbra, Instituto Superior de Engenharia de Coimbra, Coimbra, Portugal

M.J. Moreira & L.M. Castro
CIEPQPF, Departamento de Engenharia Química, Universidade de Coimbra, Coimbra, Portugal
Instituto Politécnico de Coimbra, Instituto Superior de Engenharia de Coimbra, Coimbra, Portugal

A.C. Veloso
CEB, Universidade do Minho, Braga, Portugal
Instituto Politécnico de Coimbra, Instituto Superior de Engenharia de Coimbra, Coimbra, Portugal

M.N. Coelho Pinheiro
CEFT, Faculdade de Engenharia da Universidade do Porto, Porto, Portugal
Instituto Politécnico de Coimbra, Instituto Superior de Engenharia de Coimbra, Coimbra, Portugal

ABSTRACT: The main goal of the present study was to explore textile dyeing ability of colorants extracted from winery wastes. Extractions with water at different pH and with a water/ethanol solution were performed. Cotton (natural and with a cationization pre-treatment) and wool were used in the dyeing process with two different time duration (100 and 200 min). The colorimetric and fastness properties, in terms of washing (with hot and cold water) and natural light exposure, were evaluated and the cationized cotton substrate was the one with the highest values of ΔE^* (parameter representing the difference in color for the fabric after and before dyeing) for all the conditions used to obtain the pomace grape extracts. Hue and tones obtained in the fabrics resulted in uniform colors going from the gray to brownish-gray with potential commercial acceptation, showing that the eco-valorization proposed for grape pomace wastes deserves further studies.

1 INTRODUCTION

Wine production generates large amounts of wastes. About 10–30% of the weight of grapes processed in winemaking remains as pomace containing seeds, skin, stems and pulp residues (Muhlack *et al.*, 2018; Mansour et al. 2017; Beres *et al.*, 2017).

These winery wastes are frequently used as fertilizer and soil conditioner due to the organic and nutrient content, as well as a source of natural antioxidants for the food industry (Arvanitoyannis *et al.* 2006). The valorization as raw material for bioconversions to produce biofuels and energy is a recent and promising alternative (Zacharof, 2017).

The grape pomace is rich in bioactive compounds, as phenolic compounds existing mainly in the seeds and skin of grapes. Among the most abundant groups of polyphenols are the flavonoids in which the family of anthocyanins is included. Anthocyanins are natural pigments with antioxidant properties (Drosou *et al.*, 2015; Tournour *et al.*, 2015). For that reason, extracts obtained from grape pomace containing anthocyanins deserved exploratory studies as possible source of natural dyes for textile dyeing (Baaka *et al.*, 2018; Baaka *et al.*, 2015).

Contributing to the transition for a circular economy, the present study aims to support the application of the grape pomace extracts as a natural source of dyes for the textile industry, as a novel process of winery wastes valorization. The experimental work conducted explored the effect of the solvent used to obtain the extracts and the dyeing process duration in the final color achieved in different textiles substrates (wool and cotton).

2 MATERIALS AND METHODOLOGIES

2.1 *Raw Material*

The grape pomace was provided by a local winemaker from Leiria region in the centre of Portugal. It resulted from a mixture of several grape varieties: Castelão, Tamarez and Baga for red grapes and Fernão Pires for white grapes. This solid waste was produced after winemaking from the 2018 harvest, by pressing the final residues remained from the grapes fermentation process. The grape pomace (a mixture of pressed grape skins, seeds, pulp residues and stalks) was taken to the laboratory and was stored in the freezer about one week before being used.

2.2 *Textiles substrates*

Three different textiles substrates (kindly provided by Tintex Textiles SA) were used to evaluate the colorant potential of the extracts obtained from grape pomace: wool (organic), cotton and cationized cotton. The cationized cotton fabric resulted from a previous treatment of the textile surface designated by cationization (the method is confidential information).

2.3 *Extraction Process*

Fresh grape pomace samples were extracted using a traditional extraction method with two different solvents: (distilled) water and a mixture of water and ethanol (50%, v/v). For the extraction in aqueous medium, three different pH values were used: 8, 12.2 and 2.4. All the extractions were carried out at 80°C ± 2°C during 3 h.

The extraction balloons with 500 mL of each of the desired solvent were immersed in a water thermostatized bath set to maintain the water at 80°C. When the extraction temperature was reached in the solvent, 30 g of grape pomace (66% moisture) was added into the extraction balloons equipped with condensers, in order to avoid losses of solvent by evaporation during the extraction process. Samples were collected from the extraction medium each 60 min to follow the evolution of colorants extracted by UV-vis spectrophotometry. Care was taken to collect small amounts to avoid significant changes in the extraction medium volume (in all cases less than 2%).

At the end of the extraction process, the solid fraction of extracts was separated by vacuum filtration with a glass fiber filter (Whatman GF-C, 0.47 mm diameter and 1.2 μm porosity). The liquid extracts were concentrated in a rotary vacuum evaporator before being lyophilized. The lyophilized extracts were stored until the assessment as colorant agents in textiles started.

After being weighed for extraction yield determination, the lyophilized extract was dissolved in 25 mL of distilled water and used, after dilution, in assays for textiles dyeing performance.

The extraction yield was determined as the percentage of the weight of the lyophilized extract relative to the dry weight of the grape pomace used for extraction.

2.4 *Dyeing Procedure*

The experiments implemented for textiles dyeing had different durations (100 min and 200 min). Tests were performed in closed flasks by immersion of textile samples (circles with 4.6 cm diameter) in the colored solutions (20 mL) obtained by dilution (10×) of the dissolved lyophilized extracts solutions. For each textile substrate, two flaks with the different extract solutions obtained were prepared. Similar flasks with distilled water were also prepared, to be used as controls for dyeing effectiveness of the three textiles substrates.

The dyeing and control flaks were stirred at 320 rpm in an orbital stirrer inside an oven at 50°C. After 100 or 200 min, dyeing flasks with wool, cotton and cationized cotton samples, and respective control flasks, were removed from de oven and the fabric samples were withdrawn.

2.5 *Fastness Tests*

After the coloration step, several tests were carried out to evaluate the color fastness obtained in the fabrics. Washing tests, with hot and cold water, and natural light exposure tests were performed before color assessment.

The wash color fastness was assessed by washing the colored textiles samples in a cold water (20°C) bath with agitation (320 rpm) during 10 min, followed by a similar washing step in a hot (45°C) water bath. Between the two washing steps, the fabric samples were removed from the water and wiped carefully in an absorbent paper before immersion in the next water bath.

To conclude if the color obtained in the fabrics resists fading caused by natural light exposure, the textile samples were placed in a tray, evenly distributed, with half of the surface covered with a black paper and remained like this one week. After that, it was possible to compare the sample color between the surface exposed to natural light with the one protected from light exposure.

2.6 *Color Measurement*

The color characteristics of the dyed samples was obtained measuring the color with a colorimeter (Konica Minolta, CR-200), according to the CIE (Commission Internationale de l'Eclairage) system using the $L^*a^*b^*$ coordinates. This system assessed the color in terms of lightness (coordinate L^*), redness-greenness and blueness-yellowness (coordinates a^* and b^*, respectively).

The CIE $L^*a^*b^*$ system is frequently used to compare color of two objects. The difference of color coordinates are ΔL^*, Δa^* and Δb^* and the total color difference can be stated as an unique value, designated by ΔE^*, calculated as:

$$\sqrt{(\Delta L^*)^2 + (\Delta a^*)^2 + (\Delta b^*)^2} \ .$$

In the present case $\Delta E*$ was used in the assessment of color difference between dyed fabric samples and respective control samples, between samples dyed in different conditions, between both sides of the same sample and also between the surface exposed to natural light and the surface protected from light exposure in the same sample. To address those objectives, 12 color measurements were made in each sample: 3 measurements evenly distributed in the light exposed surface and another 3 in the surface not exposed to light, in both sides of the sample.

2.7 *UV-vis Absorption Spectra*

The absorption spectras were recorded in a spectrophotometer (Thermo Scientific Evolution 201) in the wavelength range of 260-900 nm for samples collected during the extraction process.

3 RESULTS AND DISCUSSION

The yield of natural colorant extracted from pomace grapes varied from 11% to 19% (dry basis). The lower value corresponds to the extraction carried out with water with pH = 8 as solvent and the higher percentage was obtained with the ethanolic solution.

The colorant solutions obtained were evaluated by UV-vis spectrophotometry. One example for acid aqueous extraction is shown in Figure 1. Two peaks are observed and correspond to anthocyanin

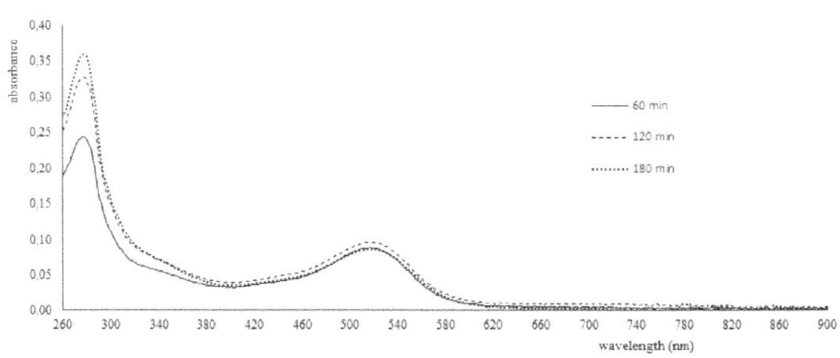

Figure 1. Absorption spectra for samples (diluted 25×) collected during pomace extraction with acid water.

Table 1. Colors (reproduced from RGB coordinates) for the textile substrates dyed with pomace grapes extracts obtained from extractions performed with different solvents and correspondents values of ΔE^* calculated using the respective controls as reference.

solvent	dyeing (min)	cotton (control / sample)	cotton cationized (control / sample)	wool (control / sample)
water (pH=8.0)	100	ΔE^*=19.58	ΔE^*=35.40	ΔE^*=23.19
	200	ΔE^*=35.36	ΔE^*=35.32	ΔE^*=25.99
water (pH=12.2)	100	ΔE^*=18.77	ΔE^*=33.48	ΔE^*=22.44
	200	ΔE^*=20.55	ΔE^*=18.63	ΔE^*=25.59
water (pH=2.4)	100	ΔE^*=20.93	ΔE^*=33.91	ΔE^*=28.62
	200	ΔE^*=26.43	ΔE^*=37.93	ΔE^*=30.58
water/etanol (50%, v/v)	100	ΔE^*=22.10	ΔE^*=31.71	ΔE^*=23.64
	200	ΔE^*=23.21	ΔE^*=31.80	ΔE^*=28.98

compounds, which have characteristics absorption peaks at wavelength regions in UV, 260–280 nm, and visible, 490–550 nm (Lombard *et al.*, 2002). An increase in absorbance with the extraction time is observed indicating an increase in the amount of colorants extracted.

Table 1 and Figure 2 present information about color assessment for all the fabric samples dyed with the extracts obtained from pomace grape treated with different solvents. The colors measured with $L^*a^*b^*$ system were convert to RGB coordinates in order to reproduce the textile sample colors before and after dyeing. The result is shown in Table 1 together with the ΔE^* value quantifying the color difference between dyed fabric samples and respective control samples. According to Mokrzycki & Tatol (2011), the following criteria can be used to establish the color difference based in the ΔE^* magnitude. If $0 < \Delta E^* < 1$ an observer does not notice difference in colors. The difference in colors can be noticed by an experienced observer if the values of ΔE^* are between 1 and 2. For values greater than 2 until 3.5 even an unexperienced observer can recognize the difference. For $\Delta E^* > 3.5$ exists a clear difference in color and when the value is greater than 5 any observer sees two different colors.

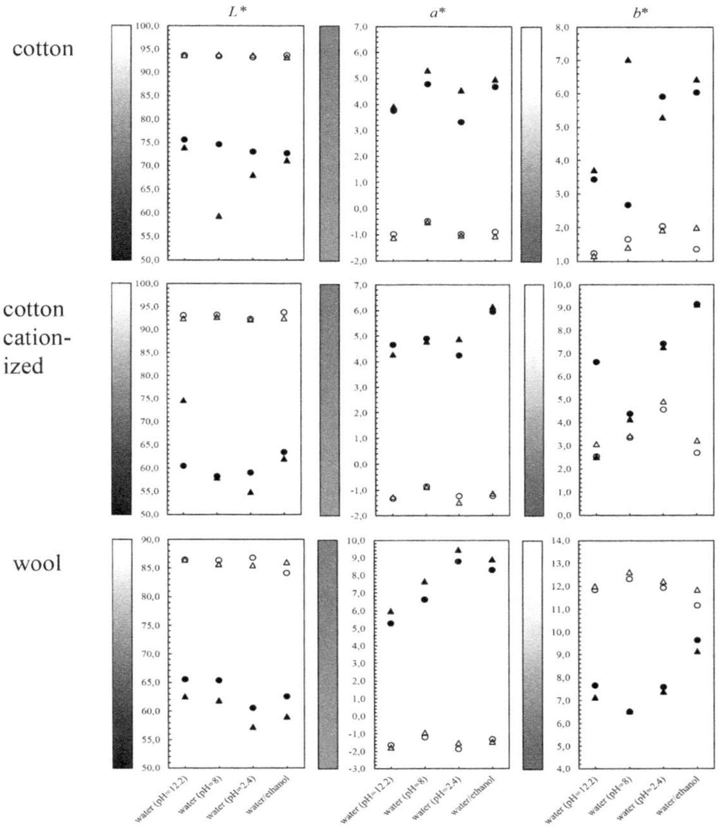

Figure 2. Coordinates $L^*a^*b^*$ measured for control samples (open symbols) and dyed samples (close symbols) for the three textile substrates used: cotton; cationized cotton and wool. The circles refer to 100 min of dyeing and the triangles correspond to 200 min of dyeing.

Regarding color difference between both sides of the same sample, all the calculated ΔE^* values were in the range [0-1] or [1-2], indicating that there are no observable color differences or only an experienced observer could noticed any difference. The same conclusion was obtained from the ΔE^* values calculated for the same sample between the surface exposed and protected from natural light, showing a good light color fastness performance of the dyed samples.

Visual comparison of colors obtained in dyed fabric samples and the respective control samples (see Table 1) allows the perception of two different colors for all the conditions used and all the textile substrates ($\Delta E^* > 5$, in all cases). Although, the more intense color was reached with the cationized cotton, which is confirmed by the highest values of ΔE^*. The exception is for 200 min of dyeing with the extract obtained with the basic aqueous solvent, where the wool showed a better coloration performance with $\Delta E* = 25.59$.

The visual inspection of Table 1, clearly shows that the time (100 or 200 min) the textile samples were immersed in the baths prepared with the pomace grape extracts obtained from the different extractions, in general, does not have influence on the final color. The same observation can be made from Figure 2 comparing the three coordinates (L^*, a^* and b^*) measured in samples with different times of dying. Two particular cases deserve mention in terms of the final tone obtained: *i*) the cationized cotton dyed with the extract resulted from the extraction with water at basic pH, presents more brightness (greater L^*) and is less yellowish (lower b^*) when it remains 200 min in the dyeing bath (Figure 2); *ii*) the cotton sample obtained with the pomace grape extract resulting from the aqueous extraction with pH = 8.0, is darker (lower L^*) and more yellow (greater b^*) when it remains more time in the dyeing bath (Figure 2).

4 CONCLUSIONS

Natural dyes obtained from wastes are an environmentally sustainable and healthy way to promote the replacement of chemical dyes used in textile industry.

The dye solutions were obtained from winery wastes using eco-friendly solvents (water with different pH and a mixture water/ethanol) as extraction agents. Samples of cotton (natural and with a cationization treatment) and wool remained immersed in the dyeing bath for 100 min and 200 min. The final color in the textile substrates was measured after washing (with hot and cold water) and natural light exposure. In general, no significant change in color intensity was obtained for samples with different dyeing times, which seems that a dyeing time of 100 min is sufficient.

The gray/brownish-gray colors obtained in the dyed fabrics appear to be interesting in a commercial point of view. Thus, the results obtained allow concluding that the valorization of winery wastes as a source of bio-dyes for textiles has an interesting industrial potential, deserving further exploratory and systematized studies to consolidate improvements in color/fastness properties.

ACKNOWLEDGMENTS

This work is supported by national funds through the Portuguese Foundation for Science and Technology (FCT) and co-financed by the European Regional Development Fund (FEDER), through the partnership agreement Portugal2020-Regional Operation Program CENTRO2020, under the project CENTRO-01-0145-FEDER-023631. The authors are grateful to Tintex Textiles S.A. for the supply of textile substrates used in this work and for technical assistance.

REFERENCES

Arvanitoyannis, I.S., Ladas D. & Mavromatis A. 2006. Potential uses and applications of treated wine waste: a review. *International Journal of Food Science and Technology* 41: 475–487.

Baaka, N., Haddar, W., Ticha, M.B. & Mhenni M.F. 2018. Eco-friendly dyeing of modified cotton fabrics with grape pomace colorant: Optimization using full factorial design approach. *Journal Natural Fibers.*

Baaka, N., Ticha, M.B., Haddar, W., Hammami, S. & Mhenni M.F. 2015. Extraction of Natural Dye from Waste Wine Industry: Optimization Survey Based on a Central Composite Design Method. *Fibers and Polymers* 16(1): 38-45.

Beres, C., Costa, G.N.S., Cabezudo, I., da Silva-James, N.K., Teles, A.S.C., Cruz, A.P.G., Mellinger-Silva, C., Tonon, R.V., Cabral, L.M.C. & Freitas, S.P. 2017. Towards integral utilization of grape pomace from winemaking process: A review. *Waste Management* 68: 581-594.

Drosou, C., Kyriakopoulou, K., Bimpilas, A., Tsimogiannis, D. & Krokida M. 2015. A comparative study on different extraction techniques to recover red grape pomace polyphenols from vinification byproducts. *Industrial Crops and Products* 75: 141–149.

Lombard, K.A., Geoffriau, E. & Peffley, E. 2002. Flavonoid quantification in onion by spectrophotometric and high performance liquid chromatography analysis. *HortScience* 37(4): 682–685.

Mansour, R., Ezzili B. & Farouk M. 2017. The use of response surface method to optimize the extraction of natural dye from winery waste in textile dyeing. *The Journal of the Textile Institute* 108(4): 528-537.

Mokrzycki W.S. & Tatol M. 2011. Color difference Delta E–A survey. *Machine Graphics and Vision* 20(4): 383-411.

Muhlack, R.A., Potumarthi, R. & Jeffery D.W. 2018. Sustainable wineries through waste valorisation: A review of grape marc utilisation for value-added products. *Waste Management* 72: 99–118.

Tournoura, H.H., Segundo, M.A., Magalhaes, L.M., Barreiros, L., Queiroz, J. & Cunha L.M. 2015. Valorization of grape pomace: Extraction of bioactive phenolics with antioxidant properties. *Industrial Crops and Products* 74: 397-406.

Zacharo, M.-P. 2017. Grape Winery Waste as Feedstock for Bioconversions: Applying the Biorefinery Concept. *Waste Biomass Valor* 8: 1011–1025.

Wastes: Solutions, Treatments and Opportunities III – Vilarinho et al. (Eds)
© 2020 Taylor & Francis Group, London, ISBN 978-0-367-25777-4

Isothermal drying of sewage sludge with eggshell for soil applications

L.A. Gomes
CIEPQPF – Center of Chemical Processes Engineering and Forest Products, Department of Chemical Engineering, University of Coimbra, Coimbra, Portugal
IFB – Federal Institute of Education, Science and Technology of Brasília – IFB, Campus Ceilândia Brasília – Federal District, Brazil

A.F. Santos & M.J. Quina
CIEPQPF – Center of Chemical Processes Engineering and Forest Products, Department of Chemical Engineering, University of Coimbra, Coimbra, Portugal

J.C. Góis
Association for the Development of Industrial Aerodynamics, Department of Mechanical Engineering, University of Coimbra, Coimbra, Portugal

ABSTRACT: The objective of this work was to analyze the effect of eggshell (ES) as adjuvant of the drying process of sewage sludge (SS) from anaerobic digestion. Small cylinders were dried in isothermal conditions at 70 and 100 °C, until complete dehydration. Two mathematical models were used to describe the experimental data, and in both cases good fitting was achieved ($R^2 > 0.98$ and RMSE < 0.02). The mixture SS:ES equal to 85:15 revealed a positive effect in the drying rate since the drying time was reduced. When ES was used, the average drying rate observed in the first period at 70°C was improved by 9.86%, while at 100°C almost 17% improvement was achieved. The diffusion coefficients (D_{eff}) were also calculated, and the improvements in this parameter were also detected. Indeed, the application of ES as adjuvant increased D_{eff} in 9% and 16,32% at 70 and 100°C, respectively.

1 INTRODUCTION

Sewage sludge (SS) is a by-product generated in wastewater treatment plants (WWTP), representing a global problem due to population growth and urbanization. It is estimated that by 2020 the SS production in the European Union (EU) will be about 13 Mt on dry basis (db) (Collard et al. 2017). In Portugal, the annual production of SS is around 300 kt (db) (LeBlanc et al. 2008). This amount will increase because new WWTP are under construction. Among the main problems associated with SS management, the following aspects stand out: i) high water content, normally above 80%; ii) release of bad odors; iii) presence of pathogenic microorganisms and iv) the possible presence of potentially toxic metals (PTM) above the limits imposed by legislation (Alvarenga et al. 2017). In the EU, the management of SS may involve agriculture applications, incineration, or landfilling. Recently, some studies highlighted the use of SS as an agriculture soil improvement agent. In fact, SS may increase organic carbon storage, promote recycling of the nutrients (e.g. N, P and K), improve water retention capacity, soil aeration and cation exchange (Alvarenga et al. 2017; Ociepa et al. 2017; Syed-Hassan et al. 2017).

Another waste produced in huge quantities in the EU is the eggshell (ES). The consumption of eggs is estimated to be around 10^{12} in worldwide (at home, restaurant, bakehouse, and industry) (Park et al. 2016). Normally, the ES wastes are neglected and discarded in landfills. However, in the last years, some researches have evaluated the potential of recycling this material as a soil amendment, agent of nutrients and PTM adsorbent (Buasri et al. 2013; Quina et al. 2017; Usman,

2010). The main constituent of ES is $CaCO_3$, which gives it the capability to be used as a buffering agent. Thus, the application of the SS mixed with ES in the soil has gained importance due to the reduced amount of organic matter (OM) in most of the Portuguese soils (less than 1%), and the acidity (pH<5.5) in about 83% of the national territory (Inácio et al. 2008; Lopes, 2017).

To mitigate some problems associated with SS, such as high water content and pathogens, thermal drying can be an efficient method. For reducing cost with energy during drying, the initial moisture content should be near to 80%. Generally, the thermal processes allow achieving final moisture of less than 30%. Several mathematical models based on the general solution of Fick's second law can be used to describe the diffusional process and to optimize the drying process. Moreover, the analysis of the kinetics can be important to obtain the parameters required to design dryer equipment. In the scope of this work and toward the circular economy, the recycling of ES was investigated as an adjuvant of the SS drying process. Thus, this study aims to analyze the drying kinetics of SS mixed with ES, and produce a material for agronomic applications.

2 MATERIALS AND METHODS

2.1 *Materials*

In this work, two wastes samples with different physicochemical compositions were studied. SS sample was obtained after mechanical dewatering by centrifugation in a WWTP, in the central region of Portugal. The WWTP valorizes primary and secondary municipal sludge by anaerobic digestion (AD). The operational capacity of the WWTP is 36,000 m^3 day^{-1} of urban effluent. The initial moisture content of SS was around 80% and the sample was kept at 4°C until further utilization. ES sample was collected in a grocery store. The sample was washed with tap water and distilled water several times, dried at room temperature, milled and sieved through a 425 μm screen.

2.2 *Physical and chemical characterization*

Moisture, total solids (TS) and OM content were determined based on EPA Method 1684. pH and electrical conductivity (EC) were measured in a 1:10 (solid:liquid) suspension (Alvarenga et al. 2016; Oleszczuk & Hollert, 2011). Total nitrogen was determined using the Kjeldahl method. The samples were dried at 105°C, ground and sieved through a ~75 μm (200 mesh) screen to determine the major elements (K, P, Ca, Mg, Mn, Si, Ti, Al, and Fe). For this determination, about 4 g of each sample was used in X-ray fluorescence (XRF) in a Nex CG Rigaku spectrometer (Healy et al. 2016). Mineralogical characterization of ES was determined by X-ray diffraction analysis (XRD) using a PANalytical X'Pert PRO diffractometer, CuKα radiation, with a scanning range from 10 to 80° (2θ).

2.3 *Drying procedures*

The drying process was investigated using small cylinders (5 mm diameter and 30 mm length) obtained through extrusion of SS (control) and SS:ES in a proportion (w/w) of 85:15 (referred as SS_ES). The mass of each cylinder was nearly 0.75 g. The drying tests were conducted placing 15 cylinders in an aluminum dish and dried in an oven with natural convection, at 70 and 100°C, until constant weight.

2.4 *Drying models*

The Fick's second law was used to describe the drying process in the first period, and Eq. (1) shows its solution for a cylinder of "infinite length", which allows the analysis along one-dimension (radial direction). To solve the Fick's second law a uniform initial moisture distribution was assumed, neglecting shrinkage and temperature gradients (Danish et al., 2016; Figueiredo et al. 2015). In this study, only the first term of the series in Eq. (1) was considered. The Henderson & Pabis model (thin-layer model), represented by Eq. (2) was used to represent the second period of drying kinetics. The modulus of drying rate |DR|, expressed in $gH_2O\ min^{-1}\ kg^{-1}SS_{wb}$ (wet basis), represents the amount of water leaving the cylinder per kg SS_{wb} min^{-1}, was determined by Eq. (3). To calculate the effect of the adjuvant in the drying process, the ES mass was discounted in the calculations.

$$MR = \frac{M(t) - M_e}{M_0 - M_e} = \sum_{n=1}^{\infty} \frac{4}{r^2 \alpha_n^2} \exp(-Deff \alpha_n^2 t) \tag{1}$$

$$MR = \frac{M(t) - M_e}{M_0 - M_e} = ae^{-kt} \tag{2}$$

$$|DR| = \frac{\partial M(r,t)}{\partial t} \approx \frac{\Delta M}{\Delta t} = \frac{M_i - M_{i-1}}{t_i - t_{i-1}} \tag{3}$$

where MR is the dimensionless moisture ratio; M_i and M(t) are the moisture content in the cylinder (g water g dry solid^{-1}) at t=0 and at any t time, respectively; M_e is the moisture content in equilibrium (g water g dry solid^{-1}); J_0 is the Bessel function of zero order ($J_0(r\alpha_n) = 0$; α_1 is the first root of the Bessel function of zero order; r is the radius of the cylinder (m); D_{eff} is the effective diffusion coefficient ($m^2\ s^{-1}$); *a* represents a parameter of the model (thin-layer model); k is the kinetic constant (min^{-1}) and *t* is the time (min).

2.5 *Statistical analysis*

The fitting quality of the mathematical models was evaluated based on the root mean square error (RMSE) and on the coefficient of determination (R^2) (Bennamoun et al. 2013; Danish et al. 2016). The length of the Period I of the drying process, where the drying rate is constant, was determined considering MR = 0.30.

3 RESULTS AND DISCUSSION

3.1 *SS and ES characteristics*

Table 1 shows the physico-chemical characteristics of the two wastes studied (SS and ES). The results found in this study are consistent with those reported in the literature.

The pH found for SS and ES are 6.7 and 9.1, respectively, and agree with the results reported in the literature. Thus, mainly ES waste can present an alternative for the correction of acid soils, such as Portuguese soils (Inácio et al. 2008). The OM and the presence of nutrients, such as N, P and K, are higher in SS comparing to ES. The SS contains some Ca, but the ES may comprise more than 90% of calcium carbonate ($CaCO_3$). The XRD analysis (spectrum not shown in this work) confirmed that ES contains mostly $CaCO_3$ in the calcite form.

3.2 *Drying kinetics*

Figure 1 presents the isothermal drying curves that correspond to the moisture ratio (MR) as a function of time, at 70 and 100°C for the control (0% of adjuvant) and SS_ES (15% of adjuvant). In addition, the experimental data was adjusted through the Fick's second law (Period I) and the thin-layer model (Period II).

Results in Figure 1 show that the theoretical models present a very good fit to the experimental data, which is confirmed by the R^2 and RMSE values (Table 2). Indeed, the drying curve can be divided into three distinct periods: heating phase (Period 0) where the temperature is rising; the

Table 1. Properties of SS and ES used in this study and in the literature.

	SS			ES		
		Literature			Literature	
Parameters	This work	(Pathak et al. 2009)	(Alvarenga et al. 2016)*	This work	(Usman et al. 2010)	(Quina et al. 2017)
pH	6.7	5.0 – 8.0	7.3	9.10	ni	8.73
OM (% TS)	63.7	30 – 88.0	70.9	4.30	ni	6.3
Moisture (%)	78	>95.0	ni	1.10	ni	1.0
EC (mS cm^{-1})	0.17	ni	0.26	0.21	ni	0.45
$N_{Kjeldahl}$(%)	3.89	ni	6.20	nd	ni	1.0
P_2O_5 (% TS)	3.83	0.8 – 11	10.20	0.29	0.15	ni
K_2O (% TS)	0.22	0.4 – 3.0	1.30	0.10	0.07	ni
MgO (% TS)	0.38	ni	1.04	0.72	0.003	ni
CaO(% TS)	5.30	ni	2.78	86.11**	91.99**	88.0**
SiO_2(% TS)	4.05	10-20	ni	0.19	0.13	ni
Al_2O_3(% TS)	1.94	ni	ni	0.19	0.14	ni
Fe_2O_3(% TS)	11.56	ni	ni	nd	0.04	ni
Na_2O (% TS)	nd	ni	0.27	nd	0.51	ni

nd – not determined; ni – not indicated; * mean of 2 sewage sludges samples. ** these values are reported in $CaCO_3$

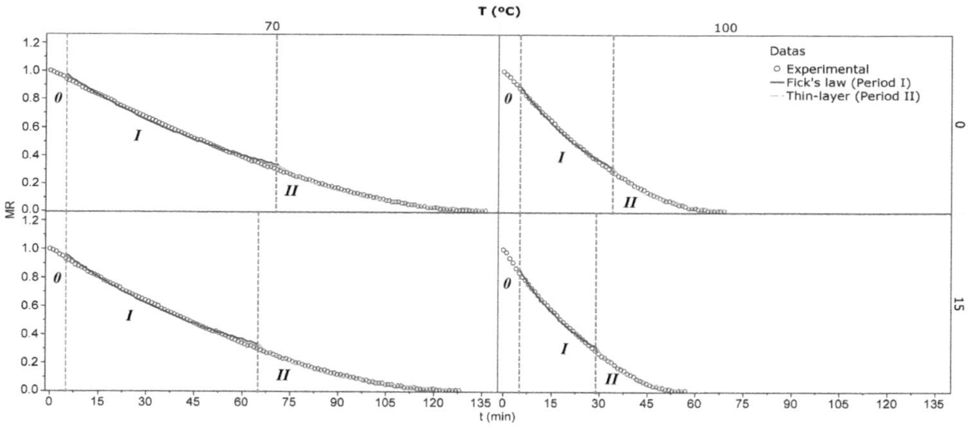

Figure 1. Moisture ratio (MR) as a function of time to Control and SS_ES samples, at 70 and 100°C. [Periods 0, I and II correspond to the rising period, the constant rate period, and falling rate period, respectively; Dashed vertical lines separate the drying phases].

constant rate phase (Period I) where the free water is evaporated; and the decay stage (Period II) where the bounded water is removed. Period 0 was assumed equal to 5 min for all the tests.

Table 2 summarizes the principal parameters that describe the drying curves and Figure 2 represents the modulus of the drying rate (DR), expressed in gH_2O kg SS_{wb} min^{-1} at 70°C and 100°C, respectively.

Through the drying curves (Figure 1), it is possible to notice that the drying profiles are dependent on the temperature. At 70°C the decay of MR is slower than at 100°C, indicating difficulties to remove both free and bound water. Moreover, it is possible to observe that at 70°C, the Period I is

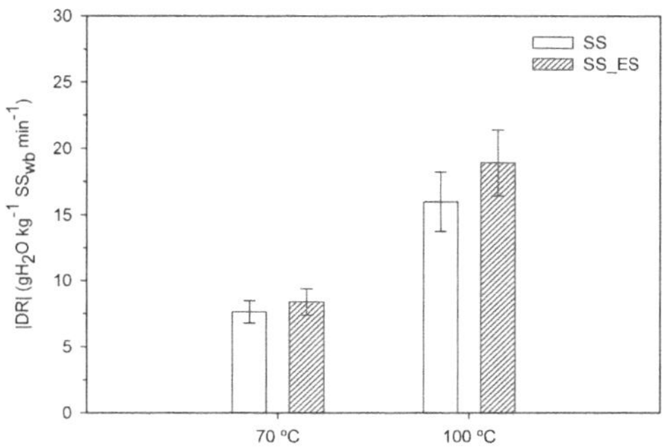

Figure 2. Drying rate |DR| expresses as gH_2O kg SS_{wb} min^{-1} at 70 and 100°C to control and SS_ES samples.

Table 2. Parameters obtained for the different periods of the drying curves.

		Fick's second law (Period I)					Thin-layer (Period II)			
T (°C)	Adj. (%)	D_{eff} (m^2 min^{-1})	R^2	t_f (min)	$\lvert\overline{DR}\rvert$ (gH_2O kg^{-1} SS_{wb} min^{-1})	RMSE	k (min^{-1})	*a*	R^2	RMSE
70	0	2.66×10^{-8}	0.992	71	7.63	0.02	0.0433	7.271	0.981	0.01
	15	2.89×10^{-8}	0.992	65	8.34	0.02	0.0446	6.054	0.984	0.01
100	0	6.01×10^{-8}	0.994	34	15.98	0.01	0.0849	5.930	0.986	0.01
	15	6.99×10^{-8}	0.995	29	18.74	0.01	0.1038	6.686	0.987	0.01

Adj. – adjuvant; t_f – time at the end of Period I; $\overline{DR}$ – average drying rate for Period I.

9% shorter in the case of SS_ES (t_f= 65min) than in the control sample (t_f= 71 min). At 100°C, the period I ends at 29 min for SS_ES, while 34 min were measured for 0% of adjuvant (Control sample). At 100°C, the D_{eff} obtained was 16.32% higher for the drying aided with adjuvant (Table 2). Besides, the values founded to D_{eff} are within the range reported in the literature (Angelopoulos et al. 2016). In terms of prediction of Period II with the Thin-layer model, the kinetic constant (k) of drying at 70 and 100°C with adjuvant increased by 3.19 and 22.26%, respectively. The average drying rate at 70°C with adjuvant, compared to the control, improved the drying process by 10%, while at 100°C almost 18% of improvement was achieved. As expected, the DR is higher at 100°C and it is possible to achieve a maximum of 18.74 gH_2O kg SS^{-1} min^{-1} in the presence of ES (Figure 2). On the other hand, only 8.34 gH_2O kg SS^{-1} min^{-1}were removed from the sample at 70°C during the Period I.

4 CONCLUSIONS

According to the results, the use of eggshell as an adjuvant to the sewage sludge drying process enhances the performance either at 70 and 100°C. In the drying period I, the addition of the adjuvant promoted an improvement in the diffusion coefficients of 9.0 and 16.3% to 70 and 100°C, respectively. Besides, the ES increased the drying rate compared to the control sample. The final product (SS mixed with 15% of ES) may be valuable for agriculture applications that require extra calcium or in case the soil requires pH correction. Finally, in the context of the circular economy, it is suggested to test other materials as SS drying agents to produce soil improvement agents with specific properties.

ACKNOWLEDGEMENTS

The authors gratefully acknowledge Federal Institute of Education, Science, and Technology of Brasília – IFB, Campus Ceilândia, for authorising the Ph.D. studies of Luciano A. Gomes.

This work was developed under the project "Dry2Value – Estudo e desenvolvimento de um sistema de secagem para valorização de lamas". Project consortium with HRV e LenaAmbiente. POCI-01-0247- FEDER-033662. Funded by Fundo Europeu de Desenvolvimento Regional (FEDER) – Programa Operacional Competitividade e Internacionalização.

REFERENCES

Alvarenga, P., Farto, M., Mourinha, C., Palma, P. 2016. Beneficial use of dewatered and composted Sewage sludge as soil amendments: behaviour of metals in soils and their uptake by plants. *Waste and biomass valorization* 7: 1189–1201.

Alvarenga, P., Palma, P., Mourinha, C., Farto, M., Dôres, J., Patanita, M., Cunha-Queda, C., Natal-da-Luz, T., Renaud, M., Sousa, J.P. 2017. Recycling organic wastes to agricultural land as a way to improve its quality: A field study to evaluate benefits and risks. *Waste Manag.* 61: 582–592.

Angelopoulos, P.M., Balomenos, E., Taxiarchou, M. 2016. Thin-Layer Modeling and determination of effective moisture diffusivity and activation energy for drying of red mud from filter presses. *J. Sustain. Metall.* 2: 344–352.

Bennamoun, L., Arlabosse, P., Léonard, A. 2013. Review on fundamental aspect of application of drying process to wastewater sludge. *Renew. Sustain. Energy Rev.* 28: 29–43.

Buasri, A., Chaiyut, N., Loryuenyong, V., Wongweang, C., Khamsrisuk, S. 2013. Application of eggshell wastes as a heterogeneous catalyst for biodiesel production, 1, 7–13.

Collard, M., Teychené, B., Lemée, L. 2017. Comparison of three different wastewater sludge and their respective drying processes: Solar, thermal and reed beds – Impact on organic matter characteristics. *J. Environ. Manage* 203: 760–767.

Danish, M., Jing, H., Pin, Z., Ziyang, L., Pansheng, Q. 2016. A new drying kinetic model for sewage sludge drying in presence of CaO and NaClO. *Appl. Therm. Eng.* 106: 141–152.

Figueiredo, R., Costa, J., Raimundo, A. 2015. *Transmissão de Calor: Fundamentos e Aplicações*, 1st ed. Lidel – Edições Técnicas, Lda, Lisboa, Portugal.

Inácio, M., Pereira, V., Pinto, M. 2008. The Soil Geochemical Atlas of Portugal: Overview and applications. *J. Geochemical Explor.* 98: 22–33.

LeBlanc, R.J., Matthews, P., Richard, R.P. 2008. *Global atlas of excreta, wastewater sludge, and biosolids management: moving forward the sustainable and welcome uses of a global resource*, Un-Habitat. Kenya.

Lopes, J.V.S. 2017. *Análise e melhoria dos processos de gestão de lamas de depuração de efluentes líquidos Analysis and improvement of the management system of sewage sludge.* Master Thesis. Universidade de Coimbra, Portugal.

Ociepa, E., Mrowiec, M., Lach, J. 2017. Influence of fertilisation with sewage sludge-derived preparation on selected soil properties and prairie cordgrass yield. *Environ. Res.* 156: 775–780.

Oleszczuk, P., Hollert, H. 2011. Comparison of sewage sludge toxicity to plants and invertebrates in three different soils. *Chemosphere* 83: 502–509.

Park, S., Choi, K.S., Lee, D., Kim, D., Lim, K.T., Lee, K.H., Seonwoo, H., Kim, J. 2016. Eggshell membrane: Review and impact on engineering. *Biosyst. Eng.* 151: 446–463.

Pathak, A., Dastidar, M.G., Sreekrishnan, T.R. 2009. Bioleaching of heavy metals from sewage sludge: A review. *J. Environ. Manage.* 90: 2343–2353.

Quina, M.J., Soares, M.A.R., Quinta-Ferreira, R. 2017. Applications of industrial eggshell as a valuable anthropogenic resource. *Resour. Conserv. Recycl.* 123: 176–186.

Syed-Hassan, S.S.A., Wang, Y., Hu, S., Su, S., Xiang, J. 2017. Thermochemical processing of sewage sludge to energy and fuel: Fundamentals, challenges and considerations. *Renew. Sustain. Energy Rev.* 80: 888–913.

Usman, A.R.A., Ok, Y.S., Jeon, W.-T., Moon, D.H., Lee, S.S., & Oh, S.-E. 2010. Application of eggshell waste for the immobilization of cadmium and lead in a contaminated soil. *Environmental Geochemistry and Health*, 33: 31–39.

Wastes: Solutions, Treatments and Opportunities III – Vilarinho et al. (Eds)
© 2020 Taylor & Francis Group, London, ISBN 978-0-367-25777-4

Optimization of the ship waste management system in the Port of Lisbon

S.A. Melón & A.M. Barreiros
ADEQ-ISEL/IPL, Lisboa, Portugal

V.C. Godinho
APL, S.A., Lisboa, Portugal

ABSTRACT: Nowadays, maritime transport is being considered the most cost-effective way to move goods and raw materials around the world. However, maritime transport introduced negative environmental impacts, not only in marine ecosystems but also in land ecosystems, such as ship generated waste. In order to prevent marine environmental pollution caused by the illegal discharge of harmful substances and waste, in 1973, an International Convention was adopted: the MARPOL protocol 73/78. Port of Lisbon follows the regulations demanded by the Directive 2000/59/CE from the EU on port reception facilities, whose main aim is to prevent waste discharge into the sea. This Directive is based on the EU policy on waste management on land. The aim of this study is to analyze and quantify the amount of ship generated waste discharged in the Port of Lisbon, identifying the existing barriers and contributing to a continuous improvement process.

1 INTRODUCTION

The international maritime industry is responsible for transportation of more than 90 % of the world commodities, and it is considered the most cost-effective way to move goods and raw materials around the world (ICS, 2019). Maritime transport introduced negative environmental impacts, not only in marine ecosystems but also in land ecosystems, such as ship generated waste (SGW) and other dangerous pollutants and air pollution.

In order to prevent marine environmental pollution caused by the discharge of illegal, harmful substances and waste, in 1973, an International Convention was adopted for the Prevention of Ship Pollution which was later modified in 1978 by the MARPOL protocol 73/78. MARPOL regulates what kind of waste can be discharged by ships into the sea and demands that States Parties make sure they offer the necessary facilities for ship reception in ports. MARPOL includes six technical Annexes providing regulations for the prevention of pollution: by "*oil*" (Annex I); by "*noxious liquid substances in bulk*" (Annex II); by packaged harmful substances (Annex III); by "*sewage from ships*" (Annex IV); by "*garbage from ships*" (Annex V) and "*the prevention of air pollution from ships*" (Annex VI) (IMO, 2019). Likewise, the European Union (EU) adopted the Directive 2000/59/CE, changed by the Directive 2015/2087, from the EU on port reception facilities (PRF), which main aims are to prevent waste discharge of ship-generated waste and cargo residues into the sea, by improving the availability and use of PRF and protection of the marine environment. The directive focuses mainly on port operations and is based on the EU policy on waste management on land. MARPOL focuses on operations at sea. In this way, the Directive is aligned with the MARPOL Convention and complements it, regulating the legal, practical and financial responsibilities in the articulation between land and sea. The Directive 2000/59/CE was transposed into national law by Decree-Law 165/2003 of 24 July change by Decree-Law 83/2017 of 18 July.

An adequate and dynamic management of ship generated waste in the Port of Lisbon (PL) is very important to minimize the discharge of waste into the sea and for that reason the PL follows the regulations demanded. The PL has implemented a Ship Generated Waste Management (SGWM) system which aims the following objectives: reducing discharges of ship-generated waste and cargo residues into the sea; reducing illegal discharges from ships using ports in the EU; improving the availability and use of PRF; protection of the marine environment with compliance with Decree-Law 165/2003 of 24 July.

PL main goals regarding SGWM are: providing adequate PRF to all and new types of waste, in safety conditions; improving the waste collection service; promoting the waste discharge in PRF; ideal with new concerns/procedures from ship-owners.

The aim of this study is to quantify the SGW discarded in the PL and apply a strengths, weaknesses, opportunities and threats (SWOT) analyze in order to identify possible management barriers and to contribute to a continuous improvement process.

2 PORT OF LISBON

2.1 Port of Lisbon characterization

PL is a natural port located in a water basin of 32 500 ha, between the Atlantic Ocean and the vast estuary of the Tagus. It is a large multifunctional European port with cargo and passenger terminals, with continuous service, which obligates a continuous monitoring and proceed with different operations related with ship-generated waste. According to the UNCTAD (1992) classification PL is a 3rd generation port.

2.2 Waste Management

The PL has an adequate and dynamic management of ship generated waste complying with legal obligations. The PL intends to be dynamic and proactive in environmental terms, in order to respond and monitor the evolution of ships, especially in waste management.

As a result, a ship waste management system has been implemented, based on the Waste Reception and Handling Plan, which joins complementary operational and administrative procedures.

The SGWM it is applied to all ships that contacts PL and should be known by the port community. The system works only with the involvement of stakeholders, such as shipping agents, waste operators, crew of ships, terminals, ship-owners and other authorities.

The crucial key for Port of Lisbon Administration action is to provide adequate and enough PRF in order to ensure the ships waste disposal needs and, consequently, to avoid illegal dumping at sea. Provide the appropriate facilities for the waste reception is a real challenge, due to the multiple variables and requirements of cargo and passenger vessels that come in contact with PL: waste type and volumes, tides, the total number of ships in the port, among others.

In order to guarantee a collection service of 24 hours a day, 365 days a year, PL established tenders/licenses with waste management operators, which detain the equipment and specialized personnel to carry out the safe collection, as well as to carry out transportation to an appropriate final destination under appropriate conditions. The PL defines the requirements and levels of service that operators have to fulfill in PL service delivery in order to prevent undue delays to ships and maximizing the recovery of the waste collected.

The PRF are selected according to the typology, hazard, physical state and volume of waste to be discharged by each ship. In the case of operations not framed in the SWMP, a specific work plan will be created.

But the availability of PRF on the pier is an almost final stage of the process, as the flow begins on board the ships. Before ship arrivals, the person responsible for the ship, such as the master or environment officer, shall forward the Waste Notification Form to the ship's agent, who in turn transcribes it to the informatics tool *Janela Única Portuária* (JUP). PL checks the required information, such as the last date and port of delivery waste, storage capacity and types of waste. In case of discharge, ships agent fills up a request in JUP considering both waste and volume to be discharged, as well as the date, hour and place to realize the service. After the analysis and

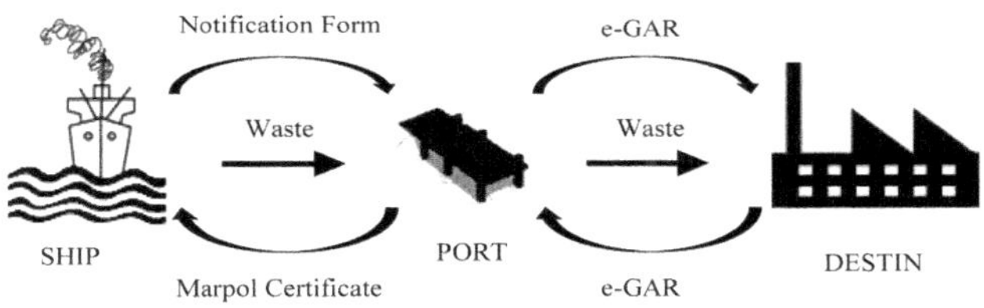

Figure 1. Scheme of waste management process in PL.

response of the operator, the PL dispatches the request so that the operation can be performed. As documentation accountability records to stay in the land, it is necessary an Electronic Waste Declaration Forms "*e-GAR*" (*Guias eletrónicas de acompanhamento de resíduos*) since leave the port terminal to waste operator. However, the MARPOL certificate is the one that has real operating value for the ship, once it is the document which proves that an unloading of the waste was carried out on land rather than at sea (Fig. 1).

As the operation wrap-up, the information of the real amounts and types of waste must be updated in the JUP. Considering the updated data and in accordance to the established in the tariff of the PL, an invoice is delivered to the ship agent, who in turn forwards it to the ship-owner. The tariffs are defined by port and must be transparent, simple, fair and non-discriminatory by all entities. It reflects the costs of equipment (including maintenance), availability, personnel (also administrative costs), transportation and acceptance at the appropriate destination.

According PL Tariffs, the waste fees are due by ship owners or their legal representatives and joins a fixed and a variable waste fee. These fees may change depending on waste collection contract service changings.

A fixed fee is defined by Directive 2000/59/CE corresponds to ship's contribution for port reception facilities costs recovery, including treatment and disposal, whether waste is discharged. It applies to all ships and vessels calling at the port and is calculated per unit of gross tonnage (GT). The variable fee is applied directly to waste discharge operations. The prices reflect the prices from tenders with operators and PL administrative costs.

At the same time, PL develops awareness-raising activities with ships that calls the port, with the purpose to: help the ships to understand the SGWM and the tariffs; help the ships to plans the operations; identify the technologies implemented on ships that call PL; incentive the use of PRF; spread the incentives regarding waste tariffs; cooperate in new environmental areas, as air emissions.

3 METHODOLOGY

In order to improve the SGWM system in the PL, a survey of the waste amount delivered per ship is carried out on the basis of the JUP records. Additionally, it will also be investigated the internal and external environmental factors that influence SGWM through a SWOT analysis (strengths, weaknesses, opportunities and threats).

4 RESULTS AND DISCUSSION

4.1 Waste collected in the Port of Lisbon

The waste collected in the PL is initially identified and quantified in volume (m^3) by the person responsible for the ship according to the MARPOL classification in the different Annex (Figs 2, 3). The PRF are available on the pier according the volume mentioned.

However, according to Decree-Law n° 73/2011 of 17 August, the legislation applied on land, the waste should be classified according to the European Waste List (EWL) established by Decision 2014/955/EU of 18 December 2014 and it is necessary to fill out an

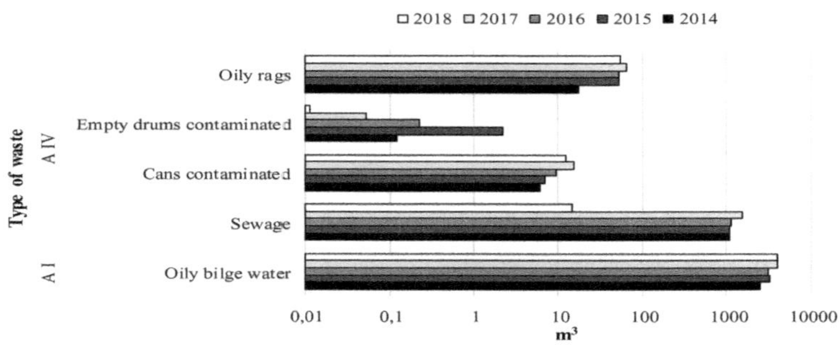

Figure 2. Evolution from 2014 to 2018 of the quantity of waste according MARPOL (Annex I and IV).

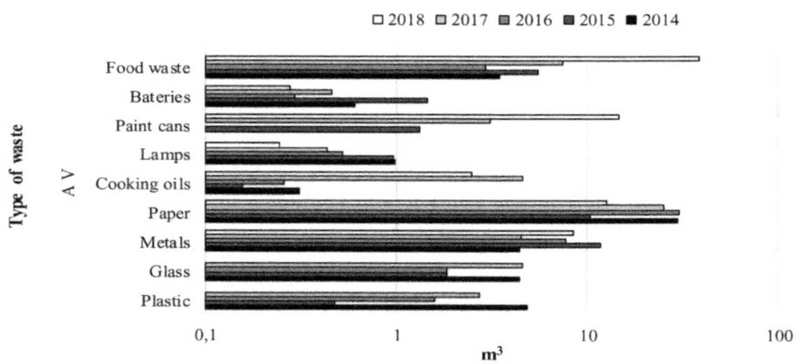

Figure 3. Evolution from 2014 to 2018 of the quantity of waste according MARPOL (Annex V).

Electronic Waste Declaration Forms to follow the transport "*e-GAR*" available on SILiAmb website (https://apoiosiliamb.apambiente.pt) for transporting waste. SILiAmb is a platform developed by the Portuguese Environmental Agency (APA) whose main objectives are the processes dematerialization and increase the communication between the APA and stakeholders. According to the user support manual of the platform, the waste quantity must be reported in kg (APA, 2018). The waste amount evolution collected at the port and classified according land legislation, in tons, is shown in Figures 4.

Due to the different requirements for the waste declaration on ships and on land, it is difficult to compare the amounts of waste. Furthermore, there is not always a perfect match between the classifications made in accordance with the MARPOL and EWL. For example, the bilge water corresponds to wastes listed in Annex I to MARPOL and the have following codes at EWL are: 13 07 03*, 15 01 10*, 15 02 02* and 16 07 08*. Wastewater is classified as waste in MARPOL Convention, but according the national legislation they are not a waste.

There are several constraints regarding the ship-generated waste management due to a sea/land mismatch.

1. Different procedures to quantify the waste – in Waste Notification send to ships agent, the ships declare waste in m^3, according to MARPOL Convention. The PRF are available on the pier according the volume mentioned. However, the legislation applied on land requires waste declared in tones.
2. Different documentation are required for ship and for land operation – on land, it is the Electronic Waste Declaration Forms (*e-GAR*). However, only the MARPOL certificate is necessary to the ship.
3. MARPOL Certificates do not reflect the waste delivered – the MARPOL certificates are extracted directly from the discharge request, and sometimes the quantity indicated initially does not correspond to the quantity delivered.

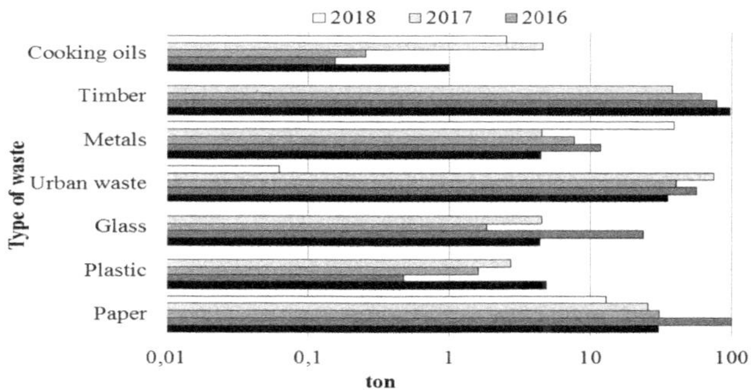

a)

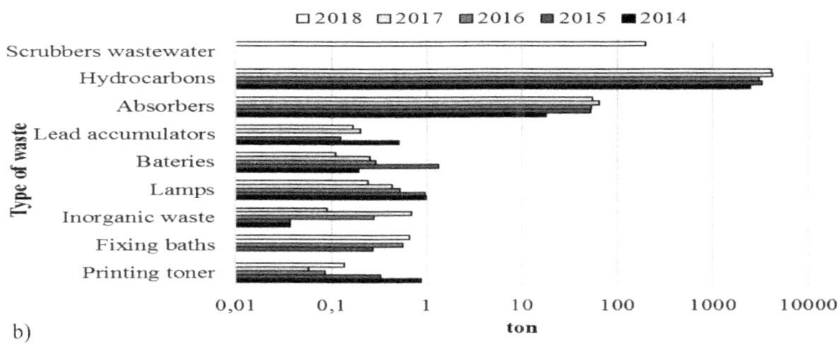

b)

Figure 4. Evolution from 2014 to 2018 of the quantity a) non-hazardous waste and b) hazardous waste.

Table 1. SWOT analysis – Internal environment.

Internal environment	
Strengths	· Good knowledge of maritime sector; · Excellent relationships and communication with port community; · Good tools to support the procedures; · Successful "inspections" strategies; · Good marketing strategies.
Weakness	· High tariffs applied to port reception facilities; · Low incentives to waste discharge; · The waste discharge is not mandatory; · The operator may not have adequate/enough port reception facilities; · Terminals have specific regulations; · Different notifications or agreements due to lack of knowledge of what is on ships; · Lack of human resources.

4. External factors to waste discharge on pier – the ship discharge operations on pier are conditioned by several factors, such as tide, food supply operations, and bunkers.
5. "Just in time" – sometimes the operations are required "just in time" before departure. Ships offload more waste than they requested or different types of waste.

4.2 SWOT analysis

The SWOT analysis is a method for investigating the internal (Table 1) and external (Table 2) environmental influencing factors in SGWM system in the PL. The aim of applying SWOT analysis is to find out the strengths and weaknesses inherent to a SGWM system in the PL as well as the opportunities and threats in the external environment.

Table 2. SWOT analysis – External environment.

External environment	
Opportunities	· Huge concern of ship-owners regarding ship generated waste; · Growth of package tourism – cruise ships; · Loyal ships; · Legislation change; · New types of ship-generated waste, as waters of scrubbers.
Threat	· Competitors ports have better tariffs and contracts; · Ships – Undeclared quantities on the volume and type of waste in Waste Notification, this implies: – Inadequate means at the dock for receiving waste, – The 3R policy (reduce, reuse and recycle) is not enforced; · Ships – Disposal of waste with a larger volume than the capacity of the container; · Very specific legislation on land, as instance to medical waste; · High fees applied by external authorities/entities, as for slops and donations; · Operator – instability of procedures; · The operator may not have adequate/enough port reception facilities; · Bad image of port reception facilities (no maintenance and language in the · containers is only in Portuguese); · Downturn in economy.

5 CONCLUSIONS

In spite of the progress towards de reduction of pollution caused by sea transport, there still remains a great challenge ahead.

The strengths and weakness of the SGWM process were clarified with a SWOT analysis. The main identified opportunities for improvement are: integration ship-port; improvement of the waste collection process; inspections to ship; regulations on emissions; compliance with the rules; maximize the number of discharge in Port; increase recycling; resource recovery; training and education; improvement of the computer system; upload the information on the waste management system to the PL website; collective vision of integrated management systems.

REFERENCES

APA 2018. *Manual de utilizador – Módulo E-GAR da plataforma SILIAMB*. Versão 3.5. Available in: https://apambiente.pt/_zdata/Politicas/Residuos/Transporte/eGAR-Manual%20de%20Utilizador-v3.5_18_01_2019.pdf. Access in 02/03/2019.

Decision 2014/955/UE Commission Decision of 18 December 2014. O.J. of the European Communities.

Decree-Law 165/2003 of 24 July, Diário da República n° 169/2003, Série I-A.

Decree-Law n° 73/2011 of 17 August, Diário da República n° 116/2011, Série I.

Decree-Law 83/2017 of 18 July, Diário da República n° 137/2017, Série I.

Directive 2000/59/EC of the European Parliament and of the Council of 27 November 2000 on port reception facilities for ship-generated waste and cargo residues. O.J. of the European Communities.

Directive (EU) 2015/2087 of the European Parliament and of the Council of 18 November 2015 amending Annex II to Directive 2000/59/EC of the European Parliament and the Council on port reception facilities for ship-generated waste and cargo residues. O.J. of the European Communities.

International Chamber of Shipping (ICS) 2019. *Shipping and World Trade – Overview*. Available in http://www.ics-shipping.org/shipping-facts/shipping-and-world-trade. Access in 02/03/2019.

International Maritime Organization (IMO) 2019. *Pollution Prevention*. Available in: http://www.imo.org/en/OurWork/Environment/PollutionPrevention/Pages/Default.aspx. Access in 02/03/2019.

United Nations Conference on Trade and Development (UNCTAD) 1992. *Port Marketing and the Third Generation Port*. TD/B C.4/AC.7/14.

Wastes: Solutions, Treatments and Opportunities III – Vilarinho et al. (Eds)
 ISBN 978-0-367-25777-4

Management of tomato waste: Biomethane production and nutrient recovery

S.R. Pinela, R.P. Rodrigues & M.J. Quina
CIEPQPF – Research Center on Chemical Processes Engineering and Forest Products, Department of Chemical Engineering, University of Coimbra, Coimbra, Portugal

ABSTRACT: In the context of biorefinery, anaerobic digestion (AD) is a well-known technology for wastes valorization. The aim of this work is the development of an integrated solution for tomato waste, which involves the production of biogas by AD and the application of the digestate as fertilizer. The AD assays were conducted using several substrates to inoculum ratios (S/I). The optimal methane production was 361 NmL CH_4 gVS^{-1}, using a $S/I = 1.0$. The digestate was separated from the liquid by centrifugation, and the characterization of the two fractions (liquid fraction – LF and solid fraction – SF) was conducted. In the LF and SF about 0.35 ± 0.08 mg TKN g TS^{-1} and 0.32 ± 0.02 mg TKN g TS^{-1} were determined, respectively. Regarding phosphorus, 0.88 ± 0.04 mgP g TS^{-1} (LF) and 113.15 ± 2.74 mgP g TS^{-1} (SF) were found. In the case of LF is diluted with a factor of 25 (v/v), no phytotoxic effect was observed.

1 INTRODUCTION

Portugal is one of the major producers of tomato in the European Union (EU) (De Cicco 2016). According to Statistics of Portugal, in 2017 there was an industrial production of 1.650 Mton of tomato (INE 2019). However, after harvesting seasoning the lack of proper management of the wastes generated are a problem that producers face.

Since 2012, the EU has a strategy for bioeconomy, which was defined as the use of biological resources and waste streams to produce high-value products (Hassan et al. 2019). In this context, organic waste streams can be used as substrates in biorefinery processes, such as extraction and bioconversion to produce added value compounds or/and energy (Higsamer & Jungmeier, 2019).

Worldwide population is growing, boosting energy and agriculture demand. Thus, to ensure sustainable development, renewable sources must be favored to produce energy and whenever possible fertilizers as well. Indeed, there are currently concerns about the depletion of P and K, which are mostly obtained from mineral deposits (Bolzonella et al. 2018).

Anaerobic digestion (AD) is a biological process that operates in the absence of oxygen, allowing the conversion of organic matter into biogas (mainly CH_4 and CO_2) and biosolids (digestate) (Li et al. 2011). AD is also a well-established technology which has been reported as an efficient method in the valorization of many substrates, where fruits and vegetables are included (Rodrigues et al. 2019). Besides biogas, this process also produces digestate, which is rich in nutrients such as N and P, and it is often considered biologically stable and relatively sanitized (Li et al. 2011; Gil et al. 2015). Although the chemical characteristics of digestate depend widely on the type of substrate and operational conditions of the process, different studies revealed that this by-product can be used in agriculture as a soil amendment or as a fertilizer (Albuquerque et al. 2012, Solé-Bundó et al. 2017). Other studies go further, reporting that digestate can be separated into two phases: N (in form of ammonia) can be recovered essentially from liquid fraction of digestate, and P can be recovered from the solid fraction (Ledda et al. 2013, Tambone et al. 2017, Bolzonella et al. 2018). While chemical properties from digestate must be analyzed to decide upon their use as a fertilizer, the phytotoxicity must be assessed to ensure the safety in agriculture applications.

Therefore, AD not only may contribute to organic waste management, but it can also reduce fossil fuel demand and the usage of mineral fertilizers in agriculture.

In the present work, an integrated solution for tomato waste was investigated. Namely, the potential for biogas production through AD was assessed; the digestate (biosolids from AD) and the liquid were separated and characterized in order to assess their potential to be used as fertilizer.

2 MATERIALS AND METHODS

2.1 *Substrate and inoculum*

Rotten tomato waste (TWR) was simulated using tomatoes (*Solanum lycopersicum*) purchased in a local supermarket. This substrate was left one week at room temperature to increase the state of maturation. The TWR was ground and sieved by a 2 mm mesh before fed the digester. The inoculum used in AD was collected in a local domestic wastewater treatment plant, in the out flow of a mesophilic anaerobic digester.

2.2 *Anaerobic digestion experimental set-up*

Anaerobic digestion (AD) assays were carried out in 5 L reactor, with a working volume of 3.75 L. The reactor was operated in batch conditions, at 37°C and manually agitated once a day. Several substrates to inoculum (S/I) ratios (0.5, 0.75, 1.0, and 1.5) were tested. The pH was adjusted to 7 by adding $NaHCO_3$. The biogas produced was measured in a gasometer filled with a solution of NaCl at 60% of the saturation and pH 2. The methane concentration in biogas was measured by washing the gas in a graduate syringe with a solution of NaOH with a concentration of 4 mol L^{-1}. After the tests of AD, the suspension was separated by centrifugation at 4000 rpm during 30 min in two phases: liquid fraction (LF) and solid fraction (SF). The two phases were collected in flasks and stored at 4°C until analysis.

2.3 *Analytic methods*

Total solids (TS), volatile solids (VS), total suspended solids (TSS) and chemical oxygen demand (COD), were determined according to standard methods (APHA 1998).

The pH was measured using the HANNA pH20 meter and electric conductivity (EC) was measured with HANNA HI2550 electric conductivity meter.

Total Kjeldahl Nitrogen (TKN) was determined by digesting 0.5 g (solid samples) or 10 mL (liquid samples) with 10 mL of 96%H_2SO_4 at 420°C during 2 h in the DKL Fully Automatic Digestion Unit from VELP Scientifica. After the digestion, all samples were distilled in the UDK Distillation Unit from VELP Scientific and then titrated with 0.1 M HCl. Total phosphorus (Total-P) was determined by a colorimetric method, EPA Method 365.3, using ascorbic acid (EPA 1978). Elemental analysis (CHNS) of TWR was determined by an Elemental Analyzer NA 2500 (NA 2500), and the SF of the digestate was measured by EA 1108 CHNS-O-Fisons equipment. The BMP was estimated based on near-infrared spectroscopy (NIR) combined with rapid calibration (FlashBMP®) developed by Ondalys (Chemometrics – Data Analytics) and commercialized by Buchi (Switzerland) (Rodrigues et al. 2019).

2.4 *Phytotoxicity assay*

Germination tests were carried out using *Lepidium sativum* L. Regarding LF, several solutions containing different percentages (v/v) of LF (100%, 20%, 10%, 4.0%, 2.0%, 1.0%, 0.5%, and 0.2%) were tested in triplicate. For testing SF, aqueous extract was obtained using dehydrated sludge with distilled water, at a solid to liquid (S:L) ratio of 1:10. The extraction lasted 2 h, at room temperature, and then centrifugation at 3500 rpm was used to obtain the supernatant. About 5 mL of liquid was used to moistening the filter paper placed in Petri dishes (9 cm diameter), over which 10 seeds of *Lepidium* was placed. The Petri dishes were incubated at 25°C in the dark for 48 h.

Table 1. Characterization of TWR used as the substrate.

Parameter	TWR
TS (%)	5.99 ± 0.46
VS (%TS)	81.08 ± 1.19
tCOD (mg O_2 gVS^{-1})	2575 ± 10
pH	4.35 ± 0.03
C (%TS)	38.9 ± 0.1
N (%TS)	1.8 ± 0.0
O (%TS)	34.1 ± 0.0
H (%TS)	6.2 ± 0.1
C/N	21.6
Empirical formula	$C_{25}H_{48}O_{17}N$

The mean of the germinated seed (N) and the length of their roots (L) was determined. Relative seed germination (RSG), relative root growth (RRG), expressed in percentage, were calculated (in respect of blank test) using Eq. 1 and Eq. 2, respectively. Germination index (GI) was calculated through Eq. 3.

$$RSG = N/N_b \times 100 \tag{1}$$

$$RRG = L/L_b \times 100 \tag{2}$$

$$GI = RSG \times RRG/100 \tag{3}$$

3 RESULTS AND DISCUSSION

3.1 *Characterization of substrate and AD*

The main physical and chemical properties of the TWR are summarized in Table 1. The percentage of VS of TWR is higher than 80%, which is very similar to values reported in the literature (Oleszek et al. 2016) and tCOD is 2575 mg g VS^{-1}. The acid nature of the substrate indicates that in order to be used in AD a correction of pH is needed. Carbon-nitrogen ratio (C/N) is 21.6, which is in the range reported as the best for methane production yield (Wang et al. 2012). Operating AD with an inappropriate C/N ratio can lead to the formation of ammonia or higher accumulation of volatile fatty acids (VFA) in the reactor, which are toxic to the microbial consortium (Li et al. 2011).

The higher experimental BMP obtained was 361 NmL CH_4 g VS^{-1} using an S/I of 1.0. For S/I ratios of 0.5, 0.75 and 1.5 the methane produced was 295 NmL CH_4 g VS^{-1}, 209 NmL CH_4 g VS^{-1}, and 60 NmL CH_4 g VS^{-1}, respectively. The higher BMP value is similar to the value obtained by NIR analysis (BMP = 371 NmL CH_4 g VS^{-1}). Based on the elemental composition ($C_{25}H_{48}O_{17}N$), the Buswell and Mueller formula leads to a prediction of 490 NmL CH_4 g VS^{-1}, which is 26% higher than the experimental value (361 NmL CH_4 g VS^{-1}). All experimental values, apart from the one obtained with S/I= 1.5, are within the range reported on literature (Calabró et al. 2015). The low BMP value obtained with an S/I ratio of 1.5 may be explained by the inhibition of microbial activity due to the excess of the substrate.

3.2 *Characterization LF and SF*

The separation by centrifugation of the two fractions of the suspension in the AD reactor, resulted in 96.81 ± 0.76% (w/w) of LF and 3.19±0.76% (w/w) of SF, in wet basis. TSS in LF represents only 2.35±0.47% of TSS of the digestate. These data show that centrifugation is an appropriate separation process, with an efficiency of 98%.

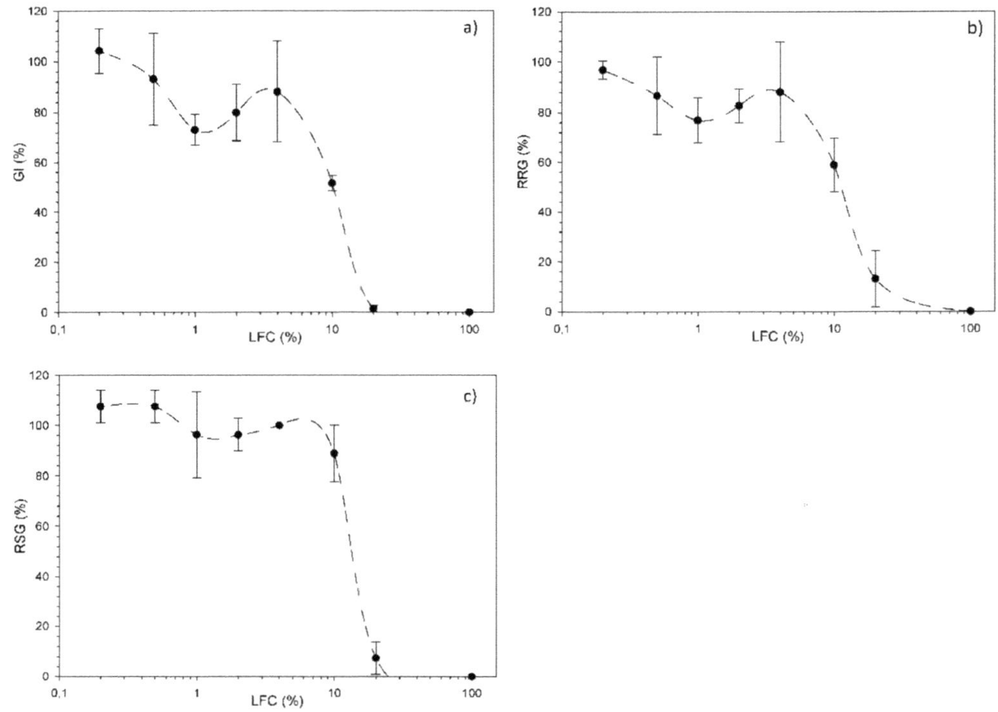

Figure 1. Phytotoxicity assays observed in different concentrations of LF: a) germination index, GI, (%), b) relative root growth, RRG, (%), and c) relative seed germination, RSG, (%).

The physical and chemical characteristics LF and SF are summarized in Table 2, all results are reported as a function of the TS of each fraction. After centrifugation, the content of TS in LF is only 0.31%. The literature reports that the TS content in digestates can vary between 1.50% to 7.75%, depending on the substrate used and the operational settings of the AD process (Coelho et al. 2018). LF contains a minor percentage of VS in comparison with SF. VS content can vary between 31.46% TS and 72.09% TS for digestates (Coelho et al. 2018). Thus, the value obtained for SF is within this range. The values of VS may vary in a large range due to several factors, such as the substrate used for methane production, initial C/N, and operational settings of the digesters (microbial activity, temperature and hydraulic retention time) (Coelho et al. 2018).

The pH of both fractions is slightly alkaline, and this parameter has a proven positive relation with N content (e.g. ammonia) and the degradation of VFAs (Albuquerque et al. 2012, Coelho et al. 2012). On the other hand, the initial correction of pH with $NaHCO_3$ may increase the buffer capacity of the digestate (Albuquerque et al. 2012). While pH is similar in the two fractions, the EC of LF is almost the double of the EC of the SF. This parameter is related to dissolved free ions (Coelho et al. 2018), and thus caution is required in the application of LF in the soil since high levels of salinity can inhibit plant growth (Albuquerque et al. 2012).

The elemental analysis obtained reveals a C/N ratio of 8.55, literature reports very similar results for the elemental analysis of anaerobic digestate from tomato waste (Gil et al. 2015).

The TKN is similar in both fractions (near 0.3 mg g TS^{-1}), and within the values found in the literature (Bolzonella et al. 2018, Ledda et al. 2013). The content of Total-P is comparable with the literature as well (Bolzonella et al. 2018). The SF contains a high concentration of Total-P, which may reach about 11.3% of the SF. Indeed, phosphorus is found mainly in SF because of the assimilation of this nutrient by the biomass in the AD reactor (Ledda et al. 2013).

Table 2. Characterisation of LF and SF.

Parameter	LF	SF
pH	8.05 ± 0.01	8.26 ± 0.02
EC (mS cm^{-1})	8.16 ± 0.06	4.26 ± 0.05
TS (%)	0.31 ± 0.09	2.21 ± 0.37
VS (%TS)	16.90 ± 1.47	59.08 ± 0.41
COD (mg O_2 g VS^{-1})	1234 ± 12	11760 ± 508
TKN (mg TKN g TS^{-1})	0.35 ± 0.08	0.32 ± 0.02
Total-P (mg P g TS^{-1})	0.88 ± 0.04	113.15 ± 2.74
C (% TS)	nd	30.87 ± 0.62
N (% TS)	nd	3.61 ± 0.05
O (% TS)	nd	17.81 ± 0.19
H (% TS)	nd	5.33 ± 0.05
S (%TS)	nd	1.46 ± 0.03
C/N	nd	8.55

nd – not determined.

Although P and TKN are not very high in LF, since they are already dissolved, the use of this fraction as a liquid fertilizer may be advantageous because it can be mixed with the irrigation water (Gil et al. 2005).

3.3 *Phytotoxicity assessment*

The phytotoxicity measured in the liquid extract from SF (1:10 S:L) revealed that none seeds germinated, meaning that a severe inhibition is observed. Thus, solids from AD should not be applied immediately to the soil. A maturation period or stabilization through composting is advised in this case. The results obtained for GI, RRG, and for RSG are shown in Figure 1a)-1c), respectively. In these figures, the RSG, RRG, and GI are expressed in function of the percentage of concentration of the liquid fraction (LFC), where 100% correspond to the application of LF without dilution and 0% correspond to distilled water. Results revealed that lower concentrations exhibited a higher RSG, RRG, and GI. Indeed, GI is higher than 60% for the dilutions 0.2%, 0.5%, 1.0%, 2.0%, and 4.0%. When GI is less than 60%, the tested material leads to a strong inhibition (Trautmann & Krasny, 1997). In this study, using a concentration of 10% of LF an RSG of $88.89 \pm 11.11\%$ was obtained. Similar results were reported by Coelho et al. (2018). The lower seed germination for the higher concentrations of the LF (20.0 and 100%) may be correlated with the high EC (8160 $\mu S\ cm^{-1}$ and 1850 $\mu S\ cm^{-1}$), since there are reports that show the inverse correlation between the germination of the seeds and the EC (Albuquerque et al. 2012). Besides salts other phytotoxins (e.g. ammonia) may be responsible for the detrimental effect of LF on *Lepidium sativum* L. These results showed that the application of the liquid fraction on soil must be diluted to a concentration lower than 10%.

4 CONCLUSIONS

Anaerobic digestion of tomato waste allows not only biogas production but also the recovery of nutrient (N and P). The S/I ratio with higher methane production was 1.0, for which a BMP of 361 NmL CH_4 g VS^{-1} was obtained. The suspension of the AD reactor was separated into two fractions (LF and SF) by centrifugation, which showed an efficiency of 98%. The physical and chemical characterization of the LF and SF showed that both have the right properties to be used as fertilizer. However, the extract of SF using a solid to liquid ratio of 1:10 was found phytotoxic, requiring a stabilization period before application. The phytotoxicity assessment of LF revealed not be phytotoxic for concentrations lower or equal to 4%. Thus, LF may be applied as a fertiliser, after dilution with irrigation water.

ACKNOWLEDGMENTS

Authors acknowledge the financial support of Fundos Europeus Estruturais e de Investimento (FEEI) através do Programa Operacional Competitividade e Internacionalização – COMPETE and Fundos Nacionais através da FCT – Fundação para a Ciência e Tecnologia, in the scope of the project POCI-01-0145-FEDER-016403.

REFERENCES

Albuquerque, J.A., Fuente, C., Ferrer-Costa, A., Carrasco, L., Cegarra, J., Abad, M., Bernal, M.P. 2012. Assessment of the fertiliser potential of digestates from farm and agroindustrial residues. *Biomass Bioenergy.* 40: 181–189.

APHA 1998. Standard Methods for the Examination of Water and Wastewater, AWWA, WPCF. 20th. Washington DC.

Bolzonella, D., Fatone, F., Gottardo, M., Frison, N. 2018. Nutrients recovery from anaerobic digestate of agro-waste: Techno-economic assessment of full scale applications. *J. Environ. Manag.* 216: 11–119.

Calabró, P., Greco, R., Evangelou, A., Komilis, D., 2015. Anaerobic digestion of tomato processing waste: Effect of alkaline pretreatment. *J. Environ. Manag.* 163: 49–52.

Coelho, J.J., Prieto, M.L., Dowling, S., Hennessy, A., Casey, I., Woodcock, T., Kennedy, N. 2018. Physical-chemical traits, phytotoxicity and pathogen detention in liquid anaerobic digestates. *Waste Manage.* 78: 8–15.

De Cicco, A. 2016. The fruit and vegetable sector in the EU - a statistical overview, *Agriculture, forestry and fishery statistics 2016 edition.*

EPA 1978. Methods for Chemical Analysis of Water and Wastewater, Washington DC.

Gil, A., Siles, J.A., Serrano, A., Martín, M.A. 2015. Mixture optimization of anaerobic co-digestion of tomato and cucumber waste. *Environ. Technol.* 36 (20): 1–9.

Hassan, S.S., Williams, G.A., Jaiswal, A.K. 2019. Lignocellulosic Biorefineries in Europe: Current State and Prospects. *Trend Biotechnol.* 37: 231–234.

Higsamer, M., Jungmeier, G. 2019. *Bioeconomy: Resource, Technology, Sustainability and Policy*, Academic Press.

Ledda, C., Schievano, A., Salati, S., Adani, F. 2013. Nitrogen and water recovery from animal slurries by a new integrated ultrafiltration, reverse osmosis and cold stripping process: A case study. *Water Res.* 4: 6157–6166.

Li, Y., Park, S.Y., Zhu, J., 2011. Solid-state anaerobic digestion for methane production from organic waste. *Rerew. Sust. Energ. Rev.* 15: 821–826.

INE 2019. Boletim Mensal de Estatística - Dezembro de 2018.

Oleszek, M., Tys, Jerzy, Wiącek, D., Król, A., Kuna, J. 2016. The Possibility of Meeting Greenhouse Energy and CO_2 Demands Through Utilization of Cucumber and Tomato Residues. *Bioenerg. Res.* 9: 624–632.

Rodrigues, R.P., Rodrigues, D.P., Klepacz-Smolka, A., Martins, R.C., & Quina, M.J. 2019. Comparative analysis of methods and models for predicting biochemical methane potential of various organic substrates. *Sci. Total Environ.* 649: 1599–1608.

Solé-Bundó, M., Cucina, M., Folch, M., Tàpias, J., Gigliotti, G., Garfí, M., Ferrer, I. 2017. Assessing the agricultural reuse of the digestate from microalgae anaerobic digestion and co-digestion with sewage sludge. *Sci. Total Environ.* 586: 1–9.

Tambone, F., Orzi, V., D'Imporzano, G., Adini, F. 2017. Solid and liquid fractionation of digestate: Mass balance, chemical characterization, and agronomic and environmental value. *Bioresour. Technol.* 243: 1251–1256.

Trautmann, N.M., Krasny, M.E. 1997. Composting in the Classroom – Scientific Inquiry for High School Students.

Wang, X., Yang, G., Feng, Y., Ren, G., Han, X. 2012. Optimizing feeding composition and carbon-nitrogen ratios for improved methane yield during anaerobic co-digestion of dairy, chicken manure and wheat straw. *Bioresour. Technol.* 120: 78–83.

Wastes: Solutions, Treatments and Opportunities III – Vilarinho et al. (Eds)

Burying solid waste problems: Sanitary landfill challenges in mainland Portugal

J. Rafael
Centro de Estudos de Geografia e Ordenamento do Território, Porto, Portugal

ABSTRACT: Solid waste is causal to pollution and contamination of soils, air and waters, as well as health and epidemiological risks. The magnitude of its consequences is capable of imprint serious changes on human genome and on the planet. It pressed climate policies – linked to demands of efficient production and sustainability precepts -, and the expansion of waste treatment and management programs to deal with the problems associated with the wastes. However, the most widely used method supporting these programs, in Portugal, is waste disposal by burial in sanitary landfills with over landscape remediation and redevelopment works. This paper argues that this method does little more than to sweep our environmental responsibility under a green carpet, and it calls for more ethical and irreverent solutions to eliminate wastes.

1 INTRODUCTION

The final disposal of solid wastes is a discomforting widespread problem in many developed and developing countries, and one of the major challenges of environmental sanitation in most countries worldwide. The rates of solid wastes generated are seen as a critical component of social, political, economic and ecological matrices: i.e. a by-product of human processes of production, construction and consumption, linked to disposability (Packard, 1960) and disorder (Douglas, 1966), inutility and neglect. Derived from improper, degraded and/or despicable (Bertolini, 1996; Eigenheer, 2003), surplus and/or excess (Strasser, 1992; MacFarlane, 2018); 2), the term 'waste' carries a negative connotation, and is commonly associated as objectively causal to pollution and contamination of soil, air and waters, as well as with potential health and epidemiological risks. This relation and connotation is not accidental. Solid wastes "exude toxic tears and boiling gases" (Engler, 2004: 81). Their agency is evaluated by the direct or indirect impact of the waste's constituent materials on the quality of human life and the environment. It emerges where and when we lose control of events and over places; through not measuring the consequences of our actions and the uses that we make of the environment, or simply because our imagination fails us (Ekberg, 2009). Waste becomes less desirable as it multiplies, accumulates, or is concentrated in the wrong places. Our approach to waste is, to a certain extent, to avoid rather than eradicate it, to cover it up and enclose it (Lynch, 1990: 1). Of waste products, many have pointed that "everyone produces them, yet no one wants to have anything to do with them, except to take them out" (Phillips, 1990: 12) – both out of sight, and out of mind.

Much of the solid waste that we produce is composed of organic matter that tends to rot, rust and ruin; other (post-industrial) waste is composed of synthetic, non-biodegradable and toxic materials that is less perishable and/or even lacks the capacity for decomposition on its own. The temporality of this later type of waste goes beyond natural and artificial boundaries. Its materiality impacts on the environment and causes serious dangers to human health and human's own existence. It induces loss of biodiversity and reduces the production or regeneration capacities of our ecosystems (Basel Convention, 2012). According to authors from the fields of science and technology studies, the spatiality of such waste is actant (Latour, 2004; Bryant, 2011), and "against the world"

(Serres, 1992); testifying to the potential misfortune brought about by social evolution (Gille, 2007; Kennedy, 2007; Neves and Mendonça, 2014), and requiring us to be more active and responsive to the impact of our human agency. For these reasons, the expansion of waste treatment and management systems to deal with the social and spatial dynamics of waste's distribution, accumulation and disposal, and the magnitude of the negative consequences that result from such processes, have become an important and integral battlefield in climate policies – linked to the demand for efficient production models and sustainability precepts (Circular Economy Package, 2015).

1.1 *Developmental history of solid waste final disposal solutions*

Since at least the nineteenth century, the solid waste we produce has motivated interdisciplinary groups of professionals and civics to organise themselves to reflect on the material foundations, functioning and structure – past and possible – of our society. These actors have sought for controlled and rational solutions for the treatment and disposal of waste produced, as well as for corrective and/or preventive public policies and reforms in defence and response to changes in social habits and attitudes, in order to reduce the production, accumulation and management of waste and/or to attempt to facilitate waste's productive recirculation. These groups denounced adverse impacts related to quantitative and qualitative changes in waste types; advocated the need for differentiated treatment options, according to the composition and decomposition features of waste matter; highlighted the socio-environmental cost of the improper disposal and accumulation of certain solid wastes; and contributed to the development of approaches to combat related problems, both in the public and private domain.

Amongst the solutions and approaches developed, and as a purpose of this study, stand the efforts of the economic-instrumental efficiency and rationality to remedy the impacts, inconveniences and problems that the accumulation of solid wastes introduce in time and space – into the environment, and above all humankind – in a pragmatic (i.e. simple, cheap and supposedly efficient) way, through spatial models for final disposal of solid waste materials by burial, such as with the sanitary landfill. Although the history of the final disposal of waste by and in landfills originated in Crete around 3000 BC (Waste Watch, 2004), through the placement of waste matter in holes in the ground to be covered by earth, the landfill space as we know it today started to be built in the 1920s, in Britain (Melosi, 2013: 704). It evolved as one of a series of responses to hygienic precepts in favour of hygenic and healthy cities, driven by regulatory requirements and based on technical demands and solutions for minimizing the harmful impacts (direct and indirect) of solid waste upon populations and their environments. From the 1970s, the landfill ultimately aimed to optimize the containment space and insulation necessary for the proper elimination of waste – layering and covering wastes to biologically, chemically and physically decompose isolated in a fully engineered space.

As a solution, known as sanitary landfill, this is based on the idea that it allows a spatially and economically efficient means of minimizing decay, pollution and the devastation of lands through local containment and concentration, by offering a better response to the risks of residual toxic leakage and obnoxious gases - such as methane, carbon dioxide, ammonium and hydrogen sulphite – created by the decomposition of the waste deposited and compacted in the landfill space, whilst also being a potential source for energy generation. Maintenance costs of sanitary landfills are high. But, particularly given its considerable affordability vis-à-vis the economic burden of alternative methods, the sanitary landfill has become the most widely used method for the final disposal of solid wastes around the world. In general, the 20*th* century landfill is construed by systems to attenuate the contamination of adjacent soils and for controlling the negative effects of the landfill's constituent waste on the overall global environment (Bagchi, 1994; Christensen, 2012), and it is subject to extensive environmental regulations, monitory and maintenance protocols. In technical terms, sanitary landfills consist of compact layers of solid waste (organic and non-organic), impermeable clays, and thick polyethylene membranes, crossed by ventilation ducts, and covered by fertile soil, following strict standards of stability, ventilation, and containment, in order to mitigate and treat waste for the protection of the surrounding environment from physical and atmospheric contamination.

In the territory, solutions for the final disposal and elimination of solid waste such as the landfill are generally found in areas where no other land use was profitable, especially on abandoned and/or neglected lands, outside cities and along borders between adjacent cities. Although many of these places have now already been absorbed by the urban fabric of populated continents and expanding cities, they were originally so located both as to restrict public concern with the epidemiological and health problems related to waste (composition and decomposition), but also to inhibit immediate public perception of waste itself. Thus, kept away from the public social space in general, banned from the population and camouflaged – most of the time by/under a vegetal mantle – the landfill also constructed a repository for the anxieties brought about by the negative connotations, qualities and quantities of the solid residues we produce and accumulate. They have become, as a result of this, common expressions of our aversion to thinking about waste: i.e. places of forgetfulness (Hird, 2013), marginalized in space and in public debate. Solutions for the final disposal and elimination of solid waste such as the landfill have encouraged scientists, engineers and landscape architects to seek better and safer measures to provide the insulation and distance required to keep waste – in reality and imagination – out of sight and thus out of mind. Most landfills in operation are being constructed with the final aim of conversion into an attractive setting for public and passive recreation (Simmons, 1999; Young, 2000) or to improve biodiversity in urbanized areas (Hobbs et al., 2006). That is, they are being built to be embellished and covered by a stable soil layer to either construct new habitats (for a diversity of animals and plants), or for the building of generic parks, making landfill compatible with ecological and recreational uses (Engler, 1995; 2003). Thus, the aim is not only for such sites to become forgotten, but, in a way, for them to become more productive.

1.2 *Burying the wastes problem*

Landscaping projects that evolve from landfill embellishment solutions are scientifically engineered and sincere in their effort to create healthier, stronger landscapes in form, space, and meaning, but further limit our interaction with the deposited waste itself. What we do forget, under these landscapes, are not only the destructive processes that they are constituted of, carving physical scars out of the planet, but also the physical, chemical, and biological processes they conceal, processes capable of imprinting serious genetic modifications on human genomes. These are processes inherent to the decomposition of waste in the landfill space, and part of the landfill's primary function; burying – physically and mentally - processes that affect our health, comfort and survival – processes that remain invisible under the vegetal layer. The idea of focusing the relegation of waste issues to marginal spaces, via the landfill, first related to the fears and environmental problems associated with the results of waste disposal in natural systems, but many of the new issues the practice currently raises are focused on the urban remediation, regeneration and redevelopment landscape projects that, lately, accompany them. Like other models of landfill (controlled and uncontrolled), the sanitary landfill has raised debate on the efficacy of the socio-technical systems and restrictions associated with the geotechnical forecasting models that are applied as a means of housing different types of waste in landfills. Such doubts have already led the EU to impose restrictions (and sanctions) on the tonnage and type of waste currently deposited in European countries' landfills, and even to consider the landfill as one of the least desirable ways of managing and treating solid wastes. This is illustrated in the Waste Hierarchy, a tool that defines preferred program priorities of waste management strategies in terms of their suitability and environmental impact, provided for in Article 4 of Directive 2008/98/EC published in the Official Journal of the European Community.

Current critical approaches recognize (following a series of incidents) an inability to provide the necessary forecast plan to effectively address all plausible worst-case scenario failures that can occur with/in the landfill reserved space (Stegmann, 2012; Christensen, 2012), as well as to effectively reduce air pollution and ground water contamination (Warith, 2002; Newman et al. 2009). Debate and research on methods to make the landfill a place of new formal, spatial and programmatic possibilities; on the idea of landfills relegating to nature the task of ensuring that waste is not a source of exposure and contamination of people and the environment – throughout the life cycle of the landfill, on the logic of entrusting the ecosystem to naturally recycle diverse matter; has not been very active. The 20th century landfill is a rather recent creation and, many bring to mind, there is no direct field experience on how long the complete post-closure phase – which extends for several decades - may last. It is uncertain if it will last the several hundred years specific wastes take to decompose, as well as if the structure is able to stand potential natural impact events - such as flooding, fires, landslides and earthquakes – that might negatively impact it. Much of the work to build a sanitary landfill is focused on the plausibility of possible outcomes, the application of better criteria, the employment of the most enduring materials and the development of better models of future (environmental and spatial) conditions. It is conducted with (not only for) the future (Rafael, 2018), a natural and certain uncertainty.

This paper posits that the feasibility and efficiency of sanitary landfills construction and maintenance (including the sorting, containment and waterproofing of soils, drainage systems and the treatment of liquids and gases), as well as the engineering, hydrology, and bioremediation services involved in the environmental remediation and reclamation work on landfills, seem to do little more than to transform landfill passive-noxious spaces from sites of waste disposal into attractive assets, and to help us sweep our responsibility towards these elements under a green carpet. Two significant challenges remain isolated and camouflaged under this green fabrication: the risks of residual toxic leakage and obnoxious gases with an impact on local populations, as well as global populations - through greenhouse gas emissions (World Bank, 2006) -; and the complex (unknown and uncertain) material exchanges that can occur on the host environment – and that have been acting without our sufficient understanding or control. This paper ultimately proffers that, as a space, the landfill serves as an architectural agent of hygene and progress that creates and supports order by disguising rather than eliminating disorder, impurity and pollution from waste. The architectural solution employed in sanitary landfills 'put a lid on' the waste but cannot possibly encompass the whole of the earth crust. It exclusively buries epidemiological and epistemological problems.

2 LANDFILLING CHALLENGES IN MAINLAND PORTUGAL

According to the latest proposal for the revision of the Urban Solid Waste Intervention Plan (PERSU 2020+) placed in public consultation by the Portuguese Environment Agency on 28 December 2018, mainland Portugal deposited in sanitary landfills, more than 50% of the approximately 4.7 million tons of urban solid waste it produced in 2017. Since the end of the 1990s these landfills have been the predominant means of treatment and integrated waste management in Portugal. Portuguese landfills receive domestic, commercial and small-scale solid waste through support programs and solutions for waste management and recycling (PERSU 2020+, 2018: 7), that put the country on the verge of failing the time-frame defined by the landfill reduction targets imposed by the European Parliament in Directive 1999/31/EC on landfill. The directive and its targets seek to reduce landfill use in favour of the prevention, recycling and recovery of waste and the (re)use of recovered materials and energy, in order to save natural resources, improve the quality of the environment and promote a more circular economy. Thus, debate and public investment in the challenges posed by the construction, remediation and re-qualification of landfill spaces in Portugal, as well as strategies to reduce our dependence on this type of space, seem not only pertinent but very urgent.

In recent years there has been a growing realization that urban solid waste management cannot be reduced to pragmatic and efficient final deposition systems (landfills), but that management can and should instead be directed to the reduction and monitoring of both the production and accumulation of waste itself, as well as the facilitation of its productive recirculation. We have seen efforts and progress related to the reduction, reuse and recycling of solid waste. Expectations of changing trends and current statistics have impacted on waste-to-resource programs, but there are few studies that incorporate and analyse the final deposition of waste in landfills, the work of enclosing these sites after use, and the complex (and uncertain) material exchanges that subsequently occur within them. The studies that there are, in Portugal, are focused mainly on a classification of the growth and critique of quantity of urban solid waste that we deposit in landfill spaces. The former secretary of state for the environment, who also owned a company responsible for the import of solid wastes and their disposal in landfill spaces noticed that the global waste management market has indeed offered the country the opportunity to "take advantage of installed capacity". Furthermore, from 2020, in line with EU directives, municipalities might start to pay close to 20 euro per tonne of waste sent to landfill spaces. It remains to be asked whether this as disciplinary measure supports efficient measures of organisation, regulation and the infrastructure for treatment, recovery or elimination of the urban solid wastes or the sector. What are the barriers to improving, if not to eliminating, the final disposal of waste in landfills?

3 CONCLUSION

The reality behind current practical solutions to dispose and eliminate urban solid wastes in landfill spaces, in Portugal and other countries, is complex and dynamic. It is dimensioned by moral, ethical and political aspects where the variables that interact are diverse. For this reason, the phenomenon encompasses qualitative aspects that must be centered on the study of epidemiological and epistemological problems as well as in the meaning, intentions, motivations and characteristic contexts of the processes that intervene in the treatment and management of urban solid waste. Since we humans and 'nature' live in spaces with more pervasive and porous artificial barriers than we often recognize, it is as important to raise awareness to the major ecological and environmental challenges that arise from our society's habits of production (and consumption) as it is to improve the current state of waste treatment and management practices and better assess the implication of how we deal with them today. Solely depending on the disposal method to solve the solid waste problem is not a means to solve the problem, the tumour remains. Our task, and thus our responsibility is held on opportunities to experiment outside the ticking clocks of an unsustainable progress to tackle the pressing solid waste problems. Thus, it is essential to tackle the problems from the root cause, i.e. to reduce the waste from being generated but also to solidly review our knowledge and practices so that they provide for the creation of replacement sites or alternative ways of dealing with the wastes and the return of waste spaces.

REFERENCES

Andrade, André Wagner Oliani. 2006. "Arqueologia do lixo: um estudo de caso nos depósitos de resíduos sólidos da cidade de Mogidas Cruzes em São Paulo". Tese de doutorado em Arqueologia, Universidade de São Paulo, São Paulo.

Bagchi, A. 1994. Design, construction, and monitoring of landfills, New York: John Wiley & Sons Inc.

Basel Convention. 2012. Vital Waste Graphics 3. Available at: http://www.basel.int?DNNAdmin/AllNews/tabid/2290/ctl/ArticleView/mim/7518/articleld/626/Vital-Waste-Graphics-3.aspx.

Beck, Ulrich. 1986. Risk Society: Towards a New Modernity. Munich: University of Munich.

Bertolini, Gérard. 1996. "Evolution des mentalités vis-à-vis des ordures ménagères". Revue de Géographie de Lyon 71 (1): 83–86. DOI: 10.3406/geoca.1996.4325.

Bouazza, A. and Kavazanjian Jr. E. 2010. "Construction on Former Landfills", Proceedings 2nd ANZ Conference on Environmental Geotechnics, Newcastle (201): 467–482.

Bryant, Levi. 2011. Democracy of Objects, Ann Arbor: University of Michigan Library.

Cahill, Cathriona and Plant, Cora. 2011. Beneficial Use of Old Landfills as a Parkland Amenity, Environmental Protection Agency (EPA), Dublin: 2011.

Directive 1999/31/EC on the landfill of waste. European Parliament and the Council of 19, April 1999. Available at: https://eur-lex.europa.eu/legal-content/EN/TXT/?uri=celex%3A31999L0031.

Directive 2008/98/EC on waste. European Parliament and of the Council of 19, November 2008. Available at: https://eur-lex.europa.eu/legal-content/EN/TXT/?uri=celex%3A32008L0098

Douglas, Mary. 1966. Purity and Danger: An Analysis of Concepts of Pollution and Taboo, London: Routledge.

Eigenheer, Emílio Maciel. 2003. Lixo, vanitas e morte: considerações de um observador de resíduos. Niterói: Universidade Federal Fluminense (UFF).
Ekberg, C.. 2009. "Waste is What is Left Behind When Imagination Fails". Sustainability: Web Journal from the Swedish Research Council Formas. Available at: http://sustainability.formas.se/en/Issues/Issues-4-December-2009/Content/Focus-articles/Waste-is-what-is-left-behind-when-imagination-fails/.
Engler, Mira. 1995. "Waste Landscape: Permissible Metaphors in Landscape Architecture". Landscape Journal 15 (1): 10–25.
Engler, Mira. 2003. Designing America's Waste Landscapes. Baltimore: John Hopkins University Press
Hird, Myra, "Waste, Landfills, and an Environmental Ethic of Vulnerability". Ethics and The Environment 18 (1): 105–124.
Latour, Bruno. 2004. Politics of Nature, How to Bring the Sciences into Democracy, Cambridge: Harvard University Press.
Lynch, Kevin, 1990. Wasting Away- An Exploration of Waste: What is, How it happens, Why we fear It, How we do it well. San Francisco: Random House.
Gille, Zsuzsa. 2007. From the Cult of Waste to the Heap of History: The Politics of Waste. Bloomington and Indianapolis: Indiana University Press.
MacFarlane, Key. 2018. "Time, Waste and the City: The Rise of the Environmental Industry". Antipode 51 (1): 225–247.
Melosi, Martin V. 2013. "The Urban Environment". Clark, Peter. The Oxford Handbook of Cities in World History. Oxford: Oxford University Press.
Neves, Fábio de Oliveria e Mendonca, Francisco. 2014. "Por uma leitura geográfico-cultural dos resíduos sólidos: reflecose para o debate na Geografia. Cuadernos de Geografía: Revista Colombiana de Geografía 25 (1): 153–169.
Packard, Vance 1960. The Waste Makers. New York: Simon and Schuster.
Philips, Patricia 1990. "Recycling Metaphors: The Culture of Garbage". The Livable City 14 (2): 12.
Rafael, Joana 2017. "450 Meters Deep into 1 Million Years Safety". Cartha Magazine. The Limits of Fiction in Architecture (1): 11–15.
Scanlan, John. 2005. On Garbage. London: Reaktion Books.
Serres, Michel 1992. The Natural Contract. Michigan: University of Michigan Press.
Simmons, Elizabeth. 2008. "Restoration of Landfill Sites for Ecological Diversity". Waste Management & Research 17(6): 511–519.
Strasser, Susan 1992. Waste and Want: The Other Side of Consumption. Oxford: Berg Publishers, Inc.
Waste Watch 2004. History of waste and recycling information sheet. United Kingdom: Waste Watch.
World Bank 2016. "Urban Population". United Nations, World Urbanization Prospects. Available at: http://data.worldbank.org/indicator/SP.URB.TOTL.IN.ZS/countries?dispay=graph.

Wastes: Solutions, Treatments and Opportunities III – Vilarinho et al. (Eds)
© 2020 Taylor & Francis Group, London, ISBN 978-0-367-25777-4

Food waste and circular economy through public policies: Portugal & Brazil

P.C. Berardi
LEPABE, Department of Metallurgical and Materials Engineering, Faculty of Engineering, University of Porto, Porto, Portugal
CELOG Centro de Excelência e Logística e Supply Chain da EAESP – FGV, São Paulo, Brazil

L.S. Betiol
FSJ, Department of Social and Legal Sciences, São Paulo School of Business Administration – FGV/EAESP, São Paulo, Brazil

J.M. Dias
LEPABE, Department of Metallurgical and Materials Engineering, Faculty of Engineering, University of Porto, Porto, Portugal

ABSTRACT: The reduction of food losses and food waste generation are a huge challenge, which demand the optimal use of food resources. This study shows how two different countries/markets face this issue through the normative aspect. Both contexts (Portugal and Brazil) present a close relationship with global agendas. However, the European Union has already established a more advanced body of rules; moreover, when demands of circular economy are taking place. It is imperative that the legal framework becomes robust and supports all social and environmental needs to promote sustainable development in the medium and long terms. Nevertheless, with a regulatory strength it will be possible to move forward across the value chain of agri-food business and considerably reduce food wastage.

1 INTRODUCTION

To feed a world population of around 9 billion is one of the greatest challenges facing the coming decades as the food demand is estimated to increase by 60% by 2050. In addition, to increase productivity and resource efficiency, it is of substantial importance to strengthen the work to reduce food loss and food waste which account for a third of all food production, in a scenario where approximately 900 million people are hungry (FAO, 2018).

Such inefficiency presents financial impacts (estimated losses of 940 billion dollars per year), as well as social and environmental impacts (FLW, 2016). Those are of course not restricted to the current generation, because the substantial loss of resources is not limited to the food itself but also to the resources which are increasingly scarce and expensive to the planet and to our society (e.g. water, energy, soil, climatic balance, supply and quality of labor and time). Thus, more risk, vulnerability and costs will be associated with food production.

To study the subject of food waste, it is necessary to take into account the multiplicity of articulations with other topics of high relevance, which confers a greater degree of complexity since it has high affinity with: agriculture, food, nutrition, food security, economic and social aspects/conditionings, the role of public and private sector, distribution and logistics, environment, waste prevention and management, eating habits, among others.

In this study, a comparison is made between the European and Brazilian efforts to advance under this topic, using as a unit of analysis the normative efforts to address the subject of food loss and food waste, as well as identifying their respective challenges.

2 EUROPEAN CONTEXT

According to the 2016 Fusions' Report, annually, it is estimated that around 90 million tons of food are wasted, which is equivalent to 173 kg per capita per year, representing 20% of all food production in the European Union (EU). The greatest attention lies in the quality of this resource, since much of it is still in conditions for human consumption (EEA, 2016). In the EU, the highest concentration of food waste is estimated to be in domestic consumption (53% of the total). However, the problem must be addressed at all stages of the value chain (Åsa Stenmarck et al., 2016).

In 2015, the EU Action Plan for the Circular Economy (APCE) was presented, with clear goals for the medium and long term, namely the 2025 target to reduce at least 30% of the waste of foodstuffs in the manufacturing, distribution and sales, trade and hotel industry and household consumption sectors, taking into account the 2005 reference values (EC, 2015).

The APCE is in alignment with the 2030 Agenda for Sustainable Development, namely with the United Nations Sustainable Development Goal 12, especially on indicator 12.3 which calls for 2030 to halve the waste production per capita (relative to 2015 as reference year) both at the global level, retail and end-consumer level and to reduce food waste over of production and supply chains; it also connects with the Paris Agreement (UN, 2015), by promoting the reduction of greenhouse gases emissions.

In May 2018, the waste Framework Directive (EU) – 2008/98/EC was revised to bring these new guidelines into practice (EP, 2018). The definition of food waste has been broadened and among the main objectives of reducing the generation of food waste is stated the priority to promote food donation and other forms for human consumption. Incentive measures were also determined, including: i) to provide incentives for the collection of unsold food at all stages of the food supply chain and for its safe redistribution, including for charity organizations; ii) to clarify consumers concerning the meaning of dates on food packaging such as "use by" and "best before" to prevent food waste generation; iii) to encourage human consumption of food surplus over animal feed and reprocessing in non-food products according to the waste hierarchy.

In order to promote cooperation between all key players in the food value chain and to accelerate EU progress towards meeting all the agendas (Agenda 2030, APCE and Paris Agreement), an EU Platform on Food Losses and Food Waste was presented (EC, 2016a) with 70 members organizations including public entities (33) and representatives of the economic operators of the food chain (37 – consumers and NGOs). The most noteworthy points of this platform are the demand for the development of indicators concerning amount of waste, working with prevention policies and implementing and monitoring actions throughout the chain.

From the Circular Economy perspective, it is necessary to rethink the flow of food, where, rather than reducing waste, it is necessary to rethink food waste as a potential resource to be absorbed as by-products and ultimately transformed into organic fertilizers or biomaterials for medicine or bioenergy, with clear reductions of negative environmental and health impacts and promotion of economic, environmental and social benefits (EMF, 2019).

The next steps to achieve the long-term goals of preventing waste generation in the EU include quantifying food waste by March 2019 and preparing a report and proposal by the end of 2023 to meet the EU's reduction target (EC, 2017).

2.1 *Characterization of the Food System in Europe*

In Europe, a great diversity of agri-food systems exists, in terms of production, scale, intensity, inputs and supply chains. Its base has a large number of small-scale family businesses, producers, retailers and restaurants operating in parallel with globalized companies (EC, 2018).

At the production side, the highest proportion of food consumed in the EU continues to be produced locally, and most EU food trade takes place between EU countries. Exports of food and beverages have a positive trade balance even though imports of commodities such as tropical fruits, coffee, tea, cocoa, soy products and palm oil are used for direct consumption or feed production. Regarding fishery products, the EU is the world's largest importer since its productive capacity has been around 45% of its demand since 2008 (EU, 2018). In what concerns biomass production, the agri-food system is estimated to generate 60% of all demand in the EU, although indicators point to annual waste production in the order of 88 Mt of food – about 20% of all food produced, with estimated costs estimated in 143 billion euros (EC, 2016b). In addition, the energy required to grow, process, pack and deliver food to European growers accounts for 17% of EU gross energy consumption, equivalent to about 26% of EU final energy (EC, 2018).

Although the agricultural sector has increased its productivity by almost 9% since 2005, while reducing greenhouse gases emission (GGE) by 21% since 1990 and reducing fertilizer use, it remains the largest source of GGE, namely of methane and nitrous oxide (EU, 2018). However, in order to consider the actions and goals of the APCE, it is still necessary to intensify the use of renewable sources of energy and reduce GGE and pollution, especially in this sector.

In agreement with IEEP (2014), to take an example of the wastage of resources in the chain, 900 Mt of paper, food and plant material are generated each year. In the fisheries sector, up to 40 Mt of fish can be discarded each year during commercial fishing. Other estimates of the waste produced in fisheries and aquaculture include amounts up to 130 Mt and annual losses of up to 50 billion dollars as a result of resource mismanagement. Every year, almost 300 Mt of domestic waste, industrial waste and wastes from other origins are generated in the EU and remain largely unexploited. Among those, 140 Mt are municipal waste, mostly generated by households (EC, 2018).

The application of the Circular Economy principles, which aim at retaining the value of different types of resources (not only biological) in the economic cycle for as long as possible, taking into account the way products and materials are designed, produced, used and disposed of, there is a diversity of demands and alternatives that must be worked out and integrated into European agri-food systems, at all stages from production to final consumption to reach such goal.

2.2 *Portugal*

As a follow-up to the European guideline in drawing up national strategies, Portugal approved the Action Plan for the Circular Economy in 2017 (PAEC 2017: 2020) (DR, 2017). It is a document of strategic growth and investment model for increased efficiency and resources recovery together with the minimization of environmental impacts, with definition of actions to be introduced and developed by 2020 by all sectors. Concerning the three levels of actions as established, the agro-industrial sector has been contemplated and presents for the next three years: macro (national) actions that consolidate some of the actions of several governmental areas; meso actions, with indication of sectorial agendas, especially for sectors more intensive in the use of resources; and micro actions, which require regional/local agendas, to be adapted to the socioeconomic specificities of each region.

In the same year, the National Strategy and Action Plan to Combat Food Waste was announced (CNCDA, 2018). The report evidences results from a Nacional study on this matter (PERDA, 2012), which estimates that in Portugal there is an annual food waste production in the order of 1 Mt (equivalent to around 17% of annual food production), which corresponds to 97 kg per capita per year, of which 32.2% is estimated to from agri-food activity; 7.5% from the industry; 28.9% in the distribution phase and the remaining 31.4% at the consumers. This strategy is built on three objectives: Prevent, Reduce and Monitor food waste. Based on these objectives, nine others were designed to support and operationalize its implementation. The plan established 14 measures to be adopted by 2021. More emphasis was given to food donation, in which it was defined that donated foods do not count into the waste metric, and food donation is considered another link of the supply chain, classified at the same level of production, transformation, trade or final consumer. In the meantime, it states that the established hygiene and safety rules for donating foodstuffs (EP, 2004) should be guaranteed.

There are already several initiatives concerning CE on this topic at different stages of the chain. At a public level, locally, the city of Porto made a roadmap in 2017 to identify flows of resources to enable the promotion of sustainable production and consumption. In the food side, direct organic collection flows were mapped at source aiming organic recovery, namely for the production of fertilizers (BCSD and 3Drivers, 2017). The "Fruta Feia" (Ugly Fruit) project, started in 2013 by a cooperative that promotes the use of food outside aesthetic marketing standards, in which every part of the chain can win: producers are able to give a proper destiny to the previously discarded production and consumers, industries, restaurants and retailers can buy and sell food products at a lower cost. In two years, a large Portuguese retailer participating in this project reported that a reduction of waste of fresh products such as vegetables in 3.4 t was achieved and another 14.3 t of food products were sold with discount price (Costa, 2016). GoodAfter, the first online Portuguese supermarket to combat food waste, started its activities in 2016. It only sells "best before" products, which is the date of minimum product durability and all products are supplied directly by the manufacturers themselves. It does not work with perishable products. In two years of activities, it managed to market 31.79 t of food and beauty products that would be wasted. The "Dose Certa" (Right Dosage) project started in 2008 by the Municipal Waste Management System of Porto – Lipor, with the Portuguese Association of Nutritionists to guide restaurants, canteens and various establishments that serve food to prepare meals without losing quality or nutritional value, but with focus on reducing waste by providing the appropriate dose. So far, it has been estimated that the impact generated by the project partners is between 30 and 35% of food waste reduction. Another interesting project is the "Embrulha" (The Wrap), a partnership between Lipor and Porto City Hall, a movement that brings together 68 restaurants that promote change in consumer behavior when meals are not fully consumed on the spot. The supply of biodegradable packages is free, and 12.72 t of food have been taken out from restaurants since 2017, which means that the emission of 2.67 t of carbon dioxide (CO_2) into the atmosphere was avoided (PCH, 2019).

3 BRAZILIAN CONTEXT

Brazil has an enormous abundance of natural resources and a unique set of physical and human attributes. These qualities are reflected in the potential of agricultural and livestock production. According to the United States Department of Agriculture (USDA), Brazil has one of the five largest agricultural production areas on the planet and is also one of the world's largest agricultural producers, namely of soy, orange, sugar, beef and chicken (USDA, 2016).

Despite the large potential for food production, there is a critical wastage of these resources at the whole chain, and a growing population migrating to unfortunately make Brazil re-appear on the world map of hunger from where the country had left in 2014 (CAISAN, 2018).

According to FAO data for 2013, in Brazil, of the 268.1 Mt of food available, 26.3 t were lost, approximately 10% of the total (CAISAN, 2018). In the fruit and vegetable sector, the scenario is even more serious. Research by EMBRAPA indicates, for example, that the loss rate varies between 30% and 50% along the productive chain of fruits and vegetables, one of the most sensitive products, given its high perishability. Currently, at the table of Brazilians, 41.6 kg of food per person each year is estimated to be wasted. Every day, every Brazilian family throws away 0.353 kg, which gives an alarming total of 128.8 kg of discarded food per family yearly (Melo et al., 2018).

In order to respond to this challenge, Brazil has been working on several paths, from its agreement to international documents that have commitments linked to the reduction of hunger and food loss and food waste, such as the UN Millennium Goals, Agenda 2030 for Sustainable Development, as well as the targets under the Paris Agreement, and have even published their National Determined Contribution. NDC embody efforts by each country to reduce *national* emissions and adapt to the impacts of climate change.

At a national level, it has worked on the elaboration and implementation of public policies focused on the topics of hunger reduction and of food loss or waste generation, starting with the Federal Constitution that indicates that the union, states and municipalities are responsible for issues related to the promotion of agricultural production and organization of food supply and the regulation of the economic order focusing on a dignified existence, as well as the insertion, since 2010, of adequate food as a social right to be guaranteed.

To address this constitutional right, the Food and Nutrition Security Policy was instituted, among other policies, through the Decree no. 7.272 of 2010, which also regulated Law no. 11.346 of 2006, which deals with the National System of Food and Nutrition Security and establishes parameters for the elaboration of the National Plan of Food and Nutritional Security (PLANSAN) (MDSA, 2017).

The PLANSAN is already in its second cycle, covering the period from 2016 to 2019, having defined nine major challenges, among them "to promote the supply and regular and permanent access of the Brazilian population to adequate and healthy food", counting on this with the fight against the food losses and food waste generation. This combat is both in the identification and mitigation of the qualitative and quantitative losses in the post-harvest of grains, as well as in the establishment of a legal framework for the reduction of losses and waste of food covering the Brazilian food banks, through the Brazilian food bank network, regulated by Ordinance 15 of May of 2017 (SNSAN, 2017).

It can be said that the legal framework of the National Policy on Food and Nutrition Security in Brazil has built a system and a policy that delimits a scope of public policy and recognizes a multisectoral agenda. But there are some challenges left. Experts believe that for the effective implementation of a policy of reducing food loss and food waste generation it is necessary to pass through the logistics of the donation, which in addition to not being simple is costly for those who wish or are obliged to do so. There is a need for, in the case of donation, a connection with food banks (Melo et al., 2018).

For the near future, what has been perceived is an increase in the importance of the theme at the National Congress. According to data from 2018, there are more than 30 law proposals distributed, covering this topic. It is possible to point out issues involving the productive chain, the role of the retail sector, to discussions about civil and criminal responsibility of those who seek to give a proper destination to foods that would be discarded before the end of their useful life. An alternative that is opening up to solve all these issues in a single document would be the elaboration of a National Consumer Food and Nutrition Education Policy, with proposals already in progress (CAISAN, 2018).

4 CHALLENGES

There are many challenges to be faced in combating and reducing food loss and food waste generation, by optimizing the use of this valuable resource. The legal basis is fundamental to enable the correct execution of the action plans. It is imperative to involve all sectors of the chain for optimal use of resources, to develop new business models, to create logistical flows to optimize the use of resources, to adopt innovation and technologies capable of ensuring food safety and quality, thus using food and waste in their totality. In the EU, the process appears to be in a more advanced stage, while in Brazil, public policy development and approvals are at an earlier stage, although all have committed themselves to the same global agendas on this matter. As the EU has an CE guidance document, there is a direct relationship with this topic whereas at Brazil, the theme of CE is still at voluntary organizational strategies level, which confers a dissociation of these topics, being another challenge to be worked on.

5 FINAL CONSIDERATIONS

The transformation of food and agricultural systems towards sustainability with healthy and resource efficient production that is circular and inclusive needs to be accelerated. It is extremely important to reduce food losses and food waste generation, namely by ensuring the optimal use of food resources. This includes waste prevention by optimizing processes throughout the supply chain, by appropriate distribution of unsold or unused food focusing at human consumption and by recovery organic by-products and wastes into valuable and safe biological products. For example, by implementing small-scale biorefineries which might help farmers and fishermen to diversify their sources of revenue and to better manage market risks, while at the same time achieving the goals of the Circular Economy, without endangering the health of the population.

ACKNOWLEDGEMENTS

This work was financially supported by project UID/EQU/00511/2019 – Laboratory for Process Engineering, Environment, Biotechnology and Energy – LEPABE funded by national funds through FCT/MCTES (PIDDAC) and Project "LEPABE-2-ECO-INNOVATION" – NORTE-01-0145-FEDER-000005, funded by Norte Portugal Regional Operational Programme (NORTE 2020), under PORTUGAL 2020 Partnership Agreement, through the European Regional Development Fund (ERDF).

REFERENCES

Åsa Stenmarck, Carl Jensen, Tom Quested & Moates, Graham 2016. Estimates of European food waste levels. *In:* Innovation, Fusions Reducing Food Waste through Social (ed.). Sweden.

BCSD, Conselho Empresarial para o Desenvolvimento Sustentável & 3Drivers 2017. Roadmap para a cidade do Porto circular em 2030. Portugal: Câmara Municipal do Porto.

CAISAN, Câmara Interministerial de Segurança Alimentar e Nutricional – 2018. Estratégia Intersetorial para a Redução de Perdas e Desperdício de Alimentos no Brasil. Brasil: Ministério do Desenvolvimento Social MDS.

CNCDA, Comissão Nacional de Combate ao Desperdício Alimentar 2018. Estratégia Nacional e Plano de Ação de Combate ao Desperdício Alimentar. Portugal: Diário da República.

Costa, Rita Marques 2016. Aproveitar o Desperdício. *Frutas, Legumes e Flores.*

DR, Diário da República – Resolução do Conselho de Ministros 2017. Plano de Ação para a Economia Circular em Portugal (PAEC 2017:2020).

EC, European Commission 2015. Communication from the Commission to the European Parliament, the Council, the European Economic and Social Committee and the Committee of the Regions Closing the loop – An EU action plan for the Circular Economy.

EC, European Commission 2016a. EU Platform on Food Losses and Food Waste. Policies, information and services ed.

EC, European Commission 2016b. Reducing food waste: the EU's response to a global challenge. *In:* Sheet, European Commission – Fact (ed.).

EC, European Commission 2017. EU actions against food waste. Policies, information and services ed.

EC, European Commission 2018. A sustainable bioeconomy for Europe: strengthening the connection between economy, society and the environment. Belgium.

EEA, Environmental European Agency 2016. Food Waste.

EMF, Ellen Macarthur Foundation 2019. Cities and circular economy for food.

EP, European Parliament 2018. Directive (EU) 2018/851 of the European Parliament and of the Council of 30 May 2018 amending Directive 2008/98/EC on waste.

EP, European Parliment 2004. Regulation (EC) no 882/2004 of the European Parliament and of the Council of 29 April 2004 on official controls performed to ensure the verification of compliance with feed and food law, animal health and animal welfare rules.

EU, European Union 2018. Agri-food trade in 2017: another record year for EU agri-food trade. *Monitoring Agri-trade Policy.*

FAO, Food and Agriculture Organization of the United Nations 2018. El estado de la seguridad alimentaria y la nutrición en el mundo. Fomentando la resiliencia climática en aras de la seguridad alimentaria y la nutrición. Italy.

FLW, Protocol Steering Committee 2016. Food Loss and Waste Accounting and Reporting Standard. Food Loss and Waste Protocol ed.: United Nations Environment Programme UNEP.

IEEP, Institute for European Environmental Policy 2014. Wasted Europe's untapped resource An Assessment of Advanced Biofuels from Wastes & Residues.

MDSA, Ministério do Desenvolvimento Social e Agrário 2017. Plano Nacional de Segurança Alimentar e Nutricional – PLANSAN 2016-2019. Brasil: CÂMARA INTERMINISTERIAL DE SEGURANÇA ALIMENTAR E NUTRICIONAL – CAISAN.

Melo, Relator: Evair Vieira de, Consultores legislativos: Rodrigo Dolabella (coordenador), Marcus Peixoto & Pinheiro, Alberto 2018. Perdas e desperdício de alimentos: Estratégias para redução. Cadernos de Trabalhos e Debates 3. Brasil: Câmara dos Deputados.

PCH, Porto City Hall 2019. Projeto Embrulha evita desperdício alimentar de 12,7 toneladas em 2018.
PERDA 2012. Do Campo ao Garfo. desperdício Alimentar em Portugal. *In:* Cestras (ed.). Lisboa.
SNSAN, Secretaria Nacional de Segurança Alimentar e Nutricional 2017. Instrução Normativa n° 1 de 15 de maio de 2017. Brasil: Diário Oficial da União.
UN, United Nations 2015. The Paris Agreement – COP 21.
USDA, Foreign Agricultural Service 2016. Agricultural Research in Brazil. USA.

Wastes: Solutions, Treatments and Opportunities III – Vilarinho et al. (Eds)
© 2020 Taylor & Francis Group, London, ISBN 978-0-367-25777-4

Energy saving potential of electronic waste management practices

B. Kuriyama & M. Madaleno
Aveiro University, Aveiro, Portugal

S. Diedler & K. Kuchta
Hamburg University of Technology, Hamburg, Germany

ABSTRACT: This study aims at estimating the energy savings that could be achieved throughout additional e-waste recycling efforts in fifteen Latin America countries, followed by a qualitative e-waste management evaluation of each considered country. The energy analysis was performed using the United States Environmental Protection Agency's Waste Reduction Model. The results indicate that up to 19,871 GJ of energy savings could have been achieved if 15% of the total e-waste generated in 2016 of each related country would have been recycled instead of landfilled. Moreover, 1,569,139 Mt of carbon dioxide equivalent emissions could have been avoided. Regarding the current status of e-waste situation in Latin America, the fifteen countries present different levels of e-waste management, and key steps have been identified to explore the energy and environmental benefits. Establishing a sound regulatory framework and foster closer cooperation between countries has proven to be essential to achieve e-waste technical standards and solutions.

1 INTRODUCTION

Entailed by the digital revolution started in the late 1970s, the present scenario of unprecedented generation of electrical and electronic waste, or e-waste for short, challenges the waste management sector worldwide (APC & HIVOS 2010). Emerging economies are those who have greater difficulty to keep up with the pace of technological advancement, and Latin America still face significant challenges to treat e-waste in an environmentally sound manner (Lundgren 2012). The region comprises fast growing cities with increasing rates of waste generation, presenting a volume of e-waste per capita higher than the world average, equivalent to 7.1 kg per inhabitant. Furthermore, in 2016, Latin American countries produced 9.4% of the total e-waste in the world, and the expected growth rate indicates that this amount will increase up to twice as fast as the global trends (GSMA 2015).

Solid waste management systems in Latin America are in the process of modernization, and only a few countries have sorting plants and employ recycling as a common practice in their municipal policies. Although many cities have initiated source-separation programs, and despite the fact that waste collection coverage is at a relative high level (nearly 85% at urban level), about 69% of total waste is still disposed in landfills and dumps (Hettiarachchi et al. 2018, Kaza S. et al. 2018).

The generation of waste represents a loss of resources in form of both materials and energy, whereby a frequently overlooked benefit of proper e-waste management is the energy savings through more efficient practices, a win-win scenario for energy consumption reduction and resource conservation (EC 2016). Once inside the e-waste stream, each stage of an electrical and electronic equipment's (EEE) lifecycle has energy impacts, and e-waste management practices such as recycling has the potential to reduce the demand for primary raw materials, lowering the energy inputs from resources extraction and processing activities, thereby saving energy (USEPA 2010, Choate et al. 2005).

This work is further structured as follows: section 2 presents the research methodology whereby in section 3 the results are discussed. The final section 4 concludes this work.

2 METHODOLOGY

This work applies the Waste Reduction Model (WARM) developed by the United States Environmental Protection Agency (EPA) to estimate the energy impacts of additional e-waste recycling efforts in fifteen countries of Latin America. Additionally, the avoided greenhouse gas (GHG) emissions are estimated, and a qualitative assessment of the different e-waste management systems is performed in order to acknowledge which areas should be the focus on in future actions (encouraging benchmarking policies and techniques).

2.1 *Energy and environmental impact assessment*

Developed to assist solid waste stakeholders in estimating GHG emissions from several different waste management practices, WARM also allows users to calculate results in terms of energy savings (USEPA 2018). Based on a lifecycle material management perspective, from a waste generation reference point, the model calculates mainly GHGs and energy implications for a baseline and alternative waste management scenarios, resulting in a comparison between both. The results represent the net emissions and energy impacts for each disposal strategy, and the benefits estimated result from the choice of one management path relative to another (Choate et. al. 2005). The analysis consisted in evaluating the change in e-waste disposal practices from landfilling to recycling. In this sense, since WARM analysis is limited to relevant lifecycle stages of a product, and the EEE composition varies considerably among product lines and categories, an approved proxy provided by EPA to represent electronics was adopted (USEPA 2015).

According to literature, although 17% of e-waste within the Americas has been documented as collected and properly recycled, it is estimated that in Latin American countries the e-waste collection rate is lower then 3% (Baldé et. al. 2007). As such, most of e-waste within the region is either sent to landfills or processed by the informal sector. In order to estimate the benefits of additional recycling efforts as a result of the change in disposal practices, it was considered that 15% of the total e-waste generated in 2016 by each studied country would be properly recycled. Table 1 presents the studied countries by amount of e-waste generated and e-waste generated per capita in 2016.

Table 1. Total amount of e-waste generated (in kt) and e-waste generated per capita (in kg/cap) in 2016 by each considered country.

	E-waste generated in 2016	
Country	Total (kt)	Per capita (kg/cap)
1. Argentina	368	8.4
2. Bolivia	36	3.3
3. Brazil	1,534	7.4
4. Chile	159	8.7
5. Colombia	275	5.6
6. Costa Rica	48	9.7
7. Ecuador	90	5.5
8. Guatemala	67	4.0
9. Honduras	19	2.3
10. Mexico	998	8.2
11. Panama	33	8.0
12. Paraguay	44	6.4
13. Peru	182	5.8
14. Uruguay	37	10.8
15. Venezuela	254	8.2

Table 2. Technology transfer performance indicators.

Indicators group	Scoring criteria
1. Regulation framework	
1.1. Ratification of Basel Convention	
1.2. Status of national waste regulation	Compliance (yes/no)
1.3. Status of national e-waste regulation	
1.4. Extended Product Responsibility (EPR)	
2. Business attractiveness	
2.1. Country Risk Index (CRI)	
2.2. Global Competitiveness Index (GCI)	Classification (good, fair, poor)
2.3. Corruption Perception Index (CPI)	
3. Public awareness	
3.1. Human Development Index (HDI)	
3.2. Illiteracy rate	Classification (good, fair, poor)
3.3 Tertiary education rate	

2.2 *E-waste management assessment*

To assess and compare the performance of different e-waste management systems, a criterion has been defined to better understand the current status of countries studied, i.e. macroeconomic factors, current recycling infrastructure, and technology transfer performance. As for the last aspect, a set of three main groups of indicators was selected, based on the areas described by the United Nations Environment Program (UNEP 2007) as challenges for a successful transfer of sustainable technologies:

(1) Regulation framework: recycling and reuse activities are directly affected by the availability and implementation of legislation. The absence of domestic laws may lead to increased informal sector, resulting in material losses and uncontrolled trade of e-waste;
(2) Business attractiveness: the interest of stakeholders can enable large-scale investments to foster the implementation of new e-waste recycling and treatment technologies;
(3) Public awareness: human development factors, in particular, education, are linked to behavior and engagement of people on environmental concerns, e.g. prevention and recycling of waste, contributing to the effectiveness of waste management systems.

Then, based on the compliance/classification grade of each indicator, for each group was assign a final adjusted score ranging from 0 to 5, with zero the worst and five the best result (Table 2). In conclusion, a comparative analysis of the overall status of the different e-waste management systems was performed.

3 RESULTS

This work endeavors to include a holistic view of e-waste management with focus on energy savings, particularly in Latin America countries. The energy savings and GHG avoided emissions were calculated based on a hypothetical scenario by adopting conservative assumptions in order to delineate an upper bound estimate. Furthermore, since the environmentally sound management of e-waste depends on the successful implementation of technological advancements and behavioral changes, a comprehensive e-waste assessment was performed to better understand the current national and regional situation on each considered country. As a final step, a comparative analysis was conducted to provide a general overview of current status of Latin America's e-waste management systems, giving an insight of which areas should be prioritized in future actions.

Table 3. WARM's outcomes.

	Energy used (GJ)			GHG emissions ($MTCO_{2eq}$)		
Country	Landfill	Recycling	Energy savings	Landfill	Recycling	Emissions avoided
1. Argentina	15.62	−1,748.99	−1,764.61	1,118	−138,226	−139,344
2. Bolivia	1.53	−171.10	−172.63	109	−13,522	−13,632
3. Brazil	65.13	−7,290.63	−7,355.75	4,661	−576,193	−580,854
4. Chile	6.75	−755.68	−762.43	483	−59,723	−60,206
5. Colombia	11.68	−1,306.99	−1,318.67	836	−103,294	−104,130
6. Costa Rica	2.04	−228.13	−230.17	146	−18,030	−18,175
7. Ecuador	3.82	−427.74	−431.56	273	−33,805	−34,079
8. Guatemala	2.84	−318.43	−321.27	204	−25,166	−25,370
9. Honduras	0.81	−90.30	−91.11	58	−7,137	−7,194
10. Mexico	42.37	−4,743.18	−4,785.56	3,032	−374,864	−377,896
11. Panama	1.40	−156.84	−158.24	100	−12,395	−12,496
12. Paraguay	1.87	−209.12	−210.99	134	−16,527	−16,661
13. Peru	7.73	−864.99	−872.72	553	−68,362	−68,915
14. Uruguay	1.57	−175.85	−177.42	112	−13,898	−14,010
15. Venezuela	10.78	−1,207.18	−1,217.97	772	−95,406	−96,178
TOTAL	175.94	−19,695.15	−19,871.08	12,590	−1,556,549	−1,569,139

Table 4. WARM's energy and emissions factors.

Factors	Unit	Landfill	Recycling
Energy use per ton of material	GJ	0.28	−31.71
GHG emission per ton of material	$MTCO_{2eq}$	0.02	−2.5

3.1 *Energy savings and GHG avoided emissions*

The environmental impacts related to energy consumption and GHG emissions were estimated as the difference between e-waste disposals practices from landfilling (baseline scenario) to recycling (alternate scenario), where negative values indicate net energy savings or avoided emissions. The model's results are presented in Table 3. Energy is shown in giga-joules (GJ), while GHG emissions in metric tons of carbon dioxide equivalent ($MTCO_{2eq}$).

As observed, the energy savings potential from e-waste management sector lies between 91.11 and 7,355.75 GJ, a total of approximately 19,871 GJ that could be achieved considering the sum of all Latin America countries studied. With regard to GHG emissions, the total reduction potential was estimated in 1,569,139 $MTCO_{2eq}$. The energy and emissions factors per ton of electronic material adopted by WARM are presented in Table 4.

The results demonstrate that the improvement of e-waste management practices can lead not only to the reduction of the total amount of e-waste generated, but also to a significant energy savings, and avoided, GHG emissions. Nevertheless, some important constraints presented in the e-waste management value-chain and also in the policy framework needs to be overcome in order to allow Latin America countries to fully explore environmental benefits of improved e-waste management practices, and so an e-waste management assessment was carried out.

3.2 *E-waste management performance*

The results of the macroeconomic analysis indicate that several trends are driving the generation of e-waste, particularly in developing countries. Higher levels of disposable income, urbanization,

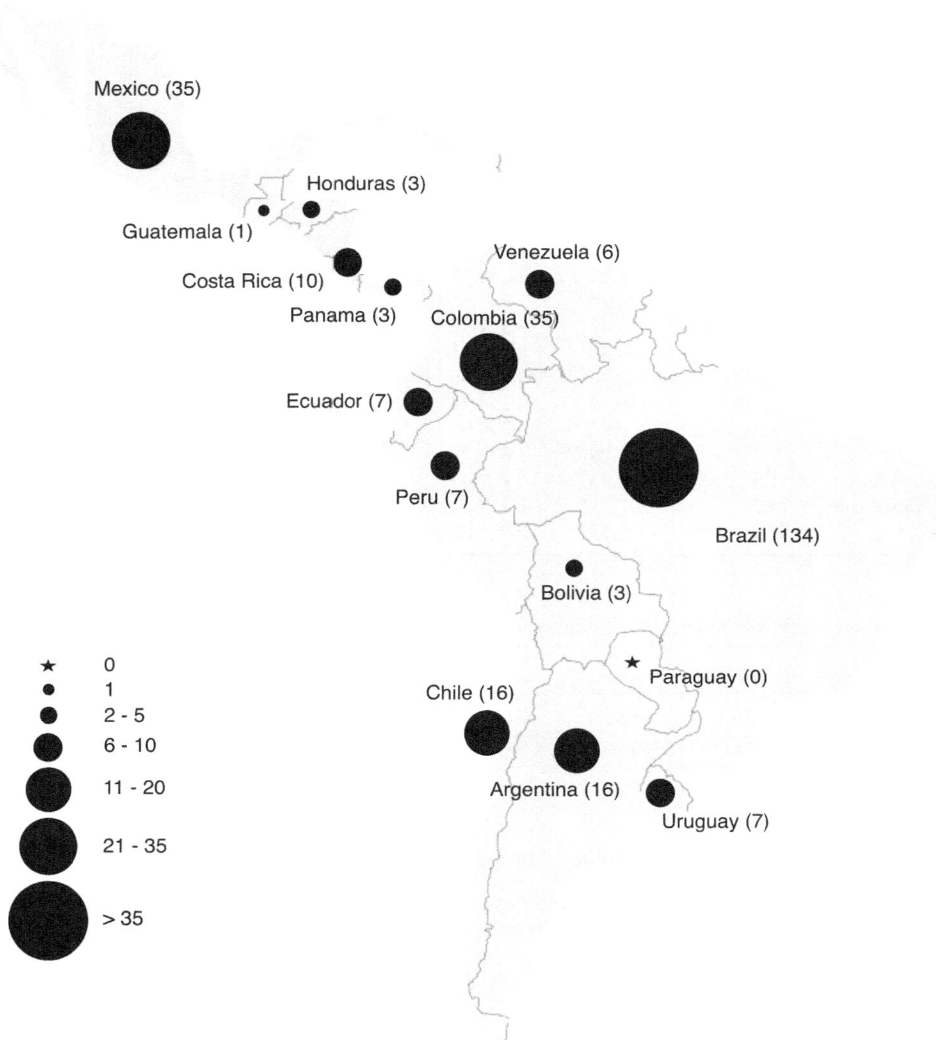

Figure 1. E-waste recycling facilities in Latin America.

and industrialization are leading to growing amounts of e-waste, and high-income countries such as Chile, Argentina and Uruguay are also those who presented a higher share of electronic products consumers Furthermore, macroeconomic indicators are also linked with the overall performance of a country in terms of financing and business, and countries with a higher levels of inflation rate, corruption and social problems are often those who face the greatest challenges to treat e-waste in an environmentally sound manner. In terms of recycling structure, 283 e-waste facilities were identified in total as current operating within the considered countries (Figure 1). Brazil leads the ranking accounting for close to 47% of the total number of facilities identified (134), followed by Colombia and Mexico, with 35 facilities each. In Paraguay, no establishment was recognized as an e-waste facility since most of companies identified were usually related to metal scrap processing or responsible for receiving e-waste for landfilling or incineration. In general, with exception of

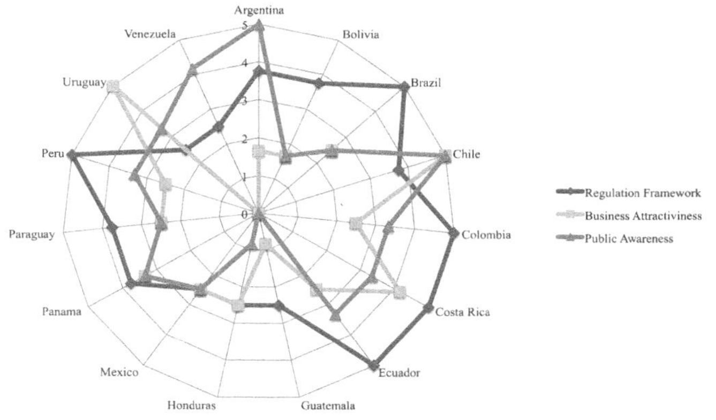

Figure 2. E-waste technology transfer performance indicators.

some specific e-waste categories such as lamps and batteries, most of e-waste facilities' activities are limited to the pre-processing stage, relying usually on manual dismantling.

Concerning the e-waste technology transfer indicators (Figure 2), results show that although all countries studied are signatories of the Basel Convention and have at least one regulation related to solid waste management or the environmental protection, only a few countries have specific regulations for e-waste management. The lack of a specific definition of key elements, principles, and responsibilities in the e-waste management processes justified why the majority of Latin American countries treat and manage e-waste under the umbrella of hazard waste legislation, resulting into a strong influence of the informal sector. In this context, a key issue is to incorporate informal recyclers into a formal system, creating opportunities in the e-waste recycling business for low-income populations.

Moreover, the economic and social indicators show that not only a regulation framework must be outlined but also a range of measures related to public awareness and business attractiveness, particularly in the education sector. In this regard, the benchmarking of e-waste data aim to support country's efforts towards improved practices, and therefore, energy savings.

4 CONCLUSIONS

E-waste is a serious problem both at local and global scales. The increase of developing countries joining the global information society coupled with the fast obsolescence of electronic products led to an unprecedented growth of e-waste, especially in emerging economies such as the majority of Latin American countries. The hazardous content of e-waste demands an effective management in order to prevent negative environmental and human health impact, contributing not only to resource efficiency but also to energy savings. In this regard, this study concludes that e-waste recycling offers a strong potential in this direction. According to WARM's results, a total of 19,871 GJ of energy savings and an environmental benefit of 1,569,130 Mt of carbon dioxide equivalent reduction could be achieved. As such, to explore all these benefits, a number of barriers need to be addressed in different stages of e-waste management value chain and also in the regulation framework, business attractiveness and public awareness.

The present scenario in Latin America indicates that the recycling industry is expected to grow in the coming years, and that the e-waste management sector presents a strong potential for improvement. In this regard, reliable and comparable e-waste data provides the necessary information to assess progresses over time, set and calculate targets and identify best practices, helping policy makers to design a strategy and assisting the industry sector in the process of making rational investments decisions. Furthermore, technological advancements are required to establish a full recycling service for e-waste in Latin America, and cooperation between countries is essential to promote technological transfer and knowledge exchange, not only for the development of new recycling systems but also for the improvement of old ones.

In conclusion, it is important to underscore that this study represents a first attempt to comprise a holistic view of e-waste management with focus on energy savings. However, due to several limitations related to the complex nature of e-waste streams, the selected model and data collected, a simplified approach was adopted in order to evaluate and compare the particularities of each considered country. In this sense, the identified gaps should be addressed in future researches to obtain a more detailed analysis.

REFERENCES

Association for Progressive Communications & Humanist Institute Cooperation with Developing Countries (APC & HIVOS). 2010. *Global information society watch 2010 Report.*

Baldé V. P. et al. 2017. *The global e-waste monitor 2017.*

Choate A. et al. 2005. *Waste management and energy savings: benefits by the numbers.*

European Commission (EC) 2016. *Study on the energy saving potential of increasing resource efficiency.* Luxembourg: Publications Office of the European Union.

Global System for Mobile Communications Association (GSMA) 2015. *E-waste in Latin America: statistical analysis and policy recommendations.*

Hettiarachchi H. et al. 2018. *Municipal solid waste management in Latin America and the Caribbean: issues and potential solutions from the governance perspective.* Recycling 2018, 3(2):19.

Kaza S. et al. 2018. *What a waste 2.0: a global snapshot on solid waste management to 2050.* Urban Developing Series. Washington, DC: World Bank Publications.

Lundgren K. 2012. *The global impact of e-waste: addressing the challenge.* Geneva: ILO.

United Nations Environment Program (UNEP). 2007. *E-waste volume I: inventory assessment manual.* New York: UNEP.

Unites States Environmental Protection Agency (USEPA) 2015. *Using WARM emission factors for materials and pathways not in WARM.* Washington, DC: USEPA.

Unites States Environmental Protection Agency (USEPA) 2018. *Documentation for greenhouse gas emission and energy factors used in the waste reduction model (WARM).* Washington, DC: USEPA.

Wastes: Solutions, Treatments and Opportunities III – Vilarinho et al. (Eds)
© 2020 Taylor & Francis Group, London, ISBN 978-0-367-25777-4

Resistance of geotextiles against mechanical damage caused by incinerator bottom ash

F. Almeida, J.R. Carneiro & M.L. Lopes
Construct-GEO, Faculty of Engineering, University of Porto, Porto, Portugal

ABSTRACT: Reuse and recycling of waste is no more a challenge, but a demand. Municipal solid waste can be treated through an incineration process that leads to the generation of a residue known as incinerator bottom ash. Previous studies addressed the possibility of using this residue as recycled aggregate in road construction, where it may be in contact with geosynthetics. In the present study, the tensile behaviour of three geotextiles (with different structures and masses per unit area) was assessed after being submitted to mechanical damage under repeated loading tests, in which different aggregates, besides incinerator bottom ash, were used (gravel 4/8, *tout-venant* and *corundum*). Results have revealed that incinerator bottom ash did not cause more damage to the geotextiles than most of the studied aggregates. This feature opens good perspectives to consider the possibility of using this residue as recycled aggregate in contact with geosynthetics.

1 INTRODUCTION

Circular economy is the paradigm that is helping a more balanced use of resources. This paradigm is based on the idea that materials and products should remain in the economy for as long as possible, minimising, for example, the generation of waste and disposal practices that induce significant environmental impacts. For achieving this purpose, actions including reuse and recycling should be carried out in order to turn the waste that is generated into a resource.

In European Union, Directive 2008/98/CE on waste sets that by 2035, 65% by weight of the municipal waste should be forwarded to valorisation actions including reuse and recycling. One of the procedures used for treating municipal solid waste is incineration, which causes the generation of a residue known as incinerator bottom ash (IBA) in considerable amounts. Over the past years, the academic community has been studying the possibility of using IBA in different situations, e.g., as cementitious material and as recycled aggregate in the manufacture of concrete and in road construction (Xuan et al. 2018). In geotechnical engineering, IBA may be in contact with geosynthetics (for example, IBA can be used as filling material in roadways or in railways infrastructures), being important to understand the degree of degradation that IBA may cause on these materials.

Geosynthetics are construction materials commonly used in engineering projects such as soil reinforcement, road construction or embankments. These materials can perform different functions, namely, reinforcement, separation, protection, fluid barrier, drainage and filtration, providing them the capability of being used in a wide range of applications.

Mechanical damage can be induced to geosynthetics during the installation on site. The damage that occurs during installation is mainly caused by the placement and compaction of aggregates and can induce, for example, cuts, abrasion and holes in the geosynthetics, which will affect their properties. The European Committee for Standardization has developed a method (EN ISO 10722 (2007)) for inducing mechanical damage under repeated loading on geosynthetics. When facing the possibility of using IBA as filling material in engineering projects, in which it can be in contact with geosynthetics, it is essential to understand if the effects caused by IBA on the mechanical behaviour of these construction materials are not greater than those imposed by natural aggregates. For instance, the nature, dimensions and shape of the constituent particles of the aggregates are features that may influence the damage induced by them to the geosynthetics.

The aim of this research was to evaluate the effects of the mechanical damage induced by IBA on the short-term tensile behaviour of geotextiles, in comparison with other aggregates. For achieving that goal, mechanical damage under repeated loading tests were carried out on three geotextiles with different structures (a woven and two nonwoven) and masses per unit area. Besides IBA, two natural aggregates (gravel 4/8 and *tout-venant*) and *corundum* (synthetic aggregate used in the method described in standard EN ISO 10722 (2007)) were used as granular materials in the mechanical damage under repeated loading tests.

2 MATERIALS AND METHODS

2.1 *Geotextiles*

Three geotextiles with different properties were used in this work: a woven geotextile manufactured with high density polyethylene (HDPE) filaments and two nonwoven geotextiles made from polypropylene (PP) fibres. The main characteristics of the geotextiles, which were supplied in the form of rolls, are exhibited in Table 1. The masses per unit area (μ_A) and thicknesses (t) of the geotextiles were determined according to standards EN ISO 9864 (2005) and EN ISO 9863-1 (2016), respectively.

The sampling and preparation of specimens for the characterisation and mechanical damage under repeated loading tests followed the instructions of standard EN ISO 9862 (2005). The specimens were collected from random positions evenly distributed over the width and length of the rolls and respecting a distance of, at least, 100 mm from the edges of the rolls.

2.2 *Aggregates*

The mechanical damage under repeated loading tests (hereinafter MD tests) were carried out with IBA (supplied by a Portuguese incineration plant (Lipor II – Maia)), and for comparison purposes, with gravel 4/8, *tout-venant* (well-graded untreated mixed aggregate) and *corundum* (synthetic aggregate from aluminium oxide used in standard EN ISO 10722 (2007)) (Fig. 1). The particle size distributions of the aggregates, which were determined by sieving according to EN 933-1 (2012), are presented in Figure 2.

Table 1. Main characteristics of the geotextiles.

Geotextile	Type	Polymer	μ_A (g.m^{-2})	t (mm)
W120	Woven	HDPE	126 (± 1)	0.54 (± 0.01)
NW150	Nonwoven	PP	164 (± 5)	1.50 (± 0.09)
NW280	Nonwoven	PP	287 (± 18)	3.08 (± 0.11)

(95% confidence intervals in brackets).

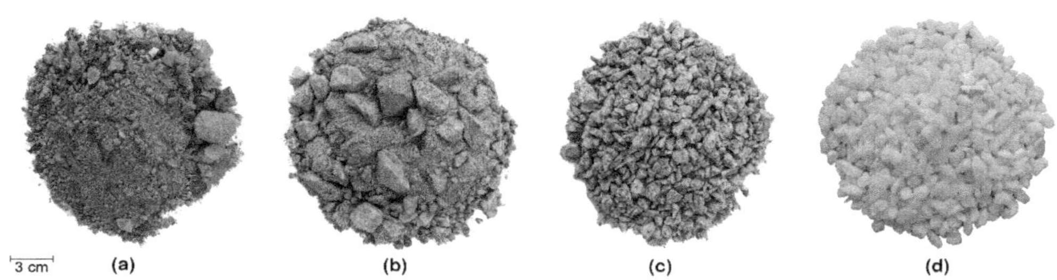

Figure 1. Aggregates used in the MD tests: (a) IBA; (b) *tout-venant*; (c) gravel 4/8; (d) *corundum*.

2.3 *Mechanical damage under repeated loading tests*

The MD tests were carried out in accordance with the procedures displayed in standard EN ISO 10722 (2007). Five specimens of each geotextile were tested with each aggregate.

The MD tests were performed on a laboratory prototype developed at the Faculty of Engineering of the University of Porto, constituted by a lower and an upper boxes (width, length and height of, respectively, 300, 300 and 75 mm), a loading plate and a compression machine (further information about the equipment can be consulted in Lopes & Lopes (2003)). Test procedures can be divided in four phases: (1) placement of a sublayer of aggregate by filling half of the lower box, followed by compaction; (2) placement of another sublayer of aggregate by filling the remaining half of the lower box, followed again by compaction; (3) placement of the specimen and upper box, followed by placement of a loose layer of aggregate (without compaction); and (4) application of a cyclic loading between (5.0 ± 0.5) kPa and (500 ± 10) kPa at a frequency of 1 Hz for 200 cycles. The compaction of the sublayers of the aggregates (placed in the lower box) was accomplished by applying a pressure of (200 ± 2) kPa for 60 s over the entire area of the box.

2.4 *Tensile tests*

The tensile tests were carried out on a *Lloyd Instruments* testing machine (model *LR10K Plus*) fitted with a load cell of 10 kN (also from *Lloyd Instruments*) at 20 mm.min^{-1}. From these tests, which were conducted according to standard EN ISO 10319 (2015), it was possible to assess the tensile strength (T, in kN.m^{-1}) and the elongation at maximum load (E_{ML}, in %) of the geotextiles. The results of the tensile properties of the geotextiles (arithmetic means of five specimens tested in the machine direction of production) are presented with 95% confidence intervals obtained by Montgomery & Runger (2010). The changes occurred in tensile strength are also expressed in terms of retained tensile strength (in %), obtained by the quotient between the tensile strengths of the damaged and undamaged samples.

3 RESULTS AND DISCUSSION

3.1 *Woven geotextile W120*

The results of the tensile tests conducted on geotextile W120 are presented in Table 2. The losses occurred in the tensile strength of geotextile W120 after the MD tests were more significant for

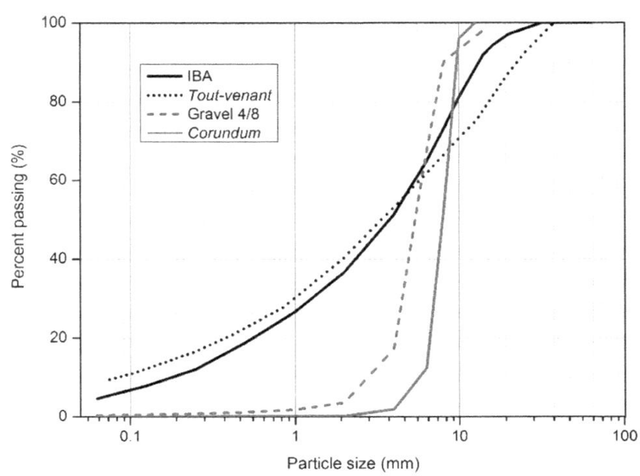

Figure 2. Particle size distributions of the aggregates.

Table 2. Tensile properties of geotextile W120 before and after the MD tests.

Mechanical damage test	T ($kN.m^{-1}$)	E_{ML} (%)
Undamaged	29.45 (± 1.59)	39.9 (± 2.4)
MD with IBA	21.58 (± 1.13)	25.1 (± 2.4)
MD with *tout-venant*	23.53 (± 2.66)	27.8 (± 2.7)
MD with gravel 4/8	15.14 (± 0.92)	19.5 (± 1.6)
MD with *corundum*	11.20 (± 1.51)	17.8 (± 1.5)

(95% confidence intervals in brackets)

Table 3. Tensile properties of geotextile NW150 before and after the MD tests.

Mechanical damage test	T ($kN.m^{-1}$)	E_{ML} (%)
Undamaged	13.94 (± 0.53)	93.7 (± 3.8)
MD with IBA	10.83 (± 1.14)	56.6 (± 6.3)
MD with *tout-venant*	9.18 (± 1.07)	51.1 (± 5.7)
MD with gravel 4/8	5.78 (± 1.32)	38.1 (± 7.0)
MD with *corundum*	5.11 (± 0.37)	33.2 (± 4.0)

(95% confidence intervals in brackets).

Table 4. Tensile properties of geotextile NW280 before and after the MD tests.

Mechanical damage test	T ($kN.m^{-1}$)	E_{ML} (%)
Undamaged	16.39 (± 1.34)	79.5 (± 2.8)
MD with IBA	15.01 (± 1.96)	66.5 (± 3.6)
MD with *tout-venant*	14.11 (± 1.84)	65.3 (± 5.5)
MD with gravel 4/8	12.75 (± 2.86)	53.8 (± 7.3)
MD with *corundum*	10.93 (± 1.51)	49.1 (± 3.0)

(95% confidence intervals in brackets)

the cases in which gravel 4/8 and *corundum* were used, being *corundum* the aggregate that led to the highest strength loss (losses of 49% and 62% after the MD tests with gravel 4/8 and *corundum*, respectively). In comparison with the aforementioned aggregates, the reductions caused by IBA and *tout-venant* were not so accented (27% and 20%, respectively). An identical trend was observed for the elongation at maximum load, as the losses were more pronounced when gravel 4/8 and *corundum* were the used aggregates, in comparison with IBA and *tout-venant*.

3.2 *Nonwoven geotextile NW150*

Table 3 shows the tensile properties obtained for geotextile NW150. For this geotextile, of a different structure but with an identical mass per unit area as geotextile W120, tensile strength losses were more evident after the MD tests in which gravel 4/8 and *corundum* were used (59% and 63%, respectively), just as observed for geotextile W120. However, and contrarily to what happened for geotextile W120, it was IBA that caused less damage to geotextile NW150 and not *tout-venant* (reductions in tensile strength of 22% and 34%, respectively). Regarding elongation at maximum load, results showed that it had the same trend observed for tensile strength: decreases were more significant after the MD tests with gravel 4/8 and *corundum*, compared to the other aggregates.

3.3 *Nonwoven geotextile NW280*

The tensile properties of geotextile NW280 can be found in Table 4. Similarly, to what was observed for geotextile NW150, IBA was the less damaging aggregate (lower deterioration of

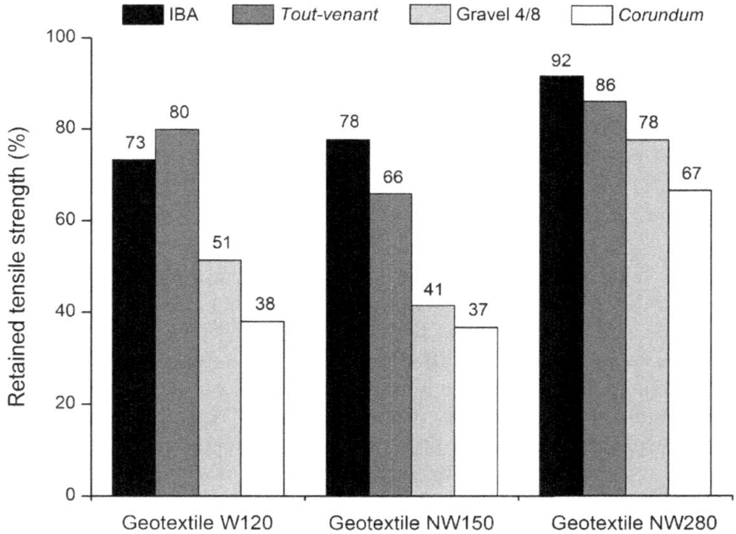

Figure 3. Comparison of the retained tensile strengths of the geotextiles after the mechanical damage tests.

tensile behaviour), while *corundum* was the most adverse one (losses on tensile strength of 8% and 33%, respectively). *Tout-venant* and gravel 4/8 caused reductions in tensile strength of, respectively, 14% and 22%. The tensile strength losses occurred in geotextile NW280 were considerably lower than those observed for geotextile NW150. The lower degree of degradation of geotextile NW280 can be explained by its higher mass per unit area. Once again, the changes in elongation at maximum load accompanied the trend observed for tensile strength (hierarchizing the aggregates in terms of induced damage).

3.4 *Comparison between the effects of the aggregates*

The comparison between the retained tensile strengths of the geotextiles is depicted in Figure 3. *Tout-venant* was one of the less damaging aggregates, not causing very considerable losses on the tensile strength of the geotextiles after the MD tests. *Tout-venant* was a well-graded aggregate with a relatively high percentage of fine particles (8.0% by weight of the particles were smaller than 0.063 mm), which allowed a relatively good spatial arrangement of individual particles and, consequently, a reduction in the volume of voids. After compaction, *tout-venant* had a flat and smooth surface (in which the geotextiles were placed), helping a better distribution of the applied loads during the MD tests. However, the particles of higher dimension and angular shape may have caused some damage in the geotextiles, which explains the losses on the tensile properties.

Gravel 4/8 was the natural aggregate that induced the most significant damage on the geotextiles during the MD tests. Gravel 4/8 was a poorly-graded aggregate whose particles, of relatively high dimension, had an angular shape. The degradation caused by this aggregate may be related to the capability of its particles to induce small cuts in the geotextiles, particularly in geotextiles W120 and NW150, which had lower masses per unit area.

The most damaging aggregate for the geotextiles was *corundum*. The rough angular particles of this poorly-graded aggregate induced cuts on the filaments or fibres of, respectively, the woven or nonwoven structures of the geotextiles. *Corundum* had also a high abrasive effect on the geotextiles.

Based on the results presented in Figure 3, it seems reasonable to state that IBA was, in general, the less damaging aggregate. Indeed, the retained tensile strengths of the geotextiles tended to be higher after the MD tests with IBA, compared to the other aggregates (exception when geotextile W120 was damaged with *tout-venant*). IBA had a relatively high amount of fine particles (4.7%

by weight of the particles were smaller than 0.063 mm) (Fig. 2), which helped fulfilling the voids and promoted the formation of a flat contact surface with the geotextiles after compaction, thereby allowing a good distribution of the applied loads (as previously described for *tout-venant*). Moreover, it is worthy to note that IBA and *tout-venant* had relatively similar particle size distributions (Fig. 2). This is an aspect in favour of IBA when considering its potential use in some applications in which *tout-venant* is commonly employed, opening good perspectives for its valorisation.

4 CONCLUSIONS

The achieved results in terms of the mechanical damage induced by IBA to geotextiles are promising to address the possibility of using this residue as filling material in engineering projects in which it may be in contact with geosynthetics. To fulfil this purpose, further research including, for example, the evaluation of the effects of IBA on the long-term behaviour of geosynthetics should be carried out. In addition, it may also be important to assess the physical, chemical and mechanical properties of IBA and compare them with those from natural aggregates, contributing to obtain additional data to decide about the suitability of using IBA in replacement of natural aggregates in the domain of civil and environmental engineering. If the employment of IBA as filling material becomes a viable option, steps are being taken in order to assign IBA a nobler role in the new economic paradigm.

ACKNOWLEDGEMENTS

This work was financially supported by: (1) project POCI-01-0145-FEDER-028862, funded by FEDER funds through COMPETE 2020 – "Programa Operacional Competitividade e Internacionalização" (POCI) and by national funds (PIDDAC) through FCT/MCTES; (2) UID/ECI/04708/2019 – CONSTRUCT – "Instituto de I&D em Estruturas e Construções" funded by national funds through FCT/MCTES (PIDDAC).

REFERENCES

Directive 2008/98/EC of the European Parliament and of the Council of 19 November 2008 on waste. *Official Journal of the European Union L312/3.*

EN 933-1. 2012. Tests for geometrical properties of aggregates – Part 1: Determination of particle size distribution – Sieving method. Brussels: CEN.

EN ISO 9862. 2005. Geosynthetics – Sampling and preparation of test specimens. Brussels: CEN.

EN ISO 9863-1. 2016. Geosynthetics – Determination of thickness at specified pressures – Part 1: Single layers. Brussels: CEN.

EN ISO 9864. 2005. Geosynthetics – Test method for the determination of mass per unit area of geotextiles and geotextile-related products. Brussels: CEN.

EN ISO 10319. 2015. Geosynthetics - Wide-width tensile test. Brussels: CEN.

EN ISO 10722. 2007. Geosynthetics - Index test procedure for the evaluation of mechanical damage under repeated loading – Damage caused by granular material. Brussels: CEN.

Lopes, M.P. & Lopes, M.L. 2003. Equipment to carry out laboratory damage during installation tests on geosynthetics. *Geotecnia (Journal of the Portuguese Geotechical Society)* 98: 7–24 (in Portuguese).

Montgomery, D.C. & Runger, G.C. 2010. *Applied Statistics and Probability for Engineers.* New York: John Wiley & Sons, Inc.

Xuan, D., Tang, P. & Poon, C.S. 2018. Limitations and quality upgrading techniques for utilization of MSW incineration bottom ash in engineering applications – A review. *Construction and Building Materials* 190: 1091–1102.

Wastes: Solutions, Treatments and Opportunities III – Vilarinho et al. (Eds)
 ISBN 978-0-367-25777-4

Impact of sewage sludge with eggshell on *Lepidium sativum* L. growth

L.A. Gomes
CIEPQPF – Center of Chemical Processes Engineering and Forest Products, Department of Chemical Engineering, University of Coimbra, Coimbra, Portugal
IFB – Federal Institute of Education, Science and Technology of Brasília – IFB, Campus Ceilândia Brasília – Federal District, Brazil

A.F. Santos & M.J. Quina
CIEPQPF – Center of Chemical Processes Engineering and Forest Products, Department of Chemical Engineering, University of Coimbra, Coimbra, Portugal

J.C. Góis
Association for the Development of Industrial Aerodynamics, Department of Mechanical Engineering, University of Coimbra, Coimbra, Portugal

ABSTRACT: The main objective of this work was to assess the possibility of recycling sewage sludge dried, using eggshell as adjuvant, for agronomic applications. The effects of these two wastes in growth parameters (e.g. root length) and soil properties (e.g. organic matter) were investigated. Germination studies indicated that sewage sludge mixed with eggshell reduce its phytotoxicity. The results from the pot experiments showed that the amendments tested improved the shoot length of the plants. Moreover, the application of sewage sludge and eggshell reduced the acidity of the soil and increased the organic matter. This study demonstrated the feasibility of recycling these two wastes in land applications with some important contributes to amend the soil characteristics.

1 INTRODUCTION

The management of sewage sludge (SS) is becoming a worldwide issue due to the growth of the population. Several SS disposal techniques are used in the European Union Countries (EU-28), such as agricultural fertilizer, incineration, composting or landfill. In Portugal, more than 80% of the total amount of SS produced is applied in agricultural land (directly or in compost form) (Alvarenga et al. 2016; Kelessidis & Stasinakis, 2012). Land application of SS has been encouraged due to its soil conditioning proprieties. The high quantities of organic matter and nutrients, especially N and P, make this residue a valuable soil amendment. However, SS may contain some potentially toxic metals (PTM), which can contaminate soil and crops as well. Consequently, the food chain could be compromised. Thus, legislation (such as Decree law n*º* 276/2009 – October 2) imposes limits to specific parameters, ensuring that the application of SS does not hamper the soil.

Nowadays, another waste that is produced in high quantities is eggshell (ES), about 700 kt were generated in 2017 by the EU-28 countries (European Comission, 2017). ES is an animal by-product that represents near to 11% of the total weight of the egg and contains about 94% of $CaCO_3$ (Baláž et al. 2016). Considering its properties, different options have been explored in the literature to add commercial value to ES (Quina et al. 2017), according to the circular economy model. Soil amendment is one of the main options to recycling ES. Since soil nutrient availability depends on its pH, when soil acidity is high (pH < 4.5) the availability of calcium is low, and the growth of the plants is hindered. Consequently, ES can be used to suppress the lack of calcium in soil because of its high concentration in $CaCO_3$. Moreover, its carbonate content promotes pH adjustment in acidic soils (raise of the pH) (Quina et al. 2017). In Portugal, the application of sewage sludge and ES in the soil can be of interest since the national territory is poor in organic matter and reveals acidic nature. Indeed, most of the Portuguese soil has less than 1% (w/w) of organic matter and more than 80% of the territory is characterized by a pH < 5.5 (Inácio et al. 2008).

The aim of this study was to evaluate the effect of SS and ES as a soil amendment on certain soil properties and plant growth. *Lepidium Sativum* L. was used as the test plant, involving both germination and growth tests.

2 MATERIALS AND METHODS

2.1 *Materials characterization*

A sandy soil from the central region of Portugal was collected from a depth of about 20 cm, dried at room temperature and then homogenized properly. Sewage sludge (SS) sample (about 5 kg) was collected after mechanical dewatering by centrifugation in a wastewater treatment plant (WWTP) from the central region of Portugal, which receives urban effluent and treats about 36,000 m^3/day. The moisture content of the collected SS was about 80%. Eggshell waste (ES) sample (about 1 kg) was collected from a grocery store. The sample was washed with distilled water to remove impurities, dried at room temperature, milled and stored until further utilization. All the materials were sieved through a 1 cm mesh size, and the oversized fraction was discarded.

The pH and electrical conductivity (EC) of the materials were measured in a 1:10 (solid:liquid) suspension (Alvarenga et al. 2016). Moisture and OM were based on EPA Method 1684. PTM and major-element oxides were determined by energy dispersive X-ray fluorescence (EDXRF), using the Nex CG Rigaku spectrometer. The nitrogen in SS was measured by the Kjeldahl method. The acid neutralization capacity (ANC) was determined by titration with HNO_3 0.1 M in a 1:10 (solid:liquid) ratio.

2.2 *Germination assays*

The phytotoxicity test was performed with *Lepidium sativum* (garden cress). For this purpose, liquid extracts were obtained for the different liquid to solid ratios (L/S), based on the standard EN 12457-2:2002, with some adaptations. The solid samples tested were SS, SS plus 15% (w/w) of ES (hereafter SS_ES), and ES. In each case, 5 mL of extract was added in a Petri dish (90 mm) along with ten seeds, which were placed over the *Whatman* filter paper. The samples stayed in an oven at 25 ± 0.1°C for 48 h, in dark conditions (Baderna et al. 2015). The pH and EC measured in the leachates are shown in Table 1. Germination Index (GI) was obtained by comparing the percentage of the number of germinated seeds and the percentage of the root lengths of the samples with respect to the control sample (Pinho et al. 2017).

2.3 *Pot experiments*

Plants (*Lepidium sativum)* in pots were grown in a chamber with light provided by a white LED light (~280 μmol/m^2.s, 400–700 nm, incidence angle of 240°) with a photoperiod of 12 h, temperature

Table 1. pH and EC measured in extracts tested in the phytotoxicity assays.

		L/S ratio (L/kg)						
Samples	Parameters	5	10	25	50	100	200	500
SS	pH	6.44	6.71	6.81	6.81	6.73	6.44	6.13
	EC (mS/cm)	4.48	1.72	1.00	0.59	0.40	0.40	0.19
SS_ES	pH	7.26	7.38	7.53	7.66	7.72	7.86	7.96
	EC (mS/cm)	3.90	2.61	1.29	0.77	0.47	0.27	0.17
ES	pH	8.86	9.10	9.10	9.50	9.41	9.33	9.01
	EC (mS/cm)	0.32	0.21	0.15	0.18	0.10	0.08	0.05

Note: The control sample (distilled water) is characterized by a pH = 6.88 and EC = 3.23 μS/cm.

control of 21 ± 0.1°C and relative humidity of 50% (Arriagada et al. 2014; Belhaj et al. 2016). Pots (90 × 75 mm) were filled uniformly to a total volume of 300 cm^3. The SS was applied to the soil at three different rates, corresponding to 6, 12 and 24 t/ha. A fourth treatment was performed with SS_ES (24 t/ha of SS). A control test using only soil was performed. At the bottom of each pot was added gravel to enable drainage of the water. About 17 seeds of garden cress were sown in each pot. Each treatment was conducted in duplicate. After 1 week, plants were harvested, remaining 10 plants in each pot (thinning phase). The experiment lasted for 4 weeks. Growth parameters were analyzed with respect to root and shoot lengths, and biomass of the plant. For determining the biomass, the plants were oven-dried at 60°C until constant weight (Mohamed et al. 2018). The OM, pH, and EC in the soil were measured at the end of the experimental period.

2.4 *Statistical analysis*

Differences between germination assays and soil treatments were performed with a Turkey HDS test ($p < 0.05$).

3 RESULTS AND DISCUSSION

3.1 *Physical and chemical characterization*

Table 2 summarizes different parameters determined in the materials under study, where data from the literature were also included. In general, the results agree with those reported in the literature. The soil used in pot tests has an acidic pH, a sandy texture and low contents in OM and P (essential nutrient). The ANC of the soil is very low, which indicates a weak buffering capacity against acidification. On the contrary, ES revealed a high ANC (19.7 meq/g). Thus, the application of SS and SS_ES will be favorable not only as pH correction agents but also as suppliers of OM and nutrients (high levels in the wastes analyzed). Additionally, the application of SS_ES can raise the buffer capacity of the soil.

All PTM concentrations measured in the soil, SS and ES are below the limit imposed by the Portuguese legislation (Decree-law n° 276/2009 – October 2). Thus, this soil can be amended not only with SS but also with the mixture of SS and ES for agronomic applications.

3.2 *Germination studies*

Germination tests with garden cress seeds were performed to evaluate the phytotoxic activity of aqueous extracts of SS, ES, and SS_ES as shown in Figure 1. Additional tests were performed with

Table 2. Characterization of the materials used.

	This work				SS – Literature	ES – Literature
Parameters	Soil	SS	ES	SS_ES	(Alvarenga et al., 2007)	(Quina et al., 2017)
Moisture (%)	na	77.9	0.84	67.8	60	1.1
pH	6.0	6.7	9.1	7.4	7.5	8.7
EC (mS/cm)	0.16	1.7	0.21	2.6	0.93	0.45
TS (%)	na	22.2	99.2	32.2	40	98.9
OM (%TS)	1.8	63.7	4.2	55.9	57	6.3
ANC_{pH4} (meq H^+/g)	0.001	0.06	19.7	0.14	na	na
$N_{kjedahl}$(%wt)	na	3.9	na	na	2.8	nd
P (%wt)	0.01	4.3	0.30	0.45	1.4	nd
K (%wt)	na	1.10	0.10	0.97	0.001	nd
Mg (%wt)	na	0.64	0.75	0.65	0.34	nd
Ca (%wt)	na	10.4	36.2	20.9	2.2	35.2

TS – total solids; na – not analyzed; nd – not detected.

Table 3. Potential toxic metals concentration (mg/kg) in materials.

					DL n° 276/2009	
PTM (mg/kg)	Soil	SS	ES	SS_ES	Soil	SS
Pb	<2	26.6	n.d	26.6	300	750
Cd	< 0.2	< QL	< QL	< QL	3	20
Cr	< 1	161.9	n.d	161.9	200	1000
Cu	7.7	251.3	17.5	229.6	100	1000
Ni	< 1	53.1	n.d	53.1	75	300
Zn	6.4	627.9	10.1	570.7	300	2500

nd – not detected; QL – quantification limit equal to 0.20 mg kg^{-1} for Cd.

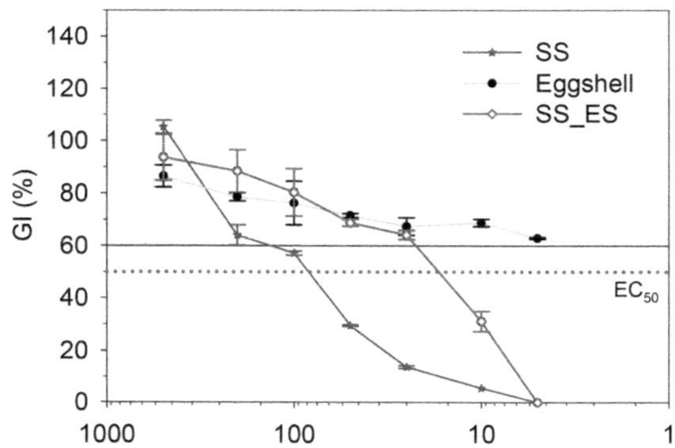

Figure 1. Germination index of SS, ES and SS_ES after 48 h of incubation.

SS plus 5% and 10% (w/w) of ES (results not shown), but phytotoxic behavior was similar when compared to SS_ES (15% of ES).

Results in Figure 1 revealed that the GI of ES is always superior to 60%, which indicates that does not exist inhibition of seed germination and growth in the different L/S ratios tested. Contrarily, SS demonstrated high inhibitory potential, except for 200 and 500 L/S ratios. The EC_{50} obtained was 105.4 L/kg, which corresponds to the concentration that induces 50% of inhibition. The SS_ES eluate only shows inhibitory effects for 5 and 10 L/S ratios. Moreover, the EC_{50} for the SS_ES was reduced to 19.2 L/kg, which represents a decrease of approximately 80% compared to the value obtained for the SS eluate. Thus, it is possible to conclude that the blend between SS and ES decreases the phytotoxic activity.

3.3 *Pot experiments*

The pot experiments were realized with a control sample (C0) and four different treatments were implemented in the soil: application of SS at 6, 12 and 24 t/ha (SS6, SS12, and SS24, respectively) and SS with 15% (w/w) of ES (SS_ES). Figure 2 presents the different treatments growths for 4 weeks of the experiment.

Figure 3(a)–(d) shows the evolution of growth parameters in terms of root length, shoot length and biomass, and organic matter in the soil, respectively. Results marked with the different letters are significantly different according to Turkey's test ($p < 0.05$).

The growth analysis revealed that the root length of the garden cress was not influenced by the treatments when compared to the control ($p < 0.05$). Contrarily, the shoot length was affected when the soil was amended with the wastes. The results show that all the treatments, except SS6, are

Figure 2. Illustration of treatments growth for 4 weeks.

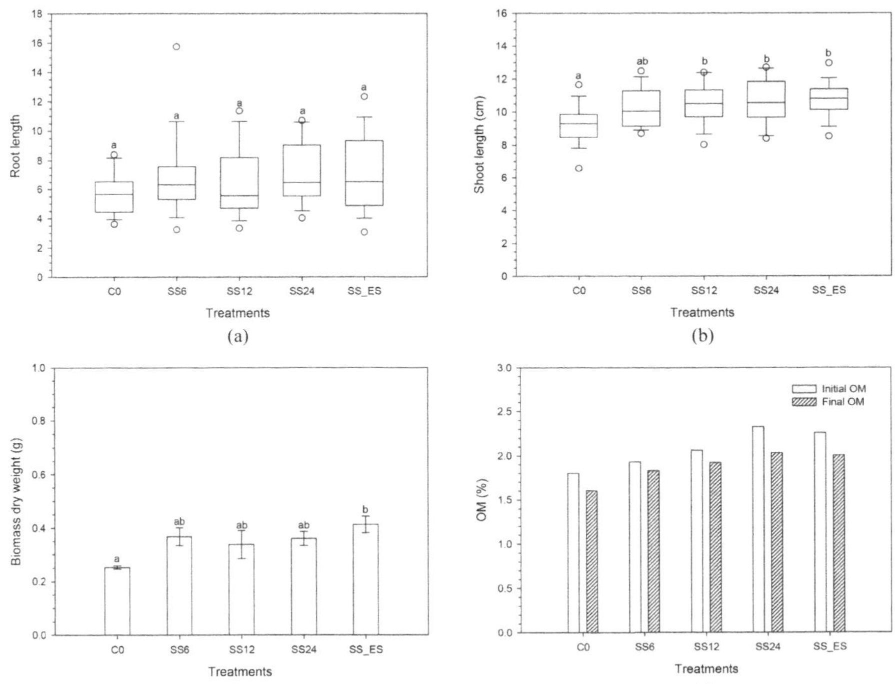

Figure 3. Effect of amendments: (a) root length; (b) shoot length; (c) biomass dry weight and (b) in soil organic matter.

statistically different in terms of shoot length ($p < 0.05$). In fact, the plants were grown in soil amended with SS12, SS24 and SS_ES treatments exhibited an increase of 14.4, 15.5 and 16.1%, respectively, in shoot length as compared with control. From Figure 3 (c), it is possible to conclude that SS_ES promotes a significant increase in biomass dry weight ($p < 0.05$). Indeed, SS_ES presents a higher biomass value (0.413 g), which represents an improvement of 63% compared to the control sample. The initial OM of the soil in all the treatments increased after amendment with SS or SS_ES. The final OM of the soil decrease in all the treatments due to the growth of the plants. Moreover, after 4 weeks of the experiment, the pH of the soil raised for values higher than 7 for the treatments SS6, SS12, SS24, and SS_ES, with a maximum of 7.80 for SS_ES. Thus, these results revealed the capability of these two wastes to act as pH correction agents.

4 CONCLUSIONS

This study aimed to assess the effect of SS and ES as soil amendments, namely for improving OM and pH conditions. The results showed that the phytotoxicity of SS tested in seeds of *Lepidium sativum* may be reduced with the addition of a small amount of ES (for example 15% w/w). Moreover, SS12, SS24, and SS_ES amendments empower the growth of plants (e.g. shoot length). Besides, SS_ES led to a higher biomass value and corrected the soil pH to 7.80. All the amendments caused an increase in soil organic matter and a better condition with respect to pH. It is important to notice that this study requires further analysis to evaluate other risks (e.g. associated with pathogens) of applying sewage sludge in the agriculture soil.

ACKNOWLEDGEMENTS

The authors gratefully acknowledge Federal Institute of Education, Science, and Technology of Brasília – IFB, Campus Ceilândia, for authorizing the Ph.D. studies of Luciano A. Gomes. This work was developed under the project "Dry2Value – Estudo e desenvolvimento de um sistema de secagem para valorização de lamas". Project consortium with HRV e LenaAmbiente. POCI-01-0247-FEDER-033662. Funded by Fundo Europeu de Desenvolvimento Regional (FEDER) – Programa Operacional Competitividade e Internacionalização.

REFERENCES

Alvarenga, P., Farto, M., Mourinha, C., Palma, P. 2016. Beneficial use of dewatered and composted sewage sludge as soil amendments: behaviour of metals in soils and their uptake by plants. *Waste and Biomass Valorization* 7: 1189–1201.

Alvarenga, P., Palma, P., Gonçalves, A.P., Fernandes, R.M., Cunha-Queda, A.C., Duarte, E., Vallini, G. 2007. Evaluation of chemical and ecotoxicological characteristics of biodegradable organic residues for application to agricultural land. *Environ. Int.* 33: 505–513.

Arriagada, C., Almonacid, L., Cornejo, P., Garcia-Romera, I., Ocampo, J. 2014. Influence of an organic amendment comprising saprophytic and mycorrhizal fungi on soil quality and growth of Eucalyptus globulus in the presence of sewage sludge contaminated with aluminium. *Arch. Agron. Soil Sci.* 60: 1229–1248.

Baderna, D., Lomazzi, E., Pogliaghi, A., Ciaccia, G., Lodi, M., Benfenati, E. 2015. Acute phytotoxicity of seven metals alone and in mixture: Are Italian soil threshold concentrations suitable for plant protection? *Environ. Res.* 140: 102–111.

Baláž, M., Ficeriová, J., Briančin, J. 2016. Influence of milling on the adsorption ability of eggshell waste. Chemosphere.

Belhaj, D., Elloumi, N., Jerbi, B., Zouari, M., Abdallah, F. Ben, Ayadi, H., Kallel, M. 2016. Effects of sewage sludge fertilizer on heavy metal accumulation and consequent responses of sunflower (Helianthus annuus). *Environ. Sci. Pollut. Res.* 23: 20168–20177.

European Comission 2017. Agriculture and Rural Development [WWW Document]. Agric. Stat. URL https://ec.europa.eu/agriculture/eggs/presentations_en (accessed 9.22.18).

Inácio, M., Pereira, V., Pinto, M. 2008. The Soil Geochemical atlas of Portugal: Overview and applications. *J. Geochemical Explor.* 98: 22–33.

Kelessidis, A., Stasinakis, A.S. 2012. Comparative study of the methods used for treatment and final disposal of sewage sludge in European countries. *Waste Manag.* 32: 1186–1195.

Mohamed, B., Mounia, K., Aziz, A., Ahmed, H., Rachid, B., Lotfi, A. 2018. Sewage sludge used as organic manure in Moroccan sunflower culture: Effects on certain soil properties, growth and yield components. *Sci. Total Environ.* 627: 681–688.

Pinho, I.A., Lopes, D. V., Martins, R.C., Quina, M.J. 2017. Phytotoxicity assessment of olive mill solid wastes and the influence of phenolic compounds. *Chemosphere* 185: 258–267.

Quina, M.J., Soares, M.A.R., Quinta-Ferreira, R. 2017. Applications of industrial eggshell as a valuable anthropogenic resource. *Resour. Conserv. Recycl.* 123: 176–186.

Wastes: Solutions, Treatments and Opportunities III – Vilarinho et al. (Eds)
© 2020 Taylor & Francis Group, London, ISBN 978-0-367-25777-4

Motivations and barriers to industrial symbiosis – the fluidized bed sands case study

I. Ferreira & H. Carvalho
UNIDEMI, Department of Mechanical and Industrial Engineering, Faculty of Science and Technology, Universidade NOVA de Lisboa, Lisboa, Portugal

M. Barreiros
CELPA – Associação da Indústria Papeleira, Lisboa, Portugal

G.M. Silva
ADVANCE/CSG, ISEG-Lisbon School of Economics and Management, University of Lisbon, Lisbon, Portugal

ABSTRACT: With an increasing population growth there is a greater pressure on natural resources. This makes imperative the transition from a traditional linear model to a circular model that contributes to a more effective and efficient waste management. This research intends to study the application of the industrial symbiosis strategy in the pulp, paper and cardboard industry. It focuses on one by-product – the fluidized bed sands, and the respective industrial symbiosis value network. A questionnaire was carried out to eight experts with the objective to find out what are the main barriers and motivations that inhibit or encourage their engagement in the industrial symbiosis value network. Their main motivations are the reduction of the waste rate, energy consumption and exploitation of natural resources. The main barriers are the logistical costs, uncertainty by the final consumer regarding the use of recyclable products and the existence of more economical "virgin" raw materials.

1 INTRODUCTION

It is increasingly evident that sustainability in the economy requires a focus on specific resources that are limited (Watkins et al., 2013). Europe loses, approximately, 600 million tons of materials that are wasteful by industries but may have the potential to be recycled or reused (Álvarez & Ruiz-Puente, 2017). Therefore, efforts in Europe and worldwide have been made for efficient and effective waste management, as well as to promote the valorization of waste. In this perspective, the concepts of circular economy and industrial symbiosis were born. To put into practice the circular economy – a model based on the recovery or reuse of waste that allows a reintroduction of materials in the supply chain – it is necessary to adopt strategies that contribute to its development. This research focuses on one of these strategies: the industrial symbiosis. The term "symbiosis" refers to the share of resources, in a mutually beneficial way, between two species. In this way, the concept of industrial symbiosis integrates the share of needs between industries or sectors, with the objective of increasing competitive advantages in a more sustainable environmental way (Fraga, 2017).

According to Chertow (2008), three types of actions can occur in an industrial symbiosis relationship: i) sharing of infrastructures: shared use of common resources such as water, energy, and wastewater; ii) sharing of common service needs: meeting common industry needs for ancillary activities, such as transportation and products supply; iii) exchange and market for by-products: exchange of materials from a specific industry to others that can use by-products as substitutes or raw materials. Whenever there is an industrial symbiosis relationship, a value network is created, comprised of a set of participants (designated by stakeholders). These participants interact with each other, exchanging value directly or indirectly, through the share of resources or exchange of information and knowledge. This research has as main objective to demonstrate using a case study, the main motivations and barriers of the different stakeholders, who constitute a specific value network, that can inhibit or promote the industrial symbiosis practice.

2 STATE OF ART

The growing popularity of the circular economy concept encourages the adoption of new strategies that promote it. Govindan & Hasanagic (2018), identify the main motivations for the circular economy implementation: policies and economics (e.g. product legislation); health (e.g. to public health); environmental protection (e.g. protection of renewable resources); social (e.g. potential job creation) and product development (e.g. increased value of products). However, there are some barriers that inhibit the adoption of a circular economic model. Kirchherra et al. (2018) identify the following ones: cultural barriers; legislative barriers; market barriers and technological barriers. Govindan & Hasanagic (2018) also point out others such as: lack of standard systems for performance evaluation; recycling policies that are ineffective; financial and economic barriers; lack of reliable information; lack of public awareness; lack of communication between entities; consumer perception of recyclable products; existence of other solutions that may be more favorable.

Among them is the industrial symbiosis strategy, which is the focus of this research. According to Álvarez & Ruiz-Puente (2017), industrial symbiosis is seen as a model of cooperation aimed at optimizing resource flows, from which collective industrial gains can be obtained greater than individual gains. There are many motivations to develop and implement industrial symbiosis relationships, according to Chertow (2008), whether they are the direct or indirect result of industries trying to achieve other goals. The environmental benefits are among the most recognized benefits in implementing the industrial symbiosis strategy. However, Chertow (2008) point out the following motivations that encourage this strategy implementation: i) traditional business reasons (as business becomes more profitable and competitive); ii) resource's sharing can decrease costs and/or increase profits; iii) industrial symbiosis can increase long-term resource security by increasing the availability of key resources. Chertow (2008) also argues that industrial symbiosis encourages a more sustainable development, therefore social, environmental and legislative motivations also exist. Among the several factors that can be considered for a symbiotic relationship, Ehrenfeld & Gertler (1997) emphasize the geographical proximity.

Chertow (2008) also identified the following barriers to the development of industrial symbiosis networks: i) technical barriers, which include the possibility that local industries do not have the potential to "unite"; ii) informational barriers, which may hinder the use of residual products without given information about their supply and potential market; iii) motivational barriers in which companies, government agencies, and other relevant local stakeholders must be willing to cooperate and commit to the process. To apply any of the strategies related to the circular economy and industrial symbiosis, communication is the key to success. Winans et al. (2017) emphasize that all communication resources, from social communication (for general society) to communication between partners, allow sharing of information between partners, community involvement, greater public awareness and accountability, and greater coverage to divulge the subject. The development of adequate information systems is indicated as a possible way to remove barriers to the implementation of industrial symbiosis (Grant et al., 2010).

3 MATERIALS AND METHODS

3.1 *Fluidized bed sands case study*

To find out about the motivations and barriers of companies to engage in an industrial symbiosis network, this research uses a case study related to the Portuguese pulp, paper, and cardboard industry, with focus on the by-product "fluidized bed sands" (FBS). FBS are formed in the biomass boilers resulting from burning the bark of the wood. In October 2017, FBS were considered and

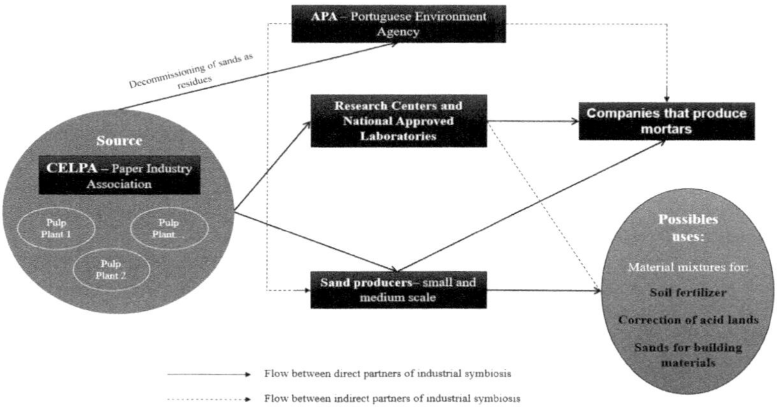

Figure 1. Value network correspondent to the industrial symbiosis case study of FBS.

classified by the Portuguese Environmental Agency (APA) as a by-product. Therefore, currently, they can be marketed to other industries and sectors. The entity leading this process is CELPA – Associação da Indústria Papeleira, which represents the pulp, paper and cardboard industry in Portugal and all its activities. Experimental tests were carried out, through partnerships between CELPA and research centers and national laboratories, to study potential FBS' applications in other industries. It was found that FBS did not have any adverse impact from an environmental and public health view. In addition to being used in the production of mortars, these sands have the potential for other uses such as: i) correction of lands for gardens/football fields/ etc.; ii) material mixtures for sands used in the construction sector; or iii) material mixtures to produce concrete or mortars.

Inherent to the industrial symbiosis process that corresponds to the FBS' commercialization, a value network is created by several stakeholders that exchange resources among them. For this specific FBS symbiosis value network, stakeholders were identified through unstructured interviews. The value network corresponding to this process is represented in Figure 1, and the description of the entities that composed it is presented in Table 1.

A questionnaire was elaborated to evaluate the stakeholders' perceptions about their main motivations and barriers in joining the FBS' value network. The questionnaire results not only from the literature review but also from unstructured interviews that were carried out to the stakeholders. Table 2 and Table 3 contain the motivations and barriers considered in the questionnaire. It was used a scale of 1 to 5, where 1 represents "nothing important" and 5 represents "extremely important". Eight participants representing the different stakeholders of the FBS' value network were inquired: two are sand producers (experts 1 and 2), a representative of the Portuguese Association of Mortars and ETICS (that represent companies that produce mortars) (expert 3), a representative of a civil construction company (expert 4), a member of CELPA (expert 5), and three representatives of CELPA associates (expert 6, 7 and 8).

3.2 *Main motivations and barriers to an industrial symbiosis network*

Table 2 presents the score attributed by the experts for the motivations to engage in an industrial symbiosis network. It is noteworthy that there are two motivations without punctuation attributed by "Expert 3" because it does not apply to the type of business developed by them.

These results allow concluding that the main motivations which participants agreed to be important are: "the contribution to sustainable development", "reduction of waste rate sent for landfill", "reduce of energy, waste and resources consumption", contribution to the company's image" and "economic growth". Motivations such as "the existence of sand scarcity" and "the possession of own fleet" are only considered important for the stakeholders who are responsible for the FBS' transportation, this is the sand producers.

Table 1. Description of the entities that composed the FBS' industrial symbiosis value network.

Description
CELPA – It represents the pulp and paper plants that produce FBS. The destination of these sands, if they are not used, recycled or reintegrated in another process is the landfill which implies that the producer's companies must pay a waste management fee.
APA – It is the Portuguese governmental entity responsible for the by-product's classification. CELPA requested to APA this classification for the FBS. Thus, after meeting the necessary conditions for the by-products' classification, later in October 2017, FBS were classified as such by APA.
Research centers and laboratories – They do the research of FBS potential use. It has been concluded that FBS could be a raw material compatible with the use of conventional sands, which can be used to produce mortars, without any additional impact on the environment.
Companies that produce mortars – Companies that produce mortars could benefit from using FBS to substitute part of the conventional sands, since they would be using a by-product, instead of increasing natural resources consumption. Due to the need of transport FBS sands, and washing and sieving before using them, the most viable alternative for the companies that produce mortars would be to buy from sand producers, mixtures of materials that integrate FBS in their composition.
Sand producers – They are responsible for the collecting, washing, sieving, transport, and bagging (if necessary) of FBS. Their main objectives are waste valorization and the reduction in the use of natural resources. The solution for the FBS commercialization is to reduce the transport logistics costs using the freight of local sand producers. After the washing and sieving treatments (using existing equipment in sand producers facilities), the FBS can be incorporated in material mixtures for diverse uses such as land correction, in the construction sector and to produce concrete or mortars.

Table 2. Motivations to the industrial symbiosis process.

	Experts score								
Description	E1	E2	E3	E4	E5	E6	E7	E8	Average
M1 – Contribution to sustainable development by using FBS, which allows the reduction in the exploitation of natural resources (sands)	5	5	5	5	5	5	2	5	4.6
M2 – Contribution to an improvement in the company's image, through ecological "flags"	5	5	5	5	5	3	3	4	4.4
M3 – Reduction of waste rate which is sent for landfill	5	5	4	5	5	4	3	5	4.5
M4 – Reduction of energy consumption, wastes, and resources consumption by increasing material efficiency and usage of renewable energy	5	5	–	4	5	4	2	5	4.3
M5 – Existence of sand scarcity in some locations that can be met by replacing by FBS	3	4	–	3	5	2	2	3	3.1
M6 – Alignment with laws and policies related to waste management	3	4	5	3	5	4	4	4	4
M7 – Promotion to networking through cooperation between companies to establish eco-industrial chains and achieve environmental improvements	3	4	5	4	5	4	3	5	4.1
M8 – Possession of own fleet to transport FBS	5	3	2	5	4	1	3	1	3
M9 – Selection of suppliers using environmental criteria	5	5	5	5	5	2	3	3	4.1
M10 – Economic growth through the implementation of the circular economy concept	4	5	5	5	5	4	4	4	4.5
M11 – Consumers pressure due to environmental awareness in the development of the circular economy concept	4	4	5	5	4	2	4	4	4
M12 – Inclusion of environmental factors in the company's internal performance	4	4	5	4	5	3	3	4	4

The score attributed by each stakeholder to the barriers is presented in Table 3. From these results, it is possible to conclude that barriers related to "lack of communication by public administration and inspection entities", "FBS' transportation", "geographical location of the industries", and "the existence of cheaper "virgin" raw materials" are the barriers that stakeholders considered

Table 3. Barriers to the industrial symbiosis process.

Description	Experts score								
	E1	E2	E3	E4	E5	E6	E7	E8	Average
B1- Inhibition of the industrial symbiosis process, due to legislation and renouncement of administrative entities	5	4	3	3	5	5	5	5	4.4
B2 – Lack of communication/information exchange by public administration and inspection entities	5	5	3	3	5	5	5	5	4.5
B3 – FBS' transportation from the origin industry to the consumer industry, resulting in high logistics costs	5	4	5	5	5	3	5	5	4.6
B4 – FBS recently introduced in the market and still little explored and known	3	4	5	5	5	3	3	3	3.9
B5 – Lack of market maturity to support the industrial symbiosis and use of FBS	3	4	4	5	5	4	3	4	4
B6 – Geographical location of the industries that produce the FBS, the industries that transport, treat and transform it, and the final consumer installations	3	5	5	5	5	4	4	5	4.5
B7 – Lack of information about FBS' use and behavior	2	4	5	5	3	2	3	3	3.4
B8 – Lack of communication between industries and top management support	3	4	4	5	4	2	4	3	3.6
B9 – Lack of standards, quantitative measures, and objectives to evaluate the company's performance	2	4	3	3	3	2	2	2	2.6
B10 – FBS' use may not be adapted to the existent production process	5	4	5	5	3	3	3	3	3.9
B11 – Existence of other solutions more efficient and profitable than the industrial symbiosis process	5	3	4	5	3	3	4	4	3.9
B12 – End-user perception of recyclable products	5	5	4	5	5	3	4	4	4.4
B13 – Existence of cheaper "virgin" raw materials	5	5	5	5	4	5	5	3	4.6
B14 – Existence of limited funds for application of the circular economy and industrial symbiosis concepts and other sustainable business models	3	4	5	5	5	4	5	3	4.3

most important. Barriers such as "FBS are still little explored and known", "lack of information about FBS' use and behavior", "lack of communication between industries and top management support", and "FBS' use may not be adapted to the existent production process" are barriers that are considered to be more important for FBS' final consumers than for industries that produce them. Barriers as "lack of standards to evaluate the company's performance" and "the existence of other solutions more efficient and profitable" are considered among the stakeholders as barriers with less importance.

4 CONCLUSIONS

Nowadays, although companies are more aware of the circular economy, they need to develop strategies and practices to promote this economic model. Thus, this research was based on a specific strategy – the industrial symbiosis and had as objective to understand the different perceptions of the stakeholders to the main motivations and barriers to participate in an industrial symbiosis network. To this end, a case study related to one of the production residues generated by the pulp and paper industry – the FBS, was carried out. FBS are currently considered as a by-product that can be commercialized and used as a secondary raw material by another industry. The study involved six stakeholders who composed the value network of this industrial symbiosis process, and a questionnaire was carried out to eight experts with the main barriers and motivations that inhibited or encouraged this strategy.

With the analysis of the stakeholders' perceptions it was possible to conclude that the stakeholders involved in this specific industrial symbiosis network participated in it with the objective of reducing the waste rate, energy consumption and exploitation of natural resources, as emphasized in the study led by Govindan & Hasanagic (2018), contributing to a sustainable development, which corroborates the study of Chertow (2008) about the motivations that lead to the industrial symbiosis practice. It was also concluded that the promotion of networking leads to this practice. Networks must be established among the various stakeholders and allow the possibility of increasing the sharing of renewable and recyclable resources. It can also be highlighted that motivations such as improving the company's image, economic growth and selection of suppliers using environmental criteria are also important, according to the stakeholders.

However, since the by-product's transportation and the geographic location between all industries demand very high logistical costs, these are very important barriers to the practice of industrial symbiosis, as emphasized by Ehrenfeld & Gertler (1997). It should also be noted that there is currently some uncertainty by the final consumer regarding the use of recyclable products and the existence of "virgin" raw materials that are more economical, as had already been highlighted in the studies by Chertow (2008) and Govindan & Hasanagic (2018). Barriers such as the existence of limited funds to support circular economy practices, inhibition of the industrial symbiosis process due to legislation, lack of communication and exchange of information were also emphasized, confirming the studies carried out by Kirchherra et al. (2018) and Govindan & Hasanagic (2018).

It should be noted that this research was limited by some factors, such as in the case study the analysis of the main barriers and motivations had been carried out by calculating an average value, inquiring eight participants. This may, therefore, causing twisted values compared to the perceptions of the different stakeholders to the practice of industrial symbiosis. Also, there were limitations regarding data availability and the availability of the participants, which limited the sample size and made not possible to extend the sample and encompass all possible FBS's final users.

ACKNOWLEDGMENTS

This work was supported by FCT (Fundação para a Ciência e a Tecnologia, Portugal) under the projects UID/EMS/00667/2019 and UID/SOC/04521/2013.

REFERENCES

Álvarez, R., & Ruiz-Puente, C. 2017. Development of the Tool SymbioSyS to Support the Transition Towards a Circular Economy Based on Industrial Symbiosis Strategies. *Waste and Biomass Valorization* 8(5): 1521–1530. https://doi.org/10.1007/s12649-016-9748-1.

Chertow, M. R. 2008. "Uncovering" Industrial Symbiosis. *Journal of Industrial Ecology* 11(1): 11–30. doi: 10.1162/jiec.2007.1110.

Ehrenfeld, J., & Gertler, N. 1997. Industrial Ecology in Practice: The Evolution of Interdependence at Kalundborg. *Journal of Industrial Ecology* 1(1): 67–79. doi: 10.1162/jiec.1997.1.1.67.

Fraga, M. 2017. *A economia circular na indústria portuguesa de pasta , papel e cartão* (Tese de Mestrado, Faculdade de Ciências e Tecnologia – Universidade Nova de Lisboa).

Govindan, K., & Hasanagic, M. 2018. A systematic review on drivers, barriers, and practices towards circular economy: a supply chain perspective. *International Journal of Production Research* 56: (1–2), 278–311. doi: 10.1080/00207543.2017.1402141.

Grant, G. B., Seager, T. P., Massard, G., & Nies, L. 2010. Information and communication technology for industrial symbiosis. *Journal of Industrial Ecology* 14(5): 740–753. doi:10.1111/j.1530-9290.2010.00273.x.

Kirchherr, J., Piscicelli, L., Bour, R., Kostense-Smit, E., Muller, J., Huibrechtse-Truijens, A., & Hekkert, M. 2018. Barriers to the Circular Economy: Evidence From the European Union (EU). *Ecological Economics* 150: 264–272. doi: 10.1016/j.ecolecon.2018.04.028.

Watkins, G., Husgafvel, R., Pajunen, N., Dahl, O., & Heiskanen, K. 2013. Overcoming institutional barriers in the development of novel process industry residue based symbiosis products – Case study at the EU level. *Minerals Engineering*, 41, 31–40. doi: 10.1016/j.mineng.2012.10.003.

Winans, K., Kendall, A., & Deng, H. 2017. The history and current applications of the circular economy concept. *Renewable and Sustainable Energy Reviews* 68: 825–833. doi: 10.1016/j.rser.2016.09.123.

Wastes: Solutions, Treatments and Opportunities III – Vilarinho et al. (Eds)
© 2020 Taylor & Francis Group, London, ISBN 978-0-367-25777-4

Magnetically responsive algae and seagrass derivatives for pollutant removal

I. Safarik
Department of Nanobiotechnology, Biology Centre, Ceske Budejovice, Czech Republic
Regional Centre of Advanced Technologies and Materials, Palacky University, Olomouc, Czech Republic

J. Prochazkova & E. Baldikova
Department of Nanobiotechnology, Biology Centre, Ceske Budejovice, Czech Republic

K. Pospiskova
Regional Centre of Advanced Technologies and Materials, Palacky University, Olomouc, Czech Republic

ABSTRACT: Microalgae, macroalgae and seagrass biomass can be considered as functional materials enabling various biotechnology and environmental technology applications. Using an appropriate magnetic modification, smart materials exhibiting a rapid response to an external magnetic field can be obtained. Such composite materials can be easily and selectively separated from desired environments. Magnetically responsive derivatives of algal and seagrass biomass have already been prepared and used as efficient biosorbents for the removal of organic and inorganic pollutants. Algal and seagrass biomass low cost and availability of waste biomass in larger amounts predetermine their utilization in large-scale processes. This chapter summarizes studies dealing with magnetic derivatives of unicellular algae, marine algae, seagrass and algae derived biochar/activated carbon, describes possibilities of their preparation and testing as efficient biosorbents of environmental contaminants, and inspires further exploration in this field of waste materials utilization.

1 INTRODUCTION

Huge amounts of waste biomaterials originating from the food, agricultural, forest, fishing and similar industries are produced all over the world. Many procedures have been developed for their valorization. In many cases biological materials can be considered as functional materials with interesting properties and broad application potential, e.g., such as biosorbents for the removal of organic and inorganic pollutants and radionuclides, or carriers enabling immobilization of various molecules and particles. Morphological and chemical properties of biomaterials can be upgraded by a broad range of modification procedures. Using an appropriate magnetic modification, smart biomaterials exhibiting a rapid response to an external magnetic field can be obtained. Such materials can be easily and selectively separated from desired environments using permanent magnets, electromagnets or appropriate magnetic separators (Safarik et al., 2018, Kanjilal and Bhattacharjee, 2018).

Biomass from unicellular algae, marine macroalgae and seagrass represents a typical example of low-cost, renewable natural material which can be obtained in large quantities. In many cases huge amounts of marine algae and seagrass can be found on beaches, thus causing problems to the tourist industry; the obtained biomass can be efficiently used as adsorbents for the removal of specific pollutants or for the production of biochar. In addition to native algae biomass also waste biomass obtained after selected industrial processes (e.g., solvent extraction of oil or colorants) can also be used for the preparation of efficient adsorbents (Delrue et al., 2016, Santos et al., 2018,

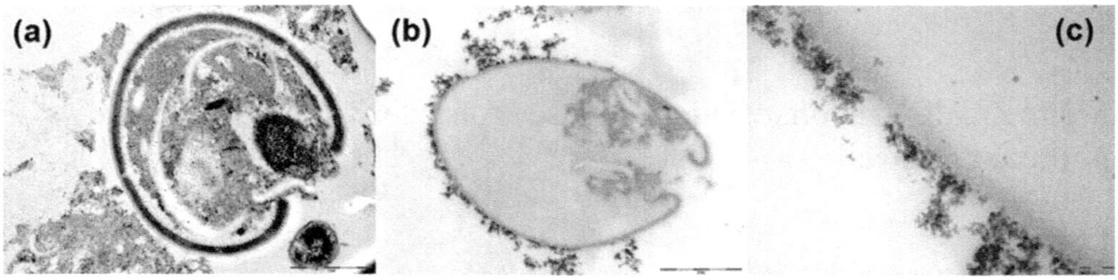

Figure 1. Transmission electron microscopy of dried *Chlorella vulgaris* cell ((a), bar line is 1 μm) and magnetic fluid modified *Chlorella* cell ((b, c), bar lines are 2 μm and 200 nm, resp.). Reproduced, with permission, from Safarikova et al., 2016c.

Safarik et al., 2018). Their low cost and availability in larger amounts predetermine their utilization in potential large-scale processes.

This short review summarizes available information about the preparation of magnetically modified algae and seagrass biomass and its subsequent application as adsorbents for the removal of both organic and inorganic pollutants.

2 MAGNETICALLY MODIFIED UNICELLULAR ALGAE

Currently, only few examples of magnetically modified unicellular algae biomass have been found in the literature. Dried *Chlorella vulgaris* cells were magnetically modified by contact with water-based magnetic fluid stabilized with perchloric acid; both isolated magnetic nanoparticles and aggregates of particles were present on the cell surface (see Fig. 1). The prepared material displayed a superparamagnetic behavior at room temperature; it was used as an inexpensive magnetic adsorbent for the removal of water-soluble dyes (aniline blue, Bismarck brown, Congo red, crystal violet, safranin O and Saturn blue LBRR). The dyes adsorption reached equilibrium in approximately 30–120 min. Langmuir isotherms were successfully used to fit the experimental data. The maximum adsorption capacities ranged between 24.2 (Saturn blue LBRR) and 257.9 (aniline blue) mg of dye per g of dried magnetically modified cells. Change of pH significantly increased the adsorption of some dyes (Safarikova et al., 2008).

Alternatively, dried biomass of *Chlorella vulgaris* was magnetically modified by magnetite nanoparticles prepared by coprecipitation reaction. The magnetically responsive composite was successfully employed in the removal of Cd(II) and Pb(II) ions from aqueous solutions. High percentage uptakes of these two toxic ions were observed in a wide range of pH and initial sorbate concentrations. The increases in background electrolyte concentrations (i.e., 0.001–0.1 mol/L $NaNO_3$) affected the uptake of Cd(II) to a greater extent whereas Pb(II) removal was almost unaffected. A simultaneous sorption study inferred that Cd(II) and Pb(II) ions were sorbed at the different binding sites available at the hybrid biomaterial; the sorption mechanism studied with FT-IR and XPS analyses has shown that Cd(II) bound with weak electrostatic forces to the dissociated carboxyl or hydroxyl groups available onto the biosorbent surface whereas Pb(II) ions were chemically bound with the amino group of magnetic *Chlorella* derivative (Lalhmunsiama et al., 2017).

Magnetically modified *Chlorella* sp. cells were also employed for the zinc ions removal. In this application, iron oxide particles prepared by coprecipitation procedure and natural magnetic clay coated with branched polyethylenimine (PEI) were successfully applied in the harvesting of *Chlorella* sp. The separated biomaterial was subsequently used for the removal of zinc ions. Magnetic clay-PEI functionalized particles increased the amount of removed metal ions after cell harvesting and biosorbent formation (Ferraro et al., 2018).

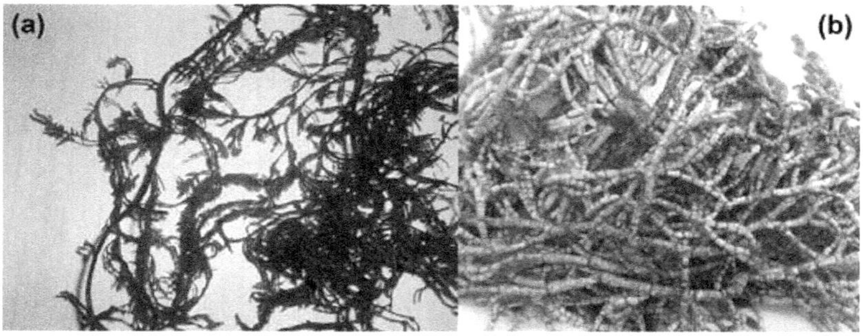

Figure 2. *Sargassum horneri* (a) and *Cymopolia barbata* (b).

3 MAGNETICALLY MODIFIED MARINE ALGAE

Marine algae can be often collected in large amounts; brown algae including *Sargassum* (Fig. 2a) can be locally very abundant (mainly in the eastern Pacific). Magnetically responsive *Sargassum horneri* was prepared by simple and rapid modification of alga biomass using microwave-synthesized iron oxide nano- and microparticles suspended in methanol. Adsorption of five water-soluble dyes of different chemical structures (acridine orange, crystal violet, malachite green, methylene blue, safranin O) from aqueous solutions was studied in a batch system under different conditions. Time necessary to reach equilibrium was less than 2 h for all tested dyes. The adsorption equilibrium data were analyzed by Langmuir and Freundlich isotherm models. The highest maximum adsorption capacity was observed for acridine orange (193.8 mg/g), the lowest one for malachite green (110.4 mg/g). The sorption kinetics could be described by pseudo-second-order model, and the thermodynamic studies indicated exothermic nature of biosorption process in the temperature range studied (Angelova et al., 2016).

Alternatively, magnetic *Sargassum* derivative was prepared using microwave synthesized magnetic iron oxide particles suspended in other water soluble organic solvents (ethanol, propanol, isopropyl alcohol, or acetone); the magnetically modified material was found to be stable in water suspension at least for 2 months (Safarik et al., 2016b).

Sargassum swartzii biomass modified with nanoscale zero-valent iron particles was evaluated for its ability to adsorb crystal violet from aqueous solutions. Significant increase in the biosorption of the tested cationic dye was observed with gradual increase of the pH value; maximum biosorption capacity was determined at pH of 8. Time necessary to reach adsorption equilibrium was ca 120 min. The Langmuir isotherm model expressed high coefficient of determination ($R^2 = 0.999$). The maximum dye uptake of 200 mg/g was reported at pH 8. Desorption was carried out with 0.1 M HCl (Jerold et al., 2017).

Brown alga *Cystoseira barbata* was coated with magnetite particles; magnetic biomaterial was used as a sorbent material for the removal of methylene blue from aqueous solution. Methylene blue adsorption capacity of this material was investigated as function of pH, contact time, initial methylene blue concentration and temperature. The equilibrium data was analyzed with Langmuir and Freundlich isotherms. The results showed that the maximum adsorption capacities were achieved in 300 min at pH 2 and reached to 5.74 and 1.08 mg/g at 25°C and 45°C respectively (Ozudogru et al., 2016).

Magnetic derivative of tropical marine green calcareous alga *Cymopolia barbata* (Fig. 2b) was used as a biosorbent for the effective safranin O removal from aqueous solutions. *Cymopolia barbata* biomass was magnetically modified using microwave-synthesized magnetic iron oxide nano- and microparticles; this modification was simple and inexpensive, without the need of drying step. The biosorption was studied in a batch system under various conditions. Time necessary to reach equilibrium was 90 min. The adsorption isotherms data exhibited best correlation to the

Figure 3. Balls of *Posidonia oceanica*.

Freundlich and Langmuir adsorption models. The maximum adsorption capacity reached the value 192.2 mg/g (dry mass). Kinetic data were best fitted to pseudo-second-order model. The adsorption process was exothermic and spontaneous (Mullerova et al., 2019).

4 MAGNETICALLY MODIFIED SEAGRASS

Posidonia oceanica (commonly known as Neptune grass) is a seagrass species that is endemic to the Mediterranean Sea. The leaf fibers of the withered Neptune grass are rolled up into balls on the bottom of the sea and washed up on Mediterranean beaches by waves and wind (Fig. 3). Magnetically modified *Posidonia oceanica* dead biomass in the form of balls was employed as an adsorbent of organic dyes. The adsorption of seven water-soluble organic dyes (methylene blue, Bismarck brown Y, safranin O, crystal violet, brilliant green, acridine orange, Nile blue A) was characterized using Langmuir adsorption model. The highest calculated maximum adsorption capacity was found for Bismarck brown Y (233.5 mg/g), while the lowest capacity value was obtained for safranin O (88.1 mg/g). The adsorption processes followed the pseudo-second-order kinetic model and the thermodynamic studies indicated spontaneous and endothermic adsorption (Safarik et al., 2016a).

5 MAGNETICALLY MODIFIED ALGAE DERIVED BIOCHAR / ACTIVATED CARBON

Biochar is a carbon-based material produced by pyrolysis of biomass in the absence of oxygen. In specific cases algal biomass can be used as biochar precursor. *Laminaria japonica*-derived activated carbon/iron oxide magnetic composites were prepared by heating powdered biomass in a horizontal electric furnace equipped with a quartz tubular reactor at 400°C under nitrogen atmosphere. After cooling to room temperature, the carbonized material was impregnated with ferric chloride and dried; subsequently the material was activated for 60–180 min at 600–800°C under nitrogen atmosphere. The prepared composite possessed a porous structure and superparamagnetic property, which substantially contributed to the effective adsorption capacity and separation from the solution using an external magnetic field. Adsorption of acetylsalicylic acid by the prepared composite was studied; the maximum adsorption capacity was ca 127 mg/g at 10°C, as fitted by Sips isotherm model. The adsorption process followed the pseudo-second-order kinetic model and was controlled by physisorption and exothermic mechanisms (Jung et al., 2019).

Brown marine macroalga was magnetically modified using a facile electro-magnetization technique. Stainless steel electrode-based electrochemical system was used to generate iron oxide particles for algal biomass modification; in the next step pyrolysis at 600°C for one hour under nitrogen atmosphere was used to prepare the biochar. Magnetite modified biochar exhibited high porosity caused by the use of electro-magnetization process, which enabled not only higher adsorption performance, but also easier separation from aqueous media using a permanent magnet. Sips model was used to quantify the adsorption of Acid orange 7; maximum adsorption capacities were 190, 297, and 382 mg/g at 10, 20, and 30°C, respectively (Jung et al., 2016).

Wasted kelp and hijikia (common seaweed found in South Korea that causes waste problems) were used to prepare magnetically modified biochar. Algae biomass was impregnated with $FeCl_3$ solution and after drying at 70°C this material was pyrolyzed in a muffle furnace at 500°C with nitrogen gas flow. The waste marine algae-based magnetic biochar exhibited great potential to remove heavy metals despite having a quite low surface area (0.97 m^2/g for kelp magnetic biochar and 63.33 m^2/g for hijikia magnetic biochar). Although magnetic biochar could be effectively separated from the solution, however, the magnetization of the biochar partially reduced its heavy metal adsorption efficiency due to the biochar's surface pores becoming plugged with iron oxide particles. The magnetic biochar's heavy metal adsorption capability was considerably higher than that of other types of biochar reported previously. Further, it demonstrated a high selectivity for copper, showing two-fold greater removal (69.37 mg/g for kelp magnetic biochar and 63.52 mg/g for hijikia magnetic biochar) than zinc and cadmium. This high heavy metal removal performance can likely be attributed to the abundant presence of various oxygen-containing functional groups (-COOH and -OH) on the magnetic biochar, which serve as potential adsorption sites for heavy metals (Son et al., 2018a).

Chitosan modified magnetic kelp biochar was successfully synthesized for efficient removal of heavy metals (Cu^{2+}) from wastewater. It was shown that this chitosan containing composite exhibited 6-times higher surface area (6.17 m^2/g) than the pristine magnetic kelp biochar described above (0.97 m^2/g). The presence of new functional groups in chitosan modified biochar improved the Cu^{2+} adsorption capacity. It was shown that the optimum pH value for the adsorption process was 6.9 (Son et al., 2018b).

The blue-green microalgae collected from Dianchi Lake, China, was also used for magnetic biochar hydrothermal production. After biomass drying at 60°C and grinding, the algal powder was mixed with $(NH_4)_2SO_4{\cdot}FeSO_4{\cdot}6H_2O$. Then the suspension was placed in a stainless steel reactor and the content was heated to different temperatures for 6 h in an electric oven. The obtained biochar possessing high amount of oxygen-containing functional groups resulted in the high adsorption performance on tetracycline. This adsorption process was fitted to Langmuir adsorption isotherm and the maximum adsorption capacity was 95.86 mg/g (Peng et al., 2014).

6 CONCLUSIONS

Magnetically responsive biological materials have already found many interesting applications, especially for the removal of various inorganic and organic pollutants and radionuclides. Although such materials have been mainly used in small-scale (laboratory) applications using model solutions, their ability to interact with external magnetic field predetermines their future applications also in large-scale environmental technology processes. Low cost, biocompatibility, high availability, and variability of algae- and seagrass-based magnetic biocomposites will enable their wide application in the near future.

ACKNOWLEDGEMENTS

The research was supported by the project LTC17020 (Ministry of Education, Youth and Sports of the Czech Republic), by the ERDF/ESF project "New Composite Materials for Environmental Applications" (No. CZ.02.1.01/0.0/0.0/17_048/0007399) and ERDF project "Development of pre-applied research in nanotechnology and biotechnology" (No. CZ.02.1.01/0.0/0.0/17_048/0007323).

REFERENCES

Angelova, R., Baldikova, E., Pospiskova, K., Maderova, Z., Safarikova, M. & Safarik, I. 2016. Magnetically modified *Sargassum horneri* biomass as an adsorbent for organic dye removal. *Journal of Cleaner Production* 137: 189–194.

Delrue, F., Álvarez-Díaz, P.D., Fon-Sing, S., Gatien Fleury, G. & Sassi, J.-F. 2016. The environmental biorefinery: Using microalgae to remediate wastewater, a win-win paradigm. *Energies* 9: Article No. 132.

Ferraro, G., Toranzo, R. M., Castiglioni, D. M., Lima, E., Vasquez Mansilla, M., Fellenz, N. A., Zysler, R. D., Pasquevich, D. M. & Bagnato, C. 2018. Zinc removal by *Chlorella* sp. biomass and harvesting with low cost magnetic particles. *Algal Research* 33: 266–276.

Jerold, M., Vasantharaj, K., Joseph, D. & Sivasubramanian, V. 2017. Fabrication of hybrid biosorbent nanoscale zero-valent iron-*Sargassum swartzii* biocomposite for the removal of crystal violet from aqueous solution. *International Journal of Phytoremediation* 19: 214–224.

Jung, K.-W., Choi, B. H., Jeong, T.-U. & Ahn, K.-H. 2016. Facile synthesis of magnetic biochar/Fe_3O_4 nanocomposites using electro-magnetization technique and its application on the removal of acid orange 7 from aqueous media. *Bioresource Technology* 220: 672–676.

Jung, K.-W., Choi, B. H., Song, K. G. & Choi, J.-W. 2019. Statistical optimization of preparing marine macroalgae derived activated carbon/iron oxide magnetic composites for sequestering acetylsalicylic acid from aqueous media using response surface methodology. *Chemosphere* 215: 432–443.

Kanjilal, T. & Bhattacharjee, C. 2018. Green applications of magnetic sorbents for environmental remediation. In Inamuddin, Abdullah M. Asiri & Ali Mohammad (eds.), Organic Pollutants in Wastewater 1. Materials Research Forum LLC, Millersville, 2018, pp. 1–41.

Lalhmunsiama, Gupta, P. L., Jung, H., Tiwari, D., Kong, S.-H. & Lee, S.-M. 2017. Insight into the mechanism of Cd(II) and Pb(II) removal by sustainable magnetic biosorbent precursor to *Chlorella vulgaris*. *Journal of the Taiwan Institute of Chemical Engineers* 71: 206–213.

Mullerova, S., Baldikova, E., Prochazkova, J., Pospiskova, K. & Safarik, I. 2019. Magnetically modified macroalgae *Cymopolia barbata* biomass as an adsorbent for safranin O removal. *Materials Chemistry and Physics* 225: 174–180.

Ozudogru, Y., Merdivan, M. & Goksan, T. 2016. Biosorption of methylene blue from aqueous solutions by iron oxide-coated *Cystoseira barbata*. *Journal of the Turkish Chemical Society A* 3: 551–564.

Peng, L., Ren, Y., Gu, J., Qin, P., Zeng, Q., Shao, J., Lei, M. & Chai, L. 2014. Iron improving bio-char derived from microalgae on removal of tetracycline from aqueous system. *Environmental Science and Pollution Research International* 21: 7631–7640.

Safarik, I., Ashoura, N., Maderova, Z., Pospiskova, K., Baldikova, E. & Safarikova, M. 2016a. Magnetically modified *Posidonia oceanica* biomass as an adsorbent for organic dyes removal. *Mediterranean Marine Science* 17: 351–358.

Safarik, I., Baldikova, E., Pospiskova, K. & Safarikova, M. 2016b. Magnetic modification of diamagnetic agglomerate forming powder materials. *Particuology* 29: 169–171.

Safarik, I., Baldikova, E., Prochazkova, J., Safarikova, M. & Pospiskova, K. 2018. Magnetically modified agricultural and food waste: Preparation and application. *Journal of Agricultural and Food Chemistry* 66: 2538–2552.

Safarik, I., Prochazkova, G., Pospiskova, K. & Branyik, T. 2016c. Magnetically modified microalgae and their applications. *Critical Reviews in Biotechnology* 36: 931–941.

Safarikova, M., Pona, B. M. R., Mosiniewicz-Szablevska, E., Weyda, F. & Safarik, I. 2008. Dye adsorption on magnetically modified *Chlorella vulgaris* cells. *Fresenius Environmental Bulletin* 17: 486–492.

Santos, S. C. R., Ungureanu, G., Volf, I., Boaventura, R. A. R. & Botelho, C. M. S. 2018. Macroalgae biomass as sorbent for metal ions. In Valentin Popa & Irina Volf (eds.), Biomass as Renewable Raw Material to Obtain Bioproducts of High-Tech Value. Elsevier, 2018, pp. 69–112.

Son, E. B., Poo, K. M., Chang, J. S. & Chae, K. J. 2018a. Heavy metal removal from aqueous solutions using engineered magnetic biochars derived from waste marine macro-algal biomass. *Science of the Total Environment* 615: 161–168.

Son, E. B., Poo, K. M., Mohamed, H. O., Choi, Y. J., Cho, W. C. & Chae, K. J. 2018b. A novel approach to developing a reusable marine macro-algae adsorbent with chitosan and ferric oxide for simultaneous efficient heavy metal removal and easy magnetic separation. *Bioresource Technology* 259: 381–387.

Wastes: Solutions, Treatments and Opportunities III – Vilarinho et al. (Eds)
© 2020 Taylor & Francis Group, London, ISBN 978-0-367-25777-4

Agribusiness residues: Water absorption and mechanical behavior of their composites

P.H.L.S.P. Domingues
State University of Rio de Janeiro, Rio de Janeiro, Brazil

J.R.M. d'Almeida
Pontifícia Universidade Católica do Rio de Janeiro, Rio de Janeiro, Brazil

ABSTRACT: The effect of water absorption on the mechanical performance of composites reinforced with chopped lignocellulosic fibers obtained from residues of other industrial processes, namely coir, sugarcane bagasse and wood fibers, was evaluated. The evaluation was performed using both destructive and non-destructive tests. In addition, since the visual perception of a panel used indoors is also an important aspect, the variation of the gloss of the composites was evaluated. The results obtained showed that coir composites absorb less water than sugarcane bagasse and wood fiber composites. A good correlation was observed between the amount of water absorbed and the variation of the properties. The observed behavior difference is displaced only with respect of the time due to the difference in the rate of water absorption between the composites. This difference in water absorption kinetics was related to the hygroscopicity of each fiber.

1 INTRODUCTION

Composite materials using lignocellulosic fibers have been used increasingly in various industrial segments. This is due to several advantages that these materials present, among which one can highlight low density and also the fact that the lignocellulosic fibers are a natural renewable and abundant resource. In addition to crops grown specifically to obtain fibers, such as jute and sisal, lignocellulosic fibers can be obtained from many sources, such as agribusiness waste. For example, a large amount of sugar cane bagasse is generated when the sugarcane is processed to obtain alcohol (Chandel *et al.*, 2012). Coconut fruits are also an abundant source of fibers. After extraction of the copra or of the liquid endosperm that fills the interior of the fruits, the fruit shell is disregarded, but fibers can be extracted from the external layer of the exocarp and from the endocarp of the fruit (Asasutjarit *et al.*, 2009). Post-processing of these wastes can generate a large amount of fibers and not impact on the current use of these materials.

Lignocellulosic fiber composites are, however, hygroscopic materials, because of the many hydroxyl groups on cellulose and hemicellulose structures. Therefore, absorption of water and the consequent fiber/matrix interface degradation, matrix swelling, and weight increase are drawbacks of using these composites.

However, to correlate the amount of water absorbed with the mechanical performance of the composites is not a simple task because of the many variables that can affect the interaction/affinity of water with the composite. Among others, one can highlight the specific affinity of the polymeric matrix to humidity, the presence or the lack of strong interfacial bonds at the fiber-matrix interface, and the fiber volume fraction.

Therefore, the aim of this work was to evaluate the effect of water absorption on the mechanical performance of chopped lignocellulosic fiber reinforced composites manufactured to be used as interior panels. The evaluation was performed using both destructive and non-destructive tests. In addition, since the visual perception of a panel used indoors is also an important aspect, the variation of the gloss of the composites also was evaluated.

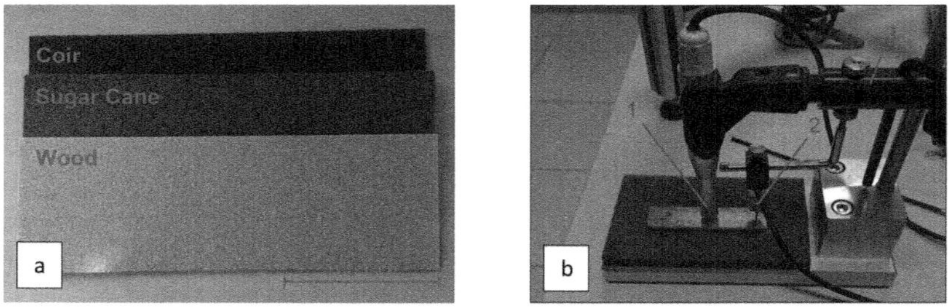

Figure 1. (a) Chopped lignocellulosic fiber-polyester composites manufactured by hot compression molding. (b) Relative position of the microphone (1) and actuator (2) in respect to the test specimen (3).

2 METHODS AND MATERIALS

2.1 *Materials*

The composites were manufactured by hot compression molding. Chopped coir (*Cocos nucifera*), sugarcane bagasse (*Saccharum sp.*) and wood fibers were used. The wood fibers were basically from eucalyptus (*Eucalyptus sp.*) chips, disregarded by furniture industry. Each chopped fiber was mixed with the polyester matrix in the appropriated ratio, and the resulting slurry was poured into the mold cavity. Plates with homogeneous appearance and with average thickness of 4 mm were produced (Figure 1a).

Flat specimens 84 mm long, 20 mm wide and 4 mm thick were machined from the plates. The actual size of each sample was determined using a caliper rule with an accuracy of 0.01 mm. Besides, the mass of each specimen was determined within ±0.0001 g. This procedure was necessary because the dimensions and the mass of each sample are input data of the non-destructive sonic test to be presented below.

2.2 *Experimental*

Aging was performed by placing the specimens in containers with tap water at room temperature, 23 ± 2°C. The vessels were kept closed and the water level was checked periodically. The samples were removed periodically from the containers and were weighted to determine the mass gain versus time behavior of each composite. These measurements were performed using the procedures recommended in ASTM D570 standard. After removing the excess of water at the surface of the specimens they were left to stand for 20 seconds and only then the mass was measured. This time was evaluated experimentally, as being sufficient to make a measurement without appreciable mass variation (±0.0005 g).

The properties of the composites were determined as a function of the immersion time by both a destructive (three-point bending) and a non-destructive (sonic test) tests. The flexural tests were performed based on the ASTM standard D790. A test speed of 5 mm/min and a span of 64 mm were used. Five samples were tested per composite, on the as-manufactured condition and after 2,496 and 4,440 hours of immersion in water (~3 and 6 months of immersion).

The sonic test was performed using an electret microphone with a detection range varying from 0.5 to 20 kHz, and an acquisition time of 0.6 s. A light actuator hits the specimen's surface every 30 seconds, and the generated signal is then processed via a dedicated software (*Sonelastic®*). Both the Young's modulus and damping factor are automatically calculated, following the procedures described at ASTM E1876 standard. The results reported are the average of 10 measurements. The test configuration used is shown at Figure 1b, where the relative position between the actuator and the microphone corresponds to the flexional-torsional boundary condition. This test was used to monitor the variation of properties along the entire time of the aging experiment. Each time the specimens were removed from immersion to be weighted they were also tested using this procedure.

Even if aging does not affect mechanical properties, the use of a material may be impaired if its surface characteristics are markedly affected. For example, if a material is used for surface finishing, gloss may be an important aesthetic feature. Therefore, the variation of the visual aspect of the composites due to aging was evaluated by measuring the gloss of the specimens. The test was performed using the ASTM D 523 standard. The measurements were made at the angles of 20°, 60°, and 85° using a T&M 268 Model Gloss Meter. This analysis was performed with the as-manufactured samples and with the samples aged for 4,440 hours.

The volume fraction of fibers of the composites was evaluated using the point count method, as described by ASTM E562-11 standard. Prior to the analysis, the cross section of each composite was prepared using the usual sanding and polishing techniques. Water sandpaper from #100 to #1200 and diamond paste with 6 μm and 3 μm were used, and the final surface obtained was found to be adequate to be analyzed. Five images were captured per sample, using a 100x magnification, and a grid with 49 points was used, as recommended by the standard.

3 RESULTS AND DISCUSSION

3.1 *Volume fraction*

The results obtained by the point count method were, respectively, for wood, sugar cane bagasse and coir fiber composites: 30.6 ± 10.7%; 37.8 ± 5.1%, and 34.7 ± 16.6%. On the average, the composites have a volume fraction of fiber ranging from 30 to 38%, which is a common value used in various lignocellulosic fiber composites. For example, Prasad et. al (1983) developed a coir-polyester composite with a volume fraction of fibers of 30% using chopped fibers, and Biwas and coworkers (2011) used approximately this same amount of chopped coir fibers.

3.2 *Water absorption behavior*

The mass gain due to water absorption is shown in Figure 2. It can be observed that the composite with coir fibers presented a much lower water up-take than the other two composites. The mass of water absorbed at saturation, M_∞, and the diffusion coefficient, D, were obtained from the non-linear fitting of the experimental points to the equation proposed by McKaque *et al.* (1976), namely:

$$M = M_\infty * \tanh\left(\frac{4}{h}\sqrt{\frac{D.t}{\pi}}\right) \quad (1)$$

where $M\%$ is the mass of water absorbed in a time t (h) and h (mm) is the thickness of the sample. The results are presented in Table 1. Considering that the matrix used was the same and that the fiber fraction of the composites is approximately equal, the lower values for the saturation value and also for the diffusion coefficient of the composite with coconut fibers can be connected to the larger lignin content that these fibers have in relation to the others. Lignin is a hydrophobic compound (Shahzad and Isaac 2015) and, therefore, the coconut fibers will absorb less moisture.

The direct comparison of the results shown in Table 1 with other composites can be misleading because both D and M_∞ are greatly affected by the amount and orientation of fibers and the specific polymer matrix used. However, the values obtained are comparable to those reported in several articles. For example, Saw and coworkers (2015) reported water saturation values varying from 9.3 to 22% when coir-luffa hybrid novolac matrix composites were studied.

3.3 *Flexural properties*

The flexural strength of the composites as a function of the immersion time show that wood and sugarcane bagasse composites degrade due to water absorption, but the flexural strength of the coir fiber composite was little affected. This result was consistent with the amount of water absorbed by each composite (Table 1). In fact, wood and sugarcane composites showed a decrease in the flexural

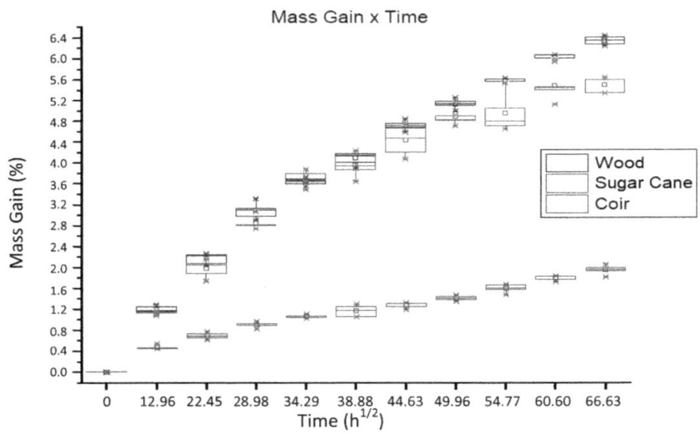

Figure 2. Mass gain as a function of the square root of the time.

Table 1. Diffusion coefficient and mass saturation value.

Composite	Diffusion coefficient (mm^2/s)	M_∞ (%)	Determination coefficient – R^2
Wood	2.32E-04	12.20	0.996
Sugar cane	3.54E-04	7.98	0.987
Coir	1.66E-04	4.82	0.994

strength of 14% and 8.2%, respectively, while the coconut fiber composite practically maintained its value (variation <1%).

Figure 3 shows a plot of the residual flexural strength as a function of the percentage of water absorbed for all the composites. The residual flexural strength was calculated as the relationship between the flexural strength of the aged specimens divided by that of the as-manufactured composite. The Pearson's correlation coefficient of the curve fitted to the experimental points has a value of −0.89, indicating a strong correlation between the variables. This behavior may indicate that the diffusion process is similar in all composites, since the points for all the composites can be satisfactorily described by a single curve. The diffusion kinetics, however, are different due to the greater or lesser affinity of the fibers by water, as inferred from Table 1.

A similar behavior was observed with respect to the maximum deformation. Coir composite was less affected than the other two composites. These showed the expected increase of deformation due to the matrix plasticization and/or fiber/matrix interface failure, as already highlighted in the introduction.

3.4 *Non-destructive sonic test*

The variation of the Young's modulus as a function of the immersion time for one of the composites is shown, as an example, in Figure 4. It can be observed that the values decreased with the time. However, the decrease of Young's modulus for coir composites was less abrupt than that observed for the other two composites. For wood, sugar cane bagasse and coir, the decrease of the Young's modulus was, respectively of −6.8%; −9.9% and −1.6%.

The variation of the damping factor as a function of the immersion time in water was similar for all the composites. However, similarly to the modulus of elasticity, there was a difference in how much this parameter was varied. Namely, an increase of 10.3%; 19.5% and 7.6% was measured for wood, sugar cane bagasse and coir composites, respectively.

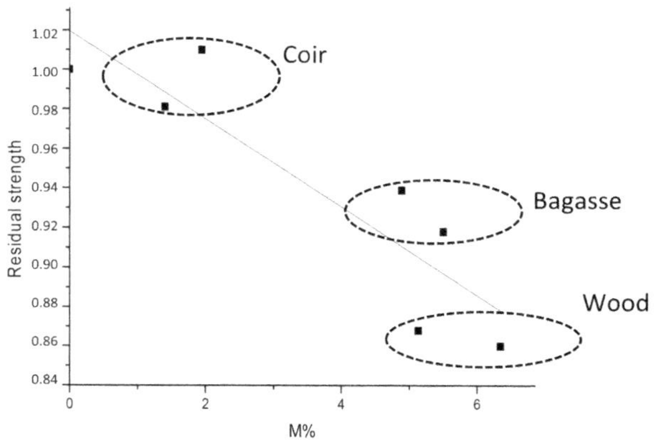

Figure 3. Variation of the residual strength vs. the amount of absorbed water.

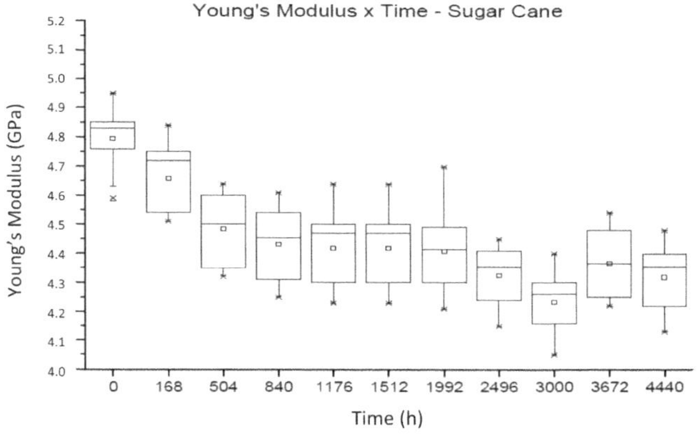

Figure 4. Sonic test. Variation of the Young's modulus as a function of the immersion time. Example for the sugar cane bagasse composite.

The variation of the damping factor is consistent with the behavior reported so far for the flexural strength and for the modulus of elasticity. It seems that the greatest variation of the results occurs only while the water is being absorbed and that, after saturation, the values of the properties are only marginally altered. A similar behavior was observed in a sponge gourd/polyester composite (Feiferis and d'Almeida, 2016). The damping factor (ʒ) reflects the attenuation of an oscillation imposed on a material and can be correlated to the internal friction due to, primarily, the fiber/matrix interfaces and to the viscoelastic response of both the polymeric matrix and lignocellulosic fibers (Chandra et al., 1999). The increase in the damping coefficient observed for all materials implies a lower amplitude of vibration in a phenomenon of resonance, and a shorter time of persistence of the vibration. This damping is consistent with the plastification of the polymer matrix by water, which increases the deformability of the polymer.

3.5 *Gloss aspect*

After 4440 hours of immersion in water the reflectance of wood fiber and sugar cane composites presented a reflectance increase, while in the coconut fiber composites the reflectance decreased. The observed behavior indicates that the wood and sugar cane composites will be brighter after moisture absorption, while the composite with coconut fibers will be more opaque than the material as-manufactured. The absolute gloss variation for coir composite was, however, lower than that of the other two composites.

4 CONCLUSIONS

The following conclusions were drawn from the experimental results obtained:

- The flexural strength of the three composites analyzed can be described by a single curve as a function of the amount of water absorbed, and varies inversely with respect to the water content absorbed. There is a strong correlation between these variables, as described by the Pearson correlation coefficient whose value was −0.89.
- Coir fiber composites absorb less water than sugarcane bagasse composites and wood fiber composites Therefore, coir fiber composites presented less variation of flexural strength and Young's modulus, and also on its damping characteristics.
- The tendency in the variation of the properties indicates that the process of water absorption and degradation of the composites were similar. The observed behavior difference is only displaced with respect to time due to the difference in the rate of water absorption between the composites. This difference in water absorption kinetics was attributed, essentially, to the hygroscopicity differences of each fiber.
- The gloss of the composites that absorbed more water (those with bagasse and wood fibers) increased with time. There was a reduction in the gloss of the composites with coir fibers, but the percentage variation of the gloss of this composites was smaller, corroborating the fact that coir composites are less susceptible to water action.

ACKNOWLEDGMENTS

The authors acknowledge the financial support from the Brazilian Funding Agency CNPq (grant number # 301299/2010-2).

REFERENCES

Asasutjarit, C., Charoenvai, S., Hirunlabh, J. & Khedari, J. 2009. Materials and mechanical properties of pretreated coir-based green composites. *Comp.: Part B* 40: 633–637.

Biswas, S., Kindo, S. & Patnaik, A. 2011. Effect of fiber length on mechanical behavior of coir fiber reinforced epoxy composites. *Fibers and Polymers* 12: 73–78.

Chandel, A.K., da Silva, S.S., Carvalho, W. & Singh, O.V. Singh. 2012. Sugarcane bagasse and leaves: foreseeable biomass of biofuel and bio-products. *J.Chem.Technol. Biotechnol.* 87: 11–20.

Chandra, R., Singh, S.P. & Gupta, K. 1999. Damping studies in fiber reinforced composites: a review. *Composite Structures* 46: 41–45.

Feiferis, A.R. & d'Almeida, J.R.M. 2016. Evaluation of Water Absorption Effects on the Mechanical Properties and Sound Propagation Behavior of Polyester Matrix- Sponge Gourd Reinforced Composite. *J. Comp. Biodegrad. Polym.* 4: 26–31.

McKaque, E.L., Reynolds, J.D. & Halkias, J.E. 1976. Moisture Diffusion in Fiber Reinforced Plastics. *J. Eng. Mater. Technol.* 98: 92–95.

Prasad, S.V., Pavithran, C. & Rohatgi, P.K. 1983. Alkali treatment of coir fibres for coir-polyester composites. *J. Mater. Sci.* 18: 1443–1454.

Saw, S.K., Sarkhel, G. & Choudhury, A. 2015. Effect of layering pattern on the physical, mechanical and acoustic properties of luffa/coir fiber-reinforced epoxy novolac hybrid composites. In V.K.Thakur (ed.), *Lignocellulosic polymer composites: processing, characterization and properties*: 369–384. Salem: Scrivener Publishing.

Shahzad, A. & Isaac, D.H. 2015. Weathering of lignocellulosic polymer composites. In V.K.Thakur (ed.), *Lignocellulosic polymer composites: processing, characterization and properties*: 327–368. Salem: Scrivener Publishing.

Wastes: Solutions, Treatments and Opportunities III – Vilarinho et al. (Eds)
© 2020 Taylor & Francis Group, London, ISBN 978-0-367-25777-4

Life cycle assessment of bio-based polyurethane foam

J. Ferreira, B. Esteves, L. Cruz-Lopes & I. Domingos
Research Centre for Natural Resources, Environment and Society (CERNAS-IPV), Instituto Politécnico de Viseu, Viseu, Portugal

ABSTRACT: The goal of this study was to understand the potential burdens associated with a new bio-based polyurethane foam using *eucalyptus globules* residues liquefied instead of an oil derivative. An LCA study was elaborated based on ISO 14040/44 series standard. Ecoinvent, was used within SimaPro to model utility process operations, transportation, and other material inputs. CML-IA (baseline) was the method chosen for Life Cycle Impact Assessment. The study considered the impact categories of abiotic depletion; abiotic depletion (fossil fuels); global warming; ozone layer depletion; human toxicity; fresh water aquatic ecotoxicity; marine aquatic ecotoxicity; terrestrial ecotoxicity; photochemical oxidation; acidification; and eutrophication. The results show that electricity and MDI Voranate (isocyanate) consumption during the foam production are the dominant sources of all potential environmental impacts. These two processes (together) play a major role in all impact categories (84–97%).

1 INTRODUCTION

Polyurethanes (PUs) are very versatile polymers with numerous applications. The main application is the furniture industry, followed by the automobile industry (Cinelli et al. 2013). In a car the PUs are present in the bumpers, seats and upholstery. In construction, PUs are used in acoustic and thermal insulation (Cinelli et al. 2013, Hu et al. 2014).

Currently, the production of PUs foam is completely dependent on petroleum, since the two main reagents necessary for its production, polyol and isocyanate, are derived from this fossil fuel (Gama et al. 2015, Hu et al. 2012). The high impacts on the environmental of this raw material have motivated the demand for renewable materials.

The LIQUERCUS project carried out by a research team of CERNAS-IPV focused on an innovative bio-based PU foam production process using *eucalyptus globules* residues liquefied instead of an oil derivative. The results of bio-based PU foam properties are shown in Table 1.

To quantify the potential environmental impacts of this new process, a Life Cycle Assessment (LCA) was done. LCA is framed by ISO (2006a, b) standards which provide comprehensive guidelines for conducting an LCA study. A technical guidance for detailed Life Cycle Assessment (LCA) studies is provided by the ILCD handbook (EC 2010). It provides the technical basis to derive product-specific criteria, guides, and simplified tools. Life Cycle Assessment (LCA) is a

Table 1. Properties of bio-based PU foam.

	PU foam (E54)	Ref. Standard
Density ($Kg.m^{-3}$)	97.2	ISO 845
Compressive strength (KPa)	107.6	ISO 844
Young's Modulus (KPa)	1691.3	ISO 844

useful methodology for examining the total environmental impact over the complete "cradle-to-grave" product or service life cycle. LCA that covers the "cradle-to-gate" stages can be applied for a material (eco-profiles) that can be used to create complete LCA for downstream uses of these materials. A life cycle perspective helps to avoid unintended environmental consequences from one life cycle phase to another during process improvement. LCA methodology has been applied in many sectors, as a direct consequence of recognition of its importance as a decision making support tool (Heijungs & Guinée 2012). Among areas where LCA has been applied are wood for energy production (Ferreira et al. 2014, 2018a, Pergola et al. 2017), wood and wood products (Ferreira et al. 2016, 2018b, Suter et al. 2017). LCA has been applied to the improvement of the PU foam technology, especially with respect to the production process and inputs (Helling & Parenti 2013) and in the PU foam eco-profiles and environmental product declarations (EPDs) (Europur 2015).

The aim of this study was to understand the potential burdens associated with a new bio-based PU foam production using *eucalyptus globules* residues liquefied instead of an oil derivative. How this may be an improvement over current technology, especially with respect to the production process and inputs are a future work.

2 MATERIAL AND METHODS

The LCA study of the bio-based PU foam was performed based on ISO 14040:2006 and ISO 14044:2006 standards. ISO defines four phases of an LCA study: Goal and Scope definition; Life Cycle Inventory (LCI); Life Cycle Impact Assessment (LCIA); and Life Cycle Interpretation.

2.1 *Goal and Scope of the study*

As referenced before, the goal of the study was to understand (from cradle-to- gate) the potential life cycle environmental impacts associated to bio-based PU foam and identify hotspots.

The target audience for this report is the CERNAS-IPV polyurethane foams R&D team. A final audience would be foam producers, for whom the information would be helpful in decisions about technology choices for new or modified production facilities.

2.1.1 *Function of the system and functional unit (FU)*

The function of the system in study is to produce bio-based PU foam to be used in different uses. Therefore, the functional unit (FU) assumed was 1 Kg of PU foam outside laboratory. The choice of this functional unit is in agreement with other PU foam systems studied from an LCA perspective (Hischier 2007).

2.1.2 *System boundary*

This was a cradle-to-grave study, so the boundaries extended upstream to materials in the earth and continued to the laboratory gate (outside). Figure 1 represents a high-level view of the life cycle stages and the primary direct inputs and emissions. Figure 2 shows a conceptual drawing of PU foam production.

Process equipment, buildings and other infrastructure was not included because typically it has minimal impact on chemical production processes (Boustead 2003).

Only transportation of raw materials was included for all inbound raw materials (except for distilled water that was produced in the laboratory), assuming 100 km of transport by truck.

2.1.3 *Allocation procedure*

The system in study is assumed to produce a single product, so no allocation is needed. All inputs and outputs related directly to the foam production operation were included.

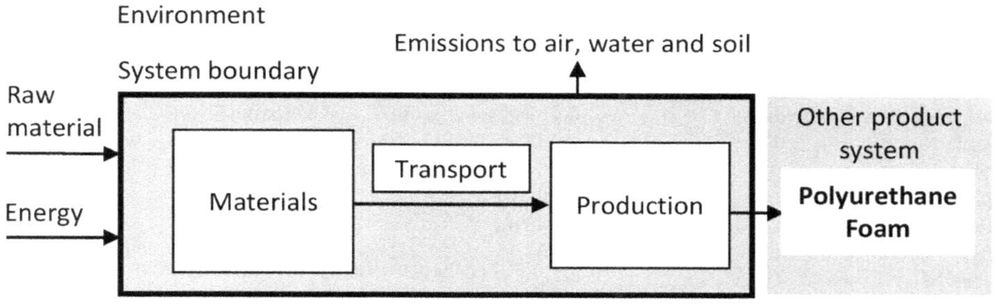

Figure 1. Life cycle stages (high level view).

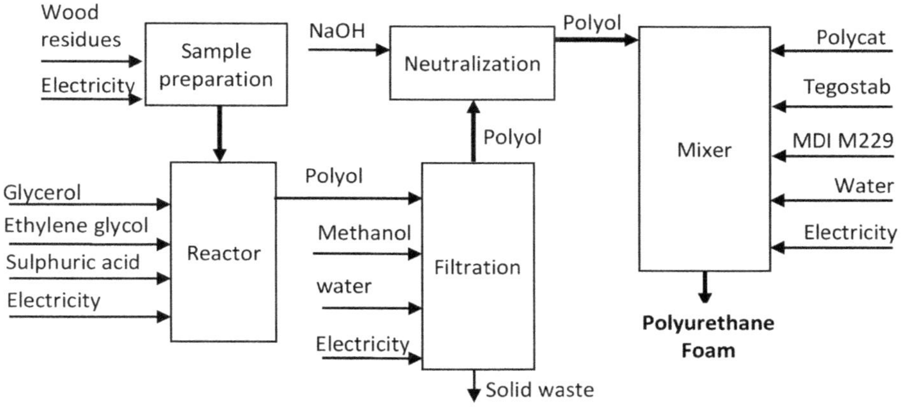

Figure 2. Conceptual flow sheet of bio-based PU foam production.

2.2 *Life Cycle Inventory Analysis*

The inventory analysis and, subsequently, the impact analysis have been performed using the LCA software SimaPro 8.5.2 (PRé 2018) and associated databases and methods.

2.2.1 *Data type/data collection*

Data were collected from LIQUERCUS project team members, belonging to the CERNAS-IPV, for the energy and material inputs. The data table for bio-based PU foam production is given in Table 2, taken from Project research team.

Data provided by project research team and other data were used directly to create process models in SimaPro. Ecoinvent v3, was used within SimaPro 8.5.2 (PRé 2018) to model utility process operations, transportation, and other material inputs as shown in Table 3.

2.3 *Life cycle impact assessment (LCIA)*

Life cycle impact assessment (LCIA) was performed using valuation systems available in Ecoinvent, and primarily that of CML-IA (baseline) V 3.05, that elaborates the problem-oriented (midpoint) approach and without any normalization or weighting (Guinée et al. 2002). CML-IA has been implemented in SimaPro 8.5.2 (PRé 2018).

Table 2. Data table for bio-based PU foam production.

Inputs	Quantity	Outputs	Quantity
Eucalyptus globules residues	0.267 g	PU foam (E54)	5.11 g
Glycerol	0.337 g	Methanol*	5.345 ml
Ethylene glycol	2.38 g	Hazardous waste	0.08 g
Sulphuric acid	0.0934 g	Air emissions:	
Methanol*	5.345 ml	VOC	5.66 g
NaOH	0.003 g	Steam water	7.724 g
Polycat 34	0.6 g		
Distilled water	8.047 g		
Tegostab B8404	0.14 g		
MDI M229 Voranate (isocyanate)	7 g		
Electricity	0.268 KWh		
Transport (truck 3.5–7.5 t)	0.00189 t.Km		

*The methanol is recycled (closed-loop)

Table 3. SimaPro models chosen to model process operations, transportation, and other material inputs.

Material/Energy/Transport/ Waste treatment	Ecoinvent model
Eucalyptus globules residues	Residual hardwood, wet {GLO} | market for | Alloc Def, U
Glycerol	Glycerine {GLO} | market for | Alloc Def, U
Ethylene glycol	Ethylene glycol {GLO} | market for | Alloc Def, U
Sulphuric acid	Sulfuric acid {GLO} | market for | Alloc Def, U
NaOH	Sodium hydroxide, without water, in 50% solution state {GLO} | market for | Alloc Def, U
Polycat 34	Trimethylamine {GLO} | market for | Alloc Def, U
Distilled water	Water, ultrapure {CA-QC} | production | Alloc Def, U
Tegostab B8404	Silicone product {GLO} | market for | Alloc Def, U
MDI M229 Voranate	Phenyl isocyanate {GLO} | market for | Alloc Def, U
Electricity	Electricity, low voltage {PT} | market for | Alloc Def, U
Transport (truck 3.5–7.5 t)	Transport, freight, lorry 3.5–7.5 metric ton, EURO6 {GLO} | market for | Alloc Def, U
Waste for treatment	Hazardous waste, for incineration (CH) / treatment of hazardous waste, hazardous waste incineration/Alloc Def, U

3 RESULTS AND DISCUSSION (LIFE CYCLE INTERPRETATION)

The life cycle impact assessment (LCIA) results (characterization) for bio-based PU foam, using CML-IA (baseline) V3.05 method are presented in Tab. 4.

Relative contribution of each process to the potential environmental results of bio-based PU foam is illustrated in Figure 3. Electricity was the main contributor (>50%) to all impact categories except to AD and OD that were mainly affected by the MDI Voranate (isocyanate). These two processes (together) play a major role in all impact categories (84–97%). The glycerin and the ethylene glycol presented a notable contribution of 15% to TET and 5% to AD (FF), respectively. Related the remaining processes they contributed with less than 2% for each impact category. The carbon dioxide (fossil) emitted into the air was responsible for 82% of GW where the electricity contributed 55% followed by the MDI (22%). Methane (fossil) contributed with 8.6% to GW which 6% is due the electricity used.

Table 4. Impact assessment results for 1 Kg of bio-based PU foam.

Impact category	Units	PU foam
Abiotic depletion	g Sb eq	0.116
Abiotic depletion (fossil fuels)	MJ	397
Global warming (GWP100a)	kg CO_2 eq	34
Ozone layer depletion (ODP)	mg CFC-11 eq	3
Human toxicity	kg 1,4-DCB[b] eq	15
Fresh water aquatic ecotox.	kg 1,4-DCB eq	18
Marine aquatic ecotoxicity	kg 1,4-DCB eq	49139
Terrestrial ecotoxicity	Kg 1,4-DCB eq	0.25
Photochemical oxidation	Kg C_2H_4 eq	0.011
Acidification	Kg SO_2 eq	0.23
Eutrophication	Kg PO_4^- eq	0.055

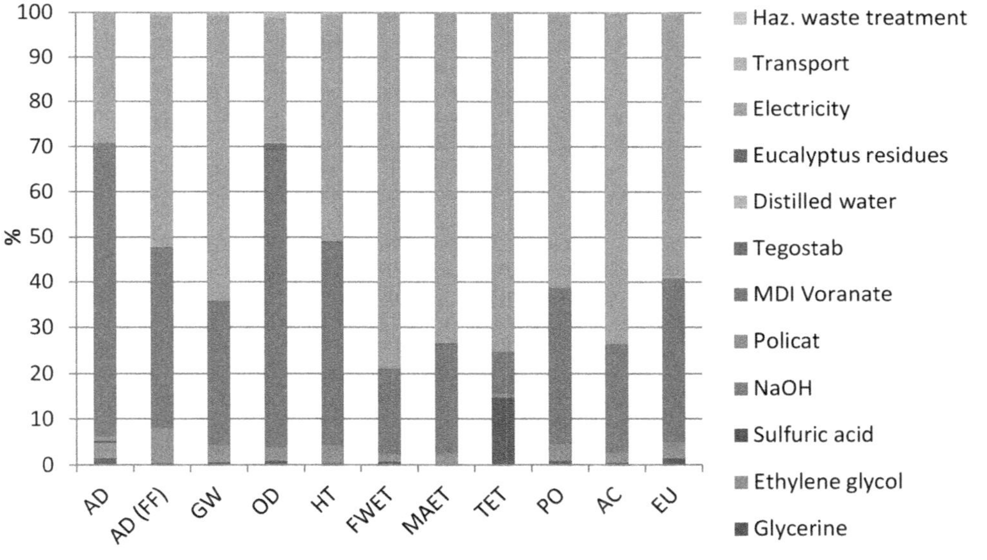

Figure 3. Relative contribution of each process/material to the potential environmental impacts of 1 Kg bio-based PU foam using CML-IA (baseline) V3.05 method. Acronyms: AD (abiotic depletion); AD(FF) (abiotic depletion (fossil fuels)); GW (global warming (GWP100a)); OD (ozone layer depletion (ODP)); HT (human toxicity); FWET (fresh water aquatic ecotoxicity); MAET (marine aquatic ecotoxicity); TET (terrestrial ecotoxicity); PO (photochemical oxidation); AC (acidification); EU (eutrophication).

4 CONCLUSIONS

A life cycle assessment (from cradle-to- gate) study was performed to understand the potential life cycle environmental impacts associated with production (in laboratory) of polyurethane foam using *eucalyptus globules* residues liquefied instead of an oil derivative.

Electricity and MDI Voranate (isocyanate) consumption during the foam production are the dominant sources of all potential environmental impacts. Together they play a major role in all impact categories (84–97%). Carbon dioxide (fossil) and methane (fossil) emitted into the air during electricity production were responsible of 55% and 6.1% to global warming respectively.

Carbon dioxide (fossil) emitted into the air during MDI production was another important source of global warming represented 22%.

REFERENCES

Boustead, I. 2003. An Introduction to Life Cycle Assessment. Boustead Consulting Ltd.

Cinelli, P., Anguillesi, I. & Lazzeri, A. 2013. Green synthesis of flexible polyurethane foams from liquefied lignin. *European Polymer Journal* 49: 1174–1184.

EC, 2010. European Commission – Joint Research Centre – Institute for Environment and Sustainability: International Reference Life Cycle Data System (ILCD) Handbook – General guide for Life Cycle Assessment – Detailed guidance. First edition March 2010. EUR 24708 EN. Luxembourg. Publications Office of the European Union.

Europur, 2015. *Flexible Polyurethane (PU) Foam*. Eco-profiles and Environmental Product Declarations of the European Plastics Manufacturers, Brussels, Belgium.

Ferreira, J., Viana, H., Esteves, B., Cruz-Lopes, L. & Domingos, I. 2014. Life cycle assessment of residual forestry biomass chips at a power plant: a Portuguese case study. *International Journal of Energy and Environmental Engineering* 5(2–3):1–7.

Ferreira, J., Esteves, B., Nunes, L. & Domingos, I. 2016. Life Cycle Assessment as a Tool to Promote Sustainable Thermowood Boards: a Portuguese Case Study. *International Wood Products Journal* 7(3): 124–129.

Ferreira, J., Esteves, B., Cruz-Lopes, L., Evtuguin, D. & Domingos, I. 2018a. Environmental advantages through producing energy from grape stalk pellets instead of wood pellets and other sources. *International Journal of Environmental Studies*, DOI: 10.1080/00207233.2018.1446646.

Ferreira, J., Herrera, R., Labidi, J., Esteves, B. & Domingos, I. 2018b. Energy and environmental profile comparison of TMT production from two different companies – a Spanish/Portuguese case study. *iForest* 11:155–161.

Gama, V., Soares, B., Freire, C., Silva, R., Neto, C., Barros-Timmons, A. & Ferreira, A. 2015. Bio-based polyurethane foams toward applications beyond thermal insulation. *Materials and Design* 76: 77–85.

Guinée J., Gorrée M., Heijungs R., Huppes G., Kleijn R., Koning A., Oers L., Sleeswijk A., Suh S., Haes H., Bruijn H., Duin R. & Huijbregts M. 2002. *Handbook on life cycle assessment. Operational guide to the ISO standards. Part III: Scientific background.* (Dordrecht: Kluwer Academic Publishers), pp. 708.

Heijungs, R. & Guinée, J.B. 2012. An Overview of the Life Cycle Assessment Method—Past, Present, and Future. In: Curran, M.A. (Ed.), Life Cycle Assessment Handbook: A Guide for Environmentally Sustainable Products, pp. 15–42. Beverly: Scrivener Publishing.

Helling, R. & Parenti, V. 2013. *Life Cycle Assessment, ENERG-ICE, a new polyurethane foam technology for the cold appliance industry*. Commissioned by the Dow Italia & Dow Polyurethanes business unit with the support of LIFE08 ENV/IT 000411.

Hischier, R. 2007. *Life Cycle Inventories of Packaging and Graphical Paper*. Ecoinvent report N° 11. Swiss Centre for Life Cycle Inventories, Dubendorf.

Hu, S., Wan, C. & Li, Y. 2012. Production and characterization of biopolyols and polyurethane foams from crude glycerol based liquefaction of soybean straw. *Bioresource Technology* 103: 227–233.

Hu, S., Luo, X. & Li, Y. 2014. Polyols and Polyurethanes from the Liquefaction of Lignocellulosic Biomass. *ChemSusChem MiniReview* 7: 66–72.

ISO. 2006a. ISO 14040:2006. Environmental management – Life cycle assessment–principles and framework. International Standard Organisation (eds), Geneva, Switzerland, pp. 28.

ISO. 2006b. ISO 14044:2006. Environmental management-Life cycle assessment-requirements and guidelines. International Standard Organisation (eds), Geneva, Switzerland, pp. 46.

Pergola, M., Gialdini, A., Celano, G., Basile, M., Caniani, D., Cozzi, M., Gentilesca, T., Mancini, I., Pastore, V., Romano, S., Ventura, G. & Ripullone, F. 2017. An environmental and economic analysis of the wood-pellet chain: two case studies in Southern Italy. *Int J Life Cycle Assessment* 23(8):1675–1684.

PRé 2018. *SimaPro Software*, version 8.5.2.0. PRé Consultants. Web site. [Online 20 October 2018] URL: https://simapro.com/.

Suter, F., Steubing, B. & Hellweg, S. 2017. Life Cycle Impacts and Benefits of Wood along the Value Chain: The Case of Switzerland. *Journal of Industrial Ecology* 21(4):874–886.

Wastes: Solutions, Treatments and Opportunities III – Vilarinho et al. (Eds)
 ISBN 978-0-367-25777-4

Stabilization of active acetylene by-product via sequestration of CO_2

M. El Gamal, A.M.O. Mohamed & S. Hameedi
Zayed University, Abu Dhabi, United Arab Emirates

ABSTRACT: The rising rates of carbide lime waste (CLW), the by-product of acetylene production, cause environmental problems and need for the repurposing of the said material. The high content of calcium in such by-products (60–80%) is the main reason for its utilization in CO_2 sequestration. This study focused on using of CLW to form simple, reliable, cost-efficient process, applying direct solid-liquid-gas carbonation reaction, which consists of bubbling CO_2 gas through an aqueous slurry of slaked CLW with different solid /water ratios. The experimental conditions were adjusted for maximum carbonation efficiency. Experimental results indicated that CLW has the potential to store CO_2 in the form of stable carbonates. Considering the total calcium content, an amount of 0.5833 kg CO_2/kg CLW was achieved. The influence of solid/water ratio on the precipitation of calcium carbonate was investigated using SEM that showed a difference in crystal carbonate sizes and morphologies.

1 INTRODUCTION

Carbide lime is the waste by-product generated in the commercial production of acetylene gas by reacting calcium carbide with water. Slurry of calcium hydroxide is generated as a by-product of acetylene manufacturing process. In a day's production, an average of 1 ton of waste is generated as per the report of middle east gases association. This lime paste is not commonly used for building or industrial purposes, and it is usually stored in ponds as a waste, thus representing a problem for the manufacturer. The final products are suspended in water. Thus, when calcium hydroxide mixes with water, it produces CLW. Once, the water is evaporated or dried, the lime slurry produced is in solid form. This makes disposal tedious and expensive. Hence, utilization of lime slurry in other industrial applications has proved to be beneficial for both the suppliers and users. Presently, the challenge is to develop and improve technologies to reduce the concentration of greenhouse gases in the atmosphere. The use of such a waste in the uptake of toxic or greenhouse gases such as CO_2 would therefore represent an alternative of significant value for its recycling. Mineral carbonation is a potentially attractive sequestration technology for the permanent and safe storage of CO_2 (Olajire, 2013).

Mineral carbonation technology is a process whereby CO_2 is chemically reacted with calcium- and/or magnesium-containing minerals to form stable carbonate materials which do not incur any long-term liability or monitoring commitments. Carbonation is already a well-known process, but a great deal of research is necessary to secure cost-effective core technologies. This requires research for the development of quick, cost-effective methods for combining carbon dioxide with magnesium or calcium compounds from rock or alkaline industrial wastes. An attempt to speed up carbonation include the use of both dry and wet methods, additives, heating and pressurizing the reactor, dividing the process into multiple steps, and pretreatment of the mineral source (Mohamed and El Gamal, 2014; El-Naas *et al.*, 2015; Mohamed and El-Gamal, 2011; Mohamed, El-Gamal and Hameedi, 2018).

Binding carbon dioxide in carbonates can be achieved through various process routes as two main methodologies: direct carbonation and indirect carbonation (Zevenhoven, Fagerlund and Songok, 2011) Direct carbonation is the simplest approach to mineral carbonation and the principal approach is that a suitable feedstock, e.g. Ca/Mg rich solid residue is carbonated in a single process step. For an aqueous process this means that both the extraction of metals from the feedstock and the subsequent reaction with the dissolved carbon dioxide to form carbonates takes place in the same reactor. Direct aqueous carbonation, in the direct aqueous mineral carbonation-route, the carbonation preformed in a single step in an aqueous solution, appears to be the most promising CO_2 mineralization alternative to date. High carbonation degrees and acceptable rates have been achieved but the process is still too expensive to be applied on a larger scale (Huijgen and Comans, 2005). Direct aqueous mineral carbonation can be further divided into two subcategories (two-step and three-step alternatives are considered under indirect carbonation routes, depending on the type of solution used. Indirect carbonation processes allow separation of silica or other by-products, such as metals and minerals, before the carbonation step (Pan, 2012). Indirect carbonation is therefore a better alternative for producing separate streams of carbonates and other materials for further recovery. If the process of mineral carbonation is divided in to several steps, it is classified as indirect carbonation. In other words, indirect carbonation means that the reactive component (usually Mg or Ca) is first extracted from the feedstock (as oxide or hydroxide) in one step and then, in another step, it is reacted with carbon dioxide to form the desired carbonates.

In this study, CLW residues generated in the manufacture of acetylene from calcium carbide have been used. A process for chemical conversion to purified precipitated calcium carbonate powder has been proposed to evaluate the capture of CO_2 by using CLW through different carbonation pathway and conditions.

2 MATERIALS

A carbide lime sludge from a "wet-process" storage pond, with 35% moisture content, is the selected CLW in the form of sludge generated in the industrial manufacture of acetylene (C_2H_2) from calcium carbide (CaC_2). The fundamental component of which is calcium hydroxide ($Ca(OH)_2$) in highly reactive formations. The raw material Carbide lime waste is a grey-black substance having an initial composition of 83% calcium hydroxide, 11% calcium carbonate and 6% silica/iron/alumina/carbon and magnesium oxide impurities. Characterization of carbide lime waste was undertaken, to estimate their mineralogical properties, average chemical composition, and particle size distribution. It was supplied by the Industrial Gas Plants, acetylene factory located in Abu Dhabi, UAE.

3 METHODOLOGY

As the carbide lime being in the form of a sludge or slurry of solid waste-product from the manufacture of acetylene, such process comprising the following steps: (i) screening carbide lime; (ii) drying the carbide lime to a moisture content of less than 0.5% by weight; (iii) fractionization of the dried powder to obtain a particle size between 300-38 microns, and (iv) dissolution of calcium hydroxide in water followed by carbonation in solid-liquid-gas carbonation system. Mineral carbonation was processed by direct carbonation through gas-solid-liquid carbonation reactions. Effective design and experimental set up were performed in a bench-scale glass reactor to achieve the optimum condition for the highest carbonation production. The effect of different parameters like solid/water ratio, flow rate, and time on the capture of CO_2 were studied to increase the carbonation efficiency and products quality. The effect of various process parameters was tested using material characterization on pre- and post-carbonated samples to enhance carbonation pathways and products. Evaluation of the degree of sequestration of CO_2 by corresponding analytical and empirical methods was undertaken. Effect of solid-liquid ratio on carbonation reaction was studied by measuring of CO_2 consumed, CO_2 captured, saturation time, pH, conductivity, TGA, XRD and SEM for generating carbonates with high economic value.

Carbide Lime samples before and after carbonation were analyzed by the following instruments: (i) Multi-parameter meter, MYRON-6PFC model to measure conductivity, total dissolved solid

Table 1. Effect of solid water ratio on the applied flow rate, carbonation time and the captured CO_2.

Run	Solid/water ratio	Flow rate [L/min]	Saturation Time [min]	CO_2 Captured [kg CO_2/kg CLW]	Loss on Ignition [Wt. %]
1	5:10	1.4	110	0.2329	30.8
2	4:10	1.2	105	0.2475	36.7
3	3:10	1.1	100	0.2929	39.5
4	2:10	1.0	95	0.3860	42.8
5	1:10	0.8	90	0.5833	34.3

(TDS) and pH (ii) X-Ray Diffractometer, XRD-6100, manufactured by Shimadzu to determine the minerals existed. (iii) California gas analyzer instrument (CAI-600 series) for detecting and measuring the concentration of the captured CO_2. It measures the reading of CO_2 gas from the carbonation reactor outlet, where it is the way to calculate the amount of CO_2 captured by the integration of CO_2 flow rate versus time curve. (vi) Inductively Coupled Plasma-Optical Emission Spectrometry (ICP-OES) ICP-OES, Avio 200, manufactured by Perkin Elmer to analyzes the chemical composition of liquid phase samples. (v) Scanning Electron Microscope, VEGA3, manufactured by TESCAN for microstructure characterization. (vii) Simultaneous Thermal Analyzer, STA 6000, Perkin Elmer, for measuring the decomposition of the sample, using heating rate range (10–20°C/min).

4 RESULTS AND DISCUSSION

Carbonation of CLW was performed by solid-liquid-gas technique, which consists of bubbling CO_2 gas through an aqueous slurry of slaked CLW with different solid /water ratios. In the carbonation process, solid $Ca(OH)_2$ is first dissolved as soluble Ca^{2+} and OH^- ions, CO_2 is absorbed in water to procedure carbonic acid, which ionized to H^+, HCO_3^- and CO_3^{2-}. After that, Ca^{2+} interact with CO_3^{2-} to form $CaCO_3$. The most significant parameter observed from these experimental was the flow rate and time needed to reach saturation. It is clearly found that, the lower solid to water ratio led to decrease in the applied flow rate due to the low viscosity and high diffusivity of CO_2 that resulted in enhanced the carbonation reaction and consequently decrease the saturation time, where no furthermore amount of CO_2 captured by the CLW (see Table 1). Same findings were reported by Thriveni et al. (2014), where they explained the important role of water in facilitating the carbonation process. (Thriveni *et al.*, 2014) The electrostatic forces of water molecules, polarize CO_2 molecules, increasing their ability to penetrate the water phase.

Carbonation process at a time during the reaction was monitored by the gas analyzer. The captured CO_2 by CLW was calculated using gas analyzer which is connected to the carbonation reactor, by plotting the instantaneous CO_2 concentration difference between inlet constant CO_2 concentration and the outlet ones versus time. Area under the curve of the plot represents numerically the amount of CO_2 consumed (captured) during carbonation reaction. as represented in Figure 1.

Different solid water ratios affect the carbonation reaction and the characteristics of produced PCC, like the particulate sizes and crystal morphology. At low initial concentrations, the solid $Ca(OH)_2$ is rapidly dissolved and exhausted in a short period of time throughout calcium carbonate formation. However, morphology and other crystal characteristic of resulted PCC would not be the identical.

Figure 2 indicates that higher calcium hydroxide contents yielded lower carbonate as compared to the concentrated one. The rate-controlling step might be the dissolution of $Ca(OH)_2$ at the water-adsorbed surface. The polymorphs of crystalline calcium carbonate particles depend mainly on precipitation conditions such as flow rate and solid concentration are the most important factors for the type of precipitated calcite (Han *et al.*, 2005).

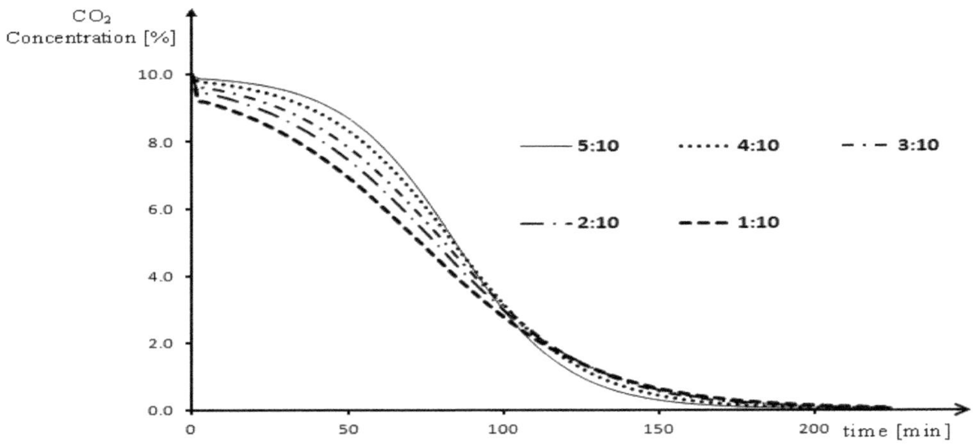

Figure 1. Percentage of CO_2 captured by CLW at different solid water ratios using Gas analyzer.

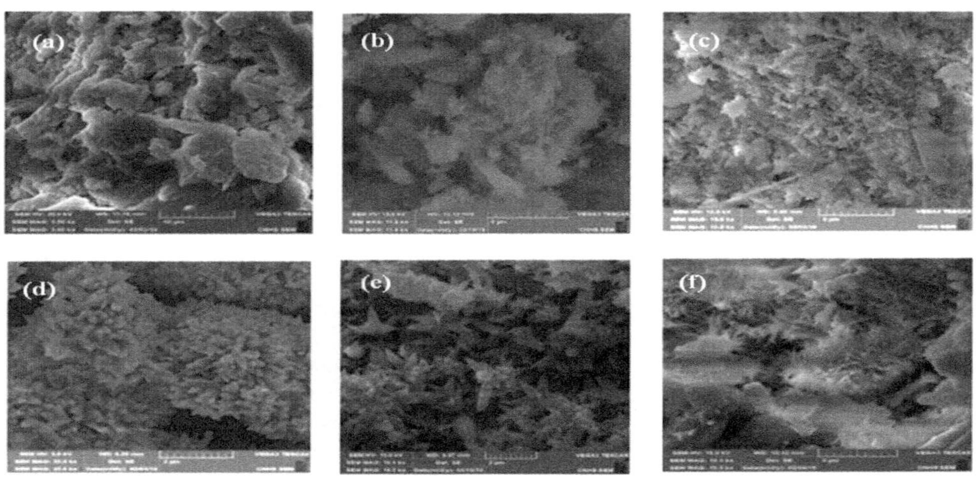

Figure 2. SEM images of (a) fresh and carbonated CLW with different solid/water ratios: (b) 5:10 (c) 4:10 (d) 3:10 (e) 2:10 and (f) 1:10.

Calcite with these morphologies is used for specific applications, the production of stable homogeneous nanosized particles.

During precipitation, the pH, conductivity and TDS values gradually decreases due to the growth of calcium carbonate. Dissolution of $Ca(OH)_2$ and CO_2 increases the concentration of ionic charges dissolved species, increase the formation of solid carbonates, consequently reduces the electro kinetic ability and caused to decrease in conductivity. The precipitation of the pure polymorphs of $CaCO_3$ has been examined by using SEM. Pure calcite formed at all concentrations which are confirmed by XRD and TGA patterns (Figures 3 and 4). To investigate the carbonation efficiency and purity of the produced solid, the qualitative comparison of x-ray diffraction spectra for fresh and carbonated CLW at different solid/water ratios were performed and the conversion of $Ca(OH)_2$ to $CaCO_3$ were revealed. The fresh sample contained mainly $Ca(OH)_2$, whereas the carbonated ones contained mainly the calcite component as indicated by the increase of the absolute intensities of calcite lines in the XRD diffractogram. This indicated that the carbonation forces the precipitation

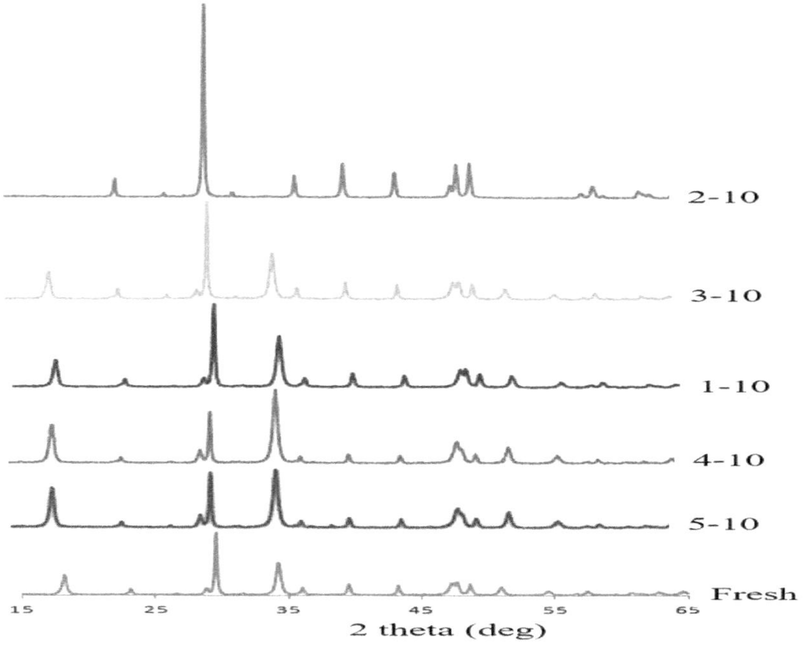

Figure 3. XRD data.

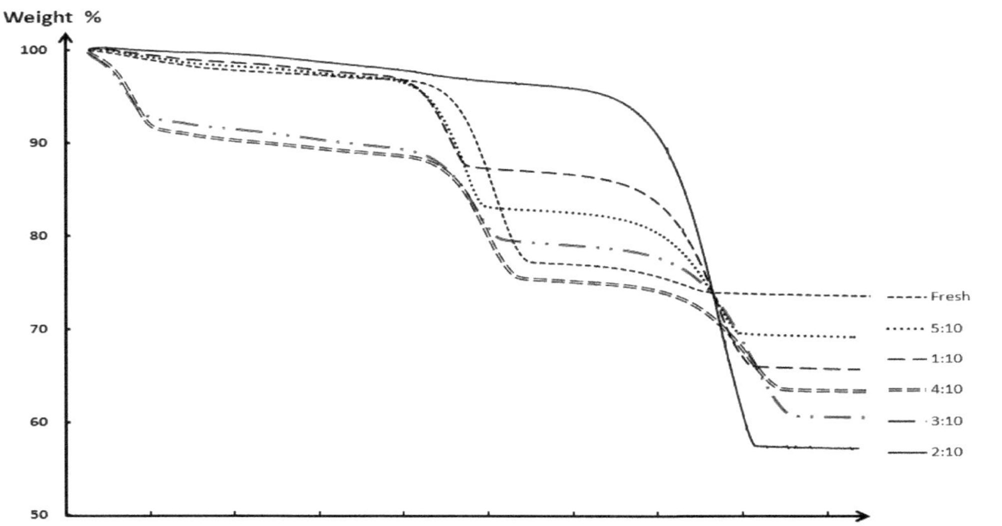

Figure 4. TGA data.

of calcium carbonates and ties up the available calcium present in the calcium carbide waste as signified in Figure 3.

The thermal behavior of the produced calcium carbonates was studied by TGA analyzer. The mass loss corresponds to the released amount of CO_2 during the heating process above the dissociation temperature of limestone (from 400–950°C) was calculated (Table 1 and Figure 4). It is indicated that higher conversion of $Ca(OH)_2$ into $CaCO_3$ is obtained at the lower solid to water ratio. This supports the idea that higher carbonic acid concentration encourages mineral dissolution and rapidly encourages precipitation of carbonates (Santos *et al.*, 2016).

5 CONCLUSION

This work investigates the CO_2 capturing capacity using CLW residues. Solid-liquid gas carbonation route has been used through bubbling CO_2 gas in an aqueous slurry of slaked CLW. The carbonation efficiency is highly affected by the carbonation condition. Different solid to water ratios were prepared and the corresponding flow rates were adjusted. carbonation time and the characteristics of the produced carbonates, like the particle sizes and crystal morphology were evaluated. The obtained results indicated that the carbonation of CLW could be a controlled system for producing fine-particles of calcite on an industrial scale. The amount of CO_2 captured per kg of CLW is 0.5833 kg, which proves the superiority of CLW for effective CO_2 sequestration. The proposed method has proven to be an efficient and productive technology for sequestration of CO_2 and stabilization of the active waste material.

REFERENCES

El-Naas, M. H., El Gamal, M., Hameedi, S. and Mohamed, A.-M. O. 2015. 'CO2 sequestration using accelerated gas-solid carbonation of pre-treated EAF steel-making bag house dust', *Journal of Environmental Management* 156: 218–224.

Han, Y. S., Hadiko, G., Fuji, M. and Takahashi, M. 2005. 'Effect of flow rate and CO2 content on the phase and morphology of CaCO3 prepared by bubbling method', *Journal of Crystal Growth* 276(3–4): 541–548.

Huijgen, W. J. J. and Comans, R. N. J. 2005. 'Mineral CO2 sequestration by carbonation of industrial residues: Literature review and selection of residue', *Energy Research Centre of the Netherlands ECN-C–05-074: Petten, The Netherlands*.

Mohamed, A.-M. O. and El Gamal, M. M. 2014. Method for treating particulate material. Google Patents.

Mohamed, A.-M. O., El-Gamal, M. and Hameedi, S. 'Advanced Mineral Carbonation: An Approach to Accelerate CO2 Sequestration Using Steel Production Wastes and Integrated Fluidized Bed Reactor'. *International Symposium on Energy Geotechnics*, Lausanne, Switzerland, 2018: Springer, 387–393.

Mohamed, A.-M. O. and El-Gamal, M. M. 2011. Method for treating cement kiln dust. Google Patents.

Olajire, A. A. 2013. 'A review of mineral carbonation technology in sequestration of CO2', *Journal of Petroleum Science and Engineering* 109(0): 364–392.

Pan, S.-Y. a. C. E. E. a. C. P.-C. 2012. 'CO2 Capture by Accelerated Carbonation of Alkaline Wastes: A Review on Its Principles and Applications', *Aerosol and Air Quality Research* 12(5): 770–791.

Santos, R. M., Thijs, J., Georgakopoulos, E., Chiang, Y. W., Creemers, A. and Van Gerven, T. 2016. 'Improving the Yield of Sonochemical Precipitated Aragonite Synthesis by Scaling up Intensified Conditions', *Chemical Engineering Communications* 203(12): 1671–1680.

Thriveni, T., Um, N., Nam, S.-Y., Ahn, Y. J., Han, C. and Ahn, J. W. 2014. 'Factors affecting the crystal growth of scalenohedral calcite by a carbonation process' *Journal of the Korean Ceramic Society* 51(2): 107–114.

Zevenhoven, R., Fagerlund, J. and Songok, J. K. 2011 'CO2 mineral sequestration: developments toward large-scale application', *Greenhouse Gases: Science and Technology* 1(1): 48–57.

Wastes: Solutions, Treatments and Opportunities III – Vilarinho et al. (Eds)
 ISBN 978-0-367-25777-4

Acetalization of glycerol over silicotungstic acid supported in silica

J.E. Castanheiro & F.I. Santos
Centro de Química de Évora, Departamento de Química, Escola das Ciências e Tecnologia, Universidade de Évora, Évora, Portugal

A.P. Pinto
Institute of Mediterranean Agricultural and Environmental Sciences, Departamento de Química, Escola das Ciências e Tecnologia, Universidade de Évora, Évora, Portugal

M.E. Lopes
HERCULES, Departamento de Química, Escola das Ciências e Tecnologia, Universidade de Évora, Évora, Portugal

ABSTRACT: The acetalization of glycerol with butanal was carried out in the presence of silicotungstic acid (HSiW) immobilized in silica, by sol-gel technique, at 70°C. The products of glycerol acetalization with butanal were (Z + E)-(2-propyl-1,3-dioxolan-4-yl) methanol and (Z + E)-2-propyl-1,3-dioxan-5-ol. Catalysts with different amount of HSiW were prepared. The catalysts were characterized by FTIR, XRD, ICP and N_2 adsorption. It was observed that the catalytic activity increases with the amount of HSiW in silica, being the HSiW2@S (with 5.7 wt. %) the most active material. All catalysts showed high values of selectivity to (Z + E)-(2-propyl-1,3-dioxolan-4-yl) methanol (about 78–83%). As the catalyst HSiW2@S exhibited the highest activity, it was also used in acetalization of glycerol with hexanal, heptanal and octanal. A decrease of catalytic activity with the length of aldehyde was observed. Catalytic stability of the HSiW2@S material was studied. After the third use, the catalytic activity stabilized.

1 INTRODUCTION

In the last years, the biodiesel production has been increased. For every 9 kg of biodiesel produced, about 1 kg of a crude glycerol is formed. In order to develop new uses for glycerol different catalytic processes, including oxidation, reforming, hydrogenolysis, etherification and esterification, acetalization reactions, have been used in the transformation of glycerol (Avhad et al. 2016, Chai et al. 2009, Deutsch et al. 2007, Lopes et al. 2015). Another catalytic process to valorization of glycerol is the condensation of glycerol with butanal, which provides a branched oxygen-containing compound and could be used as an additive in the biodiesel formulation, improving the cold properties and lowering the viscosity (Silva et al. 2010, Trifoi et al. 2016).

Traditionally, the acetalization of glycerol with butanal is carried out over mineral acids, as catalysts. However, the effluent disposal leads to environmental problems and economical inconveniences. These problems can be overcome using heterogeneous catalysts. The acetalization of glycerol has been carried out in the presence of some heterogenous catalysts (Kong et al. 2016, Cornejo et al. 2017, Carlota et al. 2018). Heteropolyacids (HPAs) are typical strong Brönsted acids, which catalyse a wide range of reactions. The major disadvantages of HPAs as catalysts are low surface area, separation problem from reaction mixtures and solubility. In order to increase the specific area of the heteropolyacids, a variety of supports have been used as support to immobilize HPAs (Patel et al. 2016, Narkhede et al. 2016).

In this work, we studied the synthesis of biofuel additives by acetalization of glycerol with butanal in the presence of silicotungstic acid (HSiW) immobilized in silica. The catalyst with the highest activity was also used in acetalization of glycerol with hexanal, heptanal and octanal.

2 EXPERIMENTAL

2.1 *Catalysts preparation*

Silicotungstic acid (HSiW) was immobilized in silica by sol-gel method, using a similar procedure developed by Y. Izumi et al. 1999. Initially, it is prepared a mixture with 2.0 mol of water with 0.2 mol of 1-butanol and different amount of HSiW. The resultant mixture is stirred during 1h at room temperature. After this period, it is added 0.2 mol of tetraethyl orthosilicate to mixture. The resultant mixture is again stirred during 3h at 80°C. The solid obtained is dried at 80°C for 2 h. This operation is performed under vacuum. All catalysts are washed with methanol using a continuous extraction (by a soxhlet apparatus) during 72 h. To conclude the preparation of catalysts, they are dried at 100°C, during 24 h. Silica was also prepared by the same process described above. However, the heteropolyacid was not added to the synthesis.

2.2 *Catalysts characterization*

BET surface area and total porous volume was determined with base on the nitrogen adsorption isotherm at 77 K, using a Micromeritics ASAP 2010 apparatus.

The amount of heteropolyacids in catalyst was determined using ICP, which was carried out in a Jobin-Yvon ULTIMA instrument.

FTIR spectra were recorded on a Perkin Elmer Spectrum FTIR spectrometer. It was prepared KBr pellets and the analysis was carried at room temperature over range of 400–4000 cm^{-1}.

A Rigaku Miniflex powder diffractometer was used to obtain the X-ray diffraction (XRD) patterns of the catalysts.

The material acidity was measured by means of potentiometric titration. It was used a Crison micropH 2001 instrument. The procedure used was developed by Pizzio et al. 2003.

2.3 *Experimental*

A 0.09 mol of aldehyde (butanal, hexanal, heptanal and octanal), 0.04 mol of glycerol and 0.3 g of catalyst was added to stirred reactor. The reaction was carried out at 70°C and atmospheric pressure. Catalytic stability was also studied. Different experiments (four consecutive experiments) was carried out.

The catalyst sample and reaction conditions used in catalytic experiments were the same. Between the catalytic experiments, the catalyst was washed with acetone. After this operation the catalyst is dried at 120°C, during 24 h.

Samples were taken periodically and analyzed by GC, using a Hewlett Packard instrument equipped with a 30 m × 0.25 mm DB-1 column.

3 RESULTS AND DISCUSSION

3.1 *Characterization of catalysts*

The textural characterization of materials is shown on Table 1. The specific surface area (S_{BET}) of the materials was determined using the BET method. The total pore volume (V_T) estimated from the value corresponding to a ratio $p/p° = 0.98$. It was observed that the surface area and total porous volume decreases with the amount of heteropolyacid immobilized in silica. Izumi et al. 1999, observed similar results. The acidity of the silica and HSiW@S materials was measurement by potentiometric titrations. The initial electrode potential (Ei) indicates the maximum acid strength of the surface sites and materials with an Ei > 100 mV can be classified with very strong sites, according to Pizzio, et al. 2003. It was observed that the Ei of materials increased with the amount of HSiW in silica (Table 1). This behavior can be due to the increases of protons amount in the material.

FTIR spectra of Silica and HSiW2@S materials are shown in Figure 1. The bands at 982, 917 and 784 cm-1, which are the same as those reported for the acid $H_4SiW_{12}O_{40}$, corresponding to

Table 1. Physicochemical characterization of HSiW immobilized in silica by sol-gel technique.

Sample	HPA load[a] (wt%)	Surface area[b] (m^2/g)	Vmic[c] (cm^3/g)	V_T^d (cm^3/g)	E_i^e(mV)
Silica	–	221	0.02	0.47	+45
HSiW1@S	0.020	495	0.27	0.27	+115
HSiW2@S	0.057	332	0.16	0.16	+137
HSiW3@S	0.072	284	0.14	0.14	+189
HSiW4@S	0.085	234	0.10	0.11	+278

(a) HSiW load determined by ICP; (b) *BET;* (c) t-method; (d) (p/p°) = 0.98; (e) Initial electrode potential

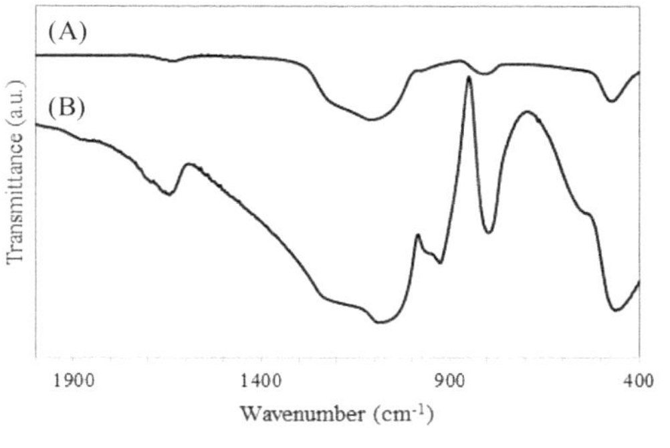

Figure 1. FTIR spectra of catalyst: (A) silica and (B) HSiW2@S.

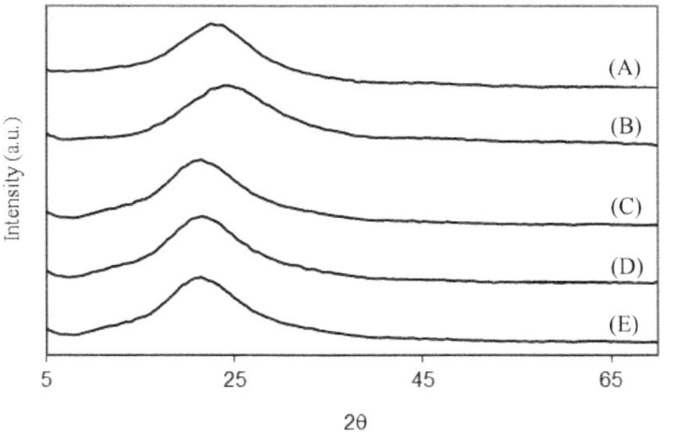

Figure 2. X-ray diffractograms of materials: (A) silica, (B) HSiW1@S, (C) HSiW2@S, (D) HSiW3@S and (E) HSiW3@S.

the vibrations W = O (terminal oxygen), Si-O and W-O-W (edge-sharing oxygen), respectively (Bielanski et al. 2003).

Figure 2 shows the X-ray diffraction patterns of catalysts. It can be observed that any crystalline phases related to HSiW are not present, which can be explained due to HSiW particles are too small. Molnár et al. 1999, observed this behavior with heteropolyacids dispersed in silica.

3.2 *Catalytic experiments*

The acetalization of glycerol with butanal was performed in the presence of HSiW immobilized in silica. The main product of glycerol acetalization was (Z + E)-(2-propyl-1,3-dioxolan-4-yl)

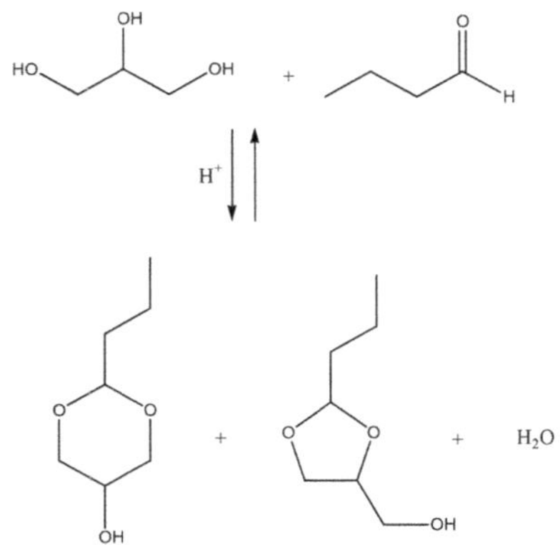

Figure 3. Acid-catalyzed acetalization of glycerol with butanal.

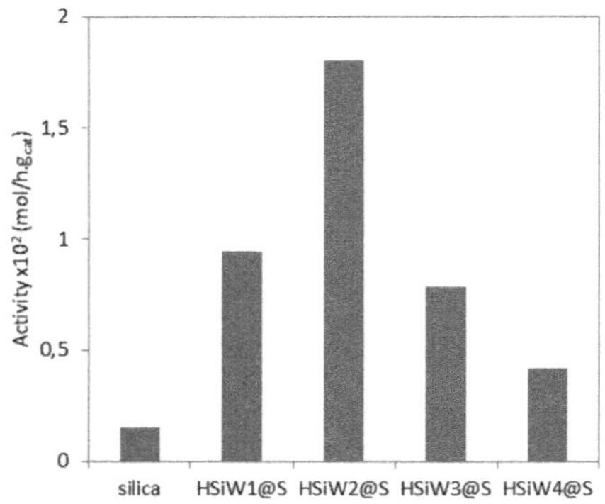

Figure 4. Acetalization of glycerol with butanal over HSiW immobilized in Silica.

methanol (five-member ring compound) being also formed and (Z + E)-2-propyl-1,3-dioxan-5-ol (six-member ring compound), which can be used as biofuel additives (Fig. 3).

Figure 4 compares the initial activity of silica, HSiW1@S, HSiW2@S, HSiW3@S and HSiW4@S. It is observed that catalytic activity increases with the amount of HSiW included in silica until a maximum, which can be explained by the increase of number of active sites (due to the increase of HSiW amount in silica). After this activity maximum, a decrease of the catalytic activity is observed when the amount of HSiW increase, which can be due to internal diffusion limitations in porous system of HSiW@S materials. In fact, it is observed a decrease of S_{BET}, microporous volume and total porous volume with the amount of heteropolyacid on silica (Table 1). Probably, all acid sites are not accessibility to the reactants.

Table 2 shows the glycerol conversion and the selectivity to the products obtained by the acetalization of glycerol with butanal over heteropolyacids in silica, after 5 hours of the reaction. The sample HSiW2@S exhibits the highest conversion. After 5 hours of reaction, the glycerol conversion was 80%, with a selectivity of 78% to five-member ring acetal and 22% to six-member ring acetal. The high selectivity to five-member ring acetal can be explained due to the five-member ring compound (1,3-dioxolane) is favored kinetically. Similar results were also observed by Umbarkar et al., 2009 and Silva et al. 2010.

The acetalization of glycerol is also performed with aldehydes of different length (hexanal, heptanal and octanal). Figure 5 compares the initial activity of HSiW2@S in different reaction of acetalization of glycerol. It is observed that the catalytic activity of HSiW2@S decrease with the increase of aldehydes' length. This behavior can be explained due to the presence of some diffusion

Table 2. Conversion and selectivity to the glycerol acetalization products with butanal over HSiW@S.

Sample	Conversion[1] (%)	Selectivity (%) five-member ring compound	Selectivity (%) six-member ring compound
Silica	22	86	14
HSiW1@S	67	80	20
HSiW2@S	80	78	22
HSiW3@S	50	81	19
HSiW4@S	38	83	17

[1] glycerol conversion after 5 h of reaction

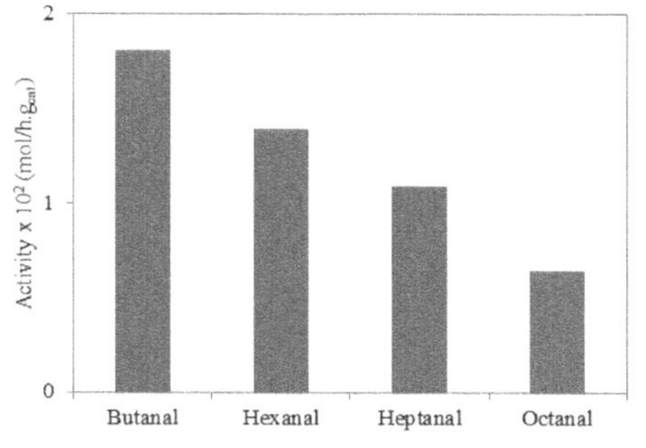

Figure 5. Acetalization of glycerol with different aldehydes HSiW2@S catalysts.

Table 3. Conversion and selectivity to the glycerol acetalization products over HSiW2@S.

Sample	Conversion[1] (%)	Selectivity (%) five-member ring compound	Selectivity (%) six-member ring compound
Butanal	80	78	22
Hexanal	67	73	27
Heptanal	59	63	37
Octanal	45	61	39

[1] glycerol conversion after 5 h of reaction

limitations inside the porous system of material. Silva et al., 2010, studied the acetalization of glycerol with different aldehydes in the presence of Amberlyst-15. It was observed, at a fixed time, the glycerol conversion decreased with the increase of aldehyde carbon chain (Silva et al., 2010). After 5 h of reaction, it was observed that the glycerol conversion (%) over HSiW2@S is 80%, 67%, 59% and 45% for the butanal, hexanal, heptanal and octanal, respectively.

Table 3 shows the selectivity to the five-member ring compound and six-member ring compound obtained by acetalization of glycerol with pentanal, hexanal, heptanal and octanal over HSiW2@S, after 5 hours of the reaction. High selectivity to five-member ring compound is also observed.

The catalytic stability of HSiW2@S was studied by consecutive batch runs with the same catalyst sample. A small decrease of the catalytic activity was observed (Fig. 6).

4 CONCLUSIONS

Silicotungstic acid immobilized in silica, with different amount of HSiW, was used as catalyst for the acetalization of glycerol with butanal. The catalyst HSiW2@S (with 0.057 g_{HSiW}/g_{silica}) showed

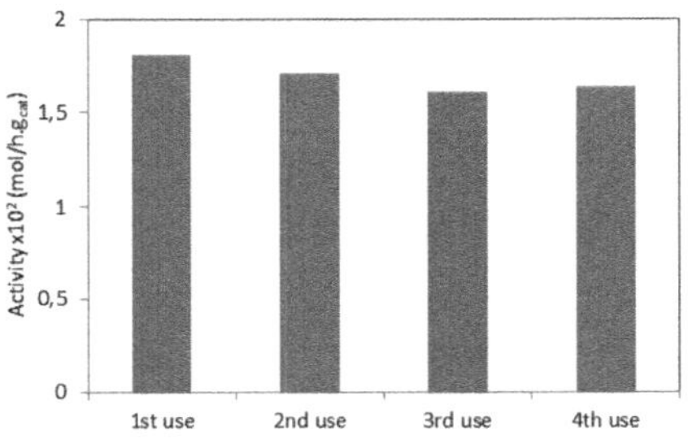

Figure 6. Catalytic activity of the HSiW2@S material on acetalization of glycerol with butanal. Study of stability.

the highest catalytic activity (1.8×10^{-2} mol/h.g_{cat}). All catalysts showed high values of selectivity to five-member ring compound (about 78–83%).

HSiW2@S catalyst was also used in acetalization of glycerol with hexanal, heptanal and octanal. It was observed that the catalytic activity decreased with the length of aldehyde's chain. HSiW2@S showed good catalytic stability. After the fourth use, it was observed that the catalytic activity decreased about 4% of its initial value.

REFERENCES

Avhad, M.R. & Marchetti, J.M. 2016. Innovation in solid heterogeneous catalysis for the generation of economically viable and ecofriendly biodiesel: a review. *Catal. Rev.* 58: 157–208.

Bielanski A., Lubanska, A., Pozniczek, J. & Micek-Ilnicka, A. 2003. The formation of MTBE on supported and unsupported $H_4SiW_{12}O_{40}$. *Appl. Catal. A:Gen.* 256: 153–171.

Cornejo, A., Barrio, I., Campoy, M., Lázaro, J. & Navarrete, B. 2017. Oxygenated fuel additives from glycerol valorization. Main production pathways and effects on fuel properties and engine performance: A critical review. *Renew. Sustainable Ener. Rev.* 79: 1400–1413.

Chai, S.-H., Wang, H.-P., Liang, Y. & Xu, B.-Q. 2009. Sustainable production of acrolein: Preparation and characterization of zirconia-supported 12-tungstophosphoric acid catalyst for gas-phase dehydration of glycerol. *Appl. Catal. A: Gen.* 353: 213–222.

Deutsch, J., Martin, A. & Lieske, H. 2007. Investigations on heterogeneously catalysed condensations of glycerol to cyclic acetals. *J. Catal.* 245: 428–435.

Izumi, Y., Hisano, K. & Hida, T. 1999. Acid catalysis of silica-included heteropolyacid in polar reaction media. *Appl. Catal. A:Gen.* 181: 277–282.

Kong, P.S., Aroua, M.K. & Daud, W.M.A.W. 2016. Conversion of crude and pure glycerol into derivatives: A feasibility evaluation. *Renew. Sustain. Energy Rev.* 63: 533–555.

Lopes, N.F., Caiado, M., Canhão, P. & Castanheiro J.E. 2015. Synthesis of bio-fuel additives from glycerol over poly(vinyl alcohol) with sulfonic acid groups. *Energ. Source Part A: Recovery, Utilization and Environmental Effects* 37: 1928–1936.

Molnár, A., Keresszegi, C. & Török, B. 1999. Heteropoly acids immobilized into a silica matrix: characterization and catalytic applications. *Appl. Catal. A:Gen.* 189: 217–224.

Narkhede, N., Singh, S. & Patel, A. 2015. Recent progress on supported polyoxometalates for biodiesel synthesis via esterification and transesterification. *Green Chem.* 17: 89–107.

Patel, A., Narkhede, N., Singh, S. & Pathan, S. 2016. Keggin-type lacunary and transition metal substituted polyoxometalates as heterogeneous catalysts: A recent progress. *Catal. Rev.* 58: 337–370.

Pizzio, L.R., Vásquez, P.G., Cáceres, C.V. & Blanco, M.N. 2003. Supported Keggin type heteropolycompounds for ecofriendly reactions. *Appl. Catal. A:Gen.* 256: 125–139.

Silva, P.H.R., Gonçalves, V.L.C. & Mota, C.J.A. 2010. Glycerol acetals as anti-freezing additives for biodiesel. *Bioresource Technol.* 101: 6225–6229.

Trifoi, A.R., Agachi, P.Ş. & Pap, T. 2016. Glycerol acetals and ketals as possible diesel additives. A review of their synthesis protocols. *Renew. Sustain. Energy Rev.* 62: 804–814.

Umbarkar, S.B., Kotbagi, T.V., Biradar, A.V., Pasrichab, R., Chanale, J., Dongare, M.K., Mamede, A.-S., Lancelot, C. & Payen, E. 2009. Acetalization of glycerol using mesoporous MoO_3/SiO_2 solid acid catalyst. *J. Mol. Catal. A: Chem.* 310: 150–158.

Wastes: Solutions, Treatments and Opportunities III – Vilarinho et al. (Eds)
© 2020 Taylor & Francis Group, London, ISBN 978-0-367-25777-4

Use of recycled C&D wastes in unpaved rural and forest roads – feasibility analysis

P.M. Pereira, F.B. Ferreira, C.S. Vieira & M.L. Lopes
CONSTRUCT-GEO, Faculty of Engineering, University of Porto, Porto, Portugal

ABSTRACT: In recent years, environmental sustainability concerns associated with over exploitation of natural resources along with the need to manage large volumes of wastes have propelled the development and implementation of waste recycling and valorisation strategies in the construction sector. In particular, Construction and Demolition (C&D) wastes have been recognised by the European Commission as a priority stream due to the large amounts generated in addition to their high potential to be reused and recycled. This paper presents the laboratory characterisation of two recycled aggregates from C&D waste to determine whether they meet the technical requirements established in the Portuguese specification LNEC E484 (2016), which would enable their application in unpaved rural and forest roads. A series of relevant physical, mechanical and chemical properties are evaluated and the fulfilment of the requirements is subsequently examined.

1 INTRODUCTION

Nowadays, waste generation and management is recognised as a key area of concern within the construction sector at international level. In fact, every year billions of tons of construction and demolition (C&W) wastes are generated worldwide from different activities, such as excavation, site preparation, construction, maintenance and demolition of buildings and other civil infrastructures. On the other hand, construction industry is responsible for as much as 50% of all materials extracted from the earth's crust (European Commission 2001). In this context, C&D waste recycling and valorisation is imperative and represents a step forward towards sustainable development, since it can contribute not only to more effective use of non-renewable natural resources, reducing the environmental impacts arising from their extraction, but also to attenuate waste disposal volumes to landfill (Arulrajah et al. 2013, Vieira & Pereira 2015).

In recent years, several studies have been conducted to assess the feasibility of using recycled C&D materials in a range of civil engineering applications, such as ground improvement works (Henzinger & Heyer 2018), backfilling of geosynthetic-reinforced structures (Santos et al. 2014, Pereira et al. 2015, Vieira et al. 2016, Pereira et al. 2017, Vieira & Pereira 2018), road construction (Arulrajah et al. 2013, Del Rey et al. 2016, Aboutalebi Esfahani 2018), pipe bedding and backfilling (Rahman et al. 2014, Vieira et al. 2017) and in concrete production (Silva et al. 2014). Most of these studies have yielded encouraging results, showing that the recycled C&D wastes can exhibit engineering properties equivalent or superior to those of typical quarry granular materials. However, the recycling rates of C&D wastes in many European countries, including Portugal, are still far below the target of 70% established by the Waste Framework Directive for 2020 (UE 2008).

Recently, the Portuguese Environment Agency in collaboration with the National Laboratory of Civil Engineering (LNEC) promoted the release of new technical specifications to regulate the use of materials from C&D waste in different civil engineering works, including the construction of rural and forest roads. In this study, a laboratory programme was carried out to evaluate a series of physical, mechanical and chemical properties of two recycled aggregates from C&D waste. The suitability of these materials as alternative aggregates for construction of unpaved rural and forest

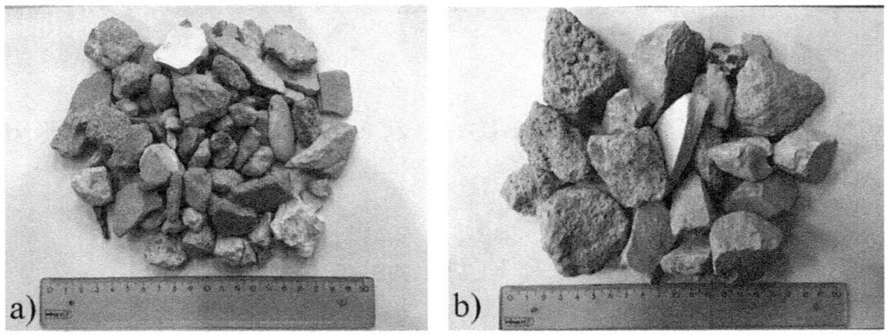

Figure 1. Recycled C&D wastes used (ruler in centimetres): a) non-selected material; b) selected material.

Table 1. Evaluation of constituents of the recycled C&D wastes.

Constituents	Non-selected C&D waste	Selected C&D waste
Concrete, concrete products, mortar, concrete masonry units, R_c (%)	31.9	43.2
Unbound aggregate, natural stone, hydraulically bound aggregate, R_u (%)	35.8	53.4
Clay masonry units, calcium silicate masonry units, aerated non-floating concrete, R_b (%)	20.7	2.0
Bituminous materials, R_a (%)	0.7	≈0
Glass, R_g (%)	0.2	≈0
Soils, R_s (%)	9.6	1.4
Other materials, X (%)	0.7	≈0
Floating particles, FL (cm^3/kg)	0.8	0.1

roads was then assessed, taking into account the requirements set out in the Portuguese specification LNEC E484 (2016).

2 LABORATORY STUDY

2.1 *General description and constituents*

Two coarse recycled aggregates from C&D waste were collected from a Portuguese recycling plant: a non-selected (mixed) aggregate (Figure 1a) and a selected aggregate from demolition of poor concrete (Figure 1b), which resulted from a selection process carried out by the supplier. The respective constituents were evaluated by hand sorting of particles, following the European standard EN 933-11 (2009). As shown in Table 1, the non-selected material consisted mainly of concrete and mortar debris, unbound aggregates and masonries, whereas the selected material was primarily composed of crushed concrete/mortar products and unbound aggregates.

2.2 *Physical and mechanical properties*

To determine the particle size distribution of the recycled aggregates, test samples were washed, dried and then sieved according to EN 933-1 (2012). Figure 2 shows the mean gradation curves (based on three tested samples) of the selected and non-selected C&D materials. The values of D_{max}, D_{50} and the percentage of fines are listed in Table 2.

The geometrical characterisation involved also the quantification of the flakiness and shape indexes, as per the EN 933-3 (2012) and EN 933-4 (2008) standards. The assessment of fines was performed through the sand equivalent test (EN 933-8:2012) as well as the methylene blue test (EN 933-9:2009), taking the 0/2 mm size fraction of the materials. The former test allows quantifying the

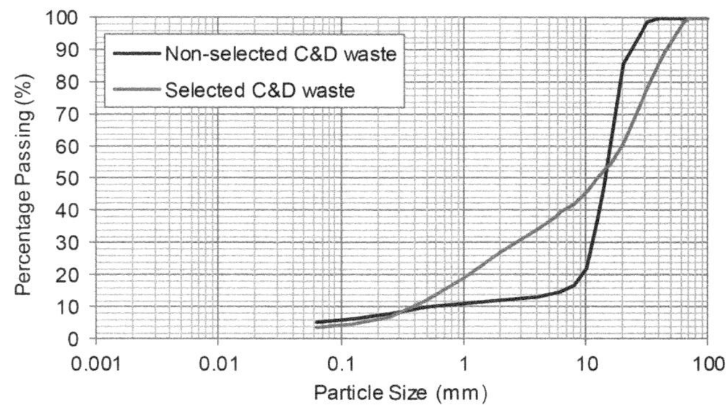

Figure 2. Particle size distribution of the recycled C&D wastes.

Table 2. Physical and mechanical properties of the recycled C&D wastes.

Parameter	Non-selected C&D waste	Selected C&D waste
D_{max} (mm)	31.5	63.0
D_{50} (mm)	14.4	12.6
Fines content (%)	5.1	3.6
Flakiness Index, FI (%)	19	9
Shape Index, SI (%)	27	9
Sand equivalent value, SE(10) (%)	19.7	47.4
Methylene blue value, MB (g/kg)	8.0 (1.0)*	0.7 (0.2)*
Los Angeles abrasion value (%)	47.6	35.7
Aggregate crushing value (%)	44.4	37.8

*The value in brackets is the value of MB multiplied by the percentage passing the 2 mm sieve ($MB_{0/D}$).

sand equivalent value (SE(10)), which gives an indication of the relative proportions of sand versus plastic fines and dust in the aggregates, whereas the latter yields the value of the methylene blue (MB), which is in function of the amount and characteristics of possibly harmful clay minerals. As shown in Table 2, a significantly higher value of MB was obtained for the non-selected C&D material (8.0 g/kg), in comparison with that for the selected C&D material (0.7 g/kg), which is directly related to the larger content of plastic fines stemming from the disintegration of clay masonry units.

The mechanical behaviour of the recycled aggregates was assessed through Los Angeles abrasion tests and aggregate crushing tests, following the recommendations of EN 1097-2 (2010) and BS 812-110 (1990), respectively. When compared with the selected C&D material, the non-selected material exhibits lower resistance to fragmentation and crushing, expressed by higher Los Angeles abrasion and crushing values (Table 2), possibly due to the presence of mortar attached to the aggregates surface and the significantly larger content of masonries. Particle density and water absorption tests were performed based on EN 1097-6 (2013), using different particle size fractions (Table 3). The values of particle density for the studied C&D materials are consistent with those typically obtained for natural aggregates. Regarding the water absorption, the values are generally higher than those of natural materials, which can be associated with the relatively high porosity of some of the constituents of the C&D materials, such as the clay masonries, mortar products and fine soils.

2.3 *Chemical properties*

The content of water soluble sulphates was estimated by spectrophotometry, according to Section 10 of the EN 1744-1 (2009). The specimens of the C&D materials were sieved through the 4 mm sieve and the retained particles were crushed to pass the same sieve. The specimens were mixed with hot water to extract water-soluble sulphate ions and barium chloride was then added so that sulphate

Table 3. Particle density and water absorption values of the recycled C&D wastes.

Parameters	Particle size fraction (mm)	Non-selected C&D waste	Selected C&D waste
Apparent particle density (Mg/m^3)		–	2.510 (0.2%)*
Oven-dried particle density (Mg/m^3)	31.5–63.0	–	2.257 (0.7%)
Saturated and surface dried particle density (Mg/m^3)		–	2.357 (0.4%)
Apparent particle density (Mg/m^3)		2.573 (0.5%)	2.623 (0.3%)
Oven-dried particle density (Mg/m^3)	4.0–31.5	2.205 (0.4%)	2.404 (0.5%)
Saturated and surface dried particle density (Mg/m^3)		2.348 (0.4%)	2.487 (0.4%)
Apparent particle density (Mg/m^3)		2.693 (0.9%)	2.619 (0.4%)
Oven-dried particle density (Mg/m^3)	0.063–4.0	2.506 (1.9%)	2.359 (0.8%)
Saturated and surface dried particle density (Mg/m^3)		2.575 (1.5%)	2.458 (0.6%)
Water absorption (%)	31.5–63.0	–	4.5 (7.7%)
	4.0–31.5	6.5 (0.9%)	3.5 (2.2%)
	0.063–4.0	2.8 (17.3%)	4.2 (4.7%)

*The value in brackets is the coefficient of variation.

Table 4. Classification of materials resulting from C&D waste (adapted from LNEC E484:2016).

	Proportion of constituents					
Material Class	$R_c + R_u + R_g$ (%)	R_g (%)	R_a (%)	$R_b + R_s$ (%)	F_L (cm^3/kg)	X (%)
CRA	No limit	≤25	No limit	No limit	≤5	≤1
CRB	≥20	≤5	≤80	≤10	≤5	≤1
CRC	≥50	≤5	≤30	≤10	≤5	≤1

R_c–Concrete and mortar; R_u–unbound aggregates; R_g–glass; R_a–bituminous materials; R_b–masonry; R_s–soil; F_L–floating particles; X–other constituents.

ions precipitate as barium sulphate. The mean values of the spectrophotometer records for the non-selected and selected C&D aggregates were 0.115% and 0.052%, respectively. Additionally, to evaluate the potential risk of ground contamination, the recycled materials were subjected to laboratory leaching tests following the standard EN 12457-4 (2002). The characterisation of the leaching behaviour of the C&D wastes (not described herein due to space constraints) showed that these materials could be accepted for inert landfill.

3 TECHNICAL REQUIREMENTS OF LNEC E484 SPECIFICATION

The Portuguese specification LNEC E484 (2016) classifies the materials resulting from C&D waste to be used in rural and forest roads into three different classes (CRA, CRB and CRC), which are defined based on the relative proportions of the constituents (Table 4).

Table 5 summarises the properties, test standards and corresponding minimum requirements that the C&D materials of each class listed in Table 4 should satisfy to be considered suitable for usage in rural and forest roads. The C&D materials, whose categories are presented in Table 5, can be applied in rural and forest roads according to the following criteria:

- Embankments for rural and forest roads – CR1, CR2, CR3 and CR4
- Subbase – CR2, CR3 and CR4
- Base course – CR3 and CR4
- Unsealed wearing course – CR4

Table 5. Properties and minimum requirements of C&D materials (adapted from LNEC E484:2016).

Compliance requirements			Material category			
			CR1	CR2	CR3	CR4
Parameters	Property	Standard	CRA, CRB, CRC	CRA, CRB, CRC	CRB, CRC	CRC
Geometry and nature	D_{max}	EN 933-1	$D_{max} \leq 180$ mm	$D_{max} \leq 80$ mm	$D_{max} \leq 40$ mm	$D_{max} \leq 40$ mm
	Fines content ($\leq$0.063 mm)	EN 933-1	–	$\leq$12%	$\leq$12%	$\leq$12%
	Assessment of fines ($MB_{0/D}$)	EN 933-9	–	$MB_{0/D} \leq 2.0$	$MB_{0/D} \leq 2.0$	$MB_{0/D} \leq 1.0$
Mechanical behaviour	Resistance to fragmentation (LA)	EN 1097-2	–	LA $\leq$ 50	LA $\leq$ 45 or MDE $\leq$ 45	LA $\leq$ 40 or MDE $\leq$ 40
	Resistance to wear (MDE)	EN 1097-1				
Chemical analysis	Water-soluble sulphates	EN 1744-1	$\leq$0.7%	$\leq$0.7%	$\leq$0.7%	$\leq$0.7%
	Release of dangerous substances	EN 12457-4	Classification as inert waste for landfill			

4 FEASIBILITY ANALYIS AND DISCUSSION

To assess the feasibility of using the studied C&D recycled aggregates in the construction of rural and forest roads, a comparison is made between the material properties (Section 2) and the corresponding requirements set out in the specification LNEC E484 (Section 3). From the comparison between the relative proportions of the constituents (Table 1) and the limits reported in Table 4, it becomes apparent that the non-selected C&D recycled aggregate belongs to the CRA class. It is worth noting that this material is excluded from CRB and CRC classes only because of the relatively high content of masonries and soil ($R_b + R_s$). On the other hand, the selected C&D material can be included in any of the three established classes (CRA, CRB and CRC).

Furthermore, taking into account the material properties described in Sections 2.2 and 2.3 and the minimum requirements summarised in Table 5, both the non-selected and selected C&D wastes can be included in category CR2, which means they can be considered as suitable aggregates for embankment construction as well as for the subbase layers of rural and forest roads. It is important to note that, if it was not for the high value of D_{max}, the selected C&D material could be accepted in the highest category (CR4), which would enable its application in any of the pavement layers, including the base course and unsealed wearing course. In the case of the non-selected C&D material, the property that hampered its classification in superior categories was the Los Angeles abrasion value. This finding is likely associated with the relatively high content of masonries, which detrimentally affected the resistance to fragmentation of this material. None of the C&D materials raised any concern in terms of chemical properties (i.e. water-soluble content and leaching behaviour).

5 CONCLUSIONS

This paper presented the laboratory evaluation of two recycled materials (selected and non-selected C&D wastes) and discussed their suitability as aggregates for construction of rural and forest roads, based on the recommendations of the specification LNEC E484 (2016). It was found that the studied C&D recycled aggregates not only presented suitable characteristics to be used in the construction of embankments, but they can also serve as alternative aggregates for the subbase layers of rural and forest roads. In fact, if it was not for the relatively high maximum particle size (D_{max}), the selected C&D waste would meet the requirements for any of the pavement layers, including the base course and unsealed wearing course of rural and forest roads. None of the C&D materials raised any environmental concerns.

ACKNOWLEDGMENTS

This work was financially supported by the Research Project CDW_LongTerm, POCI-01-0145-FEDER-030452, funded by FEDER funds through COMPETE2020 – Programa Operacional Competitividade e Internacionalização (POCI) and by national funds (PIDDAC) through FCT/MCTES and also by UID/ECI/04708/2019- CONSTRUCT – Instituto de I&D em Estruturas e Construções funded by national funds through the FCT/MCTES (PIDDAC).

REFERENCES

Aboutalebi Esfahani, M. 2018. Evaluating the feasibility, usability, and strength of recycled construction and demolition waste in base and subbase courses. *Road Materials and Pavement Design*: 1–23.

Arulrajah, A., Piratheepan, J., Disfani, M.M. & Bo, M.W. 2013. Geotechnical and geoenvironmental properties of recycled construction and demolition materials in pavement subbase applications. *Journal of Materials in Civil Engineering* 25(8): 1077–1088.

BS 812-110 1990. Testing aggregates. Methods for determination of aggregate crushing value (ACV). BSI.

Del Rey, I., Ayuso, J., Galvín, A.P., Jiménez, J.R. & Barbudo, A. 2016. Feasibility of using unbound mixed recycled aggregates from CDW over expansive clay subgrade in unpaved rural roads. *Materials,* 9(11).

EN 933-1 2012. Tests for geometrical properties of aggregates – Part 1: Determination of particle size distribution – Sieving method. CEN.

EN 933-3 2012. Tests for geometrical properties of aggregates – Part 4: Determination of particle shape. Flakiness index. CEN.

EN 933-4 2008. Tests for geometrical properties of aggregates – Part 3: Determination of particle shape. Shape index. CEN.

EN 933-8 2012. Tests for geometrical properties of aggregates – Part 8: Assessment of fines. Sand equivalent test. CEN.

EN 933-9 2009. Tests for geometrical properties of aggregates. Assessment of fines. Methylene blue test. CEN.

EN 933-11 2009. Tests for geometrical properties of aggregates – Part 11: Classification test for the constituents of coarse recycled aggregate. CEN.

EN 1097-1 2011. Tests for mechanical and physical properties of aggregates – Part 1: Determination of the resistance to wear (micro-Deval). CEN.

EN 1097-2 2010. Tests for mechanical and physical properties of aggregates – Part 2: Methods for the determination of resistance to fragmentation. CEN.

EN 1097-6 2013. Tests for mechanical and physical properties of aggregates – Part 6: Determination of particle density and water absorption. CEN.

EN 1744-1 2009. Tests for chemical properties of aggregates – Part 1: Chemical analysis. CEN.

EN 12457-4 2002. Characterisation of waste – Leaching – Compliance test for leaching of granular waste material and sludges – Part 4. CEN.

European Commission 2001. Competitiveness of the Construction Industry. A Report drawn up by the Working Group for Sustainable Construction with participants from the European Commission, Member States and Industry. European Commission, Brussels.

Henzinger, C. & Heyer, D. 2018. Soil improvement using recycled aggregates from demolition waste. *Proceedings of the Institution of Civil Engineers: Ground Improvement* 171(2): 74–81.

LNEC E484 2016. Guide for use of materials resulting from Construction and Demolition Waste in rural and forest roads. LNEC (Portuguese Laboratory of Civil Engineering). 8p. (in Portuguese).

Pereira, P.M., Vieira, C.S. & Lopes, M.L., 2015. Characterization of construction and demolition wastes (C&DW)/geogrid interfaces. 3rd Edition of the International Conference Wastes: Solutions, Treatments and Opportunities. CRC Press Taylor & Francis Group, Viana do Castelo, Portugal, pp. 215–220.

Pereira, P.M., Vieira, C.S. & Lopes, M.L., 2017 Damage induced by recycled C&D wastes on the short-term tensile behaviour of a geogrid, 4th Edition of the International Conference Wastes: Solutions, Treatments and Opportunities. CRC Press Taylor & Francis Group, Porto, Portugal, pp. 119–124.

Rahman, M.A., Imteaz, M., Arulrajah, A. & Disfani, M.M. 2014. Suitability of recycled construction and demolition aggregates as alternative pipe backfilling materials. *Journal of Cleaner Production* 66: 75–84.

Santos, E.C., Palmeira, E.M. & Bathurst, R.J. 2014. Performance of two geosynthetic reinforced walls with recycled construction waste backfill and constructed on collapsible ground. *Geosynthetics International* 21(4): 256–269.

Silva, R.V., De Brito, J. & Dhir, R.K. 2014. Properties and composition of recycled aggregates from construction and demolition waste suitable for concrete production. *Construction and Building Materials* 65: 201–217.

UE 2008. Directive 2008/98/EC of the European Parliament and of the Council of 19 November on waste and repealing certain Directives. *Official Journal of the European Union (L 312/3 of 22 November 2008).*

Vieira, C.S. & Pereira, P.M. 2015. Use of recycled construction and demolition materials in geotechnical applications: A review. *Resources, Conservation and Recycling* 103: 192–204.

Vieira, C.S. & Pereira, P.M. 2018. Use of Mixed Construction and Demolition Recycled Materials in Geosynthetic Reinforced Embankments. *Indian Geotechnical Journal* 48(2): 279–292.

Vieira, C.S., Pereira, P.M. & Lopes, M.D.L. 2016. Recycled Construction and Demolition Wastes as filling material for geosynthetic reinforced structures. Interface properties. *Journal of Cleaner Production,* 124: 299–311.

Vieira, C.S., Cristelo, N., Lopes, M.L. 2017. Geotechnical and geoenvironmental characterization of recycled Construction and Demolition Wastes for use as backfilling of trenches, 4th Edition of the International Conference Wastes: Solutions, Treatments and Opportunities. CRC Press Taylor & Francis Group, Porto, Portugal, pp. 175–181.

Wastes: Solutions, Treatments and Opportunities III – Vilarinho et al. (Eds)
© 2020 Taylor & Francis Group, London, ISBN 978-0-367-25777-4

Slow pyrolysis of oil palm mesocarp fibres: Effect of operating temperature

A.F. Almeida, D. Direito, R.M. Pilão & A.M. Ribeiro
CIETI, Instituto Superior de Engenharia do Porto, Porto, Portugal

B. Mayer
Erasmus Student, Department of Chemical Engineering, Instituto Superior de Engenharia do Porto, Porto, Portugal

ABSTRACT: This study investigated the effect of temperature on the slow pyrolysis of oil palm mesocarp fibres. The biomass was characterized in terms of proximate and ultimate analysis, and its higher heating value (HHV) was 18.51 MJ/kg. For pyrolysis temperatures from 469 to 783°C, at a heating rate of 20°C/min, bio-char yield varied between 32.7% and 25.8%. Gas and liquid phases were analysed by gas chromatography and Fourier-Transform Infrared Spectrometry (FTIR), respectively. CO_2 was the major gas produced for all temperatures (6.9 to 10.0 mol/kg of biomass) and H_2 concentration increased rapidly as the temperature rose (0.6 to 8.0 mol/kg of biomass). FTIR measurements show that the bio-oils contained alcohols, phenols, alkanes, alkenes, carboxylic acids, aldehydes and aromatic compounds. The HHV of both bio-chars (27.50 to 28.86 MJ/kg) and bio-oils (25.95 to 28.50 MJ/kg) were measured. Thermal decomposition of the fibres was also studied using thermogravimetric analysis.

1 INTRODUCTION

Increasing concern over declining fossil fuel reserves has raised interest in the search for alternative energy solutions, namely the use of lignocellulosic biomass. Biomass is generally considered to be a clean energy source because it contains negligible amounts of nitrogen, sulphur and ash when compared to conventional fossil fuels. This results in lower emissions of SO_2, NO_x and soot than those produced by conventional fossil fuels (Sembiring *et al.*, 2015).

According to the Index Mundi 2018, Indonesia, Malaysia and Thailand are the biggest palm oil producers in the world, with total productions of 41.5×10^6, 20.5×10^6 and 2.9×10^6 metric tons (MT), respectively. Significant quantities of waste material originate from oil palm tree plantation activities (trunks, fronds and leaves) and from the milling process: empty fruit bunches (EFB), oil palm mesocarp fibre (OPMF) and palm kernel shells. In addition, palm oil mill operations also produce large quantities of palm oil mill effluent (Zafar, 2019).

After arriving at the mills, fresh fruit bunches are sterilized before removing the fruits, leaving empty fruit bunches. Each fruit is composed of a hard kernel inside a shell surrounded by a fleshy mesocarp. The fruits are then pressed, releasing an edible oil, and the kernels detached from the press cake (mesocarp fibres). The detached palm kernels are then crushed and the palm kernel oil (used in the soap industry) is extracted, leaving another waste, the palm kernel shells. (Sumathi *et al.*, 2008 and Zafar, 2019). According to Hooi *et al.* (2009), each MT of processed fresh fruit bunches produces 220 kg of EFB, 67 kg of mesocarp fibre, 70 kg of shells, and 30 kg of palm kernel cake.

Residues from the palm oil industry can be treated by thermochemical processes such as combustion, gasification or pyrolysis for use as alternative energy sources. Pyrolysis is a process which uses relatively high temperatures and an inert atmosphere to convert biomass waste into bio-char, bio-oil and gases. Bio-char may be employed as a solid fuel, as a soil fertilizer or transformed into activated carbon. Further refinement of the bio-oil allows the liquid to be used as a fuel or as the basis for the extraction of other chemicals. The gas phase, mainly composed of CO_2, CO, CH_4, and H_2 can be used as an energy source for maintaining the pyrolysis process itself, in addition to applications in power generation and heat production. Several parameters, including temperature, biomass type, heating rate, reactor type, particle size, and the use of catalysts, affect pyrolysis product yields and the composition of the gas phase (Basu, 2013).

Several authors investigated the pyrolysis of various residues coming from the oil palm tree and related industry processes. EFB were studied by Sukiran *et al.* (2009), Abdullah *et al.* (2011) and Sembiring *et al.* (2015). Asadullah *et al.* (2013) presented results on the fast pyrolysis of PKS. Yang *et al.* (2006) and Abnisa *et al.* (2013) studied the pyrolysis of palm shells, EFB and OPMF. Hooi *et al.* (2009) and Chen & Lin (2016) investigated the effect of temperature on the product yields obtained during the slow pyrolysis of OPMF.

This paper presents new data on the slow pyrolysis of OPMF. The influence of temperature on the properties of bio-oils and bio-chars, and on the yield of bio-char was studied. Due to experimental set-up restrictions, it was not possible to determine the bio-oil or the gas phase yields. Thermal degradation of the palm oil wastes was followed by thermogravimetric (TG) analysis.

2 MATERIALS AND METHODS

2.1 *Biomass characterization*

As mentioned earlier, OPMF is a fibrous material obtained from the mesocarp of the oil palm fruit after oil extraction. The OPMF was characterized in terms of its proximate (moisture, volatiles, ash and fixed carbon) and ultimate (C, H, N and O) analyses, and by its higher heating value.

The moisture content of the air-dried fibres was determined according to standard EN14774. The ash content was measured following standard EN14775, and standard EN15148 was used for quantifying the volatiles. The fixed carbon was calculated by mass difference. The ultimate analysis of raw material was carried out using standard ISO 16948:2015. Measurement of the HHV of the feed material and of the chars and bio-oils obtained after pyrolysis, was performed using a Parr 6200 calorimeter. The oil content of the OPMF was determined according to standard ISO 734-1:1998.

2.2 *Thermogravimetric analysis*

The thermal degradation characteristics of the OPMF were studied using a thermal analyser type Netzsch STA 449 F3 Jupiter, with nitrogen as the carrier gas. During the experiments, nitrogen was fed to the equipment at a flow rate of 50 mL/min and the temperature was varied from 50°C to 900°C. The influence of the heating rate (HR) was studied using HR = 10, 20 and 30°C/min, and a sample mass of 10 mg. Proteus software was used to acquire, store and analyse the data.

2.3 *Experimental procedure*

Pyrolysis of the OPMF was carried out using a slow pyrolysis reactor and the procedure described by Almeida *et al.* (2017). Prior to the pyrolysis experiments, the fibre residue was cut into pieces of less than 20 mm and dried at 105°C, until a constant weight was reached. The mass of the dried fibres was approximately 40 g, and the nitrogen flow rate was set at 4.11×10^{-5} kg/s. During all experiments the maximum temperature inside the bed varied from 462 to 783°C, which corresponded to set point temperatures of 550 to 850°C. In all runs a HR of 20°C/min was used. Each experiment continued for at least 15 minutes after the final temperature was reached. All experimental conditions were repeated at least twice in order to test the reproducibility of the results.

During each run, seven gas samples were collected in Tedlar bags, from the point where the concentration of the CO started increasing until the end of the experiment.

Table 1. Proximate and ultimate analyses, oil content and HHV of the palm oil fibres.

Proximate analysis (dry basis – % w/w)	Ultimate analysis (dry basis – % w/w)	Oil content (%)	HHV (MJ/kg)
Volatiles – 74.3 Ash – 4.2 Fixed carbon – 21.5	C – 49.3 N – 1.0 H – 6.1 O – 39.4	2.40	18.51

After the reactor had cooled, the basket with the bio-char was taken out and weighed. Samples of bio-oils were taken, and after removal of the water they were analysed using a Nicolet 6700 Fourier Transform Infrared Spectrometer. The HHV of the bio-chars and bio-oils was determined using a Parr 6200 calorimeter.

The gas samples collected during each pyrolysis experiment were analysed in a Dani 1000 DPC gas chromatographer equipped with an OPT333 injector and a thermal conductivity detector (TCD OPT266). The column used was a 60/80 Carbonex 1000 with argon as the carrier gas. The gases analysed were CO_2, CO, H_2, CH_4, O_2 and N_2.

3 RESULTS AND DISCUSSION

3.1 *Characterization of the oil palm fibres*

Data concerning the proximate and elemental analyses, oil content and HHV of the palm oil residues are listed in Table 1.

For the ultimate analysis, on a dry basis, Yang *et al.* (2006) reported carbon, hydrogen, nitrogen, sulphur and oxygen contents of 50.27%, 7.07%, 0.42%, 0.63% and 36.28%, respectively. These values are close to the ultimate analysis obtained for the oil palm fibres used in this study (Table 1). The HHV of the OPMF reported by Hooi *et al.* (2009) was 18.76 MJ/kg, which is also in agreement with the value obtained in this investigation (18.51 MJ/kg).

Abnisa *et al.* (2013) reported that the lignocellulosic content of oil palm mesocarp fibres contained 23.7% of cellulose, 30.5% of hemicellulose and 27.3% of lignin. During the pyrolysis process, each of these components will decompose and contribute to the formation of volatiles and char.

3.2 *Bio-char yields*

For the conditions being studied the bio-char yield varied between 32.7% and 25.8% as the pyrolysis temperature increased from 469 to 783°C. For temperatures between 450°C and 800°C, Hooi *et al.* (2009) obtained char yield varying from 34.82% to 25.89%. This data shows a good agreement with the results obtained during these experiments, with char yield decreasing as the temperature increases. Other authors found the same behaviour using different biomass residues (Asadullah *et al.* 2013, Yang *et al.* 2006).

3.3 *Characterization of bio-chars and bio-oils*

After pyrolysis, the bio-oil and biochar samples obtained from the runs at 462 and 469°C (T_{avg1} = 466°C), 568 and 584°C (T_{avg2} = 576°C), 652, 661 and 671°C (T_{avg3} = 661°C) and 775 and 783°C (T_{avg4} = 779°C) were mixed together. Table 2 contains the higher heating values of chars and bio-oils as a function of the average pyrolysis temperature.

The HHV of the chars obtained varied from 27.50 to 28.86 MJ/kg, and for the bio-oils from 25.95 to 28.50 MJ/kg. For the conditions under study, the pyrolysis temperature does not seem

Table 2. Higher heating values of the bio-chars and bio-oils obtained during the pyrolysis experiments.

T_{avg} (°C)	HHV_{char} (MJ/kg)	$HHV_{bio\text{-}oil}$ (MJ/kg)
466	27.50	–
576	28.04	28.50
661	27.62	28.57
779	28.86	25.95

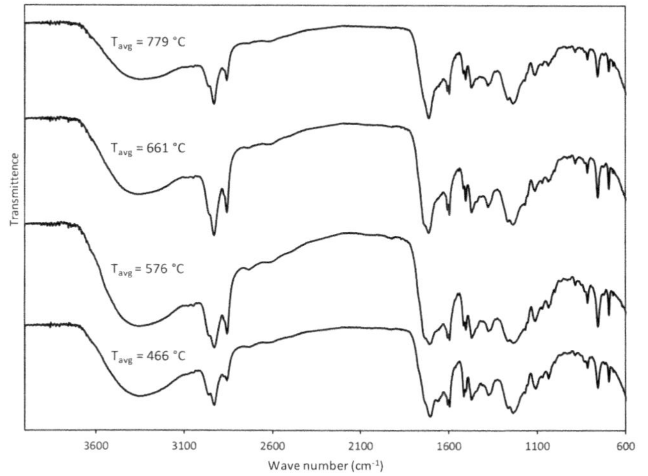

Figure 1. FTIR analysis of the bio-oils.

to affect the HHV of the chars and bio-oils obtained. It was observed that the pyrolysis products present higher HHV than that of the initial biomass (18.51 MJ/kg).

After removal of the water content from the bio-oils, they were analysed using the Nicolet 6700 Fourier Transform Infrared Spectrometer. Figure 1 shows the transmittance spectra of the bio-oils for average pyrolysis temperatures of 466°C, 576°C, 661°C and 779°C. In general, the spectra were very similar for all the bio-oils. Table 3 displays the values from the analysis of the curve referring to $T_{avg4} = 779°C$. According to Table 3, the bio-oils obtained could be a blend of alcohols, phenols, alkanes, alkenes, carboxylic acids, aldehydes and aromatic compounds. Similar results were found by Yang *et al.* (2006) for palm shell bio-oil and by Abnisa *et al.* (2013) for palm shell, EFB and OPMF bio-oils.

3.4 *Gas composition*

The gas phase produced during pyrolysis was analysed by gas chromatography (GC). Figure 2 presents the total amount of each gas produced during pyrolysis as a function of the temperature inside the reactor (462 to 783°C). CO_2 is the gas produced in the greatest quantity for all temperatures studied (6.9 to 10.0 mol/kg of biomass), followed by CO (3.8 to 8.5 mol/kg of biomass). Increasing the reactor temperature caused a rapid increase in H_2 concentration from 0.6 to 8.0 mol/kg and an increase of methane production from 1.1 to 3.2 mol/kg. In general, it can be observed that increasing the temperature favours the formation of H_2, which is also confirmed by the work of Yang *et al.* (2006). These authors mention that hydrogen production is mainly affected by the secondary pyrolysis of biomass.

Table 3. Main functional groups of the bio-oil obtained from OPMF.

Peak wave number (cm^{-1})	Wave number range* (cm^{-1})	Functional group*	Class of compounds*
3338	3600–3300	O–H stretching	alcohols, phenols
2925, 2854	3050–2800	C–H stretching	alkanes
1710	1750–1650	C=O stretching	ketones, aldehydes, carboxylic acids
1595	1650–1580	C=C stretching	alkenes
1465, 1375	1470–1350	C–H bending	alkanes
1230	1300–950	C–O stretching	primary, secondary, tertiary alcohol
756	915–650	O–H bending	phenol, esters, ethers and aromatic compounds

*from Yang *et al.* (2006)

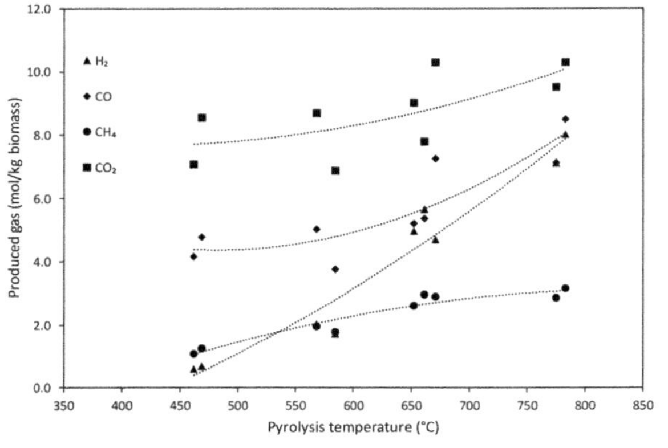

Figure 2. Influence of bed temperature on producer gas composition.

For pyrolysis temperatures from 462°C to 783°C, the HHV of the produced gas varied from 843 to 1980 kJ/(Nm^3), indicating that higher temperatures favour the production of a richer combustible gas.

3.5 *Thermogravimetric analysis of OPMF*

The thermal decomposition of OPMF was studied for temperatures in the range of 40 to 900°C, and for HR = 10, 20 and 30°C/min using an inert atmosphere. The differential weight loss curves of oil palm fibres for the different heating rates are shown in Figure 3.

The curves in Figure 3 present three different zones. The first stage of pyrolysis runs from around 40 to 120°C for HR = 10°C/min, and from 40 to 145°C for HR = 20 and 30°C/min and could be linked to the release of moisture and other volatile compounds. The second stage of pyrolysis (primary pyrolysis) extends from 188 to 369°C for HR = 10°C/min, from 182 to 395°C for HR = 20°C/min and from 174 to 386°C for HR = 30°C/min. During primary pyrolysis, decomposition of hemicellulose and cellulose occurs. The third region is due to secondary pyrolysis and is mainly concerned with the thermal degradation of lignin (Pathasarathy & Narayanan, 2014). For the first and second stages of the thermal degradation, and as the HR increases, the differential thermogravimetry (DTG) curves are shifted towards higher temperatures and the magnitude of the respective DTG values also increase. This behaviour was also found by other researchers and was justified by Quan *et al.* (2009).

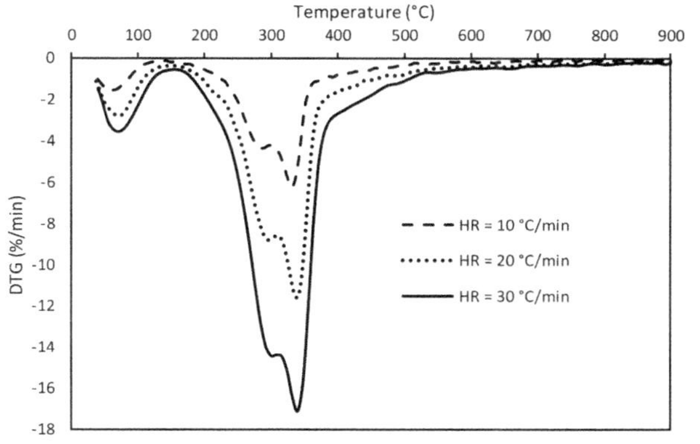

Figure 3. DTG curves for the decomposition of OPMF at HR of 10, 20 and 30°C/min.

4 CONCLUSIONS

In this work new experimental results on the slow pyrolysis of OPMF are presented. The biomass was characterized in terms of proximate and ultimate analysis, HHV (18.51 MJ/kg) and oil content (2.40%). The effect of pyrolysis temperature on the bio-char yield and on the characteristics of the produced products was studied. For bed temperatures in the range of 469 to 783°C and at a heating rate of 20°C/min, the bio-char yield varied between 32.7% and 25.8%. The HHV of the bio-chars obtained varied from 27.50 to 28.86 MJ/kg and for the bio-oils from 25.95 to 28.50 MJ/kg. They are higher than the HHV of the OPMF and do not seem to be affected by pyrolysis temperature. The GC analysis showed that CO_2 was the gas produced in greatest quantity for all the temperatures studied (6.9 to 10.0 mol/kg of biomass), followed by CO (3.8 to 8.5 mol/kg biomass). The amount of H_2 increased rapidly from 0.6 to 8.0 mol/kg as the temperature was raised. Methane production increased from 1.1 to 3.2 mol/kg as the reactor temperature increased. The FTIR measurements indicated that the bio-oils produced could be a blend of alcohols, phenols, alkanes, alkenes, carboxylic acids, aldehydes and aromatic compounds. The TG analysis carried out at HR = 10, 20 and 30°C/min showed that, within the range of temperatures tested (40 to 900°C), the DTG curves were shifted towards higher temperatures as the HR was increased.

ACKNOWLEDGEMENTS

Financial support from the FCT, under Research Project UID/EQU/04730/2019.

REFERENCES

Almeida, A.F., Pereira, I.M., Silva, P., Neto, M.P., Crispim, A.C. Pilão, R.M. 2017. Pyrolysis of leather trimmings in a fixed bed reactor. *JALCA* 112:112–120.

Abdullah, N., Sulaiman, F., Gerhauser, H. 2011. Characterisation of oil palm empty fruit bunches for fuel application. *Journal of Physical Science* 22 (1):1–24.

Abnisa, F, Arami-Niya, A., Wan Daud, W.M.A., Sahu, J.N. 2013. Characterization of bio-oil and bio-char from pyrolysis of palm oil wastes. *Bioenergy Research* 6:830–840.

Asadullah, M., Rasid, N.S.A., Kadir, S.A.S.A, Azdarpour, A. 2013. Production and detailed characterization of bio-oil from fast pyrolysis of palm kernel shell. *Biomass and Bioenergy* 59: 316–324.

Basu, P. 2013. Biomass Gasification, Pyrolysis and Torrefaction – Practical Design and Theory, Elsevier, Oxford, UK.

Chen, WH., Lin, BJ. 2016. Characteristics of products from the pyrolysis of oil palm fibre and its pellets in nitrogen and carbon dioxide atmospheres. *Energy* 94:569–578.
Hooi, K.K., Alauddin, Z.A.Z, Ong, L.K. 2009. Laboratory-scale pyrolysis of oil palm pressed fruit fibers. *Journal of Oil Palm Research* 21: 577–587.
Parthasarathy, P. & Narayanan, S. K. 2014. Determination of kinetic parameters of biomass samples using thermogravimetric analysis. *Environmental Progress & Sustainable Energy* 33 (1): 256–266.
Quan, C., Li, A. and Gao, N. 2009. Thermogravimetric Analysis and Kinetic Study on Large Particles of Printed Circuit Boards Wastes. *Waste Management* 29: 2353–2360.
Sembiring, K.C., Rinaldi, N., Simanungkalit, 2015. Bio-oil from fast pyrolysis of empty fruit bunch at various temperatures. *Energy Procedia* 65:162–169.
Sukiran, M. A., Bakar, N.K.A., Chin, C.M. 2009. Optimization of pyrolysis of oil palm empty fruit bunches. *Journal of Oil Palm Research* 21: 653–658.
Sumathi, S., Chai, S.P., Mohamed, A.R. 2008. Utilization of oil palm as a source of renewable energy in Malaysia. *Renewable & Sustainable Energy Reviews* 12: 2404–2421.
Yang, H., Yan, R., Chen, H., Lee, D.H., Liang, D.T., Zheng, C. 2006. Mechanism of palm oil waste pyrolysis in a packed bed. *Energy & Fuels* 20: 1321–1328.
Zafar, S. 2019. Biomass Energy in Indonesia, BioEnergy Consult, www.bioenergyconsult.com/tag/palm-oil-biomass/, accessed on 26/2/2019.

Wastes: Solutions, Treatments and Opportunities III – Vilarinho et al. (Eds)
© 2020 Taylor & Francis Group, London, ISBN 978-0-367-25777-4

Organic municipal solid wastes: Analyzing a system transition

A. Padrão & A. Guerner Dias
Faculty of Sciences, University of Porto, Porto, Portugal

ABSTRACT: Recycling of matter is essential to the agroecosystems that support our society, and developing an optimal system to valorize organic wastes, turning them into a sustainable source of agricultural fertilizer, is a pressing concern. However, even the systems utilized in developed countries such as Portugal are deficient, either lacking the capacity to recycle sufficient organic waste or failing to ensure the quality of the resulting compost. This paper focuses on the possibility of waste management companies adopting a system based on domestic sorting of organic waste, analyzing the strengths, weaknesses, opportunities and threats associated with such transition.

1 INTRODUCTION

In 2002 Portugal elaborated ENRRUBDA, a national strategy designed to accomplish the goal, set by the European Union's Landfill Directive (portuguese "Diretiva 1999/31/CE"), of reducing the organic municipal solid wastes (MSW) deposited in landfill to 35% of what it was in 1995 until 2020. The point of such goal would have been something like addressing the valorization of one of the most abundant (37.2% of all the waste produced in continental Portugal, according to APA 2017a) and important resource in our society. Indeed, we depend on the functioning of a huge agroecosystem, industrial and globalized, which is now supposed to feed over 7 billion people while facing the depletion of mineral fertilizer reservoirs, the hazards to public health associated with chemical fertilizers and the degradation of agricultural soil – according to PERSU 2020 (most recent strategic plan for MSW in Portugal, "Portaria n.° 187-A/2014"), portuguese soil is notable for a low fertility, with 66% of soils classified as low quality. The proper valorization of organic wastes could produce sustainable fertilizer which would be key to dealing with all these problems, but the infrastructures that are supposed to do it are dangerously underdeveloped – again, in comparison to other developed countries, Portugal is especially lacking adequate systems and infrastructures of organic valorization (Duarte 2016).

We shall consider three different types of organic valorization systems in this paper:

A. Undifferentiated collection of organic wastes, followed by a system that relies on mechanical and biological treatment (MBT) to extract the organic fraction of MSW and turn it into compost. Most companies in Portugal have adopted this system, with 77% of the national compost being produced from undifferentiated waste according to PERSU 2020; however, such compost withholds serious quality problems, such as high heavy metal content (Smith 2004).
B. Industrial sorting of organic wastes, a system that collects organic waste among big producers (gardens, restaurants, etc.) and sends them directly to composting. Being responsible for approximately 35% of the nationally produced compost (APA 2017b), it probably produces all 28% that can be applied in agriculture (PERSU 2020) as well; nevertheless, all companies that use this system require energetic valorization centers (PERSU 2020) to incinerate the large amount of organic wastes, originated by small producers, that remain in the undifferentiated waste – therefore forsaking their proper valorization.
C. Domestic sorting of organic wastes, an alternative system which entrusts the general population with the sorting of their organic waste, for posterior selective collection and composting. Its efficiency has already been proved in states with strong environmental policies (Sullivan 2011),

as the organic MSW are not contaminated by undifferentiated waste and it encompasses the organic waste of large and small producers; however, it is not yet clear whether or not it is economically viable without such policies.

The dissertation work of Padrão (2018) focuses on assessing the amount of organic waste a population can sort without financial incentives, on the premise that the domestic sorting of organic MSW is the most advantageous system. As this paper is based on that work, it will use its results, along with the data available on two representative companies, to analyze the possible effects of a transition from systems A or B to a system C.

2 METODOLOGY

2.1 *Accessing representative companies*

The MSW produced in continental Portugal are managed by 23 plurimunicipal companies (APA 2017b), each with its own different methods. The data about their activities is available in the website of the public institution *Agência Portuguesa do Ambiente*; there we have collected information about two companies, each representative of a different organic valorization system.

2.1.1 *SULDOURO and undifferentiated collection of organic wastes*

Undifferentiated collection is the main system associated with valorization of organic wastes in Portugal, as it is used by several companies – among which SULDOURO. There, the organic fraction is separated from the rest of MSW through mechanical treatment, being then exposed to biological treatment: homogenization in a suspension tank, anaerobic degradation (with biogas production) in digesters, solid-liquid separation and aerobic composting in piles (Silva 2016), with maturation of compost at the end of the process to ensure their stability (SULDOURO 2017).

Figure 1 (above) represents the waste flux in SULDOURO's MBT unit in 2016. About 90,000 tons of waste got into this unit, most of it collected undifferentiatedly. Most of the waste mass was redirected to landfill, while 21% of it has disappeared by decomposition, 2.1% was recovered for material valorization and just 1.5% was transformed into non-rejected compost (APA 2017a).

2.1.2 *LIPOR and industrial sorting of organic wastes*

Industrial sorting of organic wastes is the focus of companies such as LIPOR, which collect organic MSW in big producers (gardens, supermarkets, cemeteries and the HoReCa sector – hotels, restaurants and cafes) and directed them straight into organic valorization. There, compost is produced after the organic waste is exposed to a series of sieves, magnetic sorters, composting tunnels and stabilization floors (LIPOR 2017).

Figure 2 represents the waste flux in LIPOR's organic valorization units in 2016. About 50,000 tons of waste were directed to organic valorization, having been collected selectively in different sources. Most of the waste mass has disappeared by decomposition, while 22% was placed in the market as 'nutrimais' compost and 8.3% had other type of destiny (APA 2017a).

2.2 *Acessing the performance of a population's domestic sorting*

The work mentioned in the introduction (Padrão 2018) involved 44 families (122 people), contacted via door-to-door inquiry, as sampling of the population in the metropolitan area of Porto. All of those families were included in the study as sample elements; most of them also agreed to participate as voluntaries (while only 6 refused so, being nevertheless included as sample elements), receiving a set of plastic bags in which to sort the organic waste produced in their homes between the 5th of February and the 28th of Abril, year 2018.

Over 12 weeks, the mass of the sorted wastes were weighted, either by the voluntaries or by the authors of the study, in a total of 2020 kg, averaging 0.20 kg.person^{-1}.day^{-1}. As the municipalities served by SULDOURO and LIPOR are also included in the metropolitan area of Porto, there is

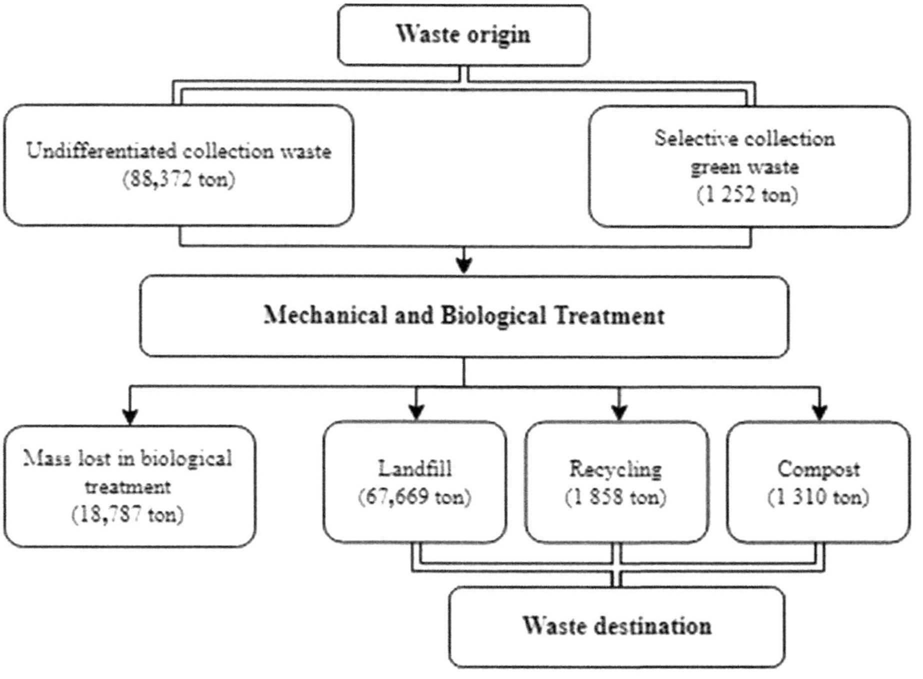

Figure 1. Origin and destiny of waste treated in SULDOURO's MBT unit in 2016 (adapted from: APA 2017a).

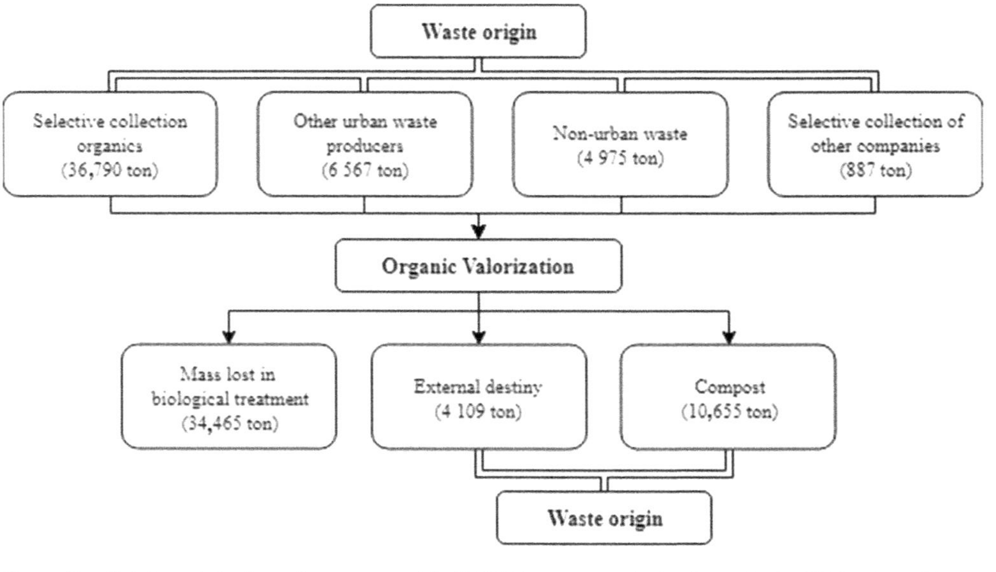

Figure 2. Origin and destiny of waste treated in LIPOR's organic valorization units in 2016 (adapted from: APA, 2017a).

every reason to expect that value to apply to these companies if they were to adopt a system based on domestic sorting of organic wastes.

3 RESULTS

Taking the example of the population served by SULDOURO, the 440,000 habitants of Santa Maria da Feira and Vila Nova de Gaia (AMP 2018), we can theorize the selective collection of 32,142 ton.year^{-1} domestically sorted organic wastes, adding to the 1252 tons of green waste already collected selectively. As for the population served by LIPOR, about 946,000 habitants in the metropolitan area of Porto (AMP 2018), we can theorize the selective collection of 69,105 ton.year^{-1} domestically sorted organic wastes, adding to the 49,229 tons of organic wastes already collected selectively.

These values are theoretical in the sense that we cannot expect any waste management company to apply a collection network over 100% of the territory it serves. Nevertheless, as we shall see, not all of the population needs to be involved in the collection network to improve the organic valorization of a company.

3.1 *Transition from undifferentiated collection to domestic sorting*

SULDOURO signals the capacity to valorize 20,000 ton.year^{-1} of organic wastes (SULDOURO 2017); as the theoretical value for its maximum domestic sorting of organic wastes surpasses that capacity by far, we can assume that SULDOURO would only need to involve 57% of its population in this system to fulfill it. Depending on the proportion of compost produced by each ton of organic waste – between 1/2 (Chynoweth et al. 1992, Batista & Batista 2007) and 1/4 (LIPOR 2017) – it would be possible to obtain between 5000 ton.year^{-1} and 10,000 ton.year^{-1} of high-quality compost.

These numbers represent a huge improvement facing the 1310 tons of low-quality compost obtained in 2016. The increase in non-rejected compost is synonymous with the decrease in compost rejected to landfill, aiding in the accomplishment of the goals set by the Landfill Directive; moreover, the increase in quality widens the possibilities of application. Not only this provides society with a sustainable source of agricultural fertilizer, it could also represent a source of income for the company that produces it. These strengths and opportunities are summarized at the first column of table 1.

Obviously, a new waste collection system requires investment in several areas (public containers, waste transportation, fuel, personal, etc.), but so did the implementation of paper, plastic and glass collection in the mid-nineties, and we can all agree it was worth it. The occupation of the MBT unit with the exclusive treatment of selectively collected organics also implies the loss of 1858 tons of waste recovered for material valorization, then again, that quantity is not particularly relevant in face of the 14,937 tons of waste collected selectively and sent directly to material valorization in 2016 (APA 2017). A threat associated with the transition to this system would be the failure of our predictions, with the population sorting fewer organic wastes than expected: a problem that would have to be confronted with environmental awareness actions. These weaknesses and threats are summarized at the second column of table 1.

3.2 *Transition from industrial sorting to domestic sorting*

LIPOR signals the capacity to valorize 60,000 ton.year^{-1} of organic wastes (LIPOR 2017); the theoretical value for its maximum domestic sorting of organic wastes surpasses that capacity, tough realistically it might not be easy for LIPOR to involve the 87% of its population required to fulfill it. Depending on the proportion of compost produced by each ton of organic waste – again, between 1/2 and 1/4 – it would be possible to obtain between 15,000 ton.year^{-1} and 30,000 ton.year^{-1} of high-quality compost.

Table 1. SWOT analysis of a system transition from undifferentiated collection to domestic sorting of organic wastes.

STRENGHTS	WEAKNESSES
Decrease in waste sent to landfill Increase in the quantity of produced compost Increase in the quality of produced compost	Investment in containers, transportation, personal, etc. Loss of waste recovered for material valorization
OPPORTUNITIES	THREATS
Use of compost as a sustainable agricultural fertilizer Profits associated to expansion of compost market	Sub-optimal population performance, demanding actions of environmental awareness

Table 2. SWOT analysis of a system transition from industrial sorting only to industrial and domestic sorting of organic MSW.

STRENGHTS	WEAKNESSES
Decrease in waste sent to incineration Increase in the quantity of produced compost Short term savings in the operation of energetic valorization units	Investment in containers, transportation, personal, etc. Investment in more organic valorization units
OPPORTUNITIES	THREATS
Profits associated to expanded compost market	Sub-optimal population performance, demanding actions of environmental awareness

These numbers look only fair when compared to the 10,655 tons of compost LIPOR appears to have produced in 2016, especially considering LIPOR also sent 4106 tons of mass to an "external destination" which might correspond to materials that should not have been sorted as organic wastes, a flaw that a system of domestic sorting would most certainly share. The quality of the compost would also be unchanged, as 'nutrimais' is already a high-quality compost. The only system transition that would make sense for a company like LIPOR would to industrial and domestic sorting of MSW: keeping both systems. As the current system already makes a good use of the current infrastructures, it would inevitably imply an investment on new organic valorization units. These weaknesses and threats are summarized at the second column of table 2.

Of course that companies like LIPOR have good reasons to invest in the construction of more organic valorization infrastructures: tough their un-composted organic waste is effectively incinerated (what copes very well with the Landfill Directive), energetic valorization units are very expensive to operate. The transitioning of waste from elimination to organic valorization would signify both environmental and economical savings. And if that investment were to happen, it would be wise to adopt the domestic sorting system instead of trying to expand the industrial sorting system, as its potential is yet untapped – indeed, pilot-projects organized by this company in Maia (LIPOR 2017) and Valongo (Jornal de Notícias 2018) may prove just that. These strengths and opportunities are summarized at the first column of table 2.

4 CONCLUSION

This paper supports the thesis of an organic valorization system based on domestic sorting of wastes being efficient, even in the absence of policies to encourage that practice. Not only that, it might be the most efficient system, allowing for the recycling of more waste than any other system and providing the high-quality compost characteristic of selective collection-based systems.

Regarding the transition of a system based on undifferentiated collection of MSW to the one mentioned above, the advantages appear to vastly outweigh the disadvantages. If SULDOURO were to arrange a collection network covering roughly two-thirds of Santa Maria da Feira and Vila Nova de Gaia, simply filling its present-day MBT unit with the domestically sorted organic waste could increase compost production up to seven times, while greatly increasing the compost's quality.

As for the transition of a system based on industrial sorting of organic wastes, the results are not as outstanding. Nevertheless, if LIPOR were to invest in new organic valorization units, complementing the collection network of big producers with key points of small producers (such as densely populated areas) might prove fruitful.

The resilience of modern ecosystems depends heavily on returning organic matter to the fields where it came from, on a scale that only the valorization of organic waste can ensure. All of this reality is connected to the concept of circular economy, which modern society needs to embrace: as scientists and citizens, we should make the best effort to make it so.

Although the sample analyzed, about 40 families, can be considered reduced, the results obtained, not only in terms of quantitative collected as the rate of adhesion to this project, are very enthusiastic.

REFERENCES

Agência Portuguesa do Ambiente 2017a. *Dados sobre resíduos urbanos.* [online] Available at: https://www.apambiente.pt/index.php?ref=16&subref=84&sub2ref=933&sub3ref=936 [Accessed 23 November 2017].

Agência Portuguesa do Ambiente 2017b. *PERSU 2020 – Relatório de Avaliação 2016.* Amadora.

AMP 2018. *Caracterização da AMP.* Available at: http://portal.amp.pt/pt/ [Accessed: 18 June 2018].

Batista, J. & Batista E. 2007. *Compostagem: Utilização de compostos em horticultura.* Angra do Heroísmo: University of Azores – CITA-A.

Chynoweth, D.P., Owens, J., O'Keefe, D., Earle, J.F.K, Bosch, G., & Legrand, R. 1992. Sequential batch anaerobic composting of the organic fraction of municipal solid waste. *Water Science and Technology* 25 (7): 327–339.

Decreto-Lei n.º 183/2009 de 10 de Agosto. *Diário Da República, 1.ª série – N.º 153.* Ministério do Ambiente, do Ordenamento do Território e do Desenvolvimento Regional. Lisboa.

Diretiva 1999/31/CE do Conselho de 26 de Abril de 1999 relativa à deposição de resíduos em aterros. *Jornal Oficial das Comunidades Europeias L 182/1.* Conselho da União Europeia. Lisboa.

Duarte, I. 2016. *Análise da recolha seletiva de resíduos urbanos em Portugal e comparação com outros países.* Masters dissertation, Faculty of Sciences and Technology of the University of Coimbra.

Jornal de Notícias 2018. *Valongo aposta na recolha de resíduos orgânicos.* [online] Available at: https://www.jn.pt/local/especial-patrocinado/interior/valongo-aposta-na-recolha-de-residuos-organicos-9121919.html [Accessed: 18 February 2018].

LIPOR 2017. *Resíduos urbanos.* [online] Available at: http://www.lipor.pt/pt/residuos-urbanos/valorizacao-organica/estrategia/a-estrategia-de-recolha-seletiva-da-fracao-biodegradavel/ [Accessed: 9 June 2017].

Padrão, A. 2018. *Separação doméstica de resíduos urbanos biodegradáveis: um sistema para otimizar a valorização orgânica.* Masters dissertation, Faculty of Sciences of the University of Porto.

Portaria n.º 187-A/2014 de 17 de setembro. *Diário Da República, 1.ª série – N.º 179 – 17 de setembro de 2014.* Ministério do Ambiente, Ordenamento do Território e Energia. Lisboa.

Silva, J. 2016. *Papel do tratamento mecânico e biológico na gestão de resíduos.* Masters dissertation, Faculty of Engineering of the University of Porto.

Smith, S.R. 2004. A critical review of the bioavailability and impacts of heavy metals in municipal solid waste composts compared to sewage sludge. *Environment International* 35 (1): 142–156.

SULDOURO 2017. *Central de valorização orgânica.* [online] Available at: http://suldouro.pt/pt/gestao-de-residuos/tratamento-mecanico-e-biologico/ [Accessed: 24 April 2018].

Sullivan, D. 2011. Zero Waste on San Francisco's Horizon. *BioCycle,* July 2011: p-p.

Wastes: Solutions, Treatments and Opportunities III – Vilarinho et al. (Eds)

Extraction of natural pigments from marine macroalgae waste

S.L. Pardilhó, M.F. Almeida & J.M. Dias
LEPABE, Departamento de Engenharia Metalúrgica e de Materiais, Faculdade de Engenharia, Universidade do Porto, Porto, Portugal

S. Machado, S.M.F. Bessada & M.B.P.P. Oliveira
REQUIMTE/LAQV, Departamento de Ciências Químicas, Faculdade de Farmácia, Universidade do Porto, Porto, Portugal

ABSTRACT: The need to replace synthetic pigments and the richness of marine macroalgae in natural pigments make their study relevant as a source of such valuable compounds. In the present study, marine macroalgae waste collected at a Northern Portugal beach was used aiming pigments extraction. *Saccorhiza polyschides* (brown algae) was the most abundant specie. The biomass (freeze-dried and ground to <1 mm) was subjected to pigments extraction using six different solvents and their amounts were evaluated through their maximum absorption using UV spectrophotometry. The results showed as promising the use of marine macroalgae waste as a natural source of pigments and acetone as the best solvent for extraction, followed by methanol. Chlorophyll *a* was found as the predominant pigment in the biomass with a maximum of 1685 $\mu g\,g^{-1}$ extracted using 90% acetone. Carotenoids and fucoxanthin could also be extracted with methanol (174 $\mu g\,g^{-1}$) and DMSO (252 $\mu g\,g^{-1}$), respectively.

1 INTRODUCTION

The inappropriate management of Marine Macroalgae Waste (MMW) accumulated at the coastal regions represents a loss of resources and an opportunity to add value to the economic sector (new products and renewable energy), considering the circular economy principles (Pardilhó et al., 2018). Moreover, MMW might lead to environmental impacts and health problems, therefore affecting the well-being of the population (Barbot et al., 2016).

Marine Macroalgae (MM), or seaweeds, are photoautotrophic organisms generally classified as red (Rhodophyta), green (Chlorophyta) and brown (Ochrophyta-Phaeophyceae), based on their pigmentation (Rodrigues et al., 2015). According to Rodrigues et al. (2015), their composition includes polysaccharides (between 4–76% dry weight), proteins (5–30% dry weight), minerals (up to 36% dry weight in some species), vitamins and lipids (1–5% of cell composition).

In the last years, MM have been considered as an important source of bioactive natural compounds, being highlighted the production of bioplastics, biostimulants, biofuels and pigments (Haryatfrehni et al., 2015, Pangestuti and Kim, 2011). Their richness in Natural Pigments (NPs) and the need to replace the synthetic pigments makes the extraction of NPs promising. There are three classes of NPs in MM, namely chlorophylls (Chls), carotenoids and phycobilins (Haryatfrehni et al., 2015, Pangestuti and Kim, 2011). Chl is the major pigment in most of photosynthetic organisms, and there are four kinds (Chl *a*, Chl *b*, Chl *c*, Chl *d*) in MM, with predominance of Chl *a* (Haryatfrehni et al., 2015). Carotenoids are considered accessory pigments, and they can be divided into carotenes and xanthophylls (Haryatfrehni et al., 2015). Fucoxanthin, a brown pigment especially present in brown algae, is one of the most abundant carotenoid (Pangestuti and Kim, 2011). Phycobilin is usually found in red algae and it has a more unstable structure, which can be easily damaged by light, heat or chemicals (Haryatfrehni et al., 2015). In short, in red MM

(Rhodophyta) the dominant pigments are Chl *a* and phycobilin, green MM (Chlorophyta) is known to contain Chl *a* and Chl *c* and brown MM (Ochrophyta-Phaeophyceae) has fucoxanthin (Fucox) and Chl *c* (Haryatfrehni et al., 2015).

A very important issue to make viable the extraction of NPs is the selection of the solvent (Henriques et al., 2007). According to Vimala and Poonghuzhali (2015), the choice of solvent for pigments extraction must take several factors into account, such as: toxicity, cost, extractions cycles and efficiency. Methanol has been the primordial solvent used to extract Chls; however, it has been replaced by other solvents (such as acetone or ethanol) due to its toxicity (Henriques et al., 2007). Acetone is known to be a very efficient extractant for Chls; however, it is not the ideal solvent essentially due to its high volatility, flammability and the fact that it is a skin irritant (Ritchie, 2006). Practical, economic and safety advantages from using ethanol for Chls extraction relatively to acetone or methanol are reported (Connan, 2015). Dimethylformamide (DMF) is a convenient solvent for Chls extraction but it is more toxic than acetone and ethanol (Connan, 2015). According to the same study, dimethylsulfoxide (DMSO) was used as extractant of Fucox, Chl *a* and Chl *c* from brown MM; for carotenoids, the extraction should be performed using methanol (Connan, 2015).

Taking into account the reported information, the present study aimed to determine the amount of NPs in MMW collected at a Northern Portugal, using different solvents, in order to evaluate their potential as a source of NPs and to establish the most adequate method to recover them.

2 MATERIALS AND METHODS

2.1 *Sample collection and preparation*

The MMW (approximately 5 kg) was collected from the Marbelo beach (Vila Nova de Gaia municipality, 41°6' 30.34" N, −8°39' 42.05" W) on the summer of 2018 (July). It was collected close to the water as it was there, transported and stored in plastic bags in the freezer without any further processing (−20°C) till analysis.

Before analysis, and in order to maintain intact as most as possible the properties of the MMW, the frozen biomass was freeze-dried at −80°C (Labconco freezone 2.5 plus), and afterwards ground to <1 mm, during 15 s at 10 000 rpm, in a laboratory mill (Retsch GM200).

2.2 *Nutritional composition*

The following physicochemical parameters were determined: residual moisture, ash, nitrogen, proteins, lipids and total carbohydrates. All the results were expressed in dry weight being further corrected for residual moisture content.

The moisture content from the freeze-dried biomass was measured by an infrared balance (Kern DBS). The ash content was determined according to AOAC (2012) in a furnace by sample calcination at 500°C up to constant weight.

The nitrogen content was obtained using the Kjeldahl method (AOAC, 2012). Samples were digested in an automatic digester (Büchi®, digestor unit model K-424 and a Büchi®, Scrubber, model B-414), followed by automatic distillation (Büchi®Kjelflex, model K-360) and volumetric titration with sulfuric acid (0.5 M).

Total lipids content was determined according to AOAC (2012), by the Soxhlet method; the result was expressed in g/100 g, on a dry basis.

The protein content was calculated according to the study of Angell et al. (2016) using a nitrogen-to-protein conversion factor of 5. The results were expressed as a percentage of dry weight.

The total carbohydrates amount was calculated by difference taking into account the results obtained for lipids, ash and protein.

All determinations were performed at least in triplicate.

2.3 *Pigments extraction*

Taking into account the literature, pigments were extracted using the following solvents: methanol (100%), ethanol (100%), methanol acid-free, dimethylsulfoxide (DMSO):water (4:1), acetone (90%) and dimethylformamide (DMF) (Connan, 2015).

To select the extraction process, different amounts of biomass (25, 50, 100, 150 mg) were firstly subjected to the following extraction conditions: i) 1 min of vortex agitation; ii) 1 min of vortex agitation, plus 1 h in a multi-rotator (Biosan, Multi RS-60); iii) 1 min of vortex agitation followed by standstill overnight; iv) 1 min of vortex agitation, plus 2 h in a multi-rotator (Biosan, Multi RS-60). Then, the sample was centrifuged (Heraeus Sepatech, Labofuge Ae) at 2500 rpm during 10 min; the supernatant was removed and centrifuged again at 5000 rpm for 5 min; the final supernatant was used for pigment determination.

Pigments amount was estimated by spectrophotometry (UV spectrophotometer, UV-1800, Shimadzu), measuring the absorbance of the extracts at selected wavelengths, corresponding to the maximum absorption of the pigment under study, and using the equations proposed by Connan (2015).

After this preliminary study, the process was then optimized for the best extraction conditions as well as the sample mass for each solvent, by making extractions in triplicate.

3 RESULTS AND DISCUSSION

The collected MMW was mainly composed by brown macroalgae, more specifically by *Saccorhiza polyschides*.

3.1 *Nutritional Characterization*

The results concerning more specifically the nutrient content of the freeze-dried MMW are presented in Table 1.

After freeze-drying, the moisture content on MMW was less than 10 % (considered adequate for the analytical determinations). The value obtained was used to correct the other parameters expressed in a dry basis.

The ash on the sample was according to the value reported in the literature, in the range 27 to 50 % for *Saccorhiza polyschides* (Jensen et al., 1985, Rodrigues et al., 2015, Rupérez, 2002, Sánchez-Machado et al., 2004). It should be highlighted that only in the study of Rupérez (2002) the biomass was freeze-dried; in the other studies the biomass was dried at temperatures between 35 and 60°C, during a few days. The value obtained shows the presence of a significant amount of inorganic materials, such as salt (NaCl) and sand.

Michalak et al. (2017) found a nitrogen content of around 1.9 wt.% for Baltic seaweeds, value in agreement with the one obtained at the present study. A nitrogen-to-protein conversion factor of 5 was used according Angell et al. (2016), who studied this conversion factor for seaweeds, using 103 algae species from different regions; they attested that this value should be used as conversion factor for seaweeds, instead of the standard 6.25 used for many materials. The protein content is also

Table 1. Characterization of Marine Macroalgae Waste (freeze-dried, size <1 mm).

Parameter	Value ($\bar{x} \pm s$)
Moisture (wt. %)	8.2 ± 0.5
Ash (wt. %)	48.7 ± 0.1
Lipids (wt. %)	0.40 ± 0.02
Nitrogen (wt. %)	1.96 ± 0.04
Protein (wt. %)	9.8 ± 0.2
Total Carbohydrates (wt. %)	41.1 ± 0.2

Results are expressed on dry basis and correspond to the sample mean value ± standard deviation.

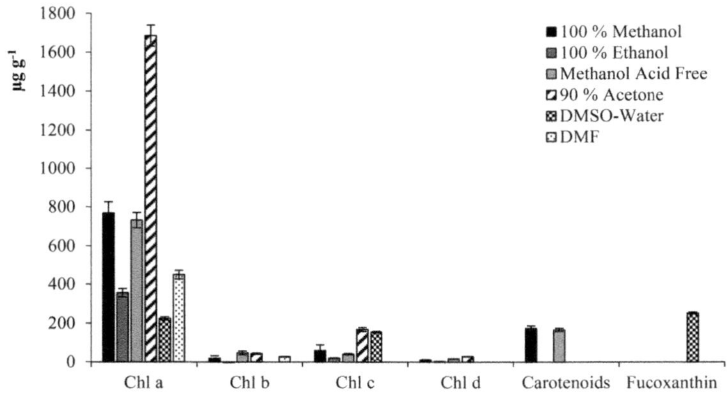

Figure 1. Pigments content extracted using several solvents at the best extraction condition (50 mg, 1 min of vortex agitation, plus 2 h in a multi-rotator). Error bars show the standard deviation. DMSO – dimethylsulfoxide; DMF – dimethylformamide; Chl – Chlorophyll.

in the range of the values reported in the literature. Sánchez-Machado et al. (2004) found a protein value of around 13 wt.% and Rodrigues et al. (2015) of around 14 wt.%. Garcia et al. (2016) studied *Saccorhiza polyschides* collected in the Barbate Estuary (Gulf of Cadiz, Spain) which presented a protein value close to 7 wt.%. Vieira et al. (2018) collected *Saccorhiza polyschides* in the North-Central coast of Portugal and obtained a protein value of 12.4 wt.%. The differences between the value obtained and those in the literature can be explained mainly by using different nitrogen-to-protein conversion factors.

The sample presents a low lipid content (less than 1 wt.%), as it was expected. Note that Maceiras et al. (2016) got a lipid content of 0.4 wt.% in *Saccorhiza polyschides* collected from Galician beaches. Also, Sánchez-Machado et al. (2004) and Rodrigues et al. (2015) obtained low lipid contents in their studies, namely 0.7 wt.% and 1.1 wt.%. The second most expressive compounds in the sample were total carbohydrates. The value obtained for them is in the range of that reported by Rodrigues et al. (2015) which was around 46 wt.%, and those from Sudhakar et al. (2018) which reported a carbohydrate content between 30 and 50 wt.% for brown seaweeds. Norton (1970) obtained a carbohydrate content around 30 wt.% for *Saccorhiza polycshides*, lower than that obtained at the present study.

3.2 *Pigments Extraction*

According to Connan (2015), DMF, acetone, methanol and ethanol are very convenient solvents for Chl extraction; in addition, DMSO was used as extractant of fucoxanthin; for carotenoids, the extraction was performed using 100 % methanol (Connan, 2015).

The preliminary extractions aimed to determine the best mass and the adequate extraction conditions. The best conditions were, therefore, those corresponding to the maximum amount of pigments extracted at the minimum amount of MMW for the established amount of solvent. When samples were kept standstill overnight (condition iii)), high amounts of pigments were observed; however, it can be due to the formation of Chl degradation compounds that severely interfere with all Chls determinations (Connan, 2015). For such reason, condition iii) was considered not advisable. The results showed that 50 mg of sample and the extraction conditions iv), with 2 h of agitation in a multi-rotator, were the best extraction scenario.

The pigments content using the best extraction conditions, for each solvent, is presented in Figure 1.

As expected, chlorophyll is the major pigment in the collected MMW sample, like in most of photosynthetic organisms, especially Chl *a* (Haryatfrehni et al., 2015, Pangestuti and Kim, 2011).

However, when DMSO-water (specific for brown seaweeds) was used, fucoxanthin (252 μg g^{-1}) together with Chl *c* (155 μg g^{-1}) appear as relevant pigments in the biomass, followed by Chl *a* (224 μg g^{-1}), findings in agreement with the literature (Haryatfrehni et al., 2015). Thus, the obtained results clearly show the relevance of using different solvents depending on the affinity to the compounds to be extracted (Chls, carotenoids, Fucox).

There are no studies in the literature focused in pigments extraction from *Saccorhiza polyschides*, and few studies were found concerning the extraction of pigments from brown algae.

Acetone was the most efficient solvent for pigments extraction, with clear evidence in the case of Chl *a* (1684.55 μg g^{-1}), followed by methanol (768.05 μg g^{-1}). These results are in agreement with those obtained by Yang et al. (2008), that extracted pigments from a brown macroalgae (*Laminaria japonica*) with six solvents. In that study, acetone was also the best extractant, with concentrations however lower than those obtained in present study and Chls as predominant pigment (Chl *a* had an amount of 174 μg g^{-1}). In the same study, methanol was the second most efficient extractant followed by ethanol with lower concentrations obtained for both solvents (methanol: 174 μg Chl *a* g^{-1}; 57 μg carot g^{-1}; ethanol: 142 μg Chl *a* g^{-1}) than those achieved at the present study (methanol: 768 μg Chl *a* g^{-1}; 174 μg carot g^{-1}; ethanol: 355 μg Chl *a* g^{-1}).

Vimala and Poonghuzhali (2015) also determined the amount of pigments in brown macroalgae and the values obtained were, in general, lower than those from the present study. Using acetone as extractant, Vimala and Poonghuzhali (2015) reached a mean value of 356 μg Chl *a* g^{-1}; using DMSO the values were 389 μg g^{-1} for Chl *a* and for fucox an amount of 8.58 μg g^{-1} was reported, respectively. Note that at the present study, 252 μg fucox g^{-1} was the value obtained. It is also important to refer that although the species used to compare were brown MM, the characteristics can differ between species; This comparation was made since there are no specific studies using *Saccorhiza polyschides* for pigments extraction.

Considering the obtained results and the literature review, the use of MMW as a source of NPs reveals to be promising. Further studies should be conducted in order to improve the selection of a solvent for making the extraction environmentally friendly, at a low cost, and with high efficiency.

4 CONCLUSIONS

The characteristics obtained for the marine macroalgae waste collected on a beach at Northern Portugal generally agreed with the values reported by the literature. Although the present work represents a preliminary study allowing the recovery of this untapped resource, it became clear that marine macroalgae waste has potential as a source of natural pigments for the replacement of synthetic pigments and the incorporation of this waste in the economic sector.

The selected extraction conditions with 1 min of vortex agitation, plus 2 h in a multi-rotator for an amount of 50 mg, ensure the non-formation of chlorophyll degradation compounds and maximum extraction. Acetone was the most efficient solvent for pigments extraction followed by methanol.

Pigments determination showed that chlorophyll is the predominant pigment in seaweeds, with a maximum of 1685 μg g^{-1} achieved when extracted with 90% acetone. Carotenoids and fucoxanthin were extracted in addition with methanol (174 μg g^{-1}) and DMSO (252 μg g^{-1}), respectively.

LIST OF ACRONYMS

Chls – Chlorophylls;
DMF – Dimethylformamide;
DMSO – Dimethylsulfoxide;
MM – Marine Macroalgae;
MMW – Marine Macroalgae Waste;
NPs – Natural Pigments.

ACKNOWLEDGEMENTS

This work was financially supported by project UID/EQU/00511/2019 – Laboratory for Process Engineering, Environment, Biotechnology and Energy – LEPABE funded by national funds through

FCT/MCTES (PIDDAC) and Project "LEPABE-2-ECO-INNOVATION" – NORTE-01-0145-FEDER-000005, funded by Norte Portugal Regional Operational Programme (NORTE 2020), under PORTUGAL 2020 Partnership Agreement, through the European Regional Development Fund (ERDF). The authors thank the financial support to the project Operação NORTE-01-0145-FEDER-000011 and the project UID/QUI/50006/2013 –POCI/01/0145/FEDER/007265. The authors also acknowledge Foundation for Science and Technology for funding Sara Pardilhó's (SFRH/BD/139513/2018) and Silvia Bessada's (SFRH/BD/122754/2016) PhD fellowships.

REFERENCES

Angell, A. R., Mata, L., de Nys, R. & Paul, N. A. 2016. The protein content of seaweeds: a universal nitrogen-to-protein conversion factor of five. *Journal of Applied Phycology* 28: 511–524.

AOAC 2012. Official Methods of Analysis. Association of Analytical Communities, USA.

Barbot, Y., Al-Ghaili, H. & Benz, R. 2016. A review on the valorization of macroalgal wastes for biomethane production. *Marine drugs* 14: 120.

Connan, S. 2015. Spectrophotometric Assays of Major Compounds Extracted from Algae. 1308, 75–101.

Garcia, J., Palacios, V. & Roldán, A. 2016. Nutritional Potential of Four Seaweed Species Collected in the Barbate Estuary (Gulf of Cadiz, Spain). *J Nutr Food Sci* 6: 2.

Haryatfrehni, R., Dewi, S. C., Meilianda, A., Rahmawati, S. & Sari, I. Z. R. 2015. Preliminary study the potency of macroalgae in yogyakarta: extraction and analysis of algal pigments from common gunungkidul seaweeds. *Procedia Chemistry* 14: 373–380.

Henriques, M., Silva, A. & Rocha, J. M. s. 2007. Extraction and quantification of pigments from a marine microalga: A simple and reproducible method. *Communicating Current Research and Educational Topics and Trends in Applied Microbiology*. A. Méndez-Vilas (Ed.), pp. 586–593.

Jensen, A., Indergaard, M. & Holt, T. 1985. Seasonal variation in the chemical composition of Saccorhiza polyschides (Laminariales, Phaeophyceae). *Botanica marina* 28: 375–382.

Maceiras, R., Cancela, Á., Sánchez, Á., Pérez, L. & Alfonsin, V. 2016. Biofuel and biomass from marine macroalgae waste. *Energy Sources, Part A: Recovery, Utilization, and Environmental Effects* 38: 1169–1175.

Michalak, I., Wilk, R. & Chojnacka, K. 2017. Bioconversion of Baltic Seaweeds into Organic Compost. *Waste and Biomass Valorization* 8: 1885–1895.

Norton, T. A. 1970. Synopsis of biological data on Saccorhiza polyschides. Roma: FAO Fisheries

Pangestuti, R. & Kim, S.-K. 2011. Biological activities and health benefit effects of natural pigments derived from marine algae. *Journal of Functional Foods* 3: 255–266.

Pardilhó, S., Duarte, R., Costa, A., Alves, R. C., Almeida, M. F., Nunes, A., Oliveira, M. B. P. P. & Dias, J. M. Characterization of Marine Macroalgae Waste aiming the production of biofuels and value added products – Preliminary studies. SUM2018 - Fourth Symposium of Urban Mining and Circular Economy, 2018 Bérgamo, Italy. CISA Publisher.

Ritchie, R. J. 2006. Consistent Sets of Spectrophotometric Chlorophyll Equations for Acetone, Methanol and Ethanol Solvents. *Photosynthesis research* 89: 27–41.

Rodrigues, D., Freitas, A. C., Pereira, L., Rocha-Santos, T. A., Vasconcelos, M. W., Roriz, M., Rodríguez-Alcalá, L. M., Gomes, A. M. & Duarte, A. C. 2015. Chemical composition of red, brown and green macroalgae from Buarcos bay in Central West Coast of Portugal. *Food chemistry* 183: 197–207.

Rupérez, P. 2002. Mineral content of edible marine seaweeds. *Food chemistry* 79: 23–26.

Sánchez-Machado, D. I., López-Cervantes, J., Lopez-Hernandez, J. & Paseiro-Losada, P. 2004. Fatty acids, total lipid, protein and ash contents of processed edible seaweeds. *Food Chemistry* 85: 439–444.

Sudhakar, K., Mamat, R., Samykano, M., Azmi, W. H., Ishak, W. F. W. & Yusaf, T. 2018. An overview of marine macroalgae as bioresource. *Renewable and Sustainable Energy Reviews* 91: 165–179.

Vieira, E. F., Soares, C., Machado, S., Correia, M., Ramalhosa, M. J., Oliva-teles, M. T., Paula Carvalho, A., Domingues, V. F., Antunes, F., Oliveira, T. A. C., Morais, S. & Delerue-Matos, C. 2018. Seaweeds from the Portuguese coast as a source of proteinaceous material: Total and free amino acid composition profile. *Food Chemistry* 269: 264–275.

Vimala, T. & Poonghuzhali, T. 2015. Estimation of pigments from seaweeds by using acetone and DMSO. *International Journal of Science and Research* 4: 1850–1854.

Yang, L., Li, P. & Fan, S. 2008. The extraction of pigments from fresh Laminaria japonica. *Chinese Journal of Oceanology and Limnology* 26: 193.

Wastes: Solutions, Treatments and Opportunities III – Vilarinho et al. (Eds)
© 2020 Taylor & Francis Group, London, ISBN 978-0-367-25777-4

Improvement of antioxidant compounds extraction by SSF from agro-food wastes

P. Leite, J.M. Salgado & I. Belo
Centre of Biological Engineering, University of Minho, Braga, Portugal

ABSTRACT: Mediterranean agro-industrial wastes are generated in huge amounts, mainly from olive mills, wineries and breweries. These wastes management poses serious environmental problems to the regions where they are generated. The objective of this work is to use the agro-industrial wastes as sources of phenolic compounds, improving its extraction from mixtures of olive pomace, brewer's spent grain and vine-shoot trimmings by solid-state fermentation (SSF) with *A. niger* and to obtain enzymes such as xylanases, cellulases and β-glucosidase. The results allowed to obtain the combination of wastes that maximized enzymes production and increased total phenolic compounds and antioxidant activity by SSF. Therefore, SSF showed to be an interesting valorization strategy to exploit agro-industrial wastes following the concept of circular economy.

1 INTRODUCTION

The industries of olive oil (2.8 Mt), wine (14.9 Mt), and beer (9.5 Mt) generate the majority of agro-industrial wastes in Mediterranean area. Mediterranean wastes represents respectively 92.8%, 51.2% and 5.3% of the total wastes generated annually by these industries in the world (FAOSTAT, 2014).

These wastes have generally no value for new applications, thus actions are being taken to change the society to a more environmentally friendly and resource-conserving. Circular economy is a concept that is gaining popularity because searches to reuse the wastes in a closed-loop (Ingrao et al., 2018; Korhonen et al., 2018).

Vegetables, fruits and beverages are the major sources of phenolic compounds in the human diet. The agricultural and food industries generate substantial quantities of phenolic-rich by-products, which could be a valuable natural sources of antioxidants (Balasundram et al., 2006). The main Mediterranean wastes are brewery spent grain (BSG) from brewery industry, olive pomace (OP) from olive oil industry, and vine-shoot trimmings (VST) from winery. Recently, it has been observed the increasing of the interest of scientific researchers for the study of biological properties of plants and active principles responsible for their therapeutic effects (Junio et al., 2011; Silva & Fernades Júnior, 2010).

Phenolic compounds can be extracted by conventional solvent extraction, such as microwave-assisted, Soxhlet, maceration, ultrasounds, high hydrostatic pressure and supercritical fluid extractions, among others (Ignat et al., 2011). However, the total recovery of them can be difficult because those compounds are present as insoluble bound to form conjugates with sugars, fatty acids or amino acids (Dey & Kuhad, 2014). In addition, the public awareness of health and environment along with safety hazards associated with the use of organic solvents in food processing and the possible solvent contamination of the final products together with the high cost of organic solvents, led to the development of a new and clean technology (Lafka et al., 2011).

Currently, enzymatic treatment for extraction of natural phenolics is a technique quite useful. Several microorganisms have the ability to produce a variety of enzymes under solid-state fermentation (SSF) (Dey & Kuhad, 2014). Besides that, SSF can be used for the production of some industrially important phenolic compounds, for the improvement of antioxidant potentials of solid substrates by increasing total phenolic compounds (TPC), and also for the bioavailability enhancement of them (Dey et al., 2016).

Filamentous fungi have the highest adaptability for SSF and are able to produce high quantities of enzymes with high commercial values (Dulf et al., 2017). *Aspergillus niger* has been used in many SSF studies. This fungi synthesize several food and industrial enzymes (cellulases, xylanase, β-glucosidase, protease, pectinase, ...) and has a significant role in the hydrolysis of phenolic conjugates (Dulf et al., 2017). The enzymes break the lignocellulosic cell walls and transform insoluble phenolics into soluble-free phenolics (Bhanja et al., 2009; Đorđević et al., 2010; Dulf et al., 2016; Wang et al., 2014; Zheng et al., 2009).

This study evaluated the use of SSF as a clean strategy to extract antioxidant phenolic compounds from mixtures of agro-industrial wastes, producing lignocellulolytic enzymes linked to the release of phenolic compounds from lignocellulosic material, such as xylanase, cellulase and β-glucosidase.

2 MATERIAL AND METHODS

2.1 *Raw material*

Two olive pomaces were used, the organic crude olive pomace (COP^{org}) and the exhausted olive pomace (EOP) that were collected from olive oil industry (Trofa, Portugal). Brewer's spent grain (BSG) was obtained from the beer industry (Vila Verde, Portugal) and vine-shoot trimming (VST) from the winery industry (Ourense, Spain) during the 2016/2017 season. These residues were dried at 65°C during 24 hours and stored at room temperature.

2.2 *Microorganisms*

Aspergillus niger CECT 2088 from CECT (Valencia, Spain) culture collection was used. It was revived on malt extract agar (MEA) plates. To obtain inoculum for SSF, the selected fungi were cultured on MEA slants, and incubated at 25°C for 6 days.

2.3 *Solid-state fermentation*

SSF process was carried out in 500 mL Erlenmeyer with 10 g of dry substrate (wastes mixtures) sterilized at 121°C for 15 minutes. Compositions of media were defined in Table 1. Moisture level was adjusted to 75% (w/w) in wet basis with distilled water and urea was added to adjust de ratio C:N to 15. The inoculation were performed following the methods described by Salgado et al. (2014).

SSF were incubated at 25°C for 7 days. The extraction of enzymes and phenolic compounds was performed with distilled water at room temperature in an L:S ratio of 5 and with agitation for 1 h. Following, extracts were centrifuged (4000 g, 15 min), filtered through Whatman N°1 filter paper and stored at 4°C. Controls of each run were performed without inoculation of fungi.

2.4 *Simplex centroid mixture design*

To evaluate the effect of mixture of agro-industrial wastes, it was implemented an experimental design (Simplex centroid mixture design). This design consists in a mixture run characterized by all one factor (all combination of two factors at equal levels and all combinations of three factors at equal levels). In addition, a center point with equal amounts of all wastes was studied. Thus, this design allowed to test four agro-industrial wastes as substrate and to evaluate the interaction effects among them in SSF (Table 1).

All experiments were performed in duplicate and in randomized order. In runs with COP^{org}, BSG, VST and EOP. The dependent variables studied were xylanase, cellulose, β-glucosidases

Table 1. Residues mixtures, obtained from Simplex centroid mixture design.

Run	A	B	C	D	COP^{org} (g)	BSG (g)	VST (g)	EOP (g)
1	1	0	0	0	10	0	0	0
2	0	1	0	0	0	10	0	0
3	0	0	1	0	0	0	10	0
4	0	0	0	1	0	0	0	10
5	0.5	0.5	0	0	5	5	0	0
6	0.5	0	0.5	0	5	0	5	0
7	0.5	0	0	0.5	5	0	0	5
8	0	0.5	0.5	0	0	5	5	0
9	0	0.5	0	0.5	0	5	0	5
10	0	0	0,5	0,5	0	0	5	5
11	0.33	0.33	0.33	0	3.33	3.33	3.33	0
12	0.33	0.33	0	0.33	3.33	3.33	0	3.33
13	0.33	0	0.33	0.33	3.33	0	3,33	3.33
14	0	0.33	0.33	0.33	0	3.33	3,33	3.33
15	0.25	0.25	0.25	0.25	2.50	2.50	2.50	2.50

activities and the increase of TPC and antioxidant capacity. A control of each experiment was performed without inoculation of fungus.

2.5 *Analysis of total phenolic compounds, antioxidant capacity and enzymes activity*

TPC, antioxidant activity by DPPH method, cellulases, xylanases and β-glucosidase activity was measured using the methods described by Leite et al., 2019.

3 RESULTS AND DISCUSSION

One of the main characteristics of SSF is the solid substrate. It acts as a physical support and source of nutrient during enzyme production. Thus the mixture of substrates can improve the enzymes growth and production, because it is difficult to acquire all the essential nutrients from the single substrate (Doriya & Kumar, 2018).

Simplex-centroid design allowed to optimize the combination of agro-industrial wastes as substrate for SSF in order to maximize the production of enzymes and the increase of TPC and antioxidant activity (Table 2).

Xylanase activity ranged from 55 to 710 U/g and the maximum value was found in 12th run consisting of 33% (w/w) of COP^{org}, 33% (w/w) of BSG and 33% (w/w) of EOP. Cellulase and β-glucosidase activity varied from 17 to 57 U/g and 43 to 262 U/g (Table 2) respectively. Maximum values were found in 5th run consisting of 50% (w/w) of COP^{org} and 50% (w/w) of BSG. On the other hand, the maximum values for the variation of antioxidant activity and TPC were found for 1st run (Table 2) in COP^{org}. Current experimental findings suggest that olive pomace (COP^{org} and EOP) and BSG have positive effects on enzymes production. COP^{org} is the best waste to obtain phenolic compounds with antioxidant activity.

Table 3 describes the optimal conditions for each dependent variable. It was concluded that the developed model can calculate the response accurately, with R^2 coefficients of 0.977 (xylanase activity), 0.960 (cellulase activity), 0.989 (β-glucosidase activity), 0.970 (variation of antioxidant activity) and 0.981 (variation of TPC).

In order to select a unique optimal substrate composition that maximize every dependent variable, it was performed an optimization of multiple response. The mixture of COP^{org} (42%, w/w), BSG (46, w/w) and EOP (12 w/w) was the optimal substrate that maximizes all dependent variables, with

Table 2. Enzymes activities, antioxidant capacity and TPC.

Run	Xylanase activity (U/g)	Cellulase activity (U/g)	β-glucosidase (U/g)	Variation of Antioxidant Activity	Variation of TPC
1	55 ± 3	25.1 ± 0.6	95 ± 5	26.6 ± 0.5	2.69 ± 0.45
2	395 ± 22	49 ± 4	214 ± 3	−3.2 ± 0.0	1,66 ± 0.06
3	83 ± 1	22 ± 2	43 ± 1	0.6 ±0.4	−0,34 ± 0.08
4	192 ± 2	20 ± 1	76 ± 4	−6.8 ± 2.2	0,73 ± 0.29
5	461 ± 6	56.2 ± 0.5	237 ± 9	4.6 ± 1.3	1,89 ± 0.07
6	82 ± 6	22 ± 1	84 ± 3	−1.7 ± 0.4	0,08 ± 0.11
7	69 ± 1	27 ± 6	84 ± 0	−9.5 ± 2.8	−0,05 ± 0.39
8	319 ± 5	57 ± 5	221 ± 11	−6.3 ± 1.2	0,91 ± 0.14
9	529 ± 19	56 ± 5	142 ± 1	1.7 ± 3.1	0,65 ± 0.32
10	63 ± 1	49.5 ± 0.0	98 ± 5	−3.6 ± 0.8	−0,02 ± 0.05
11	353 ± 38	17 ± 2	157 ± 3	−0.7 ± 0.5	−0,53 ± 0.18
12	710 ± 29	49.8 ± 0.5	262 ± 5	2.1 ± 1.4	−0,13 ± 0.52
13	104 ± 13	24 ± 0	95 ± 2	4.2 ± 3.6	0,01 ± 0.17
14	425 ± 16	54 ± 6	158 ± 8	−3.3 ± 0.8	−1,14 ± 0.23
15	652 ± 72	40 ± 3	158 ± 7	9.1 ± 4.7	−0,81 ± 0.17
16	547 ± 9	38 ± 2	148 ± 2	10.4 ± 1.2	−0,59 ± 0.06
17	649 ± 72	44.7 ± 0.3	167.4 ± 4.8	9.2 ± 0.4	−0,38 ± 0.06

Table 3. Optimum parameters for each dependent variable and statistical parameter.

	Xylanase activity (U/g) 790.47	Cellulase activity (U/g) 60.33	β-glucosidase (U/g) 273.16	Variation of Antioxidant Activity 26.66	Variation of TPC 2.70
COP^{org} (g)	0.26	0.00	0.33	1.00	0.00
BSG (g)	0.44	0.55	0.49	0.00	0.00
VST (g)	0.00	0.24	0.00	0.00	0.00
EOP (g)	0.30	0.21	0.17	0.00	1.00
R^2	0.977	0.960	0.989	0.970	0.981

Table 4. Optimization of multiple response.

	Xylanase activity (U/g)	Cellulase activity (U/g)	β-glucosidase (U/g)	Variation of Antioxidant Activity	Variation of TPC
Optimum value	667.691	57	267.12	5.925	0.968

the theoretical maximum activities on Table 4. Significant differences can be observed when are compared the maximum enzymes production of each dependent variable (Table 3) and optimizing all at once (Table 4).

This work shows that mixtures of wastes are an effective solution for the improvement of SSF by filamentous fungi, leading to the enhancement of enzymes production and to the liberation of phenolic compounds, which is in accordance with several works in the literature (Kumar et al., 2018; Ohara et al., 2017; Oliveira et al., 2017; Sousa et al., 2018).

Zimbardi et al. (2013) optimized the production of β-glucosidase, β-xylosidase and xylanase. The maximal production occurred in the wheat bran, but Sugarcane trash, peanut hulls and corncob enhanced β-glucosidase, β-xylosidase and xylanase production, respectively. Maximal levels after optimization reached 159.3 ± 12.7 U g^{-1}, 128.1 ± 6.4 U g^{-1} and 378.1 ± 23.3 U g^{-1}, respectively. Salgado et al. (2015) also observed an increase of production of the lignocellulolytic enzymes with the mixture of crude olive pomace, VST and Exhausted Grape Marc (EGM).

Cai et al. (2011), Razak et al. (2015) and Singh et al. (2010) reported that the use of filamentous fungi in SSF enhances the phenolic compound release in cereals due to the produced enzymes action.

Therefore, SSF has been proven to be an excellent process for the improvement of antioxidant properties and nutritional quality, of a great variety of vegetables and cereals, including agro-industrial wastes (Dey et al., 2016).

4 CONCLUSIONS

The mixture of wastes and its use as substrate in SSF improved the extraction of antioxidant phenolic compounds and the production of lignocellulolytic enzymes in comparison to the use of each waste alone. The simplex centroid mixture design allowed to optimize the wastes combination to maximize the extraction of antioxidant phenolic compounds or the production of lignocellulolytic enzymes. The optimization of multiple response selected only one combination of wastes that led to a maximum of all dependent variables studied. The best wastes combination was composed of COP^{org}, BSG and EOP. SSF showed to be a suitable and clean technology to extract antioxidant compounds from agro-industrial wastes.

ACKNOWLEDGEMENTS

Paulina Leite is recipient of a PhD fellowship (SFRH/BD/114777/2016) supported by the Portuguese Foundation for Science and Technology (FCT). José Manuel Salgado was supported by grant CEB/N2020 – INV/01/2016 from Project “BIOTECNORTE – Underpinning Biotechnology to foster the north of Portugal bioeconomy” (NORTE-01-0145-FEDER-000004). This study was supported by the Portuguese Foundation for Science and Technology (FCT) under the scope of the strategic funding of UID/BIO/04469/2019 unit and BioTecNorte operation (NORTE-01-0145-FEDER-000004) funded by the European Regional Development Fund under the scope of Norte2020 – Programa Operacional Regional do Norte.

REFERENCES

Balasundram, N., Sundram, K., & Samman, S. 2006. Phenolic compounds in plants and agri-industrial by-products: Antioxidant activity, occurrence, and potential uses. *Food Chemistry* 99: 191–203.

Bhanja, T., Kumari, A., & Banerjee, R. 2009. Enrichment of phenolics and free radical scavenging property of wheat koji prepared with two filamentous fungi. *Bioresource Technology* 100(11): 2861–2866.

Cai, S., Wang, O., Wu, W., Zhu, S., Zhou, F., & Ji, B. 2011. Comparative study of the effects of solid-state fermentation with three filamentous fungi on the total phenolics content (TPC), flavonoids, and antioxidant activities of subfractions from oats (Avena sativa L.). *Journal of Agricultural and Food Chemistry* 60(1): 507–513.

Dey, T. B., Chakraborty, S., Jain, K. K., Sharma, A., & Kuhad, R. C. 2016. Antioxidant phenolics and their microbial production by submerged and solid state fermentation process: A review. *Trends in Food Science & Technology* 53: 60–74.

Dey, T. B., & Kuhad, R. C. 2014. Enhanced production and extraction of phenolic compounds from wheat by solid-state fermentation with Rhizopus oryzae RCK2012. *Biotechnology Reports* 4: 120–127.

Đorđević, T. M., Šiler-Marinković, S. S., & Dimitrijević-Branković, S. I. 2010. Effect of fermentation on antioxidant properties of some cereals and pseudo cereals. *Food Chemistry* 119(3): 957–963.

Doriya, K., & Kumar, D. S. 2018. Optimization of solid substrate mixture and process parameters for the production of L-asparaginase and scale-up using tray bioreactor. *Biocatalysis and Agricultural Biotechnology* 13: 244–250.

Dulf, F. V., Vodnar, D. C., Dulf, E. H., & Pintea, A. 2017. Phenolic compounds, flavonoids, lipids and antioxidant potential of apricot (Prunus armeniaca L.) pomace fermented by two filamentous fungal strains in solid state system. *Chemistry Central Journal* 11(1): 92.
Dulf, F. V., Vodnar, D. C., & Socaciu, C. 2016. Effects of solid-state fermentation with two filamentous fungi on the total phenolic contents, flavonoids, antioxidant activities and lipid fractions of plum fruit (Prunus domestica L.) by-products. *Food Chemistry* 209: 27–36.
FAOSTAT. 2014. FAO statistical Database. Retrieved from http://faostat.fao.org
Ignat, I., Volf, I., & Popa, V. I. (2011). A critical review of methods for characterisation of polyphenolic compounds in fruits and vegetables. *Food Chemistry* 126(4): 1821–1835.
Ingrao, C., Faccilongo, N., Gioia, L. Di, & Messineo, A. 2018. Food waste recovery into energy in a circular economy perspective: A comprehensive review of aspects related to plant operation and environmental assessment. *Journal of Cleaner Production* 184: 869–892.
Junio, H. A., Sy-Cordero, A. A., Ettefagh, K. A., Burns, J. T., Micko, K. T., Graf, T. N., ... Cech, N. B. 2011. Synergy-Directed Fractionation of Botanical Medicines: A Case Study with Goldenseal (Hydrastis canadensis). *Journal of Natural Products* 74(7): 1621–1629.
Korhonen, J., Honkasalo, A., & Seppälä, J. 2018. Circular Economy: The Concept and its Limitations. *Ecological Economics* 143: 37–46.
Kumar, P., Chawla, P., & Singh, J. 2018. Food Bioscience Fermentation approach on phenolic, antioxidants and functional properties of peanut press cake. *Food Bioscience*, 22(January), 113–120.
Lafka, T. I., Lazou, A. E., Sinanoglou, V. J., & Lazos, E. S. 2011. Phenolic and antioxidant potential of olive oil mill wastes. *Food Chemistry* 125(1): 92–98.
Leite, P., Silva, C., Salgado, J. M., & Belo, I. 2019. Simultaneous production of lignocellulolytic enzymes and extraction of antioxidant compounds by solid-state fermentation of agro-industrial wastes. *Industrial Crops and Products* 137: 315–322.
Ohara, A., Santos, J. G., Angelotti, J. A. F., Barbosa, P. P. M., Dias, F. F. G., Bagagli, M. P., & Castro, R. J. S. 2017. A multicomponent system based on a blend of agroindustrial wastes for the simultaneous production of industrially applicable enzymes by solid-state fermentation, 1–7.
Oliveira, F., Salgado, J. M., Abrunhosa, L., Pérez-Rodríguez, N., Domínguez, J. M., Venâncio, A., & Belo, I. 2017. Optimization of lipase production by solid-state fermentation of olive pomace: from flask to laboratory-scale packed-bed bioreactor. *Bioprocess and Biosystems Engineering* 40(7): 1123–1132.
Razak, D. L. A., Rashid, N. Y. A., Jamaluddin, A., Sharifudin, S. A., & Long, K. 2015. Enhancement of phenolic acid content and antioxidant activity of rice bran fermented with Rhizopus oligosporus and Monascus purpureus. *Biocatalysis and Agricultural Biotechnology* 4(1): 33–38.
Salgado, J. M., Abrunhosa, L., Venâncio, A., Domínguez, J. M., & Belo, I. 2014. Screening of winery and olive mill wastes for lignocellulolytic enzyme production from Aspergillus species by solidstate fermentation. *Biomass Convers. Biorefinery* 4: 201–209.
Salgado, J. M., Abrunhosa, L., Venâncio, A., Domínguez, J. M., & Belo, I. 2015. Enhancing the Bioconversion of Winery and Olive Mill Waste Mixtures into Lignocellulolytic Enzymes and Animal Feed by Aspergillus uvarum Using a Packed-Bed Bioreactor. *Journal of Agricultural and Food Chemistry* 63(42): 9306–9314.
Silva, N. C. C., & Fernades Júnior, A. 2010. Biological properties of medicinal plants: a review of their antimicrobial activity. *Journal of Venomous Animals and Toxins Including Tropical Diseases* 16(3): 402–413.
Singh, H. B., Singh, B. N., Singh, S. P., & Nautiyal, C. S. 2010. Solid-state cultivation of Trichoderma harzianum NBRI-1055 for modulating natural antioxidants in soybean seed matrix. *Bioresource Technology* 101(16): 6444–6453.
Sousa, D., Venâncio, A., Belo, I., & Salgado, J. M. 2018. Mediterranean agro-industrial wastes as valuable substrates for lignocellulolytic enzymes and protein production by solid-state fermentation. *Journal of the Science of Food and Agriculture* 98(14): 5248–5256.
Wang, C. Y., Wu, S. J., & Shyu, Y. T. 2014. Antioxidant properties of certain cereals as affected by food-grade bacteria fermentation. *Journal of Bioscience and Bioengineering* 117(4): 449–456.
Zheng, H., Hwang, I. W., & Chung, S. K. 2009. Enhancing polyphenol extraction from unripe apples by carbohydrate-hydrolyzing enzymes. *Journal of Zhejiang University SCIENCE B* 10(12): 912.
Zimbardi, A. L. R. L., Sehn, C., Meleiro, L. P., Souza, F. H. M., Masui, D. C., Nozawa, M. S. F., ... Furriel, R. P. M. 2013. Optimization of β-Glucosidase, β-Xylosidase and Xylanase Production by Colletotrichum graminicola under Solid-State Fermentation and Application in Raw Sugarcane Trash Saccharification. *International Journal of Molecular Sciences* 14: 2875–2902.

Wastes: Solutions, Treatments and Opportunities III – Vilarinho et al. (Eds)
© 2020 Taylor & Francis Group, London, ISBN 978-0-367-25777-4

Systematic approach in waste management and recycling

A. Adl
Value Oriented Lda, Lisbon, Portugal

ABSTRACT: Waste is called to an unusable thing required to be disposed and thrown away. So, if we start changing our approach and point of view to these things, we would achieve remarkable improvements in waste management and recycling. This paper would draw our attention to the stage before waste disposal and recycling and discuss about somehow big advantages of having different new approach at this stage which would result in less amount of wastes, more recycling, upcycling, reuse and recovery rates and respectively creating more values for available resources than before. We would also rarely hear from the author as an investor entrepreneur about his programs for training and empowerment of the society for having this new approach and for doing research and producing valuable things from wastes as secondary available resources.

1 INTRODUCTION

Since 1996 while I started my final B.Sc. thesis project in "10% recycling of mill scales in the steel production industry", I fell in love with waste management and recycling and so far, I was part of some unique projects and experiences: reuse of steel mill scales in other industries, producing Strontium Sulfate from rare impure resources, recovery of cracked four meter diameter of mill cap in cement industry, producing foam concrete using foam agent from slaughter houses.

I was trained very well during being employee career, so I have started my own business since 2007 and now I am running my fourth company here in Portugal in waste management and recycling. My partner and I started planning for our current business by attending second world congress and expo on recycling in 2016 in Berlin and then Wastes 2017 in Porto and now we are in the beginning of building production, training and research center for correct consuming, recycling and re-use of materials.

Recently during our studies and research for our new business, we found out by applying systematic approach, it would be much easier to accomplish Waste Management Plans and meet declared targets.

Investigating literatures and reviewing conferences proceedings were not enough so far to find any direct mention of anybody discussing systematic, process and standard approach for preventing or reducing wastes in the scope of waste management and recycling. It has not been seen so far discussing the effect of cultural informative programs, optimum and correct consumption of resources and the attitude of thinking of adding value to seemingly unusable things before disposing them in the territory of waste management and recycling.

It may seem so simple and this article is going to invite all people on earth to get back few steps, think outside the box and try a kind of simple new approach to whatever they call "waste" and it would result in a huge deduction in waste amounts around the world.

2 WHAT IS SYSTEMATIC APPROACH?

Systematic approach is one of concepts I discuss in my business coaching classes as a part of strategic approach. Very simple, it is methodical approach repeatable and learnable through a step by step procedure (business dictionary, in press), focus on whole big picture, asking WH questions, combining information and intuition, thinking creatively outside-the-box and step out of our shoes and into the shoes of interested parties to find system-wide focus and gain systemic insights into complex situations and problems (Kotelnikov, in press).

3 WHY SYSTEMATIC APPROACH IN WASTE MANAGEMENT AND RECYCLING?

When we approach to waste management and recycling systematically and we find the system-wide focus and systemic insight into it, literally and genuinely we would find some practical missing rings which need more attention and play key roles in meeting planned targets.

In my opinion, while we are ignoring these key indicators because of whatever, all seminars, conferences, speeches, gatherings, etc. would lose their efficiencies. During my more than twenty years career, I was not lucky enough to witness many follow-up presentations of a research project to announce practical pilot and industrial tests.

In this presentation, I am going to mention some of those missing rings and declare our current business goals to return them to the mainstream which may play a role as well in meeting National Waste Management targets for 2020 and beyond (PERSU 2020).

Since Last July which we started our business Deming PDCA (Plan-Do-Check-Act) cycle (Landesberg 1999), we explored following facts as missing rings and key indicators that need to be taken care seriously to speed up making close of real results to the designed plan as much as possible.

Investigating literatures and reviewing conferences proceedings were not enough so far to find any direct mention of anybody discussing systematic approach for preventing or reducing wastes in the scope of waste management and recycling.

4 IMPORTANCE OF INFORMATION IN WASTE MANAGEMENT AND RECYCLING

There is a psychological concept in business coaching that explains when you are pointing out to a cause, as a matter of fact it is just one finger showing that cause and three others show yourself as the main cause. This fact simply notifies that we should take care of our responsibility as the main attitude while doing cause and effect analysis.

So, let's start from ourselves: how many bins do we have in our living place to dispose our wastes? how many of us look for the right bin to dispose our wastes? How often are we thinking of the waste we are disposing as a secondary available material and resource for making something valuable? Why should we do that? What is in it for us if we dispose our wastes correctly? Who would be beneficiary of this much sensitivity?

During our visit of some urban waste collection and preliminary processing sites in Portugal, it could clearly be seen that people mostly mix different kinds of wastes at the disposal bins and before going to disposal bins at their living places as well.

Many efforts have been done and is being done to inform families about the concepts of separation at the origin and correct disposal into the correct bins, but we explored that they are not enough, and we need more cultural informative programs to assure that every member of a family has been informed about benefits could be achieved for themselves and for the society by fulfilling abovementioned concepts.

5 OPTIMUM AND CORRECT CONSUMPTION

Some of us may think what it is, some of us may say that: "consumption is consumption, we consume something and that's all, what is optimum consumption?" I would say that optimum consumption shows our responsibly and respect to resources. I would like to share my experience as an example in this regard. Fortunately, by means of technology, the rate of fuel consumption can be monitored for most of the cars, I did a small study with my hybrid car that by controlling our foot pressure on the accelerator, we can reach an optimum rate of fuel consumption specially in a bit long trips with uphill downhill roads. Consuming fuel in its optimum rate, will cause the car lives longer and the least rate of depreciation, this in return will save a lot of money in maintenance costs which means less wastes of car parts, less carbon footprint and more capability for investment opportunities.

Upcycling is another example for optimizing consumption rates. It just needs to reconsider broken things, appliances, etc. one more time before throwing them out. *UPCAST OY*, a Finnish company, which has developed a production line for 100% recycling of copper scraps, has published that: "*Even a 10% reduction in energy consumption means a remarkable amount of money in production costs*" (upcast, in press).

Clearly, it can be understood that the first direct benefit of optimized consumption returns to the person who is doing it.

There is another point of view: correct consumption/use. This is another attitude that decreases the amount of wastes very easily and importantly. You might have experienced this: using knife tip for opening small screws or paint and adhesive cans. That is right, mostly you would have regretted yourself, because you either bend the tip or scratch the plastic base of the small screw. Using appliances according to their standard manuals is another approach of correct use which would result in decreasing waste quantity. For example, if we use our washing machine every time in either less or more weight of its standard weight described in its manual, we should repair or replace it sooner than its normal working life.

In my opinion, for using and consuming in the optimum and correct way, we need to believe and stick to quality definition and principles.

In this regard, we have to refer to the last version of ISO (the International Organization for Standardization)'s quality management definitions. According to this international standard quality is the degree to which a set of inherent characteristics (distinguishing features) of an object (anything perceivable or conceivable) fulfils requirements (needs or expectations that are stated, generally implied or obligatory) (ISO 9000:2015). In Joseph Juran's words quality is fitness for use (Defeo 2017, Landesberg 1999). Very simple, quality in each process elements (Fig. 1) is avoiding accepting bad materials as inputs, avoiding interrelated or interacting bad activities (procedures) and finally avoiding delivery of bad products or services as outputs and this is the definition when I was being trained as QMS (quality management system) consultant.

In other words, according to quality principles (ISO 9000:2015), if each person, family and enterprise gets engaged in the correct consumption activities and has a process approach in accomplishing those activities, the desired result i.e. less disposed wastes, would be achieved more efficiently and effectively.

6 ADDING VALUE

While we are practicing for process approach in our lifestyle, we will respectively experience another important and valuable attitude and that is thinking of adding value to seemingly unusable things before disposing them. If every human being gets used to this attitude, most of waste problems around the world would be resolved. Practicing this attitude would result in developing secondary available resources and new products from unusable things and wastes.

Restoration, repair, producing new things are kinds of adding value to reusable items as available resources. Nowadays many of different ways of restoration, repair, recycling, upcycling, recovery and reuse of unusable items could easily be learnt on the worldwide web.

Figure 2 shows a very simple example of adding value to disposable egg trays and toothpicks, I made it the first days I had arrived in Portugal in July 2018. It is simply made of a piece of used cardboard egg tray, six pieces of used toothpicks, few drops of hot melted glue and a little paint to turn it to a brand-new egg cup.

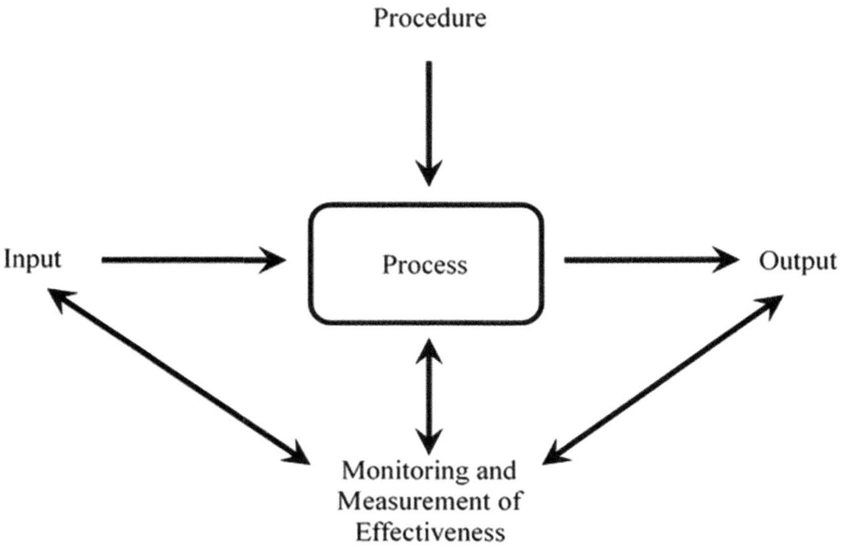

Figure 1. Process interaction diagram (ISO 9001:2015).

Figure 2. Upcycling used cardboard egg trays and toothpicks to make a brand-new value-added egg cup.

I also shred our all plastic wastes at home to make them ready for our business which is described in the following clause.

7 R2V – RECYCLE TO VALUE

This brand has been born in 2016 in Denmark while we were investigating European territory to find our favorite state to start our new business. Currently, we are following regulations to build production, training and research center for correct consuming, recycling and reuse of materials in the industrial zone of Montalvo in Constância, district of Santarém, not with concrete and normal construction but with recycled materials like used shipping containers and plastic wastes.

As it can be realized from the name of the business, we are going to train young students, young graduates, industry staff and public society about optimized and correct consuming, reuse, recycling and upcycling of used items and materials to look at these things not as wastes to be disposed but as secondary available resources.

8 CONCLUSIONS

This paper tried to introduce a simple new approach at one level before something to be called waste, being disposed and managed for recycling.

Practicing systematic and process approach in life by every human being would increase our responsibility and respect to our available primary and secondary resources.

Holding more cultural informative programs about more efficient and effective separation at the origin and correct disposal in the correct bins would increase efficiency and effectiveness of waste management and recycling plans.

Consuming resources correctly at their optimum rate would decrease a lot of payable costs for all of us at first and for fulfilling waste management and recycling plans.

Process approach would cause thinking twice and reconsidering seemingly unusable items as valuable secondary available resources to be restored, repaired and produced new added-value products.

Taking seriously whatever has been discussed in this paper would result in having more joyful environment and Value Oriented Lda would apply these simple attitudes in its business to train people for correct and optimum consuming and reuse, recycle and upcycle of available secondary resources in the center region of Portugal.

ACKNOWLEDGMENT

I would like to thank the municipality of Constância from the reception lady at the entrance who welcomes all visitors with a happy and smiley face, to the mayor, Mr. Sérgio Oliveira who has run an efficient and effective system for attracting helpful innovative businesses in Constância.

REFERENCES

Defeo, Joseph (7) 2017. *Juran's quality handbook: The Complete Guide to Performance Excellence.* New York:McGraw-Hill.

http://www.businessdictionary.com

http://www.kotelnikov.biz

https://www.upcast.com

ISO 9000:2015. *Quality management systems – Fundamentals and vocabulary.*

ISO 9001:2015. *Quality management systems – Requirements.*

Landesberg, Phil 1999. In the Beginning, There Were Deming and Juran, *The Journal for Quality & Participation*: 59–61.

PERSU 2020. *National Waste Management Plans.*

Wastes: Solutions, Treatments and Opportunities III – Vilarinho et al. (Eds)
© 2020 Taylor & Francis Group, London, ISBN 978-0-367-25777-4

Valorization of bamboo wastes for the production of particleboards

A.C.C. Furtini, C.A. Santos, D.L. Faria, A.P.L. Vecchia, I.C. Fernandes,
L.M. Mendes & J.B. Guimarães Júnior
Federal University of Lavras, Lavras, Brazil

ABSTRACT: This work aimed to verify the potential of bamboo wastes and pinus wood to produce particleboards. The boards were produced with 9% urea- formaldehyde adhesive, with press cycle with temperature 160°C, pressure 4.00 MPa and time of 8 minutes. The nominal density established for the boards was 0.59 g cm^{-3}. Particleboards of bamboo, pinus and boards with 50% of bamboo and 50% of pinus were produced. Were evaluated the modulus of elasticity (MOE) and modulus of rupture (MOR) in static bending, internal bond (IB), nominal density (ND), thickness swelling (TS) and water absorption (WA). Based on the results it can be concluded that the production of the bamboo particleboards are viable.

1 INTRODUCTION

The s particleboards are plates of wood particles or lignocellulosic materials bonded together with synthetic adhesive (Moslemi, 1974; Tsoumis, 1991). According to Guimarães Júnior et al. (2011), conventional agglomerated boards are by concept the possibility of using lower quality raw materials such as forest or agro-industrial waste in their production.

With the growth of the wood boards industry, the demand for raw material also increases, making necessary not only the increase of areas of plantations with currently used species of the genus pinus and eucalyptus, but also the search of new options of raw materials (Arrapo et al., 2014; Mendes et al., 2014).

According to Mendes et al. (2010), the production of particleboards from non-wood lignocellulosic materials provides added value to this raw material, meeting the growing demand of the wood boards industry, in addition to allowing its expansion, reducing the use of wood and consequently, pressure on forests, and reduce production costs of the boards, making them even more competitive in the economic scenario.

An alternative of lignocellulosic biomass with potential to be used for the production of boards is bamboo, which is a fibrous plant belonging to the family *Gramineae* or *Poaceae*, composed of approximately 45 genera and 1300 species worldwide. Bamboo has been widely used as a source of fibers for particleboards because of its excellent physical and mechanical properties. Characteristic perennial, renewable, fast growing and high yield per area, low cost and diversity in use as well as being an excellent carbon kidnapper (Gatto et al., 2007).

In this sense, the objective of this work was to verify the potential of bamboo biomass and pinus wood to produce low densification particleboards.

2 MATERIAL AND METHODS

2.1 *Chemical characterization*

The chemical characterization of the residues particles of *Pinus oocarpa* and *Dendrocalamus Giantus* were realized after the reduction of sawdust particles with the help of a Willey type mill.

After milling, the sawdust was separated in overlapping sieves of 0.420 and 0.250 mm, using for analysis only the fraction that was retained in the sieve of 0.250 mm opening. After air conditioning at a temperature of 22 ± 2°C and 65 ± 5% relative humidity, the characterization was initiated by means of moisture analysis according to ASTM E871-82 (2013), analysis of total extractives according to with NBR 14853 (2010), insoluble lignin content according to the norm NBR 7989 (2010), ash content according to norm NBR 13999 (2017) and holocellulose content obtained by the difference between the proportions.

2.2 *Production of Boards*

The particles of wood and bamboo used in the manufacture of the boards came respectively from the *Pinus* tree *oocarpa* and *Dendrocalamus Giantus* located on the campus of the Federal University of Lavras, in Lavras, Minas Gerais, Brazil.

Sliver particles of pinus and bamboo were obtained from hammer mill processing. After the processing, the resulting material was sieved, and selected the one retained between the 10 and 30 mesh sieves to produce the boards, which were oven dried with forced air circulation, with an initial temperature of 50 ± 3°C, then passed to 80 ± 3°C, to the moisture content of 3% on the dry basis.

Subsequently, the particles were mixed manually, taking into account the proportions of the treatments defined in Table 1, with 9% urea-formaldehyde adhesive. The particulate mattress was then pre- pressed at 0.5 MPa for 5 minutes at room temperature. Subsequently, the boards passed the pressing cycle with a temperature of 160°C and a specific pressure of 4.00 MPa, for a period of 8 minutes, thus obtaining the homogeneous boards.

Three boards were produced for each treatment, the first being produced only with particles of Pinus, the secon 50% Pinus and 50% bamboo and the third only with bamboo particles whose dimensions and nominal density were, respectively, 250 mm × 250 mm × 15 mm and 0.59 g cm^{-3}.

To determine the physico-mechanical quality of the boards, four est bodies per boards were extracted and used to determine the apparent density, water absorption and swelling in thickness after 24 hours of immersion; 4 test specimens per boards for determination of perpendicular tensile strength, according to procedures of ASTM D1037 (2006). In order to determine the strength and stiffness of the static flexion test, two test pieces per boards were extracted, and the test procedure was performed according to ASTM D1037 (2006). Preliminarily to the tests, the boards were air-conditioned at 20 ± 3°C and relative humidity of 65 ± 5%.

2.3 *Quantitative methodologies for potential boards analysis*

For the evaluation of the physical and mechanical properties, it was followed the norm CS 236-66 (COMMERCIAL STANDARD – CS, 1968). In relation to density, the standard classifies as low density boards values less than 0.60 g cm^{-3}; medium density between 0.60 and 0.80 g cm^{-3} and high density higher than 0.80 g cm^{-3}.

For swelling in thickness, after 24 hours of immersion in water, the maximum values s required by the standard are 30% for low density boards, 35% for medium density boards.

In mechanical tests the standard establishes minimum values for low density boards for MOE, MOR and perpendicular traction of respectively 1029.7 MPa; 5.6 MPa and 0.14 MPa, for medium density 2401.96 MPa;10.98 MPa; 0.42 MPA.

3 RESULTS AND DISCUSSION

3.1 *Chemical characterization of lignocellulosic materials*

Knowledge of chemical composition of the material is of fundamental importance for the understanding of its influence on the characteristics of the composite material and also on the choice

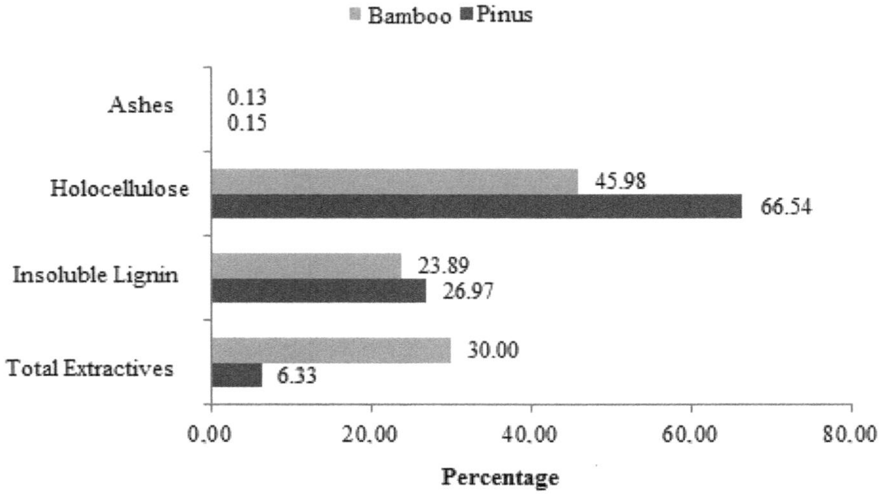

Figure 1. Chemical characterization of lignocellulosic materials.

of chemical treatments (Clemons; Caulfield, 2005). Figure 1 shows the results obtained with the chemical analysis for total extractives, insoluble lignin, holocellulose and ash.

The bamboo presented high content of total extractives compared to the studied pinus wood. The extractives are a group of heterogeneous chemical substances related to the defense mechanisms of the plant and influenced by genetic and edaphoclimatic factors (Soares et al., 2017).

As mentioned by Bufalino et al. (2012), species with high extractives contents can generate boards whose collage is less efficient and of inferior quality in relation to those with low contents. The extractives present in the lignocellulosic material may migrate to the surface during the drying process and/or pressing of the sheet and hence inactivate the surface and impede wettability of wood and penetration of the adhesive (Frihart; Hunt, 2010; Bufalino et al., 2012).

As for the lignin content, bamboo presents a small disadvantage compared to pinus wood. The high levels of lignin can contribute to the improvement of the physical properties of the boards produced. According to Araújo Júnior (2014), in the temperature range between 180°C and 220°C, the lignin fills filling the voids present on the surface and throughout the boards, which will make it difficult for water to enter the material, thus reducing the physical absorption properties of water and swelling in thickness.

According to Iwakiri et al. (2005), higher lignin contents are desirable for the production of reconstituted wood boards since lignin may serve as a natural adhesive. Soon it is expected that plates produced withlignocellulosic materials with higher lignin content have higher water resistance and better mechanical properties (Khedari et al., 2004).

The ash content of the two materials was very low. According to Kniess et al. (2015) low ash content is an advantage because high values can negatively influence the physical and mechanical properties of these boards. The presence of larger amounts of minerals and some apolar extractives may result in the blocking of chemical groups reactive for adhesion with polar adhesives, thus affecting the quality of the bonding and the mechanical performance of the reconstituted sheets (Ndazi et al., 2007).

3.2 *Physical-mechanical quality of agglomerated boards*

Table 1 shows the apparent density results of the boards. For this property no differences were observed with the inclusion of bamboo in particleboards. In addition, the densities of the boards were lower than the nominal value (0.60 g/cm^3). This can be attributed to the specificity of the

Table 1. Physical and mechanical properties of the boards.

Treatments	DA (g/cm^3)	AA (%)	IE (%)	MOE (MPa)	MOR (MPa)	TP (MPa)
Wood boards	0.55 A	85.88 A	9.59 A	1954 A	15.28 A	0.44 A
Wood-Bamboo Boards	0.57 A	92.15 A	16.28 B	1624 A	11.23 B	0.39 A
Bamboo boards	0.52 A	118.18 B	19.92 C	1214 B	11.31 B	0.33 B

laboratory conditions in relation to the production, with loss of materials during the handling of the particles in the steps of adhesive application, formation of the mattress and pressing of the boards.

According to CS 236-66 (COMMERCIAL STANDARD – CS, 1968) all boards produced were classified as low density ($<0.60\ g\ cm^{-3}$).

In the water absorption and in the thickness swelling (TS), it is observed that the boards produced exclusively with bamboo fibers presented higher values for this property. This phenomenon may have occurred due to the fact that lignocellulosic materials with lower density present in general higher water absorption indices because they present greater release of the compressive stresses imposed during the pressing Iwakiri et al. (2005).

According to Soares et al. (2017), the particleboards of low density and with the addition of sugarcane bagasse had an average value of 132.8% for absorption after 24 hours, showing to be superior to the present work. Scatolino et al. (2017), reported water absorption values after 24 hours of water immersion of 138.1%, respectively, for low density boards produced with eucalyptus wood, showing higher values than those observed in this study. Thus, it is assumed that the higher water absorption in boards of lower density occurred because there was a greater volume of voids that could be occupied by water.

Standard CS 236-66 (COMMERCIAL STANDARD, 1968) requires boards of swelling in thickness after 24 hours of immersion in water, of a maximum of 30% for boards of low density. In this case, both boards met the normalization.

With the insertion of bamboo in the boards, it was observed reduction in MOE and MOR. This behavior is probably due to the extractive content of bamboo being larger than wood. The presence of these chemical components may hinder the bonding process, generating more fragile glue lines, leading to lower strength and stiffness (Soares et al., 2017).

This may have occurred because bamboo presented low lignin values and high extractive values. For Tsoumis (1991) to decrease in the amount of extractives causes an improvement in the bonding of the particles, due to the lower probability of interference of these substances in adhesive cure. Already Iwakiri et al. (2005) the relative increase of the amount of lignin, can act as an adhesive, helping in the approximation and interaction between the particles.

The internal bond (IB) is a property that evaluates the bonding relationship between the particles. It was observed that the presence of bamboo caused a decrease of this property. This behavior can be explained by the increase of the extractive content in the boards, one that such chemicals act to hinder or render unfeasible the bonding. The extractives present in the lignocellulosic material may migrate to the surface during the drying and / or pressing process of the sheet and, consequently, inactivate the surface and hinder the wettability of the sheet. wood and adhesive penetration (FRIHART; HUNT, 2010; BUFALINO et al., 2012a, 2012b).

The standard CS 236-66 (CS 1968) establishes minimum values pray for MOE, MOR and internal bond of respectively 1029.7 MPa; 5.6 MPa and 0.42 MPa. In this sense all the treatment met the normative requirements.

4 CONCLUSION

According to the results, it is concluded:

- The production of boards exclusively made of bamboo and with a part being pinus, are viable.

- There was increased water absorption and thickness swelling (TS) for boards produced with bamboo.
- There was a decrease in the mechanical properties of MOE, MOR and internal bond (IB) for boards produced with bamboo.
- Given the low density of the boards enables study indicates its application as insulating thermal acoustic. However, more specific tests should be carried out on the material to verify its efficiency when applied under these conditions.
- The association of bamboo and pine wood for the production of chipboard boards has been shown to be a viable alternative and can be applied in internal environments according to the adhesive used, with application on table tops, side doors and cabinets, racks, partition and side shelves.

ACKNOWLEDGMENTS

We thank FAPEMIG, CNPq, CAPES, RELIGAR NETWORK and PPG BIOMAT / UFLA.

REFERENCES

American Society for Testing Materials. *ASTM E871-82:* Standard Test Method for Moisture Analysis of Particulate Wood Fuels, ASTM International, West Conshohocken, PA; 2013.

American Society for Testing and Materials, "Standard test methods for evaluating properties of wood fiber-base and particle panel materials", *ASTM D1037-06*, Philladelphia, 2006.

Araújo Júnior, CP de. *Fibreboards made from the green coconut shell without the addition of binder resins.* 2014. 82 p. Dissertation (Master in Engineering and Materials Science) – Federal University of Ceará, Fortaleza.

Brazilian Association of Technical Standards. *NBR 13999*: Paper, paperboard, cellulosic pulp and wood: determination of the residue (ash) after incineration at 525°C. Rio de Janeiro; 2017.

Brazilian Association of Technical Standards. *NBR 14853:* Wood – determination of the material soluble in ethanol-toluene and in dichloromethane and acetone. Rio de Janeiro, 2010a. 3 p.

Brazilian Association of Technical Standards. *NBR 7989*: cellulosic pulp and wood: determination of acid-insoluble lignin. Rio de Janeiro, 2010b. 6 p.

Bufalino, L.; Albino, Vcs; Sá, Va; Corrêa, Arr; Mendes, Lm; Almeida, NA Particle boards made from Australian red cedar: processing variables and evaluation of mixed species. Journal of Tropical Forest Science, v. 24, n. 2, p.162–172, 2012b.

Bufalino, L.; Protasio, Tpp; Couto, AM; Nassur, Oac; Sá, Va; Harvest, Pf; Mendes, Lm *Chemical and energetic characterization for the utilization of the veneer wood and thinning of Australian cedar.* Brazilian Forest Research, v. 32, n. 70, p. 129–137, 2012.

Clemons, CM; Caulfield, DF *Natural fibers: Functional fillers for plastics,* Wiley-VCH Verlag: Weinheim, 2005, chap. 11.

Commercial Standard. *Mat formed wood particleboard. CS* 236–66. In: Book of Commercial Standards. Wallingford, 1968.

Farrapo, CL; Mendes, RF; Guimarães Jr., JB; Mendes, LM *Use of wood from Pterocarpus violaceus in the production of agglomerated panels.* Scientia Forestalis, Piracicaba, v. 42, n.103, p. 329–335, set. 2014.

Frihart, CR; Hunt, CG *Adhesives with wood materials: bond formation and performance*. In: Forest Products Laboratory. Wood handbook: wood as an engineering material. Washington, DC: USDA, 2010. p. 10 / 1–10 / 24.

Gatto, D; Calegari, L.; Haselein, CR; Scaravelli, TL; Santini, EJ; Stangerlin, DM; Trevisan, R. *Physical-mechanical performance of panels made of bamboo (bambusa vulgaris Schr .) In combination with wood.* Cerne, v. 13, n. 1, p. 57–63, 2007.

Guimarães Junior, J.B.; Mendes, LM; Mendes, RF; Mori, FA *Agglomerated wood panels from laminates from different origins of Eucalyptus grandis, Eucalyptus saligna and Eucalyptus cloeziana.* Cerne, v. 17, n. 4, p. 443–452, 2011.

Iwakiri, S.; Andrade, AS; Cardoso Júnior, AA; Chipanski, ER; Prata, JG; Adriazola, MK *O. Production of high densification agglomerate panels using melamine – urea – formaldehyde resin*. However, 11, n. 4, p. 323–328, Oct./Dec. 2005.
Khedari, J.; Nankongnab, N.; Hirunlabh, J.; Teekasap, S. *New low-cost insulation particleboards from the mixture of durian peel and coconut coir. Building and Environment*, v. 39, n. 1, p. 59–65, 2004.
Kniess, D. C.; Vieira, HC; Bourscheid, CB; Grubert, W.; Cordova, F.; Zulianello, V.; Cunha, AB; Rios, PD. *Physical properties of agglomerated panels produced with fibers and shavings of Pinus spp.* In: II Brazilian Congress of Science and Technology of Madeira, 2015, Belo Horizonte.
Mendes, RF; Mendes, LM; Abranches, RAS; Santos, RC; Guimarães JR., JB *Agglomerated panels produced with sugarcane bagasse in association with eucalyptus wood*. Scientia forestalis, Piracicaba, v. 38, n. 86, p. 285–295, Jun.2010.
Mendes, RF; Mendes, LM; Mendonça, LL; Guimarães JR., JB; Mori, FA *Quality of homogeneous agglomerated panels produced with wood from clones of Eucalyptus urophylla*. However, 20, n. 2, p. 1–10, jan./mar. 2014.
Moslemi, AA *Particleboard*. London: Southern Illinois University Press, 1974. 245 p.
Ndazi, BS; Karlsson, S.; Tesha, JV; Nyahumwa, CW *Chemical and physical modifications of rice husks for use as composite panels*. Composites Part A: Applied Science and Manufacturing, v.38, n. 33, p.925–935, 2007.
Scatolino, MV; Costa, AO; However, Protasio, TP; Mendes, RF; Mendes, LM *Eucalyptus wood and coffee parchment for particleboard production: physical and mechanical properties*. Science and Agrotechnology, v. 41, n. 2, p.139–146, 2017.
Soares, SS et al. *Valuation of sugarcane bagasse in the production of low density agglomerated panels*. Science of the Wood, Pelotas, v. 8, n. 2, p. 64–73, 2017.
Tsoumis, G. *Science and technology of wood: structure, properties and utilization*. New York: Chapman & Hall, 1991. P. 309–339–494.

Wastes: Solutions, Treatments and Opportunities III – Vilarinho et al. (Eds)
 ISBN 978-0-367-25777-4

The potential of biodiesel production from WWTP Wastes

R.M. Salgado
ESTS-CINEA/IPS, Setúbal, Portugal
LAQV-REQUIMTE/FCT-UNL, Caparica, Portugal

A.M.T. Mata
ESTS-CINEA/IPS, Setúbal, Portugal
iBB-IST/UL, Lisboa, Portugal

L. Epifâneo
ESTS-CINEA/IPS, Setúbal, Portugal

A.M. Barreiros
ADEQ-ISEL/IPL, Lisboa, Portugal

ABSTRACT: The biofuel incorporation in fossil fuels is a measure that promotes energy sustainability and reduction of fossil fuels dependency. The European Union established a target of incorporating 14% renewable energy in the transport sector by 2030. The biodiesel is one of the most widely used biofuel in Europe mainly from vegetable oils. Land use to grow biodiesel feedstock may have negative environmental impacts. It is necessary to look for an alternative feedstock for biodiesel production that consider the sustainability, and which can contribute to the decarbonisation. The wastewater contains fat, oil and grease (FOG) that can be used in the biodiesel production to reduce the carbon footprint and to promote the circular economy. This work analyses the potential of biodiesel production from FOG collected in WWTP. Considering the quantity and quality of FOG separated, it can be verified that this may be an important alternative feedstock for biodiesel production.

1 INTRODUCTION

Oil and grease (O&G) can be found in wastewater and they need to be removed in wastewater treatment plants (WWTP) by a separation process. The storage can release odours to the atmosphere and the final treatment of this fraction is still a problem for the WWTP management. The floating mixture, scums, composed of inert and fat oil and grease (FOG) is removed for a storage device and seldom go for a specific treatment at WWTP. For many wastewater treatment facilities, FOG and scums, preceded or not by a concentration step, are mixed and treated with activated sludge in an anaerobic digestion process. Other wastewater treatment facilities will treat FOG and scum as a waste product, either incinerating it or sending it to landfill.

These solutions, apart from the incorporation of the FOG and scums in the anaerobic digestion, represent economic costs which include transportation, energy (pumping and aeration), labor, storage and final destination. Regarding the incorporation of FOG and scums into the solid phase treatment line, although the economic costs may be relatively small but present disadvantages. The presence of FOG in the affluent to anaerobic processes can be a problem for two reasons: inhibition of methanogenesis by long chain fatty acids and sludge flotation due to hydrophobicity and lower specific mass of these compounds in relation to water. The inhibition of methanogenesis may be associated with the adsorption of long chain fatty acids on the cell wall or membrane of the organisms responsible for methanogenesis, interfering with the functions of substrate transport. To maintain a stable operation of an anaerobic treatment system and promote the degradation of FOG a loading of less than 1.0 kg $m^{-3}d^{-1}$ is needed (Tchobanoglous et al. 2014). Moreover, to the drawbacks of FOG and scum incorporating into anaerobic digestion, in a recent work, Kumar et al. (2019) suggests that scum to biodiesel was 9.6 times more energetically favorable than scum to methane.

New sources of O&G to produce biodiesel are one of the challenges nowadays to avoid reduce greenhouse gas emissions from diesel, so the reutilization of the FOG waste separated at WWTP to produce biodiesel is extremely interesting. The potential of this waste as a raw material should be evaluated.

To reduce greenhouse gas emissions and dependence on fossil fuels and promotes energy sustainability, the European Union established in the Renewable Energy Directive a target of introducing 10% renewable energy (such as biodiesel) in the transport sector by 2020 (Directive 2009/28/EC) and increased the percentage to 14% in the revision of this Directive in 2018 (Directive (EU) 2018/2001).

Biodiesel is considered one of the best solutions as an additive to the diesel to reduce the CO_2 emissions and particles to the atmosphere.

Biodiesel can be used directly in the vehicle at 100% or mixed with diesel. Its biodegradability, non-toxicity and being free of sulfur and aromatics makes it advantageous over the conventional petrol diesel. It emits less air pollutants and greenhouse gases. In addition, it is safer to handle because it is less toxic and easy to store than petroleum (Gebremariam & Marchetti, 2018). However, despite these environmental advantages, biodiesel cannot be extensively applied as a complete substitute fuel for conventional diesel. The main reason is its higher cost of production, which is about one and a half times more expensive than petroleum diesel fuel. The reasons are the price of raw materials and the high cost of production.

Biodiesel is a mixture of fatty acid alkyl esters (FAME) that can be produced from different lipid feedstock or alcohol with or without the presence of a catalyst (Abdullah et al. 2017). Depending on the nature of the lipid feedstock the reaction is called transesterification or esterification (Diamantopoulos et al. 2015). The selection of an appropriate catalyst to the type of reaction and the lipid feedstock depends on the amount of free fatty acids in the oil (Abdullah et al. 2017; Diamantopoulos et al. 2015), and can be homogeneous (acid or base) or heterogeneous (acid or base, or enzymatic).

Vegetable oils, edible or non-edible, are the main source of biodiesel, according to a review carried out by Sajjadi et al. (2016). The same authors indicate that more than 95% of the world biodiesel is produced from edible vegetable oils, being rapeseed (84%) the main source of edible oil, followed by sunflower (13%), palm (1%), soybean and others (2%). The most commonly used edible oil in EU Member States is rapeseed oil (70.2%), but oil from soybean (5.8%), palm (5.0%), are also used (Ecofys, 2014). The use of raw material waste like animal fat waste (Kirubakaran & Selvan 2018), waste cooking oil from the restaurant industry (Abed, et al. 2018), microalgae (Chen et al. 2018a, Shomal et al. 2019) have been extensively investigated. Biodiesel production from wastes like cooking oil or animal fats is gaining a foothold, and more capacity is expected in the coming years. In European Union 11.4% of biodiesel is produced from cooking oil and 4.6% from animal fat (Ecofys, 2014). The increased demand for animal fats to produce biodiesel has resulted in the price of animal fats increasing significantly and there are also difficulties in obtaining the enough waste cooking oil to ensure production based on this waste source.

A life cycle assessment and economic analysis of the scum-to-biodiesel done by Mu et al. (2016) shows that the technology provides great benefits to the environment by reducing fossil resource depletion, CO_2 emissions, and N and SO_2 discharge to the environment. When comparing the impact of biodiesel from scum to conventionally produced biodiesel from soybean and vegetable oil, scum as a feedstock has less environmental impacts per kg diesel produced in all impact categories, fossil fuel use, GHG emissions, eutrophication, and acidification.

The aim of this study was to evaluate the potential of biodiesel production from scums and FOG removed in Portuguese WWTP. Extraction of FOG from the separated fraction was studied in Portugal by extraction the raw material for the biodiesel production and produce biodiesel using this O&G source.

2 MATERIALS AND METHODS

2.1 *Chemicals*

Methanol, *n*−Hexane and sulfuric acid were purchased from Panreac (Panreac, Portugal) and sodium hydroxide was purchase from Merck (Merck, Portugal).

2.2 *Solid sample collection in WWTP and FOG extraction*

The area selected for this study is in the metropolitan area of Lisbon (Lisbon and Setúbal districts), where an important populated and industrialized region is responsible for the existence of several wastewater treatment plants. Ten WWTP were selected and all the WWTP are activated sludge processes, with exception to WWTP n°8 which is a biofiltration process. All the WWTP have aerated channels for the FOG waste removal, with exception of WWTP n°5. Total annual solid waste fraction collected in degrease separators (ton/year) was kindly provided by WWTP management.

Wet solid samples were collected in PET bags and preserved at +4°C. Samples were homogenized and dried in an oven at 100°C. This dried fraction was then used for FOG extraction and to determine the inert and non-oil fraction.

The extraction of FOG was carried out by soxhlet, where 3.0 g of sample were extracted with n-Hexane for 3 h. Extracts were concentrated under vacuum at 69°C with a rotary evaporator. The FOG extract was completely dried in an oven at 50°C and the mass of FOG was measured. The FOG extract previously pre-treated with an extraction step as clean-up process was then used for the biodiesel production. The extraction was carried out in duplicates.

2.3 *Biodiesel production process*

The acid or basic catalysis selection for the transesterification reaction was defined according to the acid value determination of the FOG extracts. The acid value determination was carried out according to the ASTM D 6751-08 standard. Acid value of the extracts less than 4 mg g^{-1} KOH suggests that the transesterification should be under alkaline catalysis, if higher the catalysis should be acidic (Nair et al. 2012). FOG extracts of the WWTP separators were in the range of 7 to 15 mg g^{-1} KOH so the transesterification reaction was catalyzed under acidic conditions for the biodiesel production from WWTP FOG extracts.

The acid catalyzed process was carried out by taking a WWTP FOG extract sample and dilute with n-Hexane in the proportion of 1:20 (w/w), adding 20 times of stoichiometric amount of methanol and 3.6% (w/w) of sulfuric acid and leaving the mixture to react for 1 h at 65°C (Wang et al. 2006). Then, excess of methanol was recovered under vacuum at 50°C with a rotational evaporator, and the mixture was left to settle for separation into two layers. The upper oil layer was the FAME mixture of biodiesel and some unreacted triglyceride, and the down layer was sulphuric acid and glycerol. After this period, the mixture was transferred to a settling device to separate the biodiesel of the glycerin and the pH of the biodiesel fraction was neutralized with NaOH. The biodiesel was washed by 10% of water at 80°C to remove soap which was produced by reaction of the alkali and the free fatty acid (FFA). The wet biodiesel was dried under vacuum at 70°C with rotational evaporator for 1 h.

3 RESULTS AND DISUSSON

3.1 *FOG recovered in the WWTP*

The solid waste removed in the WWTP degrease system was composed of wastewater, inert material and FOG. Combining the information of reported annual solid waste fraction collected in degrease separators with the percentage of FOG found in these samples (black dashed line), Figure 1 was constructed. As seen in Figure 1, in some WWTP the total solid removed was mainly FOG (e.g.

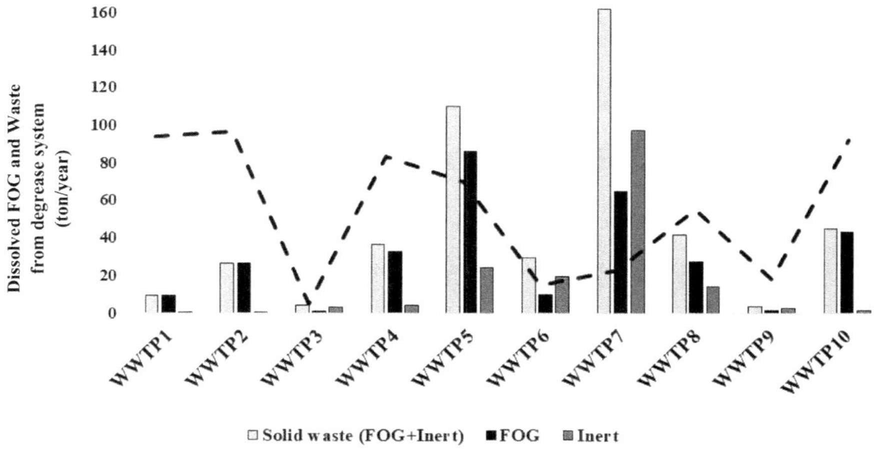

Figure 1. Total amount of FOG recovered from the solid waste in the degrease system ($t.y^{-1}$). Black dashed line is % of FOG in the solid waste recovered from the WWTP separator.

WWP 1, 2, 4, 5, 8 and 10) and in others only a little fraction of FOG was found suggesting that these WWTP were not interesting sources of FOG for biodiesel production (e.g. WWTP 3 and 9).

In the ten WWTP, it is possible to recover 469 ton per year (dry weight) of total solid where 303 ton per year correspond to FOG and only 166 t y^{-1} to inert material.

At least in seven WWTP, the total solids are composed mostly by FOG (e.g. WWTP 1, 2, 4, 5, 7, 8 and 10). The FOG and water content of the WWTP 3, 6 and 9 suggests that the separation system is less efficient. A large amount of wastewater is present in the sample, so it contains a very little percentage of FOG.

The FOG percentages found were higher than 80% for five of the ten WWTP studied (e,g 1, 2, 4, 5 and 10). The potential source of FOG collected in the degrease separators of WWTP can give an important source contribution for biodiesel production when compared with other sources. The amount of FOG found WWTP degrease separators in this study was higher than the one found in grease trap waste (GTW) (Almeida et al. 2016, Hums et al. 2016, Montefrio et al. 2010), or in other WWTP wastes like sewage sludge (Chen et al. 2018b, Demiirbas 2017, Melero et al., 2015, Pastore et al., 2013) or in scum from other systems (Almeida et al. 2016, Anderson et al. 2016a, b, Anderson et al. 2018, Bi et al. 2015, Bitonto et al. 2015, Kumar et al. 2019, Pastore et al., 2014).

According to our knowledge one of the highest FOG content obtained with scum was a value of 60% (Bi et al. 2015), this indicates the exceptional potential of some of the WWTP studied as sources of biodiesel raw material, since half of them have 80% or higher of FOG content. This study reveals that the FFA content of FOG derived from grease interceptors did not exceed 8% (w/w) due to constant influx of fresh FOG from wastewater. However, if the FOG can hydrolyze without dilution, the FFA content can reach 15% (w/w) in more than 20 days (Montefrio et al. 2010).

Nevertheless, not all WWTP have degrease separators, or even having them, cannot be considered as feedstock for biodiesel production, since according to Olkiewicz et al. (2012) the amount of extracted lipids for primary sludge was 25.3% compared to 21.9%, 10.1% and 9.1% (dry wt) for blended, stabilized and secondary sludge, respectively, and the FAMEs yield obtained for primary, blended, secondary and stabilized sludge were 13.9%, 10.9%, 2.9% and 1% (dry wt), respectively, showing that an important amount of O&G are mixed in the sludge fraction.

In can be concluded that in the metropolitan area of Lisbon FOG from WWTP degrease system can be a very interesting possibility to increase the amount of O&G for biodiesel production,

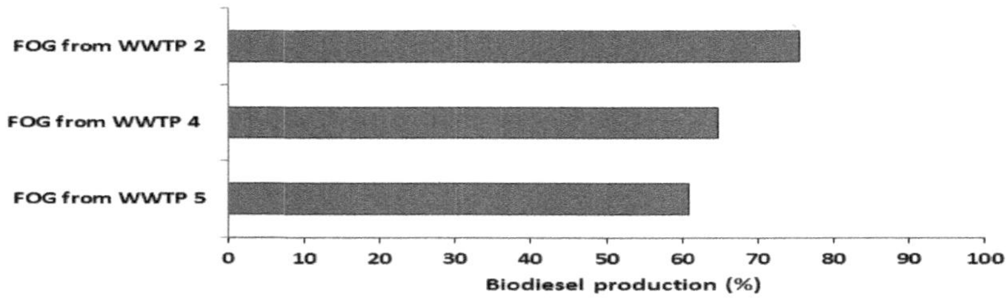

Figure 2. Percentage of biodiesel obtained from WWTP FOG.

ensuring both a more sustainable supply of this feedstock and a way to reuse this waste and reduce its treatment and disposal costs in the WWTP.

3.2 *Biodiesel production*

The FOG extracts of the WWTP 2, 4, and 5 were selected to produce biodiesel due to their high FOG content and results are presented in Figure 2.

The biodiesel yield percentage varied from 61 to 75% using the WWTP FOG extracts as raw material for the production. The production was carried out by diluting the biodiesel with n-Hexane as solvent to be comparable to the use of diesel fuel or diesel oil as solvent according to Kojima et al. (2009). The n-Hexane was only used as a solvent to support the occurrence of the reaction for the biodiesel production.

The results of the biodiesel produced from the WWTP agree with the biodiesel produced from a small community with high content of animal fat where they obtain also yields in biodiesel production from 74 to 92% (Phalakornkule et al. 2009). In other hand, when other FOG sources are used such as feather meal, only 7–11% of biodiesel can be obtained (Kondamudi et al. 2009).

Pokoo-Aikins et al. (2009) extracted lipids from sewage sludge using different organic solvents, and obtained lipids yields close to 25 wt%. A two-step production of FAME from municipal waste water sludge using n-Hexane in acidic ambient followed by methanolysis with sulfuric acid allowed FAME yields between 12 and 22 wt%. Huynh et al. (2010) reported the in-situ production of FAME, from untreated wet activated sludge under subcritical water and methanol conditions with sulfuric acid. Additionally, Mondala et al. (2009) investigated the feasibility of using homogeneous acid catalysts to produce biodiesel from primary and secondary sewage sludge by in-situ transesterification process, with sulfuric acid and n-Hexane to improve the solubility in the reaction mixture; obtaining FAME yields close to 15 wt% from primary sludge and 3 wt% from secondary sludge (Melero et al., 2015). According to these studies, it can be concluded that biodiesel produced by FOG from degrease separation process at the entrance of the WWTP have an exceptional potential when compared with other studies and sources of O&G from wastewater processes.

4 CONCLUSION

The total amount of FOG available in WWTP is much higher compared with other sources as feather meal, algae or biomass lipids of primary and secondary sludge. The average yield of biodiesel production from FOG extracts of WWTP obtained was higher than 60%. This work suggests that production of biodiesel should be considered as an option for O&G WWTP disposal, avoiding the problems associated with the addition of O&G to anaerobic digestion, and reinforcing a circular economy with energy produced from waste and reduction of CO_2 emissions to the atmosphere. This work provides an important contribution, revealing not only a new raw material source of O&G but also the feasibility of biodiesel production from these waste material. The amount of the feedstock available and the final characteristics of biodiesel produced justify further investments in this process.

REFERENCES

Abdullah, S.H.Y.S., Hanapi, N.H.M., Azid, A. Umar, R., Juahir, H., Khatoon, H., Endut, A. 2017. A review of biomass-derived heterogeneous catalyst for a sustainable biodiesel production. *Renewable and Sustainable Energy Reviews* 70: 1040–1051.

Abed, K.A., El Morsi, A.K., Sayed, M.M., El Shaib, A.A., Gad, M.S. 2018. Effect of waste cooking-oil biodiesel on performance and exhaust emissions of a diesel engine. *Egyptian Journal of Petroleum* 27: 985–989.

Almeida, H.S., Corrêa, O.A., Eid, J.G., Ribeiro, H.J., Castro, D.A.R., Pereira, M.S., Mâncio, A.A., Santos, M.C., Souza, J.A.S, Borges, L.E.P., Mendonça, N.M., Machado, N.T. 2016. Production of biofuels by thermal catalytic cracking of scum from grease traps in pilot scale. *Journal of Analytical and Applied Pyrolysis* 118: 20–33.

Anderson, E., Addy, M., Chen, P., and Ruan, R. 2018. Development and operation of innovative scum to biodiesel pilot-system for the treatment of floatable wastewater scum. *Bioresource Technology* 249: 1066–1068.

Bi, C., Min, M., Nie, Y., Xie, Q., Lu, Q., Deng, X., and Ruan, R. 2015. Process development for scum to biodiesel conversion. *Bioresource Technology* 185: 185–193.

Bitonto, L., Lopez, A., Mascolo, G. Mininni, G. 2016. Efficient solvent-less separation of lipids from municipal wet sewage scum and their sustainable conversion into biodiesel. *Renew. Energy* 90: 55–61.

Chen, J., Li, J., Dong, W., Zhang, X., Tyagi, R.D., Drogui, P., Surampalli, R.Y. 2018a.The potential of microalgae in biodiesel production. *Renewable and Sustainable Energy Reviews* 90: 336–346.

Chen, J., Tyagi, R.D., Li, J., Zhang, X. Drogui, P. Sun, F. 2018b. Economic assessment of biodiesel production from wastewater sludge. *Bioresource Technology* 253: 41–48.

Choi, O.K., Song J.S., Cha, D.K. Lee, J.W. 2014 Biodiesel production from wet municipal sludge: Evaluation of in situ transesterification using xylene as a cosolvent. *Bioresource Tech.* 166:51–56.

Demirbas, A. 2017. Biodiesel from municipal sewage sludge (MSS): Challenges and cost analysis. *Energy Sources, Part B: Economics, Planning, and Policy* 12 (4): 351–357.

Diamantopoulos, N., Panagiotaras, D., Nikolopoulos, D. 2015. Comprehensive Review on the Biodiesel Production using Solid Acid Heterogeneous Catalysts. *J. Thermodyn. Catal* 6(1): 143.

Directive 2009/28/EC of the European Parliament and of the Council, of 23 April 2009. Renewable Energy Directive. *Official Journal of the European Union.*

Directive (EU) 2018/2001 of the European Parliament and of the Council, of 11 December 2018. Renewable Energy Directive. *Official Journal of the European Union.*

Ecofys, 2014. Renewable energy progress and biofuels sustainability.

Gebremariam, S.N., Marchetti, J.M. 2018. Economics of biodiesel production: Review. *Energy Conversion and Management* 168: 74–84.

Hums, M.E., Cairncross, R.A., Spatari, S. 2016. Life-Cycle Assessment of Biodiesel Produced from Grease Trap Waste. *Environ. Sci. Technol.* 50: 2718–2726.

Huynh, L. Kasim, N., Ju, Y. 2010. Extraction and analysis of neutral lipids from activated sludge with and without sub-critical water pre-treatment, *Bioresource Technology* 101(22):8891–6

Kirubakaran, M. & Selvan, V.A.M. 2018. A comprehensive review of lowcost biodiesel production from waste chicken fat. *Renewable and Sustainable Energy reviews* 82: 390–401.

Kojima, S., Du, D., Sato, M., Park, E.Y. 2004. Efficient production of fatty acid methyl ester from waste activated bleaching earth using diesel oil as organic solvent, *J. Biosci. Bioeng.* 98 (6): 420–424.

Kondamudi, N., Strull, J., Misra, M., Mohapatra, S.K. 2009. A green process for producing biodiesel from feather meal, *J. Agric. Chem.* 57: 6163–6166.

Kumar, L.R., Yellapu, S.K., Zhang, X., Tyagi, R.D. 2019. Energy balance for biodiesel production processes using microbial oil and scum. *Bioresource Technology* 272: 379–388.

Melero, J.A., Sánchez-Vázquez, R., Vasiliadou, I.A., Castillejo, F.M., Bautista, L.F., Iglesias, J. Morales, G., Molina, R. 2015. Municipal sewage sludge to biodiesel by simultaneous extraction and conversion of lipids. *Energy Conversion and Management* 103: 111–118.

Montefrio, M.J., Xinwen, T., Obbard, J.P. 2010. biodiesel production. *Appl. Energ.* 87(10): 3155–3161

Mu, D., Addy, M., Anderson, E., Chen, P., and Ruan, R. 2016. A life cycle assessment and economic analysis of the Scum-to-Biodiesel technology in wastewater treatment plants. *Bioresource Technology* 204: 89–97.

Nair, P., Singh, B., Upadhyay, S.N., Sharma, Y.C. 2012. Synthesis of biodiesel from low FFA waste frying oil using calcium oxide derived from mereterix as a heterogeneous catalyst, *Journal of Cleaner Production* 2012, 29–30, 82–90.

Olkiewicz, M. Fortuny, A., Stüber, F., Fabregata, A. Fonta, J. Bengoaa, C. 2012. Evaluation of different sludges from WWTP as a potential source for biodiesel production. *Procedia Engineering* 42:634–643.

Pastore, C., Lopez, A., Lotito, V. Mascolo, G. 2013. Biodiesel from dewatered wastewater sludge: A two-step process for a more advantageous production. *Chemosphere* 92: 667–673.

Phalakornkule, C., Petiruksakul, A., Puthavithi, W., 2009. Biodiesel production in a small community: Case study in Thailand, Resources Conservation and Recycling 53(3): 129–135.

Pokoo-Aikins, G. Nadim, A., El-Halwagi, M.M., Mahalec, V. 2009. Design and analysis of biodiesel production from algae grown through carbon sequestration, *Clean Technol. Environ.* 12(3): 239–254.

Sajjadi, B., Raman, A.A.A., Arandiyan, H. 2016. A comprehensive review on properties of edible and non–edible vegetable oil-based biodiesel: Composition, specifications and prediction models. *Renewable and Sustainable Energy Reviews* 63: 62–92.

Shomal, R., Hisham, H., Mlhem, A., Hassan, R. Al-Zuhair, S. 2019. Simultaneous extraction–reaction process for biodiesel production from microalgae. *Energy Reports* 5: 37–40.

Tchobanoglous, G., Stensel, H.D., Tsuchihashi, R., Burton, F., Abu-Orf, M., Bowden, G., Pfrang, W. (5th) 2014. *Wastewater engineering – Treatment and resource recovery*. Metcalf & Eddy AECOM.

Tu, Q., Wang, J., Lu, M., Brougham, A., Lu, T. 2016. A solvent-free approach to extract the lipid fraction from sewer grease for biodiesel production. *Waste Management* 54: 126–130.

Yi, W., Sha, F., Xiaojuan, B., Jingchan, Z., Siqing, X. 2016. Scum sludge as a potential feedstock for biodiesel production from wastewater treatment plants. *Waste Manag.* 47: 91–97.

Wastes: Solutions, Treatments and Opportunities III – Vilarinho et al. (Eds)
 ISBN 978-0-367-25777-4

Application of industrial wastes in substrates for ecological green roofs

V. Pinheiro & A. Ribeiro
CVR – Centre for Waste Valorisation, Guimarães, Portugal

P. Palha
Neoturf, Espaço Verdes, Lda., Matosinhos, Portugal

J. Almeida
Itecons, Coimbra, Portugal

T. Teixeira & J. Ribeiro
W2V, SA, Guimarães, Portugal

I. Lima, M. Abreu, I.A.P. Mina & F. Castro
University of Minho, Braga, Guimarães, Portugal

ABSTRACT: Green roofs are systems that can help to solve some urban environmental problems. A multi-layer green roof has five layers: from waterproof membrane till substrate and vegetation. Substrate is arguably the most important component of green roofs but there is still much to optimize and exploit. Also, world needs to avoid uncontrolled dispersal of waste. This new approach can be applied to substrates production. The main objective of this study is to compare and evaluate 8 substrates incorporating industrial wastes with a commercial substrate. Employed plants were *Armeria welwitschii, Festuca glauca* and *Sedum sediforme*. Alongside, the rhizosphere eukaryotic biocenosis (protists communities) were observed in optical microscope. Preliminary results revealed that some industrial wastes may replace commonly used materials on green roofs substrate increasing plant growth. With regards to eukaryotic biocenosis, the performed observations, point out that, these organisms can be good bioindicators for green roof technology.

1 INTRODUCTION

A green roof also referred as a living, vegetated or eco-roof is a building roof that is partially or completely covered with vegetation planted in a growing medium (substrate) over a waterproof membrane (Mickovski et al., 2013). The concept was designed and developed to promote the growth of various plants forms on buildings top and thereby provide aesthetically as well as environmental and economic benefits (Dunnett & Kingsbury, 2004; Oberndorfer et al., 2007; Vijayaraghavan, 2016).

In the 19th century, Scandinavian engineers started using birch bark and sod on rooftops, to ensure thermal insulation under wet and cold climates (Carson et al., 2012). The present green roofs may be constructed according the ancient technique' concept, but technological advances make them much more efficient, practical and beneficial than their homologous ancestors (Vijayaraghavan, 2016). The growing concern with the environment that started in 1970s, especially on Germany' urban areas, introduced opportunities for progressive environmental thought, policy and technologies which has led to the adoption by Germany, in 1982, of the first professional rules for green roofing (FLL, 2008).

The need for greater functional diversity led to the definition of different typologies of green roofs - intensive, semi-intensive and extensive - mainly based on their depth, substrate type for the growth layer (and therefore additional load), plants type and requirements for its maintenance. Intensive green roofs have a thick substrate layer (20-30 cm) and so a more weight, a wide variety of plants, and high maintenance and price. Due to increased soil or substrate depth, the plants selected may include shrubs and small trees. Extensive green roofs are characterized by a thin substrate layer (4-15 cm depending on climate conditions and plant type), low weight and minimal maintenance and so, they are cheaper (Lata et al., 2018). Owing to its thin substrate layer, extensive roofs can accommodate only limited type of vegetation, like grasses, moss and a few succulents. An extensive green roof system is commonly used in situations where no additional structural support is needed. Semi-intensive green roofs are those with a moderate to thick substrate layer (12-30 cm), may accommodate small herbaceous plants, grasses and small shrubs (Lata et al., 2018). These roofs require frequent maintenance and high prices. So, extensive green roofs are the more common all over the world, due to their weight restrictions, building costs, and maintenance.

Common multi-layer green roofs include five layers: a waterproof and roots resistant membrane, a drainage layer, a filtering one, a growth substrate (medium) and vegetation on top (Berndtsson, 2010). Previous studies have reported that green roofs substrates properties have a great influence on the water retention capacity (VanWoert et al., 2005; Dunnett et al., 2008) as well as on water purification (Hunt et al., 2006; Berghage et al., 2008;). Green roofs substrates should be lightweight to meet the weight constraints of buildings structures, provide good drainage to channel excess water from the roof, have low organic content to prevent decomposition and collapse of the growth layer, and should be able of ensuring an efficient plants growth (Emilsson et al., 2007; Morgan et al., 2013). Usually, green roofs' substrates comprise a single layer composed of a mixture of organic matter (e.g., matured and stabilized humus, blonde peat and coco peat) and inorganic matter (e.g., perlite, crushed bricks, volcanic rocks and expanded clay).

Generally, the substrate has to play the role of a soil for plant growth (Sutton, 2015) providing to the plants, moisture, nutrients and physical support. Some research on new substrate materials (Molineux et al., 2009; Solano et al., 2012), their biological properties (Kolb et al., 1982) and influence on vegetation growth (Emilsson, 2008; Farrell et al., 2012; Kotsiris et al., 2012; Nagase and Dunnett, 2011; Rowe et al., 2006) has already been done. But, the effects of individual substrate components (e.g. mineral content, type of organic matter, artificial additives, mixing ratios) upon the growth and physiological performance of the vegetation are little known (Dvorak & Volder, 2010; Ouldboukhitine et al., 2012).

There are already some papers on green roofs biodiversity (Berndtsson, 2010), but there are still gaps on its research. For example, there are no information on the study of rhizosphere eukaryotic biocenosis, i.e., eukaryotic microbial communities on roots neighborhood. Protists are the most diverse eukaryotes, and are keystone organisms of soil ecosystems, regulating essential processes of soil fertility such as nutrient cycling and plant growth. Despite this, protists have received little scientific attention in soil studies, especially compared to bacteria, fungi and nematodes (Geisen et al., 2017).

Organic and inorganic components of substrates can be replaced by recycled materials (Chen et al., 2018), for example, compost or green manure, sewage sludge, waste bricks and tiles, waste paper, and carbonated limestone (Molineux et al., 2009) or also industrial wastes like foundry sands, slags, furnace refractories and sludges. However, previous studies have focused on the effects on plant growth but not on drained water quality (Chen et al., 2018). The substrates showed to influence the water quality that cross them, and therefore, the use of recycled materials should consider not only its effects on the plants, but also on the leachate quality. An optimal substrate is expected to benefit plant growth and simultaneously produce few contaminants (Chen et al., 2018).

The aim of the present study was to produce green roof substrates with industrial wastes as replacement for inorganic and organic components, to evaluate their effects on plants' growth and on leachate quality and look into rhizosphere eukaryotic biocenosis - identify and quantify the eukaryotic organisms on substrates to evaluate its relationship with plants growth and water quality.

Table 1. Composition (%) of the different tested substrates.

Substrates	Standard inorganic Substrate	Standard organic substrate	Inorganic industrial waste	Organic industrial waste
0N	50	50	0	0
0L	50	50	0	0
1	50	40	10	0
2	50	30	20	0
3	50	49	1	0
4	50	48	2	0
5	50	49	1	0
6	50	48	2	0
7	25	50	0	25
8	37.5	50	0	12.5

Notes: Substrates – 0N, commercial; 0L – lab made; 1 to 8, with industrial wastes.

Figure 1. a) Pot with a substrate planted with *Armeria welwitschii* coupled to a plastic container; b) Planted pots' layout, over the outside cover.

2 MATERIAL AND METHODS

Employed substrates were made with defined amounts of industrial wastes, to replace components usually used, like volcanic rock or expanded clay and peat. Industrial wastes were searched in different types of industry, like the ceramic sector and foundries. Usual organic substrates were replaced by a compost made from municipal solid wastes. Eight different mixtures were formulated (Table 1), and about 500 g of each were placed in small pots. Three pots with each of the substrates were planted with *Armeria welwitschii*, *Festuca glauca* and *Sedum sediforme* and placed in makeshift containers for water drainage (Fig. 1a). These plant species are all autochthonous and were selected from different botanic families, because its widely used on extensive green roofs, and has different appearance and morphology – *A. welwitshii* from Plumbaginaceae family have purple flowers; *F. glauca* is an herbaceous perennial ornamental grass from Poaceae and *S. sediforme* is a succulent species from Crassulaceae.

The tests were performed on an outer cover (Fig. 1b) and, as controls, a commercial substrate (0N) and a lab' made substrate (0L) prepared according to the commercial one, were used. One pot with each of the different substrates without plants was used to estimate rainwater retention during the experimental period which began December 12, 2018.

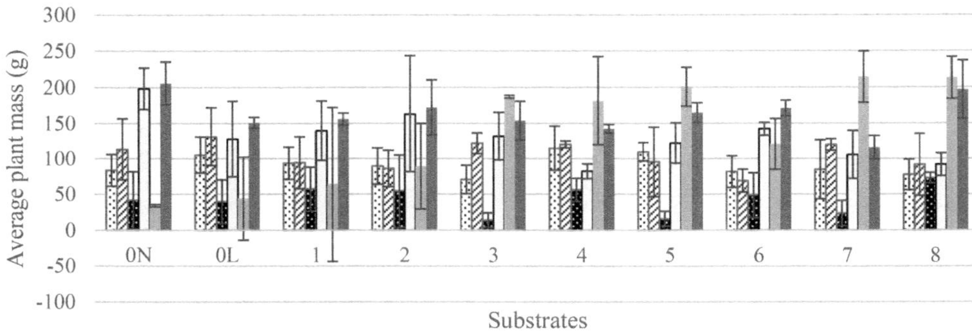

Figure 2. Average mass (g) ± standard deviation (n = 3) of *Armeria welwitschii* (Aw), *Festuca glauca* (Fg) and *Sedum sediforme* (Ss) planted on 3 pots with the tested substrates; at the beginning (Aw; Fg; Ss) and after 10 weeks (Aw□; Fg ; Ss■). Notes: Substrates – 0N, commercial; 0L – lab made; 1 to 8, with industrial wastes.

Evaluation of plant's growth was made by weight change regular measurements. Weekly, each pot was weighed on a balance with 0.01 g precision. After the experimental period of 10 weeks, change of plant weights were determined, making the necessary adjustments for rain water (considering the change of weight in control pots without plants).

From drainage water of some selected pots, 25 mL samples were withdrawn to 50 mL polypropylene conical tubes for further microscopic analysis. Subsamples (50 μL) withdrawn from the conical part of these tubes were placed over microscope slides covered with, 24x24 mm small slides. Light field microscopic observations were performed on an OLYMPUS BX63 microscope equipped with cellSens imaging software.

3 RESULTS

Figure 2 presents the average of 3 samples for the plant's weight at the beginning of the experiment and after 10 weeks.

Globally the plant species with the best growth during the experimental period were *S. sediforme*. A mean mass increase of about 3 x were observed, although the plants in substrates 3 and 5 have increased about 10 times. *A. welwitschii* and *F. glauca* had a similar average mass increase (respectively 1.5x and 1.3x) with very little variation between pots. But, the highest growth of *A. welwitschii* was observed in 0N substrate (2.4 x) while mass increase of *F. glauca* double, only in substrate 5.

Microscopic analyses of the observed samples evidenced the presence of a variable abundance of bacteria, protozoa, mostly ciliates (Fig. 3a) and amoeboids (Fig. 3b) and even some metazoans.

Although the microscopic observations are still few and preliminary, differences between the "pots' biocenosis", seem to be more obvious between samples from pots with different plants than between samples from pots with different substrates but with the same plant' species. For example, the biocenosis observed on samples from substrates 0N and 8 were more similar in the samples from pots with the same plant' species than from pots with the same substrate planted with different species.

4 CONCLUSIONS

Up to the present, although preliminary, our results seem promising. They point out that some of the mixture of industrial wastes used in substrates formulation may replace the conventional inorganic products, such as volcanic rock, expanded clay or peat, with expected lower costs. While

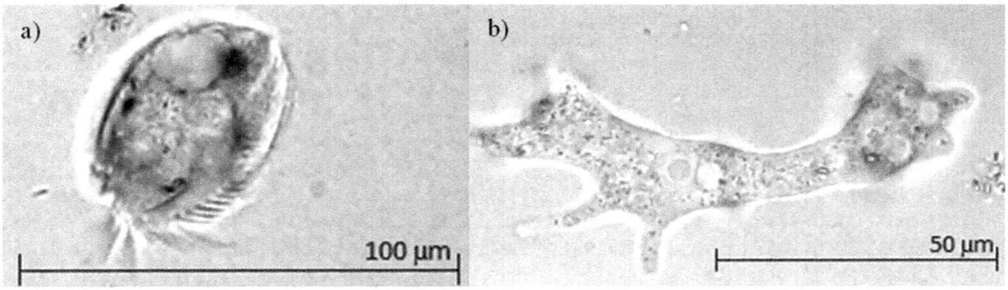

Figure 3. Protozoa microphotography in bright field: a) ciliate, magnification 200x; b) amoeboid, magnification 400x.

A. welwitschii just doubled its mass on 0N substrate, *F. glauca* grew better on substrates 5 and 8 – the first one supplemented with foundry wastes rich in zinc oxide and the latter one containing compost.

The higher growth of *S. sediforme* was enhanced on substrates 3 and 5 (ca. 10 times increase). On substrate 3, that contains ceramic wastes and aluminum anodizing sludge (rich in aluminum hydroxide), the successful growth can be explained, such as on substrate 5, by the increase of ion exchange properties related to metal hydroxides. These results indicate the viability of the use of these materials as alternatives to produce ecological and economic substrates, for green roofs.

In what concerns eukaryotic biocenosis the performed observations, point out that, these organisms can be good bioindicators for green roof technology. As they seem associated with specific plants' rhizosphere their proliferation may contribute to more efficient plant growth. The study of these particular ecosystems will provide an understanding of its structure and functioning, "paving the way to a new era in soil biology" (Geisen et al, 2017).

ACKNOWLEDGEMENTS

This work has been co-financed by Compete 2020, Portugal 2020 and the European Union through the European Regional Development Fund – FEDER within the scope of the project EGR - EcoGreenRoof: Desenvolvimento de eco-materiais para coberturas verdes (POCI-01-0247-FEDER-033728).

REFERENCES

Berghage, R., Wolf, A., Miller, C. 2008. Testing green roof media for nutrient content. Proceedings of the Greening Rooftops for Sustainable Communities, Baltimore, MD 2008.

Berndtsson, J.C. 2010. Green roof performance towards management of runoff water quantity and quality: a review. *Ecological Engineering* 36: 351–360.

Carson, T. B., Hakimdavar, R., Sjoblom, K. J., Culligan, P. J. 2012. Viability of recycled and waste materials as Green Roof substrates, in Geotechnical Special Publication 226 (225): 3644–3653.

Chen, C-F., Kang, S-F., Lin, J-H. 2018. Effects of recycled glass and different substrate materials on the leachate quality and plant growth of green roofs. *Ecological Engineering*. 112: 10–20.

Dunnett, N.P., Kingsbury, N. 2004. Planting Green Roofs and Living Walls. Portland (OR), Timber Press.

Dunnett, N., Nagase, A., Booth, R., Grime, P. 2008. Influence of vegetation composition on runoff in two simulated green roof experiments. *Urban Ecosyst*. 11: 385–398.

Dvorak, B., Volder, A. 2010. Green roof vegetation for North American ecoregions: a literature review. Landsc. *Urban Plan*. 96: 197–213.

Emilsson, T., Czemiel Berndtsson, J., Mattsson, J.E., Rolf, K. 2007. Effect of using conventional and controlled release fertilizer on nutrient runoff from various vegetated roof systems. *Ecol. Eng*. 29: 260–271.

Emilsson, T. 2008. Vegetation development on extensive vegetated green roofs: influence of substrate composition, establishment method and species mix. *Ecol. Eng.* 33: 265–277.
Farrell, C., Mitchell, R.E., Szota, C., Rayner, J.P., Williams, N.S.G. 2012. Green roofs for hot and dry climates: interacting effects of plant water use, succulence and substrate. *Ecol. Eng.* 49: 270–276.
FLL 2008 Guidelines for the Planning, Construction and Maintenance of Green Roofing – Green Roofing Guidelines. Forschungsgesellschaft Landschaftsentwicklung Landschaftsbau e.V., Bonn, Germany.
Geisen, S., Mitchell, E. A. D., Wilkinson, D. M., Adl, S., Bonkowski, M., Brown, M. W., Fiore-Donno, A. M., Heger, T. J., Jassey, V. E. J., Krashevska, V., Lahr, D. J. G., Marcisz, K, Mulot, M., Payne, R., Singer, D., Anderson, O. R., Charman, D. J., Ekelund, F., Griffiths, B. S., RØnn, R, Smirnov, A., Bass, D., Belbahri, L., Berney, C., Blandenier, Q., Chatzinotas, A., Clarholm, M., Dunthorn, M., Feest, A., Fernandez, L. D., Foissner, W., Fournier, B., Gentekaki, E., Hajek, M., Helder, J., Jousset, A., Koller, R., Kumar, S., La Terza, A., Lamentowicz, M., Mazei, Y., Santos, S. S., Seppey, C. V. W., Spiegel, F. W., Walochnik, J., Winding, A., Lara, E. 2017. Soil protistology rebooted: 30 fundamental questions to start with. *Soil Biology & Biochemistry.* 111: 94–103.
Hunt, W., Hathaway, A., Smith, J., Calabria, J. 2006. Choosing the right green roof media for water quality. Proceedings of the Greening Rooftops for Sustainable Communities, Boston, MA 2006.
Kotsiris, G., Nektarios, P.A., Paraskevopoulou, A.T. 2012. Lavandula angustifolia growth and physiology is affected by substrate type and depth when grown under Mediterranean semi-intensive green roof conditions. *Hort Science.* 47: 311–317.
Lata, J. C., Dusza, Y, Abbadie, L., Barot, S., Carmignac, D., Gendreau, E., Kraepiel, Y., Mériguet J., Motard, E., Raynaud, X. 2018. Role of substrate properties in the provision of multifunctional green roof ecosystem services. *Applied Soil Ecology.* 123: 464–468.
Mickovski, S.B., Buss, K., McKenzie, B.M., Sökmener, B. 2013. Laboratory study on the potential use of recycled inert construction waste material in the substrate mix for extensive green roofs. *Ecological Engineering* 61: 706–714.
Molineux, C. J., Fentiman, C. H., Gange, A. C. 2009. Characterising alternative recycled waste materials for use as green roof growing media in the U.K. *Ecological Engineering* 35: 1507–1513.
Morgan, S., Celik, S., Retzlaff, W. 2013 Green roof storm-water runoff quantity and quality. *J. Environ. Eng.* 139: 471–478.
Nagase, A., Dunnett, N. 2011. The relationship between percentage of organic matter in substrate and growth in extensive green roofs. Landsc. *Urban Plan.* 103: 230–236.
Oberndorfer, E., Lundholm, J., Bass, B., Coffman, R.R., Doshi, H., Dunnett, N., Gaffin, S., Kohler, M., Liu, K.K.Y., Rowe, B. 2007. Green roofs as urban ecosystems: ecological structures, functions, and services. *Bioscience* 57 (10): 823–833.
Ouldboukhitine, S. E., Belarbi, R., Djedjig, R. 2012. Characterization of green roof components: measurements of thermal and hydrological properties. *Build. Environ.* 56: 78–85.
Rowe, D.B., Monterusso, M.A., Rugh, C.L. 2006. Assessment of heat-expanded slate and fertility requirements in green roof substrates. *Hort Technology.* 16: 471–477.
Solano, L., Ristvey, A.G., Lea-Cox, J., Cohan, S.M. 2012. Sequestering zinc from recycled crumb rubber in extensive green roof media. *Ecol. Eng.* 47: 284–290.
Sutton, K.R., 2015. Green roof ecosystems. In: Sutton, K.R. (Ed.), Ecological Studies 223. Springer International Publishing, Switzerland. http://dx.doi.org/10.1007/978-3- 319-14983-7_1.
VanWoert, N.D., Rowe, D.B., Andresen, J.A., Rugh, C.L., Xiao, L. 2005. Watering regime and green roof substrate design affect sedum plant growth. *Hort. Sci.* 40: 659–664.
Vijayaraghavan, K. 2016. Green roofs: A critical review on the role of components, benefits, limitations and trends. *Renewable and Sustainable Energy Reviews* 57: 740–752.

Wastes: Solutions, Treatments and Opportunities III – Vilarinho et al. (Eds)
© 2020 Taylor & Francis Group, London, ISBN 978-0-367-25777-4

Agroindustrial residues as cellulose source for food packaging applications

D.J.C. Araújo
Institute for Polymers and Composites/I3N, University of Minho, Guimarães, Portugal
CVR – Centre for Waste Valorization, Guimarães, Portugal

M.C.L.G. Vilarinho
Mechanical Engineering and Resources Sustainability Centre, University of Minho, Guimarães, Portugal

A.V. Machado
Institute for Polymers and Composites/I3N, University of Minho, Guimarães, Portugal

ABSTRACT: Food packaging materials available in the market are mainly made of synthetic plastic derived from fossil resources. However, growing economic and environmental threats related to their life cycle have fostered the search for alternative sustainable raw-materials to produce biodegradable products. In this sense, cellulose extracted from agroindustrial residues, as an available, renewable and low-cost polymer, is a suitable feedstock to produce bio-based packaging. Although research in this field is still limited, cellulose in the form of nanostructures, regenerated products, derivatives and fibers have shown potential to produce packaging components with enhanced properties. This review summarizes the up-to-date developments/applications of cellulosic materials obtained from agroindustrial residues in bio-based packaging, including some prospective applications resulting from the introduction of smart and active functionalities.

1 INTRODUCTION

Polymer-based packaging play an essential role throughout the food distribution and storage chain. In addition to ensuring food safety and quality, packaging must prevent environmental degradation. In 2015, almost 115 million tons of synthetic polymers, predominantly consisting of polyolefins, were used for packaging (Geyer et al. 2017). Despite the versatile properties of polyolefins there are growing concerns over economic and environmental issues related to them. Namely oil market price fluctuations, finiteness of fossil raw materials and pollution of earth's compartments have become nucleation supports for new sustainable thoughts on polymers' life cycle. At the research level, great efforts are currently being focused on the search for alternative renewable resources to produce nonpersistent plastic packaging.

In this scenario, agroindustrial residues have emerged as a source of low-cost raw material, widely available and suitable to produce bio-based and biodegradable plastics as they are composed almost entirely of natural polymers. Their valorization not only avoids the disposal of huge amounts of residues but also does not generate competition with food production. Within the natural polymers, cellulose stands out due to its wide availability and properties, including renewability, low cost, biodegradability, as well as its ability to form derivatives. Recently, cellulose fibers, cellulose derivatives, regenerated cellulose and nanocellulose obtained from agroindustrial residues have been successfully applied to produce food packaging materials with remarkable properties. Besides that, some improvements and multifunctionality have been achieved by introducing additives to the cellulose-based packaging. As far as we know, few reviews and book chapters have covered the use of agroindustrial residues as cellulose source for food packaging applications. Therefore, this review concentrates in summarizing the up-to-date developments/applications on this field and briefly discuss related topics, namely agroindustrial residues, cellulose structure and properties and fractionation techniques.

2 AGROINDUSTRIAL RESIDUES: SOURCE, COMPOSITION AND PRETREATMENT

The term agroindustrial residues spans over two main categories, depending on the generation source: (1) crop residues and (2) industry processing residues. The crop residues are generated during the harvesting phase and comprise straw, branches, stover, leaves, stalks, roots, trimmings and pruning. On the other hand, crops such as sugarcane, grape, orange, potato, coffee, cocoa and coconut are particularly interesting to obtain processed products. From processing food industry, residues in the form of bagasse, pomace, husk, hull and peels are commonly generated.

The attractiveness of agroindustrial residues as a promising sustainable source of biomaterials comes from their availability and chemical composition. They are mainly composed by cellulose (50%), hemicellulose (20–40%) and lignin (10–40%) linked together by covalent crosslinks, which results in their recalcitrant property (Hassan et al. 2018). In a recent publication, Araújo et al. (2018) provided the chemical composition and availability of various crop and industry processing residues. Cellulose, a high molecular weight homopolymer generated from repeating D-glucopyranose ring units linked by $\beta-1$,4-glycosidic bonds (cellobiose unit), stands out as the most abundant renewable resource in nature (Habibi et al. 2010). The drawback of cellulose high hydrophilicity and low thermoplasticity can be solved by its modification in the form of derivatives, nanoparticles, regenerated products, or blending with other materials. However, when sourced from agroindustrial residues, cellulosic fibers must be extracted through fractionation techniques prior to their processing.

The fractionation techniques, also named as pretreatment methods, are carried out to disrupt the compact structure of biomass and fractionate it into its main components. They can be classified basically into physical, chemical, physicochemical or biological treatments. Currently, the pretreatment step is considered one of the main bottlenecks to the cost-effective valorization of agroindustrial residues. Its application may demand high energy and material consumption and generate hazardous wastes (Bhutto et al. 2017). Hence, recently, the use and development of green pretreatments and fine-tuned green solvents, such as ionic liquid, deep eutectic solvents, liquid hot water, steam explosion and biological have been increasingly encouraged (Farrán et al. 2015; Yoo et al. 2017; Hassan et al. 2018).

3 CELLULOSE IN FOOD PACKAGING APPLICATIONS

The development of cost-effective and sustainable methods to produce cellulose food packaging from residues is still challenging and the most innovative products released in the market are based on high-quality cellulose from wood pulp or cotton. The production of new bio-based materials has to meet market requirements and keep up with innovative trends on food packaging, that appears as the enhancement of properties by the introduction of smart, intelligent and active functionalities, obtained via the intentional embedment of supplementary components including inorganic and organic materials/nanomaterials (Ghoshal 2018; Huang et al. 2018). Recent developments and applications on this field are presented in the following subsections, which are themed into the main application forms of cellulose. It's worth to emphasize that we only covered applications that used cellulose emerged from agroindustrial residues. Moreover, it is not the propose of this paper to make fundamental explanations on design, synthesis, preparation methods or concepts regarding cellulose-based materials/derivatives, but rather to present the latest applications.

3.1 *Regenerated cellulose in packaging applications*

Regenerated cellulose refers to a class of materials prepared directly via the dissolution of cellulose into a solution, followed by its shaping and regeneration process (Wang et al. 2016). By changing the regeneration parameters (*e.g.* coagulating anti-solvent, time and temperature), regenerated cellulose with different shapes (such as powder, fibers, films, hydrogels, aerogels and spheres) and properties can be obtained (Wang et al. 2016). Many green solvents have been used to prepare regenerated cellulose, including N,N-dimethylacetamide/lithium chloride (DMAc/LiCl), ionic liquids (ILs) and NaOH/urea aqueous solution (Wang et al. 2016; Li et al. 2018). Regenerated films with mechanical and barrier properties comparable with those of conventional films were also prepared by simple dissolution in IL and regeneration of pretreated sugarcane bagasse (Vanitjinda et al. 2019) and borassus fruit fibers (Reddy et al. 2017).

A fairly new concept of materials, the so-called all-cellulose composite (ACC), have been developed from the application of cellulose regeneration techniques and have shown potential application in food packaging (Li et al. 2018). A homogeneous and transparent ACC film, with tensile strength (TS) reaching 67 MPa, was produced in an IL medium by using corn husk as the regenerated matrix and partially dissolved cellulose microcrystalline (MCC) as filler (Zhang et al. 2017). Using an opposite approach, Wei et al. (2016) employed an alkali-treated straw as the reinforcing phase in a MCC regenerated matrix and produce an ultra-higher tensile strength (568 MPa) ACC.

Currently, hydrogels and aerogels are an attractive subject of research in food packaging, mainly due to their liquid absorption capacity (Batista et al. 2019; Oliveira et al. 2019). However, the development of regenerated cellulose-based hydrogels and aerogels from agroindustrial residues are still limited, and their application in food packaging has not been reported yet. Notwithstanding, some published works present results of great interest to the concerned application.

Hydrogels are three-dimensional networks of polymeric chains, randomly crosslinked by physical or chemical bonds, characterized by the ability to absorb large amounts of fluids (Batista et al. 2019). Hydrogels with high water absorption were obtained by homogenization of aqueous dispersion of PVA and cellulosic samples extracted from rice and oat husk, followed by freezing and thawing (Oliveira et al. 2017). Liu et al. (2017) added graphene oxide on a tea residue cellulose/1-allyl-3-methylimidazolium chloride solution, and by coagulation in distilled water produced composite hydrogels with enhanced thermal stability and textural properties.

In opposite to hydrogels, aerogels are highly porous solids that contain gas instead of a liquid phase within their pores. They are typically made through a solvent exchange and freeze-drying process of hydrogels, in a way that its three-dimensional polymeric network remains intact (Gan et al. 2017). Highly porous lignocellulose aerogels, with great water adsorption and swelling properties, were synthesized by dissolving ethylenediamine-pretreated soybean straw in LiCl/DMSO, followed by coagulation, solvent replacing and freeze-drying (Liu et al. 2019). Wheat straw (Li et al. 2016) and cotton stalk (Mussana et al. 2018) were also used as a cellulose source to produce aerogels with high specific surface area and absorption capacity.

3.2 *Nanocellulose in packaging applications*

In recent years, nanostructured cellulose has attracted research attention due to its potential to maximize mechanical, structural, thermal and barriers properties of plastic-based packaging (Azeredo et al. 2017). Depending on how the nanocellulose is extracted from biomass, two main nanostructures can be obtained: cellulose nanocrystals (CNCs) or nanofibrils (CNFs). The nanocellulose is generally applied as a reinforcement filler for food packaging applications, and improvements of composite properties (namely mechanical and barriers) are intrinsically related to the source and characteristics of nanostructures (Silvério et al. 2013; Rhim et al. 2015; Oun and Rhim 2016; Asad et al. 2018). Nanocellulose extracted from corn cob (Silvério et al. 2013), onion skin (Rhim et al. 2015), wheat straw, rice straw and barley straw (Oun and Rhim, 2016) have already been applied in bio-based packaging.

Usually physical and chemical surface treatments are necessary to disperse the cellulose nanostructures in polymeric matrix. For example, 4-acetamido-TEMPO/NaBr/NaCLO and ultrasonic treatment were employed to promote concurrently the isolation and oxidation of cellulose nanocrystals from oil palm empty fruit bunches. Composite films were produced and the oxidized CNCs presented better dispersibility in the polyvinyl alcohol (PVA) matrix (Asad et al. 2018).

Recently, kiwi pruning residues were used as precursors for the extraction of CNCs through a standard treatment (Luzi et al. 2017). The nanostructures were used as a reinforcement element in a PVA matrix blended with chitosan and carvacrol. The addition of CNCs increased the mechanical properties of composite and their combination with carvacrol improved the shelf-life of perishable food products, by maintaining low moisture migration and bacterial activity.

Another relatively new application in food packaging is the use of nanocellulose-based aerogels and hydrogels. In Ooi et al. (2016), CNCs extracted from rice husk were simply mixed with gelatin and by an evaporative concentration method a pH-responsive CNC-gelatin hydrogel, with potential to be applied in active food packaging, was produced. Similar to regenerated cellulose-based aerogels, nanocellulose-aerogels are prepared via freeze-drying or critical point drying of hydrogels. In Oliveira et al. (2019), an aqueous suspension of CNCs extracted from rice and oat husk was homogenized with PVA and freeze-thawing to form hydrogels. Aerogels with high water absorption capacity (maximum WAC of 402.8% for oat husk CNC-aerogel) were produced by freeze-drying the hydrogel. Shamskar et al. (2016) produced CNC-only aerogels with high specific surface area and mesoporous structure using cotton stalk as nanocellulose precursor.

3.3 *Cellulosic fibers and derivatives in packaging applications*

Similar to the nanostructured cellulose, the cellulosic fibers are usually applied as fillers in polymeric matrix, mainly due to their non-thermoplastic property. In Safont et al. (2018), mechanically ground almond shell (AS) was submitted to a standard purification (alkali and bleach treatment) and blended with poly(3-hydroxybutyrate) (PHB) to produce a fully compostable biocomposite packaging. Compared to the neat PHB, the mechanical properties of the produced composite were enhanced and its total disintegration under composting conditions occurred after 35 days. Using a different and greener approach, Asgher et al. (2017) applied a bacterial cellulose-assisted method to delignify and modify wheat straw fibers and applied them in a PVA matrix. The obtained fibers showed a bacterial cellulose integration with the cellulose microfibrils leading to a significant improvement in the mechanical and water absorption properties of the composite.

Interesting biodegradable biocomposite films made up of carrot processing waste, hydroxypropyl methylcellulose and high pressure microfluidized cellulose fibers were produced without any pretreatment other than mechanical ground (Otoni et al. 2018). The authors extended the production of biocomposites to a pilot scale through continuous casting approach and were able to process 1.56 m^2 of biocomposite film per hour. Recently, by applying an easy and industrial suitable approach, Bilo et al. (2018) produced a cellulose-based bioplastic from rice straw with promising dual-shaping memory.

While natural cellulosic fibers cannot be processed directly into plastic packaging, their chemical modification may result into derivatives that are able to be used in thermoplastic processing. Among the most important derivatizing techniques are etherification and esterification, particularly because they can result in high value-added products. Due to its broad range of applications and easy processability, the modification of cellulosic biomass into cellulose acetate (CA) has been the most commonly explored approach, and current research focuses mainly on the development of more sustainable processing methods (Daud and Djuned 2015; Camiscia et al. 2018). Recently, a fully bio-based and transparent ACC film was fabricated by the simple aqueous blending of water-soluble CA synthesized from waste cotton fabrics (WCFs) and nanocelluloses (Cao et al. 2016a). The same research group (Cao et al. 2016b) used the synthesized water-soluble CA to produce Ag Nps@CA nanohybrid films with good antibacterial property. Great focus has also been devoted to carboxymethyl cellulose (CMC) as it is currently also finding an increasing number of applications. Sugar beet residues and sugarcane bagasse were used as cellulosic fiber sources and carboxymethylated to produce high strength and stiffness film (Šimkovic et al. 2017). Other residues such as corn husk and rice stubble have also been applied as a cellulose source to produce CMC (Mondal et al. 2015; Rodsamran and Sothornit 2017).

4 CONCLUSIONS

Cellulose-based packaging materials emerged from agroindustrial residues are a promising alternative to replace the conventional polymers. Besides to present suitable physicochemical properties, they are biodegradable and can be produced on a large scale due to the raw material availability. Recently, the embedment of natural additives and nanoparticles has been shown to be effective in bringing out functionalities capable to enhance the expected performance of packaging. Notwithstanding, the valorization of agroindustrial residues towards an innovative, acceptable and commercial packaging is still inching and their use must be increasingly encouraged.

REFERENCES

Araújo, D. J. C. *et al.* 2018 'Availability and Suitability of Agroindustrial Residues as Feedstock for Cellulose-Based Materials: Brazil Case Study', *Waste and Biomass Valorization*. Springer Netherlands, p. 16. doi: 10.1007/s12649-018-0291-0.

Asad, M. *et al.* 2018 'Preparation and characterization of nanocomposite films from oil palm pulp nanocellulose/poly (Vinyl alcohol) by casting method', *Carbohydrate Polymers*. Elsevier, 191(February), pp. 103–111. doi: 10.1016/j.carbpol.2018.03.015.

Asgher, M. *et al.* 2017 'Bacterial cellulose-assisted de-lignified wheat straw-PVA based bio-composites with novel characteristics', *Carbohydrate Polymers*. Elsevier Ltd., 161, pp. 244–252. doi: 10.1016/j.carbpol.2017.01.032.

Azeredo, H. M. C. *et al.* 2017 'Nanocellulose in bio-based food packaging applications', *Industrial Crops & Products*. Elsevier B.V., 97, pp. 664–671. doi: 10.1016/j.indcrop.2016.03.013.

Batista, R. A. *et al.* 2019 'Hydrogel as an alternative structure for food packaging systems', *Carbohydrate Polymers*. Elsevier, 205(September 2018), pp. 106–116. doi: 10.1016/j.carbpol.2018.10.006.

Bhutto, A. W. *et al.* 2017 'Insight into progress in pre-treatment of lignocellulosic biomass', *Energy*. Elsevier Ltd, 122, pp. 724–745. doi: 10.1016/j.energy.2017.01.005.

Bilo, F. *et al.* 2018 'A sustainable bioplastic obtained from rice straw', *Journal of Cleaner Production*. Elsevier Ltd, 200, pp. 357–368. doi: 10.1016/j.jclepro.2018.07.252.

Camiscia, P. *et al.* 2018 'Comparison of soybean hull pre-treatments to obtain cellulose and chemical derivatives: Physical chemistry characterization', *Carbohydrate Polymers*. Elsevier, 198(July), pp. 601–610. doi: 10.1016/j.carbpol.2018.06.125.

Cao, J. *et al.* 2016a 'Water-soluble cellulose acetate from waste cotton fabrics and the aqueous processing of all-cellulose composites', *Carbohydrate Polymers*. Elsevier Ltd., 149, pp. 60–67. doi: 10.1016/j.carbpol.2016.04.086.

Cao, J. *et al.* 2016b 'Homogeneous synthesis of Ag nanoparticles-doped water-soluble cellulose acetate for versatile applications', *Inter. J. of Biol. Macrom.*. Elsevier B.V., 92, pp. 167–173. doi: 10.1016/j.ijbiomac.2016.06.092.

Daud, W. R. and Djuned, F. M. 2015 'Cellulose acetate from oil palm empty fruit bunch via a one step heterogeneous acetylation', *Carbohydrate Polymers*. Elsevier Ltd., 132, pp. 252–260. doi: 10.1016/j.carbpol.2015.06.011.

Farrán, A. *et al.* 2015 'Green Solvents in Carbohydrate Chemistry: From Raw Materials to Fine Chemicals', *Chemical Reviews*, 115(14), pp. 6811–6853. doi: 10.1021/cr500719h.

Gan, S. *et al.* 2017 'Highly porous regenerated cellulose hydrogel and aerogel prepared from hydrothermal synthesized cellulose carbamate', PLOS ONE pp. 1–13.

Geyer, R., Jambeck, J. R. and Law, K. L. 2017 'Production , use , and fate of all plastics ever made', Science Advances (July), pp. 25–29.

Ghoshal, G. 2018 'Recent Trends in Active, Smart, and Intelligent Packaging for Food Products', in *Food Pack. and Pre.*. Academic Press, pp. 343–374. doi: //doi.org/10.1016/B978-0-12-811516-9.00010-5.

Habibi, Y. *et al.* 2010 'Cellulose nanocrystals: Chemistry, self-assembly, and applications', *Chemical Reviews*, 110(6), pp. 3479–3500. doi: 10.1021/cr900339w.

Hassan, S. S. *et al.* 2018 'Emerging technologies for the pretreatment of lignocellulosic biomass', *Bioresource Technology*. Elsevier, 262(April), pp. 310–318. doi: 10.1016/j.biortech.2018.04.099.

Huang, Y. *et al.* 2018 'Recent Developments in Food Packaging Based on Nanomaterials', *Nanomaterials*, 8(830), pp. 1–29. doi: 10.3390/nano8100830.

Li, J. *et al.* 2016 'Fabrication of cellulose aerogel from wheat straw with strong absorptive capacity', *Front. Agr. Sci. Eng*, 1(1), pp. 46–52. doi: 10.15302/J-FASE-2014004.
Li, J. *et al.* 2018 'All-cellulose composites based on the self-reinforced effect', *Composites Communications*. Elsevier, 9(April), pp. 42–53. doi: 10.1016/j.coco.2018.04.008.
Liu, Z. *et al.* 2017 'Enhanced properties of tea residue cellulose hydrogels by addition of graphene oxide', *J. of Molec. Liq.*. Elsevier B.V., 244, pp. 110–116. doi: 10.1016/j.molliq.2017.08.106.
Liu, Z. *et al.* 2019 'Characterization of lignocellulose aerogels fabricated using a LiCl/DMSO solution', *Ind. Crops & Prod.*. Elsevier, 131(February), pp. 293–300. doi: 10.1016/j.indcrop.2019.01.057.
Luzi, F. *et al.* 2017 'Cellulose nanocrystals from Actinidia deliciosa pruning residues combined with carvacrol in PVA CH films with antioxidant/antimicrobial properties for packaging applications', *International Journal of Biological Macromolecules*. Elsevier B.V., 104, pp. 43–55. doi: 10.1016/j.ijbiomac.2017.05.176.
Mondal, I. H. *et al.* 2015 'Preparation of food grade carboxymethyl cellulose from corn husk agrowaste', *Inter. J. of Biol. Macro.*. Elsevier B.V., 79, pp. 144–150. doi: 10.1016/j.ijbiomac.2015.04.061.
Mussana, H. *et al.* 2018 'Preparation of lignocellulose aerogels from cotton stalks in the ionic liquid-based co-solvent system', *Industrial Crops & Products*. Elsevier, 113(January), pp. 225–233. doi: 10.1016/j.indcrop.2018.01.025.
Oliveira, J. *et al.* 2017 'Cellulose fibers extracted from rice and oat husks and their application in hydrogel', *Food Chemistry*. Elsevier Ltd, 221, pp. 153–160. doi: 10.1016/j.foodchem.2016.10.048.
Oliveira, J. P. *et al.* 2019 'Cellulose nanocrystals from rice and oat husks and their application in aerogels for food packaging', *International Journal of Biological Macromolecules*. Elsevier B.V., 124, pp. 175–184. doi: 10.1016/j.ijbiomac.2018.11.205.
Ooi, S. Y. *et al.* 2016 'Cellulose nanocrystals extracted from rice husks as a reinforcing material in gelatin hydrogels for use in controlled drug delivery systems Gelatin/CNC hydrogel Gelatin hydrogel', *Industrial Crops & Products*. Elsevier B.V., 93, pp. 227–234. doi: 10.1016/j.indcrop.2015.11.082.
Otoni, C. G. *et al.* 2018 'Optimized and scaled-up production of cellulose-reinforced biodegradable composite fi lms made up of carrot processing waste', *Industrial Crops & Products*. Elsevier, 121(November 2017), pp. 66–72. doi: 10.1016/j.indcrop.2018.05.003.
Oun, A. A. and Rhim, J. 2016 'Isolation of cellulose nanocrystals from grain straws and their use for the preparation of carboxymethyl cellulose-based nanocomposite films', *Carbohydrate Polymers*. Elsevier Ltd., 150, pp. 187–200. doi: 10.1016/j.carbpol.2016.05.020.
Shamskar, K. R. *et al.* 2016 'Preparation and evaluation of nanocrystalline cellulose aerogels from raw cotton and cotton stalk', *Industrial Crops and Products*. Elsevier B.V., 93, pp. 203–211. doi: 10.1016/j.indcrop.2016.01.044.
Reddy, K. O. *et al.* 2017 'Preparation and characterization of regenerated cellulose films using borassus fruit fibers and an ionic liquid', *Carbohydrate Polymers*. Elsevier Ltd., 160, pp. 203–211. doi: 10.1016/j.carbpol.2016.12.051.
Rhim, J. W. *et al.* 2015 'Isolation of cellulose nanocrystals from onion skin and their utilization for the preparation of agar-based bio-nanocomposites films', *Cellulose*, 22(1), pp. 407–420. doi: 10.1007/s10570-014-0517-7.
Rodsamran, P. and Sothornvit, R. 2017 'Rice stubble as a new biopolymer source to produce carboxymethyl cellulose-blended films', *Carbohydrate Polymers*. Elsevier Ltd., 171, pp. 94–101. doi: 10.1016/j.carbpol.2017.05.003.
Safont, E. L. S. *et al.* 2018 'Biocomposites of different lignocellulosic wastes for sustainable food packaging applications', *Composites Part B*. Elsevier, 145(January), pp. 215–225. doi: 10.1016/j.compositesb.2018.03.037.
Silvério, H. A. *et al.* 2013 'Extraction and characterization of cellulose nanocrystals from corncob for application as reinforcing agent in nanocomposites', *Industrial Crops and Products*. Elsevier B.V., 44, pp. 427–436. doi: 10.1016/j.indcrop.2012.10.014.
Šimkovic, I. *et al.* 2017 'Composite films prepared from agricultural by-products', 156, pp. 77–85. doi: 10.1016/j.carbpol.2016.09.014.
Vanitjinda, G. *et al.* 2019 'Effect of xylanase-assisted pretreatment on the properties of cellulose and regenerated cellulose fi lms from sugarcane bagasse', *Intern. J. of Biol. Macrom.*. Elsevier B.V., 122, pp. 503–516. doi: 10.1016/j.ijbiomac.2018.10.191.
Wang, S., Lu, A. and Zhang, L. 2016 'Recent advances in regenerated cellulose materials', *Progress in Polymer Science*. Elsevier Ltd, 53, pp. 169–206. doi: 10.1016/j.progpolymsci.2015.07.003.
Wei, X. *et al.* (2016) 'All-cellulose composites with ultra-high mechanical properties prepared through using straw cellulose fiber', *RSC Advances*, 6(96), pp. 93428–93435.

Yoo, C. G., Pu, Y. and Ragauskas, A. J. 2017 'Ionic liquids: Promising green solvents for lignocellulosic biomass utilization', *Current Opinion in Green and Sustainable Chemistry*. Elsevier B.V., 5, pp. 5–11. doi: 10.1016/j.cogsc.2017.03.003.

Zhang, J. *et al.* 2017 'Directly Converting Agricultural Straw into All-Biomass Nanocomposite Films Reinforced with Additional in Situ-Retained Cellulose Nanocrystals', *ACS Sustainable Chem. Eng*, 5(6), pp. 5127–5133. doi: 10.1021/acssuschemeng.7b00488.

Wastes: Solutions, Treatments and Opportunities III – Vilarinho et al. (Eds)
© 2020 Taylor & Francis Group, London, ISBN 978-0-367-25777-4

Valorization of *Caryocar brasiliense* Camb. waste for the production of particleboards

D.L. Faria, G.O. Chagas, K.M. Oliveira, A.C.C. Furtini, C.A. Santos, L.M. Mendes, C.R. Andrade, K.C. Lozano & J.B. Guimarães Júnior
Federal University of Lavras, Lavras, Minas Gerais, Brazil

ABSTRACT: The aim of this work was to evaluate the physical properties of particleboards produced with *Pinus oocarpa* and *Caryocar brasiliense* Camb. For this, boards containing 0, 5, 10 and 15% of particles of *Caryocar brasiliense* Camb. in substitution of *Pinus oocarpa* wood and 8% of urea-formaldehyde adhesive. The boards were produced with nominal density of 0.7 $g.cm^{-3}$, pressing cycle with temperature of 160°C, pressure of 3.92 MPa and time of 8 minutes. The properties of water absorption (WA), thickness swelling (TS), both after 2 and 24 hours, nominal density (ND), compaction ratio and moisture were evaluated. The results of the evaluated properties indicated that with the increase of the particle content of the fruit of *Caryocar brasiliense* Camb. resulting in boards with better dimensional stability on contact with water. In general, the use of particles of the fruit of *Caryocar brasiliense* Camb. is feasible for the production of particleboards.

1 INTRODUCTION

Environmental issues, now perceived by humanity, instigate the search for new alternatives for global improvement. One of the sectors that needs care is the construction industry, which generates waste and often presents no solutions to its consequences. In this case, construction systems and building materials, such as wood, are also used and are often not planned for disposal (Almeida et al., 2012).

Just as planted forest areas are being used as alternative sites to minimize environmental hardship, new materials also appear on the market as a way to alleviate current problems. In these cases the boards that take advantage of wood processing wastes for their production fit together. Such proposed and already consolidated materials in the market have also generated worries, because, although there is a lot of supply, its demand is increasing every year, with its production/consumption ratio very close. These materials present certain constraints to the manufacturing, which implies the need of research to take advantage of its benefits (Almeida et al., 2012).

Pinus and eucalyptus wood constitute the raw material base for the production of particleboards and MDF boards in Brazil. In the last two decades, industries of the reconstituted wood boards industry have made great investments in the implantation of new productive units, besides the increase in new areas of forest plantations to assure the supply of wood (Iwakiri et al., 2012). The particleboards industries in Brazil are located in the south and southeast regions, focusing on furniture centers located mainly in the States of eucalyptus are the base of supply of wood for these industries. Although the northern and mid-western regions of the country have extensive native tropical forest areas available, few studies have been conducted on the feasibility of using tropical timber as well as non-timber wastes for the production of particleboards.

Among the species that stand out in the Cerrado biome is the pequiary, *Caryocar brasiliense* Camb., which is known for its economic and nutritional value, besides its regional importance due to the high consumption of fruits and derivatives by the population (Côrrea et al., 2008; Moura et al., 2013). In this way, the Pequi tree deserves special attention due to its occurrence, the volume of fruits commercialized in the region and the organoleptic characteristics of its fruit.

In view of the above, this work has the objective of evaluating the physical quality of particleboards produced with *Pinus oocarpa* wood and different particle additions of the *Caryocar brasiliense* Camb. Fruit, in order to add value to this waste.

2 MATERIAL AND METHODS

Two trees of the species *Pinus oocarpa*, approximately 25 years old, were cut on the campus of the Federal University of Lavras, located in Lavras, Minas Gerais, Brazil. After cutting and sectioning into logs of 60 cm in length, the logs were placed in a tank with water at 70°C for approximately 24 hours until the wood was soft to obtain laminas in a rolling mill. The laminas obtained were oven dried for 24 hours at a temperature of 105°C to 3% moisture content, and subsequently the laminas were ground in a hammer mill to obtain "*sliver*" particles. The fruits of *Caryocar brasiliense* Camb. were obtained from a farm located in the municipality of Jataí, located in the state of Goiás, Brazil. The fruits were dried in a greenhouse with forced air circulation, and later they were ground in a knife mill, obtaining the "*sliver*" particles.

Before the production of the particleboards, the basic density and chemical constituents of the particles of *Pinus oocarpa* and *Caryocar brasiliense* Camb. were evaluated. The basic density was determined according to the NBR 11941 (Brazilian Association of Technical Standards, 2003). The quantification of the total extractive content was done in triplicate using the fraction of the biomass retained between the sieves of 40 and 60 mesh, according to the norm NBR 14853 (ABNT, 2010). Samples were subjected to a toluol-ethanol (2:1, 5 h), ethanol (4 h) and hot water (2 h) sequence. The determination of the insoluble lignin content was carried out using a procedure described in standard NBR 7989 (ABNT, 2010), using samples of approximately 1 g free of extractives. The solvent used was 72% sulfuric acid kept cooled. For the quantification of the ashes contents, the methodology established in the norm NBR 13999 (ABNT, 2017) was considered. Holocellulose content was obtained by difference in relation to the other molecular and mineral components of the biomass.

The boards were produced with dimensions of 300 × 300 × 15 mm and nominal density of 0.600 g.cm^{-3}, being evaluated three substitutions of wood by the residue of "*Pequizeiro*": 0, 5, 10 and 15%. Each treatment had three replicates, totaling 12 boards. The adhesive used was urea-formaldehyde, which had a solids content of 60%, viscosity of 200.95 cP, pH of 8.45 and gel timer of 7.85 min. The boards were pressed at a temperature of 160°C and a specific pressure of 3.92 MPa for eight minutes. After pressing, the boards were packed in a climatic chamber at a temperature of 20 ± 3°C and relative humidity of 65 ± 5% until equilibrium moisture of approximately 12% was reached and the solidification process of the adhesive was terminated.

The compaction ratio of each particleboard was calculated by the ratio of the basic density of the wood of *Pinus oocarpa* and the fruit of *Caryocar brasiliense* Camb. by the nominal density of the panel. Standard D 1037 (ASTM, 2006) was used for the evaluation of the water absorption properties after 2 and 24 h of immersion (WA2h and WA24h) and thickness swelling after 2 and 24 h (TS2h and TS24h). The nominal density of the boards was determined with the average density of each of the test pieces used to evaluate the physical properties. The moisture content of the particleboards was determined using surplus specimens, using the mass before and after the permanence of the test specimens in a greenhouse until there was no change in mass on the order of 0.01 g. To evaluate the properties of the boards studied, the data were submitted to analysis of variance (ANOVA) and Scott-Knott test, both at 5% significance. The data was processed in *Sisvar* software.

3 RESULTS AND DISCUSSION

3.1 *Physical and chemical characterization of lignocellulosic materials*

The mean apparent density of *Pinus oocarpa* wood was 0.477 g.cm^{-3}, a result close to that obtained by Mattos et al. (2011), whose authors obtained apparent density ranging from

Table 1. Chemical composition of the wood of *Pinus oocarpa* and *Caryocar brasiliense* Camb.

Material	Extractives (%)	Lignin (%)	Holocellulose (%)	Ashes (%)
Pinus oocarpa	4.03 ± 0.17 B	26.08 ± 2.50 B	69.63 ± 2.63 B	0.26 ± 0.02 A
Caryocar brasiliense	59.60 ± 0.44 A	10.79 ± 0.20 A	23.76 ± 0.89 A	2.66 ± 0.07 B

0.350 to 0.450 g.cm^{-3} for *Pinus oocarpa* wood. Already the *Caryocar brasiliense* Camb. presented mean values for apparent density of 0.520 g.cm^{-3}.

Regarding the chemical properties of the species under study (Table 1), there was a statistically significant difference for all the constituents.

When evaluating the average levels of extractives, it was observed that there was a statistical difference between the studied species. The fruit of *Caryocar brasiliense* Camb. presented the highest extractive contents (59.60%) followed by *Pinus oocarpa* wood (4.03%). According to Fonte; Trianoski (2015) boards produced with species with high extractive values may exhibit deficiency in bonding. Depending on the species and the drying conditions of the wood, due to the migration and excessive concentration of extractives, a so-called "inactive or contaminated surface" may occur, damaging the adhesive-wood contact (Iwakiri, 2005). High levels of this component may influence the cure time of the adhesive as well as the quality of adhesion (Boa et al., 2014). The extractives are worthy of note, since the high content of certain types of extractives may impair the quality of particleboards. According to Iwakiri (2005), the presence of a great amount of extractives in the material used may incur problems on the consumption of adhesives, decrease of mechanical resistance, besides the occurrence of air bubbles during pressing. Another problem generated by the extractives is the weak moistening of the particles, since the extractives are hydrophobic and concentrate on the surface after high temperatures. In this way, the adhesives may have an impaired penetration.

The lignin content of *Pinus oocarpa* wood was 26.08%, where as for the fruit of *Caryocar brasiliense* Camb. was 10.79%. Lignin is a natural adhesive, being excellent for the production of reconstituted wood panels, giving better mechanical properties and resistance to water (Khedari et al., 2004; Bufalino et al., 2012).

For holocellulose contents, it was observed that *Pinus oocarpa* wood had the highest average contents (69.63%) in relation to the fruit of *Caryocar brasiliense* Camb. (23.76%). For the content of holocellulose present in *Pinus oocarpa* wood, the results are close to those of the literature, as observed in Mendes et al. (2014), where they evaluated the quality of particleboards produced with *Pinus oocarpa* wood and obtained results for holocellulose content of 66.50%. Holocellulose values can affect the physical properties of particleboards due to the hygroscopicity of these structures because they refer to free hydroxyl groups that can adhere to water (Iwakiri, 2005).

The ash contents were 0.26% for *Pinus oocarpa* wood and 2.66% for fruit of *Caryocar brasiliense* Camb. The presence of higher amounts of minerals and some apolar extractives may result in the blocking of chemical groups reactive towards adhesion, thus affecting the quality of the bonding and the mechanical performance of particleboards (Ndazi et al., 2007). With this, it is observed that the fruit of *Caryocar brasiliense* Camb. may present problems in bonding, due to the presence of more than 1% of ash in its constitution.

3.2 *Physical quality of particleboards*

Table 2 shows the average values of nominal density (ND), compaction ratio (CR) and moisture (M) of the particleboards produced.

The values obtained for nominal density of the produced boards varied statistically by the Tukey average test at 5% of significance, presented values varying from 0.551 to 0.614 g.cm^{-3}. According

Table 2. Mean values of nominal density, compaction ratio and moisture for the particleboards produced.

Treatments	ND (g.cm^{-3})	CR	M (%)
T1	0.551 ± 0.03 A	1.15 ± 0.07 D	7.37 ± 0.44 A
T2	0.579 ± 0.02 B	1.21 ± 0.05 C	7.30 ± 0.43 A
T3	0.614 ± 0.01 B	1.28 ± 0.02 B	6.73 ± 0.45 A
T4	0.582 ± 0.03 B	1.22 ± 0.06 B	6.99 ± 0.47 A

Values followed by the same letter do not differ from each other by the Tukey test at 5% significance.

Table 3. Mean values of water absorption and thickness swelling for the particleboards produced.

Treatments	WA2h (%)	WA24h (%)	TS2h (%)	TS24h (%)
T1	103.89 ± 11.14 A	104.85 ± 19.64 A	18.15 ± 1.70 B	18.92 ± 1.58 B
T2	108.84 ± 7.94 A	110.71 ± 8.15 A	19.60 ± 1.55 B	20.17 ± 2.64 B
T3	96.33 ± 9.92 A	96.61 ± 10.16 A	14.60 ± 1.40 A	14.94 ± 1.25 A
T4	96.47 ± 6.18 A	98.23 ± 7.19 A	14.63 ± 1.95 A	14.67 ± 2.29 A

Values followed by the same letter do not differ from each other by the Tukey test at 5% significance.

to NBR 14810-2 (ABNT, 2018), the boards can be classified as medium density, as they have an nominal density greater than 0.551 g.cm^{-3}.

For the compaction ratio, there was an increase whereas the amount of particles of *Caryocar brasiliense* Camb. in the boards, ranging from 1.15 to 1.28. None of the evaluated boards were within the range considered ideal by Maloney (1993), ranging from 1.3 to 1.6.

The moisture varied from 6.734 to 7.37% and there was no statistical difference between the treatments. The Brazilian standard NBR 14810-2 (ABNT, 2018) stipulates moisture values between 5 and 11%, thus, all treatments fall within the cited standard.

Table 3 shows the average values of water absorption (WA) after 2 and 24 hours, thickness swelling (TS) after 2 and 24 hours of the particleboards produced.

No significant statistical difference was observed between the treatments for the water absorption property after 2 and 24 hours immersion in water. There were only significant statistical differences for the property of thickness swelling after 2 and 24 hours of immersion in water. As particles of *Caryocar brasiliense* Camb. were added to the boards, there was a reduction of the thickness swelling. This behavior is due to the higher specific mass of this species and consequently increasing the compaction ratio of the boards, facilitating the entry of water due to the greater porosity of the boards.

Behavior similar to this work was observed by Andrade et al. (2018), whose authors studying the physical and mechanical properties of agglomerated medium density panels of *Eucalyptus sp.* obtained 53, 78, 108 and 130% water immersion contents after 24 hours for boards produced with 0, 25, 50 and 75% cellulose pulping waste.

The results obtained in this study for swelling in thickness showed the same tendency of reduction of those verified by Trianoski et al. (2013) when studying the use of *Cryptomeria japonica* wood in different proportions for the production of particleboards in pure composition and in mixture with *Pinus spp*. The authors verified mean values of 5.71 to 3.55% after 2 hours and from 18.61 to 7.85% after 24 hours of immersion in water.

Moslemi (1974) reports that an increase in the compaction ratio of the boards results in greater swelling in thickness, fact observed in several works of the area, but was not verified in this research. In addition, the presence of this constituent may have inhibited swelling (Iwakiri et al., 2018).

Only the treatments containing 10 and 15% of particles of the fruit of the *Caryocar brasiliense* Camb. met the requirement of EN 312 (EN, 2003) of 15% for swelling in thickness after 24 hours of immersion in water.

4 CONCLUSION

In view of the obtained results, it can be concluded:

- The increase in the particle contents of the *Caryocar brasiliense* Camb. fruit increased the nominal density of the particleboards, giving higher compaction ratio values.
- The higher extractive contents of the *Caryocar brasiliense* Camb. fruit gave better physical properties to the particleboards, reflecting in lower percentages of water absorption (WA) and thickness swelling (TS), both after 2 and 24 hours of immersion in water.
- According to the current regulations, only particleboards produced 10 and 15% of particles of the fruit of the *Caryocar brasiliense* Camb. met the minimum requirement for thickness swelling (TS) after 24 hours of immersion in water.

ACKNOWLEDGMENTS

We thank FAPEMIG, CNPq, CAPES, RELIGAR NETWORK and PPG BIOMAT / UFLA.

REFERENCES

American Society for Testing and Materials. 2006. "Standard test methods for evaluating properties of wood fiber -base and particle panel materials", *ASTM D1037-06*, Philladelphia.

Almeida, J.E., Logsdon, N.B., Jesus, J.M.H. 2012. Wood panels produced with sawdust and expanded polystyrene. *Floresta* 42: 189–200.

Andrade, L.M.F., Scatolino, M.V., Faria, D.L., César, A.A., Mendes, L.M., Guimarães Júnior, J.B. 2018. Inclusion of cellulose pulping waste for production of medium density particleboards. *Scientia Forestalis* 46: 120.

Boa, A.C., Gonçalves, F.G., Oliveira, J.T.S., Paes, J.B., Arantes, M.D.C. 2014. Eucalypts timber wastes glued with urea formaldehyde resin at room temperature. *Scientia Forestalis* 42: 279–288.

Brazilian Association ofTechnical Standards. 2017. *NBR 13999* : Paper, paperboard, cellulosic pulp and wood: determination of the residue (ash) after incineration at 525° C. Rio de Janeiro.

Brazilian Association ofTechnical Standards. 2003. *NBR-11941*: Wood - Determination of basic density. São Paulo.

Brazilian Association of Technical Standards. 2018. *NBR 14810*: Medium density particleboards Part 2: Requirements and test methods.

Brazilian Association ofTechnical Standards. 2010a. *NBR 14853:* Wood – determination of the material soluble in ethanol-toluene and in dichloromethane and acetone.

Brazilian Association of Technical Standards. 2010b. *NBR 7989*: cellulosic pulp and wood: determination of acid-insoluble lignin.

Bufalino, L., Protásio, T.P.P., Couto, A.M., Nassur, O.A.C., Sá, V.A., Trugilho, P.F., Mendes, L.M. 2012. Chemical and energetic characterization for utilization of thinning and slab wood from Australian red cedar. *Pesquisa Florestal Brasileira* 32: 129–137.

Côrrea, G.C. et al. 2008. Physical determinations in fruit and seeds of baru (*Dipteryx alata* Vog.), Cajuzinho (*Anacardium othonianum* Rizz.) and pequi (*Caryocar brasiliense* Camb.) aiming genetic breeding. *Bioscience Journal* v.24: 42–47.

European Standard. EN 312: particleboards: specifications. British Standard Institution, London, 2003. 22p

Fonte, A.P.N., Trianoski, R. 2015. Effect of grammage on the bonding quality of glue side of Tectona grandis wood, *Lages* 14:224–233.

Iwakiri, S. Panels, wood reconstituted. FUPEF. Curitiba, 2005.

Iwakiri, S., Trianoski, R., Raia, R.Z., Keinert, A.C., Paula, C.R.P., Protzek, G.R., Kobylarz, R., SChweitzwer, V.R. 2018. Production of particleboard of *Hevea brasiliensis* (Clone RRIM 600) in mixture with three species of Eucalyptus used by São Paulo's industries. *Scientia Forestalis* 46: 31–39.
Iwakiri, S., Vianez, B.F., Weber, C., Trianoski, R., Almeida, V.C. 2012. Evaluation of the properties of particleboard made from sawmill waste of nine tropical wood species of Amazon. *Acta Amazonica* 42: 59–64.
Khedari, J., Nankongnab, N., Hirunlabh, J., Teekasap, S. 2004. New low-cost insulation particleboards from mixture of durian peel and coconut coir. *Building and Environment* 39: 59–65.
Maloney, T. M. 1993. Modern particleboard and dry-process fiberboard manufacturing. 2. Ed. São Francisco: Miller Freeman.
Mattos, B.D., Gatto, D.A., Stangerlin, D.M., Calegari, L., Melo, R.R., Santini, E.J. 2011. Axial variation of wood basic density three gymnosperms species. *Revista Brasileira de Ciências Agrárias* 6: 121–126.
Mendes, R. F., Mendes, L.M., Mendonça, L.L., Guimarães Júnior, J.B., Mori, F.A. 2014. Quality of homogeneous particleboard produced with *Eucalyptus urophylla* clone wood. *Cerne* 20: 329–336.
Moslemi, A.A. 1974. *Particleboard.* London: Southern Illinois University Press, 245 p.
Moura, N.F. et al. 2013. Variability between provenances and progenies of Pequizero (*Caryocar brasiliense* Camb.). *Scientia Forestalis* 41: 103–112.
Ndazi, B.S., Karlsson, S., Tesha, J.V., Nyahumwa, C.W. 2007. Chemical and physical modifications of rice husks for use as composite panels. *Composites Part A: Applied Science and Manufacturing* 38: 925–935.
Trianoski, R., Iwakiri, S., Matos, J.L.M., Chies, D. 2013. Use of Cryptomeria japonica wood for the production of particleboard. *Scientia Forestalis* 41: 057–064.

Wastes: Solutions, Treatments and Opportunities III – Vilarinho et al. (Eds)
© 2020 Taylor & Francis Group, London, ISBN 978-0-367-25777-4

Char valorization in construction materials

F. Andreola, L. Barbieri, I. Lancellotti, P. Pozzi & V. Vezzali
Department of Engineering "Enzo Ferrari", University of Modena and Reggio Emilia, Modena, Italy

ABSTRACT: Char coming from gasification of biomasses was mixed with red clay to create both lightweight aggregates (LWAs) for green roofs applications and bricks and with polyoils and other reactants to obtain polyurethane panels for insulating purpose. Application in LWAs by substituting the clay with char leads to a weight-lightening of the material. To optimize LWAs pH, spent coffee grounds (SCG) were added. A greater decrease in weight and pH values in the neutrality range were observed. Adding biochar on bricks led to a significant reduction of materials bulk density (from 2 to 1.5 g/cm^3). Polyurethane panels show a decrease of thermal conductivity by increasing the content of char. The addition of char increases the maximum applicable load related to stress relaxation tests and improves the recovery of the load after 10 minutes. Higher elastic modulus values are seen after compressive tests. Electrical conductivity increases in samples containing char.

1 INTRODUCTION

Biomass from river maintenance is not widely used for bioenergy production. In fact, the high moisture, the high ash content and the variable size distribution decrease the quality of these feedstock which it is less exploitable into power plants compared to wood chips or wood pellets (IEA, 2012). However, the wide availability of biomass per hectare of riverbed justifies the cost of power plants optimized for these kind of feedstocks. Over the energy convenience of the conversion processes, river maintenance is a vital operation to ensure the safety of banks and to prevent ruinous floods. The by-product of this process is char: a fine-grained vegetable carbon extracted from the bottom of the gasifier. If the use of char as soil or substrate improver has been studied a lot in the last 15 years, while the application of this substance in construction materials or other composites is starting to gain more attention recently. Indeed, in addition to the great advantage of carbon sequestration, the use of char can reduce the energy associated with the production process of such materials, by decreasing firing temperature, as well as the consumption of raw materials. Furthermore, in many applications, for example cement or other materials, an improvement of various physical properties is found when char is added (Gupta et al., 2017), (Choi et al., 2012). (Khushnood et al.,2016). (Ahmad et al., 2015). Other recent uses of biochar in the building sector can be identified in the field of asphalts. (Zhao et al., 2014).

The introduction of char into a production process of bricks, lightweight aggregates or polyurethane panels can overcome the problem of natural resources consumption.

In this work char with grain size under 4 mm coming from the gasification reactor was characterized. Chemical, thermal and physical analyses were made to fully characterize the materials and understand their nature and behaviour. Char was then applied in various construction materials such as bricks and LWAs (lightweight aggregates) for green roofs and polyuethane panels for insulating.

2 MATERIALS AND METHODS

2.1 *Char characterization*

Char was examined by scanning electron microscope (SEM) "ESEM-Quanta 200 FEI", to evaluate particle structure and surface morphology of the materials. Energy Dispersive Spectroscopy (EDS),

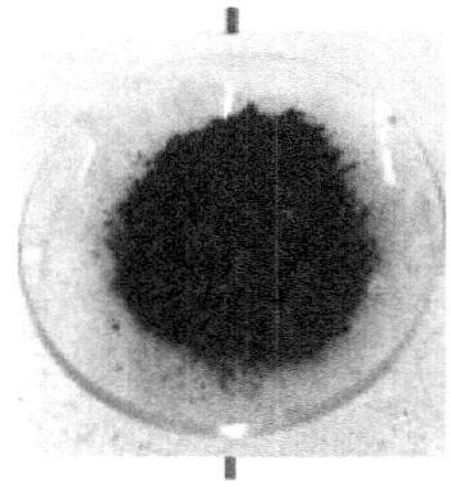

Figure 1. Photo of Char.

Figure 2. Photo of lightweight aggregates.

coupled with SEM, was also carried out to identify the atomic elements in these materials and their relative proportions (%). Content of C, H, N, S was measured on an elemental analyser "CHNS-O Thermo Finnigan Elementary Analyzer Flash EA 1112". pH and conductivity were measured following UNI EN 13037:2012 and UNI EN 13038:2012. Crystalline phases (X-ray diffractometer "Phillips PANalyticalPW3710") and carbonates ("Dietrich-Fruhling" calcimeter) were also analyzed.

2.2 *LWA samples production*

For the production of LWAs, spent coffee grounds (SCG) mixed with char, both sieved under 250 μm, was used. The ratio in this case was 85% red local clay b-15% biochar/SCG. The aggregates were subjected to a firing treatment in a static oven at 1000°C for 1h. Some important parameters were assessed for the use of aggregates in green roofs: weight loss during drying and firing; pH value; specific electrical conductivity (EC); static absorption in water over 24h; surface microstructure by SEM.

2.3 *Bricks samples production*

Several cylindrical specimens (40mm Ø x 4mm) with char ratios from 5 to 40% and a red local clay were obtained by pressing at 22 bar after 7% humidification of the mixtures. Bricks were subjected to a firing treatment in a static oven at 950°C for 30 minutes. Weight loss in drying and firing, drying and firing shrinkage and apparent density of the fired samples were evaluated.

2.4 *Polyurethane panel preparation and characterization*

Char dried and ground <125μm was mixed with two different polyols, isocianate, water, catalyst and silane agents. The mixture was poured in a mould, maintained at 40°C, with dimensions of

Figure 3. Photo of polyurethane panel containing char.

20x20x2 cm. Samples containing 0.5, 1 e 2 wt% of char with respect to polyol were characterized by: thermal conductivity, stress relaxation, compressive test and electrical resistivity measurements.

By means of the thermal conductivity test, λ_m [W/(m*K)] was determined by "Heat Flow Meters HFM 436 Lambda". In order to investigate the visco-elastic behaviour of materials containing char compared to the standard, stress relaxation test was performed. The test applies a constant deformation on the material (0.1 mm/s) and after maximum load is reached (5% of deformation) the relaxing behaviour after 10min is studied. Compressive strength tests (5567 Instron with maximum load 30 kN) were performed in agreement with the norm UNI EN ISO 604. Electrical resistivity measurements were conducted by the instrument "Keithley 6517a".

3 RESULTS AND DISCUSSION

3.1 *Char characterization*

The pH measurements of char are strongly alkaline; indeed, the high temperatures reached during the gasification process strongly influence the pH of the final products, causing alkalinity. This parameter is important because a very alkaline pH could influence the final material in the application that will be chosen, altering its acceptability in a certain range of values. The values of specific electrical conductivity of char are in the range 5-6 mS/cm. The analysis with a calcimeter shows a significant presence of calcium carbonates, later confirmed by the EDS and XRD analyses data. Yuan et al. found a linear correlation between alkalinity and carbonate content, although also carboxyl and hydroxyl groups may contribute to the pH of chars (Yuan et al., 2011). The EDS chemical analysis shows the total carbon from 45% to 85%. These values are in agreement with literature studies on various types of biochars from different feedstocks and different processes. The values obtained from elemental analysis, referred to the red spruce char, fall within the ranges identified by EDS analysis (C 54%, H 1%, N 0.33%, S 0%). The X-ray diffractometry patterns show that char has predominantly amorphous nature, the only crystalline phases present can be identified as calcite ($CaCO_3$), with traces of quartz (SiO_2).

3.2 *Lightweight Aggregates*

All the samples of LWA prepared were characterised by a good feasibility in processing and showed a good level of final aggregation. No samples showed any breakage during the firing process. The biochar gives a darker colour to the aggregates but, after the firing process, this is lost with a final standard red colour. In Table 1 the results of the characterization tests are reported. The addition of 15% of char (ROC) inside the clay (RO), leads to an increase in weight loss during firing and a consequent weight-lightening of the material (13.64% of the ROC compared to 9.08% of the RO). The alkaline pH of the char gives at the final product a value beyond the optimal range of plant

Table 1. Characterization results on LWAs samples.

	RO	ROC	ROCC
W.L. (%) 105°C 24h	22.9	27.1	26.2
W.L. (%) 1000°C 1h	9.1	13.6	17.8
pH	6.7	9.3	7.9
Specific electrical conductivity (mS/cm)	0.32	0.90	0.41
Static water absorption (%) 24h	4.7	13.6	12.5

Table 2. Characterization results on brick samples.

	% W.L. (%) 950°C 30 min	Drying shrinkage (%)	Firing shrinkage (%)	Apparent density (g/cm^3)
RO	3.34	0.10	0.62	1.99
ROC5	4.82	0.03	0.07	1.85
ROC10	6.58	0.05	0.32	1.75
ROC20	10.25	0.00	0.57	1.47

comfort (6-8). The value of EC is less than 2 mS/cm (optimal range of plant growth). To optimize these parameters, aggregates were created with spent coffee grounds (SCG). The addition involves, as expected, a greater loss in firing and therefore a further weight-lightening (sample ROCC). This is because this type of waste is organic, has a LOI of about 98% and has a high theoretical high calorific value. The low pH value of the coffee that remains in the aggregates after firing, gives encouraging results that fall within the optimum values of pH and conductivity.

3.3 *Bricks*

The samples with only clay and with various percentages from 5 to 20 of char show the same red-clay-colour, typical of ceramic bricks. Table 2 shows weight loss after firing up to 10% with a corresponding lightening of the product. The shrinkage of the samples both during the drying phase and the firing phase, remains below 1%. This value can be considered negligible and in line with values relating to industrial processes. Considering the apparent density measured on the samples, it can be seen how the introduction of char makes possible to reach values around 1.5 g/cm^3, starting from about 2 g/cm^3 of the clay alone. Minor densities of 1.2 g/cm^3 – 1.3 g/cm^3, which can be considered very interesting for the market and a future industrial use, are achieved only by adding 40% of char in the samples. This, however, weakens the material, which crumbles quite easily to the touch and does not exhibit good mechanical strength.

3.4 *Polyurethane panels*

Thermal conductivity of polyurethane panels decreases by increasing the content of char. Sample containing char up to 2% show values in the range 0,037–0,038 W/(m*K), while reference sample shows value of 0,044 W/(m*K). The variation is significant and it is due to the presence of char which behaves as graphite acting as insulating material. In stress relaxing test reference sample shows a maximum load (5% of total deformation) of 53 kg with a load recovery of 22 kg. This specimen recovers much less than specimen which has the same formulation but with the addition of 1% biochar and which has a load of 132.6 kg and a load recovery of 44 kg. After compressive tests elastic modulus was measured and has been found increased in samples with additions up to 0.5-1% of biochar, but not in samples with addition of 2%. With additions of char higher than 1% the polyurethane has a rapid loss of structural properties. Probably the complete polymerization of

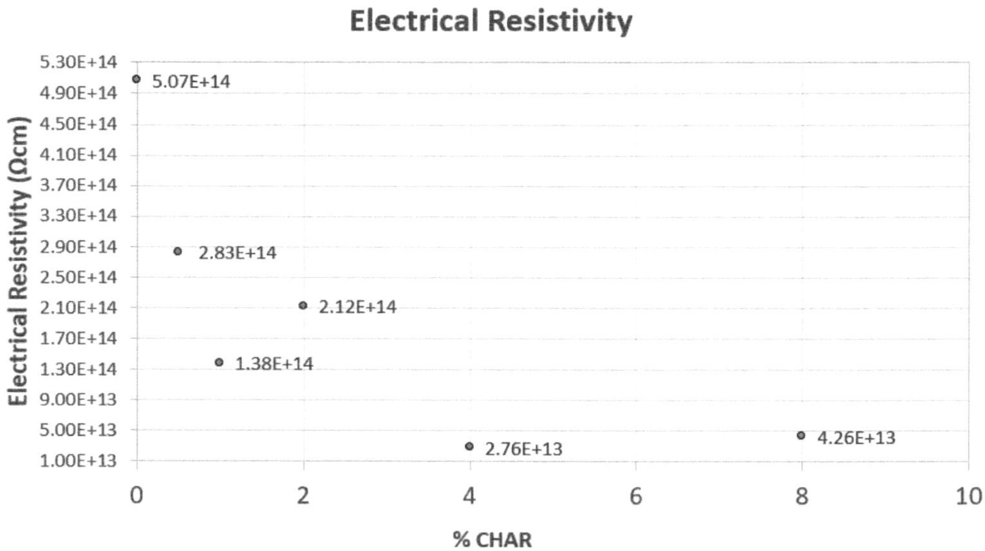

Figure 4. Electrical conductivity related to char percentage in polyurethane samples.

the material does not occur causing a decrease in mechanical properties. For electrical conductivity tests, samples containing char has higher values with respect to the reference. The resistivity values vary from $5*10^{14}\Omega*cm$ for reference to $2*10^{14}$ (0,5% char), $1{,}5*10^{14}$ (1% e 2% char) and values around $3*10^{13}$ are measured for percentages over 4%. These samples are not further investigated because shows problems during melting. The improvement of electrical behaviour is in line with the similar property between graphite and char.

4 CONCLUSIONS

Application on lightweight aggregates for green roofs by substituting 15%wt of the raw clay leads to an increase in weight loss during firing and a consequent weight-lightening of the material. The alkaline pH of the char gives the final product a value beyond the optimal range of plant comfort (6-8) and, to optimize this parameter, aggregates were prepared with addition of spent coffee grounds. This involves, as expected, a greater loss in firing and therefore a further weight-lightening, because this type of waste is organic and has a high calorific value. For instance, another interesting residue usefull in this context could be represented by cherries seeds, the waste resulting from cherries industrial processes, taking into account that sweet cherry cv "Moretta di Vignola", is a Protected Geographical Indication (PGI) product from Modena surrounding area, within the km 0 concept. The low pH value of the coffee that remains in the aggregates after firing, gives encouraging results that fall within the optimum values of pH and conductivity. Application on bricks leads to a weight-lightening effect as well, with a significant reduction in the bulk density of the materials. The insertion of char is feasible up to a percentage of 20/25%wt. The positive results about char ashes reusing can make more sustainable the gasification process by extending its use in different application fields. The polyurethane specimen with the addition up to 1% biochar mixed with polyol has good mechanical characteristics, in terms of increase of elastic modulus and visco-elastic recovery after the application of a constant deformation. It also presents an interesting value of thermal conductivity, lower than reference sample and a significantly lower electrical conductivity with respect to the same specimen without char. This could lead to benefits such as avoiding the accumulation of electrostatic charges. An added advantage in this application is the carbon sequestration effect (carbon negative technology). Storing char in a stable matrix such as polyurethane and, generally, in construction materials, can lead to a decrease in atmospheric carbon dioxide; a good practice to counter global warming.

AKNOWLEDGMENT

The authors thank "Regione Emilia Romagna" POR-FESR REBAF (2016-2018), FAR UNIMORE 2017 "La valorizzazione degli scarti agroindustriali tra diritto e scienza: processi innovativi dalla sperimentazione all'industrializzazione nel contesto legale"and FAR UNIMORE 2018 "Prunus avium L. cherries and other red fruits as new sources of neuroprotective compounds: a multidisciplinary study" for financial support.

REFERENCES

Ahmad S., Khushnood R.A., Jagdale P., Tulliani J.-M., Ferro G.A. 2015. High performance self-consolidating cementitious composites by using micro carbonized bamboo particles, *Materials & Design* 76: 223–229.

Choi W.C., Yun H.D., Lee J.Y. 2012. Mechanical properties of mortar containing biochar from pyrolysis, *Journal of the Korea institute for structural maintenance and inspection* 16: 67–74.

Gupta S., Kua H.W. 2017. Factors determining the potential of biochar as a carbon capturing and sequestering construction material: a critical review, *Journal of Materials in Civil Engineering* 29.

IEA, 2012, Technology Roadmap - Bioenergy for Heat and Power. Technical report, On line at: https://webstore.iea.org/technology-roadmap-bioenergy-for-heat-and-power

Khushnood R.A., Ahmad S., Restuccia L., Spoto C., Jagdale P., Tulliani J.-M., Ferro G.A. 2016 Carbonized nano/microparticles for enhanced mechanical properties and electromagnetic interference shielding of cementitious materials, *Frontiers of Structural and Civil Engineering* 10: 209–213.

Yuan J., Xu R. and Zhang H. 2011. The forms of alkalis in the biochar produced from crop residues at different temperatures, *Bioresource Technology*. Elsevier Ltd. 102(3): 3488–3497.

Zhao S., Huang B., Shu X., Ye P., 2014. Laboratory investigation of biochar-modified asphalt mixture, *Transportation Research Record: Journal of the Transportation Research Board* 2445:56–63.

Wastes: Solutions, Treatments and Opportunities III – Vilarinho et al. (Eds)
© 2020 Taylor & Francis Group, London, ISBN 978-0-367-25777-4

Struvite quality assessment during electrodialytic extraction

V. Oliveira
Polytechnic Institute of Coimbra & CERNAS, College of Agriculture, Coimbra, Portugal
CICECO, University of Aveiro, Aveiro, Portugal

G.M. Kirkelund
Department of Civil Engineering, Technical University of Denmark, Lyngby, Denmark

J. Labrincha
CICECO, University of Aveiro, Aveiro, Portugal

C. Dias-Ferreira
Polytechnic Institute of Coimbra & CERNAS, College of Agriculture, Coimbra, Portugal
Universidade Aberta, Lisboa, Portugal

ABSTRACT: Struvite recovered from anaerobically digested sludge can be conducted by a two-step process. The first step consists of electrodialytic extraction of phosphorus and heavy metals from the waste. The second step comprises the chemical precipitation of phosphorus as struvite. This work studied if the quality of phosphorus recovered as struvite is affected by the reduction of energy on the electrodialytic extraction step. Four electrodialytic experiments were carried out to assess the effect of the energy reduction through stirring time. This was followed by four struvite-precipitation experiments. The efficacy of the electrodialytic extraction of heavy metals is not affected by reducing the stirring from continuous mode to just 25% of the time. XRD analysis confirmed that the precipitates were constituted by pure struvite, while no significant accumulation of heavy metals was found in produced struvite. The struvite meets the requirements for use as phosphorus-based fertiliser.

1 INTRODUCTION

In the European Union, each person generates an average of 487 kg of municipal solid waste per year (Eurostat 2019). When the organic waste fraction is anaerobically digested the resulting sludge contains a high content of phosphorus (P) (Kalmykova & Fedje 2013). P is a vital element for all living beings; however P is also a finite resource whose reserves are declining, which makes crucial the search for alternative sources. Anaerobically digested sludge may be considered a potential raw material to be used for the production of fertilisers instead of the nonrenewable phosphate rocks. Nevertheless, this sludge also contains heavy metals (European Commission 2014) limiting its direct application in agriculture. Heavy metals may accumulate in soils and then be transferred to the food chain. Therefore, the real challenge on the valorisation of anaerobically digested sludge is the separation of P from the heavy metals to ensure the production of a safe and clean fertiliser.

Struvite is a slow-release fertiliser widely used in agriculture to increase the contents of P, nitrogen and magnesium in soil for plant uptake. The production of struvite as a second generation P-based fertiliser was previously conducted using the anaerobically digested sludge. The first attempt was carried out by Oliveira et al. (2016). In that work, 90% of P was firstly released from the waste using nitric acid as extractant ($1.2 < pH < 1.5$) followed by chemical precipitation of struvite. Although P has been successfully extracted from the waste, the formation of struvite did not occur due to the presence of large amounts of calcium. Afterwards, in the work of Oliveira et al. (2018), an electrodialytic (ED) process was conducted to separate P and heavy metals from the waste before the struvite production. The results showed that the P extracted was successfully transformed into struvite; however, the presence of zinc in the struvite did not allow its utilisation in agricultural applications. In the following work of Oliveira et al. (2019a), the contamination of struvite with zinc was avoided and the production of a high-quality struvite was attained; in that work, the ED process achieved an extraction of 90% of P. However, ED extraction of P is an energy-consuming process; the generation of an electric field and the stirring of waste are required to promote the extraction of P and separation of heavy metals from the waste. Oliveira et al. (2019b) investigated the possibility of reducing energy use during ED extraction of P from anaerobically digested sludge through the replacement of continuous stirring by pulse stirring and the utilisation of pulse electric current as an alternative to a constant current. The results revealed that about 70% of energy could be saved during the ED extraction process by operating the stirrer during 25% of the time without compromising the P extraction yields (close to 90%). However, the reduction of stirring time during the ED extraction might affect the separation of heavy metals from P and interfere with struvite formation. Thus, it is also crucial to evaluate the effects of the reduction of stirring time on the separation of heavy metals from the waste.

The main objective of this work is to assess the quality of the produced struvite after the reduction of energy usage during ED extraction of P and heavy metals from anaerobically digested sludge.

2 MATERIALS AND METHODS

2.1 *ED experiments*

Four ED experiments were carried out for 7 d, to evaluate the extraction of heavy metals from anaerobically digested sludge (Table 1), before the struvite production. The anaerobically digested sludge used in the experiments presented a pH between 7.67 – 7.71 and the following P and heavy metal concentrations (dry weight): 7.5 ± 0.5 g P kg^{-1}; 351.4 ± 74.4 mg Zn kg^{-1}; 158.6 ± 14.6 mg Pb kg^{-1}; 148.1 ± 21.9 mg Cu kg^{-1}; 34.0 ± 2.1 mg Cr kg^{-1}; 29.3 ± 3.6 mg Ni kg^{-1} and 1.2 ± 0.05 mg Cd kg^{-1}. A more detailed physico-chemical characterisation of the anaerobically digested sludge can be found in Oliveira et al. (2019a).

All experiments were conducted in a three chamber ED cell setup (Fig. 1), which is fully described in Oliveira et al. (2019). Briefly, a two-step ED extraction was used; in the first step, the anode (+) was placed directly in the waste chamber to enhance the acidification reactions occurring in the ED process. During this step, the negatively charged ions (e.g. PO_4^{3-}, Cl^-) are solubilised from the waste and kept in the waste chamber, while the positively charged ions (e.g. Cu^{2+}, Zn^{2+}) are solubilised and electromigrate into the cathode chamber. In the second step, the anode (+) was moved from the waste chamber to the anode chamber which further allows the separation of the negative charged species from the waste towards into the anode chamber.

In all experiments, 35 g of waste and 350 mL of distilled water (L:S = 10 mL g^{-1}) were placed into the waste chamber and stirred using a plastic flap fastened to a glass spatula connected to an overhead stirrer (VWR VOS). 500 mL of 0.01 M $NaNO_3$ was circulated in the anode and cathode chambers. A power supply (E3612A, from Hewlett Packard) was used to generate an electric current of 50 mA.

During the experiments, the electric current, voltage and pH in the waste suspension were registered every day. The catholyte's pH was maintained below 2 by addition of 1:1 HNO_3.

At the end of the experiments, heavy metals were measured in: i) liquid and solid phase of the waste, after filtration using quantitative filter paper and drying in an oven (105 °C), overnight; ii) soaking electrode solutions (in 5 M HNO_3, during 24 h); iii) ion-exchange membrane cleaning solutions (in 1 M HNO_3, 24 h); and iv) anolyte and catholyte solutions, after their volumes were recorded.

Table 1. Experimental conditions tested in ED experiments.

Experiment	Stirring	Electric current
A	Continuous	50 mA
B	Pulse: 45 min ON + 15 min OFF	50 mA
C	Pulse: 30 min ON + 30 min OFF	50 mA
D	Pulse: 15 min ON + 45 min OFF	50 mA

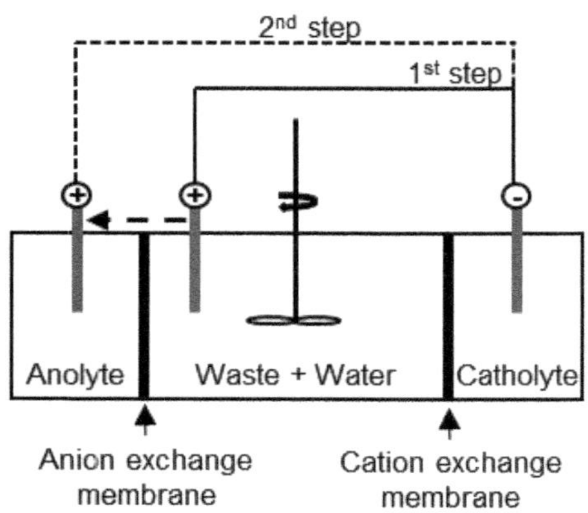

Figure 1. ED apparatus used in all experiments. The three chambers are made in Plexiglas® (D = 8 cm): i) anode and cathode chambers have L = 5 cm and, ii) waste chamber has L = 10 cm. Ion exchange membranes are from IONICS (204 SZRA B02249C; CR67 HUY N12116B) and the electrodes (+) and (-) are platinum coated titanium bars from Permascand® (D = 3 mm and L = 5 cm).

2.2 *Struvite production*

The anolytes obtained at the end of the four ED experiments were used as P concentrate solutions and were used to produce struvite through chemical precipitation. The detailed procedure is described in Oliveira et al. (2018). The structure of the harvested precipitates was analysed by X-ray diffraction (XRD, Rigaku Geigerflex (JP) with a Cu anode, operating at 20 kV and 40 mA. The precipitates were also analysed for the contents of P and heavy metals.

2.3 *Analytical procedures*

Heavy metals concentrations in the solid samples resulting from the ED experiments and the struvite precipitates were measured after a pre-treatment described in Danish Standard DS259 (Danish standard 2003). The digested samples were filtered by vacuum through a 0.45 μm filter and diluted to 50 mL. The concentrations of the liquid samples resulting from ED experiments (after filtration with a 0.45 μm syringe filter) and the digested samples were quantified by ICP-OES (Varian 720-ES). All concentration units are given in dry weight. The pH measurements during ED experiments were made using Radiometer electrodes.

3 RESULTS AND DISCUSSION

The presence of heavy metals in the P rich solutions used in the struvite-precipitation experiments can affect the recovery of P as struvite. Thus it is important to firstly assess what are the effects of the reduction of energy usage on the efficacy of ED process to separate heavy metals from the

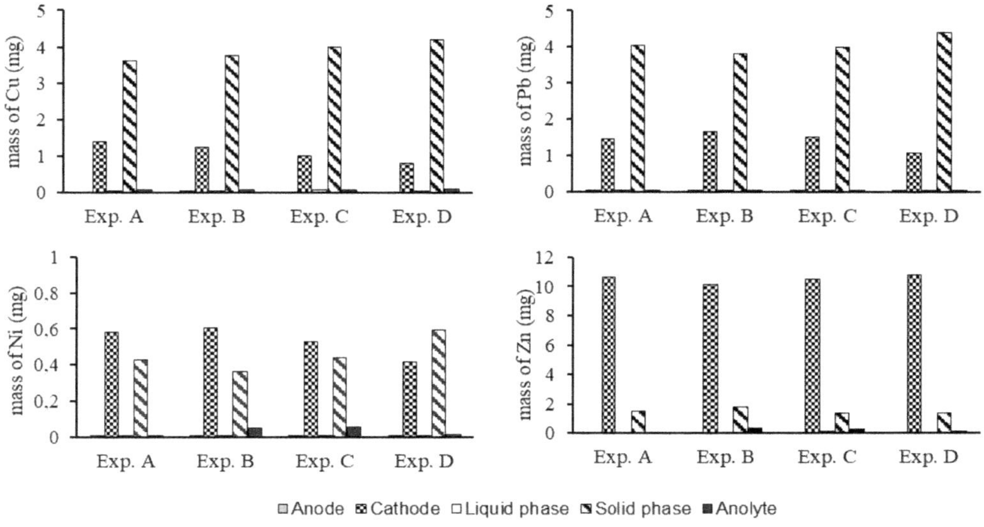

Figure 2. Distribution of Cu, Ni, Pb and Zn in different parts of the ED cell at the end of extraction experiments.

waste (section 3.1). Then, the produced struvite is analysed in terms of its structure and elemental composition in order to assess its applicability as fertiliser (section 3.2).

3.1 *Effects of the reduction of stirring time on the efficacy of the ED extraction of heavy metals*

The heavy metal concentrations (Cu, Cd, Cr, Ni, Pb and Zn) in the P concentrate solutions should be inexistent or low to avoid the production of a contaminated struvite which would make unsuitable its further utilisation for agriculture applications. The ED process will thus be efficient if the heavy metals are moved from the waste into the cathode chamber and/or if the heavy metals are not present in the P concentrate solution. Figure 2 shows the amount (mg) of heavy metals found in the different parts of the ED cell at the end of the extraction experiments. The heavy metals were mostly found in the solid phase of the waste and/or moved into the cathode chamber; only a very small amount of heavy metals was found in the anolyte. More than 93% of Cr remained in the solid phase (data not shown) as well as the Cu and Pb, but in a smaller extent, while Zn and Ni were mainly moved into the cathode chamber. Overall, similarly to P (Oliveira et al. 2019b), the reduction of stirring during the ED extraction of heavy metals from continuous to 25% of stirring does not negatively affect the separation of heavy metals (Fig. 2).

3.2 *Quality of recovered struvite*

3.2.1 *P and heavy metal removal from solution during struvite precipitation*

P and heavy metal removals (Table 2) at the end of struvite-precipitation experiments was defined as the difference between the initial and the final concentration in relation to the initial concentration. P and heavy metal removals were similar in all four struvite-precipitation experiments which indicates a good repeatability of the experimental conditions. Circa 99% of P was removed from the anolyte solutions as a solid white powder indicating that struvite was successfully produced. Only 1% P was left in solution, allowing the recovery of the maximum amount of struvite. Regarding the heavy metal removal, it was found that under the experimental conditions tested herein, Zn and Pb had the highest removal during precipitation (≈92%) followed by Cr (≈50%) and Ni (≈30%). Cu was the element less susceptible to be precipitated (≈3%). Despite the higher heavy metal removals, a

Table 2. P and heavy metals removals (%) at the end of struvite-precipitation experiments. In parenthesis are shown the initial concentration (mg L^{-1}) of the element in anolyte solution.

Element	Exp. A	Exp. B	Exp. C	Exp. D
P	99 (530)	99 (542)	99 (547)	99 (559)
Cr	43 (0.05)	58 (0.06)	52 (0.06)	43 (0.06)
Cu	10 (0.15)	2 (0.12)	0 (0.14)	1 (0.17)
Ni	54 (0.03)	29 (0.1)	15 (0.10)	23 (0.05)
Pb	76 (0.02)	100 (0.02)	100 (0.02)	92 (0.03)
Zn	86 (0.10)	98 (0.89)	95 (0.78)	90 (0.40)

Table 3. Contents of P and heavy metal in the recovered struvite and limits for fertiliser application in agriculture in two European countries. (*) values retrieved from Ministério da Economia (2015). (**) (Ministerio De La Presidencia 2005).

Precipitate	P (g/kg)	Zn (mg kg^{-1})	Cu (mg kg^{-1})	Pb (mg kg^{-1})	Ni (mg kg^{-1})	Cr (mg kg^{-1})	Cd (mg kg^{-1})
Struvite A	132.6	26.4	2.4	8.3	<0.02 mg L^{-1}	6.7	<0.02 mg L^{-1}
Struvite B	130.4	169.5	2.2	8.4	3.0	8.1	<0.02 mg L^{-1}
Struvite C	127.6	154.9	1.5	6.4	<0.02 mg L^{-1}	7.5	<0.02 mg L^{-1}
Struvite D	129.5	80.9	1.4	8.2	<0.02 mg L^{-1}	7.3	<0.02 mg L^{-1}
Portugal(*)	–	500	200	150	100	150	1.5
Spain (class A)(**)	–	200	70	45	25	70	0.7

pure struvite is expected, because the concentrations of heavy metals in the anolyte solutions were very low (<1 mg L^{-1}) compared to the P concentrations.

3.2.2 *Confirmation of struvite formation*

The XRD analysis of the precipitates obtained from the P concentrate solutions from the ED experiments are shown in Figure 3. The XRD pattern confirms the appearance of struvite crystals. In all precipitates (struvite A – D), the position of the diffraction peaks well matched the ones of the standard (Fig. 3 – vertical lines) and no traces of impurity phases were identified, which validates the purity degree of the produced struvite. No difference was distinguished between the XRD pattern of struvite produced from P concentrate solution obtained from ED process with continuous stirring and with only 25% of the stirring time.

3.2.3 *Heavy metal concentrations in the struvite*

Table 3 presents the contents of P and heavy metals in the struvite. Overall, a similar P content was measured in all struvite precipitates (around 13%). Cd and Ni, except for one sample, were not detected in the struvite. Similarly to P, the Cu, Cr and Pb concentrations in the struvite precipitates were very similar. In contrast, the Zn concentration was highest in the struvite produced from P concentrate solutions obtained after the pulse stirring ED process. The Zn concentration was about 3 to 6 times higher in these experiments compared to the struvite produced from the P concentrate solution obtained from the ED experiment with continuous stirring. Nevertheless, the Zn concentration still meets the requirements established by Portuguese and Spanish legislation for fertilisers application. These results highlighted that there is no major effect of the energy reduction on the ED process in the quality of the recovered struvite.

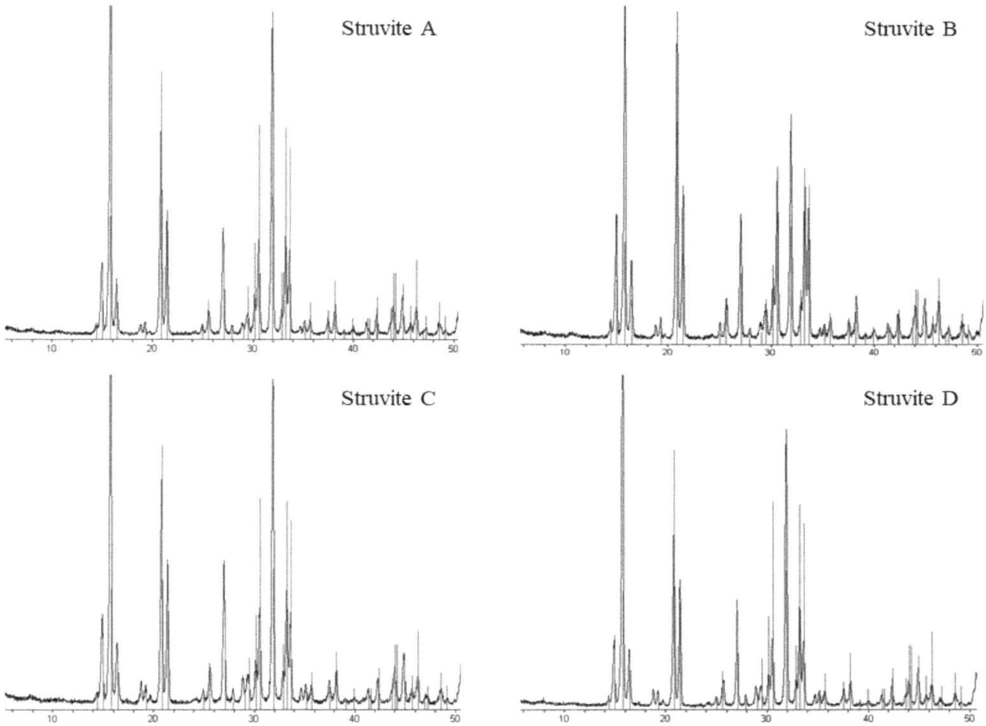

Figure 3. XRD pattern of struvite produced from ED P concentrate solutions (the vertical lines represent the struvite standard: 00-015-0762).

4 CONCLUSIONS

Struvite recovered from phosphorus concentrate solutions obtained after electrodialytic extraction of phosphorus and heavy metals from anaerobically digested sludge presents the required quality to be considered as phosphorus-based fertiliser for agriculture. Even if the tested strategy of energy reduction during the electrodialytic process caused an increase of the zinc in the phosphorus concentrate solutions used in struvite-precipitation experiments, the concentration of this species in the struvite is below the guideline limit. Overall, the recovery of phosphorus from anaerobically digested sludge as struvite was not affected by the energy reduction of the electrodialytic process.

AKCNOWLEDGMENTS

The authors would like to thank Sabrina Hvid for carrying out ICP analyses. This work has been funded by Portuguese National Funds through FCT – Portuguese Foundation for Science and Technology under CERNAS (UID/AMB/00681/2013) and by project 0340-SYMBIOSIS-3-E co-funded by FEDER "Fundo Europeu de Desenvolvimento Regional" through Interreg V-A España-Portugal (POCTEP) 2014-2020. C. Dias-Ferreira and V. Oliveira have been funded through FCT "Fundação para a Ciência e para a Tecnologia" by POCH – Programa Operacional Capital Humano within ESF – European Social Fund and by national funds from MCTES (SFRH/BPD/100717/2014; SFRH/BD/115312/2016).

REFERENCES

Danish standard 2003. *Determination of metals in water, sludge and sediments – General guidelines for determination by atomic absorption spectrophotometry in flame.*

European Commission 2014. *End-of-waste criteria for Biodegradable waste subjected to biological treatment (compost & digestate): Technical proposals.* Final Report, December 2013.

Eurostat, 2019. *Municipal waste statistics* [WWW Document]. URL https://ec.europa.eu/eurostat/statistics-explained/index.php/Municipal_waste_statistics#Municipal_waste_treatment (accessed 3.4.19).

Kalmykova, Y. & Fedje, K.K. 2013. Phosphorus recovery from municipal solid waste incineration fly ash. *Waste Manag.* 33: 1403–1410. https://doi.org/10.1016/j.wasman.2013.01.040

Ministério da Economia 2015. *Decreto-Lei nº 103/2015 de 15 de Junho.* Diário da República, 1ª série – Nº 114 (In Portuguese).

Ministerio De La Presidencia 2005. *REAL DECRETO 824/2005, de 8 de julio, sobre productos fertilizantes.* BOE 171 (In Spanish).

Oliveira, V., Dias-Ferreira, C., Labrincha, J. & Kirkelund, G.M. 2019a. Testing new strategies to improve the recovery of phosphorus from anaerobically digested organic fraction of municipal solid waste. *J. Chem. Technol. Biotechnol.* (Submitted).

Oliveira, V., Kirkelund, G.M., Horta, C., Labrincha, J. & Dias-Ferreira, C., 2019b. Improving the energy efficiency of an electrodialytic process to extract phosphorus from municipal solid waste digestate through different strategies. *Appl. Energy* (Submitted).

Oliveira, V., Labrincha, J. & Dias-Ferreira, C. 2018. Extraction of phosphorus and struvite production from the anaerobically digested of organic fraction of municipal solid waste. *J. Environ. Chem. Eng.* 6: 2837–2845. https://doi.org/10.1016/j.jece.2018.04.034

Oliveira, V., Ottosen, L.M., Labrincha, J. & Dias-Ferreira, C. 2016. Valorisation of Phosphorus Extracted from Farm Yard Slurry and Municipal Solid Wastes Digestates as a Fertilizer. *Waste and Biomass Valorization* 7: 861–869. https://doi.org/10.1007/s12649-015-9466-0

Wastes: Solutions, Treatments and Opportunities III – Vilarinho et al. (Eds)
© 2020 Taylor & Francis Group, London, ISBN 978-0-367-25777-4

Effects of struvite recovered from wastes on crop cultivation

V. Oliveira
Polytechnic Institute of Coimbra & CERNAS, College of Agriculture, Coimbra, Portugal
CICECO, University of Aveiro, Aveiro, Portugal

C. Horta
Polytechnic Institute of Castelo Branco & CERNAS, College of Agriculture, Castelo Branco, Portugal

J.L. Rocha
Polytechnic Institute of Coimbra & CERNAS, College of Agriculture, Coimbra, Portugal

C. Dias-Ferreira
Universidade Aberta, Lisboa, Portugal
Polytechnic Institute of Coimbra & CERNAS, College of Agriculture, Coimbra, Portugal

ABSTRACT: Struvite recovered from wastes has been considered a promising second generation fertiliser. The main objective of this work is to evaluate the effects of struvite fertilisation on i) rye (*Secale cereale* L.) cultivation (through a 45-d pot experiment), in terms of macronutrients, micronutrients and heavy metals and ii) on soil fertility (through a 30-d incubation experiment) in terms of available phosphorus and exchangeable bases (Ca^{2+}, Mg^{2+} and K^{+}). Struvite-fertilisation increased the soil available phosphorus resulting in improvement of the soil fertility class from low to medium class. In terms of exchangeable bases, the soil fertility class did not change. Furthermore, the struvite-fertilisation led to the highest concentration of total phosphorus in the plant. No relevant effects were found regarding the concentrations of Ca, Mg and K in the plant as well as the micronutrients. The levels of heavy metals on crop did not increase after soils fertilisation with struvite.

1 INTRODUCTION

Ensuring access to safe, nutritious and sufficient food for all people is one of the Sustainable Development Goals of the 2030 Agenda (FAO, 2006). There are several factors affecting it such as Global Water Crisis, Climate Change, Land Degradation and Greedy Land Deals. Focusing on Land Degradation, intensive farming has caused a serious loss of soil fertility and declining of agricultural yields (Disabled World, 2015). For instance, Vance et al. (2003) mentioned that more than 40% of the world arable soils have a low phosphorus (P) content. P is one of the essential nutrients for all living beings. Consequently, to maximise the agricultural productivity, the needs of fertilisation are increasingly. The increasing food demand for a rising human world population has led to an increase of the demand on P fertilisers (IPW, 2019). P fertilisers are manufactured from non-renewable rock phosphate deposits and its overconsumption has contributed to its rapid decline, which may put at risk the supply of P for the next generations. Alternative sources of P have been exploited. One of them is P from waste streams. Afterwards, the extracted P is used for the manufacturing of the fertilisers.

In a previous work, Oliveira et al. (2018) produced struvite from anaerobically digested organic fraction of municipal solid waste. Struvite ($MgNH_4PO_4 \cdot 6H_2O$) is a slow-release fertiliser and can be an effective source of several essential nutrients to crops: magnesium, nitrogen and P (Karunanithi et al., 2015; Rahman et al., 2014; Xu et al., 2012). The presence of contaminants in struvite during

Table 1. Classification of available P and exchangeable bases in soil. (*) Egnér-Riehm method.

Soil fertility class	Available P ($mg\ kg^{-1}$)$^{(*)}$	Ca^{2+} ($cmol_c\ kg^{-1}$)	Mg^{2+} ($cmol_c\ kg^{-1}$)	K^+ ($cmol_c\ kg^{-1}$)
Very low	≤11	≤2.0	≤0.5	≤0.1
Low	12–22	2.1–5.0	0.6–1.0	0.1–0.25
Medium	23–44	5.1–10.0	1.1–2.5	0.26–0.50
High	45–87	10.1–20.0	2.6–5.0	0.51–1.0
Very high	>87	>20.0	>5.0	>1.0

its production was previously assessed in Oliveira et al. (2018), whereas the agronomic efficiency of struvite was already evaluated in Oliveira et al. (2019). The results showed that the fertilisation with struvite almost doubled the plant biomass production compared with an unfertilised soil. In spite of struvite having been shown promising as a P fertiliser, the effects of other nutrients and of phytotoxic heavy metals on the crop production were not evaluated. The separation of P from heavy metals is a challenge when the waste streams are used for sourcing of P. Heavy metals may be toxic to the crops and consequently can affect the fertilising value of struvite. Similarly to P, the shortage of exchangeable bases in soil can also affect the plant growth and yield and could even cause plant death. The ability of the soil to provide nutrients to the plants from its organic and mineral reserves is closely related to the soil exchangeable complex (LQARS, 2006). The exchangeable bases held in this soil exchangeable complex are mineral nutrients for plants, such as calcium (Ca^{2+}), magnesium (Mg^{2+}) and potassium (K^+). The available P and the exchangeable bases in soil are indicators of the soil class fertility (Table 1).

The current work intends to assess the bioaccumulation of heavy metals in the plant when struvite derived from anaerobically digested organic fraction of municipal solid waste is used as fertiliser. It aims also to assess the effects of struvite fertilisation on the presence of micro- and macro- nutrients in the plant and on soil fertility. The effect of struvite was compared with that of a commercial fertiliser (single superphosphate, SSP). The substitution of fertilisers produced from P natural resources by a new generation of recycled P fertilisers will contribute to assure the sustainability of the agricultural sector as well as contribute to the circular economy goals.

2 MATERIALS AND METHODS

Struvite and SSP were the P-based fertiliser used in soil incubation experiments and in the plant pot experiment. Struvite was produced from anaerobically digested organic fraction of municipal solid waste according to the procedure described in Oliveira et al. (2018). SSP is a commercial fertiliser which was used as a standard mineral fertiliser.

The soil sample was taken from the layer 0–0.20 m and was air dried and sieved in a 2.0 mm-mesh sieve, before its use. The soil was classified as acid sandy loam soil ($pH_{H2O} = 5.1$) with low content of Ca^{2+} (1.2 $cmol_c\ kg^{-1}$), Mg^{2+} (0.3 $cmol_c\ kg^{-1}$) and K^+ (0.2 $cmol_c\ kg^{-1}$) and an available P content of 12.9 mg kg^{-1}.

Three treatments were set: i) with struvite, ii) with SSP and iii) without P fertiliser (control). Four replicates of each treatment were prepared using 1.5 kg of soil mixed with 19.6 mg of P fertiliser ($\approx$ 13.1 mg P per kg of soil), with exception of the control treatment in which no P fertiliser was added. Then, the 12 pots were watered to 70% of field capacity and placed in an incubator during 30 d at 25°C in the dark (maintaining field capacity). At the end of experiment, the available P (using the Egnér-Riehm method) and the exchangeable bases (Ca^{2+}, Mg^{2+} and K^+) (using ammonium acetate buffered at pH $\approx$ 7) were analysed in the air-dried soils (40°C; 48 h).

Afterwards, a new set of 12 pots was prepared using 150.0 g of soil taken from each incubated pot mixed with 50.0 g of purified sand. In each pot, three seeds of rye (*Secale cereale* L.) were sown (after sprouting only one plant was left in every pot) and then the pots were placed in a growth climate chamber at a temperature of 24°C daytime and 15°C night-time, 75% of relative humidity and 16 h of daylight (90 μmol photons m^{-2} s^{-1}). During the experiment, the pots were watered with the half-strength Hoagland solution (without P) to avoid nutrient deficiencies. After 6 weeks, the plants were harvested and shoots were separated from roots. Then, shoots and roots were air-dried at 65°C for 48 h and weighed, and the plant dry matter was determined. The dried plants were grounded in a ball mill, sieved at a 0.5 mm-mesh sieve and then were placed in a muffle furnace at a temperature of 480°C for 16 h to obtain the ashes. Total P, calcium (Ca), magnesium (mg), potassium (K), copper (Cu), zinc (Zn), cadmium (Cd), chromium (Cr), niquel (Ni) and lead (Pb) were measured after digestion of the ashes with hydrochloric acid solution (HCl 20%, v/v). Total P was quantified by spectrophotometry at a wavelength of 470 nm while the other elements were quantified by atomic absorption spectrometry.

One-way ANOVA analyses were performed to identify differences between treatments (control, struvite and SSP) on available P and exchangeable bases in soil at the end of incubation experiment, and total elements in the dried plants and nutrients extracted by plant, at the end of pot experiment. Tukey's test was used to compare means at 0.05 probability level. Different letters (a, b and c) were used to identify groups of results which are statistically different, at confidence level of 95%. All statistical analyses were conducted using IBM SPSS Statistics software (version 23).

3 RESULTS AND DISCUSSION

3.1 *Struvite effects on soil fertility: available P and exchangeable bases*

The contents of available P and some exchangeable bases (Ca^{2+}, Mg^{2+} and K^{+}) of the soil at the end of incubation experiment are shown in Table 2. Available P in soil depended on the fertiliser. Soils fertilised with struvite presented a significantly increase of the available P when compared with SSP and control (no P) ($p < 0.001$). The fertilisation with struvite increased about 52% of content of the available P in soil while with SSP only about 36%. This means struvite is more effective in supplying available P to the soil than SSP. In addition, the soils fertilised with struvite or SSP did not present any statistical difference between the content of available P measured at the beginning and at the end of incubation experiment (data not shown). This result demonstrates that after incorporation of the fertilisers in soil, an immediate source of P is provided for uptake by crop.

Regarding the amount of exchangeable bases, Ca^{2+} was the dominant cation in soil and it ranged from 1.39 to 1.50 $cmol_c$ kg^{-1}. Soils fertilised with SSP registered a significantly higher value than the soils fertilised with struvite ($p < 0.05$), but no difference was found when compared with unfertilised soils. This suggests that the amount of exchangeable Ca^{2+} in soil was not dependent on the fertilisation with either struvite or SSP. Exchangeable Mg^{2+} was the second cation most abundant, with a concentration between 0.39 to 0.57 $cmol_c$ kg^{-1}. The fertilisation with struvite resulted in increase of about 45% of Mg^{2+} compared to SSP and control. As like P, struvite fertiliser also provided an immediate source of Mg for plant uptake. The amount of exchangeable K^{+} was not statistically different among fertilised and unfertilised treatments ($p \geq 0.05$).

The soil fertility class previous to the addition of fertilisers is low, according to the classification by LQARS (2006). Overall, struvite-fertilisation improved the soil fertility in terms of available P; the struvite-fertilised soils can be included in the medium fertility class instead of low fertility class. Contrarily, in terms of exchangeable bases, the soil class fertility did not change.

3.2 *Struvite effects on crop*

In a previous work, it was confirmed that the application of struvite as a P fertiliser has a significantly effect on the shoot and root biomass production, increasing it (Oliveira et al. 2019). Consequently,

Table 2. Amount of available P and exchangeable bases in soil after 30 d-incubation experiment (different small letters denote significant differences among treatments, at confidence level of 95%).

Treatments	Available P (mg kg^{-1})	Ca^{2+} (cmol$_c$ kg^{-1})	Mg^{2+} (cmol$_c$ kg^{-1})	K^+ (cmol$_c$ kg^{-1})
Control	14.8 ± 0.6 c	1.41 ± 0.05 ab	0.39 ± 0.04 b	0.32 ± 0.03
SSP	23.3 ± 0.6 b	1.50 ± 0.05 a	0.40 ± 0.04 b	0.29 ± 0.01
Struvite	28.4 ± 0.9 a	1.39 ± 0.01 b	0.57 ± 0.04 a	0.30 ± 0.01
Significance level	$p < 0.001$	$p < 0.05$	$p < 0.001$	$p \geq 0.05$

it was referred that the nutrition of plant is strongly affected by the P fertiliser; but the nutrient contents in the plant were not measured. This analysis is carried out in sections 3.2.1, by analysing those nutrients required by the plant in large quantities ("macronutrients"), and 3.2.2, in where are presented those nutrients needed in only very small quantities (micronutrients"). In addition, as struvite was produced from a waste, it is also relevant to determine if its application has led to an increase of heavy metals content in the plant (section 3.2.3).

3.2.1 *Macronutrients: P, Ca, Mg and K*

Table 3 presents the concentration of P as well as Ca, Mg and K in the plant. Both shoots and roots obtained from soils fertilised with struvite registered a higher total P (by 21% at shoot level and 41% at root level) than the obtained from soils fertilised with SSP ($p < 0.01$). This is in accordance with the higher P extraction by the plant registered in soils fertilised with struvite than with SSP (Table 4), for a similar plant biomass production. This suggests that the dissolution of P provided by struvite was higher than by SSP. P was taken up by plant more effectively after struvite-fertilisation than with SSP. It should also be noted that in spite of the concentration of total P being similar in the control and SSP treatments, the biomass production was strongly affected when no fertiliser was used.

The concentrations of Ca, Mg and K in shoots were not significantly different among treatments ($p \geq 0.05$) (Table 3). It is interesting to observe that Mg concentration in shoots obtained from soils fertilised with struvite was not significantly higher than that obtained from soils fertilised with SSP or even from unfertilised soils. Our results are not in accordance with reported by Ryu & Lee (2016) which referred that Mg concentration in dried lettuce obtained from soils fertilised with struvite derived from swine wastewater was almost double than the obtained from soils fertilised with commercial fertilisers. Regarding the Ca, Mg and K extracted by plant (Table 4), it was found a significant difference between the fertilised and unfertilised treatments, because a higher shoot biomass production was obtained in fertilised soils than in unfertilised soils.

Although the Mg and K concentrations in the roots obtained from soils fertilised with struvite have been lowest (Table 3), the root biomass production was not affected, which demonstrates these nutrients are not the major contributors for the root development during the plant growth. Ca concentration was similar in roots obtained in all treatments ($p \geq 0.05$); however, the struvite has no Ca in its composition, which means that all the Ca contained in the roots was provided by the raw soil, not by the fertiliser. Nevertheless, the amount of Ca extracted by roots obtained in soils fertilised with struvite was significantly higher than in unfertilised soils (Table 4).

Overall, the results indicate that the application of struvite as fertiliser led to the highest concentration of total P in the plant and a higher uptake of Ca, Mg and K.

3.2.2 *Micronutrients: Cu, Zn and Ni*

Like macronutrients, micronutrients are also key elements without which the plants cannot grow. For micronutrients only minor amounts are uptaken by plants. If these micronutrients are unavailable, the plant functions are limited which may cause plant abnormalities and/or reduce their growth. However, it should also be noticed that when the micronutrients are in higher quantities than

Table 3. Plant biomass production and concentration of nutrients in dried plants obtained from soil fertilisation with SSP and struvite (different letters denote significant differences among treatments, at confidence level of 95%). (*) values from Oliveira et al. (2019).

Crop	Treatments	Plant biomass ($g\ kg^{-1}$ soil)(*)	Total P ($g\ kg^{-1}$)	Total Ca ($g\ kg^{-1}$)	Total Mg ($g\ kg^{-1}$)	Total K ($g\ kg^{-1}$)
Shoots	Control	0.83 b	2.6 b	7.5	2.5	60.0
	SSP	1.69 a	2.8 b	6.0	2.9	45.6
	Struvite	1.73 a	3.4 a	6.7	2.6	49.4
	Significance level	$p < 0.001$	$p < 0.01$	$p \geq 0.05$	$p \geq 0.05$	$p \geq 0.05$
Roots	Control	1.16 b	2.3 b	3.6	1.4 a	11.4
	SSP	1.57 a	2.1 b	3.3	1.1 b	10.9
	Struvite	1.88 a	3.0 a	3.8	1.1 b	8.7
	Significance level	$p < 0.001$	$p < 0.01$	$p \geq 0.05$	$p < 0.05$	$p \geq 0.05$

Table 4. Amount of macronutrients extracted at the end of pot experiment (different letters denote significant differences among treatments, at confidence level of 95%).

Crop	Treatments	P extracted ($mg\ pot^{-1}$)	Ca extracted ($mg\ pot^{-1}$)	Mg extracted ($mg\ pot^{-1}$)	K extracted ($mg\ pot^{-1}$)
Shoots	Control	0.32 c	0.93 b	0.30 b	7.33 b
	SSP	0.71 b	1.53 a	0.74 a	11.52 a
	Struvite	0.88 a	1.71 a	0.65 a	12.61 a
	Significance level	$p < 0.001$	$p < 0.01$	$p < 0.01$	$p < 0.05$
Roots	Control	0.41 b	0.63 b	0.25	1.99
	SSP	0.50 b	0.79 ab	0.26	2.55
	Struvite	0.85 a	1.07 a	0.31	2.41
	Significance level	$p < 0.01$	$p < 0.05$	$p \geq 0.05$	$p \geq 0.05$

those required by crops they may also cause toxicity to the plants, similarly to the heavy metals. Figure 1 shows the concentration of some micronutrients in the plant. Overall, the concentration of micronutrients was significantly higher in roots than in shoots ($p < 0.05$). This trend is opposite to the observed for macronutrients, which demonstrates that in rye micronutrients accumulate mainly in roots. The concentrations of Cu and Ni in the roots obtained from soils fertilised with struvite and SSP are significantly lower than in the unfertilised soils ($p < 0.001$); however, the production of root biomass was not affected in the fertilised soils. At shoot level, the struvite fertilisation did not increase significantly the concentrations of Cu and Ni in the plant ($p < 0.01$). Regarding the concentration of Zn, no significant differences were found among treatments, which indicates that struvite does not limit Zn concentration in the plant.

3.2.3 *Heavy metals: Cd, Cr and Pb*

The levels of some heavy metals on dry plants should be analysed when struvite recovered from wastes is applied as fertiliser, because its presence is prejudicial to plant growth due to their toxicity. The concentrations of Cr (Fig. 2), Cd and Pb were measured in the shoots and roots obtained at the end of pot experiment. Cd and Pb were not detected in plants. Overall, there were significant differences of the Cr concentration in the shoots ($p < 0.01$) and roots ($p < 0.001$); it was significantly higher in roots than in shoots ($p < 0.05$). In addition, it was also found that Cr concentration in plants grown in fertilised soils was higher than in unfertilised soils.

At shoots level, Cr accumulation in the struvite-fertilised soils was significantly smaller than in the SSP-fertilised soils and also than unfertilised soils ($p < 0.05$). At roots level, Cr concentration in

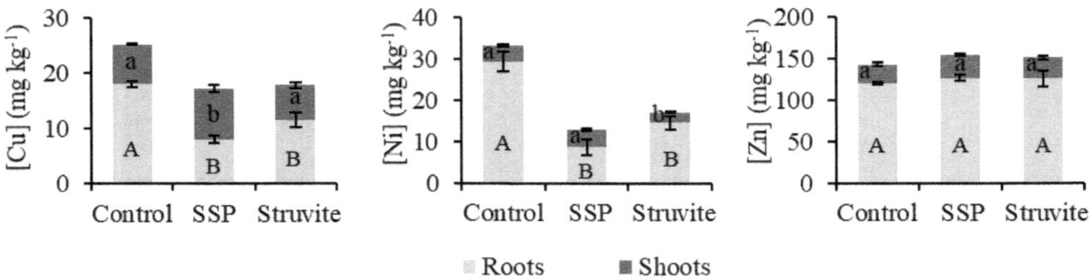

Figure 1. Micronutrients (mg kg^{-1}) in dried plant at the end of pot experiment. Different capital letters denote significant differences among treatments in roots and different small letters denote significant differences among treatments in shoots, at confidence level of 95%. Error bars represent standard error.

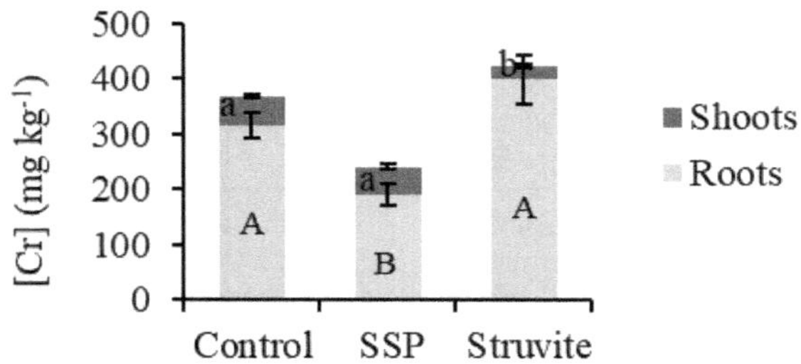

Figure 2. Cr concentration (mg kg^{-1}) in dried plant at the end of pot experiment. Different capital letters denote significant differences among treatments in roots and different small letters denote significant differences among treatments in shoots, at confidence level of 95%. Error bars represent standard error.

struvite-fertilised soils was significantly higher than in SSP-fertilised soils being that concentration was similar to those of unfertilised soils. Rye et al. (2012) reported similar levels of heavy metals in Chinese cabbage which was fertilised with struvite recovered from wastewater.

Based on the above results, the use of struvite recovered from anaerobically digested of organic fraction of municipal solid waste does not increase the levels of heavy metals in the plant, which indicates that struvite could be used as a P nutrient fertiliser.

4 CONCLUSIONS

The effects of struvite derived from the anaerobically digested organic fraction of municipal solid waste were evaluated on cultivation of rye (*Secale cereale* L.). Under the experimental conditions tested herein, the main findings of this work are: i) struvite has the ability of increasing the available P content of the soil immediately after its incorporation; ii) total P is higher in rye grown in struvite treatment than in SSP; however, shoot concentration of Ca, Mg and K were not significantly affected by struvite application; and iii) the levels of micronutrients and of heavy metals in rye were not negatively affected by the fertilisation with struvite.

ACKNOWLEDGMENTS

The authors would like to thank Marta Solipa Batista, for laboratorial support and analyses. This work has been funded by Portuguese National Funds through FCT – Portuguese Foundation for Science and Technology under CERNAS (UID/AMB/00681/2013) and by project 0340-SYMBIOSIS-3-E co-funded by FEDER "Fundo Europeu de Desenvolvimento Regional" through Interreg V-A España-Portugal (POCTEP) 2014-2020. C. Dias-Ferreira and V. Oliveira have been funded through FCT "Fundação para a Ciência e para a Tecnologia" by POCH – "Programa Operacional Capital Humano" within ESF – European Social Fund and by national funds from MCTES (SFRH/BPD/100717/2014; SFRH/BD/115312/2016).

REFERENCES

Disabled World 2015. *Food security: definitions & general information* [WWW Document]. URL https://www.disabled-world.com/fitness/nutrition/foodsecurity/#docs (accessed 2.19.19).

FAO 2006. *Food security – Policy Brief* [WWW Document]. URL http://www.fao.org/fileadmin/templates/faoitaly/documents/pdf/pdf_Food_Security_Cocept_Note.pdf (accessed 2.18.19).

IPW 2019. *Pant nutrition* [WWW Document]. URL http://www.plantnutrition.ethz.ch/ipw9/program.html (accessed 2.18.19).

Karunanithi, R., Szogi, A.A., Bolan, N., Naidu, R., Loganathan, P., Hunt, P.G., Vanotti, M.B., Saint, C.P. & Ok, Y.S. 2015. Phosphorus recovery and reuse from waste streams. Adv. Agron. 131, 173–250. https://doi.org/10.1016/bs.agron.2014.12.005

LQARS 2006. *Fertilisation guidelines for agricultural crops*. Instituto Nacional de Investigação Agrária, Lisboa.

Oliveira, V., Horta, C. & Dias-Ferreira, C. 2019. Evaluation of a phosphorus fertiliser produced from anaerobically digested organic fraction of municipal solid waste. *J. Clean. Prod.* IN PRESS.

Oliveira, V., Labrincha, J. & Dias-Ferreira, C. 2018. Extraction of phosphorus and struvite production from the anaerobically digested of organic fraction of municipal solid waste. *J. Environ. Chem. Eng.* 6, 2837–2845. https://doi.org/10.1016/j.jece.2018.04.034

Rahman, M.M., Salleh, M.A.M., Rashid, U., Ahsan, A., Hossain, M.M. & Ra, C.S. 2014. Production of slow release crystal fertilizer from wastewaters through struvite crystallization – *A review. Arab. J. Chem.* 7, 139–155. https://doi.org/10.1016/j.arabjc.2013.10.007

Ryu, H.D. & Lee, S.I. 2016. Struvite recovery from swine wastewater and its assessment as a fertilizer. *Environ. Eng. Res.* 21, 29–35. https://doi.org/10.4491/eer.2015.066

Ryu, H.D., Lim, C.S., Kang, M.K. & Lee, S.I. 2012. Evaluation of struvite obtained from semiconductor wastewater as a fertilizer in cultivating Chinese cabbage. *J. Hazard. Mater.* 221–222, 248–255. https://doi.org/10.1016/j.jhazmat.2012.04.038

Vance, C.P., Uhde-stone, C. & Allan, D.L. 2003. Phosphorus acquisition and use: critical adaptations by plants for securing a nonrenewable resource. *New Phytol.* 157, 423–447. https://doi.org/10.1046/j.1469-8137.2003.00695.x

Xu, H., He, P., Gu, W., Wang, G. & Shao, L. 2012. Recovery of phosphorus as struvite from sewage sludge ash. *J. Environ. Sci. (China)* 24, 1533–1538. https://doi.org/10.1016/S1001-0742(11)60969-8

Wastes: Solutions, Treatments and Opportunities III – Vilarinho et al. (Eds)
 ISBN 978-0-367-25777-4

Development of recycled geotextiles towards circular economy

J.R. Carneiro, F. Almeida & M.L. Lopes
Construct-GEO, Faculty of Engineering, University of Porto, Porto, Portugal

ABSTRACT: Legislators, researchers and the distinct industrial sectors are carrying out efforts to adopt proper procedures towards circular economy. Solutions involving the reuse, reduction and recycling of generated waste have to be put into practice, otherwise the amounts of natural resources currently available are going to attain critical levels. It is essential to find solutions in which end-of-use materials and products may play the role of raw materials. Geotextiles are construction materials (mostly made from synthetic polymers) to which the concept of circular economy can be improved. If properly treated, textile and plastic waste may be possible sources of raw materials for manufacturing recycled geotextiles. This work discusses the feasibility of developing recycled geotextiles from textile and plastic waste, pointing out production procedures, concerns, drawbacks and possible applications, which have to be in accordance with the final properties of the recycled products.

1 INTRODUCTION

The 2030 Agenda for Sustainable Development adopted by more than 150 world leaders at the 2015 United Nations Summit defined the path that must be followed towards a more prosperous and balanced World. To achieve this, the 2030 Agenda includes seventeen Sustainable Development Goals within the framework of social, economic and environmental aspects. The Goal 12 is related to sustainable consumption and production patterns and includes targets in the domains of the exploitation of natural resources and waste management (UN 2015).

The linear model of economic development "take-make-dispose" is no longer appropriate and should be replaced by a circular paradigm in which materials and products have their lifetime extended for as long as possible. To achieve this purpose, actions including reuse, reduction and recycling of waste should be carried out in order to implement a more efficient energy consumption and a more sustainable way to cope with the natural resources.

In European Union, Directive 2008/98/EC on waste sets targets in terms of preparation for reuse and recycling of municipal solid waste and construction and demolition waste. In relation to the latter, if selective demolition of old structures is carried out, a path is created to boost the reuse and recycling of construction and demolition waste, in which are included construction materials such as geotextiles, which are commonly used in engineering projects.

The Directive 2008/98/EC also gives attention to the waste generated in other industrial sectors. Concerning textile waste, it is defined that by 1 January 2025, Member States shall promote separate collection for textiles and that by 31 December 2024 targets referring to the preparation for reuse and recycling of textile waste should be established. Considering that in the textile sector it seems to be a lack of plans clearly defining how the waste generated can be used or recycled, it is time to cogitate about what strategies can be implemented in order to achieve the aforementioned goals.

The waste generated throughout the production and use chain of textile materials and products can be labelled as pre-consumer or post-consumer textile waste (Roznev et al. 2011). The pre-consumer textile waste is generated in the industry during the development and manufacture of the final products. On the other hand, the discarded textile products that were previously placed on the market for consumption are considered post-consumer textile waste. If properly treated, polyester,

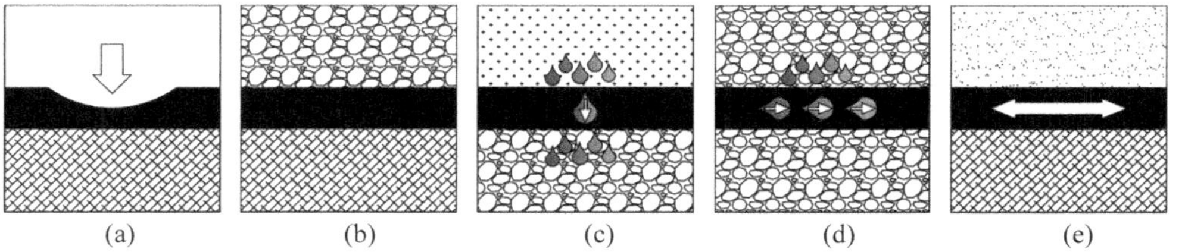

Figure 1. Functions of geotextiles in civil and environmental engineering applications: (a) protection; (b) separation; (c) filtration; (d) drainage; (e) reinforcement.

Figure 2. Geotextiles: (a) woven; (b) nonwoven needle-punched; (c) nonwoven thermally bonded.

nylon, cotton or wool, which are examples of raw materials employed in the textile industry, can be reused in the development of recycled products and, hence, promoting circularity.

Plastic waste is also a matter of great concern. In European Union, according to data from 2016, approximately 15.8 million tonnes of plastic packaging waste are generated every year, being about 40% of that amount sent for recycling (Eurostat 2018). Therefore, actions should be also carried out to promote the development of solutions involving the reuse, reduction and recycling of plastic packaging waste, as well as for other types of plastic waste.

What if some of the textile and plastic waste could be forwarded to the development of recycled geotextiles? Is it possible to convey confidence in the use of recycled geotextiles? And is it feasible to create connections between the textile, plastic and construction industries? These are key questions whose answers may help to open the door to increase circularity in and between these industrial sectors.

2 GEOTEXTILES

Geotextiles are polymeric materials (most often synthetic) commonly used in civil and environmental engineering applications. These materials can perform many different functions, such as separation, protection, drainage, filtration or reinforcement (Fig. 1). Due to their high versatility, high efficiency, relative low cost and ease of installation, geotextiles can be applied, for example, in roads, railways, waste landfills, embankments, stabilisation of slopes, erosion control or coastal protection structures. Polypropylene, polyethylene, polyesters and polyamides are the most common polymers used for manufacturing geotextiles. Chemical additives, such as antioxidants and/or ultraviolet (UV) stabilisers are often added to the base polymers of the geotextiles. The polymeric mixtures are converted into filaments, fibres or strips, which are used as components to produce woven or nonwoven geotextiles (Fig. 2).

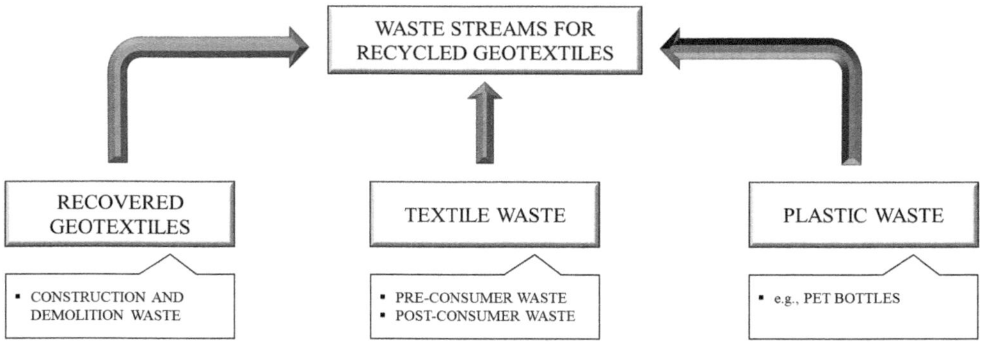

Figure 3. Waste streams for obtaining recycled raw materials for manufacturing geotextiles.

3 RECYCLED GEOTEXTILES

3.1 *Production*

Geotextiles are mostly manufactured from virgin polymers, which leads to a consumption of natural resources. Developing strategies in which waste materials can be used to manufacture geotextiles is a challenging task. It is important to mention that aesthetic requirements are generally not important regarding geotextiles, since these construction materials are usually buried or covered by liquids. This non-aesthetic need is an advantage, increasing the possible waste sources for manufacturing recycled geotextiles and facilitating sorting and pre-treatment steps. For example, and taking into consideration that colour is not an issue, unsorted post-consumer textile waste can be a valuable raw material for manufacturing nonwoven geotextiles.

Recycled geotextiles can be produced with 100% recycled materials (synthetic or natural) or by mixing recycled with virgin materials. The addition of virgin raw materials is intended to improve the properties of the final products. For example, the fibres obtained from recycling textile waste (many times physically damaged by the pre-treatment steps) often have poor mechanical properties when compared to virgin fibres.

The definition of proper waste streams for manufacturing recycled geotextiles is an important and demanding task. Nevertheless, three possible sources of waste raw materials have been identified: recovered geotextiles (construction and demolition waste), textile waste (pre-consumer and post-consumer waste) and plastic waste (for example, PET bottles) (Fig. 3).

There are possibilities for the use of geotextiles recovered from the selective demolition of old structures. However, for the time being, the recycling of geotextiles is not developed. There are two reasons that may explain this circumstance: geotextiles are relatively recent construction materials and the structures in which they are inserted are often designed to last for long periods of time. Therefore, not many geotextiles have yet been removed during the deconstruction of old structures.

For being possible the employment of pre-consumer and post-consumer textile waste as raw materials in the manufacture of nonwoven geotextiles, pre-treatment actions are needed. These actions include, for example, sorting, cutting or shredding of textile products. This way, recycled fibres can be obtained and, depending on their properties and on the desired properties for the final product, virgin fibres may be also incorporated in the nonwoven structure. The fibres are then arranged in a web and bonded in order to manufacture the nonwoven products. The bonding process may be carried out by mechanical, chemical or thermal procedures, depending on the characteristics of the recycled fibres (and on the expected characteristics of the final product). The bonding process is a key step in manufacturing recycled geotextiles since it will have a high influence on the physical, mechanical and hydraulic properties of the final products (Fig. 4).

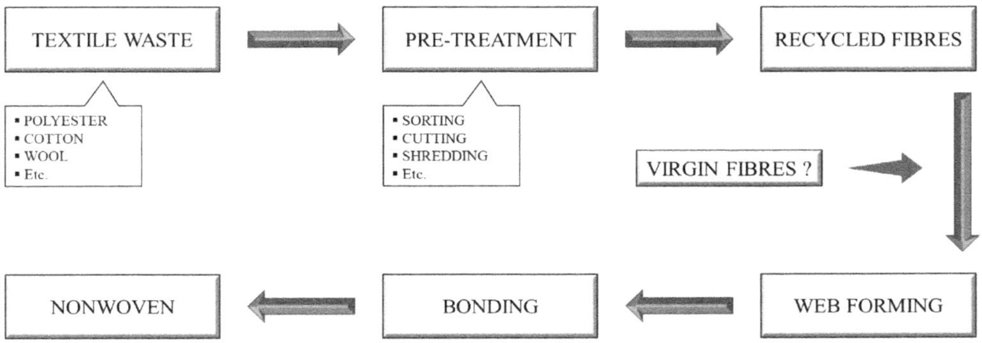

Figure 4. Process to manufacture recycled nonwoven geotextiles from textile waste.

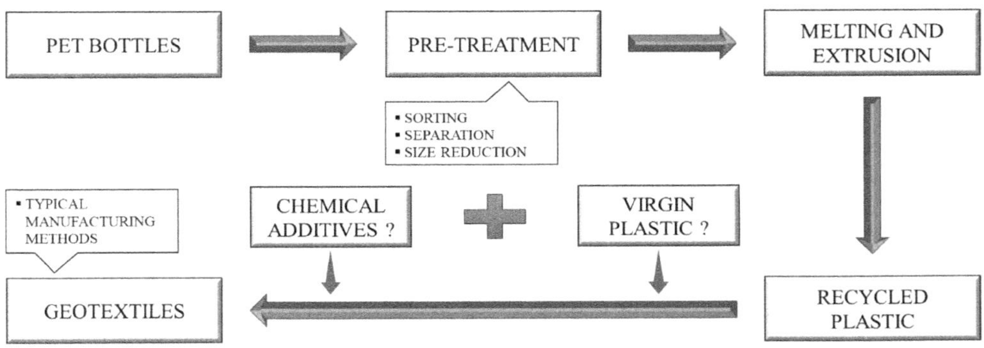

Figure 5. Process to manufacture recycled geotextiles from PET bottles.

The other source of raw materials for manufacturing recycled geotextiles is plastic waste. Geotextiles are typically manufactured from the same polymers (such as polyolefins or polyesters) used in plastic packaging and other plastic products. Polyethylene terephthalate (PET) bottles are a type of plastic waste generated in high amounts and a possible source to manufacture fibres or strips for the production of, respectively, nonwoven or woven geotextiles. The use of PET bottles for manufacturing geotextiles also requires some pre-treatment actions, such as sorting, separation and reduction in size. The bottle paper labels must be removed in order to avoid contamination of the PET raw material. The treated PET bottles can be then melted and extruded, thereby obtaining the recycled plastic. Once again, the addition of virgin plastic to improve the properties of the recycled geotextiles has to be considered. Moreover, it has also to be considered the addition of chemical additives (such as UV stabilisers or antioxidants) to protect the recycled geotextiles from degradation, improving their long-term performance (Carneiro & Lopes 2017). Finally, the recycled geotextiles can be developed through typical manufacturing methods (Fig. 5).

3.2 *Concerns and drawbacks*

Launching recycled materials or products into the market involves, inevitably, a discussion about reliability. The consumers need to have guarantees that these materials or products will properly accomplish the tasks for which they were designed. The comparison with the long-established materials or products will always be present.

Naturally, there are concerns and drawbacks related to the use of recycled geotextiles that must be discussed. The first issue that has to be addressed is the drawback of not being possible a

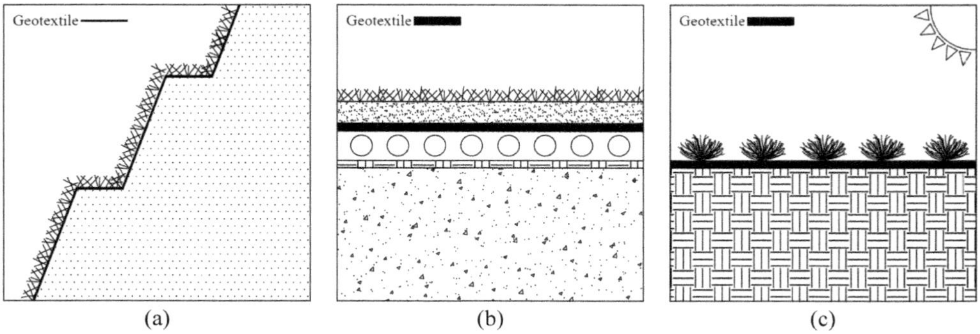

Figure 6. Examples of possible applications of recycled geotextiles: (a) erosion control; (b) rooftop gardens; (c) agrotextiles.

full traceability of the raw materials over the entire lifespan, which prevents from knowing the degradation state of the polymers and the possible existence of potential contaminants in those raw materials. The degradation state of the recycled raw materials may be higher or lower depending on the environmental conditions to which the original products have been exposed.

When textile waste is employed as raw material, there is no guarantee that the properties of the recycled geotextiles will be uniform along their width and length. Since textile waste can be from many different sources (high heterogeneous waste stream), it is difficult to ensure a final product with regular properties. Regarding mechanical properties, the recycled geotextiles made from textile waste are expected to have lower strengths in comparison with conventional geotextiles. This is a consequence of using recycled fibres that are very likely to be physically damaged (the pre-treatment actions, such as cutting or shredding, promote the occurrence of such damage).

If plastic waste (like PET bottles) is employed, the properties of the recycled geotextiles are expected to be more homogenous and more reproducible compared to the use of textile waste. Indeed, the manufacture process of recycled geotextiles using plastic waste is more uniform and less random (due to the production of new components by melting and extrusion) and the waste stream is by itself much more homogenous (compared to textile waste). Nevertheless, the recycled geotextiles manufactured from plastic waste are also expected to have poor mechanical performance in comparison with geotextiles obtained from virgin polymers. This may be explained by the previous degradation suffered by the polymers and by the presence of contaminants.

Durability is a cross-cutting issue in all construction materials, arousing even greater prominence when their compositions incorporate recycled raw materials. Long-term performance is of utmost importance in geotextiles, since they may be used in structures with an expected lifetime of 50 years of more. Polymer waste may have already experienced degradation, depending on the conditions that those materials had to face during their service life, which may adversely compromise the durability of the recycled geotextiles. Taking into consideration that the use of polymer waste may have serious repercussions on the long-term performance of the recycled geotextiles, it is not advisable the use of such construction materials in rigorous environments or in long-term applications (unless it is proven that they will last and correctly perform their functions over time).

3.3 *Applications*

Considering the concerns and drawbacks associated with recycled geotextiles, it seems to be reasonable to encourage the development of solutions for these materials involving temporary and/or less-demanding applications. Therefore, recycled geotextiles may be used, for example, in erosion control, rooftop gardens or in agricultural applications (Fig. 6).

The use of recycled geotextiles in erosion control may be a feasible alternative. In this application, geotextiles can be used to protect the soil, providing a barrier against rainfall and/or wind. The growth of vegetation will allow to cover the geotextiles, minimizing the environmental impacts. If made from natural polymers (such as cotton or wool), the recycled geotextiles will be biodegradable, being an advantage in some cases. It is worthy to mention that recycled textile waste has recently been used in a study related with erosion protection. Indeed, Broda et al. (2017) designed a solution in which meandrically arranged ropes made from textile waste were used in the development of geotextiles for the protection of the banks of a deep drainage ditch and for the reinforcement of a roadside ditch.

In rooftop gardens, a recent solution aiming the increase of "green lungs" in the cities, recycled geotextiles can be employed to accomplish the functions of filtration, retaining the soil particles and allowing the free flow of water. Finally, recycled geotextiles may be suitable to carry out the functions usually assigned to agrotextiles, which can be applied in agriculture, horticulture or gardening. These products are intended to help protecting the soil and, among other competences, they can offer shade and allow preservation of soil humidity (Ajmeri & Ajmeri 2016).

4 CONCLUSIONS

The manufacture of recycled geotextiles may be a good opportunity to boost the use of textile and plastic waste as raw materials. This will help improving the recycling rates of these waste streams (contributing for their rational management) and simultaneously will allow a reduction on the exploitation of natural resources. The use of recycled geotextiles can be considered as a valuable contribution for a more sustainable construction.

Considering that the properties of the recycled geotextiles are expected to be not as good as those from conventional ones, it is reasonable to suggest the use of these recycled construction materials in low-demanding applications, such as erosion control, rooftop gardens or as agrotextiles. Nevertheless, there seems to be conditions to deeply explore the development of innovative geotextiles in accordance with the premises of circular economy.

ACKNOWLEDGEMENTS

This work was financially supported by: (1) project POCI-01-0145-FEDER-028862, funded by FEDER funds through COMPETE 2020 – "Programa Operacional Competitividade e Internacionalização" (POCI) and by national funds (PIDDAC) through FCT/MCTES; (2) UID/ECI/04708/2019 – CONSTRUCT – "Instituto de I&D em Estruturas e Construções" funded by national funds through FCT/MCTES (PIDDAC).

REFERENCES

Ajmeri, J.R. & Ajmeri, C.J. 2016. Developments in nonwovens as agrotextiles. In KELLIE, G. (ed.), *Advances in Technical Nonwovens*: 365-384. Woodhead Publishing.

Broda, J., Gawlowski, A., Laszczak, R., Mitka, A., Przybylo, S., Grzybowska-Pietras, J. & Rom, M. 2017. Application of innovative meandrically arranged geotextiles for the protection of drainage ditches in the clay ground. *Geotextiles and Geomembranes* 45(1): 45–53.

Carneiro, J.R. & Lopes, M.L. 2017. Natural weathering of polypropylene geotextiles treated with different chemical stabilisers. *Geosynthetics International* 24(6): 544–553.

Directive 2008/98/EC of the European Parliament and of the Council of 19 November 2008 on waste. *Official Journal of the European Union L312/3*.
Eurostat. 2018. *How much plastic packaging waste do you produce?* [Online]. Available: https://ec.europa.eu/eurostat/en/web/products-eurostat-news/-/EDN-20180422-1 [Accessed 1 March 2019].
Roznev, A., Puzakova, E., Akpedeye, F., Sillstén, I., Dele, O. & Ilori, O. 2011. *Recycling in textiles*. HAMK University of Applied Sciences.
Transforming our world: the 2030 Agenda for Sustainable Development – Resolution adopted by the General Assembly on 25 September 2015. *United Nations A/RES/70/1*.

Wastes: Solutions, Treatments and Opportunities III – Vilarinho et al. (Eds)
© 2020 Taylor & Francis Group, London, ISBN 978-0-367-25777-4

Environmental product declaration of recycled aggregates: A LCA approach

A. Pires, A. Gomes, M. Ramos, P. Santos & G. Martinho
MARE, School of Sciences and Technology, NOVA University of Lisbon, Caparica, Portugal

ABSTRACT: This study used the life cycle assessment (LCA) methodology embedded in environmental product declaration (EPD) to quantify the environmental impact related to the production of recycled aggregates from construction and demolition waste. The software Umberto LCA+ was used, and the impact assessment was made using several impact categories established in EPD norm for construction materials from DAP Habitat system (EPD for Portuguese construction materials). The results showed that 1 t of recycled aggregates have impacts on the environment, resulting mostly from transportation of waste aggregates into the recycling plant. More research is needed to improve the environmental performance of the recycled aggregates, like comparing them with virgin aggregates. The landfilling of waste aggregates must be compared with the recycled aggregates to demonstrate the environmental benefits of such circular economy process. Other destinations of the recycled aggregates instead of the use at road construction should also be addressed in future research.

1 INTRODUCTION

1.1 *Construction and demolition waste in Europe*

The construction sector in Europe is responsible for the consumption of half of the total resources extracted, being also responsible for the generation of one-third of all waste (European Commission, 2011, 2018). The recycling of construction and demolition waste (CDW) could replace the consumption of natural resources as aggregates. The aggregates sector is one of the largest amongst the on-energy extractive industries, where the demand for aggregates is 5 tonnes per capita per year, leading to the consumption of 2.7 billion tonnes per year, representing an annual turnover of an estimated €15 billion (UEPG, 2017). The consumption of recycled aggregates is 8%, where it is needed more efforts to bring such materials into the market (UEPG, 2017). In Portugal, the material productivity of the sector is one of the lowest at European level. Despite the efforts to improve the market value of CDW, it is also true that such effort is restricted to leading companies of the sector, which is a niche at the entire Portuguese construction sector which is mainly composed by small and medium-sized enterprises, representing 87.7% of total workforce of the sector (ECSO, 2018).

One of the instruments used in the construction sector to impulse the consumption of recycled materials is sustainable certification. Certification of buildings raises from the need to define methodologies capable of measuring how sustainable the building is. Buildings sustainability level, so-called green rating systems, have been developed in the last years namely LEED (USA), Green Star (Australia), BREEAM, Arge TQ, Minergie, Green Globes (Mattoni et al., 2018).

1.2 *The aim of the study*

This study intends to develop an LCA to recycled aggregates generated by a private company in Portugal. By using the Product Category Rules norm from DAP Habitat (the Environmental Product Declaration (EPD) system for construction materials in Portugal) (CentroHabitat, 2018) which is based on EN 15804:2012+A1:2013 (CEN, 2013) is expected to verify the advantages and disadvantages of such streamlined LCA method.

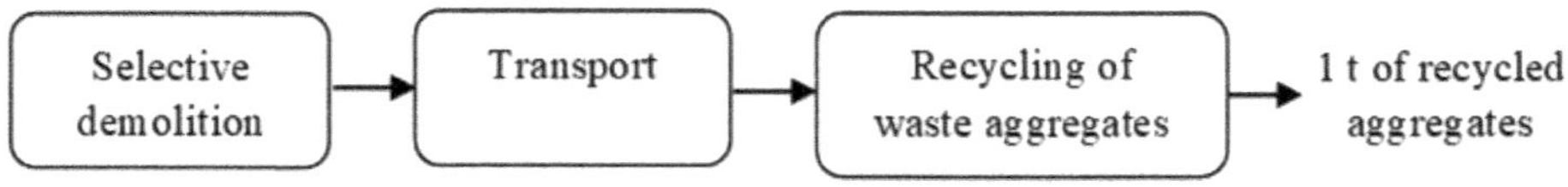

Figure 1. Frontiers of the system.

2 MATERIAL AND METHODS

2.1 *Materials*

The recycled aggregates analyzed in this paper are resulting from a demolition work occurring Póvoa de Lanhoso, Portugal. The recycled aggregates produced respects the European Union Regulation n° 305/2011, with CE mark, by the NP EN 13108-1:2011 standard.

2.2 *Goal and scope of the LCA*

The goal of this LCA study is to determine the environmental impact resulting from the production of recycled aggregates, according to the EN 15804:2012+A1:2013 (CEN, 2013), the norms elaborated to help on the development of EPD. The functional unit used as a reference is 1 t recycled aggregate. The environmental burdens generated during the recycling were determined in order to identify the critical recycling phases.

The system boundaries of the recycled aggregates include the extraction and processing of raw materials from the building (the CDW), transport into the recycling plant, and recycling. The phases related to the transport to the final consumer of the recycled aggregates, construction use, and demolition after use have not been taken into account. Thus, the LCA is a cradle to gate level type. The CDW is to be burden-free type (also named zero-burden). The impacts of the raw material from CDW has been attributed to the construction itself, while the impact caused during the demolition, transport, and recycling processes has been allocated to the production of recycled aggregates. Recycling of materials during the recycled aggregates production occurs, the avoided burdens due to the production of a secondary material as a partial replacement of the primary can be considered. However, that was not the case. Benefits are resulting from the use of recycled material like avoiding the landfilling of the waste. Also, the recycled aggregates use to save the use of virgin aggregates, and those benefits were excluded by the EPD norms.

The recycled aggregates production consists of deconstruction or selective demolition, transportation, and recycling of CDW, is described in Fig. 1. In the case of selective demolition, it is necessary to carry out a careful dismantling of the structure, in order to promote the possible reuse of materials or their recovery. The waste resulting from the demolition process is separated according to its typology for later referral to waste management operators.

In general, the demolition process is agreed between the contractor and the supervisor/contractor and in the first phase consists of the removal of non-structural elements (such as window frames, doors, windows) and a later stage in the demolition of the concrete structure armed.

The residues resulting from the demolition process are stored in the existing separation media, identified with the description and the European Waste Code. When the containers reach their storage capacity, they are referred to as the waste management operator. Electronic waste monitoring guides accompany this transport. The work under study provides for the deconstruction of the following structures:

- demolition of existing buildings identified in design;
- demolition of floors and interior walls;
- demolition of sidewalks in concrete, demolition of the bituminous floor for passage of new infrastructures and construction of new elements;
- demolition of the metal walkway;
- disassembling of aluminum frames, separating aluminum from glass for correct separation of waste;

Table 1. Mass and energy balance of selective demolition phase.

Inputs	Amount	Outputs	Amount
CDW	1,538,700 kg	CDW for recycling	907,020 kg
Diesel	2520 L	CDW for disposal	631,680 kg
Electric energy	2.61E10 J		
Lubricant oil	<1%		

Table 2. Mass and energy balance of recycling phase.

Inputs	Amount	Outputs	Amount
CDW source separated	907,020 kg	Recycled aggregates	770,970 kg
Diesel	454.5 L	Fines	136,050 kg
Lubricant oil	<1%		

- disassembling of interior wood vents, separating the wood from glass and fittings for correct separation of waste;
- removal of electrical infrastructures, water supply, drainage of rainwater and wastewater;
- removal of fiber cement roofs.

The consumption of energy and raw materials were considered during selective demolition. The ones considered were electric energy and diesel. Raw materials and energy consumptions are presented in Table 1.

The transportation A2) phase at EPD norm of the recyclable fraction of CDW was made with a vehicle of EURO4, lorry superior to 32 t. The distance between the demolition place and the recycling unit was 23 km.

The recycling process A3) phase at EPD norm is where the recycled aggregates are produced. Table 2 present the consumption of raw materials and energy during A3). The recycled aggregates production line begins with the primary crushing where the aggregates of 0/40 mm dimension are produced. Aggregates are sent to a secondary crushing where the particles are reduced again and then routed to a sieve where different sizes separate them. These sizes can always be adjusted to meet the desired end product needs by just changing the dimensions of the screen sieves. The consumption occurring in this phase is only diesel. In summary, the scheme is as follows:

- receipt of waste (inert and bituminous);
- storage by types;
- crushing and screening;
- characterization of the final product.

2.3 *Life cycle inventory data quality and data collection*

The collection of the data required for the recycled aggregates life cycle inventory was dealt by the team of owners of the demolition site. The data collected included the amount of CDW recovered to be recycled by type, the CDW which is to be landfilled without recovery, the equipment used in the selective demolition process and in the recycling process, the description of selective demolition and recycling processes, means of transport and distances, and resulting products and waste. The Ecoinvent v.2.2 database (Ecoinvent, 2010) was used as background data for generic materials, including temporal, geographical, and technological representativeness.

2.4 *Methodology and selection of impact categories*

The impact assessment was conducted using impact categories recommended in EN 15804:2012+A1:2013 (CEN, 2013) concerning the sustainability of construction works for

Table 3. Characterization results of recycled aggregates production (per 1 t).

		Phases			
Impact category	Units	A1	A2	A3	Total
GW	kg CO_2eq.	1.62E-2	2.86E0	2.12E0	5.00E0
ODP	kg CFC-11 eq.	1.73E-9	4.70E-7	2.63E-7	7.35E-7
A	kg SO_2 eq.	1.55E-4	1.00E-2	2.00E-2	3.02E-2
E	kg PO_4^{3-} eq.	1.77E-4	2.00E-2	3.00E-2	5.02E-2
POF	kg C_2H_4 eq.	2.21E-6	3.02E-4	3.74E-4	6.78E-4
Ade	kg Sb eq.	1.15E-4	2.00E-2	1.00E-2	3.01E-2
Adf	J	2.40E5	4.61E7	3.11E7	7.74E7

construction products and services. These impact categories are abiotic depletion of elements (ADe), abiotic depletion of fossil fuels (ADf), global warming (GW), ozone layer depletion (ODP), photochemical oxidation (POF), acidification (A), and eutrophication (E). For these categories, the characterization factors of the CML-IA method (Guinée et al., 2002) were used as established in EN 15804:2012+A1:2013 (CEN 2013). The other impact categories considered were related to resource use: renewable primary energy resources used as energy carrier (RPEE); renewable primary energy resources used as raw materials (RPEM); total use of renewable primary energy resources (TPE); non-renewable primary energy resources used as energy carrier (NRPE); non-renewable primary energy resources used as materials (NRPM); total use of non-renewable primary energy resources (TRPE); use of secondary materials (SM); use of renewable secondary fuels (RSF); use of non-renewable secondary fuels (NRSF); use of net freshwater (W). Those impact categories were obtained from different impact assessment models. RPEE, RPEM, TPE, NRPE, NRPM, and TRPE were obtained from cumulative energy demand calculated by Umberto LCA+. SM, RSF, and NRSF were obtained from EDIP 2013 LCIA method (Hauschild & Potting, 2005). The water use was calculated directly by Umberto LCA+. Hazardous waste disposed (HW), non-hazardous waste disposed (NHW), and radioactive waste disposed (RW) were also determined using EDIP 2013. Concerning components for reuse (CR), materials for recycling (MR), materials for energy recovery (MER), exported electric energy (EEE), and exported thermal energy (ETE), all were calculated from the life cycle inventory.

The data collected during the inventory phase were loaded into the Umberto LCA+ version 10.0 software (ifu Hamburg, 2018) and processed using CML-IA method (Guinée et al., 2002) and EDIP 2003. The results of characterization were presented for the impact categories proposed by the EPD norm.

3 RESULTS AND DISCUSSIONS

The results showed that the production of recycled aggregates generates positive values in all impact categories. The phase with higher impact is the transportation of CDW streams to recycling for GW, ODP, Ade and Adf. The results are also similar to the ones reached by Cuenca-Moyano et al. (2019) for recycles aggregates case study in Spain, where transportation is also the main contributor to the environmental impacts.

Concerning the results from energy resources impact categories, the transport is, again, the phase responsible for the higher positive values. In Table 4, the use of non-renewable energy to transport CDW streams into the recycling unit is significant.

The last group of impacts calculated was related to the amount of waste generated. In Table 5 are calculated the waste generated hazardous, non-hazardous and radioactive send for landfill. The phase with a higher amount of waste generated has been transportation. The waste generated is related to the production of the diesel to be used during the transport of CDW into recycling unit.

Table 4. Characterization results of recycled aggregates production: energy resources and water consumption (per 1 t).

		Phases			
Impact category	Units	A1	A2	A3	Total
RPEE	J	1.00E4	6.11E5	1.20E5	7.41E5
RPEM	J	0	0	0	0
TPE	J	1.00E4	6.11E5	1.20E5	7.41E5
NRPE	J	2.80E5	4.88E7	3.17E7	8.08E7
NRPM	J	0	0	0	0
TRPE	J	2.80E5	4.88E7	3.17E7	8.08E7
SM	kg	0	0	0	0
RSF	J	0	0	0	0
NRSF	J	–	–	–	–
W	m^3	3.91E-1	1.00E-2	3.96E-3	4.05E-1

Table 5. Characterization results of recycled aggregates production: waste send for landfilling (per 1 t).

		Phases			
Impact category	Units	A1	A2	A3	Total
HW	kg	2.00E-7	5.11E-5	1.53E-5	6.66E-5
NHW	kg	2.00E-3	3.88E-1	2.00E-2	4.10E-1
RW	kg	9.00E-8	3.63E-5	8.81E-6	4.52E-5

Components send for reuse, for recycling, materials for energy recovery, and energy exported were all null. Although the recycled aggregates result from a recycling process, no waste neither materials are resulting from this life cycle send for those destinations.

4 CONCLUSION

This study has used LCA procedure developed in EPD norm to assess the environmental performance of recycled aggregates generated from specific selective demolition work. Findings of the LCA have been presented openly and comprehensively, making possible to others to compare the results obtained with other EPD existing on the market, having in mind the assumptions made in this study. Detailed data from inventories was possible to include in the LCA.

The advantage of the EPD norm – a streamlined LCA – is the easiness of getting data for the inventory, few data needed from the owner of the recycled aggregates process, the ability to compare with other products following the same EPD norm, and the interpretation of the results. For a market product is important that environmental impacts can be provided in a fast and understandable way by non-LCA experts, and the LCA defined by the EPD norm is doing it. The drawbacks of conducting this LCA is the inadequacy to improve the environmental performance of the product. Forward LCA simulations must be conducted to test different scenarios to improve the product before submitting results to the EPD process.

The limitations of this study will be addressed in future research (i) uncertainty and sensitivity analysis, (ii) life cycle phases alterations, (iii) scenarios of landfilling of waste aggregates, (iv) extension of life cycle phases, (v) LCA of virgin aggregates to compare with recycled aggregates. Future works will include uncertainty and sensitivity analysis, where different scenarios and change of parameters will be made to assess the reliability of the LCA. Uncertainty and sensitivity analysis are not mandatory in terms of the EPD certification system in Portugal (the DAP Habitat). However, the results obtained from the uncertainty and sensitivity analysis may be used to improve the LCA. More, changes will be proposed to the recycler to improve their environmental performance, before the LCA results could be submitted to the DAP Habitat. Scenarios concerning the landfilling of waste aggregates should also be made to verify the environmental benefit of recycling such waste instead of landfilling.

Possible future developments into the existing recycled aggregate product should consider the extension of the life cycle system to include the use and end-of-life phases. Recycled aggregates are mostly used in road construction, where lixiviation of substances may occur during the use and where the end-of-life destination like recycling may not be possible.

Another crucial future work is to develop the LCA of the virgin aggregate used in the market and compare the two DAP Habitat profiles. With the comparison of both aggregates, recycled and virgin, more information could be obtained from the process, understand where the recycled aggregates need to be improved to face the concurrence.

ACKNOWLEDGMENTS

This work had the financial support from Fundo Ambiental through the call: *Apoiar a Economia Circular do Setor da Construção (CIRCULAr – Construção), aviso n.º 5573/2018, de 24 de abril,* with the support of Domingos da Silva Teixeira, S.A. and also supported by Fundação para a Ciência e Tecnologia (FCT), through the strategic project UID/MAR/04292/2013 granted to MARE (Marine and Environmental Sciences Centre).

REFERENCES

CEN. 2013. *EN 15804:2012 + A1:2013 – Sustainability of construction works - Environmental product declarations - Core rules for the product category of construction products.*

CentroHabitat. 2018. *Sistema de declarações ambientais de produto. dapHabitat* [Online]. Available: https://daphabitat.pt/ [Accessed 10-01-2018 2018].

Cuenca-Moyano, G.M., Martín-Morales, M., Bonoli, A. & Valverde-Palacios, I. 2019. Environmental assessment of masonry mortars made with natural and recycled aggregates. *The International Journal of Life Cycle Assessment 24*(2): 191–210.

ECSO. 2018. *Country profile Portugal. European Commission* [Online]. Available: https://ec.europa.eu/growth/sectors/construction/observatory_en [Accessed 21-02-2019 2019].

European Commission. 2011. *Roadmap to a resource efficient Europe (COM 571).* European Commission [Online]. Available: https://eur-lex.europa.eu/legal-content/EN/TXT/?uri=CELEX:52011DC0571 [Accessed 21-02-2019 2019].

European Commission. 2018. *Construction and demolition waste (CDW).* European Commission [Online]. Available: http://ec.europa.eu/environment/waste/construction_demolition.htm [Accessed 21-02-2019 2019].

UEPG. 2017. *A sustainable industry for a sustainable Europe – Annual Review 2016–2017* [Online]. Available: http://www.uepg.eu/uploads/Modules/Publications/uepg-ar2016-17_32pages_v10_18122017_pbp_small.PDF [Accessed 21-02-2019 2019].

Ecoinvent. 2010. *Ecoinvent v.2.2 database.* Ecoinvent. https://www.ecoinvent.org/database/database.html [Accessed 10-01-2019 2019].

Guinée, J.B., Gorrée, M., Heijungs, R. et al. 2002. *Handbook on life cycle assessment, operational guide to the ISO standards, I: LCA in perspective, IIa: guide, IIb: operational annex, III: scientific background.* Dordrecht: Kluwer Academic Publishers.

Hauschild, M.Z. & Potting, J. 2005. *Spatial differentiation in LCA impact assessment – The EDIP 2003 methodology.* Copenhagen: Danish Environmental Protection Agency.

ifu Hamburg. 2018. *Umberto LCA+, version 10.0.* ifu Hamburg. Available: https://www.ifu.com/en/umberto/lca-software/ [Accessed 10-12-2018 2018].

Mattoni, B., Guattari, C., Evangelisti, L., Bisegna, F., Gori, P. & Asdrubali, F. 2018. Critical review and methodological approach to evaluate the differences among international green building rating tools. *Renewable and Sustainable Energy Reviews 82*: 950–960.

Wastes: Solutions, Treatments and Opportunities III – Vilarinho et al. (Eds)
© 2020 Taylor & Francis Group, London, ISBN 978-0-367-25777-4

Anaerobic digestion in the fruit waste disposal and valorization

M.M. Baumguertner & C. Kreutz
Federal University of Technology, Paraná, Brazil

R.J.E. Martins
Polytechnic Institute of Bragança, Bragança, Portugal
LSRE – LCM, Porto, Portugal

ABSTRACT: Due to their characteristics, a significant portion of fruit produced on a large scale, especially kiwi, is wasted as organic waste. This problem has created an opportunity to evaluated the potential for biogas generation using kiwi waste as a substrate by the anaerobic process. Eight distinct assays were performed in two batch reactors (R1 and R18). Different substrate/inoculum ratios, addition of $NaHCO_3$ and with and without nutrients addition were tested. The most satisfactory results were obtained with: kiwi waste (8.4 g), anaerobic digester slurry (192 mL), sodium bicarbonate (500.4 mg), ammonium chloride (453.0 mg) and potassium phosphate monobasic (106.0 mg) added. In this assay, 759.1 mL biogas/g VS was produced and methane quality of 60%, indicating that kiwi fruit waste has potential for biogas production. It is also concluded that the quality of the inoculum and substrate/inoculum ratio influences the biogas production.

1 INTRODUCTION

Kiwi (*Actinidia deliciosa*) is a fruit of Chinese origin that was introduced in western agriculture in the 19th century. In 1973, it began its cultivation in Portugal and today the country is the 11th largest producer in the world (FAO, 2013, Franco, 2008, Silveira et al. 2012). Fruits are perishable materials and for this reason are considered as a source of organic waste generation. It is estimated that approximately 25% of the kiwi fruits produced chain are lost, according to Coelho (2015). The most common destination for these wastes is landfilling, which is an inexpensive option but can lead to a number of environmental problems (Sanjaya et al. 2016). One of the ways of exploiting this organic waste is its use as a substrate in the anaerobic digestion (AD) process.

Organic materials, such as fruit waste, are mainly constituted of carbohydrates and proteins, which can be degraded to simpler compounds by microorganisms in an oxygen-free environment, by means of a complex biologic treatment process called Anaerobic Digestion (AD), used to stabilize organic matter while producing biogas, a mixture formed mainly of methane and carbon dioxide (Hagos et al. 2017, Paritosh et al. 2017). So, the AD process is an efficient method of biochemical degradation of biomass and provides the treatment and energetic recovery of many organic waste (Abbasi et al. 2012, Yan et al. 2017).

According Zhang et al. (2018) each kilogram of food waste (included fruit waste) can generate up to 0.1 m^3 of methane gas. So, methane has a high calorific value of 17 to 25 MJ/m^3, which can be burnt to release heat energy or converted to electricity using internal combustion engines. About 15% of this gas is captured for beneficial use or flaring and the remainder converts into fugitive greenhouse gas emission from landfilling of food waste, that could amount to 3.1 gigatons CO_2-eq/year based on the global figure of 1.6 gigatons of food waste each year (Nghiem et al. 2017).

Biofuels, the category that encompasses biogas, accounted for a 1.9% share of the total electricity generated in the world in 2015. In the same year, the total capacity of electricity generation from

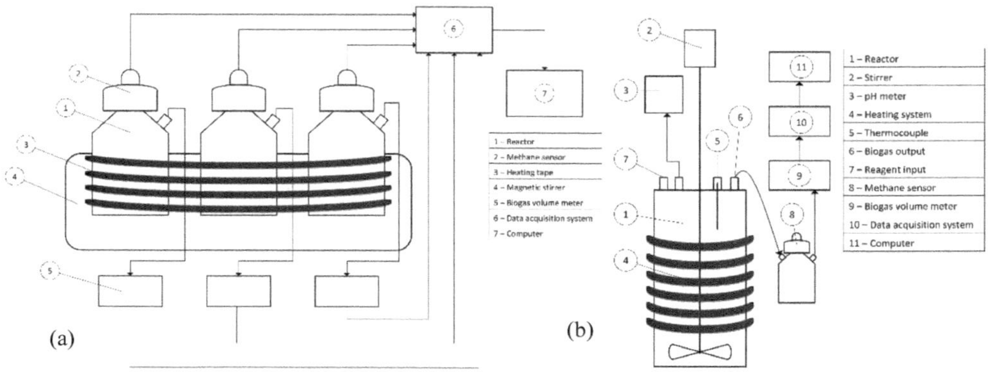

Figure 1. Schematic drawing of the anaerobic batch system: (a) R1 and (b) R18.

renewable sources in the world was of 5534 TWh (IEA, 2017a, b). According to the IEA (2015) estimates for the period 2014–2020, the renewable electricity generation capacity is expected to grow 40% reaching about 7150 TWh compared to 5420 TWh produced in 2014.

Whereas world energy production rose from 6131 TWh in 1973 to 24,255 TWh in 2015 and the share of renewable energy sources, excluding water, increased from 0.6% to 7.1% in the same period, the use of the kiwi becomes a viable alternative and a socioeconomic and environmental interest to society. The aim of this work was to investigate the recovery of kiwi waste through its conversion to biogas with the anaerobic digestion process, and to solved a waste disposal problem.

2 MATERIAL AND METHODS

2.1 *Anaerobic digestion system*

To evaluate the biogas generation potential, the AD system was carried out on two different reactors. The R1 (Figure 1a), an anaerobic glass vial with a useful volume of 200 mL and R18 (Figure 1b), an acrylic cylinder with 17 L useful volume.

The reactors temperature control was carried out by means of a heating tape coupled to a control unit (Electrothermal MC242) and kept under mesophilic conditions (37°C). Intermittent mechanical stirring was performed with a magnetic stirrer (IKA RO 5P) in R1, connected in a timer (Aslo ASTH) to effect 15 minutes of agitation and 15 minutes with-out agitation. In R18 the stirring was carried out with an internal stirrer (CAT R100C), with 12 hours of agitation in a day.

The data acquisition of methane percentage, of both reactors, was carried out in a continuous mode, through the BlueSens BCS-CH4 biogas model methane sensors, in a 20 seconds' interval time. This data was then sent to the data acquisition system (BlueSensBACCom12) and transmitted to BlueSens® BACVis Version 7.6.0.2 software. The biogas production, in terms of volume, was measured by a flowmeter Ritter®MGC-1 V3.0 model, that was coupled in R1. At R18, the volumetric method was used, in which the biogas was washed in NaOH solution, and the methane volume was measured through a cylinder glass graduated.

2.2 *Inoculum and substrate*

The inoculum used in the tests consisted of anaerobic sludge from the anaerobic digesters of the Wastewater Treatment Plant (WWTP) of Bragança, Portugal. After collection, the inoculum was stored and refrigerated at 4°C until use. The substrate used in the tests was kiwi waste collected from a food distribution unit in Bragança city. The kiwi was cut and then crushed until a pasty and

Table 1. Physico-chemical characteristics of inoculum and substrate.

Parameter	Inoculum	Substrate
TS (g/L)	14.0	176.3
VS (g/L)	10.7	165.3
FS (g/L)	3.3	10.9
COD (g O_2/L)	–	226.3
Mass density (g/cm^3)	–	1.15

Table 2. Operational conditions and experimental time of trials.

Assay	Reactor	Substrate (g)	Inoculum (mL)	$NaHCO_3$ (mg)	NH_3Cl (mg)	KH_2PO_4 (mg)
1.1	R1	2.2	198	301.9	–	–
1.2		2.0	198	301.0	–	–
1.3		2.1	198	302.3	–	–
2.1		4.2	196	501.5	226.0	52.9
2.2		8.4	192	500.4	453.0	106.0
2.3		20.3	180	503.0	1130.2	260.0
3		8.2	192	503.9	453.3	106.1
4	R18	775.5	17000	45200.0	40700.0	9400.0

homogeneous mass is obtained. After being ground the substrate was packed in a glass vial and stored at 4°C until use. The physico-chemical characteristics of inoculum and substrate are shown in Table 1, performed according to the procedures described in APHA (1998).

2.3 *Operating procedures and process monitoring*

The physico-chemical parameters used for inoculum and substrate characterization were pH, Total Solid (g/L), Fixed Solid (g/L), Alkalinity (mg $CaCO_3$/L) and Chemical Oxygen Demand (g O_2/L), performed according to the procedures described in APHA (1998).

Eight distinct assay, batch type, were performed, with triplicate tests in each one. The substrate mass and inoculum volume were varied. In the last series, the addition of salts of nutrients (nitrogen and phosphorus) was tested. The operating conditions and the assay duration are shown in Table 2.

For the calculation of the Kiwi theoretical methane production ($PCH_{4\ theoretical}$), the equation presented by Raposo et al. (2011) was used, which allows to calculate the methane volume (mL CH_4) produced by each gram of volatile solid (g VS) of substrate. The data needed to perform the calculations were obtained from the USDA (2018) database.

3 RESULTS AND DISCUSSION

According the inoculum characterization, the VS/TS ratio obtained was 0.76, as reporting by Andreoli et al. (2015) the ideal range would be 0.75–0.80. This value is indicative of undigested sludge, that is, that there is unstable organic matter and consequently the active biomass presence. The substrate (kiwi) characterization shown a TS content of 176.3 g/L and organic matter concentration of 226.3 g/L (expressed as COD), Table 1. Similar results were found by Dias (2014) for

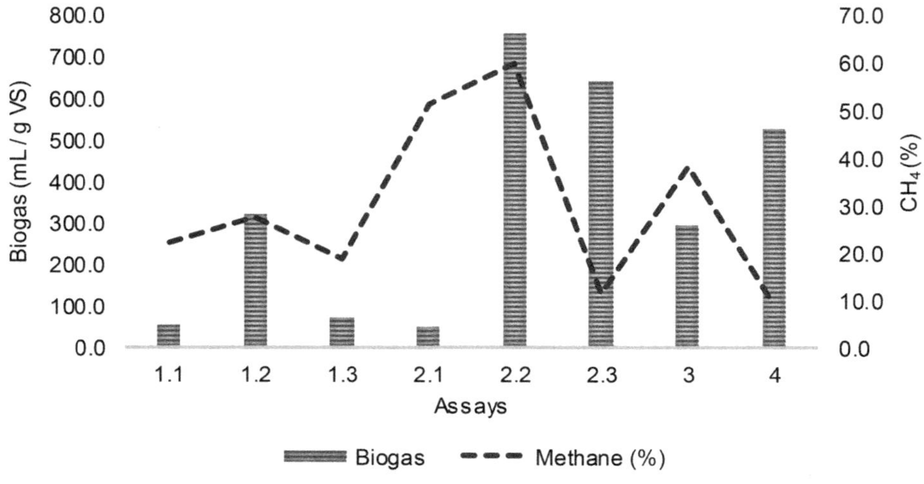

Figure 2. Biogas and methane yield in the experiments.

the pear, another fruit used as anaerobic digestion substrate, with COD of 203.4 g/kg and TS of 196.3 g/kg.

The results showed a significantly difference between the assays, in terms of biogas produced, as can be seen in Figure 2. In assay 2.1 was obtained the lowest value, with 46.7 mL_{biogas}/g VS while the assay 2.2 produced the highest value, 759.1 mL_{biogas}/g VS, the maximum biogas produced in all trials. The optimal conditions were observed in assay 2.2 that was biogas contained approximately 60% of methane. The worst methane yield was observed in assay 4, which presented values of 9%. This result may be associated by the scale-up of the reactor, from 200 mL (R1) to 17 L (R18) of useful volume.

Gonçalves (2016) performed eight similar experiments, varying only the amount of $NaHCO_3$added, that was 300 mg against the 500 mg in this study. The author showed results that ranging between 241.8 and 437.4 mL CH_4/g SV. When using kiwi as a substrate, Menardo and Balsari (2012) reported concentration of 371 mL CH_4/g SV, that represent 46% of CH_4 present in the biogas.

Calabrò and Panzera (2018) evaluated the anaerobic digestion for methane production from orange peel waste (OPW) to analyze the effects of storage, called ensiling, of OPW. The results of their batch experiments showed the highest production was registered for samples of OPW ensiled for 37 days, with a value of 365 Nml CH_4/g VS.

The results of methane production and theoretical methane production referring to the experimental data of this research is showed in Figure 3.

According to the results showed in Figure 2, in all assays, with the exception of 1.2, the methane yield obtained was lower than the theoretical potential, that was 448 mL CH_4/g SV. Odedina et al. (2017) investigated the theoretical methane potential (TMP) in comparison to biochemical methane potential (BMP) of tropical fruit wastes. Rambutan waste has the highest theoretical methane yield of 0.49 L CH_4/g VS, followed by longan waste having 0.45 L CH_4/g VS and then lady-finger banana peel having 0.41 L CH_4/g VS, while the BMP produced is 0.193, 0.234 and 0.268 L CH_4/g VS, respectively.

Taking the values obtained in this study as a basis for comparison with recently published studies, in which other types of organic matter are used for the biogas production with the DA process, it can be observed that the use of kiwi waste is a viable option in terms of methane production.

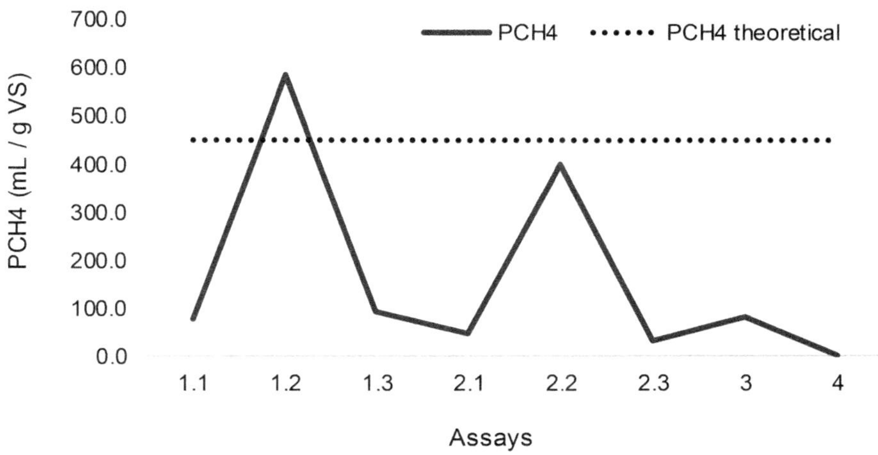

Figure 3. Methane production for each assay and theoretical production.

4 CONCLUSION

According to the results, it can be concluded that the kiwi waste has the potential to be used as a substrate in the anaerobic digestion, presenting values close to those already reported in other studies that use other organic waste as substrates.

It was possible to concluded that the substrate/inoculum ratio influence the anaerobic digestion process efficiency, in terms of methane yield. The best results showed 759.1 mL biogas/g VS with methane quality about 60%.

REFERENCES

Abbasi, T., Tauseef, S. M., & Abbasi, S. A. 2012. *Biogas Energy*. New York, NY: Springer New York.

Andreoli, C. V., von Sperling, M., & Fernandes, F. 2015. Sludge Treatment and Disposal. *Water Intelligence Online 6*: 237.

Calabrò Paolo S; Panzera, Maria F. 2018. Anaerobic digestion of ensiled orange peel waste: Preliminary batch results. *Thermal Science and Engineering Progress 6*: 355–360.

Coelho, R. A. 2015. *Obtenção de óleo de sementes de quiuí (Actinidia deliciosa) utilizando extração com solvente pressurizado e extração assistida com ultrassom*. Curitiba: Universidade Federal do Paraná.

Dias, T. R. D. 2014. *Co-digestão como solução para a valorização energética de resíduos de fruta e legumes*. Lisboa: Universidade de Lisboa.

FAO. 2013. *Countries by commodity*. Rome: Food and Agriculture Organization of the United Nations.

Franco, J. 2008. História e Desenvolvimento comercial. In *Kiwi, da produção à comercialização*. In M. D. Antunes (Ed.), Ciências da Terra: Universidade do Algarve.

Gonçalves, B. C. 2016. *Eliminação/Valorização de Resíduos de Frutas (kiwi) por Digestão Anaeróbia*. Bragança: Instituto Politécnico de Bragança.

Hagos, Kiros; Zong, Jianpeng; Li, Dongxue; Liu, Chang; Lu, Xiaohua. 2017. Anaerobic co-digestion process for biogas production: Progress, challenges and perspectives. *Renewable and Sustainable Energy Reviews 76*: 1485–1496.

IEA. 2015. *Renewable Energy – Medium-Term Market Report 2015*. International Energy Agency.

IEA. 2017a. *Key world energy statistics*. International Energy Agency.

IEA. 2017b. *Renewables information: Overview*. International Energy Agency.

Menardo, S., & Balsari, P. 2012. An Analysis of the Energy Potential of Anaerobic Digestion of Agricultural By-Products and Organic Waste. *BioEnergy Research 5*(3): 759–767.

Nghiem, Long D; Koch, Konrad; Bolzonella, David; Drewes, Jörg E. 2017. Full scale co-digestion of wastewater sludge and food waste: Bottlenecks and possibilities. *Renewable and Sustainable Energy Reviews 72*: 354–362.

Odedina, Mary Jesuyemi; Charnnok, Boonya; Saritpongteeraka, Kanyarat; Chaiprapat, Sumate. 2017. Effects of size and thermophilic pre-hydrolysis of banana peel during anaerobic digestion, and biomethanation potential of key tropical fruit wastes. *Waste management 68*: 128–138.

Paritosh, Kunwar; Kushwaha, Sandeep K.; Yadav, Monika; Pareek, Nidhi; Chawade, Aakash; Vivekanand, Vivekanand. 2017. Food Waste to Energy: An Overview of Sustainable Approaches for Food Waste Management and Nutrient Recycling. *BioMed Research International* 2017: 19.

Raposo, F., Fernández-Cegrí, V., de la Rubia, M. A., Borja, R., Béline, F., Cavinato, C., de Wilde, V. 2011. Biochemical methane potential (BMP) of solid organic substrates: Evaluation of anaerobic biodegradability using data from an international interlaboratory study. *Journal of Chemical Technology and Biotechnology 86*(8): 1088–1098.

Sanjaya, A. P., Cahyanto, M. N., & Millati, R. 2016. Mesophilic batch anaerobic digestion from fruit fragments. *Renewable Energy 98*: 135–141.

Silveira, S. V. da, Anzanello, R., Simonetto, P. R., Gava, R., Garrido, L. da R., Santos, R. S. S. dos, & Girardi, C. L. 2012. *Aspectos Técnicos da Produção de Quivi*. Bento Gonçalves: Embrapa Uva e Vinho.

USDA National Nutrient Database for Standard Reference. 2018. *Full Report (All Nutrients) 09148, Kiwifruit, green, raw*. U.S. Department of Agriculture.

Yan, H., Zhao, C., Zhang, J., Zhang, R., Xue, C., Liu, G., & Chen, C. 2017. Study on biomethane production and biodegradability of different leafy vegetables in anaerobic digestion. *AMB Express 7*(1): 27.

Zhang, Le; Loh, Kai-Chee; Zhang, Jingxin. 2018a. Enhanced biogas production from anaerobic digestion of solid organic wastes: Current status and prospects. *Bioresource Technology Reports 5*: 280–296.

Wastes: Solutions, Treatments and Opportunities III – Vilarinho et al. (Eds)
© 2020 Taylor & Francis Group, London, ISBN 978-0-367-25777-4

Rhodobacter and pigments in anaerobic digestion of brewery effluent

A. Neves, L. Ramalho, I.P. Marques & A. Eusébio
LNEG-Unit of Bioenergy, Lisboa, Portugal

ABSTRACT: Brewery wastewater (BWW) was digested anaerobically in mesophilic conditions and batch mode. The presence of a reddish pigmentation associated to the better removal capacity of the BWW digestion (64%), indicates that the treatment and energetic valorisation of an organic effluent can occur at the same time and inside the same unit as the production of a photosynthetic pigment. The reddish pigmentation found in this experiment was mainly attributed to bacteriochlorophyll *a*, and to carotenoids pigments of the spirilloxantin series (characteristics of purple non-sulfur bacteria). Microbial identification through Next-generation sequencing of 16S rRNA genes showed the presence of bacterial genus *Rhodobacter* in the inoculum and in anaerobic digestion of BWW.

1 INTRODUCTION

The brewing sector has an important impact on the Portuguese economy, with beer production increasing by 11% in 2017 (EUROSTAT 2018). However, for each liter of beer produced, 3–10 liters of highly polluted wastewater are generated (Simate et al. 2011). Due to their organic load (sugars, soluble starch, ethanol and volatile fatty acids), suspended solids content and the presence of phosphorus and nitrogen (ammonia and/or nitrate) (Raposo et al. 2010), they have to be treated before being thrown into the environment.

Anaerobic digestion (AD) has been used as a sustainable and environmentally friendly method for converting organic content wastes into a renewable energy and a flow for agricultural use. Moreover, it was found the presence of a ubiquitous group of anoxyphototroph microorganisms – the purple non-sulfur bacteria (PNSB) – which can produce photosynthetic pigments in anaerobic conditions, like carotenoids and bacteriochlorophyll *a* (Zhang et al. 2002, Soto-Feliciano et al. 2010), which gave them a reddish pigmentation. The members of this group belongs to α and β-Proteobacteria and are photoautotrophic, photoheterotrophic, and chemoheterotrophic (Soon et al. 2014). They show an important role in wastewater treatment processes loaded with high concentrations of acetate and lower fatty acids (Okubo et al. 2005).

The objective of this work is to characterize a pigmented population detected in the liquid medium during the anaerobic digestion process of BWW and evaluate possible conjugation of the effluent treatment and the simultaneous production of the pigment during the same process.

2 MATERIAL AND METHODS

2.1 *Substrate and inoculum*

Brewery wastewater (BWW) was collected from the Sociedade Central de Cervejas e Bebidas brewery (SCC, Vialonga, Portugal) after a primary treatment stage. Biological solids, collected from an anaerobic digester plant (SIMLIS, Leiria, Portugal), were used as inoculum (I).

Table 1. Analytical characterization of brewery wastewater (BWW) and control.

Materials	pH	COD [g/L]	TS [g/L]	VS [g/L]	Acetic acid [g/L]	Total VFA [g/L in acetic acid]
BWW	5.1	7.37 ± 0.00	3.6 ± 0.09	1.3 ± 0.09	2.27	3.13
Control	7.4	17.55 ± 0.38	12.5 ± 0.06	9.6 ± 0.04	1.09	1.53

2.2 *Anaerobic digestion experimental set-up*

The AD assay was performed in transparent batch vials of 71.5 mL total volume (31.5 mL headspace), digesting a substrate to inoculum ratio of 70%BWW+30%I (v/v). Digestion units were kept permeable to sunlight for the experiment duration. The trial was conducted in triplicate, under mesophilic conditions (37 ± 1°C) for 34 days. A control assay was also carried out using the same inoculum concentration but without substrate addition.

2.3 *Analytical and chromatograph methods*

Performance of the process was monitored by analytical characterizations and by the volume and quality of the obtained biogas. Total and volatile solids (TS, VS), chemical oxygen demand (COD), and pH were assayed according to Standard Methods (APHA 2012). Biogas production was monitored daily with a pressure transducer while the gas composition and volatile fatty acids (VFA) (acetate, propionate, butyrate, isobutyrate, iso-valerate and valerate) were analysed weekly by gas chromatographic techniques (Varian 430-GC, TDC; HP-5890, FID), according to ASTM Standard Method (D1946–90 2000). All gas volumes were corrected to STP conditions (Standard Temperature and Pressure: 1 bar, 0°C). The characterization of the substrate and control is shown in Table 1.

2.4 *Microscopy*

A sample was collected at the final of the experiment, and cells were observed under optical microscopy (Olympus BX51), with a 40x ocular making a total amplification of 400x.

2.5 *Analysis of pigments*

The presence of photosynthetic pigments like bacteriochlorophyll *a*, and carotenoids pigments, was assessed by spectrometry. An aliquot of culture medium was collected after the experiment and was diluted in water. Then, the absorption spectrum of intact cells was measured within a range of 380–900 nm (Shimadzu UV – 2401PC).

2.6 *Molecular analysis*

Aliquots were centrifuged at 12.900 g for 20 min to obtain total genomic DNA from samples collected from batch experiments according to Eusébio et al. (2011). The extracted DNA was pooled, quantified and checked for purity using Qubit^{TM} (Thermo Fisher Scientific) prior to storage at −20°C. Next-generation Sequencing (NGS) of DNA molecules was performed at STAB VIDA facilities (Lisbon, Portugal). V3 and V4 regions of bacterial and archaeal 16S rRNA gene were amplified with universal primers 515F – 806R. Library construction was performed using the Illumina 16S Metagenomic Sequencing Library preparation protocol. The generated DNA fragments (DNA libraries) were sequenced with MiSeq Reagent Kit v3 in the Illumina MiSeq platform, using 300 bp paired-end sequencing reads.

Table 2. Performance in terms of removal capacity [percentage].

Assay	COD	TS	VS	Total VFA [in acetic acid]
70%BWW+I	63.5	18.8	33.3	70.91
Control	35.4	0	5.9	15.93

The bioinformatics analysis of the generated raw sequence data was carried out at STAB VIDA facilities (Lisbon, Portugal) using the Quantitative Insights Into Microbial Ecology (QIIME2, version 2018.11) (Caporaso et al. 2010). The reads were denoised using the Divisive Amplicon Denoising Algorithm 2 (*DADA2)* plugin (Callahan et al. 2016), where the following processes were applied: Trimming and truncating low quality regions; dereplicating the reads; filtering chimeras. After denoising, the reads were organized in features, which are operational taxonomic units (OTUs) and a feature table was generated using the plugin feature-table (https://github.com/qiime2/q2-feature-table), with each feature being represented by exactly one sequence. After applying the plugins Alignment (Katoh & Standley 2013), Phylogeny (Price et al. 2010), a pre-trained sk-learn classifier based on the SILVA (Glöckner et al. 2017) (release 132 QIIME) with a clustering threshold of 97% similarity was applied to generate taxonomy tables. Taxonomic classification was achieved by using plugins Feature-classifier (https://github.com/qiime2/q2- feature-classifier) and Taxa (https://github.com/qiime2/q2-taxa) where only OTUs containing at least 10 sequence reads were considered as significant.

3 RESULTS AND DISCUSSION

BWW is an acidic effluent (pH 5) with a low concentration of organic materials (7 g/L COD, 4 g/L TS and 1 g/L VS), as shown in Table 1. It is rich in VFA compounds, mainly acetic acid (73% of total VFA), which were efficiently converted. Anaerobic digestion of BWW exhibited a good removal capacity, presenting values of 63% (COD removal) and 71% (VFA removal (Table 2) and providing a methane yield of 0.298 m^3 CH_4/Kg COD removed.

During the AD process, the liquid medium became reddish and a sample was taken at the final of the experiment. According to literature, the color change resulted from an increase of phototrophic PNSB populations in anaerobic conditions (Zhang et al. 2002, Soto-Feliciano et al. 2010). The pigmented cells were observed under light microscopy (Fig. 1), and typical reddish-like clusters were observed.

The analysis of photosynthetic pigments of [70%BWW+I] sample was done by spectrometry and the absorption spectrum is presented in Figure 2. The absorbance maxima of whole cells were found at 862, 806, 592, 528, and 490 nm, indicating the presence of bacteriochlorophyll *a*, and carotenoids pigments of the spirilloxantin series (characteristics of purple non-sulfur bacteria) (Okubo et al. 2006, Soto-Feliciano et al. 2010).

The microbial composition of samples collected from inoculum (I), BWW and before [70%BWW+I in] and after [70%BWW+I out] the AD process, was analyzed by NGS of 16S rRNA genes and relative abundance of major phyla is presented in Figure 3a. There was found 12 phyla including Actinobacteria, Armatimonadetes, Bacteroidetes, Chloroflexi, Firmicutes, Fusobacteria, Planctomycetes, Proteobacteria, Saccharibacteria, Spirochaetae, Synergistetes and Tenericutes. The dominant phyla during anaerobic digestion process and in the inoculum were Chloroflexi, Firmicutes and Proteobacteria. In BWW, the dominant phylum is Bacteroidetes (53% relative abundance). Molecular analysis of all samples revealed high diversity of populations inside each phyla (data not shown). The important role of Firmicutes as fermentative bacteria producing extracellular enzymes for hydrolysis (Wang et al. 2017) and Chloroflexi in degradation of organic matters could explain their dominance in AD (Cheng et al. 2018).

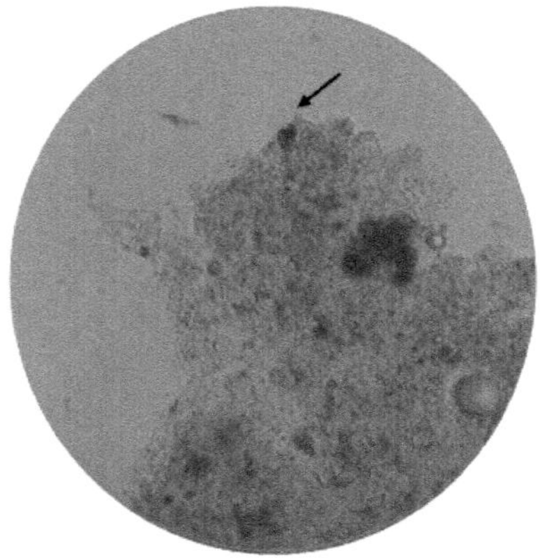

Figure 1. A liquid sample of culture medium analyzed under light microscopy (amplification 400x). The black arrow shows the reddish-like clusters that contain pigmentation.

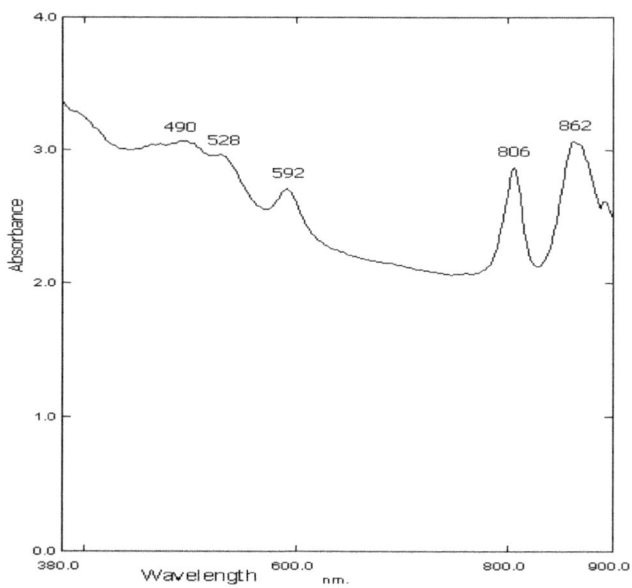

Figure 2. Absorption spectrum of the whole cells sample. Wavelength (nm) of the absorption maxima are shown in the top of the peaks.

Concerning the presence of PNSB, 16S rRNA gene sequences affiliated with genus *Rhodobacter* (97% similarity), showing the presence of this population in the inoculum with relative abundance of 0.73% relative to all detected populations (Fig 3b) and was maintained in the AD process of BWW. In the raw BWW, *Rhodobacter* was not detected (relative abundance of 0%). The high concentration of acetate in the beginning of the process (1.3 g/L), together with anaerobic process occurring in glass reactors under light intensity could be an explanation for the favorable conditions for the growth of *Rhodobacter* populations, and production of red pigments.

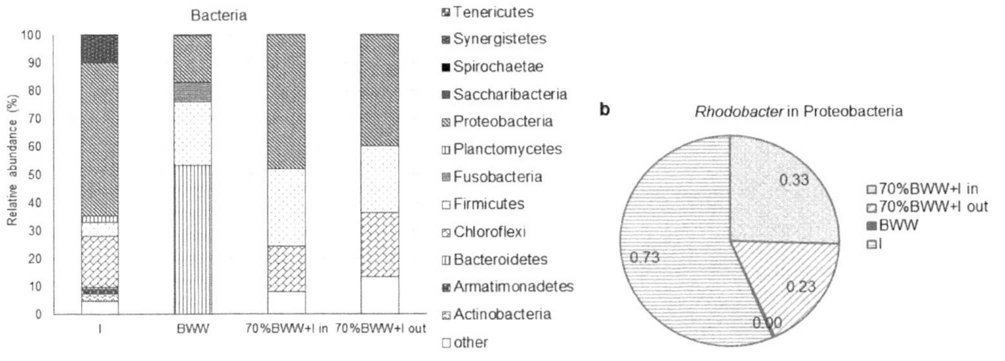

Figure 3. Relative abundance of bacterial taxonomic groups in samples collected from inoculum (I), BWW and before [70%BWW+I in] and after [70%BWW+I out] the anaerobic digestion process. The taxonomic classification of bacterial reads at phylum (a) and *Rhodobacter* genus (b) levels are shown. Bacterial groups accounting for less than 1% of all classified sequences were gathered in the category "other".

ACKNOWLEDGEMENTS

This work was carried out under the Project Phoenix (MSCA/RISE Contract number 690925) and was financed by national funds through the FCT – Fundação para a Ciência e a Tecnologia, I.P. under the project ERANETLAC/0001/2014, GREENBIOREFINERY – Processing of brewery wastes with microalgae for producing valuable Compounds. The authors would like to thank Sociedade Central Cervejas e Bebidas (SCC, Portugal) for wastewater, and Natércia Santos for laboratory assistance.

REFERENCES

APHA-American Public Health Association. 2012. Standard Methods for examination of water and wastewater, Washington DC.

ASTM D1946-90. 2000. Standard Practice for Analysis of Reformed Gas by Gas Chromatography. ASTM International, West Conshohocken, PA.

Callahan BJ,.McMurdie PJ, Rosen MJ, Han AW, Johnson AJA, Holmes SP. 2016. DADA2: High-resolution sample inference from Illumina amplicon data. *Nat. Methods 2016*. doi:10.1038/nmeth.3869.

Caporaso J.G., Kuczynski J., Stombaugh J., Bittinger K., Bushman F.D., Costello E.K., et al. QIIME allows analysis of high-throughput community sequencing data. Nat Methods. 2010; 7(5):335–336. doi:10.1038/nmeth0510-335.

Cheng C, Zhoua Z, Pang H, Zheng Y, Chen L, Jiang L-M, Zhao X. 2018. Correlation of microbial community structure with pollutants removal, sludge reduction and sludge characteristics in micro-aerobic side-stream reactor coupled membrane bioreactors under different hydraulic retention times. Bioresource Technology, 260: 177–185

Eurostat. 2018. Happy Beer day! Access in 27 feb 2019, available: https://ec.europa.eu/eurostat/en/web/products-eurostat-news/-/EDN-20180803-1.

Eusébio A., Tacão M., Chaves S., Tenreiro R., Almeida-Vara E. 2011. Molecular assessment of microbiota structure and dynamics along mixed olive oil and winery wastewaters biotreatment. Biodegradation 22: 773–795. DOI 10.1007/s10532-010-9434-0.

Glöckner F.O., Yilmaz P., Quast C., Gerken J., Beccati A., Ciuprina A., Bruns G., Yarza P., Peplies J., Westram R., Ludwig W. 25 years of serving the community with ribosomal RNA gene reference databases and tools. Journal of. Biotechnology. 2017. doi:10.1016/j.jbiotec.2017.06.1198.

Katoh, K., Standley, D.M.. MAFFT multiple sequence alignment software version 7: Improvements in performance and usability. Mol. Biol. Evol. (2013). doi:10.1093/molbev/mst010.

Okubo Y, Futamata H, Hiraishi A. 2005. Distribution and capacity for utilization of lower fatty acids of phototrophic purple nonsulfur bacteria in wastewater environments. Microbes Environ. 20:135–143.

Okubo Y, Futamata H, Hiraishi A. 2006. Characterization of Phototrophic Purple Nonsulfur Bacteria Forming Colored Microbial Mats in a Swine Wastewater Ditch. Applied Environmental Microbiology,72: 6225–6233.

Price, M. N., Dehal, P. S. & Arkin, A. P. FastTree 2 – Approximately maximum-likelihood trees for large alignments. PLoS One (2010). doi:10.1371/journal.pone.0009490.

Raposo M.F.J., Oliveira S.E., Castro P.M., Bandarra N.M., Morais R.M. 2010. On the Utilization of Microalgae for Brewery Effluent Treatment and Possible Applications of the Produced Biomass. J. Inst. Brew. 116: 285–292.

Simate G.S., Cluett J., Iyuke S.E., Musapatika E.T., Ndlovu S., Walubita L.F., Alvarez A.E. 2011. The treatment of brewery wastewater for reuse: State of the art. Desalination 273: 235–247.

Soon TK, Al-Azad S, Ransangan J. 2014. Isolation and Characterization of Purple Non-Sulfur Bacteria, *Afifella marina*, Producing Large Amount of Carotenoids from Mangrove Microhabitats. J. Microbiol. Biotechnol., 24(8), 1034–1043. DOI:10.4014/jmb.1308.08072.

Soto-Feliciano K, De Jesus M, Vega-Sepulveda J, RiosVelazquez. 2010. Isolation and characterization of purple nonsulfur anoxyphototropic bacteria from two microecosystems: tropical hypersaline microbial mats and bromeliads phytotelmata, pp. 109–116. In Mendez-Vilas A (ed.). Current Research, Technology and Education Topics in Applied Microbiology and Microbial Biotechnology. Formatex Research Center, Badazoz, Spain.

Wang M, Zhang X, Zhou J, Yuan Y, Dai Y, Li Y, Li Z, Liu X, Yan Z. 2017. The dynamic changes and interactional networks of prokaryotic community between co-digestion and mono-digestions of corn stalk and pig manure. *Bioresource Technology* 225: 23–33

Zhang D, Yang H, Huang Z, Zhang W, Liu S-J. 2002 *Rhodopseudomonas faecalis* sp. nov., a phototrophic bacterium isolated from an anaerobic reactor that digests chicken faeces. *International Journal of Systematic and Evolutionary Microbiology* 52: 2055–2060. doi: 10.1099/ijs.0.02259-0.

Wastes: Solutions, Treatments and Opportunities III – Vilarinho et al. (Eds)
© 2020 Taylor & Francis Group, London, ISBN 978-0-367-25777-4

e-Plastics as partial replacement for aggregates in concrete

L.A. Parsons & S.O. Nwaubani
School of Civil and Environmental Engineering, Faculty of Engineering and the Built Environment, University of the Witwatersrand, Johannesburg, South Africa

ABSTRACT: This paper describes an investigation aimed at producing structural strength concrete (>25 MPa) which incorporates different types of waste electrical and electronic plastics (e-Plastic), as replacement for natural aggregates. The concrete specimens studied were made with polycarbonate and acrylonitrile butadiene styrene (PC/ABS), high impact polystyrene (HIPS) and ABS as well as blends of the aforementioned e-Plastic in a ratio of 1:1:1. Granulated e-Plastic was used to replace both the natural coarse and fine crushed aggregates in proportions of 5%, 10%, 20% and 30% by volume of the natural aggregates. Tests carried out in the fresh state showed that as the percentage replacement of e-Plastic increased, the workability decreased. Consequently, a superplasticising admixture was needed to maintain a similar slump value for all the mixtures. Compressive strength after curing for 3, 7, 28 and 90-days showed that the concrete incorporating the e-Plastic did meet the minimum requirements for structural application of concrete.

1 INTRODUCTION

The e-Waste Association of South Africa defines e-Waste as electronic wastes derived from information technology equipment, consumer electronics, small household appliances and large household appliances (eWasa, 2018). The United Nations University (UNU) – an autonomous organisation of the UN General Assembly operating through the worldwide network and coordinated by UNU Center in Tokyo, estimate that South Africa (5.7 kg/inh) and Algeria (6.2 kg/inh) generates the largest amount of e-Waste on the continent per inhabitant. This is much compared to Africa's average e-Waste consumption of 1.9 kg/inh (Baldé, et al., 2017). E-Waste recycling in South Africa focuses mainly on the recovery of ferrous and non-ferrous metal fractions, while approximately 80% of its e-Plastic are exported to other countries (Lydall, et al., 2017).

The e-Plastic considered in this study are those discarded plastics derived from electrical and electronic plastics (EEP). EEP may be used for insulation, noise reduction, sealing, housing, structural parts, functional parts and electronic parts in electrical and electronic equipment (EEE). EEP is the second largest component of EEE which contributes to on average 30% by weight (wt%) (Mahesh, et al., 2012). However, less than 25 wt% of e-Plastic is recycled globally (Anon., 2011).

The recycling of e-Plastic is difficult due to numerous amounts of additives incorporated into EEP and that not all EEP parts used in EEE are labelled according to their respective plastic types among other. Moreover, it is well known that in both its use and recycling phases, polymers consume antioxidants which may lead to unpredictable and inadequate properties in recycled plastics (Stenval, 2013).

However, one of the greatest concerns pertaining to the recycling of e-Plastic is the presence of flame-retardants (FR). Approximately 25% of all EEP contain flame retardants and are found mainly in printed circuit boards, component connectors and plastic covers. FR are deemed necessary for fire safety reasons. They are added to plastics to increase the products flammability protection, reduce or prevent flame related injury and property destruction (Mahesh, et al., 2012). Flame retardants supresses ignition, limits the spread of fire or slows combustion down due to halogen formation. These halogens capture free radicals, and are added to plastics where fire resistance is required (Stenval, 2013).

Table 1. Oxide composition of AfriSam cement (CEM 1, 52.5R).

Oxide	*MgO*	Al_2O_3	*CaO*	Fe_2O_3	SiO_2
Percentage	1.64	9.54	61.85	5.46	21.51

Table 2. Physical properties of aggregates.

Physical properties	Coarse aggregates	Fine aggregates	PC/ABS	HIPS	ABS
Specific gravity	2890	2580	1190	1390	940
Bulk density (kg/m^3)	1660	2060	580	580	520
Fineness modulus	6.2	3.4	6	6	6
Dust content (%)	2.1	8.3	0	0	0

Table 3. Technical and dosage data for Sika®ViscoCrete®-3088 superplasticiser.

Density	pH value	Dosage	Effects on setting
1.06 kg/L	5.5 ± 0.5	0.2% – 2.0% wt cement	Retarding according to dosage

Brominated flame retardants (BFR's) have been found in numerous environments and organisms, such as: household dust, sewage, air, soil, water, human and wildlife tissue. This is of great concern as these BFR's can negatively affect memory and learning, thyroid function and even increase hyperactivity (Mahesh, et al., 2012).

Manjunath investigate the *Partial replacement of E-Plastic Waste as Coarse-aggregate in Concrete* and found a drop in strength between concrete made without e-Plastic and with 30% e-Plastic replacement of 47.18 MPa and 22.15 MPa (Manjunath, 2016). Similar results were found by Kumar and Baskar who studied *the Recycling of E-plastic Waste as Construction Material in Developing Countries* (Kumar & Baskar, 2015). There is a clear strength drop when substituting e-Plastic into concrete. However, little research has been conducted in evaluating the strength differences for concrete made using different types of e-Plastic as partial replacement for aggregates.

Clearly, there is a need to find innovative ways of eliminating the negative impact of this form of waste materials in the environment or to convert them into safer usage. Concrete offers a unique solution to encapsulate such toxic materials, whilst helping to solve the possible environmental pollution problem created by the careless disposal of e-Plastic in different communities. Furthermore, it may also reduce the demand for natural aggregates and thus, save natural resources. This study is therefore timely in seeking to address this issue in the South African context.

2 EXPERIMENTAL

2.1 *Materials used*

The cement used was ordinary Portland cement (CEM 1, 52.5R), donated by AfriSam in South Africa. The superplasticiser used was Sika® ViscoCrete®-3088, donated by Sika South Africa. Table 1 shows the main oxide composition of the cement used. Table 2 shows the Physical properties of aggregates used. Table 3 shows the main technical and dosage data.

Fig. 1 shows the grading curve for the e-Plastic used.

2.2 *Specimen preparation and curing*

The specimens were cast in 100 mm × 100 mm × 100 mm plastic cube molds and compacted using a vibrating table. The specimens were covered with a plastic sheet in the laboratory for 24 hours before demolding. Curing was continued by placing the specimens in a water curing tank with water temperatures maintained at 23°C±2°C.

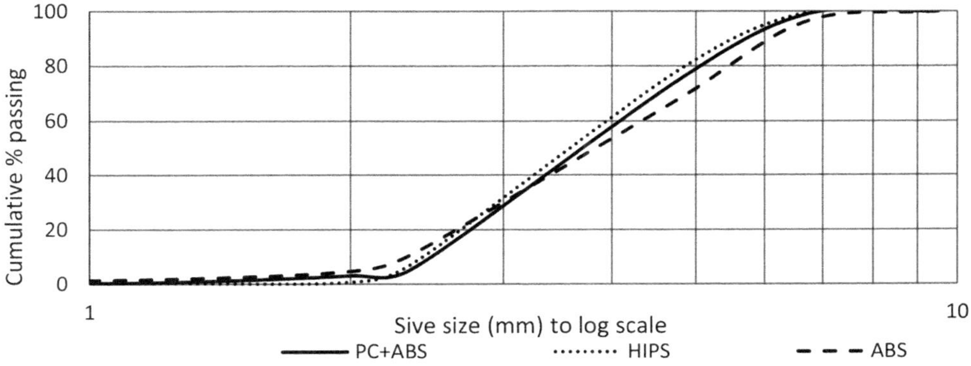

Figure 1. Grading curve for granulated e-Plastic.

Table 4. Mix proportions of concrete mixes (kg/m^3).

Mix design	Fine aggregates	Coarse aggregates	e-Plastic	Cement	Water
S	931	706	0	444	231
S (5P)	908	688	18	444	231
S (10P)	884	671	36	444	231
S (20P)	838	635	72	444	231
S (30P)	791	600	108	444	231
S (5H)	908	688	21	444	231
S (10H)	884	671	42	444	231
S (20H)	838	635	84	444	231
S (30H)	791	600	126	444	231
S (5A)	908	688	14	444	231
S (10A)	884	671	28	444	231
S (20A)	838	635	57	444	231
S (30A)	791	600	85	444	231
S (5B)	908	688	18	444	231
S (10B)	884	671	36	444	231
S (20B)	838	635	71	444	231
S (30B)	791	600	107	444	231

2.3 *Concrete mix design*

One mix design was used for all e-Plastic replacements using the Cement and Concrete Institute (C&CI) method, which is based on the American Concrete Institute standard (ACI 211.1-91). A water-cement ratio of 0.52 was used in preparing all specimens using a 13.2 mm unwashed crushed aggregate and fine crusher sand. Both the coarse and fine aggregates were replaced simultaneously as trail mixes indicated that this replacement yielded higher compressive strengths, compared to replacing only fine or coarse aggregates. Furthermore, PC/ABS, HIPS, ABS and a combination of the three e-Plastics in a ratio of 1:1:1 was used to replace crushed aggregates by volume (5, 10, 20 and 30%).

Table 4 presents the mix proportions for each mix design whereby S is the standard mix, S (5/10/20/30) is the percentage preplacement and S (P/H/A/B) are replacements with PC/ABS, HIPS, ABS or a blend of the three e-Plastics.

Table 5. Compressive strength results for relevant mix designs (average values, MPa).

Mix design	3 Days	CoV(%)	7 Days	CoV(%)	28 Days	CoV(%)	90 Days	CoV(%)
S 0	37.5	2.2	46.4	0.9	57.6	1.7	60.1	3.0
S (5P)	37.4	3.2	43.5	2.5	50.4	2.2	53.1	1.4
S (10P)	36.3	1.1	40.7	1.6	46.5	1.1	46.6	0.6
S (20P)	34.2	1.5	37.5	5.8	41.6	1.8	43.8	3.5
S (30P)	33.2	6.6	36.4	4.2	40.5	3.8	40.9	2.5
S (5H)	37.0	2.5	42.2	5.9	50.5	1.8	51.1	1.0
S (10H)	34.0	0.5	39.5	0.9	46.1	0.2	46.7	3.1
S (20H)	32.3	2.2	35.5	1.8	39.8	4.3	41.1	1.8
S (30H)	34.0	2.0	38.5	4.3	39.5	2.1	40.7	4.7
S (5A)	35.2	6.9	42.6	3.2	49.7	2.2	53.3	2.2
S (10A)	31.5	2.8	37.8	1.7	44.2	1.1	49.2	3.5
S (20A)	28.8	1.9	34.0	1.7	38.1	1.8	42.1	1.1
S (30A)	19.1	1.6	22.1	6.3	25.1	3.8	27.0	6.1
S (5B)	37.1	0.3	42.9	0.6	49.4	1.9	53.3	2.6
S (10B)	34.9	2.5	41.4	0.4	46.5	1.4	49.9	2.4
S (20B)	31.3	0.9	36.8	0.7	42.0	2.1	42.5	1.0
S (30B)	31.6	1.6	34.9	1.6	37.5	3.1	37.5	5.1

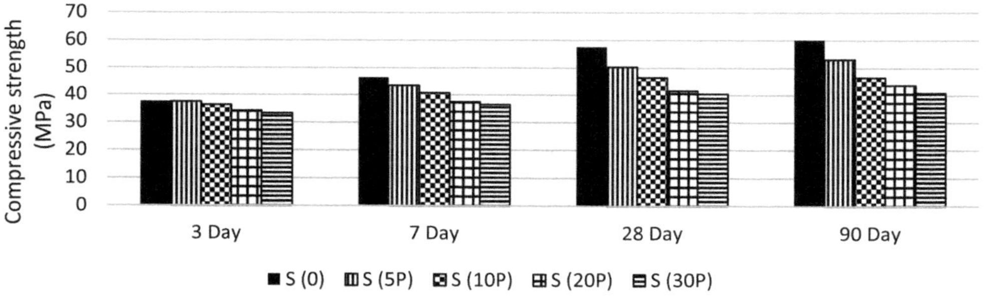

Figure 2. Compressive strength of PC/ABS e-Plastic in concrete.

3 TEST PROCEDURE AND RESULTS

3.1 *Workability*

The standard mix was designed to have a slump of 80 mm. However, it was found that the addition of e-Plastic decreased the slump. Hence, the aforementioned superplasticiser was used to maintain a slump of 80 mm ± 25 mm. A slump test cone was used to perform the slump test according to SANS:2862-1.

3.2 *Compressive strength test*

The compressive strength for the aforementioned cubes were tested according to SANS 5863. An automatic cube press-foote test machine was used with a load rate of 150 kN/min. The cubes were tested at 3, 7, 28- and 90 days. Table 5 provides all the results for the average compressive strengths of the above-mentioned mix designs and testing dates.

Fig. 2 – 5 provides a comparison for the 3, 7, 28 and 90-day compressive strength results for the standard mix design and mixes with e-Plastic replacement. Fig. 6 provides a comparison for the different e-Plastic at 30% replacements, at the mentioned dates.

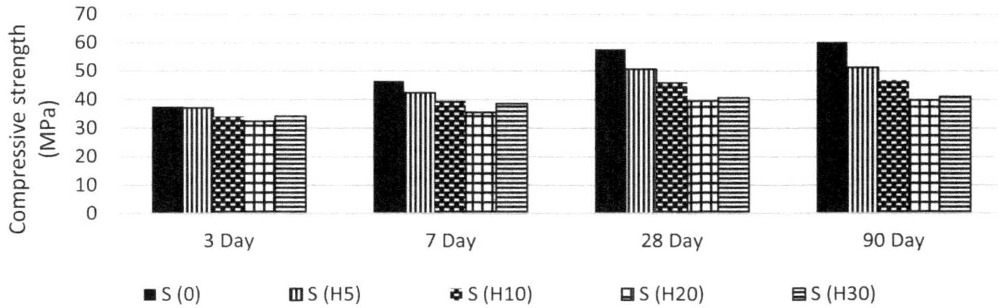

Figure 3. Compressive strength of HIPS e-Plastic in concrete.

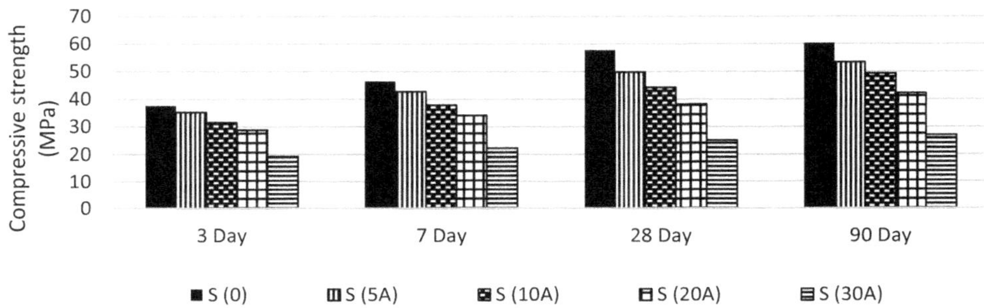

Figure 4. Compressive strength of HIPS e-Plastic in concrete.

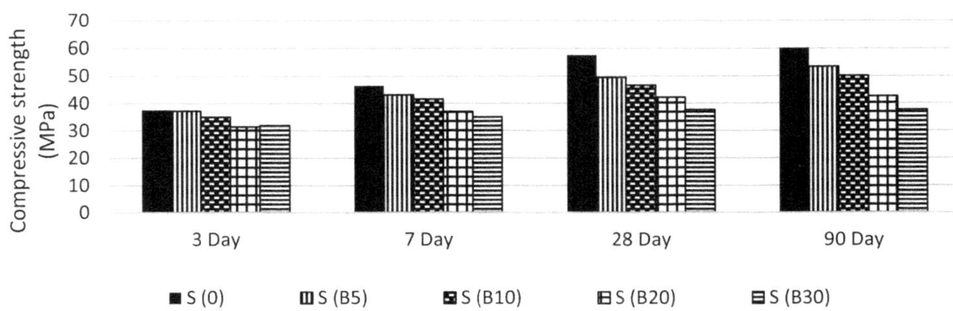

Figure 5. Compressive strength of Blended e-Plastic in concrete.

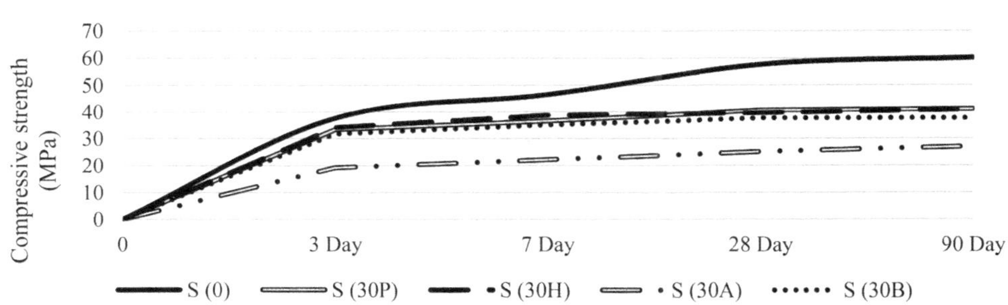

Figure 6. Compressive strength of HIPS e-Plastic in concrete.

4 DISCUSSION

The compressive strength result clearly shows that the control mix containing no plastic substitution of the natural aggregate had the highest strength at all ages. The compressive strength of the mixtures containing waste plastics was observed to decrease as the percentage of substitution increased. At 28 days of curing, the mixtures containing 30% plastics had compressive strength in the following order:

	S	>	SP	>	SH	>	SB	>	SA	
i.e,	(57.6		40.5		39.5		37.5		25.1)	MPa.

At 90 days of curing, the mixtures containing 30% plastics had a very marginal increase in compressive strength over that obtained at 28 days, in the following order:

	S	>	SP	>	SH	>	SB	>	SA	
i.e,	(60.1		40.9		40.7		37.5		27.0)	MPa.

Another important observation is that even at 30% of substitution with e-Plastic, the compressive strength was above 25 MPa which is the minimum requirement for structural applications.

These results suggest that replacing the natural aggregates with e-Plastic decreased its compressive strength. However, the final strength obtained may depend on what type of e-Plastic is used in the replacement. SP replacement produced concrete with a much higher 28-day compressive strength compared to the mixture incorporating SB and SH. It is however clear that the compressive strength of specimens incorporating SH was only slightly lower than that incorporating SP.

5 CONCLUSION

Based on the results obtained from this study, the following conclusions can be drawn:

- The compressive strength of hardened concrete incorporating e-Plastic reduced as the quantity of replacement increased;
- For mixture containing e-Plastic, SP substitution produced concrete with the highest compressive strength at 28 and 90 days.
- The results show that e-Plastic substitution of up to 30% can be used to produce structural strength concrete.
- Using e-Plastic may reduce the demand for natural or crushed aggregates and reduce the amount of waste landing up in landfills.

REFERENCES

Anon. 2011. Recycling and Disposal of Electronic Waste. *Oxford,* p. 6417.

Baldé, C. et al., 2017. The Global e-Waste Monitor. *United Nations University, International Telecommunication and International Solid Waste Association.*

eWasa, 2018. *e-Waste Association of South Africa.* [Online] Available at: https://ewasa.org

Kumar, S. & Baskar, K. 2015. Recycling of e-Plastic Waste as a Construction Material in Developing Countries. *Journal of Material Cycles and Waste Management* 718–724.

Lydall, M., Nyanjowa, W. & James, Y. 2017. Mapping South Africa's Waste Electrical and Electronic Equipment (WEEE) Dismantling, Pre-Processing and Processing Technology Landscape. In: Johannesburg(Gauteng): Mintek.

Mahesh, P., Rajankar, P. & Kumar, V. 2012. *Improving Plastic Management in Delhi. A Report on WEEE Plastic Recycling.* New Delhi: Toxics Link.

Manjunath, A. 2016. Partial Replacement of e-Waste as Coarse Aggregate in Concrete. *Science Direct* 731–739.

Stenval, E. 2013. *Electronic Waste Plastics Characterisation and Recycling by Melt-processing.* Gothenburg (Sweden): Department of Materials and Manufacturing Technology: Chalmers University of Technology.

Wastes: Solutions, Treatments and Opportunities III – Vilarinho et al. (Eds)

Production of bioactive compounds by solid-state fermentation of oilseed cakes

D.F. Sousa, J.M. Salgado & I. Belo
Centre of Biological Engineering, University of Minho, Braga, Portugal

M. Cambra-López
Institute of Science and Animal Technology, Universitat Politècnica de València, Valencia, Spain

A.C.P. Dias
Centre of Molecular and Environmental Biology, University of Minho, Braga, Portugal,
Centre of Biological Engineering, University of Minho, Braga, Portugal

ABSTRACT: Vegetable oils are an important part of human diet. By-products from vegetable oil's industry are a reliable source of protein and fat, with variable amounts of fibre, and a source of phenolic compounds. However, their digestibility by some animals is difficult due to polysaccharides and lignin content. This work aims to assess the use of sunflower cake (SFC), rapeseed cake (RSC) and soybean cake (SBC) as substrate in solid-state fermentation (SSF) with filamentous fungi *Rhyzopus oryzae* and *Aspergillus ibericus*, for the production of animal feed additives and bioactive substances such as lignocellulolytic enzymes and antioxidant phenolic compounds. Results showed the highest cellulase and xylanase activities were achieved with *A. ibericus* using SFC and SBC as substrates. Highest β-glucosidase activity was observed in SSF with *R. oryzae* using RSC as substrate. *R. oryzae* release the maximum amount of total phenols and SSF improved antioxidant capacity of RSC and SBC extracts.

1 INTRODUCTION

Global population is increasing and it is expected to reach more than 9 billion people by 2050 (Food and Agriculture Organization, 2009). Food demand will increase over the next decades due to population growth and rising incomes by the families.

Agro-food wastes are a consequence of food industry production, which can be used in biotechnology industry. The use of agro-food wastes as raw materials for the production of added-value compounds using biotechnological process has economic and environmental impact (Lio and Wang, 2012; Horita *et al.*, 2015). Over the last decade, initiatives such as ZeroWaste Europe have been established to promote alternatives for a positive valorization of wastes instead of their incineration or landfill disposal. Biotechnological eco-friendly approaches for valorization of wastes, besides their environmental and economic benefits, also creates new business opportunities and workplaces (Zero Waste Europe, 2011). Production of value-added compounds from wastes is a key point to circular economy.

Nowadays, despite their industrial applications, vegetable oils play an important role in human diet. Major sources of vegetable oils are soybean, rapeseed and sunflower seeds. Oilseed cakes are the main by-products obtained after the extraction of these vegetable oils. These byproducts are mainly composed by proteins, fibers, carbohydrates, minerals and vitamins (Sunil *et al.*, 2016). These by-products are mainly used as animal feed in ruminant diets or to complement monogastric animal rations and also as fuel in thermal power stations (Matthäus, 2002; Lomascolo *et al.*, 2012;

Sunil *et al.*, 2016). These by-products, however, can contain antinutritional factors that can be harmful to animals and do not match their nutritional needs (Ajila *et al.*, 2012).

Soybean, sunflower and rapeseed are natural sources of antioxidants (Pandey *et al.*, 2000). During extraction of vegetable oils natural antioxidants present in the oilseeds are separated into liposoluble and hydrophilic fractions (Schmidt and Pokorný, 2005). Despite this fact, oilseed cakes still have large amounts of antioxidants that have not been damaged or removed during oil extraction process. Antioxidant activity of these meals is mainly due to the presence of phenolic compounds such as phenolic acids, flavonoids and lignans (Schmidt and Pokorný, 2005).

Oilseed cakes are lignocellulosic materials, suitable to be used as substrate in biotechnological process like solid-state fermentation (SSF). SSF is defined as a fermentation process that occurs in the absence or near-absence of free water to which is applied a natural or inert substrate used as solid support. The solid support must contain enough moisture to support the growth and metabolism of microorganisms. This fermentation reproduces the conditions close to the natural habitat of microorganisms; and filamentous fungi are the ones that better adapt to SSF (Pandey, 1992; Pandey *et al.*, 2000).

This study assessed the production of animal feed additives such as lignocellulolytic enzymes (cellulases, xylanases and β-glucosidase) and bioactive compounds such as antioxidant extracts, using sunflower, rapeseed and soybean cakes as substrate in SSF.

2 METHODOLOGY

2.1 *Raw material*

Sunflower cake (SFC), rapeseed cake (RSC) and soybean cake (SBC) were collected from companies related with the vegetable oils industry from Portugal. Residues were dried at 65°C for 24 hours and stored at room temperature.

2.2 *Microorganisms*

Rhyzopus oryzae 10.260 and*Aspergillus ibericus* 03.113 were obtained from Micoteca of University of Minho, Braga, Portugal (MUM). Fungi were cultivated in potato dextrose agar (PDA). In order to obtain inoculum for SSF, the selected fungi were subcultured in PDA slants and incubated at 25°C for 7 days.

2.3 *Solid-State Fermentation*

SFC, RSC and SBC were used as substrate in SSF assays to evaluate the two fungi. SSF process was carried out in 500 mL Erlenmeyer flasks with 10 g of dried residue with moisture level adjusted to 75% (w/w) in wet basis. Erlenmeyer flasks with the solid substrate were sterilized at 121°C for 15 minutes. Inoculation process was performed following the method described by Sousa et al. (2018). The extraction of enzymes was performed at the end of 7 days of fermentative process, with water, at 4°C in a solid/liquid ratio of 1:5, and 1 h of stirring, at 150 r.p.m. Following, extracts were filtered through a net. The liquid fraction was centrifuged (4 000 r.p.m.). All SSF were performed in duplicate.

2.4 *Enzymes activity, antioxidant capacity and total phenols of fermented extracts*

Cellulases and xylanases activity was quantified using 2% carboxymethylcellulose (CMC) and 1% xylan as substrate in 0.05 M citrate buffer (pH 4.8), respectively. The enzymatic reaction was carried at 50°C for 30 and 15 min respectively. The reducing sugars released during the enzymatic reaction were quantified by 3,5-dinitrosalicylic (DNS) method and measured at 540 nm. β-glucosidase activity was quantified using the method described by Leite *et al.*, (2016). Antioxidant capacity of fermented extracts was quantified using the 2,2-Diphenyl-1-picrylhydrazyl (DPPH) radical. 200 μL of fermented extracts were placed in a 96 well microplate and 100 μL of DPPH (0.3 mM, in methanol) was added. The mixture reacted in the dark for 30 min. Then, the variation of absorbance was measured at 520 nm. All samples were performed in duplicate. Methanol was used as blank solution and DPPH methanolic solution was used as control. Known amounts of Trolox were used to build a standard curve. Scavenging activity of fermented extracts was expressed as micromole of Trolox equivalents per g of dry solid substrate ($\mu mol \cdot g^{-1}$).

Total phenols were determined by the Folin – Ciocalteu method (Commission Regulation (EEC) No. 2676/90).

3 RESULTS

3.1 *Enzymes activity*

SFC, RSC and SBC were fermented with *R. oryzae* and *A. ibericus*. Both fungi were able to grow in the different residues. It is possible to observe the production of cellulase, particularly by *A. ibericus* (Figure 1). For this species, higher production of cellulase was found with SFC and SBC than with RSC. Contrary, it was for RSC that more cellulase was produced by *R. oryzae*, however at lower values than 20 U/g.

Figure 2 shows xylanase activity obtained after analysis of crude extracts. It is possible to observe that residues fermented with *A. ibericus* achieved higher values of xylanase activity when compared to residues fermented with *R.* oryzae.

In the case of SSF with *A. ibericus* there are no significant differences among obtained values of xylanase activity. Regarding the fermentation performed with *R. oryzae*, it was observed a maximum of xylanase activity using RSC as substrate, but not statistically different from the values obtained with SFC.

β-glucosidase activity was also detected in fermented (Figure 3) where the highest β-glucosidase activity was achieved in the fermentation with *R. oryzae*, using RSC as substrate. No statistically significant differences were found between β-glucosidase activity in fermentation with both fungi, using SFC and SBC as substrate.

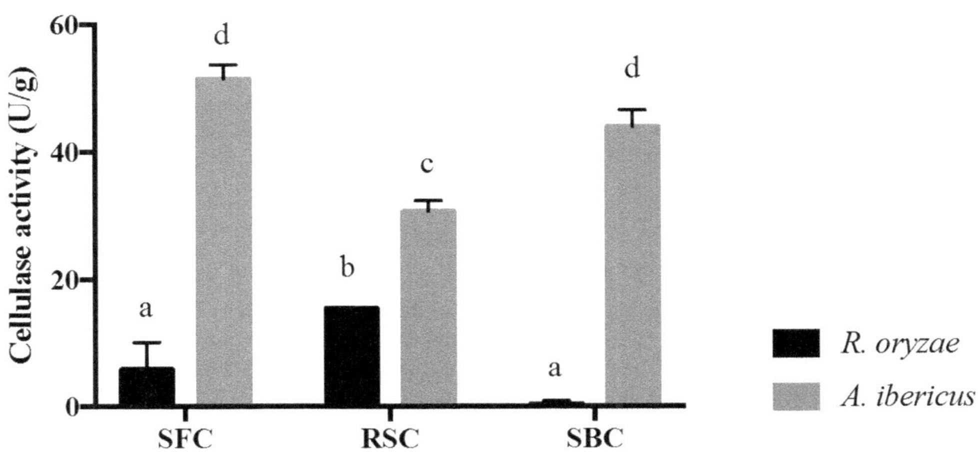

Figure 1. Cellulase activity of crude extracts obtained after SSF. Results represent the average of two independent fermentation experiments and error bars represent standard deviation. Bars with equal letters are not significantly different (Tukey test; $P < 0.05$). SFC, sunflower cake; RSC, rapeseed cake; SBC, soybean cake.

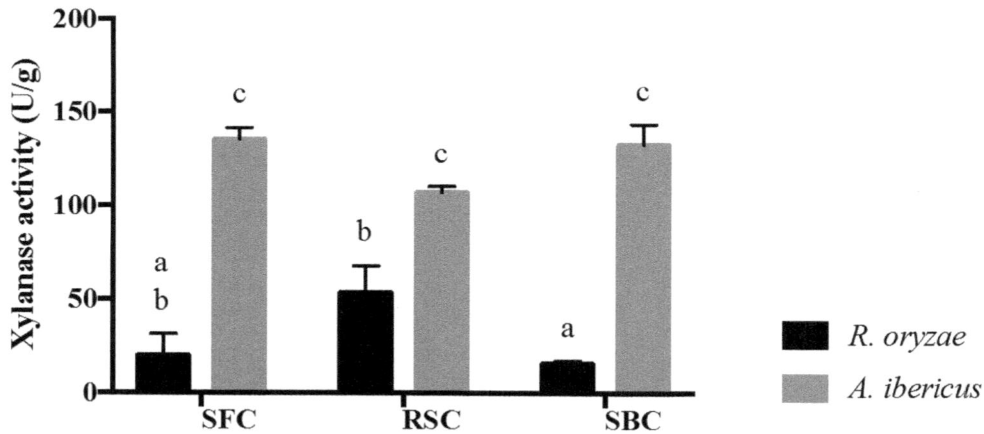

Figure 2. Xylanase activity of crude extracts obtained after SSF. Results represent the average of two independent fermentation experiments and error bars represent standard deviation. Bars with equal letters are not significantly different (Tukey test; $P < 0.05$). SFC, sunflower cake; RSC, rapeseed cake; SBC, soybean cake.

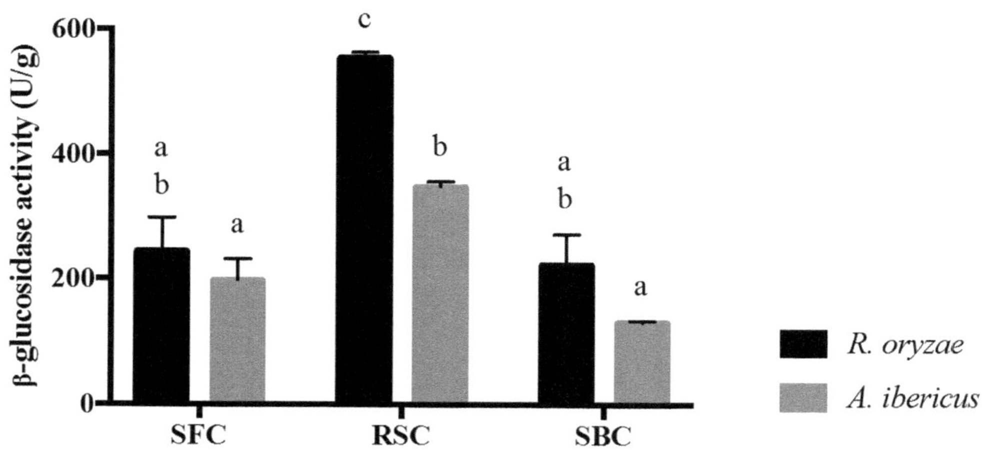

Figure 3. β-glucosidase activity of crude extracts obtained after SSF. Results represent the average of two independent fermentation experiments and error bars represent standard deviation. Bars with equal letters are not significantly different (Tukey test; $P < 0.05$). SFC, sunflower cake; RSC, rapeseed cake; SBC, soybean cake.

3.2 *Total phenols and antioxidant activity*

Total phenols content of extracts from the residues studied are presented in Figure 4a. Generally, the phenolic content of the analyzed extracts increased during the fermentative process, except for SFC fermented by *A. ibericus*. The higher content of total phenols was achieved in fermentations using *R. oryzae* but no statistically significant differences were observed among residues. These results showed that during SSF there was an increase of total phenols, which may be related with carbohydrate-hydrolyzing enzymes. As stated before, the higher activity of β-glucosidase in fermented extracts was achieved with *R. oryzae*, using RSC as substrate. These enzymes have been reported to be involved in mobilization of phenolic compounds during SSF (Vattem and Shetty, 2002; Bhanja *et al.*, 2009).

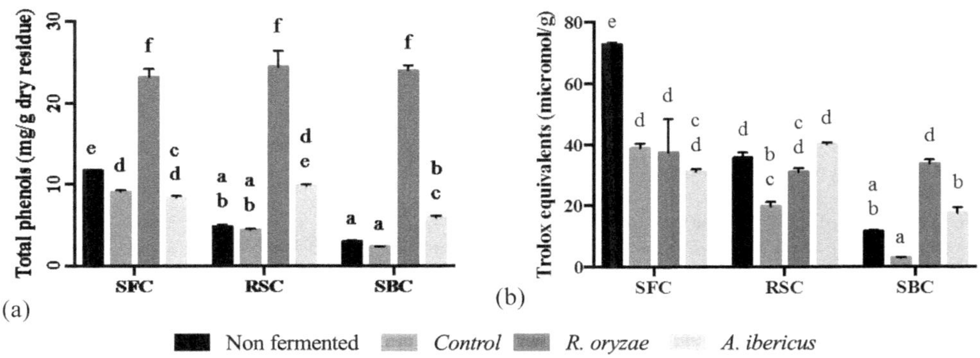

Figure 4. a) Total phenols content of initial substrate (non-fermented), control (solid after sterilization) and residues after SSF. b) Antioxidant capacity of initial substrate (non-fermented), control (solid after sterilization) and residues after SSF. Bars with equal letters are not significantly different (Tukey test; $P < 0.05$). SFC, sunflower cake; RSC, rapeseed cake; SBC, soybean cake.

Antioxidant capacity of non-fermented and fermented extracts is depicted in Figure 4b. The higher antioxidant capacity was obtained in non-fermented extract of SFC. In the case of RSC and SBC it was possible to observe an increase of antioxidant capacity after SSF when compared to control.

Extracts from *A. ibericus* contain smaller amounts of phenolic compounds rather than extracts from *R. oryzae*. Despite this fact, it was possible to observe that regarding to SFC and RSC, extracts from SSF by different fungi do not show statistically significant differences. In the case of SBC there was a clear difference between antioxidant capacity of fermented extracts.

4 CONCLUSIONS

SFC, RSC and SBC proved to be suitable substrates for SSF. During SSF lignocellulolytic enzymes were produced that can be used as feed additives. *A. ibericus* induced a higher activity of cellulases using SFC and SBC as substrate, and also the higher activity of xylanases was achieved with *A. ibericus*. β-glucosidase activity was highest when RSC was fermented with *R. oryzae*. These fungi enhanced the extraction of phenolic compounds in every substrate and this may be related to the higher activity of β-glucosidase.

SSF by *A. ibericus* improved the antioxidant capacity of RSC and SBC extracts. Phenolic compounds present in extracts from SSF by *R. oryzae* did not show the same antioxidant capacity as phenolic compounds present in extracts from SSF by *A. ibericus*.

AKNOWLEDGEMENTS

Daniel Sousa was supported by FCT-Portuguese Foundation for Science and Technology (PD/BD/135328/2017), under the Doctoral Programme "Agricultural Production Chains – from fork to farm" (PD/00122/2012), under the project UID/AGR/04033/2019. José Manuel Salgado was supported by grant CEB/N2020 – INV/01/2016 from Project "BIOTECNORTE – Underpinning Biotechnology to foster the north of Portugal bioeconomy" (NORTE-01-0145-FEDER-000004). This study was supported by the Portuguese Foundation for Science and Technology (FCT) under the scope of the strategic funding of UID/BIO/04469/2019 unit and BioTecNorte operation (NORTE-01-0145-FEDER-000004) funded by the European Regional Development Fund under the scope of Norte2020 – Programa Operacional Regional do Norte.

REFERENCES

Ajila, C. M., Brar, S. K., Verma, M., Tyagi, R. D., Godbout, S. & Veléro, J. R. 2012. Bio-processing of agro-byproducts to animal feed., *Critical Reviews in Biotechnology*, 32(January), pp. 1–19. doi: 10.3109/07388551.2012.659172.

Bhanja, T., Kumari, A. & Banerjee, R. 2009. Enrichment of phenolics and free radical scavenging property of wheat koji prepared with two filamentous fungi., *Bioresource Technology*. Elsevier, 100(11), pp. 2861–2866.

Food and Agriculture Organization 2009. *How to Feed the World in 2050.*

Horita, M., Kitamoto, H., Kawaide, T., Tachibana, Y. & Shinozaki, Y. 2015. On-farm solid state simultaneous saccharification and fermentation of whole crop forage rice in wrapped round bale for ethanol production., *Biotechnology for biofuels*, 8(1), p. 9. doi: 10.1186/s13068-014-0192-9.

Leite, P., Salgado, J. M., Venâncio, A., Domínguez, J. M. & Belo, I. 2016. Ultrasounds pretreatment of olive pomace to improve xylanase and cellulase production by solid-state fermentation., *Bioresource technology*. Elsevier, 214, pp. 737–746.

Lio, J. & Wang, T. 2012. Solid-state fermentation of soybean and corn processing coproducts for potential feed improvement., *Journal of Agricultural and Food Chemistry*, 60(31), pp. 7702–7709. doi: 10.1021/jf301674u.

Lomascolo, A., Uzan-Boukhris, E., Sigoillot, J. C. & Fine, F. 2012. Rapeseed and sunflower meal: a review on biotechnology status and challenges., *Applied microbiology and biotechnology*. Springer, 95(5), pp. 1105–1114.

Matthäus, B. 2002. Antioxidant activity of extracts obtained from residues of different oilseeds., *Journal of Agricultural and Food Chemistry*. ACS Publications, 50(12), pp. 3444–3452.

Pandey, A. 1992. Recent process developments in solid-state fermentation., *Process Biochemistry*, 27(2), pp. 109–117. doi: 10.1016/0032-9592(92)80017-W.

Pandey, A., Soccol, C. R., Nigam, P. & Soccol, V. T. 2000. Biotechnological potential of agro-industrial residues. I: Sugarcane bagasse., *Bioresource Technology*, 74(1), pp. 69–80. doi: 10.1016/S0960-8524(99)00142-X.

Pandey, A., Soccol, C. R. & Mitchell, D. 2000. New developments in solid state fermentation: I-bioprocesses and products., *Process biochemistry*. Elsevier, 35(10), pp. 1153–1169.

Schmidt, S. & Pokorný, J., 2005. Potential application of oilseeds as sources of antioxidants for food lipids–a review., *Czech J. Food Sci*, 23(3), pp. 93–102.

Sunil, L., Prakruthi, A., Prasanth Kumar, P. K. & Gopala Krishna, A. G. 2016. Development of Health Foods from Oilseed Cakes., *Journal of Food Processing and Technology*. OMICS International, 7(11), pp. 1–6.

Vattem, D. A. & Shetty, K. 2002. Solid-state production of phenolic antioxidants from cranberry pomace by Rhizopus oligosporus, *Food Biotechnology*. Taylor & Francis, 16(3), pp. 189–210.

Zero Waste Europe, 2011. Available at: http://zwia.org/ (Accessed: 2 February 2019).

Wastes: Solutions, Treatments and Opportunities III – Vilarinho et al. (Eds)

Mechanical damage of geotextiles caused by recycled C&DW and other aggregates

D.M. Carlos, J.R. Carneiro & M.L. Lopes
CONSTRUCT-GEO, Faculty of Engineering, University of Porto, Porto, Portugal

ABSTRACT: The use of recycled materials in construction contributes significantly to the protection of the environment and to the reduction of waste. This work evaluates the effect of a recycled aggregate (construction and demolition waste) and, for comparison, a natural aggregate (gravel 14/20) and a synthetic aggregate (*corundum*) on the mechanical damage under repeated loading suffered by two nonwoven geotextiles with different masses per unit area. The damage occurred in the geotextiles (in the laboratory mechanical damage tests) was evaluated by visual inspection and by monitoring changes in their short-term tensile and puncture behaviours. Results showed that the recycled aggregate induced lower damage to the geotextiles (lower deterioration of their mechanical properties) than gravel 14/20 or *corundum*. This way, and in terms of mechanical damage induced to geotextiles, there are good perspectives for the use of the recycled aggregate in civil engineering applications.

1 INTRODUCTION

The use of recycled construction and demolition waste (C&DW) as an alternative to construction materials obtained from natural sources has been increasing over the last years. After proper treatment, these wastes can be recycled into aggregates that can be used in different construction applications (examples can be found in Vieira & Pereira (2015)). In some applications, the recycled aggregates may be in contact with geosynthetics (polymeric materials applied in many civil engineering structures). Geosynthetics and aggregates can be used together, for example, in embankments, base layers for transportation infrastructures, retaining walls, waste landfills or drainage structures.

The installation on site of the geosynthetics can lead to changes in their physical, mechanical and/or hydraulic properties. Installation damage is often induced by handling the geosynthetics and by the placement and compaction of aggregates over them. In some cases, the geosynthetics can be submitted to higher stresses during installation than during service (Shukla & Yin 2006). The most common effects of installation damage include cuts in components, holes, tears, punctures and abrasion.

The damage occurred in the geosynthetics during the installation process can be evaluated by laboratory tests (that try to reproduce installation damage) or by field tests (installation under real conditions). Damage assessment is often performed by monitoring changes in the mechanical, hydraulic and/or interface properties of the geosynthetics (Carneiro et al. 2013, Dias et al. 2017). For inducing mechanical damage on geosynthetics, the European Committee for Standardization developed a standard procedure (EN ISO 10722 (2007)). This procedure has been used by many authors to evaluate the damage that occurs during the installation of geosynthetics, while others tried to correlate it with field installation conditions.

The available studies about the use of recycled aggregates in contact with geosynthetics are relatively few and mostly related with interface properties (Vieira & Pereira 2016). The mechanical damage of geosynthetics induced by recycled aggregates has also been assessed (Carneiro et al. 2018). However, there are no comparisons available between the effects of recycled and natural aggregates on the geosynthetics. This is a relevant issue when considering the use of recycled aggregates as filling materials in contact with geosynthetics.

In this work, two geotextiles were submitted to mechanical damage under repeated loading tests (for simplification, hereinafter MD tests) with different confinement materials: a recycled aggregate (C&DW) and, for comparison, a natural aggregate (gravel 14/20) and a synthetic aggregate (*corundum* – aggregate used in the procedure described in EN ISO 10722 (2007)). The main aims of the work included: (1) evaluation of the effects of the MD tests with the recycled aggregate in the tensile and puncture properties of the geotextiles, (2) comparison of the damage induced by the recycled and natural aggregates and (3) evaluation if the recycled aggregate can be used as filling material in contact with geotextiles (in terms of mechanical damage).

2 MATERIALS AND METHODS

2.1 *Geotextiles*

This work used two nonwoven needle-punched polypropylene (PP) geotextiles (designated by NW100 and NW300) with different masses per unit area. These geotextiles can be used to perform functions like filtration or separation in many engineering applications. The main characteristics of the geotextiles can be found in Table 1. The sampling process was carried out according to the guidelines of EN ISO 9862 (2005).

2.2 *Mechanical damage tests*

The MD tests were carried out according to EN ISO 10722 (2007) (exception for the use of different aggregates other than *corundum*) in a prototype equipment developed at the Faculty of Engineering of the University of Porto. The equipment was formed by a test container (rigid metal box where the geotextiles and aggregates were placed), a loading plate and a compression machine (full description of the equipment in Lopes & Lopes 2003). The geotextiles (specimens with a width of 250 mm and a length of 500 mm) were placed between two layers of aggregate (each with a height of 75 mm) and subjected to cyclic loading between 5.0 ± 0.5 kPa (minimum) and 500 ± 10 kPa (maximum) at a frequency of 1 Hz for 200 loading cycles.

2.3 *Aggregates*

The MD tests were performed using a recycled aggregate (C&DW formed by about 25% ceramics, 30% concrete and 45% natural soil/aggregates), gravel 14/20 (natural granite aggregate) and *corundum* (synthetic aggregate of aluminium oxide) (Fig. 1). Table 2 presents some parameters (D_{10} – effective particle size, D_{50} – particle size corresponding to 50% passing and D_{Max} – maximum particle size) related to the particle size distributions of the aggregates illustrated in Figure 1d (determined according to EN 933-1 (2012)).

Table 1. Main characteristics of the geotextiles.

Geotextile	Type	Polymer	μ_A^* (g.m^{-2})	t^{**} (mm)
NW100	Nonwoven	PP	116 (±7)	0.97 (±0.06)
NW300	Nonwoven	PP	325 (±11)	3.83 (±0.09)

(95% confidence intervals in brackets calculated according to Montgomery & Runger (2010))
*Mass per unit area (determined according to EN ISO 9864 (2005)).
**Thickness (determined according to EN ISO 9863-1 (2016)).

2.4 *Evaluation of the damage occurred in the geotextiles*

The damage suffered by the geotextiles (in the MD tests) was evaluated qualitatively by visual inspection and quantitatively by tensile tests (according to EN ISO 10319 (2015)) and static puncture tests (according to EN ISO 12236 (2006)). These tests were carried out on a *Lloyd Instruments* equipment (model LR 10K Plus). The experimental conditions used in the tensile and puncture tests can be found in Table 3.

The mechanical properties determined in the tensile tests (mean values of 5 specimens in the machine direction of production) included tensile strength (T, in kN.m^{-1}) and elongation at maximum load (E_{ML}, in %). Puncture strength (maximum push-through force) (F_P, in kN) and push-through displacement at maximum force (h_P, in mm) were the properties determined in the puncture tests (also mean values of 5 specimens). The properties of the geotextiles are presented with 95% confidence intervals calculated according to Montgomery & Runger (2010). The changes occurred in tensile and puncture strengths are also presented as retained strengths (respectively, $T_{Retained}$ and $F_{P\ Retained}$, in %), obtained by dividing the strength (tensile or puncture) of the damaged samples by the respective resistance of the reference samples (undamaged).

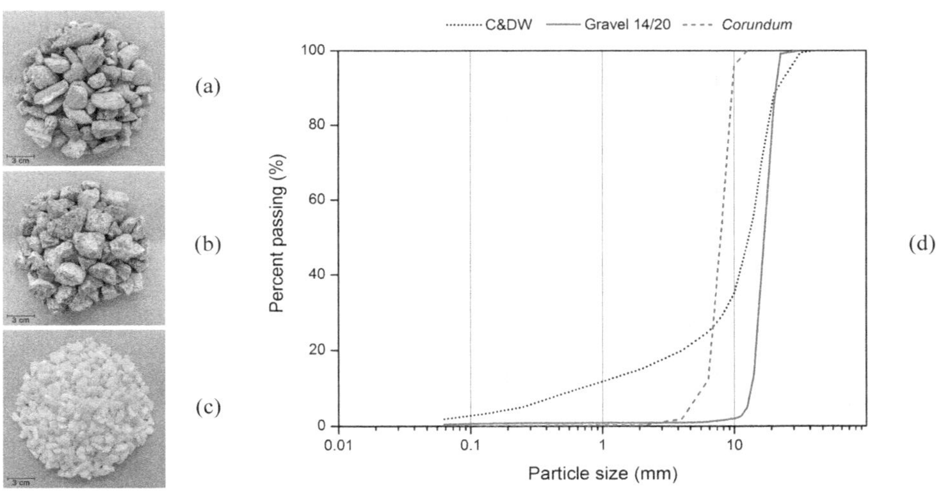

Figure 1. Aggregates used in the MD tests: (a) C&DW; (b) gravel 14/20; (c) *corundum*; (d) particle size distributions.

Table 2. Characterization of the particle size distribution of the aggregates.

Aggregate %	<0.063 mm	D_{10} (mm)	D_{50} (mm)	D_{Max} (mm)
C&DW	1.89	0.71	12.85	22.4
Gravel 14/20	0.53	13.37	16.92	22.4
Corundum	0.06	5.77	7.91	10.0

Table 3. Experimental conditions of the tensile and puncture tests.

Test	Standard	Specimens type	Specimens size	N	Test speed
Tensile	EN ISO 10319 (2015)	Rectangular	100 mm* × 200 mm	5	20 mm.min^{-1}
Puncture	EN ISO 12236 (2006)	Circular	150 mm**	5	50 mm.min^{-1}

(N – number of specimens; *length between grips; **diameter between grips)

3 RESULTS AND DISCUSSION

3.1 *Geotextile NW100*

The MD tests induced different defects in geotextile NW100, namely cuts in fibres, formation of holes and punctures. The occurrence of cuts in fibres and formation of holes were more significant after the MD tests with *corundum* (holes in large amount, but small in size, about 2–3 mm). The MD tests with the recycled aggregate and gravel 14/20 caused less holes compared to the tests with *corundum*, but with larger dimensions (about 5–6 mm). The defects mostly found in geotextile NW100 after these MD tests were punctures.

The tensile and puncture properties of geotextile NW100, before and after the MD tests, can be found in Table 4. The tensile strength of geotextile NW100 suffered considerable reductions after the MD tests, reflecting the defects present in the nonwoven structure. The reduction occurred in tensile strength was higher after the MD tests with *corundum* ($T_{Retained}$ of 43.7%) than after the MD tests with the recycled aggregate or gravel 14/20 ($T_{Retained}$ of 59.3% and 54.9%, respectively). As for tensile strength, the reduction observed in puncture strength was also more pronounced after the MD tests with *corundum* ($F_{P\ Retained}$ of 34.3%) and less significant after the MD tests with the recycled aggregate ($F_{P\ Retained}$ of 67.9%). With exception to the MD tests with *corundum*, the losses occurred in tensile strength were higher than those observed in puncture strength. The elongation at maximum load and push-through displacement at maximum force of geotextile NW100 also suffered reductions after the MD tests (once again, more relevant reductions after the MD tests with *corundum*).

3.2 *Geotextile NW300*

Similarly to what happened for geotextile NW100, the MD tests also provoked cuts in fibres, formation of holes and punctures in geotextile NW300. Yet, the defects were less abundant and less pronounced when compared to geotextile NW100.

The tensile and puncture strengths of geotextile NW300 also suffered relevant reductions after the MD tests (Table 5). However, and compared to geotextile NW100, the reductions were less significant. For instance, the MD tests with the recycled aggregate induced losses in tensile and puncture strengths of, respectively, 10.8% and 15.0% (in geotextile NW100, losses of 40.7% and 32.2%, respectively). This can be explained by the higher mass per unit area of geotextile NW300, which provided a better resistance against the damaging actions.

Comparing the different aggregates, *corundum* was again the most damaging, while the recycled aggregate induced the lowest reductions in the tensile and puncture strengths of geotextile NW300. The reductions occurred in puncture strength were slightly higher than those observed in tensile strength. The elongation at maximum load and push-through displacement at maximum force of geotextile NW300 also decreased after the MD tests.

3.3 *Effect of the type of aggregate on the mechanical damage of the geotextiles*

The deterioration of the tensile and puncture behaviours of geotextiles NW100 and NW300 can be related with the characteristics of the aggregates used in the MD tests. The comparison between the retained tensile and puncture strengths of the geotextiles can be observed in Figure 2. The highest

Table 4. Tensile and puncture properties of geotextile NW100 before and after the MD tests.

Mechanical damage test	T ($kN.m^{-1}$)	E_{ML} (%)	F_P (kN)	h_P (mm)
Undamaged	8.61 (±0.71)	53.7 (±3.5)	1.40 (±0.12)	46.1 (± 0.9)
MD with C&DW	5.11 (±0.40)	34.1 (±2.9)	0.95 (±0.10)	41.7 (±1.1)
MD with gravel 14/20	4.73 (±0.37)	32.0 (±1.4)	0.93 (±0.11)	41.5 (±1.9)
MD with *corundum*	3.76 (±0.42)	30.1 (±1.2)	0.48 (±0.06)	35.0 (±3.4)

Table 5. Tensile and puncture properties of geotextile NW300 before and after the MD tests.

Mechanical damage test	T (kN.m^{-1})	E_{ML} (%)	F_P (kN)	h_P (mm)
Undamaged	23.44 (±1.33)	138.4 (±13.8)	4.66 (±0.15)	65.6 (±4.1)
MD with C&DW	20.90 (±1.61)	100.2 (±5.5)	3.96 (±0.32)	60.5 (±1.2)
MD with gravel 14/20	17.40 (±1.51)	87.6 (±6.7)	3.06 (±0.31)	53.0 (±5.6)
MD with *corundum*	16.36 (±1.79)	81.0 (±7.6)	3.03 (±0.22)	54.4 (±2.4)

(95% confidence intervals in brackets)

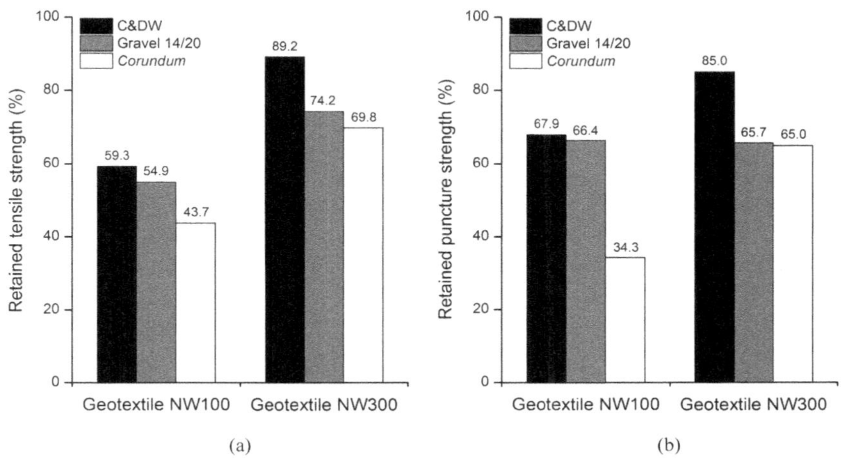

Figure 2. Comparison of the mechanical resistances of geotextiles NW100 and NW300 after the MD tests: (a) retained tensile strength; (b) retained puncture strength.

reductions in tensile and puncture strength occurred after the MD tests with *corundum*. Despite having lower D_{50} and D_{Max} than the recycled aggregate or gravel 14/20, *corundum* was formed by rough and angular particles with a high abrasive effect. This explains why *corundum* was the aggregate that provoked more cuts and holes in the nonwoven structures (and, consequently, a higher deterioration of the mechanical behaviour of the geotextiles). Compared to *corundum*, the damage induced by gravel 14/20 was less pronounced. Although gravel 14/20 was constituted by larger particles than *corundum* (as can be seen in Figure 1), they were less angular, rough and abrasive.

Finally, the recycled aggregate was the least damaging to the geotextiles (lower deterioration of tensile and puncture behaviours). This aggregate had also large particles, but in comparison with *corundum* and gravel 14/20 (uniformly graded aggregates), it had a more extended particle size distribution (Fig. 1d), which resulted in a flatter and smoother contacting surface with the geotextiles. Indeed, the large particles were surrounded by smaller ones, reducing the voids and creating a larger contact area with the geotextiles, thereby allowing a better distribution of the applied loads (in the MD tests) and minimizing the occurrence of damage. In addition to the particle size distribution, other characterisation tests (such as LA abrasion tests or determination of shape indexes) must be carried out for the recycled aggregate. The results of those tests may provide additional useful information for explaining the lower deterioration caused by the recycled aggregate to the geotextiles when compared to the other aggregates.

4 CONCLUSIONS

The MD tests with different confinement materials (recycled C&DW, natural and synthetic aggregates) induced relevant changes in the tensile and puncture properties of two nonwoven geotextiles.

These changes depended on the mass per unit area of the geotextiles and on the characteristics of the confinement materials.

The geotextile with higher mass per unit area had a better resistance (lower deterioration of tensile and puncture properties) against mechanical damage. The damage provoked by the recycled aggregate on the geotextiles was less significant than that caused by gravel 14/20 (natural aggregate) or *corundum* (synthetic aggregate used in the procedure described in EN ISO 10722 (2007)). In terms of mechanical damage induced to geotextiles, this opens good perspectives for using the recycled aggregate in geotechnical works as a viable alternative to filling materials obtained from natural sources, thereby contributing for a more sustainable construction. However, before application, further studies are needed to evaluate the physical, mechanical and hydraulic properties of the recycled aggregate. In addition, its environmental impact is also a relevant issue to be addressed.

ACKNOWLEDGEMENTS

This work was financially supported by: (1) project POCI-01-0145-FEDER-028862, funded by FEDER funds through COMPETE 2020 – "Programa Operacional Competitividade e Internacionalização" (POCI) and by national funds (PIDDAC) through FCT/MCTES; (2) UID/ECI/04708/2019 – CONSTRUCT – "Instituto de I&D em Estruturas e Construções" funded by national funds through FCT/MCTES (PIDDAC).

REFERENCES

Carneiro, J.R., Morais, L.M., Moreira, S.P. & Lopes, M.L. 2013. Evaluation of the damages occurred during the installation of non-woven geotextiles. *Materials Science Forum* 730–732: 439–444.

Carneiro, J.R., Lopes, M.L. & da Silva, A. 2018. *Mechanical damage of a nonwoven geotextile induced by recycled aggregates*, Wastes-Solutions, Treatment and Opportunities II, CRC Press/ Balkema, Leiden, Netherlands.

Dias, M., Carneiro, J.R. & Lopes, M.L. 2017. Resistance of nonwoven geotextiles against mechanical damage and abrasion. *Ciência e Tecnologia dos Materiais* 29 (1): 177–181.

EN 933-1. 2012. Tests for geometrical properties of aggregates – Part 1: Determination of particle size distribution – Sieving method. Brussels, Belgium: CEN.

EN ISO 9862. 2005. Geosynthetics. Sampling and preparation of test specimens. Brussels, Belgium: CEN.

EN ISO 9863-1. 2016. Geosynthetics. Determination of thickness at specified pressures. Part 1: single layers. Brussels, Belgium: CEN.

EN ISO 9864. 2005. Geosynthetics. Test method for the determination of mass per unit area of geotextiles and geotextile related products. Brussels, Belgium: CEN.

EN ISO 10319. 2015. Geosynthetics. Wide-width tensile test. Brussels, Belgium: CEN.

EN ISO 10722. 2007. Geosynthetics. Index test procedure for the evaluation of mechanical damage under repeated loading. Damage caused by granular material. Brussels, Belgium: CEN.

EN ISO 12236. 2006. Geosynthetics. Static puncture test (CBR test). Brussels, Belgium: CEN.

Lopes, M.P. & Lopes, M.L. 2003. Equipment to carry out laboratory damage during installation tests on geosynthetics. *Geotecnia (Journal of the Portuguese Geotechnical Society)* 98: 7–24 (in Portuguese).

Montgomery, D.C. & Runger, G.C. 2010. *Applied Statistics and Probability for Engineers*. New York: John Wiley & Sons, Inc.

Shukla, S.K. & Yin, J.-H. 2006. *Fundaments of Geosynthetics Engineering*. Taylor & Francis/ Balkema, Leiden, Netherlands.

Vieira, C.S. & Pereira, P.M. 2015. Use of recycled construction and demolition materials in geotechnical applications: A review. *Resources, Conservation and Recycling* 103: 192–204.

Vieira, C.S. & Pereira, P.M. 2016. Interface shear properties of geosynthetics and construction and demolition waste from large-scale direct shear tests, *Geosynthetics International* 23(1): 62–70.

Wastes: Solutions, Treatments and Opportunities III – Vilarinho et al. (Eds)
© 2020 Taylor & Francis Group, London, ISBN 978-0-367-25777-4

Influence of heavy metals fraction on quality of composts

M.E. Silva & I. Brás
Departamento de Ambiente, Escola Superior de Tecnologia e Gestão and CI&DETS, Instituto Politécnico de Viseu, Viseu, Portugal

A.C. Cunha-Queda
LEAF-Linking Landscape, Environment, Agriculture and Food, Instituto Superior de Agronomia, Universidade de Lisboa, Lisboa, Portugal

O.C. Nunes
LEPABE – Laboratory for Process Engineering, Environment, Biotechnology and Energy, Faculty of Engineering, University of Porto, Porto, Portugal

ABSTRACT: In this study the fractionation of the heavy metals in four commercial composts was assessed. The heavy metals mobility that most influenced the chemical composition of humic-like substances (HS-like) and the physicochemical, maturity and stability properties of the composts were identified. The heavy metals fractionation was performed following the modified sequential BCR (Community Bureau of Reference) extraction, which permitted to quantify the elemental concentration in the exchangeable/bioavailable, reducible, oxidisable and residual fractions. Cu, Ni, Cr and Pb were most abundant in the immobile phases, while Zn showed no dominant chemical phase. The HS-like chemical composition was influenced negatively by the residual fraction of Cu, Pb, and mainly Cr. The type or mobility of the heavy metals did not show any influence on composts maturity. However, mobile forms of Zn and Pb were correlated with high moisture content and low stability, respectively as well as with the increase of compost phytotoxicity.

1 INTRODUCTION

Composting is one of the methods mostly used for organic solid wastes recycling, yielding the compost. Compost is a mixture of inorganic and organic matter and the later can be divided into two classes, non-humic and humic substances (HS). Composts are, thus, a source of humic-like substances (HS-like), contributing to increase soil fertility when applied in soils (Smidt et al., 2008). However, both non-humic and humic substances of composts may interact with different pollutants, such as heavy metals, which may cause noxious effects in the environment when composts are used for land application as soil amendments. The heavy metals distribution in composts is influenced by their release during organic matter (OM) mineralization occurring during the composting process, metal solubilization by the decrease of pH, metal biosorption by the microbial biomass or metal complexation with the newly formed humic substances (Haroun et al., 2009; Liu et al., 2007). The concentration and mobility of heavy metals depends also on the total metal content in the raw material (Liu et al., 2007), and on the sulfur content, since the heavy metals may form sulfides (Fuentes et al., 2004). Chemical sequential extraction techniques have been used to remove heavy metals bound into different operationally defined phases of composts (Miaomiao et al., 2009; Singh & Kalamdhad, 2012). This process enables the assessment of their bioavailability and, thus, suitability for land application (Cai et al., 2007). On the other hand, heavy metals are known to be toxic to most organisms when present in excessive concentrations (Adejumo et al., 2018). This may cause a negative influence on the maturation/stabilization of OM, decreasing the humification degree and the stability parameters of the compost. Therefore, it is important to understand which are the heavy metals fraction (exchangeable, reducible, oxidisable and residual) that influence the maturation/stabilization of the composts. In the present study the heavy metals fractionation in four commercial composts was assessed. The mobility of the heavy metals that influenced most the chemical composition of HS-like substances and the physicochemical, maturity and stability properties of the composts were identified.

2 MATERIALS AND METHODS

2.1 *Total content and fractionation of heavy metal on composts*

Four commercial composts, two from non-separated municipal solid waste (MSW1, MSW2), one from source-separated municipal solid waste (MSW3) and one produced from poultry litter (P1) were used to carry out this study. Heavy metals content (Zn, Cu, Ni, Cd, Cr and Pb) were analysed by flame atomic absorption spectrophotometry (Perkin Elmer model AAnalyst 300) after sample digestion with "aqua regia", followed by filtration (EN 13650, 2001). Metal fractionation was performed using the modified BCR (Community Bureau of Reference) sequential extraction procedure, which has been applied to a variety of matrices, including composts (Bogusz & Oleszczuk 2018; Fuentes et al., 2004; Jamali et al., 2009; Rauret et al., 2000) and municipal solid waste (Farrell & Jones, 2009). The extraction procedure was performed on aliquots of 1 g air-dried composts samples. The exchangeable, water soluble and bound to carbonates fraction, the reducible fraction (fraction associated with Fe and Mn oxides), the oxidisable fraction (fraction bound to organic matter) and residual fraction was studied. The data obtained from the quantification of the heavy metals in each fraction (exchangeable, reducible, oxidisable and residual) constituted the heavy metals profile.

2.2 *Physical, chemical, maturity and stability properties of the composts*

Silva et al. (2013) characterized the composts used in this study previously. In general, P1 and MSW2 showed the lowest and highest organic matter content, respectively. MSW2 showed the lowest total N content, while the highest value was registered for MSW3. The highest and lowest stability degrees were observed, respectively, in MSW3 (stability degree V) and MSW2 (stability degree I). In opposition, the highest and lowest respiration activity values were registered in MSW2 and MSW3, respectively. MSW1 registered the highest humic acid content, in opposition to P1. MSW1 showed highest values of humification indices, while, MSW2 presented the lowest values.

2.3 *Statistical analysis*

Data were subjected to one-way analysis of variance (ANOVA) and the Newman-Keuls test was used to separate the means. All statistics analyses were carried out using Statistica 6.0 software. Canonical Correspondence Analyses (CCA) were carried out in order to assess the influence of heavy metals profiles of the analysed composts on their HS-like chemical composition, and physicochemical, maturity and stability properties (previous described by Silva et al., 2013). These analyses were performed with the software package CANOCO version 4.5. The significance of the relationships between the HS-like chemical composition (principal matrix – species matrix) and the heavy metals profiles (second matrix – environmental matrix); and the compost physicochemical, maturity and stability properties (principal matrix – species matrix) and the heavy metals profiles (second matrix – environmental matrix) was tested by Monte Carlo permutations test ($n = 499$). Explanatory variables included in CCA analysis were selected by manual forward selection including the Monte Carlo permutations test.

3 RESULTS AND DISCUSSION

Metals in the exchangeable and reducible fractions of composts or soil are relatively labile and potentially bioavailable, and consequently potentially toxic to the native microbiota or plants (Bogusz & Oleszczuk 2018). On the other hand, metals in the oxidisable fractions (organically bound) and residual fractions are relatively immobile and may not be readily bioavailable (Fuentes et al., 2004), and thus, are less toxic. In this study, the potential of mobility of the heavy metals as well as their possible influence on the HS-like chemical composition, stability and maturity of four commercial composts were assessed.

3.1 *Distribution of heavy metals fractions on composts*

In general, MSW1 and MSW3 composts showed the highest and lowest content in heavy metals, respectively. Among the heavy metals analysed, Zn and Cd were, respectively, those detected in highest (132–861 mg kg^{-1} d.m.) and lowest (0.8–3.0 mg kg^{-1} d.m.) concentrations (Table 1). Zinc, the heavy metal with the highest total concentration (132–861 mg kg^{-1} d.m.) among the analysed composts, was majorly distributed in the exchangeable (9–23%) and reducible (36–50%) forms for all composts, a result also reported before (Cai et al., 2007; Fuentes et al., 2004). In opposition, more than 95% of the total Cu was associated with the oxidisable and residual forms, in all the composts, suggesting that it was associated with strong organic ligands (Cai et al., 2007), and probably occluded in minerals like quartz, and/or feldspars. The distribution of Pb, Cr and Ni among the fractions was related with their concentration in the composts. Cd concentration did not exceed 3 mg kg^{-1} d.m., and was not detectable by the sequential extraction process. In composts MSW3 and P1, where the concentration of these heavy metals was lower than in composts MSW1 and MSW2, all the extractable Ni, Cr and Pb were found in the residual fraction (100%). In composts MSW1 and MSW2, these heavy metals were mainly associated with the residual (40–73%) and oxidisable (15–60%) fractions. Among the analysed composts, only MSW2 contained Cr in its reducible fraction (8%). Overall, the MSW2 compost showed the highest metals content in bioavailable fraction (present in the exchangeable and reducible fraction), which may represent a potential risk of soil contamination if it used as soil amendment.

3.2 *Assessment of the influence of heavy metals fraction on the hs-like chemical composition*

The correlation between the heavy metal content in each fraction on the distribution of the composts based on their HS-like profiles was assessed. The CCA analysis could explain 90.1% of the total variation among the HS-like profiles of the composts. As described before, the HS-like profile of compost MSW2 was the most distinct, being separated from all the others over axis 1. The highest content of residual Cr in this compost (Table 2) contributed to distinguish its HS-like profile ($p < 0.05$; species-environment correlations with axis 1 of 0.98) (Fig. 1).

The Cu content contributed to separate MSW1 from MSW3 and P1 over axis 2. Beside Cu, also Pb contributed to this differentiation (Fig. 1), but only those present in the residual fraction ($p < 0.05$; species-environment correlations with axis 2 of 0.97, both). Indeed, the residual Cu and Pb contents (103 and 168 mg kg^{-1} d.m., respectively) were highest in compost MSW1.

These results suggest that even at an immobile form, Cr may have had a negative impact on the microbiota and/or the chemical processes involved on the condensation of the organic matter, contributing to a HS-like profile with a higher content of alkylic compounds and polysaccharides in MSW2 than in other analysed composts (Silva et al., 2013). On the other hand, the influence of relatively immobile Cu and Pb, and Cr on the HS-like composition of, respectively, composts MSW1 and MSW2 may have also been due to co-extraction. Indeed, a higher concentration of these heavy metals, particularly of Cu, in the humic- and fulvic-like acid fractions of these composts than in MSW3 and P1 was previously observed (Silva et al., 2013).

Table 1. Heavy metals fractionation of compost.

Metal (mg kg^{-1} d.m.)	Fraction	MSW1	MSW2	MSW3	P1
Zn	Total*	861 ± 25a**	389 ± 4b	132 ± 1c	178 ± 7d
	Exchangeable	49.0 ± 5.6a	52.3 ± 5.2a	9.9 ± 0.3b	26.1 ± 0.8c
	Reducible	185 ± 11a	164 ± 13b	42.4 ± 0.9c	47.4 ± 2.6c
	Oxidisable	82.8 ± 1.8a	71.7 ± 18.1a	40.3 ± 4.0b	14.9 ± 0.3c
	Residual	142 ± 18a	40.8 ± 1.3b	20.3 ± 1.2b	23.7 ± 3.5b
Cu	Total*	749 ± 27a	141 ± 3b	41 ± 4c	77 ± 4d
	Exchangeable	<LD	<LD	<LD	<LD
	Reducible	12.0 ± 1.3	<LD	<LD	<LD
	Oxidisable	159 ± 17a	84.4 ± 2.7b	30.9 ± 1.4c	22.9 ± 1.1c
	Residual	103 ± 4a	25.3 ± 3.8b	8.3 ± 0.9c	9.9 ± 0.4c
Ni	Total*	74 ± 7a	31 ± 3b	18 ± 1c	15 ± 1c
	Exchangeable	<LD	<LD	<LD	<LD
	Reducible	<LD	<LD	<LD	<LD
	Oxidisable	11.3 ± 1.6a	8.0 ± 0.5b	<LD	<LD
	Residual	16.3 ± 0.8a	5.4 ± 0.4b	2.6 ± 0.3c	2.8 ± 0.4c
Cr	Total*	85 ± 6a	180 ± 3b	46 ± 4c	89 ± 1a
	Exchangeable	<LD	<LD	<LD	<LD
	Reducible	<LD	11.6 ± 1.8	<LD	<LD
	Oxidisable	21.8 ± 2.9a	62.5 ± 3.3b	<LD	<LD
	Residual	36.8 ± 3.9a	76.8 ± 1.7b	40.3 ± 1.7a	40.6 ± 2.6a
Cd	Total*	3.0 ± 0.1a	1.3 ± 0.1b	1.0 ± 0.1c	0.8 ± 0.0d
	Exchangeable	<LD	<LD	<LD	<LD
	Reducible	<LD	<LD	<LD	<LD
	Oxidisable	<LD	<LD	<LD	<LD
	Residual	<LD	<LD	<LD	<LD
Pb	Total*	243 ± 15a	143 ± 3b	36 ± 6c	21 ± 0c
	Exchangeable	<LD	<LD	<LD	<LD
	Reducible	29.6 ± 1.4a	28.3 ± 0.1b	<LD	<LD
	Oxidisable	34.2 ± 1.7a	32.2 ± 2.0a	<LD	<LD
	Residual	168 ± 8a	49.0 ± 2.7b	25.3 ± 0.1c	12.3 ± 0.6d

*data from Silva et al. (2013); ** values in a row followed by different letters are statistically different ($p < 0.05$); d.m. – dry matter; LD – limit of detection (limits of detection: Zn, 1.3 mg kg^{-1}; Cu, 2.0 mg kg^{-1}; Ni, 0.3 mg kg^{-1}; Cr, 2.3 mg kg^{-1}; Cd, 0.3 mg kg^{-1}; Pb, 3.7 mg kg^{-1}).

3.3 *Assessment of the influence of heavy metals fraction on physicochemical, maturity and stability properties of composts*

The influence of the different forms of the heavy metals on the physicochemical properties, maturity and stability of the composts was assessed (Fig. 2). The CCA analysis could explain 83.5% of the total variation among the properties of the composts.

The lowest HS-like content (HAC, FAC), humification indices (DP, HR, HI, PAH) as well as the highest pH and the NH_4^+-N content of compost P1 (Silva et al., 2013) permitted its separation from the other composts, mainly MSW1, over axis 1. Oxidisable Zn ($p < 0.05$; species-environment correlation with axis 1 of 0.81), which was present in a higher content in composts MSW than in P1 (Table 2), contributed most to this differentiation. The highest moisture content and lowest germination index of MSW2 were the main properties responsible for its separation from the other composts, over axis 2 (Fig. 2). The highest oxidisable Cr content in MSW2 contributed to its differentiation ($p < 0.05$; species-environment correlation with axis 2 of 0.92). These results suggest that none of the forms of the heavy metals influenced negatively the degree of humification of the composts, since the highest degree of humification was found in composts with highest heavy metals content (MSW). However, when only the mobile heavy metals were included in the analysis,

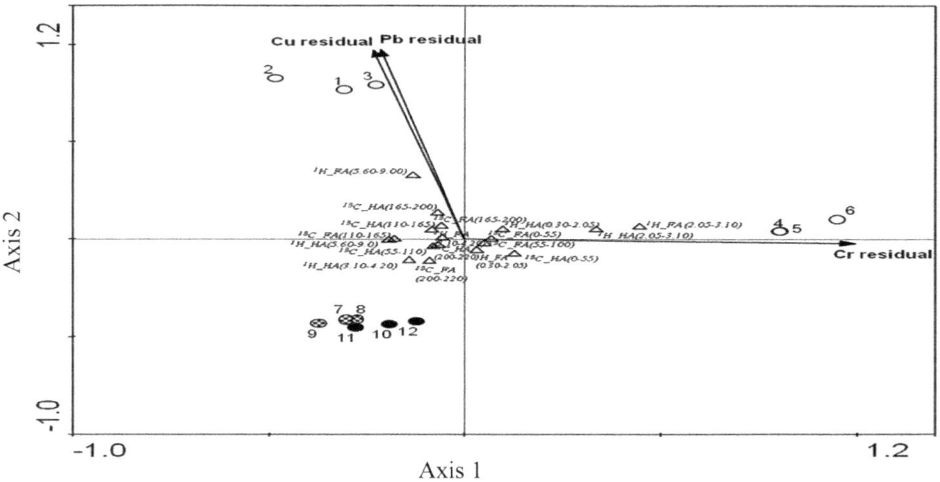

Figure 1. Canonical correspondence analysis biplot of HS-like chemical composition [Humic-like acid (HA) and Fulvic-like acid (FA) fractions] (67.6% could be explained by axis 1 and 22.5% by axis 2) in function of the heavy metals profiles of the respective composts (Cr residual, $p = 0.002$; Cu residual, $p = 0.002$ and Pb residual, $p = 0.008$). The species-environmental correlations for axis 1 and 2 were, respectively, 0.987 and 0.999. Only the variables significantly ($p < 0.05$) explaining the observed HS-like chemical composition variation are shown.

^{1}H-FA (0.30–2.05) – ^{1}H-NMR assignment to terminal CH_3 and CH_2, CH of methylene chains of FA fraction; ^{1}H-FA (2.05–3.10)– ^{1}H-NMR assignment to CH_3 and CH_2, CH proton α to aromatic or carboxyl groups of FA fraction; ^{1}H-FA (3.10–4.20) – ^{1}H-NMR assignment to protons on carbon α to oxygen, carbohydrates of FA fraction; ^{1}H-FA (5.60–9.00) – ^{1}H-NMR assignment to aromatic protons of FA fraction; ^{1}H-HA (0.30–2.05) – ^{1}H-NMR assignment to terminal CH_3 and CH_2, CH of methylene chains of HA fraction; ^{1}H-HA (2.05–3.10) – ^{1}H-NMR assignment to CH_3 and CH_2, CH proton α to aromatic or carboxyl groups of HA fraction; ^{1}H-HA (3.10–4.20) – ^{1}H-NMR assignment to protons on carbon α to oxygen, carbohydrates of HA fraction; ^{1}H-HA (5.60–9.00) – ^{1}H-NMR assignment to aromatic protons of HA fraction; ^{13}C-FA (0–55) – ^{13}C-NMR assignment to alkylic C of FA fraction; ^{13}C-FA (55–110) – ^{13}C-NMR assignment to O- or N-substituted alkylic C of FA fraction; ^{13}C-FA (110–165) – ^{13}C-NMR assignment to aromatic and olefinic C of FA fraction; ^{13}C-FA (165–200) – ^{13}C-NMR assignment to derivate of carboxylic C of FA fraction; ^{13}C-FA (200–220) – ^{13}C-NMR assignment to carbonyl C of FA; ^{13}C-HA (0–55) – ^{13}C-NMR assignment to alkylic C of HA fraction; ^{13}C-HA (55–110) – ^{13}C-NMR assignment to O- or N-substituted alkylic C of HA fraction; ^{13}C-HA (110–165) – ^{13}C-NMR assignment to aromatic and olefinic C of HA fraction; ^{13}C-HA (165–200) – ^{13}C-NMR assignment to derivate of carboxylic C of HA fraction; ^{13}C-HA (200–220) – ^{13}C-NMR assignment to carbonyl C of HA fraction.

although the distribution of the composts in the CCA biplot was maintained (Fig. 3), different parameters were the main responsible for their distribution. In this analysis, which explained 77.8% of the total variation among the properties of the composts, besides pH and the NH_4^+-N content, the humification ratio (HR) and the respiration activity after 4 days (AT_4) were the parameters that contributed most to distinguish P1 from composts MSW over axis 1. Although the moisture and the germination index (GI) were, again, the parameters that contributed most to differentiate MSW2 from the others. The higher content of exchangeable Zn and reducible Pb ($p < 0.05$; species-environment correlations with axes 1 and 2 of 0.23 and 0.65, and 0.60 and 0.59, respectively) in MSW2 and MSW1 than in the other composts, and mainly the presence of Cu ($p < 0.05$; species-environment correlation with axis 1 of 0.64) in the reducible fraction of MSW1 explained the variation found. High loads of reducible Pb and Cu were correlated with low stability (high AT_4 values) and high HR (HEC/TC) indices (high HEC content, but low HAC and high FAC contents). These results suggest that the presence of the mobile forms these heavy metals affected negatively

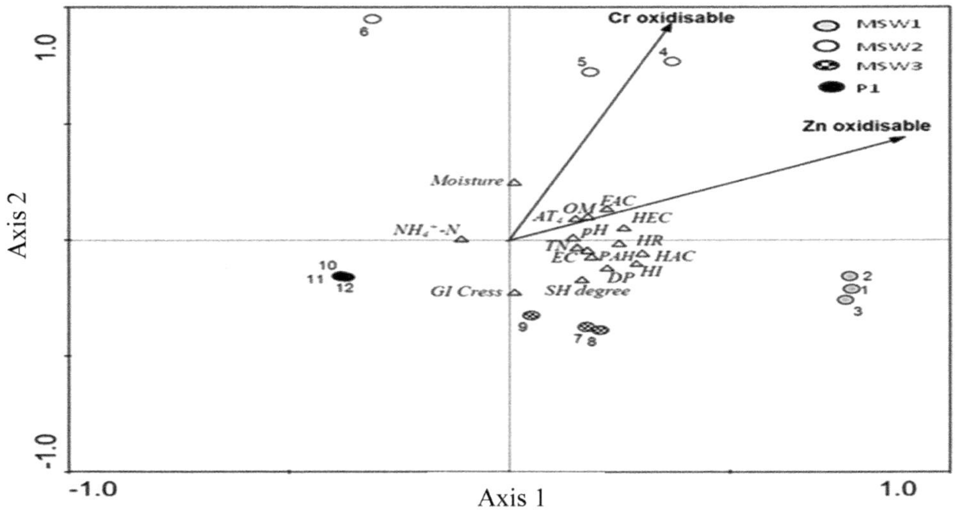

Figure 2. Canonical correspondence analysis biplot of physicochemical, maturity and stability properties (56.9% could be explained by axis 1 and 26.6% by axis 2) in function of the heavy metals profiles of the respective composts (Zn oxidisable, $p = 0.002$; Cr oxidisable, $p = 0.002$). The species-environmental correlations for axis 1 and 2 were, respectively, 0.901 and 0.993.

EC – Electrical Conductivity; OM – organic matter; GI – germination index; SH – self-heating; AT_4 – respiration activity after 4 days; HEC – carbon content of humic-like substances extract; HAC – carbon content of humic-like acids; FAC – carbon content of fulvic-like acids; DP – degree of polymerization (HAC/FAC); HI – humification index (HAC/TC); HR – humification ratio (HEC/TC); PAH – percentage of humic acid (HAC/HEC).

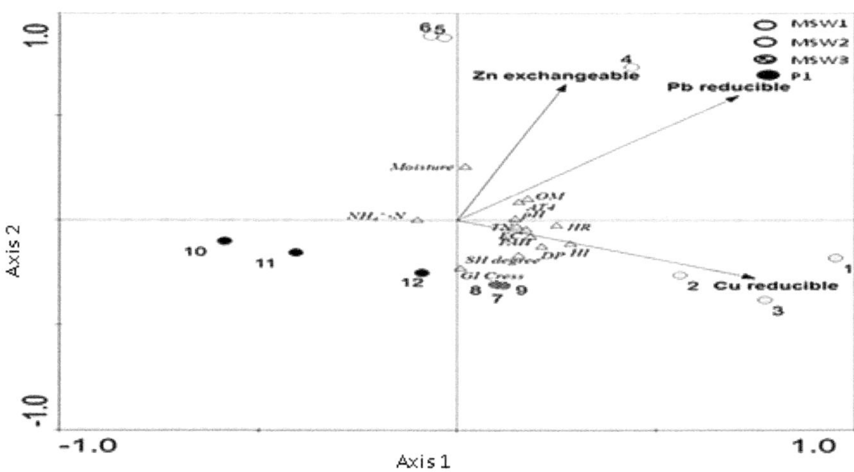

Figure 3. Canonical correspondence analysis of physicochemical, maturity and stability properties (50.2% could be explained by axis 1 and 27.6% by axis 2) in function the heavy metals mobile fraction of respective composts (Zn exchangeable, $p = 0.022$; Cu reducible, $p = 0.026$; Pb reducible, $p = 0.046$). The species-environmental correlations for axis 1 and 2 were, respectively, 0.854 and 0.992.

EC – Electrical Conductivity; OM – organic matter; GI – germination index; SH – self-heating; AT_4 – respiration activity after 4 days; HEC – carbon content of humic-like substances extract; HAC – carbon content of humic-like acid; FAC – carbon content of fulvic-like acid; DP – degree of polymerization (HAC/FAC); HI – humification index (HAC/TC); HR – humification ratio (HEC/TC); PAH – percentage of humic-like acids (HAC/HEC).

the activity microbiota involved on the transformation of the biodegradable organic carbon into stabilized humic-like acid substances. On the other hand, the simultaneous high moisture and Zn and Pb contents may increase the mobility of these heavy metals in composts and contribute to their low stability and high phytotoxicity.

4 CONCLUSION

The results obtained after applying the sequential extraction scheme indicate that Cu, Ni, Cr and Pb were most abundant in the immobile phases, while Zn showed no dominant chemical phase. The HS-like chemical composition was influenced negatively by the residual fraction of Cu, Pb and, mainly, Cr. It seems that the type of heavy metals or their mobility do not influence the maturity of compost. Pb and Cu mobile phase may have a negative influence on the microbial activity, involved on the formation of stabilized humic-like substances. Mobile forms of Zn and Pb associated with high moisture content may contribute to the increase of compost phytotoxicity.

ACKNOWLEDGEMENTS

This work was financially supported by the Portuguese Foundation for Science and Technology (FCT) through the projects UID/EQU/00511/2019 – Laboratory for Process Engineering, Environment, Biotechnology and Energy (LEPABE), through the Grant FRH/BD/43807/2008 and through Instituto Politécnico de Viseu by the Center for Studies in Education, Technologies and Health (CI&DETS).

REFERENCES

Adejumo, S.A., Ogundiran, M.B. & Togun, A.O. 2018. Soil amendment with compost and crop growth stages influenced heavy metal uptake and distribution in maize crop grown on lead-acid battery waste contaminated soil. *Journal of Environmental Chemical Engineering* 6: 4809–4819.

Bogusz, A. & Oleszczuk, P. 2018. Sequential extraction of nickel and zinc in sewage sludge- or biochar/sewage sludge-amended soil. *Science of the Total Environment* 636: 927–935

Cai, Q-Y., Mo, C-H., Wu, Q-T., Zeng, Q-Y. & Katsoyiannis, A. 2007. Concentration and speciation of heavy metals in six different sewage sludge-composts. *Journal of Hazardous Materials* 14: 1063–1072.

EN 13650 2001. *Soil Improvers and Growing Media – Extraction of aqua regia soluble elements*. European Committee for Standardization, Technical Committee CEN/TC 223.

Farrell, M & Jones, DL. 2009. Heavy metal contamination of a mixed waste compost: Metal speciation and fate. *Bioresource Technology* 100: 4423–4432.

Fuentes, A, Lloréns, M, Sáez, J, Soler, A, Aguilar, MI, Ortuño, JF, Meseguer, VF. 2004. Simple and sequential extractions of heavy metals from different sewage sludges. *Chemosphere* 54: 1039–1047.

Haroun, M., Idris, A. & Omar, S. 2009. Analysis of heavy metals during composting of the tannery sludge using physicochemical and spectroscopic techniques. *Journal of Hazardous Materials* 165: 111–119.

Jamali, M.K., Kazi, T.G., Arain, M.B., Afridi, H.I., Jalbani, N., Kandhro, G.A., Shah, A.Q. & Baig, J.A. 2009. Speciation of heavy metals in untreated sewage sludge by using microwave assisted sequential extraction procedure. *Journal of Hazardous Materials* 163: 1157–1164.

Liu, Y, Ma, L, Li, Y & Zheng, L. 2007. Evolution of heavy metal speciation during the aerobic composting process of sewage sludge. *Chemosphere* 67: 1025–1032.

Miaomiao, H., Wenhong, L., Xinqiang, L., Donglei, W. & Guangming, T. 2009. Effect of composting process on phytotoxicity and speciation of copper, zinc and lead in sewage sludge and swine manure. *Waste Management* 29: 590–597

Rauret, G., López-Sánchez, J-F., Sahuquillo, A., Barahona, E., Lachica, M., Ure, A.M., Davidson, C.M., Gomez, A., Lück, D., Bacon, J., Yli-Halla, M., Muntau, H. & Quevauviller, P. 2000. Application of a modified BCR sequential extraction (three-step) procedure for the determination of extractable trace metal contents in a sewage sludge amended soil reference material (CRM 483), complemented by a three-year

stability study of acetic acid and EDTA extractable metal content. *Journal of Environmental Monitoring* 2: 228–233.
Silva, M.E.F., Lemos, L.T., Bastos, MMSM, Nunes, O.C. & Cunha-Queda, A.C. 2013. Recovery of humic-like substances from low quality composts. *Bioresource Technology* 128: 624–632.
Singh, J. & Kalamdhad, A.S. 2012. Concentration and speciation of heavy metals during water hyacinth composting. *Bioresource Technology* 124:169–179.
Smidt, E., Meissl, K., Schmutzer, M. & Hinterstoisser, B. 2008. Co-composting of lignin to build up humic substances – Strategies in waste management to improve compost quality. *Industrial Crops and Products* 27: 196–201.

Wastes: Solutions, Treatments and Opportunities III – Vilarinho et al. (Eds)

Dark fermentation sludge as nitrogen source for hydrogen production from food waste

J. Ortigueira & P. Moura
LNEG I.P., Laboratório Nacional de Energia e Geologia, Unidade de Bioenergia, Lisboa, Portugal

C. Silva
Instituto Dom Luiz, Faculdade de Ciências, Universidade de Lisboa, Lisboa, Portugal

ABSTRACT: The biological conversion of food waste (FW) into hydrogen (H_2) by anaerobic fermentation is associated with high production costs and complex supplementation requirements. The present study focused on the simplification of the H_2 production through dark fermentation (DF) by reusing its residual solid fraction, herein referred as DF-sludge, as nitrogen source for a subsequent FW conversion. The non-sterile FW fermentation with addition of *C. butyricum* as H_2-producing microorganism and supplemented with two nitrogen sources was compared: ammonium chloride (NH_4Cl) or DF-sludge. The maximum biogas productivity, H_2 production yield and H_2 cumulative production were obtained with the DF sludge supplementation, reaching values of 433.3 ± 34.3 mL biogas $(L\ h)^{-1}$, 194.2 ± 24.4 mL $H_2\ g_{sugar}^{-1}$ and 3.2 ± 0.0 L $H_2\ L^{-1}$, respectively. The use of DF sludge improved the fermentation efficiency on H_2 production by 40%, underlining the impact of nutrient recycling in *C. butyricum* fermentative performance.

1 INTRODUCTION

Over the past decade, the scientific community has emphasized the environmental and economic impact of wasted food products (Scherhaufer et al. 2018). The act of throwing out food due to inefficient production and consumption practices, reaches farther than the mere gesture. It represents the loss of all resources required for its production as well as those necessary for its proper treatment and disposal. According to the Food and Agriculture Organization of the United Nations, approximately 88 billion tonnes of food waste (FW) were discarded in the 28 countries of the European Union in 2013, a wastage that represents up to 186 Mt CO_{2-eq} carbon emissions (Scherhaufer et al. 2018). While FW prevention is a required target in the majority of legislation packages dealing with this problem (Corrado & Sala 2018), it is not feasible to assume that it will be possible to reduce or eliminate this type of waste in a nearby future. Therefore, several studies focus the possibility of FW valorisation.

Food waste is a highly heterogeneous mixture of chemical components, containing water, carbohydrates, proteins, fat, among others. This composition makes it an interesting substrate for dark fermentation (DF), a biological process which consists on the anaerobic conversion of carbohydrates into hydrogen (H_2), organic acids, such as butyrate and acetate, and compost (Ortigueira et al. 2019). Hydrogen is considered to be an extremely interesting bioenergy carrier as it has a considerably high energy density (120 MJ kg^{-1}), is storable at $-253°C$ in the liquid form or in the gaseous form at high pressures of 300–700 bar, and its combustion is not associated to carbon or sulphur emissions (Dutta et al. 2014). However, the low production yields which are normally associated with biological conversion systems tend to inflate the production costs and difficult the implementation at industrial scale. One strategy commonly used for costs reduction involves the minimisation of the nutrient requirements in the culture media and/or its replacement by cheaper alternatives. These alternatives should be, in principle, easily attainable, readily available in large quantities, renewable and environmentally friendly (Han et al. 2016). Dark fermentation sludge can be defined as the solid residue obtained after completion of the biological conversion process, being composed by substrate leftovers that remain in the fermentation sludge, produced metabolites possibly adsorbed to the solid residues, and cellular biomass (Moser-Engeler et al. 1998). Chemically, DF-sludge is composed by a considerable carbon fraction and also nitrogen. The latter is an essential nutrient for bacterial growth and necessary for the correct conversion performance through DF. In fact, the compost that is traditionally produced from DF sludge after maturation and stabilisation has a high concentration in nitrogen (Wilson & Novak 2009). This study analysed the effect of replacing NH_4Cl as nitrogen source by DF nitrogen-rich sludge on the fermentative conversion of FW to H_2. The main objective was not only to reduce process costs, but also to promote the reincorporation of nutrients and achieve disposal savings.

2 MATERIALS AND METHODS

The FW used for this study was collected from a local restaurant during a period of 8 hours, located in the southern area of metropolitan Lisbon (Trafaria, Almada, November 2016). The chemical characterisation was performed after bones removal, mashing and homogenisation of the samples. The processed samples were then stored at −20°C prior to characterisation and fermentation. The FW samples were characterised in terms of water, total sugars, crude protein, total fat and ash. Moisture and ash were determined by oven drying (100°C, 12 hours) followed by organic component volatilisation (550°C, 6 hours) according to standard methods (Horwitz & Latimer 2005). The protein content was estimated by the Kjeldahl method using 6.25 as the conversion factor of total nitrogen into crude protein (Horwitz & Latimer 2005). Total fat was determined after ether extraction in a Soxhlet system (Sukhija & Palmquist 1988). Total sugars were determined through an adapted anthrone method as published elsewhere (Ortigueira et al. 2018). The FW fermentation assays were divided in two series, referred to as NH_4Cl assay and DF-sludge assay, according to the nitrogen source used. Three independent batch fermentations were performed for each experimental condition, and where prepared as follows:

NH_4Cl assay – 60.8 g FW were submitted to microwave pretreatment (550 W for 4 mins) (Ortigueira et al., 2019). The pretreated sample was suspended up to a total volume of 500 mL in Minimum Mineral Medium (MMM), containing per litre of 100 mM phosphate buffer (pH 6.8): NH_4Cl, 12 g, 3.3 mg $FeSO_4.7H_2O$, 0.56 g cysteine-HCl.H_2O and 1 mg resazurin (Ortigueira et al. 2018).

DF sludge assay – 60.8 g FW were mixed with 14 g of DF-sludge obtained from a previous non-sterile, *C. butyricum* bioaugmented, FW fermentation. The mixture was submitted to microwave pretreatment as described above, and then suspended in phosphate buffer, pH 6.8, up to total volume of 500 mL.

The initial concentration of total sugars was 20 g L^{-1} for both NH_4Cl and DF sludge assays. Both NH_4Cl and DF sludge assay were undertaken under non-sterile conditions. The bacterial culture *C. butyricum* DSM 10702 from the German Collection of Microorganisms and Cell Cultures (DSMZ, Braunschweig, Germany) was used as additional biocatalyst. *C. butyricum* was pre-cultured in Reinforced Clostridial Medium (RCM) (Difco laboratories, Le Pont de Claix, France) and the cells in exponential growth phase were inoculated at 5% (v v^{-1}) in each fermentation medium that was contained in a 1.65 L bioreactor to be operated at pH 5.5, 37°C and 150 rpm. The produced biogas at the outlet of the bioreactor was routed through a gas washing bottle containing NaOH 250 mM, for CO_2 stripping, after which was continuously quantified by means of a gas flowmeter (μflow, Bioprocess Control, Lund, Sweden) and collected in gas sampling bags (SKC sample bags, 263-03, SKC, PA, USA). The produced biogas in the sampling bags, herein expressed as cumulative biogas, and the biogas in the bioreactor headspace at the end of the fermentation were characterised by gas chromatography (GC). A chromatographer (Varian 430-GC) equipped with a thermal conductivity detector (TCD) and a fused silica column (Select Permanent Gases/CO2-Molsieve 5A/Borabound Q Tandem #CP 7430) was used in the following operational conditions:

Table 1. Proximate composition of the FW samples.

Component	% d.w.
Total sugars	62.1 ± 0.1
Crude protein	10.4 ± 0.2
Total fat	26.3 ± 2.2
Ash	1.2 ± 0.1

the injector and column were operated at 80°C and the detector at 120°C with argon as the carrier gas, at a flow rate of 32.4 mL min^{-1}. The GC column was kept at 30–60°C, the injector at 60°C and the TCD at 150°C. The liquid fermentation samples were processed for oil removal by hexane extraction, using a 1:2 sample/hexane ratio, and the organic acids were subsequently quantified by high performance liquid chromatography (HPLC). The HPLC system was equipped with an Aminex HPX-87H column (Bio-Rad Laboratories, USA) and a refraction index (RI) detector (LaChrom L-7490). The temperature of the column and the RI detector were kept constant at 50°C and 45°C, respectively, and the samples were eluted using H_2SO_4 5 mM at a flow rate of 0.5 mL min^{-1}. The total H_2 production was calculated as the sum of the H_2 volume in the biogas sampling bags and the volume of H_2 in the biogas that remained inside the bioreactor headspace at the end of each fermentation assay. The H_2 productivity was estimated from the graphical representation of the cumulative biogas production (L L^{-1}) versus time (hours), as the slope of the exponential production period, and using the percentage of H_2 (% vol.) in the total volume of biogas produced. The volumetric H_2 yield was defined as the ratio between the total H_2 volume (mL) produced and the mass of FW volatile solids (g_{VS}) supplied to the culture medium in each assay. Total sugars consumption was defined as the percentage of initial sugars consumed.

3 RESULTS AND DISCUSSION

3.1 *Food waste and dark fermentation-sludge characterisation*

The use of FW as a substrate for DF is highly dependent on an appropriate chemical composition, more specifically the carbohydrate composition and its chemical form. Highly polymerised components such as cellulose, hemicelluloses and starch might be more recalcitrant for direct bioconversion and usually require an additional pretreatment and/or saccharification stage prior to fermentation. Additionally, a high nitrogen content in the FW may circumvent the need for media supplementation. The chemical characterisation of the collected FW is presented in Table 1.

While the chemical variability of FW is one of the main characteristics of this kind of samples, the FW collected for this work was considered to be a good representative of the typical summer Portuguese diet. It was composed by residues of cooked vegetables, fish skins/bones/remains and a visibly high fraction of starch-rich leftovers, such as bread, rice and potato scraps. The high concentration of total sugars suggests the sample to be quite adequate for DF conversion. Furthermore, the carbon:nitrogen molar ratio was 2.9:1 which was comparable to the ratio of 3:1 of FW samples used in previous fermentations (Ortigueira et al. 2019). However, in that study it was found that not all the crude protein supplied by the substrate was easily converted, which led to the need of supplementing the fermentation medium.

The DF-sludge used in the present study corresponded to the solid fraction obtained at the end of a FW non-sterile fermentation with addition of *C. butyricum* as H_2-producing microorganism (Ortigueira et al. 2019). The concentration of total sugars and nitrogen in the DF-sludge fraction totalised approximately 14 ± 4.7 % and 58.2 ± 0.9 % (w w^{-1}). It contained mainly residual carbohydrates that were not metabolised by the microbial population, and organic acids, particularly butyric and acetic acid, that were present in the liquid fermentation and were possibly adsorbed to the remnant solids. The DF-sludge was separated from the liquid fermentation throughcentrifugation,

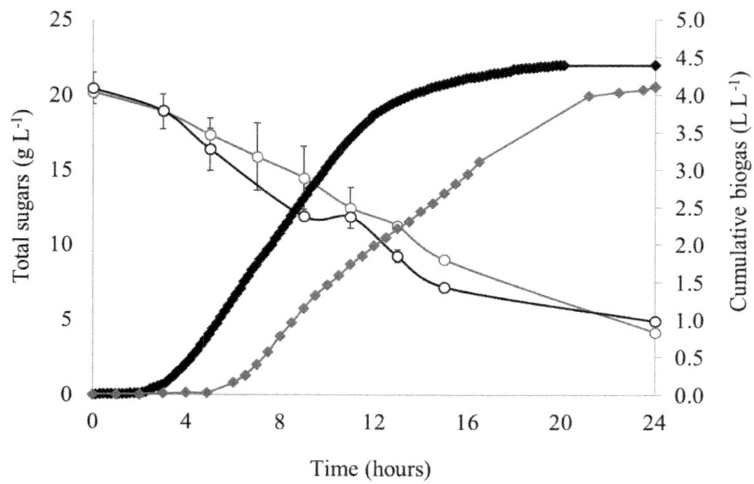

Figure 1. Non-sterile FW mesophilic fermentation, at pH 5.5, 37°C and 150 rpm, with addition of *C. butyricum* as biocatalyst and NH_4Cl or DF-sludge supplementation: time-course of the cumulative biogas production and sugar consumption (NH_4Cl assay: ◆ – biogas; ○ – total sugars) (DF-sludge assay: ◆ – biogas; o – total sugars).

kept wet to minimise additional drying costs and then rerouted into the DF sludge assay (see section 2). The microwave pretreatment could, theoretically, breakdown the celular material, reduce contamination and degrade further carbohydrate components which were not previously converted.

3.2 *Comparative FW fermentation with NH_4Cl or DF-sludge supplementation*

The fermentability of the FW samples for H_2 production was evaluated in a series of batch experiments conducted in a bench-scale bioreactor with pH control. The fermentation results of the NH_4Cl and DF-sludge assays are shown in Figure 1.

The change in the nitrogen source altered visibly the fermentation start-up. The lag phase for biogas production in the DF-sludge assay decreased to 3 hours, while in the NH_4Cl assay it started only 5 hours after the inoculation of *C. butyricum*. The maximum biogas productivity was achieved between 3 and 12 hours of operation in the DF-sludge assay, reaching a maximum value of 433.3 ± 34.3 mL biogas $(L\ h)^{-1}$ (Table 2). This result represents an increase of almost 64% when compared to the NH_4Cl assay that achieved a productivity of 264.4 ± 13.8 mL biogas $(L\ h)^{-1}$ from 7 to 16 hours of fermentation. It was suggested that both the decrease in the lag phase and the higher productivity obtained in the DF-sludge assay were likely due to the presence of enzymatic material in the sludge, which enabled the bacterial population to use the polymeric components more readily than in the NH_4Cl assay. The improvement in both these fermentation parameters led to a shorter conversion time in the DF-sludge assay that is confirmed by the negligible biogas production, *i.e.* bellow 10 mL $(Lh)^{-1}$, after 18 hours of fermentation.

The replacement of NH_4Cl by DF-sludge impacted significantly in the H_2 concentration of the produced biogas (Table 2) and increased the total H_2 production by 41%. This can be explained by a faster adaptation of *C. butyricum* to the DF-sludge composition because adsorbed organic acids may exert an inhibitory effect over the FW native microbial community. As there were no significant differences in both the initial sugar concentration (Figure 1) and the total sugars consumption (Table 2) between the assays, it was suggested that the metabolic behaviour of the microbial population in the DF-sludge assay was more efficient. This pattern was reflected upon analysis of the H_2 yield of the same assay, which reached a maximum of 194.2 ± 24.4 mL $g_{total\ sugars}^{-1}$ (Table 2). It was not possible in any fermentation condition to completely convert the carbohydrates

Table 2. Results of the non-sterile FW fermentation with addition of *C. butyricum* and NH_4Cl or DF-sludge supplementation.

Assay	Total sugars consumption [%]	Total biogas production [$L L^{-1}$]	Max. biogas productivity [$mL (L h)^{-1}$]	H_2 concentration [% vol]	H_2 production yield [$mL g_{VS}^{-1}$]	H_2 production yield [$mL g_{total\ sugars}^{-1}$]
NH_4Cl	79.3 ± 0.0	4.1 ± 0.1	264.4 ± 13.8	55.6 ± 0.1	77.9 ± 4.1	139.1 ± 2.6
DF-sludge	75.9 ± 0.0	4.4 ± 0.1	433.3 ± 34.3	73.0 ± 0.0	111.9 ± 1.6	194.2 ± 24.4

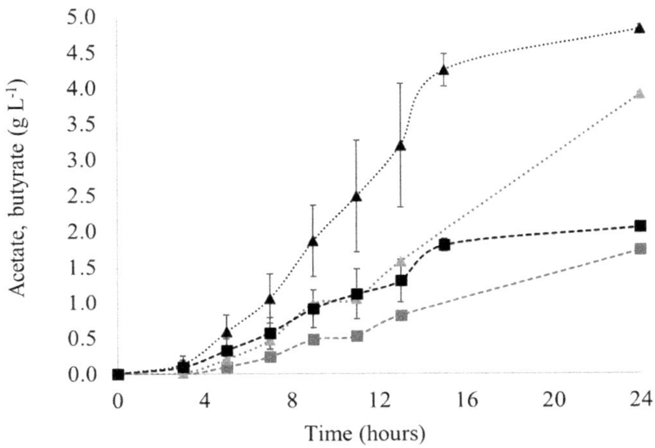

Figure 2. Non-sterile FW fermentation, at pH 5.5, 37°C and 150 rpm, with addition of *C. butyricum* as biocatalyst and NH_4Cl or DF-sludge supplementation: time-course of acetic and butyric acid production (NH_4Cl assay: ■ – acetate; ▲ – butyrate) (DF-sludge assay: ■ – acetate; ▲ – butyrate).

supplied, and the final concentration of total sugars was approximately $5\,g\,L^{-1}$ (Figure 1). This may indicate the presence of polymeric components that *C. butyricum* in association with the FW native microbiota cannot metabolise without additional FW pretreatment and/or saccharification apart from the exposure to MW as performed in the present study.

3.3 *Production of organic acids*

Hydrogen production was accompanied by the production of organic acids, particularly acetate and butyrate, which started in the first 3 h (DF-sludge assay) or 5 h (NH_4Cl assay) after the inoculation of *C. butyricum* (Figure 2). Apparently, there was no microbial inhibition by the presence of adsorbed organic acids in the sludge biomass. Once again, the concentration of both acids was higher in the DF-sludge assay, where the production of butyric acid reached a maximum of $4.8 \pm 0.1\,g\,L^{-1}$. The butyrate-to-acetate molar ratio in the DF-sludge assay was accordingly higher ($1.49 \pm 0.0\,mol\,mol^{-1}$), which illustrates a situation of high H_2 partial pressure in the bioreactor headspace. In fact, in that fermentation, the concentration of H_2 in the produced biogas reached 62% vol. and 73% vol. before and after CO_2 stripping, respectively (Table 2). Lactate was produced at a maximum concentration of $0.5\,g\,L^{-1}$ throughout the DF-sludge assay runtime up to 12 hours (data not shown), point after which it steadily decreased to $0.06\,g\,L^{-1}$ until the end of the experiment. This behaviour was not mirrored in NH_4Cl assay, in which the lactate concentration increased steadily up to the end of the fermentation, reaching a maximum concentration of $0.4\,g\,L^{-1}$.

According to the obtained results, the DF sludge has the correct characteristics to serve as a potential nitrogen source for the FW conversion through DF. Furthermore, the rerouting of this solid effluent stream to a new fermentation process has two main positive impacts: reducing the need of using NH_4Cl as external resource in the H_2 and/or organic acids production stage, and reducing the cost and environmental burden associated with the conversion of the fermentation sludge into compost, including the required stabilisation, drying and transportation stages prior to its use as fertiliser. Therefore, it is likely that the replacement of the nitrogen source by a cheaper and environmentally friendly option can lead to a significant decrease in the overall production and operation costs in scalable FW bioconversion processes.

4 CONCLUSIONS

The replacement of NH_4Cl by DF-sludge in non-sterile FW fermentations bioaugmented with *C. butyricum* was especially effective in terms of process productivity. The biogas production lag phase was shortened from 5 to 3 hours and the maximum biogas productivity was increased by 64%. The H_2 concentration achieved 73% vol. in the produced biogas and yielded 194.2 ± 24.4 mL H_2 g_{sugar}^{-1}, which represents a 40% increase when compared to the NH_4Cl assay. Butyric acid was the major cometabolite and exhibited a production profile similar to that of biogas. The present study shows that, in compliance with circular economy practices, the recycling of nutrients in non-sterile FW fermentations for H_2 production increases the overall process efficiency.

FUNDING AND ACKNOWLEDGEMENTS

This work is integrated in the project CONVERTE, supported by POSEUR (POSEUR-01-1001-FC-000001) under the PORTUGAL 2020 Partnership Agreement. Joana Ortigueira acknowledges FCT for the PhD grant SFRH/BD/107780/2015. The authors would also like to thank the technical assistance of Céu Penedo (biomass characterisation) and Paula Costa (GC analysis). Publication supported by FCT- project UID/GEO/50019/2019 – Instituto Dom Luiz.

REFERENCES

Corrado, S. & Sala, S. 2018. Food waste accounting along global and European food supply chains: State of the art and outlook. *Waste management,* 79, 120–131.

Dutta, S. 2014. A review on production, storage of hydrogen and its utilization as an energy resource. *Journal of Industrial and Engineering Chemistry, 20*(4), 1148–1156.

Han, W., Yan, Y., Gu, J., Shi, Y., Tang, J. & Li, Y. 2016. Techno-economic analysis of a novel bioprocess combining solid state fermentation and dark fermentation for H_2 production from food waste. *International Journal of Hydrogen Energy, 41*(48), 22619–22625.

Horwitz W. & Latimer G.W. 2005. A.O.A.C. Official methods of analysis. 18th ed. Gaithersburg MD, USA: Association of the Official Analytical Chemists (A.O.A.C.) International. ISBN-13: 978e0935584783.

Moser-Engeler, R., Udert, K. M., Wild, D. & Siegrist, H. 1998. Products from primary sludge fermentation and their suitability for nutrient removal. *Water Science and Technology, 38*(1), 265–273

Ortigueira, J., Silva, C. & Moura, P. 2018. Assessment of the adequacy of different Mediterranean waste biomass types for fermentative hydrogen production and the particular advantage of carob (Ceratonia siliqua L.) pulp. *International Journal of Hydrogen Energy, 43*(16), 7773–7783.

Ortigueira, J., Martins, L., Pacheco M., Silva, C. & Moura, P. 2019. Improving the non-sterile food waste bioconversion to hydrogen by microwave pretreatment and bioaugmentation with *C. butyricum. Waste management, 88*, 226–235.

Scherhaufer, S., Moates, G., Hartikainen, H., Waldron, K., & Obersteiner, G. 2018. Environmental impacts of food waste in Europe. *Waste management, 77*, 98–113.

Sukhija, P. S. & Palmquist, D. L. 1988. Rapid method for determination of total fatty acid content and composition of feedstuffs and feces. *Journal of Agricultural and Food Chemistry, 36*(6), 1202–1206.

Wilson, C. A. & Novak, J. T. 2009. Hydrolysis of macromolecular components of primary and secondary wastewater sludge by thermal hydrolytic pretreatment. *Water research, 43*(18), 4489–4498.

Wastes: Solutions, Treatments and Opportunities III – Vilarinho et al. (Eds)
© 2020 Taylor & Francis Group, London, ISBN 978-0-367-25777-4

Development of methodology for quantification of unsorted WEEE in scrap metal mixtures

M. Luízio
Electrão – Associação de Gestão de Resíduos, Lisboa, Portugal

P. Costa
TÜV Rheinland Portugal, Lisboa, Portugal

ABSTRACT: Electrão, a take back scheme in Portugal, has defined and implemented the first auditable methodology to reliably quantify the amount of unsorted WEEE collected by waste management operators together with scrap metal. This methodology intends to address the well-known problem of unaccounted quantities of unsorted WEEE collected together with scrap metal in Europe by ensuring the proper reporting of these quantities. This methodology was carried out in a number of national waste management operators, which resulted in the identification of about 20 000 tonnes of WEEE in 2017 otherwise unaccounted for. The implementation of this methodology has allowed the reliable quantification and reporting of this waste stream thus improving the environmental performance in Portugal.

1 INTRODUCTION

1.1 *Motivation*

The publication of the European Directive 2012/19/EU on waste electrical and electronic equipment (WEEE) has resulted in a significant increase on the collection targets. According to the Directive, each Member State must ensure that, from 2016, the minimum collection rate is 45%, calculated on the basis of the total weight of WEEE collected in a given year, expressed as a percentage of the average weight of electrical and electronic (EEE) placed on the market in the three preceding years. It was also established that from 2019, each Member State must achieve an annual average weight of EEE placed on the market of 65%, or alternatively 85% of WEEE generated.

The European Commission's report to the European Parliament and the Council regarding the review of the scope of the Directive and the re-examination of the deadlines for reaching the collection targets (European Commission, 2017) concluded that some Member States may not reach the 2019 collection targets if the current pace of progression and collection practices are maintained. The report identified one of the possible drivers for this outcome to be the high rate of collection that is unaccounted for in WEEE collection statistics, particularly when collection occurs outside the framework of extended producer responsibility compliance systems and is mislabelled as metal or plastic scrap or when WEEE is not handled by authorised recyclers.

This is mostly due to the lack of knowledge of the consumers, the producers of WEEE, on its value chain, namely their misclassification as metal scrap which results in sending this waste stream to scrap metal operators outside the producer compliance scheme. The education of the consumer regarding this aspect constitutes a considerable limitation on the waste management model that has proved difficult to overcome.

Due to these limitations, significant quantities of WEEE continue to appear in scrap metal collectors and shredders which is why it is crucial to quantify and report them.

The need for accounting this waste flow is backed by the European Commission that supported the development of an information system that can trace WEEE flows from retailers and scrap metal collectors in order to account for the complementary WEEE collection that occurs (European Commission, 2014).

The activities of these operators who do not report their WEEE remain unaccounted for their treatment. With the methodology developed by Electrão, it is possible to have a reliable quantification and reporting of otherwise unaccounted amounts, thus providing a tool for Portuguese authorities to monitor the entire WEEE amounts generated and processed in the country.

Many producer compliance schemes have presented estimates for the concentration of WEEE in metal scrap, however they are not based on an adequate sampling methodology, with values varying significantly and appearing to be conservative (Huisman, J. et al., 2015; ADEME, 2013).

Given the more demanding collection targets set by the European Commission and the limitations of the current WEEE management system that led to significant quantities of WEEE being collected and treated by waste management operators outside producer compliance schemes, Electrão found an opportunity to address this challenge and provide a trailblazing solution for the quantification of this waste flow. Electrão has worked alongside TUV Rheinland Portugal to address this issue and develop a methodology that ensures that the quantification of unsorted WEEE is done in a systematically and statistically representative way, which can be audited and verified by an independent third party.

1.2 *Objectives and scope*

This paper describes the research and development performed by Electrão take back scheme in Portugal, to define and implement the first auditable methodology to quantify and characterise the input of unsorted WEEE collected by waste management operators (WMO) together with scrap metal.

This methodology was defined in a procedure which is followed by the responsible technician of the Treatment and Recovery Unit (TRU) for the technical audit monitoring. This procedure summarises the objectives and the methodology of the works to be carried out and defines the technical and human resources that the TRU should provide.

2 METHODOLOGY

The methodology designed by Electrão set out to characterise a representative sample amount of scrap metal, with sorting and weighing of large and small waste electrical and electronic equipment (LWEEE and SWEEE) (Figure 1).

2.1 *Characterisation events*

This was achieved by organising characterisation events in each waste management operator that manages scrap metal. These events had the purpose of separating the Waste electrical and electronic equipment (large and small) and their components from a certain amount of scrap metal (Figure 2).

These separated flows were then weighed and classified according to the following categories:

- Large waste electrical and electronic equipment;
- Large waste electrical and electronic equipment components;
- Small waste electrical and electronic equipment;
- Small waste electrical and electronic equipment components.

Other WEEE categories were also separated but not weighted. These can be classified as:

- WEEE components (this flow refers to components that can be identified as being part of any WEEE);

Figure 1. Characterisation of scrap metal sample with sorting of large and small waste electrical and electronic equipment.

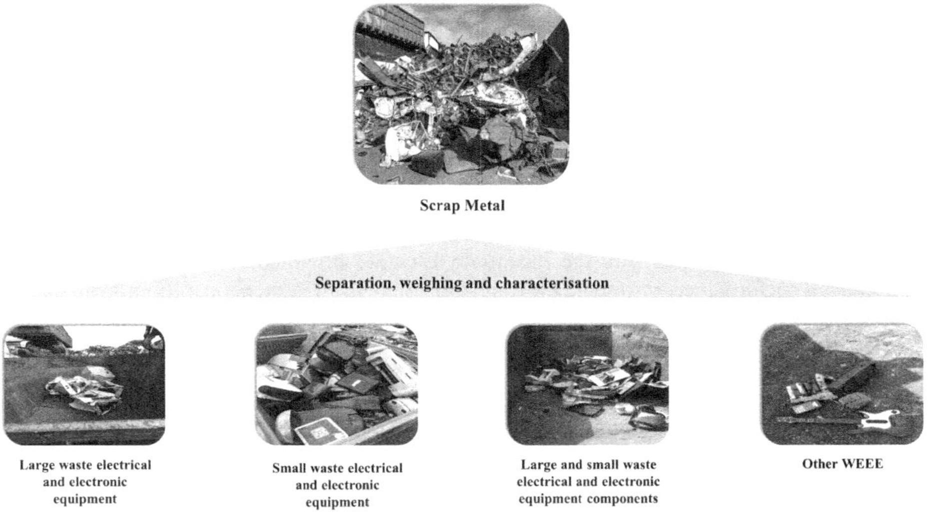

Figure 2. Separation and classification of WEEE flows from scrap metal.

- Other WEEE (this flow refers to equipment other than LWEEE and SWEEE, for example, CRT televisons and monitors, refrigerators, lamps, and other).

This methodology can either be applied to a stack, which is equivalent to a day's worth of processing, or to all the loads of scrap metal that are received and processed in a TRU in a day's work (without significant stops). It was established that the minimum amount of material to be sampled should equal the daily processing capacity of the least efficient operator in the Electrão system, which in 2017 was 30 tonnes. This represents approximately 63% of the average sample amount in the 2017 events.

It is the TRU's responsibility to provide the mechanical means to move the material, such as a mechanical gripper or a forklift, in addition to the adequate weighing equipment. The TRU must also provide the necessary human resources to move, separate and weigh the materials.

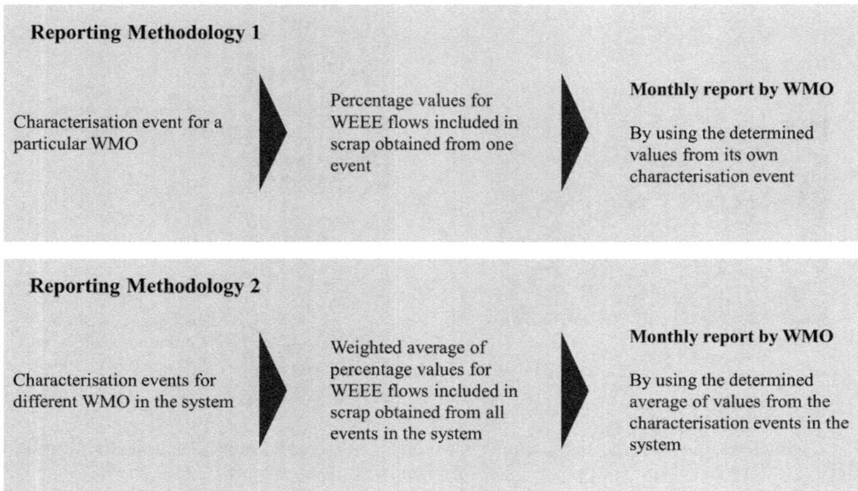

Figure 3. Description of the two WEEE reporting methodologies for WMO.

Electrão, on the other hand, is responsible for the definition of the methodology to be implemented in each characterisation, as well as the recruitment of the auditor that will conduct the works. The WEEE management system is also be responsible for the development of a characterisation report that should be made available to interested parties.

It is also important to highlight that Electrão simultaneously assessed the WMO's environmental performance, as it is the organisation's standard procedure to all operators included in its system. These assessments, together with the incentives Electrão provided for operators with the best environmental performance, promoted the improvement of the treatment operations in all operators.

2.2 *Reporting Methodologies*

The development of characterisation events resulted in percentage values of WEEE in a sample of scrap metal, and their classification in flow LWEEE and flow SWEEE for a WMO that manages this waste stream. Electrão proposed two different methodologies for the WMO to report their quantities of unsorted WEEE resorting to these values (Figure 3).

In the first methodology, the WMO provides a monthly report of their collected WEEE quantities equivalent to the results of their characterisation event (with an interval of $\pm 1\%$). In the second methodology, each WMO provides a monthly report resorting to the weighted average values of all characterisation events organized by Electrão for several WMO (with an interval of $\pm 1\%$).

Since this was a new process, the first methodology was perceived as the most adequate because the used values are easily recognised by the WMO who were more willing to use them in their monthly reports. Additionally, given that these events were still low numbered and hardly representative of the system, resorting to average values was expected not to add further value to the reporting.

3 RESULTS

Between November of 2016 and July of 2018, 12 characterisation events were organised by Electrão in different WMO resulting in the intervals of values for WEEE quantities in scrap metal presented in Table 1, according to reporting methodology 1. In these events, individual samples with around

Table 1. Results of the characterisation events conducted between 2016 and 2018 according to reporting methodology.

WMO	WEEE	LWEEE	SWEEE
1	[12.5–14.5]%	[9.7–11.7]%	[1.8–3.8]%
2	[13.9–15.9]%	[8.8–10.8]%	[4.1–6.1]%
3	[14.3–16.3]%	[12.9–14.9]%	[0.4–2.4]%
4	[30.8–32.8]%	[18.0–20.0]%	[11.7–13.7]%
5	[12.2–14.2]%	[11.5–13.5]%	[−0.4–1.6]%

Table 2. Results of the characterisation events conducted between 2016 and 2018 according to reporting methodology 2.

	WEEE	LWEEE	SWEEE
2016	[12.6–14.6]%	[7.7–9.7]%	[3.8–5.8]%
2017	[22.5–24.5]%	[14.9–16.9]%	[6.6–8.6]%
2018	[28.8–30.8]%	[14.4–16.4]%	[13.4–15.4]%
System's Results	[21.3–23.3]%	[12.4–14.4]%	[7.9–9.9]%

Figure 4. Weighted average results of the implementation of the characterisation events from 2016 to2018.

50 tonnes of scrap metal were analysed, for the representative daily scrap metal processing capacity of a TRU. In all events, the methodology involved the extraction of sample from the scrap. This method was representative because the stack includes scrap metal from weeks of input of material and from different origins.

If a WMO chooses to use reporting methodology 2 to report its quantities of WEEE, then he should resort to the values presented in Table 2.

The quantification of unsorted WEEE from scrap in 2017 has resulted in the accounting of over 20 000 tonnes of this waste flow, representing 50% of the total collection that year, thus proving the significance of the unreported quantities of WEEE in the country.

Overall, the events performed from 2016 to 2018 produced a result of WEEE present in scrap on average of 17.6%, with 13.3% corresponding to large equipment and the remaining 4.3% corresponding to small equipment (Figure 4).

4 CONCLUSIONS

The problem of unreported quantities of unsorted WEEE collected together with scrap metal (and incorrectly classified as such) is well known and observed in many Member States. Electrão has designed and implemented an auditable methodology in a number of national waste management operators in Portugal, which allowed the reliable quantification of about 20 000 tonnes of otherwise unaccounted WEEE, representing 50% of the collected quantities in 2017.

With this paper, Electrão has proved that it is possible for countries to monitor the full amounts of WEEE flows. By sharing this tool and the acquired knowledge from this experience, Electrão provides the opportunity for Authorities and other entities to establish a comprehensive and reliable reporting of the whole WEEE amounts managed in the country, thus contributing to improve its environmental performance. This is an important tool for Member States to quantify and monitor all the WEEE amounts, particularly in the context of the new and more demanding collection targets set by the Directive on WEEE.

REFERENCES

European Commission. 2017. Report from the Commission to the European Parliament and the Council on the review of the scope of Directive 2012/19/EU in waste electrical and electronic equipment (the new WEEE Directive) and on the re-examination of the deadlines for reaching the collection targets referred to in Article 7(1) of the new Directive and on the possibility of setting individual collection targets for one or more categories of electrical and electronic equipment in Annex III to the Directive. COM(2017) 171 final, Brussels.

European Commission. 2014. Study on Collection Rates of Waste Electrical and Electronic Equipment (WEEE). Possible measures to be initiated by the Commission as required by Article 7(4), 7(5), 7(6) and 7(7) of Directive 2012/19/EU on Waste Electrical and Electronic Equipment (WEEE).

Huisman, J. et al. 2015. Countering WEEE Illegal Trade (CWIT) Summary Report, Market Assessment, Legal Analysis, Crime Analysis and Recommendations Roadmap.

ADEME. 2013. Study on the Quantification of Waste of Electrical and Electronic Equipment (WEEE) in France – Household and similar WEEE arising and destinations.

Wastes: Solutions, Treatments and Opportunities III – Vilarinho et al. (Eds)
© 2020 Taylor & Francis Group, London, ISBN 978-0-367-25777-4

Adding value to soapstocks from the vegetable oil refining: Alternative processes

M. Cruz, M.F. Almeida & J.M. Dias
LEPABE, Departamento de Engenharia Metalúrgica e de Materiais, Faculdade de Engenharia, Universidade do Porto, Porto, Portugal

M.C. Alvim-Ferraz
LEPABE, Departamento de Engenharia Química, Faculdade de Engenharia, Universidade do Porto, Porto, Portugal

ABSTRACT: Soapstock is a major by-product from vegetable oil refining, conventionally recovered by acidification with a strong acid to obtain an acid oil which might be further used. In the present study, acidification with HCl was performed and compared to an alternative process of enzymatic hydrolysis with hexane as organic solvent to enable the direct recovery of soaps and lipids aiming at further integration within a biofuel production process. The best condition achieved in the first case was 10 wt.% of HCl allowing to obtain FFA content of around 60 wt.% and mass yield of 43 wt.%; and, considering the moisture content of the soapstock (47 wt.%), the soaps were converted into FFA efficiently. Regarding the enzymatic hydrolysis using 10 wt.% of enzyme and 1:8 of soapstock: hexane mass ratio, a final FFA content of around 36 wt.% was obtained, showing the hydrolysis of glycerides present in the soapstock.

1 INTRODUCTION

The Food and Agriculture Organization of the United Nations and the Organization for Economic Cooperation and Development expect that the global oilseeds production will expand at around 1.5% per year, for the ten-year period of 2018-27, a rate below that observed during the last decade. It is also expected that the demand for vegetable oil grows more slowly due to the slower growth in per capita food use in developing countries and also due to the projected stagnation concerning its demand as feedstock for biodiesel (OECD, 2018).

Crude oil can be refined using several processes to remove undesirable compounds, before making it acceptable for human consumption. The refining processes generate by-products and/or waste products that contain lipids, such as olive oil pomace, resulting from olive oil production, soapstocks (SS), deodorizer distillates, acid oil (AO) and acid water (Dijkstra, 2017). The first refining step involves the removal of phospholipids using the degumming process. The best known and the most widely used process is that which employs an acid followed by sodium hydroxide (Dumont and Narine, 2007). The reaction of the alkaline solution with the free fatty acids (FFA) leads to the formation of soaps. The SS created is then continuously separated from crude oil by centrifugation. The elimination of FFA from oils by distillation (steam refining) is known as physical refining. Although being an environmentally friendly process (no SS to be treated) it is more sensitive to the crude oil quality and stability, once experience has shown that it only leads to acceptable results when good quality starting oils are used (Talal et al., 2013, Cvengros, 1995).

SS are mostly wet lipidic mixtures. Their physicochemical properties tend to change with vegetable oil source, seed processing, as well as handling and storage conditions. SS is generated at a rate of 6% of the volume of the crude oil produced, or in some cases up till 20% of the crude oil (Laoretani et al., 2017, Haas, 2005). At the same time, the SS price is only one-tenth of the refined oil cost (Wang et al., 2007). The by-product consists in a heavy alkaline aqueous emulsion of lipids, containing about 50% water, with the balance made up of FFA, phosphoacylglycerols, mono, di and triglycerides, pigments and other minor nonpolar components (Basheer and Watanabe, 2016). The process widely adopted to add value to the SS is the obtention of AO. This process is of easy implementation and it allows to obtain a very versatile product. AO can be used as feedstock for animal feed, raw material for biodiesel production, boiler fuel or can also be used as a source of fatty acids for industrial applications (Haas, 2005).

The most conventional way to add value to SS involves their acidification at high steam temperature (80–95°C) using excess of a strong concentrated acid. However, the acidification of SS is one of the undesirable steps in an integrated facility, due to the difficulty to perform it effectively because it requires high energy and high effluent treatment cost (Piloto-Rodríguez et al., 2014, Laoretani et al., 2017). The problems associated with the acidification of SS are well known, namely the corrosive nature of the process, and the difficult separation between AO and acid water phases, which leads to high losses of lipids and, consequently, high contamination of wastewater. The use of heterogeneous catalysts, such as biological catalysts (biocatalysts) is an appealing alternative. In addition, enzymatic reactions in organic solvents provide numerous industrially attractive advantages, such as increased solubility of non-polar substrates and reduction of viscosity when compared to aqueous systems (Doukyu and Ogino, 2010, Haas et al., 1995).

The aim of this study was to evaluate an alternative process to conventional acidification of SS. In that way, the study includes analysis of the acidification with hydrochloric acid as well as the recovery by enzymatic hydrolysis with organic solvents. Hydrochloric acid was used as an alternative to sulfuric acid for comparison, considering that products aim integration in the biodiesel production process and sulfur contamination makes less viable product use with such purpose.

2 MATERALS AND METHODS

2.1 *Materials and analytical methods*

The SS from the vegetable oil refining (mixture of seeds) was kindly provided by the Portuguese company Frabioleo S.A.. Hexane (LabChem ≥90%) was the organic solvent used. The acid used was the concentrated (37%) hydrochloric acid, and the enzyme was the lipase from *Thermomyces lanuginosus* (Lipolase 100 L, activity ≥100,000 U/g), purchased from Sigma-Aldrich. All the other reagents were of analytical grade.

The SS physicochemical properties determined were moisture and FFA contents. Taking into account the expected high value for moisture content, it was determined by weight loss at $T = 105 \pm 2$°C (oven method), until constant weight, according to EN 12880 (2000); results are expressed as weight percentage, in wet basis. The FFA content was determined according to NP EN ISO 660 (titrimetric method) and the results are expressed as the weight percentage in terms of oleic acid (molar mass of 282 g mol^{-1}).

2.2 *HCl acidification*

As previously mentioned, SS is commonly acidified with sulfuric acid to break the soaps. According to Haas et al. (2003), the acid is added in excess (relative to soaps) to the SS and the mass is boiled for 2 to 4 h. The products are then settled, and the acid water layer separated. The AO is finally water-washed by adding 25 to 50% water to remove mineral acid in excess. The acidic wastewater from this treatment is a significant part of the environmental pollution resulting from such refineries.

Taking into account the information from literature, in this study the SS were treated with hydrochloric acid (Tripathi and Subramanian, 2017, Haas et al., 1995, Morshed et al., 2011), aiming minimum acid consumption. Although being less used industrially, due to the higher price compared to sulfuric acid, the process was evaluated with the advantage of being performed at room temperature and allowing to obtain a sulfur free product to incorporate in the biodiesel industry.

The reaction was carried out with 50 g at room temperature and constant magnetic stirring, with hydrochloric acid (37%) at different amounts relatively to SS (1 wt.%, 2.5 wt.%, 5 wt.% and 10 wt.%) and variable contact time (7 h and 14 h). After acidification, the mixture was centrifuged at 3500 rpm for 15 minutes, allowing to separate the phases and subsequent remove the aqueous phase by decantation. The oily phase was washed twice with hot water (1:1 v/v of oil:water per wash) followed by settling and phase separation to eliminate excess of mineral acid; samples of around 0.5 mL were used to measure the FFA content. All experiments were performed in duplicate and the scatter was evaluated in terms of the relative percentage difference to the mean (RPD), in all cases less than 10%.

2.3 *Enzymatic hydrolysis in organic solvent*

The use of lipases in organic media is of growing importance because of enhanced solubility of the reaction products, better yields and possibly to have higher enzyme stability (Meng et al., 2013).

The reactions were conducted at 42°C in 250 mL Erlenmeyer flasks, stirred at 300 rpm in an orbital shaker, during 20 h, according to Haas et al. (1995). The mass ratio of soapstock:hexane was evaluated using four conditions (1:1, 1:4, 1:8 and 1:16), and the amount of enzyme (10 wt.% and 20 wt.%), according to (Haas et al., 1995). No additional water was added due to the high moisture content of SS. Following incubation, the solvent was removed in a rotary evaporator at about 70°C and samples of around 0.5 mL were used to measure the FFA content. All experiments were performed in duplicate and the scatter was evaluated in terms of the RPD, in all cases less than 10%.

3 RESULTS AND DISCUSSION

3.1 *Soapstock characterization*

SS is described in the literature as an alkaline raw material with a high moisture content (Piloto-Rodríguez et al., 2014). This high water content is due to the neutralization step of the crude oil which depends on the centrifugation step, FFA content of the crude oil, the amount of sodium hydroxide used and the presence of phospholipids in the crude oil (Santos et al., 2014).

The SS here used had moisture of about 47.2 ± 0.1 wt.%; the washing water presented a pH around 8.2 which is a quite low value compared to those found in the literature. Haas et al. (1995) studied a SS containing 45% of water, 12% of neutral oil, 32% of phospholipids and 10% of fatty acids, having pH of 10.3. Knothe et al. (2010) described a SS with approximately 12% of acylglycerols, 10% FFA and 8% phospholipids. It also contained nearly 50% water and it was quite alkaline (typically pH > 9).

In agreement, Wang et al. (2007) described a SS with 47% of water and total fatty acid of about 35%; and, after acidification, they obtained AO with 50.0% of FFA, 3.1% of monoglycerides, 6.9% of diglycerides and 15.5% of triglycerides. Santos et al. (2014) reported a canola SS with moisture of 42 wt.%, corn SS with 53 wt.% and sunflower SS with 30 wt.%.

3.2 *Acidification with HCl*

The acidification was performed (Figure 1) and the effect of the hydrochloric acid load (1, 2.5, 5 and 10 wt.%) was evaluated by determining the acid value. The reaction was performed during 7 h or 14 h, for all the conditions tested. The increase of the reaction time did not bring any benefit, either in terms of mass yield or of acidity increase. As expected, the lowest amounts of hydrochloric acid showed lowest conversions to AO, with final FFA content of 8.1 ± 0.2 wt.%, corresponding to a yield of 11.3 ± 1.2 wt.% for 1 wt.% of HCl at 7 h. The increase of the hydrochloric acid concentration led to an increase of the FFA content. The best condition achieved was 10 wt.% of HCl, which allowed obtaining FFA content of 61.4 ± 0.6 wt.% with a moisture content of about

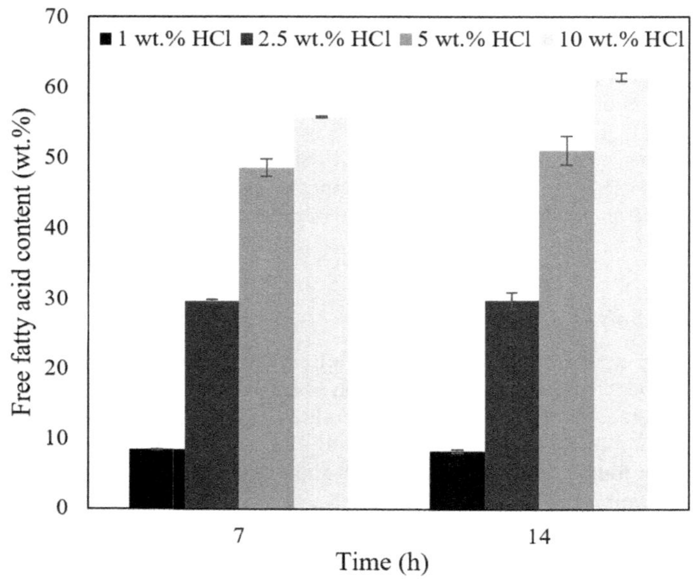

Figure 1. Free fatty acid content during the acidification with hydrochloric acid at different concentra-tions (50 g, room temperature, constant magnetic stirring). Relative percentage differences always lessthan 10% of the mean. Error bars show maximum and minimum values.

2.7 ± 0.3 wt.%, corresponding to a yield of 43.3 ± 1.2 wt.%. Considering the moisture content of the SS (47.2 ± 0.2 wt.%) and according to Wang et al. (2007), the results showed that the process was effective to convert soaps into FFA.

Wang et al. (2007) described the recovery of AO from SS with a sulfuric acid solution (100 g SS, 60 g deionized water, magnetic stirrer) under the room temperature (25 ± 2°C). The recovery yield of AO was about 97% (w/w) based on the total fatty acids of the SS containing around 50% FFA.

3.3 *Enzymatic hydrolysis in organic solvent*

Figure 2 shows the FFA content under the different conditions studied (soapstock: hexane mass ratio and enzyme concentration). In enzymatic reactions, the high price of lipase is a crucial factor to ensure the sustainability of the process (Dave et al., 2010). Although the conversion of FFA increased with increasing catalyst concentration, the differences obtained do not justify the use of 20 wt.% of enzyme.

The highest FFA content was obtained using 20 wt.% of enzyme and 1:8 of soapstock: hexane mass ratio, which afforded a final FFA content of 40.2 ± 1.0 wt.%. However, considering the enzyme and solvent costs and compared to the acid cost and amounts used, the best condition was 10 wt.% of enzyme and 1:8 of soapstock: hexane mass ratio, which allowed to reach a final FFA content around 36.1 ± 0.2 wt.%. Regarding the FFA content of AO described in literature, and according to SS composition described by Haas et al. (2001) and Knothe et al. (2010), the final FFA content achieved indicates that the hydrolysis of tri–, di– and/or monoglycerides that were present in SS may have occurred. Despite such results, they are considered interesting since enzymes are known to be very sensitive to the pH, as it is the case of this raw-material with alkaline pH. Haas et al. (1995) showed that an alkaline SS (pH 10.3) could suffer enzymatic hydrolysis only when neutralized with HC1 before being dissolved in hexane. Using 1 wt.% of HCl and employing the enzymatic process with the solvent, the final FFA content was 47.9 ± 0.5 wt.% after 20 h, which reveals both the hydrolysis and the soaps break by the acid.

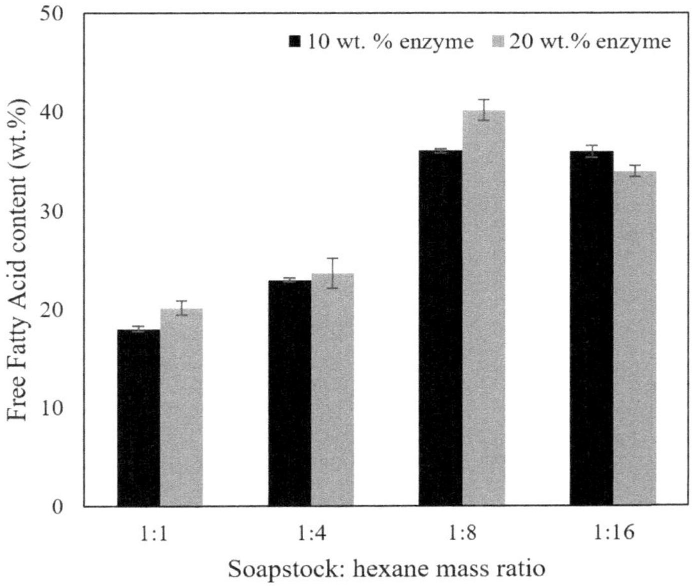

Figure 2. Effect of the soapstock:hexane mass ratio as well asenzyme concentration (lipase from Ther-momyces lanuginosus) in the free fatty acid content of the product after enzymatic process (10 g, 42°C, 300 rpm during 20 h). Relative percentage differences always less than 10% of the mean. Error bars show maximum and minimum values.

Considering the best condition achieved with 10 wt.% of HCl, the enzymatic hydrolysis in organic solvent was tested with 10 wt.% of enzyme and 1:8 of soapstock: hexane mass ratio (this could make possible to obtain both the maximum FFA from soaps break plus that resulting from tri, di and or monoglycerides hydrolysis). After 20 h the FFA content remained constant (~60 wt.%) which may be explained by the inefficient water-washing after acidification at the higher acid concentration, which led to inhibition of the enzyme (Cruz et al., 2018). Further studies will intend to obtain the optimum combination of the evaluated processes as well as a more detailed characterization of the raw material and products.

4 CONCLUSION

Soapstock acidification with HCl was successfully conducted. An acid oil with FFA content of 61.4 ± 0.6 wt.% with mass yield of 43.3 ± 1.2 wt.% was obtained, with 10 wt.% of HCl, indicating effective conversion of soaps into free fatty acids. Under the conditions tested, enzyme concentration only slightly influenced the hydrolysis with organic solvent, whereas the solvent amount presented more relevance. The soapstock:hexane mass ratio of 1:1 was insufficient to allow conversion into FFA. Increasing hexane amount, FFA increased too. The best conditions established were 10 wt.% of enzyme and 1:8 of soapstock:hexane mass ratio, which allowed to obtain a final FFA content of 36.1 ± 0.2 wt.%. The additional test of adding HCl to the reactional mixture showed an increment of FFA to 47.9 ± 0.5 wt.%.

ACKNOWLEDGMENTS

This work was financially supported by project UID/EQU/00511/2019 – Laboratory for Process Engineering, Environment, Biotechnology and Energy – LEPABE funded by national funds through FCT/MCTES (PIDDAC) and Project "LEPABE-2-ECO-INNOVATION" – NORTE-01-0145-FEDER-000005, funded by Norte Portugal Regional Operational Programme NORTE 2020), under PORTUGAL 2020 Partnership Agreement, through the European Regional Development Fund (ERDF).

REFERENCES

Basheer, S. & Watanabe, Y. 2016. Enzymatic conversion of acid oils to biodiesel. *Lipid Technology,* 28, 16–18.

Cruz, M., Pinho, S. C., Mota, R., Almeida, M. F. & Dias, J. M. 2018. Enzymatic esterification of acid oil from soapstocks obtained in vegetable oil refining: Effect of enzyme concentration. *Renewable Energy,* 124, 165–171.

Cvengros, J. 1995. Physical refining of edible oils. *Journal of the American Oil Chemists' Society,* 72, 1193–1196.

Dave, D., E. Ghaly, A., S. Brooks, M. & S, B. 2010. *Production of Biodiesel by Enzymatic Transesterification: Review.*

Dijkstra, A. J. 2017. Questions begging for answers (Part V): Fatty waste. *Lipid Technology,* 29, 3–5.

Doukyu, N. & Ogino, H. 2010. Organic solvent-tolerant enzymes. *Biochemical Engineering Journal,* 48, 270–282.

Dumont, M.-J. & Narine, S. S. 2007. Soapstock and deodorizer distillates from North American vegetable oils: Review on their characterization, extraction and utilization. *Food Research International,* 40, 957–974.

Haas, M. J. 2005. Improving the economics of biodiesel production through the use of low value lipids as feedstocks: vegetable oil soapstock. *Fuel Processing Technology,* 86, 1087–1096.

Haas, M. J., Cichowicz, D. J., Jun, W. & Scott, K. 1995. The enzymatic hydrolysis of triglyceride-phospholipid mixtures in an organic solvent. *Journal of the American Oil Chemists' Society,* 72, 519–525.

Haas, M. J., Michalski, P. J., Runyon, S., Nunez, A. & Wagner, K. 2003. *Production of FAME from acid oil, a by-product of vegetable oil refining.*

Haas, M. J., Scott, K. M., Alleman, T. L. & McCormick, R. L. 2001. Engine performance of biodiesel fuel prepared from soybean soapstock: A high quality renewable fuel produced from a waste feedstock. *Energy and Fuels,* 15, 1207–1212.

Knothe, G., Krahl, J. & Gerpen, J. V. 2010. *The Biodiesel Handbook.*

Laoretani, D. S., Fischer, C. D. & Iribarren, O. A. 2017. Selection among alternative processes for the disposal of soapstock. *Food and Bioproducts Processing,* 101, 177–183.

Meng, Y., Yuan, Y., Zhu, Y., Guo, Y., Li, M., Wang, Z., Pu, X. & Jiang, L. 2013. Effects of organic solvents and substrate binding on trypsin in acetonitrile and hexane media. *Journal of Molecular Modeling,* 19, 3749–3766.

Morshed, M., Ferdous, K., Khan, M. R., Mazumder, M. S. I., Islam, M. A. & Uddin, M. T. 2011. Rubber seed oil as a potential source for biodiesel production in Bangladesh. *Fuel,* 90, 2981–2986.

OECD 2018. *OECD-FAO Agricultural Outlook 2018-2027,* OECD Publishing.

Piloto-Rodríguez, R., Melo, E. A., Goyos-Pérez, L. & Verhelst, S. 2014. Conversion of by-products from the vegetable oil industry into biodiesel and its use in internal combustion engines: a review. *Brazilian Journal of Chemical Engineering,* 31, 287–301.

Santos, R. R. d., Muruci, L. N. M., Santos, L. O., Antoniassi, R., Silva, J. P. L. d. & Damaso, M. n. C. T. 2014. Characterization of Different Oil Soapstocks and Their Application in the Lipase Production by Aspergillus niger under Solid State Fermentation. *Journal of Food and Nutrition Research,* 2, 561–566.

Talal, E. M. A. S., Jiang, J. & Liu, Y. 2013. Chemical refining of sunflower oil: Effect on oil stability, total tocopherol, free fatty acids and colour. *International Journal of Engineering Science and Technology,* 5, 449–454.

Tripathi, S. & Subramanian, K. A. 2017. Experimental investigation of utilization of Soya soap stock based acid oil biodiesel in an automotive compression ignition engine. *Applied Energy,* 198, 332–346.

Wang, Z.-M., Lee, J.-S., Park, J.-Y., Wu, C.-Z. & Yuan, Z.-H. 2007. Novel biodiesel production technology from soybean soapstock. *Korean Journal of Chemical Engineering,* 24, 1027–1030.

Wastes: Solutions, Treatments and Opportunities III – Vilarinho et al. (Eds)
© 2020 Taylor & Francis Group, London, ISBN 978-0-367-25777-4

Production of polyurethane foams from *Betula pendula*

L. Cruz-Lopes, I. Domingos, J. Ferreira, L. Teixeira de Lemos & B. Esteves
Research Centre for Natural Resources, Environment and Society (CERNAS-IPV), CI&DETS, Instituto Politécnico de Viseu, Viseu, Portugal

P. Aires
Department of Environment, Superior School of Technology and Management of Viseu, Viseu, Portugal

ABSTRACT: The production of flexible foams from liquefied bio-wastes is a theme that has been recently approached by various groups of investigators. In this work we have used polyols from liquefied *Betula pendula* bio-wastes for the production of flexible foams. During that process we have made combinations, varying the isocyanate, the expansion agent (water), the catalyst and surfactant, in order to optimize the process. The production, physical and mechanical properties were determined to evaluate their quality ca. density, compression modulus and resistance to compression. Those parameters were determined based on the mass percentage of each reagent in the foam, evaluating the impact that they had in their proprieties.

1 INTRODUCTION

Over the past few decades, the extensive use of fossil resources, in particular the oil in many industries, has led to major shortages of this raw material (Zou et al. 2012). Therefore, industry and research groups have tried to solve this problem with the search for renewable alternatives to the production of various products: liquefied soy wool (Hu et al. 2012; Esteves et al. 2015); liquefied cork (Gama et al. 2014; Esteves et al., 2014); *Quercus cerris* bark (Cruz-Lopes et al. 2016); sugar cane bagasse (Hakim et al. 2011); cork rich barks from *Pseudotsuga menziesii* and *Quercus cerris* (Esteves et al. 2017; Cruz-Lopes 2016) among others.

The most abundant renewable resources are lignocellulosic materials, which can serve as a substitute for oil in many applications. Efforts have been made to convert this feedstock into commercially attractive products by solvent liquefaction (Pan 2011). In solvent liquefaction, this raw material is converted into a liquid through a complex sequence of chemical changes that will lead to smaller molecules (Demirbas, 2000; Hu & Li, 2014; Strezov, 2015). These molecules are unstable and reactive, and can re-polymerize into oily compounds (bio-oil) with a wide range of molecular weight distribution.

Currently the polyurethane (PU) foam production is completely dependent on oil, since the two primary reagents needed for its production, the polyol and the isocyanate, derive from this fossil fuel (Cruz-Lopes et al. 2016; Hu et al. 2012). The choice of reagents for the synthesis of a PU foam, as well as the adequate ratio between them is essential to produce high quality foams that can make them occupy an important position on the world market of high performance synthetic polymers (Thomson, 2005). Commercially, the PU foams are those with greater importance and can be classified into flexible, semi-rigid and rigid, according to the specific mass and mechanical characteristics (Cinelli et al. 2013).

The production of flexible polyurethane foams results mainly from the competing of two distinct reactions, the gelling reaction and the expansion reaction. The polymerization of the PU foams consists of a reaction between an isocyanate and a polyol, with the formation of a urethane group (gelling reaction). At the same time the isocyanate reacts with water and generates an intermediate product, carbamic acid, which due to its instability decomposes in amine and carbon dioxide (expansion reaction). The decomposition of carbamic acid leads to the expansion of the material through the diffusion of gaseous bubbles, forming the PU foam (Mahmood et al. 2016).

Isocyanate is known to give rigidity to the foam while the diol group gives plasticity (Duarte et al. 2017) a balanced proportion of each agent has to be used. There are catalysts acting in the gelling reaction like organometallic compounds (bismuth, iron, mercury and cobalt) or for example 4-diazabicyclo [2, 2, 2] octane (DABCO), and there are others acting in the expansion reaction like tertiary amines (Yan et al. 2008). The blowing agent is responsible for the expansion of the foam. The surfactant, usually a silicone, is responsible for the emulsification of the components, stabilization of the cell structure and adjustment of the size of the cells (Mahmood et al. 2016).

The aim of this work was to test the influence of the proportion of isocyanate, catalyst, surfactant and blowing agent in physical and mechanical properties of the PU foams using liquefied material obtained from *Betula pendula* biopolyol.

2 MATERIAL AND METHODS

2.1 *Material*

The biopolyol used in the production of polyurethane foams was obtained from the liquefaction process of *Betula pendula* bark waste. The 10 g of the material (grain size of 80 mesh), were liquefied in a double walled reactor with heated oil using glycerol as solvent catalysed by potassium hydroxide 6% based on the solvent mass. A ratio of 10:1 (solvent/lignocellulosic material) was used. The liquefaction process was conducted at 200°C for 60 min.

2.2 *Preparation of PU foams*

Polyurethane foams were produced with the neutralized liquefied material. Approximately 3 g of the polyol was mixed in a polypropylene glass, with the catalyst (Polycat 34), the blowing agent (water) and the surfactant (Tegostab B8404). This mixture was stirred for about 30 s at 750 rpm, in an IKA Ost Basic mixer and then added the polymeric isocyanate (MDI Voranate M229). After the addition of the isocyanate, the mixture was stirred again, for a few seconds at 750 rpm until the chemical reaction starts. This procedure was repeated for the preparation of all the PU foams. Several amounts of each chemical were tested in order to determine the effect of different quantities of isocyanate, catalyst, surfactant and blowing agent on the properties of the foam. The results were presented in percentage relative to polyol.

2.3 *Physical and mechanical properties of PU foams*

The density of PU foams was determined by the ratio between the mass and the volume of a cylindrical sample. The mass was measured on a digital scale, and the linear dimensions of the sample were assessed with a digital caliper, Mitutoyo Absolute CD – 15DCX, with an uncertainty of ±0.01 mm.

The determination of the mechanical properties was held in a universal test machine Servosis I – 405/5 according to ISO 844 (Kim et al 2008) standard with some modifications. Each PU foam slice (cylindrical shape) was placed on the machine and compressed at a rate of 10 % of its original thickness per minute.

Since in most of the foams it is impossible to find a maximum compressive strength, compressive stress at 10% deformation was determined. Equation 1 presents the calculations made to determine compressive stress at 10% (σ_10)

$$\sigma_{10} = \frac{F_{10}}{A_0} \quad \text{(MPa)} \tag{1}$$

F_{10} is the applied force for 10% of the sample deformation (N);
A_0 is the area of the base of the cylindrical specimen (mm^2).

Equation 2 presents the calculations made to determine Young's modulus.

$$Young's\ modulus = \frac{\Delta F/_{\Delta x} \times h_0}{A_0} \text{ (MPa)} \tag{2}$$

$\Delta F/\Delta x$ is the slope of the linear zone of the stress vs. deformation curve (N/mm);
h_0 is the average height of the cylindrical specimen (mm);
A_0 is the area of the base of the cylindrical sample (mm^2).

3 RESULTS AND DISCUSSION

3.1 *Influence of isocyanate mass*

Figure 1 presents a typical stress-strain curve where we can see three regions. The first one corresponds to the elastic region where after removing the stress the material will return to its original shape. In the second region there is a large plateau where the foam presents plastic behaviour and the final region is when the cells are crushed and there is the densification of the material.

Figure 2 presents the influence of isocyanate percentage (relative to polyol) in density, uniaxial compression modulus (Young's modulus) and compressive strength of the produced foams.

Results show that with the increase of isocyanate the foam density decreases, from 65.2 kg/m3 to 28.5 kg/m3, when the quantity of isocyanate mass added to the sample was about 8 grams, decreasing afterwards. Regarding the compression modulus, it was found that at first when the isocyanate mass increased the compression modulus decreases, showing later an increase at higher percentages of isocyanate. The compressive strength presented a similar variation to the compression modulus.

The highest values of density and compression modulus 65, 2 kg/m3 and 487 kPa, respectively, were obtained when 167% of isocyanate were used. Some studies show that an increase of isocyanate leads to an increase in density, however this relationship is not linear, and from a certain value of isocyanate mass, the density begins to decrease (Hakim et al. 2011, Yan et al. 2008). This decrease of density is due to the fact that the isocyanate mass, when in excess, reacts with urethane and urea groups, forming allophanate and biuret which form three-dimensional networks (Hakim et al. 2011, Yan et al. 2008). In this study 167%, 233%, 367% and 433% of isocyanate mass were tested. The different variations in the properties with increasing isocyanate, is because the properties of the foams depend on a balance between the various additives, essentially the isocyanate content, the blowing agent and the catalyst. This balance requires a comprehensive study to be conducted for each case. The decrease in density and mechanical properties with increasing isocyanate is due to the increased growth of the foam, thus leading to a decrease in density. For higher values of

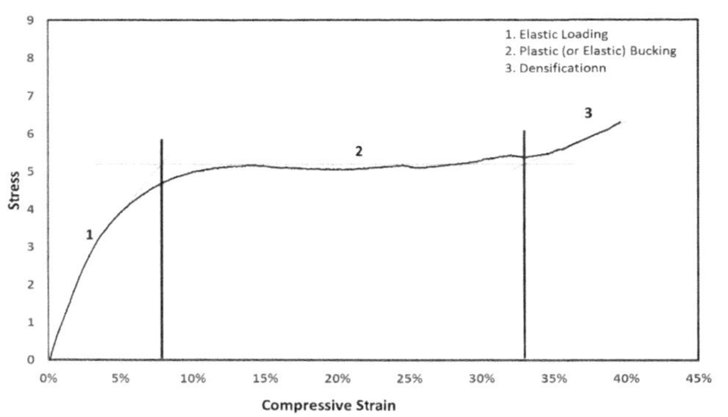

Figure 1. Typical stress-strain curve for the foam.

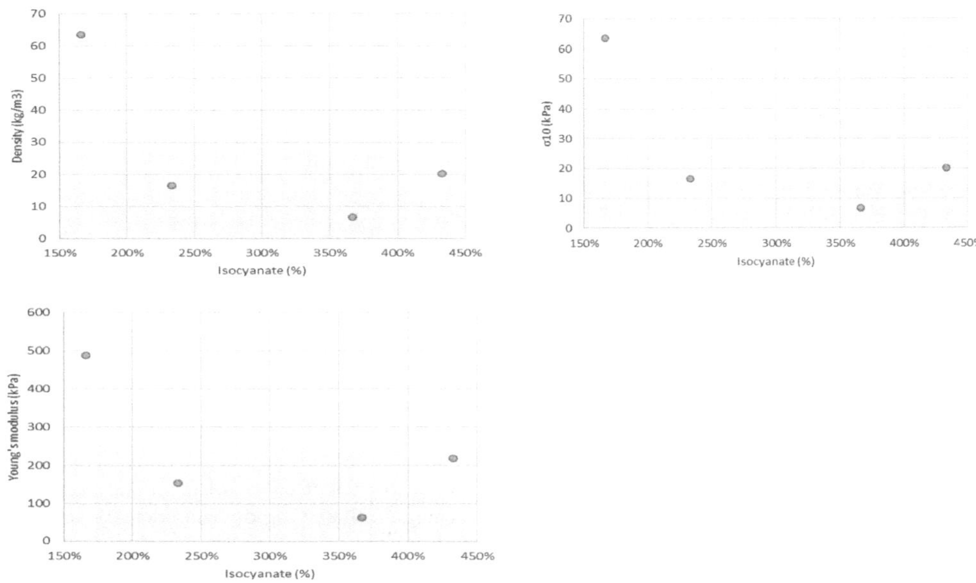

Figure 2. Density, compressive stress (σ_{10}) and Young's modulus as a function of isocyanate percentage.

isocyanate mass an increase of density is expected, when it is impossibility for the foam to grow more, as showed by the values of mechanical properties for values up to 400%.

3.2 *Influence of catalyst mass*

Figure 3 presents the influence of catalyst percentage in density, compression modulus (Young's modulus) and compressive strength of the produced foams.

Figure 3 shows that the use of 8.84% g of catalyst lead to the production of foams with the highest values in almost all physical properties, 37.5 kg/m3, 25.3 kPa and 263 kPa for density, compression stress and Young's modulus, respectively. It seems that an increase in the mass of catalyst, leads to more compression resistant foam. The reagent used catalyses both the reaction of expansion (reaction with water) and formation (reaction with the isocyanate) of the foam, accelerating both reactions. The increase in the speed of the expansion reaction leads to foams with lower density and less resistance to compression, while if there is an increase in the speed of formation reaction the foams show an increase on the density and resistance to compression.

3.3 *Influence of blowing agent mass*

Figure 4 presents the influence of blowing agent percentage in density, compression modulus (Young's modulus) and compressive strength of the produced foams.

Figure 4 analysis allows us to conclude that the best compression modulus and compressive strength was achieved using 7.64% of blowing agent. It can be verify that with the increase in blowing agent mass, compressive strength presents an oscillating variation. According to Choe et al. (2004), when the agent is just water, it reacts with the isocyanate and producing di-substituted urea that then reacts with isocyanate, originating biuret and establishing additional networks. This is the reason why density and the resistance to compression remains constant (Choe et al. 2004). Increasing this agent usually drives to a decrease of density which is not always the case. This is

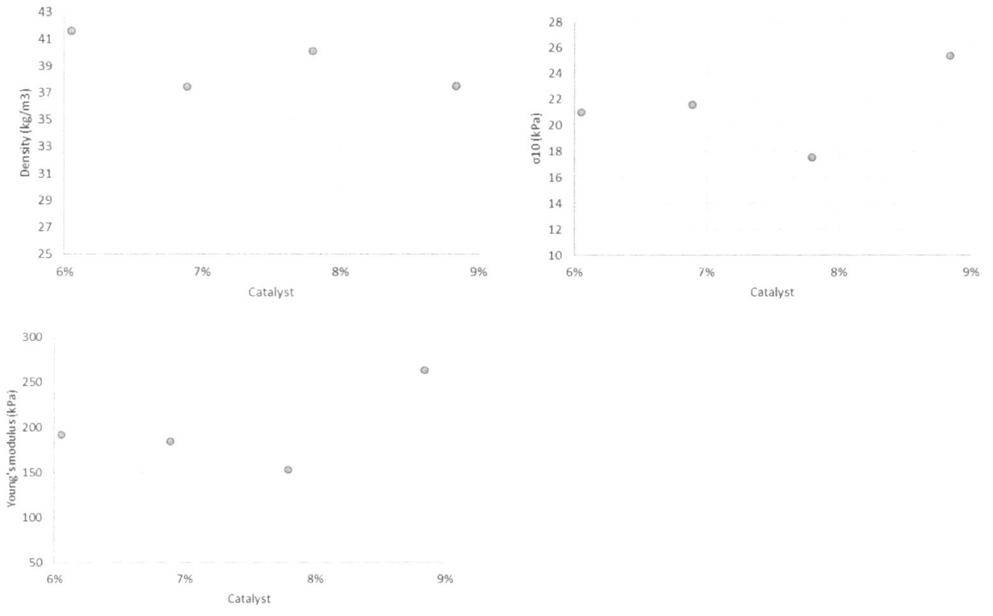

Figure 3. Density, compressive stress (σ_{10}) and Young's modulus as a function of catalyst percentage.

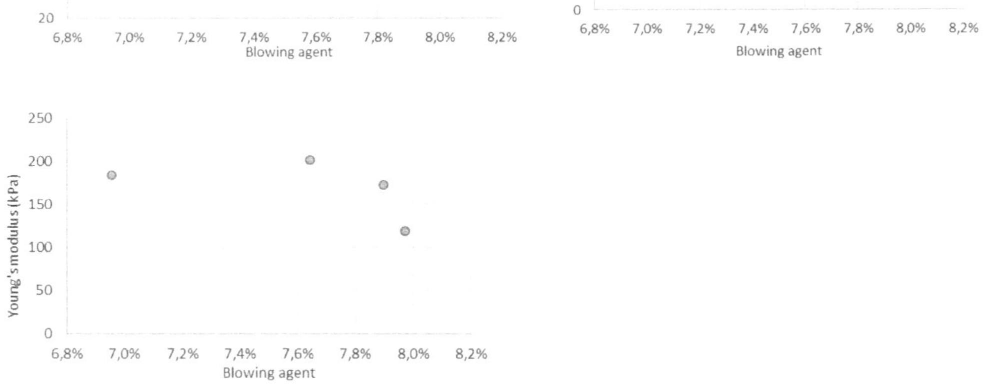

Figure 4. Density, compressive stress (σ_{10}) and Young's modulus as a function of blowing agent percentage.

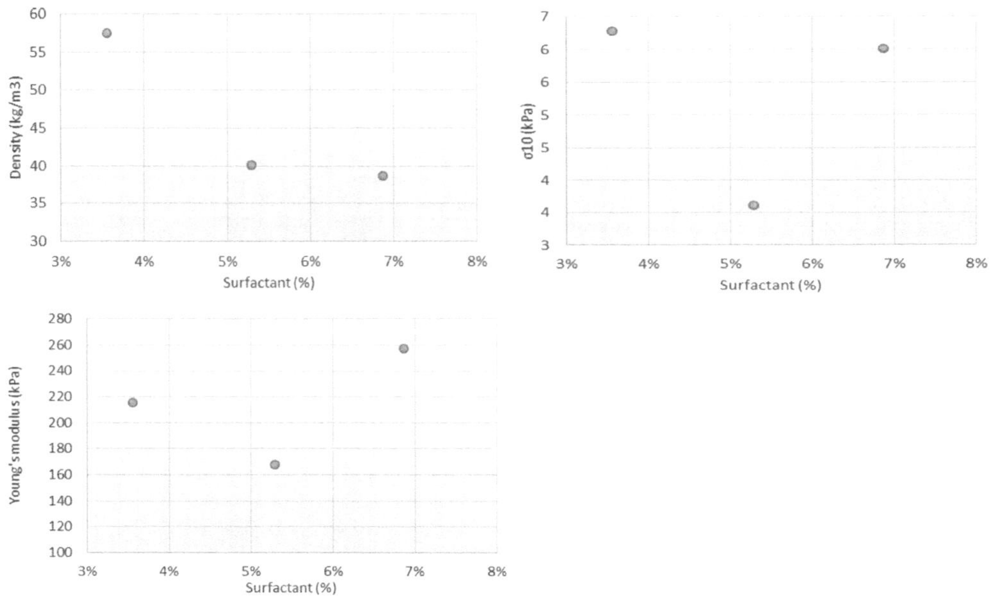

Figure 5. Density, compressive stress (σ_{10}) and Young's modulus as a function of surfactant percentage.

due to the balance with other additives or the fact that the maximum of the expansion has been reached and the further increase of water does not improve the reaction.

3.4 *Influence of surfactant mass*

Figure 5 presents the influence of surfactant percentage in density, compression modulus (Young's modulus) and compressive strength of the produced foams.

The influence of the increase of surfactant mass in function of compressive strength is very variable (Figure 5). The highest value obtained in terms of compressive strength was respectively 28.7 kPa (using a 3.56% of surfactant). Published studies, regarding the effect of surfactant on compressive strength, are divergent. According to Yan et al., who studied the effect of surfactant, the compressive strength of PU foams obtained from liquefied biomass, proved that this property displays nonlinear behavior in relation to the increase of surfactant. In this investigation the curve of the compressive strength was represented as a function of the mass of surfactant used, and the curve showed that increasing mass of surfactant, initially the resistance increases, up to a maximum, then there is a decrease, but continuing to increase the mass of surfactant, the resistance increases again (Yan et al. 2008).

4 CONCLUSIONS

The results showed that the properties of the foam could be tailored by a careful choice of the additives used in the foam formation. It is possible to convert liquefied lignocellulosic wastes of *Betula pendula* into PU foams, a product with potential value for various applications. The PU foams which registered better mechanical properties were obtained using 167% of isocyanate, 7.83% blowing agent, 7.80% of catalyst and 5.29% of surfactant. The properties obtained have a density, a compression modulus (Young) and a compressive strength of 65.2 kg/m^3, 63.5 kPa and 487 kPa, respectively.

ACKNOWLEDGMENTS

This work is financed by national funds through FCT – Fundação para a Ciência e Tecnologia, I.P., under the project UID/Multi/04016/2016. Furthermore we would like to thank the Instituto Politécnico de Viseu and CI&DETS for their support.

REFERENCES

Choe, K. H., Lee, D. S., Seo, W. J. & Kim, W. N. 2004. Properties of Rigid Polyurethane Foams with Blowing Agents and Catalysts. Polymer Journal 36(5): 367–374.

Cinelli, P., Anguillesi, I. & Lazzeri, A. 2013. Green synthesis of flexible polyurethane foams from liquefied lignin. European Polymer Journal: 1174–1184.

Cruz-Lopes, L., Silva, H.C., Domingos, I., Ferreira, J., Lemos, L.T. & Esteves, B. 2016. Optimization of Quercus cerris Bark Liquefaction. Int. J. Chem. Mol. Nucl. Mater. Metall. Eng. 10(8): 1060–1063.

Cruz-Lopes, L., Rodrigues, L., Domingos, I., Ferreira, J., Lemos, L. & Esteves, B. 2016. Production of polyurethane foams from bark wastes. Int. J. Chem. Mol. Nucl. Mater. Metall. Eng. 10(8): 1056–1059.

Demirbas, A. 2000. Mechanisms of liquefaction and pyrolysis reactions of biomass. Energy Convers Manage 41:633–646.

Duarte, J., Cruz-Lopes, L., Dulyanska, Y., Domingos, I., Ferreira, J., de Lemos, L.T. & Esteves, B. 2017. Orange peel liquefaction monitored by FTIR. J. Int. Sci. Publ. 5: 309–313.

Esteves, B., Cruz-Lopes, L., Ferreira, J., Domingos, I., Nunes, L. & Pereira, H. 2017. Optimizing Douglas-fir bark liquefaction in mixtures of glycerol and polyethylene glycol and KOH. Holzforschung. 72: 25–30.

Esteves, B., Dulyanska, Y., Costa, C., Ferreira, J.V., Domingos, I., Pereira, H., Lemos, L.T. & de, Cruz-Lopes, L. 2017. Cork Liquefaction for Polyurethane Foam Production. BioResources. 12: 2339–2353.

Esteves, B., Martins, J., Martins, J., Cruz-Lopes, L., Vicente, J. & Domingos, I. 2015. Liquefied wood as a partial substitute of melamine-urea-2 formaldehyde and urea-formaldehyde resins 3. Maderas-Cienc Tecnol. 17: 3.

Gama, N. V., Soares, B., Freire, C. S.R., Silva, R., Brandão, I., Neto, C. P., Barros-Timmons, A. & Ferreira, A. 2014. Rigid polyurethane foams derived from cork liquefied at atmospheric pressure. Polymint 64: 250–258.

Hakim, A.A.A., Nassara, M., Emamb, A. & Sultana, M. 2011. Preparation and characterization of rigid polyurethane foam prepared from sugar-cane bagasse polyol. Materials Chemistry and Physics 129: 301–307.

Hu, S. & Li, Y. 2014. Two-step sequential liquefaction of lignocellulosic biomass by crude glycerol for the production of polyols and polyurethane foams. Bioresource Technology: 410–415.

Hu, S. & Wan, C., Li, Y. 2012. Production and characterization of biopolyols and polyurethane foams from crude glycerol based liquefaction of soybean straw. Bioresource Technology 103: 227–233.

Kim, S.H., Kim, B.K. & Lim, H. 2008. Effect of isocyanate index on the properties of rigid polyurethane foams blown by HFC 365mfc. Macromol. Res. 16: 467–472.

Mahmood, N., Yuan, Z., Schmidt, J. & Xu, C. 2016. Depolymerization of lignins and their applications for the preparation of polyols and rigid polyurethane foams: A review. Renew. Sustain. Energy Rev. 60: 317–329.

Pan, H. 2011. Synthesis of polymers from organic solvent liquefied biomass: A review. Renewable and Sustainable Energy Reviews: 3454–3463.

Strezov, V. & Evans, T.J. 2015. Biomass Processing Technologies: 1–12.

Thomson, T. 2005. Polyurethanes as Specialty Chemicals – Principles and Applications, 1a Ed. CRC Press.

Yan, Y., Pang, H., Yang, X., Zhang, R. & Liao, B. 2008. Preparation and characterization of water blown polyurethane foams from liquefied cornstalk polyol. J. Appl. Polym. Sci. 110: 1099–1111.

Zou, X., Qin, T., Wang, Y., Huang, L., Han, Y. & Li, Y. 2012. Synthesis and properties of polyurethane foams prepared from heavy oil modified by polyols with 4,4-methylene-diphenylene isocyanate (MDI). Bioresource Technology: 654–658.

Wastes: Solutions, Treatments and Opportunities III – Vilarinho et al. (Eds)
© 2020 Taylor & Francis Group, London, ISBN 978-0-367-25777-4

Recovery of byproducts from the rice milling industry for biofuel production

E.T. Costa, M.F. Almeida & J.M. Dias
LEPABE, Department of Metallurgical and Material Engineering, Faculty of Engineering, University of Porto, Porto, Portugal

C. Alvim-Ferraz
LEPABE, Department of Chemical Engineering, Faculty of Engineering, University of Porto, Porto, Portugal

ABSTRACT: In this study, the efficiency of enzymatic esterification and homogenous acid esterification for the reduction of Free Fatty Acid (FFA) content of acid rice bran oil (47 mg KOH g^{-1}) was studied. Further, alkaline transesterification of the pretreated oil mixed with sunflower oil (to achieve FFA content ≤2%) was conducted to maximize fatty acid methyl esters (FAME) production. Homogeneous acid esterification was conducted at respectively 65°C and 45°C, using 2 wt.% and 4 wt.% of H_2SO_4 as catalyst and 9:1 methanol:oil molar ratio. Enzymatic esterification was performed at 35°C using a 6:1 molar ratio of methanol:oil and 5 wt.% of catalyst. Homogenous acid esterification allowed to achieve the highest acid value reduction (92%), although enzymatic pretreatment alone allowed to obtain the highest FAME content (86 wt.%) compared to acid catalyzed process (41 wt.%). The obtained results show potential of this alternative raw material for bioenergy purposes.

1 INTRODUCTION

Fossil fuels are the main resource used for energy production in developing and developed countries, representing more than 80 % of the world primary energy consumption (Mohr et al., 2015). Several problems are associated to this kind of fuels, namely, air pollution, global warming and direct and indirect health problems (Cruz et al., 2018). In addition, the instability of related markets and prices reduce the attractiveness of fossil fuels. For this reason, the production of energy from renewable sources have been adopted as alternative to the conventional fuels in Europe; biofuels are one of the more relevant alternatives to replace fossil fuels in the transport sector (Directive 2009/28/EC). Several strategies have been adopted to encourage the use of biodiesel especially that produced from feedstock which does not compete with food market (Directive 2009/28/EC).

Rice is the second most consumed cereal in the world, only exceed by wheat. In Portugal, in 2017, the area of rice sown was around 30 000 ha corresponding to an annual production around 180 kt (INE, 2018). It should be noted that the national consumption per capita is in average 15.3 kg, three times more than the European average. Crambe oil, jatropha oil, castor oil or waste raw materials and byproducts, e.g. fats and used cooking oil might be used as more sustainable feedstocks for biodiesel production, compared to the traditional food oils (Costa et al., 2019). According to economic indicators, more than 80% of the biodiesel production costs are related to the feedstock and for this reason alternative low cost raw materials should be studied (Knothe and Razon, 2017). Byproducts such as rice bran might be seen as a non-conventional and low cost feedstock for biodiesel production (El Khatib et al., 2018). This byproduct results from the milling process of rice grain. The processing consists in the removal of the husk (dehulling) and bran layers, resulting the white rice product at the end. Dehulling process generates rice husk equivalent to 20% of the unprocessed paddy rice (rice harvested) (Chen et al., 2019). Rice husk is usually used for energy/heat production in the industrial process (Chen et al., 2019). Bran residue corresponds to 10 % of the paddy rice (Sharif et al., 2014). Several studies showed that bran residue has high oil content (between 15 wt.% and 20 wt.%) and for this reason could be used for biofuels production (El Khatib et al., 2018, Raja Rajan and Gopala Krishna, 2009). After the milling process, the quick deterioration of the bran makes this byproduct unsuitable for human consumption (El Khatib et al., 2018); in agreement, rice bran is mostly used for animal feed, namely due to its high protein content (Raja Rajan and Gopala Krishna, 2009). However, it is estimated that only 10% of rice bran oil is used as edible oil, due to the high content of free fatty acids (FFA) (El Khatib et al., 2018). In Portugal, rice bran is mostly valued for animal feed corresponding to an annual consumption of 8122 t in 2017 higher than that of 4592 t for 2016 (INE, 2018). In 2017, the use of rice bran for animal feed corresponded to approximately 44% of the total amount of rice bran produced in Portugal. For this reason, it is essential to study alternatives for the recovery of the remaining byproduct, which from the gathered information, does not have a clear application. At an industrial scale, biodiesel is conventionally produced by a chemical process which includes a homogeneous alkaline transesterification reaction. Alkaline catalysis cannot however be employed when the raw material has a high FFA content, since it reacts leading to soap formation and the limit to apply the conventional process is 1wt.% FFA (Costa et al., 2019). To overcome this drawback, the pretreatment of the oil is necessary, namely by acid or enzymatic esterification. Depending on the conditions, both processes might enable both the esterification and the transesterification, although at a slower reaction rates than homogeneous alkali processes (Moazeni et al., 2019). Esterification reaction leads to the production of fatty acid methyl esters (biodiesel) and water. From the transesterification reaction, biodiesel and glycerol are obtained.

The present work evaluated the pretreatment of acid rice bran oil (RBO) extracted from rice bran byproduct by enzymatic and acid esterification in order to establish alternatives for viable recovery of this byproduct aiming integration in the biofuels industry.

2 EXPERIMENTAL

2.1 *Materials and methods*

The rice bran by-product was provided by the company Novarroz S.A (North of Portugal). Commercial sunflower oil was blended with RBO for transesterification reaction. Methanol (Fischer Scientific $\geq$99%) was used as the acyl acceptor in the chemical reactions. The biocatalyst used was lipase *Thermomyces lanuginosus* (Lipolase 100 L, activity $\geq$100,000 U/g), purchased from Sigma-Aldrich. For homogenous acid esterification, sulfuric acid was employed with concentration of 98% (Fischer Chemical). The other reagents were of analytical grade.

2.2 *Analytical methods*

The raw material and biodiesel were characterized taking into account several parameters. The oil content of the rice bran was determined using a 100 mL Soxhlet extractor, following NP EN ISO 659 (2002). Crude oil quality was evaluated for the following parameters: (i) oxidation stability by accelerated oxidation using an automatic equipment (873 Biodiesel Rancimat from Metrohm); (ii) viscosity at 40°C, measured according to ISO 3104 using capillary viscometers immersed in a thermostatic bath; and (iii) acid value (AV) by volumetric titration according to NP EN ISO 660. For biodiesel characterization, fatty acid methyl esters content parameter was measured by gas chromatography (GC) according to EN 14103. GC analysis were conducted using a Dani Master GC with a DN-WAX capillary column of 30 m, 0.25 mm internal diameter and 0.25μm of film thickness. The temperature program was: 120°C as the starting temperature, rising at 4°C per minute, up to 220°C, held for 10 min.

2.3 *Oil extraction*

The oil extraction was performed using a Soxhlet extractor (1 L). The raw material was introduced inside a thimble and dipped in petroleum ether solvent for 6 h (equivalent to 14 turns of the solvent in the extractor). After, the solvent was removed in a rotary evaporator at 70°C. Several extraction cycles were conducted until obtaining the necessary amount of oil for the study. Crude oil was after submitted to centrifugation at 3500 rpm during 10 min and the lipid fraction – Rice Bran Oil (RBO) – was used in the following studies.

2.4 *Pretreatment of rice bran oil*

Regarding the reduction of FFA content, two different pretreatments were performed, namely, heterogeneous (enzymatic) and homogenous (acid) esterification.

2.4.1 *Pretreatment of RBO by heterogeneous/enzymatic esterification*

For enzymatic reaction, the oil (15 g) was added to a batch reactor in an orbital shaking incubator (Agitorb 200IC), with constant stirring of 200 rpm, during 24 h at 35°C according to the literature (Cruz et al., 2018). The temperature of reaction adopted was the optimum temperature suggested by the enzyme supplier. The methanol:oil molar ratio adopted (6:1) was according to that referred in several studies (Cruz et al., 2018, Arumugam and Ponnusami, 2017). After reaching the reaction temperature (35°C), 5 wt.% of enzyme was added to the reactional mixture (Cruz et al., 2018). The total amount of methanol was added to the oil in 5 different steps with the objective to avoid enzymatic inhibition. The FFA content was assessed for the following instants: 0 min; 15 min, 30 min, 60 min, 120 min, 240 min, 480 min and 24 h. At the end of the reaction, the excess of methanol was removed in a rotary evaporator and the remaining fraction centrifuged during 5 min at 3500 rpm. The AV of the oil was then evaluated in the resulting product. This procedure was performed in duplicate.

2.4.2 *Pretreatment of RBO by homogeneous/acid esterification*

For the acid esterification reactions, the following reference conditions were selected according to the literature (Dias et al., 2009, Costa et al., 2017). Two different temperatures were tested (45°C and 65°C) and the amount of catalyst (sulfuric acid) were 2 wt.% and 4 wt.% (in relation to the weight of the oil). For all experiments, the methanol:oil molar ratio used was 9:1, referred in the literature as the condition which allows the highest FFA reduction (Costa et al., 2017).

The tests were performed in a batch reactor and FFA content of the mixture was measured at: 0 min; 15 min, 30 min, 60 min, 120 min, 240 min and 480 min. The amount of sample (RBO) used for each experiment was also 15 g. At the end of the esterification reactions, the pretreated oils were washed with 50 vol.% of water to remove the excess of catalyst and dried using a rotary evaporator. The final product was evaluated in terms of AV and fatty acid methyl ester (FAME) content. This procedure was performed in duplicate for the condition, which, in the preliminary experiments, led to a higher FFA reduction.

2.4.3 *Transesterification*

Synthesis of biodiesel was performed, with the best result in terms of acidity reduction, by homogenous acid esterification. The product of homogenous acid esterification (FFA content >2 wt.%) was previously mixed with sunflower oil (SFO), to obtain a raw material with suitable acidity for alkaline transesterification. In the presence of acid catalyst, both transesterification and esterification might occur; however, low FAME content is expected due to the slow reaction rates. For this reason, transesterification reaction was only performed for the pretreated oil obtained by acid esterification, which presented a low FAME content. The alkaline transesterification was performed in a batch reactor at 65°C during 1 h. The amount of catalyst was 1 wt.% NaOH and the methanol:oil molar ratio was 6:1 (Costa et al., 2019). Acid/water washings and drying procedures were also performed according to previously published procedures (Costa et al., 2019).

The weight percentage of each oil (pretreated RBO and sunflower oil) mixed to obtain a blend with acidity close to 2 wt.% was calculated (virgin oil presented an acidity of 0.39%) according to equation 1.

$$Acidity_{mixture} = (Acidity_{RBO} * X) + (Acidity_{SFO} * (1 - X)) \quad (1)$$

3 RESULTS AND DISCUSSION

3.1 *Raw material properties*

The oil content of the rice bran was 18.9 ± 0.5 wt.%, suitable for energy production and similar to previously published values (Raja Rajan and Gopala Krishna, 2009). The properties of the RBO are presented in Table 1.

As referred in several studies, RBO shows high content of FFA due to natural degradation of the oil (Raja Rajan and Gopala Krishna, 2009, Ihoeghian and Usman, 2018). The FFA content was in agreement to that presented in others studies (Ihoeghian and Usman, 2018). This oil is recognized as having high viscosity, close to that mentioned in the study of Diamante and Lan (2014). For oxidation stability few information is available in the literature. The value obtained for the RBO oxidation stability (4.77 h) was higher than that found in other study for rice bran oil with 1 month of storage (Maszewska et al., 2018). Further studies should be developed to identify the presence of natural antioxidants. For a commercial biodiesel, the oxidation stability has to be higher than 8 h, value very near to the found for RBO crude oil. It is expected that with the processing of the RBO the oxidation stability decreases due to the removal of natural antioxidants, namely during the required washing steps.

3.2 *Enzymatic esterification pretreatment*

The application of enzymatic esterification had as main goal the reduction of the FFA content of RBO until acceptable levels (2 mg $KOH.g^{-1}$), aiming its subsequent application for biodiesel production. For the beginning of the reaction, the AV obtained were different from the acid value mentioned previously in section 3.1, because acid value of the mixture is affected by the presence of methanol and enzyme (dilution effect). Figure 1 shows the evolution of FFA content at 35°C, after addition of methanol in five steps.

The results showed that the conversion of FFA occurs very quickly, and after 60 min the reaction tends to stabilize (around a final value of 10 mg $KOH.g^{-1}$). After centrifugation and evaporation of methanol to purify the oil, the AV of the RBO was of 7.2 mg $KOH.g^{-1}$. Enzymatic esterification was also performed with addition of methanol in only one step; however, the AV of the mixtures remained unchanged, thus confirming the inhibition of the enzyme by the alcohol as reported in different studies (Moazeni et al., 2019, Cruz et al., 2018). For pretreated oil, the FAME content was 86.2 wt.%. As initial FFA content represent only 23 wt.% of oil's mass that is possible to convert into FAME, it is possible to conclude that both esterification and transesterification occurred. Similar results were previously published (Veljković et al., 2006, Cruz et al., 2018).

Table 1. Physicochemical properties of acid RBO.

	Acid value (mg KOH g^{-1})	Viscosity at 40°C ($mm^2.s^{-1}$)	Oxidation Stability (h)
Result	46.5	29.0	7.6
RPD* (%)	0.3	5	5

*Relative Percentage Difference to the mean.

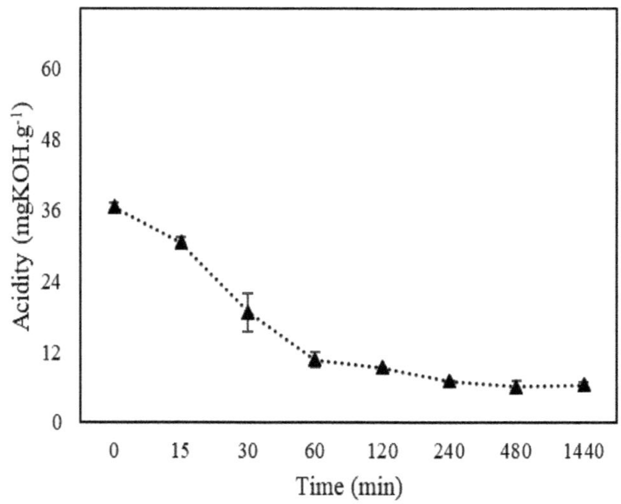

Figure 1. Evolution of the acid value during enzymatic esterification of RBO (35°C, 5 wt.% of enzyme, 6:1 methanol:oil molar ratio, stepwise addition). Mean values presented; errors bars show maximum and minimum values.

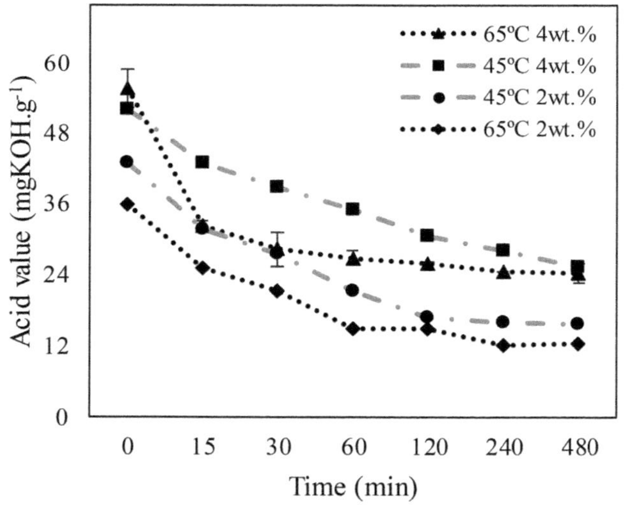

Figure 2. Evolution of the acid value during the homogenous acid esterification of the acid oil (45°C and 65°C). Mean values presented; errors bars show maximum and minimum values.

3.3 *Homogenous acid esterification pretreatment*

In Figure 2 is depicted the evolution of the acid value present in the mixture during homogenous acid esterification.

As shown in Figure 2, for a temperature of 65°C the reaction progresses faster than at 45°C (for the same catalyst concentration). Although at the end, the acid value of the mixtures were very similar (using 4 wt.% of catalyst at 45°C or 65°C led to AV around 25 mg $KOH.g^{-1}$, whereas using 2 wt.% of catalyst at 45°C or 65°C led to AV around 14 mg $KOH.g^{-1}$). For 65°C, after 60 min the reaction tends to stabilize, although for 45°C the reduction of acid value is only reached at

Table 2. Characterization of the RBO after homogenous acid esterification pretreatment.

	65°C 4 wt.%	65°C 2 wt.%	45°C 4 wt.%	45°C 2 wt.%
Acid value (mg $KOH.g^{-1}$)	3.69	5.35	6.82	5.83
FAMES (wt.%)	40.7	30.4	28.2	23.7

more than 120 min for the highest concentration or at the end of 480 min at the lowest. It should be noted that the AV are not comparable in the graph for both concentrations since the mineral acid affects the acid value of the mixture. Table 2 presents the AV after purification of the mixture and removal of the mineral acid.

According to the results, homogenous acid esterification performed at 65°C with 4 wt.% of sulfuric acid was chosen as the best condition for pretreat RBO since it allowed the highest FAME content and lowest AV.

3.4 *Alkaline transesterification of oil mixtures*

The RBO pretreated by homogenous acid esterification at 65°C with 4 wt.% of catalyst (best condition) was submitted to alkaline transesterification. The AV of this oil was 3.7 mg $KOH.g^{-1}$ and for this reason alkaline transesterification can't be performed directly using this raw material. As consequence, pretreated RBO was blended with sunflower virgin oil (low acidity) to produce a mixture with an AV lower than 2 mg $KOH.g^{-1}$, as previously referred (section 2.4.3). The final product presented a FAME content of 88 wt.%, which although high, is lower than that required for use as automotive fuel (>96.5 wt.%). This biodiesel was mostly constituted by C18:1, oleic acid methyl ester (47 wt.%), and C18:2, linoleic acid methyl ester (41 wt.%).

4 CONCLUSION

In this study, rice bran was used as source of acid oil (18.9 ± 0.5 wt.%). Enzymatic esterification and homogenous acid esterification proved to be effective to reduce the free fatty acid content of rice bran oil (from 47 mg $KOH.g^{-1}$ until 4 mg $KOH.g^{-1}$). Homogenous acid esterification was the pretreatment which allowed the highest AV reduction (92%), although enzymatic pretreatment alone allowed to obtain high FAME content (86 wt.%) compared to homogeneous acid esterification (41 wt.%). During homogeneous esterification, catalyst amount and temperature clearly influenced the rate of the reaction. The best conditions established were 65°C and 4 wt% of catalyst. The alkaline transesterification of the pretreated RBO by the homogenous process allowed to obtain 88 wt.% of FAME content; although high, it is still below the value set by the standard requirements for biodiesel use. Independently, RBO showed a clear potential as alternative/complementary raw material for bioenergy production. In addition, the enzymatic process seems to be promising aiming the one step recovery of such raw material.

AKNOWLEDGEMENTS

This work was financially supported by project UID/EQU/00511/2019 – Laboratory for Process Engineering, Environment, Biotechnology and Energy – LEPABE funded by national funds through FCT/MCTES (PIDDAC) and Project "LEPABE-2-ECO-INNOVATION" – NORTE-01-0145-FEDER-000005, funded by Norte Portugal Regional Operational Programme (NORTE 2020), under PORTUGAL 2020 Partnership Agreement, through the European Regional Development Fund (ERDF). The authors also acknowledge Foundation for Science and Technology for funding Emanuel Costa's (PD/BD/114312/2016) PhD fellowship.

REFERENCES

Arumugam, A. & Ponnusami, V. 2017. Production of biodiesel by enzymatic transesterification of waste sardine oil and evaluation of its engine performance. *Heliyon* 3: e00486.

Chen, W., Oldfield, T. L., Katsantonis, D., Kadoglidou, K., Wood, R. & Holden, N. M. 2019. The socio-economic impacts of introducing circular economy into Mediterranean rice production. *Journal of Cleaner Production* 218: 273–283.

Costa, E., Almeida, M. F., Alvim-Ferraz, C. & Dias, J. M. 2019. The cycle of biodiesel production from Crambe abyssinica in Portugal. *Industrial Crops and Products* 129: 51-58.

Costa, E., Cruz, M., Alvim-Ferraz, C., Almeida, M. F. & Dias, J. 2017. Acid esterification vs glycerolysis of acid oil soapstock for FFA reduction. 287-292; *WASTES–Solutions, Treatments and Opportunities II*. CRC Press.

Cruz, M., Pinho, S. C., Mota, R., Almeida, M. F. & Dias, J. M. 2018. Enzymatic esterification of acid oil from soapstocks obtained in vegetable oil refining: Effect of enzyme concentration. *Renewable Energy* 124: 165-171.

Diamante, L. M. & Lan, T. 2014. Absolute viscosities of vegetable oils at different temperatures and shear rate range of 64.5 to 4835 s- 1. *Journal of food processing,* 2014.

Dias, J. M., Alvim-Ferraz, M. C. M. & Almeida, M. F. 2009. Production of biodiesel from acid waste lard. *Bioresource Technology* 100: 6355–6361.

El Khatib, S. A., Hanafi, S. A., Barakat, Y. & Al-Amrousi, E. F. 2018. Hydrotreating rice bran oil for biofuel production. *Egyptian Journal of Petroleum* 27: 1325–1331.

Ihoeghian, N. A. & Usman, M. A. 2018. Exergetic evaluation of biodiesel production from rice bran oil using heterogeneous catalyst. *Journal of King Saud University – Engineering Sciences*.

INE; Statistics Portugal for 2018.

Knothe, G. & Razon, L. F. 2017. Biodiesel fuels. *Progress in Energy and Combustion Science* 58: 36–59.

Maszewska, M., Florowska, A. Dłużwska, E., Wroniak, M., Marciniak-Lukasiak, K. & Żbikowska, A. 2018. Oxidative Stability of Selected Edible Oils. *Molecules (Basel, Switzerland)* 23: 1746.

Moazeni, F., Chen, Y.-C. & Zhang, G. 2019. Enzymatic transesterification for biodiesel production from used cooking oil, a review. *Journal of Cleaner Production* 216: 117–128.

Mohr, S. H., Wang, J., Ellem, G., Ward, J. & Giurco, D. 2015. Projection of world fossil fuels by country. *Fuel* 141: 120–135.

Raja Rajan, R. & Gopala Krishna, A. 2009. Refining of high free fatty acid rice bran oil and its quality characteristics. *Journal of food lipids* 16: 589–604.

Sharif, M. K., Butt, M. S., Anjum, F. M. & Khan, S. H. 2014. Rice bran: A novel functional ingredient. *Critical reviews in food science and nutrition* 54: 807–816.

Veljković, V. B., Lakićević, S. H., Stamenković, O. S., Todorović, Z. B. & Lazić, M. L. 2006. Biodiesel production from tobacco (Nicotiana tabacum L.) seed oil with a high content of free fatty acids. *Fuel* 85: 2671–2675.

Wastes: Solutions, Treatments and Opportunities III – Vilarinho et al. (Eds)
ISBN 978-0-367-25777-4

Olive mill wastewater treatment by electro-Fenton with heterogeneous iron source

A.H. Ltaïef, A. Gadri & S. Ammar
Electrochemistry, Water and Environment, Faculty of Sciences of Gabes, Erriadh City, Gabes, Tunisia

A. Fernandes, M.J. Nunes, L. Ciríaco, M.J. Pacheco & A. Lopes
FibEnTech-UBI and Department of Chemistry, Universidade da Beira Interior, Covilhã, Portugal

ABSTRACT: The production of olive oil is an added value for the Mediterranean countries, although it presents a negative impact to the environment, due to olive mill effluents, with recalcitrant properties. In this study, the degradation of olive mill wastewater (OMW) from Portuguese and Tunisian olive mills was performed by utilizing the electro-Fenton (EF) process, using as iron source commercial iron salts and mined pyrite and chalcopyrite. The EF process proved to be suitable to treat OMWs and, utilizing the most favorable experimental conditions, after 8 h assay, a chemical oxygen demand removal of 22.7 g L^{-1} was attained. Pyrite proved to be a more efficient iron source than chalcopyrite. The use of iron-containing solid catalysts, pyrite and chalcopyrite, when compared with iron sulfate or chloride, showed to promote identical treatment results, with the advantage that they can be recovered and reutilized, since they are kept in suspension.

1 INTRODUCTION

The production of olive oil is a very important activity for the economy of the Mediterranean countries and accounts for approximately 95% of the worldwide olive oil production. It is a major agro-industrial activity for countries, such as Tunisia and Portugal. However, olive mills wastewater (OMW) is generated during the olive oil extraction and the Mediterranean countries are confronted with the problem of OMW elimination. Wastewaters from olive processing industry present high organic load due to the high concentrations in organic acids, sugars, tannins and aromatic compounds, which are very difficult to eliminate (Khoufi et al. 2006). They represent a serious environmental contamination source, since they are often discharged untreated into water streams and soil, or stored in evaporation ponds. Thus, there is great interest in developing appropriate technologies for the treatment of this effluent. Studies have been conducted to treat these effluents by conventional processes (Beltran et al. 2008, Pulido 2016, Flores et al. 2018). Most of these technologies are insufficient to destroy toxic and biorecalcitrant micropollutants and others are expensive or generate sludge.

OMW organic pollutants can be removed using advanced oxidation processes (AOPs), which are able to non-selectively destroy most organic and organometallic contaminants until their mineralization. They are all based on the *in situ* production and utilization of the hydroxyl radical ($^{\bullet}OH$), a powerful oxidizing agent. Although studies involving the hydroxyl radical to treat OMW were already conducted (Barbosa et al. 2016, Bellakhal et al. 2006) more research is necessary to improve the treated effluent and to decrease treatment costs.

Nowadays, electro-Fenton (EF) technique is under considerable development (Brillas et al. 2009). In this process, an acid solution, containing Fe^{2+}, is continuously supplied with hydrogen peroxide, formed by reduction of oxygen gas injected into a carbon cathode, generally carbon felt (Eq. 1). The H_2O_2 formed is decomposed to $^{\bullet}OH$ by the Fenton reaction (Eq. 2), being Fe^{2+} oxidized. Fe^{3+} formed is reduced at the cathode, regenerating Fe^{2+}, the whole process resulting in a higher degradation rate of organic pollutants than the traditional Fenton process (Ammar et al. 2015, Bellakhal et al. 2006).

$$O_2 + 2H^+ + 2e^- \rightarrow H_2O_2 \quad (1)$$

$$Fe^{2+} + H_2O_2 \rightarrow Fe^{3+} + OH^- + {}^{\bullet}OH \quad (2)$$

Boron-doped diamond (BDD) electrode may be used as anode, being the oxidation power greatly improved by the *in situ* production of H_2O_2 and, thus, by the production of heterogeneous hydroxyl radicals BDD($^{\bullet}$OH) on the surface of this anode (Eq. 3):

$$BDD + H_2O \rightarrow BDD({}^{\bullet}OH) + H^+ + e^- \quad (3)$$

The EF process with BDD anode and carbon felt cathode has shown efficiency to treat real wastewaters (Fernandes et al. 2017). Iron sulfate or chloride are the most common iron sources, but pyrite (FeS_2) and chalcopyrite ($CuFeS_2$) have also been used as catalysts for (Barhoumi et al. 2015). They allow self-regulation of the solution pH by releasing H^+, acidifying the solution, while regenerating Fe^{2+} (Eqs. 4 to 9) (Ammar et al. 2015).

$$2FeS_2 + 7O_2 + 2H_2O \rightarrow 2Fe^{2+} + 4SO_4^{2-} + 4H^+ \quad (4)$$

$$2FeS_2 + 15H_2O_2 \rightarrow 2Fe^{3+} + 4SO_4^{2-} + 14H_2O + 2H^+ \quad (5)$$

$$FeS_2 + 14Fe^{3+} + 8H_2O \rightarrow 15Fe^{2+} + 2SO_4^{2-} + 16H^+ \quad (6)$$

$$CuFeS_2 + 4O_2 \rightarrow Cu^{2+} + Fe^{2+} + 2SO_4^{2-} \quad (7)$$

$$CuFeS_2 + 4H^+ + O_2 \rightarrow Cu^{2+} + Fe^{2+} + 2S^0 + 2H_2O \quad (8)$$

$$CuFeS_2 + 16Fe^{3+} + 8H_2O \rightarrow Cu^{2+} + 17Fe^{2+} + 2SO_4^{2-} + 16H^+ \quad (9)$$

The aim of this work was to evaluate the catalytic activity of different iron sources, iron sulfate or chloride, mined pyrite (FeS_2) and chalcopyrite ($CuFeS_2$) for the oxidation of Portuguese and Tunisian OMWs by EF, using a BDD anode and a carbon-felt cathode.

2 MATERIALS AND METHODS

2.1 *Analytical determinations*

Degradation tests were followed by chemical oxygen demand (COD), dissolved organic carbon (DOC), dissolved inorganic carbon (DIC) and total dissolved nitrogen (TDN), which were performed according to standard procedures (Eaton et al. 2005). COD determinations were made using the closed reflux titrimetric method. DOC, DIC and TDN were measured in a Shimadzu TOC-VCPH analyzer combined with a TNM-1 unit. Before these determinations, samples were filtered through 0.45 μm membrane filters. Total dissolved iron and dissolved iron (II) concentrations were determined using a spectrometric method with 1,10-phenanthroline according to ISO 6332 (1988). H_2O_2 concentration was determined by colorimetric metavanadate method (Nogueira et al. 2005). Suspended solids analysis was performed following the analytical procedure (Eaton et al. 2005). The pH was measured using a HANNA pH meter (HI 931400). The conductivity was determined using a Mettler Toledo conductivity meter (SevenEasy S30K).

2.2 *Olive mill wastewater samples*

The OMW samples used in this study were collected at Portuguese and Tunisian olive processing industries, before being submitted to any treatment, and they were kept refrigerated. Before the EF treatment, samples were filtered (500 μm mesh) to remove larger suspended solids. Samples characterization, raw and after filtration, is presented in Table 1. Portuguese OMW presents much higher COD and suspended solids than Tunisian OMW and lower conductivity. Moreover, dissolved iron (II) was found in Tunisian OMW, whereas in Portuguese OMW it was not present. After

Table 1. Physicochemical characteristics of the Portuguese and Tunisian OMW samples, raw and after filtration through a 500 μm mesh.

	Portuguese OMW		Tunisian OMW	
Parameter	Raw	After filtration	Raw	After filtration
COD/g L^{-1}	61 ± 9	41 ± 4	39 ± 9	31 ± 3
DOC/g L^{-1}	13.2 ± 0.8	12.4 ± 0.4	11.1 ± 0.9	10.5 ± 0.3
Dissolved iron (II)/mg L^{-1}	n.d.[a]	n.d.[a]	44 ± 3	43 ± 2
Total dissolved iron/mg L^{-1}	n.d.[a]	n.d.[a]	45 ± 4	43 ± 3
Suspended solids/mg L^{-1}	41.8 ± 0.5	21.2 ± 0.2	2.7 ± 0.2	1.3 ± 0.1
pH	5.13 ± 0.01	4.99 ± 0.09	4.85 ± 0.01	4.8 ± 0.1
Conductivity/mS cm^{-1}	5.1 ± 0.2	5.3 ± 0.1	9.8 ± 0.3	9.9 ± 0.1

[a]n.d. – not detectable.

filtration, COD reductions of 36% and 21% in Portuguese and Tunisian OMWs were observed, respectively.

2.3 *Electro-Fenton experiments*

EF experiments were conducted in batch mode, with stirring, in an undivided cylindrical glass cell containing 200 mL of OMW sample. A 110 cm^2 carbon-felt piece (Carbone Loraine) with a thickness of 0.5 cm was used as cathode and a 10 cm^2 BDD electrode (Adamant Technologies) was used as anode. The anode was centered in the cell and surrounded by the cathode, which covered the inner wall of the cell. Continuous O_2 saturation at atmospheric pressure was ensured by bubbling compressed air through a fritted glass diffuser at 1 L min^{-1}, starting 10 min before the electrolysis, to reach a steady O_2 concentration that allowed H_2O_2 electrogeneration (Eq. 1). The imposed current intensity was 1.0 A, with a DC power supply GW, Lab DC, model GPS-3030D (0–30 V, 0–3 A). Different iron suppliers were evaluated: $Fe_2(SO_4)_3 \cdot 5H_2O$, $FeCl_3 \cdot 6H_2O$, pyrite or chalcopyrite. When iron salts were utilized, they were added to the OMW 10 min before electrolysis to ensure complete dissolution. $Fe_2(SO_4)_3 \cdot 5H_2O$ and $FeCl_3 \cdot 6H_2O$ were analytical grade and were used without further purification. Pyrite and chalcopyrite were mined from Jendouba (Tunisia). They were milled with a ceramic mortar and sieved (<80 μm). To remove surface impurities, the powder was sonicated (45 kHz, 200 W) in 95% ethanol for 5 min, washed with 1 M HNO_3, rinsed with deionized water and further with 95% ethanol, finally being air-dried at 30°C (Ammar et al. 2015). The iron content in pyrite and chalcopyrite is 39.0% and 30.6%, respectively, and minerals characterization is described elsewhere (Ltaïef et al. 2018). For the OMW degradation assays with $Fe_2(SO_4)_3 \cdot 5H_2O$ and $FeCl_3 \cdot 6H_2O$, pH adjustment to an initial value of 3 was done by adding concentrated H_2SO_4 solution. EF assays had 8 h duration and were performed at least in duplicate. The values presented for the parameters used to follow the assays are mean values.

3 RESULTS AND DISCUSSION

A first set of experiments was conducted with the Portuguese OMW using as iron supplier $Fe_2(SO_4)_3 \cdot 5H_2O$ (total iron concentrations of 50, 100 and 200 mg L^{-1}) and $FeCl_3 \cdot 6H_2O$ (total iron concentrations of 50 and 200 mg L^{-1}). Figure 1a presents the relative COD removal for these EF assays, performed at an initial pH value of 3. From the results for the assays performed with iron sulfate, it can be seen that the highest COD removal rate was achieved for an iron initial concentration of 100 mg L^{-1}. Apparently, for an iron initial concentration of 200 mg L^{-1}, the excess

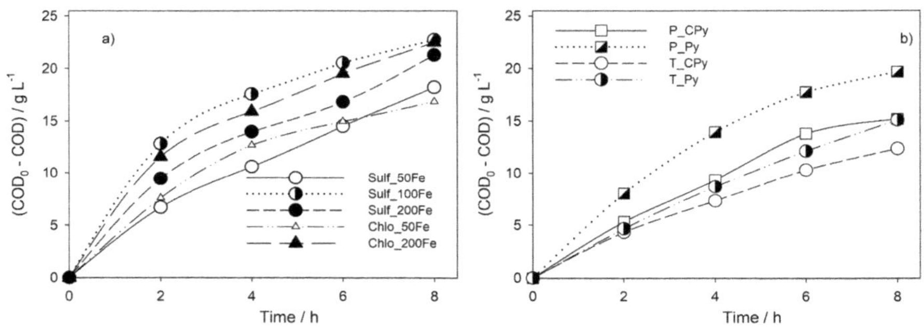

Figure 1. COD removal for the assays with: (a) Portuguese OMW at different iron concentration, using $Fe_2(SO_4)_3 \cdot 5H_2O$ and $FeCl_3 \cdot 6H_2O$ as iron source, at initial pH = 3; (b) Portuguese and Tunisian OMWs using pyrite and chalcopyrite as iron source. P – Portuguese; T – Tunisian; xFe – total iron initial concentration; iron sources – iron sulfate (Sulf), iron chloride (Chlo), pyrite (Py) and chalcopyrite (CPy).

of ferrous ions can lead to the consumption of hydroxyl radicals through a competitive reaction between hydroxyl radicals and ferrous ions (Eq. 10).

$$^{\bullet}OH + Fe^{2+} \rightarrow Fe^{3+} + OH^- \qquad (10)$$

Results for the EF oxidation process in the presence of iron chloride are also presented in Figure 1, and after 8 h, COD removal in the presence of iron chloride is lower than the one observed with iron sulfate for similar initial total iron concentration, indicating that the presence of chloride ions in such concentration could be detrimental for the EF process. This can be explained by the formation of organochloride compounds as by-products, more difficult to undergo degradation. In fact, in the first stage of the degradation, solutions containing chloride present higher COD decay than those in the presence of sulfate. However, in the presence of chloride, COD decay slows down, which may be due to the formation of more recalcitrant by-products. Another reason is the possible formation of side reactions between HOCl (Eqs. 11–12) with the ferrous ion (Eq. 13) that will consume iron and hypochlorous acid (Labiadh et al. 2016), particularly when the more easily degradable compounds are already consumed.

$$2Cl^- \rightarrow Cl_2 + 2e^- \qquad (11)$$

$$Cl_2 + H_2O \rightarrow HOCl + H^+ + Cl^- \qquad (12)$$

$$2Fe^{2+} + 3HOCl + 3H_2O \rightarrow 2Fe(OH)_3 + 3Cl^- + 3H^+ \qquad (13)$$

Regarding DOC removals for the EF assays with Portuguese OMW and addition of iron salts (Table 2), with sulfate, DOC removals follow similar trend to COD removals, i.e., the best results were obtained for $100\,mg\,L^{-1}$ iron total initial concentration. However, with iron chloride, DOC removal decreased with the increase in iron total initial concentration. This indicates that by-products are more oxidized than the initial molecules but carbon is not being eliminated.

Using Portuguese and Tunisian OMWs, a second set of EF experiments was performed with $3\,g\,L^{-1}$ of pyrite or chalcopyrite as iron source. COD (Fig. 1b) and DOC (Table 2) removals were higher with pyrite than with chalcopyrite, which must be due to lower iron concentration in chalcopyrite and the competition of iron and copper ions for reduction at the cathode. The most unexpected results are COD and DOC decays for Portuguese and Tunisian OMWs, either with pyrite or chalcopyrite. In fact, Tunisian OMW present lower COD decays, which could be explained by the lower initial COD, but it presents higher DOC decays than the Portuguese OMW. This means that the mineralization degree is much higher for the Tunisian wastewater, with pyrite or chalcopyrite as iron source. An explanation is the different composition of the Portuguese and Tunisian OMWs, particularly the molecular size of the constituents.

Table 2. Results of dissolved organic carbon removal and pH for the different EF assays performed with Portuguese and Tunisian OMWs.

Sample	Iron source and mass added $g\ L^{-1}$	Dissolved iron $mg\ L^{-1}$	DOC removal $g\ L^{-1}$	$pH_{initial}/pH_{final}$	
Portuguese	$Fe_2(SO_4)_3 \cdot 5H_2O$	0.44	56	3.2	2.8/2.4
		0.87	130	4.3	2.7/2.4
		1.32	209	3.7	3.0/2.3
	$FeCl_3 \cdot 6H_2O$	0.48	57	3.7	2.9/2.6
		1.50	203	3.5	2.9/3.1
	Chalcopyrite	3.0	4	2.0	5.0/4.2
	Pyrite	3.0	5	3.4	4.9/4.2
Tunisian	Chalcopyrite	3.0	60	2.7	4.7/4.0
	Pyrite	3.0	53	4.4	4.7/3.8

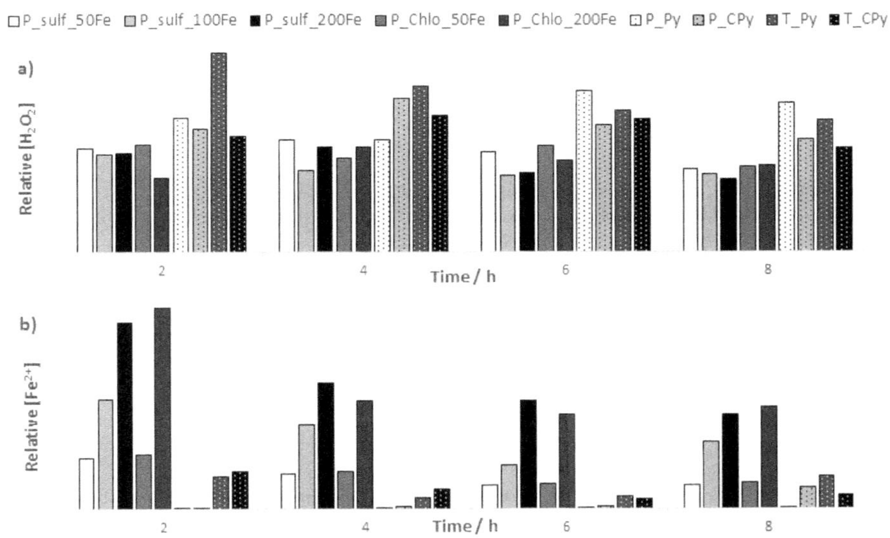

Figure 2. (a) H_2O_2 and (b) Fe^{2+} relative concentrations at the 2nd, 4th, 6th and 8th hours of the different EF assays. P – Portuguese OMW; T – Tunisian OMW; Fe – total iron initial concentration; iron sources – iron sulfate (sulf), iron chloride (chlo), pyrite (Py) and chalcopyrite (CPy).

Figure 2, a and b, presents the H_2O_2 and Fe^{2+} variation during the 8 h assay, respectively. For the assays performed with the Portuguese OMW using iron sulfate, H_2O_2 concentration presented the lowest values for the iron initial concentration of 100 mg L^{-1}, showing that this is the iron concentration ideal to consume the formed hydrogen peroxide. In fact, a further increase in the Fe^{2+} concentration did not correspond to an increase in the H_2O_2 consumption, which can be explained by Equation 10. Results for H_2O_2 and Fe^{2+} variations during the assays performed with iron chloride are similar to those obtained with iron sulfate, showing that the small differences in COD decays for the different iron sources must be related with Equations 11 to 13 and recalcitrant by-products. Regarding EF assays with pyrite and chalcopyrite, H_2O_2 concentrations are anomalously high during the assays probably because iron is being slowly released from the mineral source. The highest Fe^{2+} concentration for Tunisian OMW is due to the iron in initial sample.

During all the EF assays, there was a small decrease in the pH of the solution (Table 2) that has been previously reported and explained by the carboxyl acids formed during the degradation process and by the oxygen evolution reaction (Labiadh et al. 2016).

4 CONCLUSIONS

The results obtained with the electro-Fenton degradation assays of olive mill wastewaters, performed with different iron sources proved that this technology can be an alternative to solve the problems raised with the olive oil industry in the Mediterranean countries. For the Portuguese OMW at the best experimental conditions applied, after 8 h assay, EF process accomplished a COD removal of 22.7 g L^{-1}. The use of iron-containing solid catalysts, mined pyrite and chalcopyrite, when compared with traditional iron sulfate and iron chloride showed to promote identical treatment results. Pyrite is a better source of Fe^{2+} than chalcopyrite, leading to higher COD removal rates for equal initial amount of both minerals. The results indicate that mined pyrite and chalcopyrite can be effectively used as heterogeneous EF catalysts for the degradation of OMW, with the advantage of being recovered and reused.

ACKNOWLEDGMENTS

Fundação para a Ciência e a Tecnologia, FCT, projects UID/Multi/00195/2013, SFRH/BPD/103615/2014 (A. Fernandes) and SFRH/BD/132436/2017 (M.J. Nunes). Tunisian authors acknowledge financial support from the University of Gabes (Tunisia).

REFERENCES

Ammar, S., Oturan, M.A., Labiadh, L., Guersalli, A., Abdelhedi, R., Oturan, N. & Brillas, E. 2015. Degradation of tyrosol by a novel electro-Fenton process using pyrite as heterogeneous source of iron catalyst. *Water Research* 74: 77–87.

Barbosa, J., Fernandes, A., Ciríaco, L., Lopes, A. & Pacheco, M.J. 2016. Electrochemical treatment of olive processing wastewater using a BDD anode. *Clean – Soil, Air, Water* 44: 1242–1249.

Barhoumi, N., Labiadh, L., Oturan, M.A., Oturan, N., Gadri, A., Ammar, S. & Brillas, E. 2015. Electro-Fenton oxidation of reverse osmosis concentrate from sanitary landfill leachate: Evaluation of operational parameters. *Chemosphere* 141: 250–257.

Bellakhal, N., Oturan, M.A., Oturan, N. & Dachraoui, M. 2006. Olive oil mill wastewater treatment by the electro-Fenton process. *Electrochemistry for the Environment* 3: 345–349.

Beltran, J., Gonzalez, T. & Garcia, J. 2008. Kinetics of the biodegradation of green table olive wastewaters by aerobic and anaerobic treatments. *Journal of Hazardous Materials* 154: 839–845.

Brillas, E., Sirés, I. & Oturan, M.A. 2009. Electro-Fenton process and related electrochemical technologies based on Fenton's reaction chemistry. *Chemical Reviews* 109: 6570–6631.

Eaton, A., Clesceri, L., Rice, E., Greenberg, A. & Franson, M.A. (eds) 2005. *Standard methods for examination of water and wastewater*. 21st ed. Washington, DC: American Public Health Association.

Fernandes, A., Labiadh, L., Ciríaco, L., Pacheco, M.J., Gadri, A., Ammar, S. & Lopes, A. 2017. Electro-Fenton oxidation of reverse osmosis concentrate from sanitary landfill leachate: Evaluation of operational parameters. *Chemosphere* 184: 1223–1229.

Flores, N., Brillas, E., Centellas, F., Rodríguez, R. M., Cabot, P.L., Garrido, J.A. & Sirés, I. 2018. Treatment of olive oil mill wastewater by single electrocoagulation with different electrodes and sequential electrocoagulation/electrochemical Fenton-based processes. *Journal of Hazardous Materials* 347: 58–66.

ISO 6332, 1988. In *Water Quality – Determination of Iron – Spectrometric Method Using 1,10-Phenanthroline*.

Khoufi, S., Aloui, F. & Sayadi, S. 2006. Treatment of olive oil mill wastewater by combined process electro-Fenton reaction and anaerobic digestion. *Water Research* 40: 2007–2016.

Labiadh, L., Fernandes, A., Ciríaco, L., Pacheco, M. J., Gadri, A., Ammar, S. & Lopes, A. 2016. Electrochemical treatment of concentrate from reverse osmosis of sanitary landfill leachate. *Journal of Environmental Management* 181: 515–521.

Ltaïef, A.H., Pastrana-Martínez, L.M., Ammar, S., Gadri, A., Faria, J.L. & Silva, A.M.T. 2018. Mined pyrite and chalcopyrite as catalysts for spontaneous acidic pH adjustment in Fenton and LED photo-Fenton-like processes. *Journal of Chemical Technology and Biotechnology* 93: 1137–1146.

Nogueira, R.F.P., Oliveira, M.C. & Paterlini, W.C. 2005. Simple and fast spectrophotometric determination of H_2O_2 in photo-Fenton reactions using metavanadate. *Talanta* 66: 86–91.

Pulido, J.M.O. 2016. A review on the use of membrane technology and fouling control for olive mill wastewater treatment. *Science of the Total Environment* 563: 664–675.

Wastes: Solutions, Treatments and Opportunities III – Vilarinho et al. (Eds)
© 2020 Taylor & Francis Group, London, ISBN 978-0-367-25777-4

Microbial conversion of oily wastes to methane: Effect of ferric nanomaterials

V.R. Martins, G. Martins, A.R. Castro, L. Pereira, M.M. Alves & A.J. Cavaleiro
Centre of Biological Engineering, University of Minho, Braga, Portugal

O.S.G.P. Soares & M.F.R. Pereira
Laboratory of Separation and Reaction Engineering – Laboratory of Catalysis and Materials (LSRE-LCM), Faculty of Engineering, University of Porto, Porto, Portugal

ABSTRACT: Petroleum-based oily wastes are generated by the oil industry and can be treated/valorized by anaerobic microbial conversion to methane. However, this process is generally slow. Conductive nanomaterials were reported to accelerate the interspecies electron transfer in anaerobic communities and therefore their addition to anaerobic processes treating hydrocarbons may also be advantageous. In this work, two ferric nanomaterials (magnetite and carbon nanotubes impregnated with 2% iron) were tested in microcosms amended with hexadecane and 1-hexadecene. Assays were also made with palmitate, acetate and H_2/CO_2, which are intermediates of hydrocarbons biodegradation. With the exception of hexadecane, methane was produced at close-to-stoichiometric amounts for each of the substrates tested. Methane production rates were similar with and without the nanomaterials, possibly due to the inability of the microorganisms to receive/transfer electrons to the materials in this microbial community, suggesting that electron transfer occurred indirectly via soluble electron shuttles (*e.g.* hydrogen or formate).

1 INTRODUCTION

The intense activity of the oil industry generates substantial amounts of petroleum-based oily wastes (Hu *et al.*, 2013; Zheng *et al.*, 2016). The management of these wastes is a concerning issue in the Oil&Gas sector, currently restricted by stringent regulations. Anaerobic treatment of these oily wastes is an attractive option, since it can couple organic treatment with the recovery of bioenergy through methane production. Hydrocarbons biodegradation to methane is performed by complex microbial communities, where different groups of microorganisms interact in a series of metabolic steps that culminate in methane production. After hydrocarbons activation, these compounds are converted into smaller molecules such as short-chain fatty acids, alcohols or H_2 by fermentative bacteria. Further degradation of these intermediates to methane is restricted by thermodynamics, and only becomes feasible when hydrogenotrophic methanogens are present, decreasing the hydrogen partial pressure (Jiménez *et al.*, 2016). Therefore, a close syntrophic relationship between bacteria and methanogenic archaea is essential (Jiménez *et al.*, 2016). The overall syntrophic reactions yield extremely low Gibbs free energy, and thus methanogenic hydrocarbon degradation typically proceeds at very low rates (Jiménez *et al.*, 2016).

Conductive ferric minerals have been reported to accelerate the syntrophic conversion to methane of diverse substrates, such as volatile fatty acids (acetate, propionate and butyrate) and benzoate (an intermediate of benzene degradation), as reviewed recently by Martins *et al.* (2018). However, the role of these materials on syntrophic hydrocarbons biodegradation to methane has never been studied. The mechanisms underlying the stimulating effects of ferric minerals are not clearly understood, and appear to depend on the crystallinity and conductivity of the Fe(III) form. Several authors suggested that conductive ferric minerals may stimulate direct interspecies electron transfer in syntrophic processes, by conducting the electrons between electron donating and electron accepting microorganisms (Kato *et al.*, 2012; Cruz Viggi *et al.*, 2014). A similar effect was also described for some conductive carbon materials, namely granular activated carbon, biochar and carbon cloth (Martins *et al.*, 2018), but recently Salvador *et al.* (2017) showed that conductive multi-walled carbon nanotubes (CNT) can directly stimulate methanogens when grown in pure culture, which may constitute an alternative or complementary mechanism of methanogenesis stimulation. Preliminary results in our research group also showed higher methane production rates in pure cultures of methanogens amended with magnetic CNT (CNT impregnated with 2% of Fe – CNT@2%Fe).

The potential of two different conductive ferric nanomaterials (magnetite and CNT@2%Fe) to accelerate hydrocarbons biodegradation to methane was investigated in this work. The two nanomaterials chosen present magnetic properties, which facilitate their recovery in waste treatment processes by applying a magnetic field (Pereira *et al.*, 2017).

2 MATERIALS AND METHODS

2.1 *Ferric nanomaterials*

Magnetite (97%, Sigma-Aldrich, 50-100 nm particle size) and CNT impregnated with 2% Fe (CNT@2%Fe) were tested in this study. The CNT@2%Fe were synthesized at the University of Porto, and their characterization was published in Pereira *et al.* (2017).

2.2 *Effect of ferric nanomaterials on anaerobic hydrocarbons degradation*

Anaerobic microcosms were prepared in 120 mL bottles containing the nanomaterial (magnetite or CNT@2%Fe, $1\,g\,L^{-1}$) and anaerobic bicarbonate-buffered mineral salt medium (50 mL, as described in Paulo *et al.*, 2017). Sludge from a real treatment plant performing *ex situ* bioremediation of hydrocarbon-contaminated groundwater was mixed with granular sludge from the sludge digester of a brewery wastewater treatment plant (1:3 v/v), and used as inoculum (5 g) at a final volatile solids (VS) content of $5\,g\,L^{-1}$ in the microcosms. The presence of hexane-extractable hydrocarbons in the sludge from the real treatment plant was confirmed by CG analysis (Figure 1), corresponding to a total petroleum hydrocarbons (TPH) content of $0.08 \pm 0.03\,g\,g^{-1}$ (wet weight). The bottles were sealed with Vitton rubber stoppers and aluminum crimp caps, and the headspace

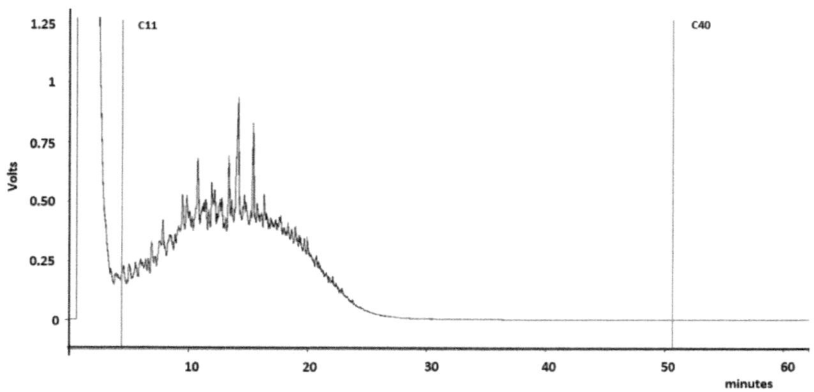

Figure 1. GC chromatogram of TPH extracted from 20 g of sludge obtained from a groundwater treatment plant performing *ex situ* bioremediation. Retention time for undecane (C11) and tetracontane (C40) are 5 and 50.2 minutes, respectively.

was flushed with N_2/CO_2 (80:20% v/v), at a final pressure of 1.7×10^5 Pa. Before incubation, the medium was reduced with $Na_2S.9H_2O$ (0.8 mmol L^{-1}).

Hexadecane (99%, Sigma-Aldrich, 1 mmol L^{-1}) or 1-hexadecene (99%, Sigma-Aldrich, 1 mmol L^{-1}) were added as model hydrocarbons. Assays amended with sodium palmitate (≥98.5%, Fluka, 1 mmol L^{-1}), H_2/CO_2 (80:20% v/v, 1.7×10^5 Pa final pressure) or acetate (99%, Sigma-Aldrich, 20 mmol L^{-1}) were also performed. In the assays with sodium palmitate, after 54 days of incubation the added substrate was completely degraded, and thus a second addition of this electron donor was performed. For that, the headspace of the bottles was flushed and pressurized with N_2/CO_2 (80:20% v/v, 1.7×10^5 Pa) under sterile conditions before new substrate addition. Control assays without nanomaterials and blank assays without any added substrate were also prepared. All tests were performed in triplicate. The bottles were incubated upside down at 37°C, with shaking (120 rpm) in the dark. Methane production was measured along the time; volatile fatty acids (VFA) and hydrocarbons concentrations were quantified at the end of the experiments.

2.3 *Analytical methods*

Methane concentration in the headspace of the bottles was measured using a gas chromatograph (Shimadzu, Kyoto, Japan) equipped with a flame ionization detector (FID) and a PoraPak Q column (80/100 mesh column, 2 m × 1/8 in, 2 mm, stainless steel) with N_2 and argon as carrier gases at 30 mL min^{-1} and 5 mL min^{-1}, respectively. The injector, column and detector temperatures were 110°C, 35°C and 220°C, respectively. VFA were quantified in liquid samples after centrifugation and filtration by high-performance liquid chromatography (HPLC Jasco equipment) with UV detection, as described by Paulo *et al.* (2017). Hydrocarbon were analyzed in the liquid and solid phases of the microcosms, which were separated by decantation. The liquid samples were sequentially extracted three times with hexane using separatory funnels. The solid samples were shaken with hexane for 4 hours at 120 rpm in Schott flasks (Siddique *et al.*, 2006). All the extracts were cleaned using Sep-Pak Florisil® cartridges (Waters, Milford, MA) and evaporated in TurboVap® LV (Biotage, Uppsala, Sweden). Hydrocarbons were quantified in a gas chromatograph with a FID, as detailed in Paulo *et al.* (2017).

3 RESULTS AND DISCUSSION

Anaerobic conversion of hydrocarbons to methane has been described as a syntrophic process. Therefore, the presence of an active methanogenic community is essential for extensive biodegradation of these compounds. The methanogenic activity of the inoculum used in this study was evaluated in microcosms supplemented with acetate or with a mixture of H_2/CO_2, either in the presence or absence of the ferric nanomaterials (Figure 2).

In the absence of nanomaterials, the cumulative methane production stabilized after 35–45 hours and after 5 hours of incubation in the assays supplemented with acetate and H_2/CO_2, respectively (Figure 2a and 2b), corresponding to a methanogenic activity of 9.2 ± 0.6 and 125.2 ± 19 mmol L^{-1} day^{-1}. The addition of CNT@2%Fe accelerated acetoclastic methanogenesis up to 13.7 ± 0.4 mmol L^{-1} day^{-1}, representing a 1.5 times faster methane production comparatively to the assays with magnetite or in the absence of the materials. The addition of the nanomaterials did not stimulate the methane production from H_2/CO_2 (Figure 2b), possibly due to the good initial activity of the hydrogenotrophic methanogenic community.

High methane production (16 ± 2 mmol L^{-1}) was measured in the blank assays during the first 19 days of incubation (Figures 3 and 4), either in the presence or in the absence of nanomaterials. The residual substrate was composed by hydrocarbons (Figure 1) and probably by other more easily biodegradable compounds. After this period, methane production proceeded at a much slower rate, probably deriving from more recalcitrant compounds or from the mineralization of dead cells (endogenous respiration).

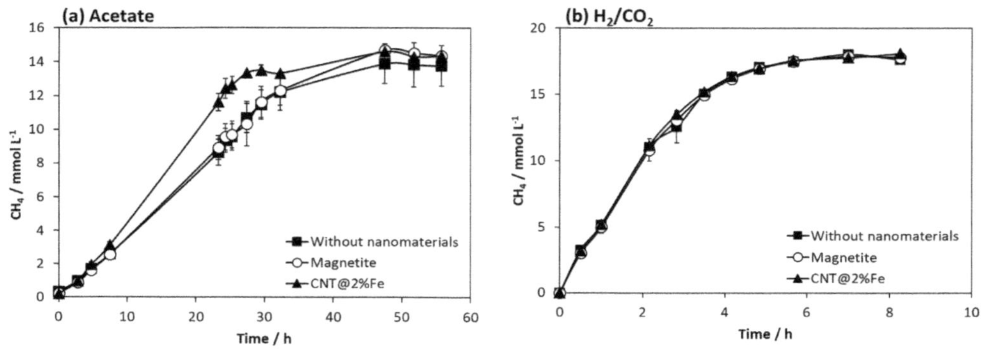

Figure 2. Cumulative methane production in the assays amended with acetate (a) or H_2/CO_2 (b).

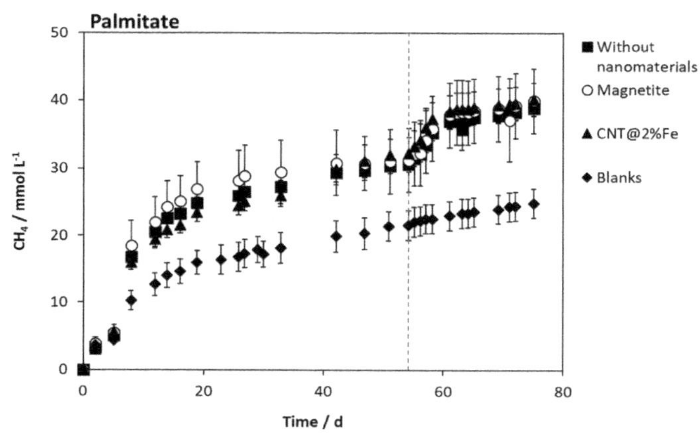

Figure 3. Cumulative methane production in the assays amended with palmitate and in the blank assays. For the blanks, the values shown represent the average of all the assays performed (in the presence and in the absence of nanomaterials). The vertical dotted line represents the moment of second palmitate addition.

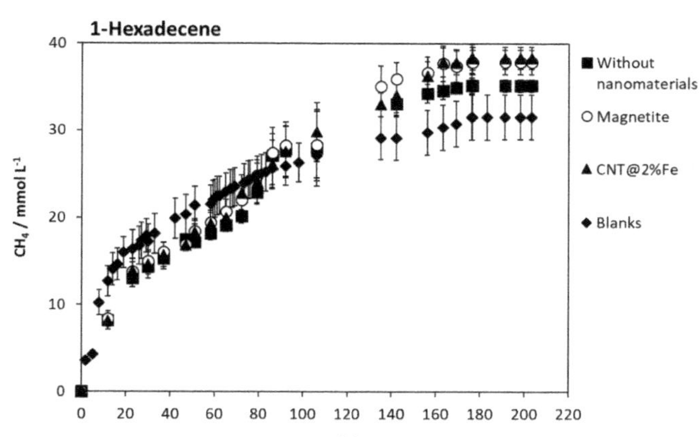

Figure 4. Cumulative methane production in the assays amended with 1-hexadecene and in the blank assays. For the blanks, the values shown represent the average of all the assays performed (in the presence and in the absence of nanomaterials).

In the assays with palmitate (Figure 3), the cumulative methane production measured after 19 days (discounting the methane produced in the blanks in the same time period) accounted for the degradation of 64–95% of the added substrate, considering the stoichiometry of palmitate conversion to methane (Equation 1).

$$C_{16}H_{31}O_2^- + 7\ H_2O + H^+ \rightarrow 11.5\ CH_4 + 4.5\ CO_2 \quad (1)$$

Moreover, more than 50% of the palmitate had been degraded to methane already after 8 days of incubation. These results show that the microbial community present was capable of effectively degrading LCFA and performing efficiently the necessary syntrophic relationships between the LCFA-degrading bacteria and the methanogens. Considering that the presence of residual substrate could influence the observed methane production rates, a second palmitate addition (marked with the vertical dotted line in Figure 3) was made, that confirmed the previous observations. No VFA were detected in the medium before the second substrate addition, neither at the end of the incubations. This efficient conversion of intermediates to methane suggests an efficient electron transfer and is probably the reason why the addition of the nanomaterials did not improve the methane production from palmitate (Figure 3).

In the assays amended with 1-hexadecene, the cumulative methane production measured in the first 80 days of incubation was similar or slightly lower than the values recorded in the blanks (Figure 4). This suggests that the added hydrocarbon was not being degraded and even slightly inhibited the degradation of the residual substrate. After that and until day 163 of incubation (*i.e.*, in 84 days) methane was produced in accumulated amounts that match the theoretical value expected from the degradation of the added 1-hexadecene (Equation 2).

$$C_{16}H_{32} + 8\ H_2O \rightarrow 12\ CH_4 + 4\ CO_2 \quad (2)$$

The almost complete 1-hexadecene biodegradation was confirmed by GC analysis, which showed the absence of this compound in the liquid and solid phases of all the assays, with the exception of the solid phase of CNT@2%Fe assays, where it could be detected in low amounts (Figure 5). Moreover, no VFA could be detected in the medium at the end of the experiments.

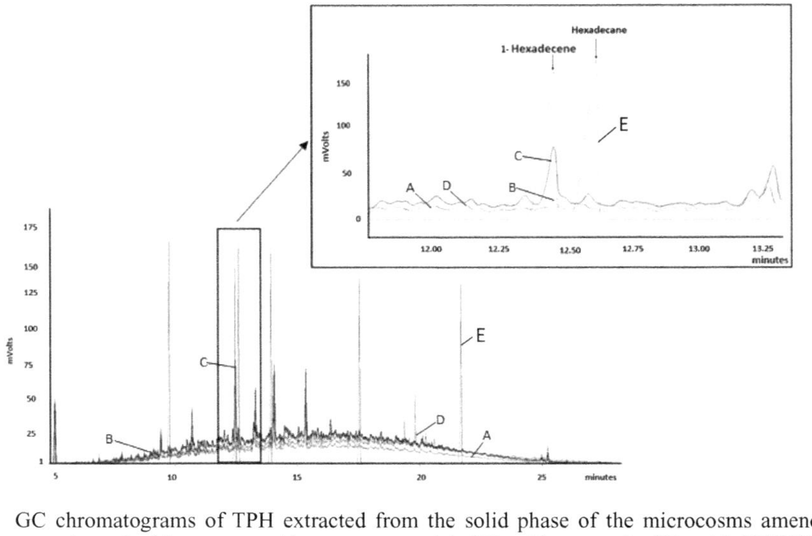

Figure 5. GC chromatograms of TPH extracted from the solid phase of the microcosms amended with 1-hexadecene, at the end of the assays: without nanomaterials (A), with magnetite (B), with CNT@2%Fe (C) and blank (D). Retention time for undecane (C11) and tetracontane (C40) are 5 and 50.2 minutes, respectively. The chromatogram of an alkanes' mixture, containing 1-hexadecene and hexadecane at 100 mg L^{-1} each, was also included, for comparison (E).

These are interesting results, since methanogenic degradation of 1-hexadecene is reported as a slow process, generally requiring more than 100 days for the complete conversion of 1–2 mmol L^{-1} to methane (Schink, 1985, Paulo *et al.*, 2017). Again, the observed absence of stimulatory effect of the nanomaterials on the methanogenesis can possibly be due to the occurrence of indirect interspecies electron transfer via soluble electron shuttles (*e.g.* hydrogen or formate). Moreover, the microorganisms present in the community may not be able to transfer/receive the electrons to/from the materials. Not all microorganisms are able to perform direct interspecies electron transfer (DIET), and the research available suggests that this may be the case for the majority of the anaerobic microorganisms. In fact, DIET was only clearly demonstrated in co-cultures of *Geobacter metallireducens* with *Methanosaeta harundinacea* or *Methanosarcina barkeri*, and frequently *Geobacter* sp. (the most well know electroactive microorganisms) are not detected in improved methane production driven systems (Martins *et al.*, 2018). In addition, experimental evidences point out for the inability of some *Syntrophomonas* (Salvador *et al.*, 2017) and *Pelobacter* species (Rotaru *et al.*, 2012) to participate in DIET.

In the assays with hexadecane, methane production was similar in all the experiments, including in the blanks, which suggests the absence of hexadecane biodegradation. This is probably related with the fact that hexadecane is highly stable and inert, and other authors have reported very long incubation periods (*e.g.* 810 days, Zengler *et al.*, 1999) as necessary for the conversion of hexadecane to methane in batch assays similar to the ones performed in this experiment.

4 CONCLUSIONS

The methanogenic community studied exhibited good syntrophic and methanogenic activity, being capable of performing palmitate and 1-hexadecene conversion to methane in less than 13 and 84 days, respectively. A stimulating effect of the nanomaterials on the metabolism of this microbial community was not observed, possibly due to the occurrence of indirect IET via soluble electron shuttles. A more detailed knowledge on the mechanisms underlying the effect of conductive material on methanogenesis, and of the microorganisms involved in these processes, is essential for the development of new strategies targeting the conversion of oily wastes to methane.

AKNOWLEDGEMENTS

The authors acknowledge Sara Escadas for the collaboration in the experimental work. This study was financially supported by the Portuguese Foundation for Science and Technology (FCT) under the scope of project MORE (PTDC/AAG-TEC/3500/2014; POCI-01-0145-FEDER-016575), Project SAICTPAC/0040/2015 (POCI-01-0145-FEDER-016403), of the strategic funding of UID/BIO/04469/2019 unit and BioTecNorte operation (NORTE-01-0145-FEDER-000004) funded by the European Regional Development Fund under the scope of Norte2020 – Programa Operacional Regional do Norte.

REFERENCES

Cruz Viggi, C., Rossetti, S., Fazi, S., Paiano, P., Majone, M. & Aulenta, F. 2014. Magnetite particles triggering a faster and more robust syntrophic pathway of methanogenic propionate degradation. *Environmental Science and Technology* 48(13): 7536–7543.

Hu, G., Li, J. & Zeng, G. 2013. Recent development in the treatment of oily sludge from petroleum industry: A review. *Journal of Hazardous Materials* 261: 470–490.

Jiménez, N., Richnow, H. & Vogt, C. 2016. Methanogenic hydrocarbon degradation: Evidence from field and laboratory studies. *Journal of Molecular Microbiology and Biotechnology* 26: 227–242.

Kato, S., Hashimoto, K. & Watanabe, K. 2012. Methanogenesis facilitated by electric syntrophy via (semi)conductive iron-oxide minerals. *Environmental Microbiology* 14(7): 1646–1654.

Martins, G., Salvador, A.F., Pereira, L. & Alves, M.M. 2018. Methane production and conductive materials: A critical review. *Environmental Science and Technology* 52(18): 10241–10253.
Paulo, A.M.S., Salvador, A.F., Alves, J.I., Castro, R., Langenhoff, A.A.M., Stams, A.J.M. & Cavaleiro, A.J. 2017. Enhancement of methane production from 1-hexadecene by additional electron donors. *Microbial Biotechnology* 11: 657–666.
Pereira, L., Dias, P., Soares, O.S.G.P., Ramalho, P.S.F., Pereira, M.F.R. & Alves, M.M. 2017. Synthesis, characterization and application of magnetic carbon materials as electron shuttles for the biological and chemical reduction of the azo dye Acid Orange 10. *Applied Catalysis B: Environmental* 212: 175–184.
Rotaru, A.E., Shrestha, P.M., Liu, F., Ueki, T., Nevin, K., Summers, Z.M. & Lovley, D.R. 2012. Interspecies electron transfer via hydrogen and formate rather than direct electrical connections in cocultures of *Pelobacter carbinolicus* and *Geobacter sulfurreducens*. *Applied and Environmental Microbiology* 78(21): 7645–7651.
Salvador, A.F., Martins, G., Melle-Franco, M., Serpa, R., Stams, A.J.M., Cavaleiro, A.J., Pereira, M.A. & Alves, M.M. 2018. Carbon nanotubes accelerate methane production in pure cultures of methanogens and in a syntrophic coculture. *Environmental Microbiology* 19(7): 2727–2739.
Zheng, J., Chen, B., Thanyamanta, W., Hawboldt, K., Zhang, B. & Liu, B. 2016. Offshore produced water management: A review of current practice and challenges in harsh/Arctic environments. *Marine Pollution Bulletin* 104(1–2): 7–19.
Zhuang, L., Tang, J., Wang, Y., Hu, M. & Zhou, S. 2015. Conductive iron oxide minerals accelerate syntrophic cooperation in methanogenic benzoate degradation. *Journal of Hazardous Materials* 293(808): 37–45.
Schink, B. 1985. Degradation of unsaturated hydrocarbons by methanogenic enrichment cultures. *FEMS Microbiology Ecology* 31: 69–77.
Siddique, T., Rutherford, P.M., Arocena, J.M., & Thring, R.W. 2006. A proposed method for rapid and economical extraction of petroleum hydrocarbons from contaminated soils. *Canadian Journal of Soil Science* 86(4): 725–728.

Wastes: Solutions, Treatments and Opportunities III – Vilarinho et al. (Eds)
© 2020 Taylor & Francis Group, London, ISBN 978-0-367-25777-4

Preliminary evaluation of the potential performance of a future PAYT system in Portugal

V. Sousa
CERIS, DECivil, IST-University of Lisbon, Lisboa, Portugal

J. Dinis, A. Drumond, M.J. Bonnet & P. Leal
EMAC, Cascais Ambiente, Cascais, Portugal

I. Meireles
RISCO, Department of Civil Engineering, University of Aveiro, Aveiro, Portugal

C. Dias-Ferreira
Universidade Aberta, Lisboa, Portugal
Aveiro Institute of Materials (CICECO), University of Aveiro, Aveiro, Portugal
CERNAS, Instituto Politécnico de Coimbra, Coimbra, Portugal

ABSTRACT: The use of Pay-As-You-Throw systems is a reality in several cities throughout the globe, in some cases for several year by now. In Portugal, as well as many other countries around the Mediterranean, there have been only a few experiences and limited information exists regarding their performance. In this contribution, the results of a first stage towards an experimental PAYT system recently implemented in the municipality of Cascais are detailed. The strategy involved the increase of collection points for packaging waste and the use of gamification as positive incentive. Following the infrastructure improvements, the segregated waste collection doubled from 10% to 20–21%. With the gamification, the proportion of waste segregation increased to nearly 30%.

1 INTRODUCTION

Waste management is an ever growing concern for the waste utilities, communities, countries and, ultimately to the society as whole. Regardless of the vast technological advances that enable the substantial reduction of landfilling, there are still countries throughout the world where the amounts of waste are increasing and the source separation into specific waste streams is limited.

Pay-As-You-Throw (PAYT) is far from being a novel concept (e.g., Stevens 1977), consisting simply in the transposition of the environmental management principle of polluter-payer to the waste management sector. Benefits from PAYT schemes include waste reduction (Canterbury 1994; Van Houtven & Morris 1999), increased recycling rates (Dijkgraaf & Gradus 2004; EC 2012), and, in some cases, also financial savings (Karagiannidis et al. 2008). It is possible to interpret a PAYT system from different perspectives, namely: i) from a technological point of view, e.g., the type of technological solution supporting the PAYT (Madureira & Dias-Ferreira, 2019a, b); ii) from a financial point of view (e.g., the tariff system to ensure the service continuity); and iii) from an educational point of view (e.g., the change in the waste producer behaviour).

Herein, the later is adopted and it will be assumed that, depending on the design of the PAYT system, the education perspective can be divided into two aspects related to the waste producer's behaviour: i) the total amount of waste produced; and ii) the degree of waste separation. In fact, a correctly designed PAYT system, should contribute to "educate" into reduce the total amount of waste they produce and the degree to which the waste is separated.

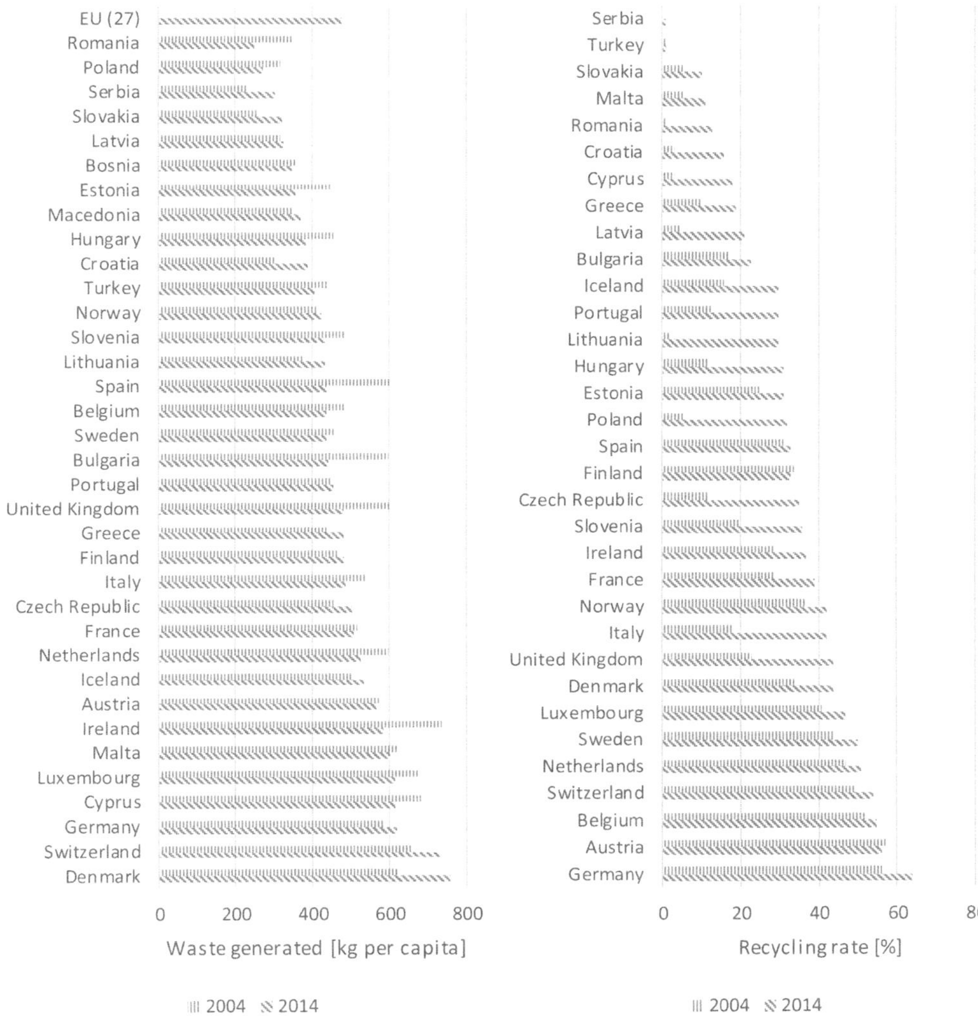

Figure 1. Performance of waste management in European countries in terms of waste generation (left) and recycling rate (right) (EEA 2016).

According the European Environmental Agency (EEA 2016), PAYT schemes are an effective instrument to boost the recycling rates since all countries with recycling rates above 45% employ them, whereas most countries with recycling rates below 20% do not (Fig. 1).

However, crossing the information of the amount of waste not recycled and the recycling rate the conclusions become mixed (Fig. 2). There are several countries with high recycling rates (e.g., Denmark, Switzerland, and Luxembourg) among the group with the highest amount of non-recycled waste per capita.

Considering Figures 1 and 2, there is an indication that while the PAYT schemes seem effective in increasing the recycling rates, their efficiency in reducing the amount of waste produced is not so evident. This raises the question if the recycling rate increase with PAYT schemes is driven by the waste tariffs (the fact that a PAYT scheme is truly implemented) or from the more developed separate waste collection service that is usually required. In practice, a PAYT scheme tends to require the waste collection to have similar levels of service for all waste streams, while more

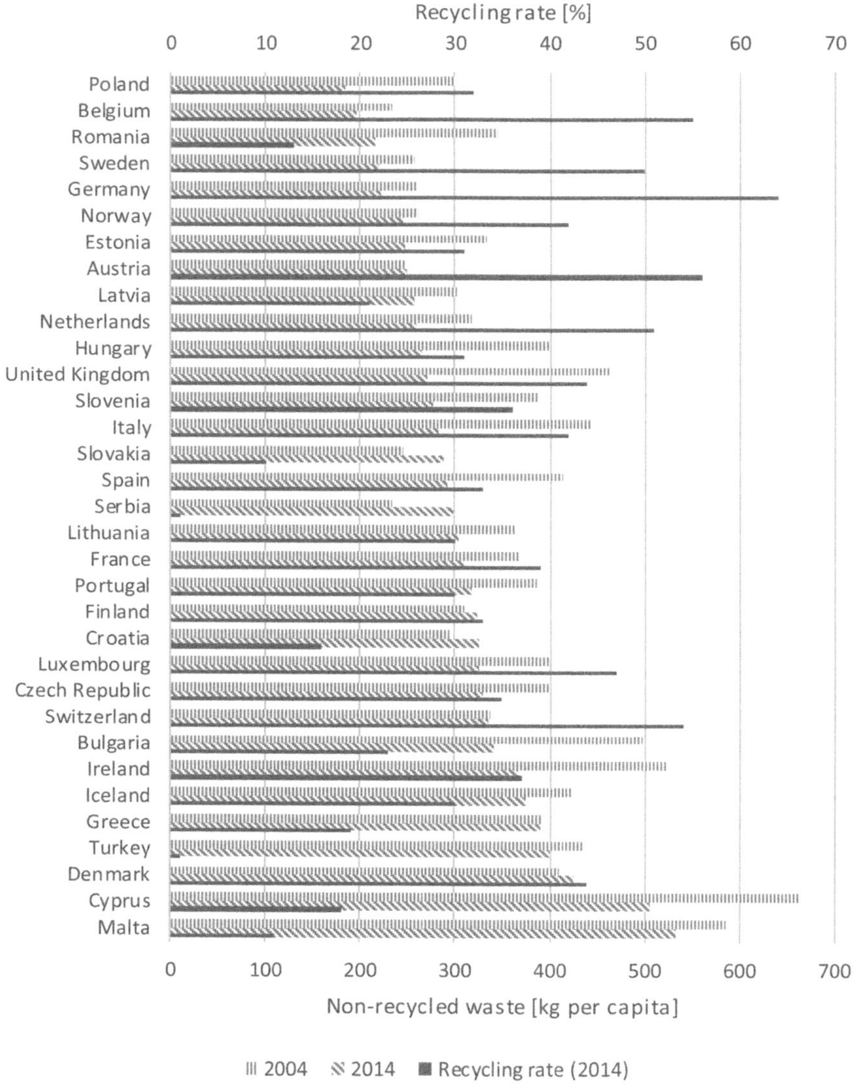

Figure 2. Comparison of the amount of non-recycled waste with the recycling rate.

traditional collection models present distinct levels of service (e.g., coverage) for the mixed and segregated wastes.

The present communication reports the results of the first stage towards the implementation of PAYT system in the municipality of Cascais (Portugal). In this first stage, only the infrastructures and public engagement were implemented, but not a varying tariff.

2 CASE STUDY

Cascais is located about 30 km West of Lisbon, the capital of Portugal. The municipality covers an area of almost 100 km^2, with a population of 211 300 inhabitants in 2017, and is divided into

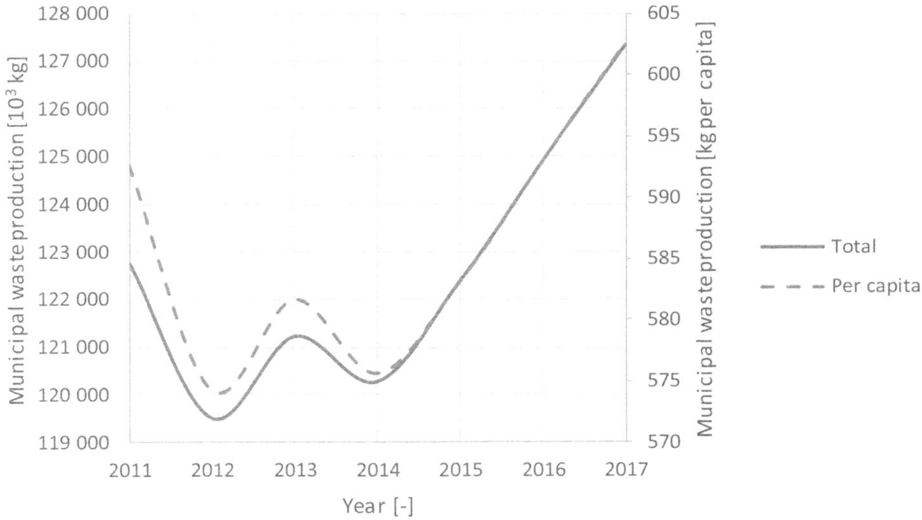

Figure 3. Total and per capita municipal waste production evolution in Cascais.

4 parishes, with the majority of the population concentrated in the two southern parishes along the coastline. The waste collection in the Cascais municipality is amongst the responsibilities of EMAC, Cascais Ambiente, a municipal company.

In average, the total and per capita solid waste produced is approximately 122×10^6 kg/year and 586 kg/per capita.year, but with some inter-annual variations (Fig. 3). The waste produced is collected through 4 services for different waste streams: i) residual waste; ii) source segregated waste, namely paper, plastic and glass; iii) parks and garden waste; and iv) bulky waste. In addition to the collection, the residual and segregated waste services include the containerization (curbside and underground containers). The residual waste service has the highest number of collection points. Contrariwise, the segregated waste service has fewer containers, and consequently superior distances between collection points. Garden and bulky waste services provide an activity subjected to variable constraints, such as public scheduling requests or evidences detected by operators of other types of waste collection services. Especially the bulky waste service has longer travel distances per vehicle due to the uncertainty arising from abandonment of waste on public roads.

Overall, the recycling rate (waste collected separately / total amount of waste collected) did not vary significantly between 2011 and 2017, recording an average value of 29% (Fig. 4). However, excluding the garden and bulky wastes that do not tend to come in the mixed residual waste, the recycling rate drops to 10%.

The PAYT system is being implemented in three neighbourhoods in the southeast of Cascais municipality: i) Quinta de São Gonçalo; ii) Lombos Sul; and iii) Bairro da Torre. Quinta de São Gonçalo is the most recent of the three neighbourhoods, comprising mostly 5 floor buildings in an orthogonal mesh. Lombos Sul grew in the 1970's and is more heterogeneous, with a mix of 3 (dominant) up to 8 floor multifamily buildings in the north and 2 floor single family buildings in the south, with various streets with no exit. Bairro da Torre is the smallest and is predominantly composed of social households managed by the municipality of Cascais. With this social character, Bairro da Torre residents have the strongest sense of community (Fig. 5).

Up to date, two stage towards the implementation of the PAYT system were concluded. The first stage consisted on harmonizing the waste collection infrastructure in the neighbourhoods, ensuring the uniform level of service in terms of bring-banks for all waste streams (unsorted waste, paper, plastic and glass). This required the installation of 13 additional collection points and the upgrade

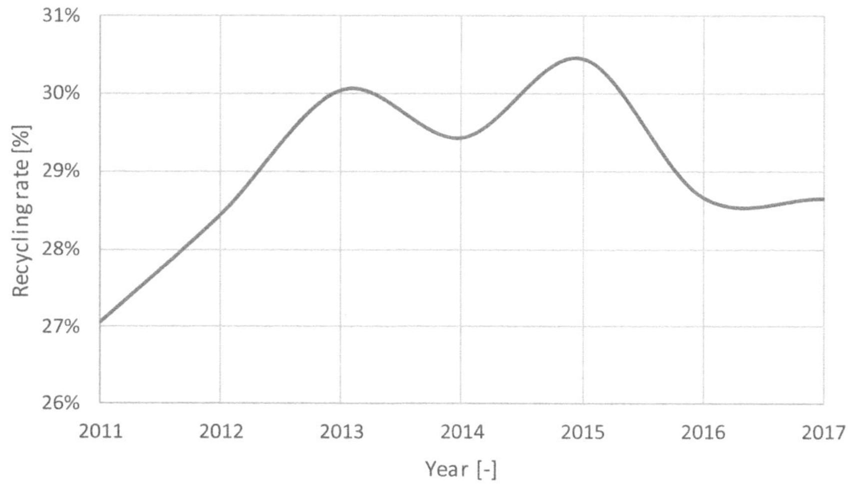

Figure 4. Annual recycling rate evolution in Cascais.

Figure 5. Location of the neighbourhoods.

of one of the 9 existing collection points. In the second stage, a gamification system was launched. The gamification is based on the attribution of points based on the users' waste performance that can be used in services provided by the municipality (e.g., cultural spaces entries, leisure and sport activities, environmental products, educational activities). The gamification was targeted to promote the use of the identification cards to access all the containers and increasing the degree of separation.

3 RESULTS AND DISCUSSION

Disregarding the garden and bulky waste, during the months following the infrastructure improvements, the segregated waste collection doubled from 10% to 20–21%. With the gamification, the proportion of waste segregation increased to nearly 30%.

Tables 1 to 3 report some of the results of the gamification system implemented. It is interesting to notice in table 3 that the number of uses per month of the participants reduces with the socio-economic characteristics of the neighbourhood, with Bairro da Torre showing a substantially lower number of uses per participant per month.

Table 1. Number of individuals enrolled (top) and proportion relatively to the number of families (bottom) in each neighbourhood.

	Month							
Neighbourhood	Jul	Aug	Sep	Oct	Households	Population	Buildings	Families
Lombos Sul	144	181	194	201	651	1069	149	469
	31%	39%	41%	43%				
Bairro da Torre	26	44	44	54	231	150	17	102
	25%	43%	43%	53%				
Quinta de São Gonçalo	102	241	270	252	801	1512	94	651
	16%	37%	41%	39%				
Total	272	466	508	507	1683	2731	260	1222
	22%	38%	42%	41%				

Table 2. Number of individuals enrolled participating (top) and proportion relatively to the number of individuals' enrolled (bottom) in each neighbourhood.

	Month			
Neighbourhood	Jul	Aug	Sep	Oct
Lombos Sul	70	116	133	130
	49%	64%	69%	65%
Bairro da Torre	9	23	23	21
	35%	52%	52%	39%
Quinta de São Gonçalo	36	137	166	150
	35%	57%	61%	60%
Total	115	276	322	301
	42%	59%	63%	59%

Table 3. Number of waste depositions (top) and ratio by the number of individuals' participating (bottom) in each neighbourhood.

	Month			
Neighbourhood	Jul	Aug	Sep	Oct
Lombos Sul	684	2266	2578	2396
	10	20	19	18
Bairro da Torre	43	261	298	281
	5	11	13	13
Quinta de São Gonçalo	145	1718	2783	2525
	4	13	17	17
Total	872	4245	5659	5202
	8	15	18	17

Presently, the system does not allow to know the amount of waste deposited in each use, but assuming that the waste containers in the households have similar sizes it is possible to infer that wealthier individuals produce more waste. This is consistent with the average per capita waste pattern in European countries, with the richer countries presenting higher values.

4 CONCLUSIONS

The stages already implemented of the PAYT scheme designed for the pilot neighbourhoods highlight that a substantial improvement in the recycling rate derive from the level of service available and the motivation/sensibilization of the waste producers. The third stage will attempt to implement a full PAYT system, by defining variable tariffs depending on the type and amount of waste produced. This will require a system for measuring the amount of waste (the most viable solution considering the existing infrastructure is by volume) and locking the access to the containers without the use of the access cards. As any other PAYT scheme in general, and for multifamily buildings in particular, there is no warranty that increasing the mixed waste collection cost will not result in a diversion of the mixed waste towards another waste stream that is cheaper.

REFERENCES

Canterbury, J. 1994. *Pay-As-You-Throw: Lessons Learned About Unit Pricing of Municipal Solid Waste*. EPA Office of Solid Waste. Report# EPA530.

Dijkgraaf, E. & Gradus, R.H.J.M. 2004. Cost savings in unit-based pricing of household waste: the case of The Netherlands. *Resource and Energy Economics* 26(4): 353–371.

EC 2012. *Use of economic instruments and waste management performances*. European Commission (EC), Sustainable Production and Consumption. http://ec.europa.eu/environment/waste/pdf/final_report_1004 2012.pdf. Acessed 15 march 2019.

EEA 2016 *Municipal waste management across European countries*. European Environment Agency (EEA), Resource efficiency and waste. https://www.eea.europa.eu/themes/waste/municipal-waste/municipal-waste-management-across-european-countries Acessed 15 march 2019.

Madureira, R.C. & Dias-Ferreira, C. 2019a. Tag & sensoring solutions overview for smart cities waste management. This issue (submitted).

Madureira, R.C. & Dias-Ferreira, C. 2019b. Data communication solutions overview for smart cities waste management. This issue (submitted).

Karagiannidis, A., Xirogiannopoulou, A. & Tchobanoglous, G. 2008. *Full cost accounting as a tool for the financial assessment of Pay-As-You-Throw schemes: a case study for the Panorama municipality, Greece*. Waste Management 28(12):2801-2808

Stevens, B. 1977. *Pricing Schemes for Refuse Collection Services: The Impact on Refuse Generation*. Research Paper # 154, Columbia University Graduate School of Business.

Van Houtven, G.L. & Morris G.E. 1999. Household behavior under alternative pay-as-you-throw systems for solid waste disposal. *Land Economics* 75(4):515–537.

Wastes: Solutions, Treatments and Opportunities III – Vilarinho et al. (Eds)
© 2020 Taylor & Francis Group, London, ISBN 978-0-367-25777-4

Development of a methodic process to recycle spent alkaline batteries

C. Ribeiro, C. Correia, M.F. Almeida & J.M. Dias
LEPABE, Departamento de Engenharia Metalúrgica e de Materiais, Faculdade de Engenharia, Universidade do Porto, Porto, Portugal

ABSTRACT: Selective dismantling of spent alkaline batteries was conducted, followed by neutral leaching of the black mass at room temperature and gravity concentration to recover KOH and metallic Zn, respectively. A counter current washing process at two different S/L ratios of 1:3 and 1:5 was evaluated and a Wilfley shaking table was employed aiming metallic Zn recovery. Neutral leaching could be effectively conducted at a 1:5 S/L ratio (17 rpm, 20 min, room temperature), using four cycles with fresh water, leading to less than 1% of the black mass K in the final water. Although less effective in terms of K recovery, the use of 1:5 S/L ratio or lower seems more technically feasible for future application of the methodic washing (34 g KOH/L in the washing water). Gravity concentration requires further optimization to effectively concentrate metallic Zn and Mn appears to be present in the fractions of higher caliber (>850 μm).

1 INTRODUCTION

In the last years, the consumption of batteries and accumulators has increased, following the trends of the electronics market. Every year, 160 000 t of batteries are placed on the European market and around 60% of them are alkaline or saline (Leite et al., 2019, Tran et al., 2018). This kind of batteries is largely used and runs out rapidly, quickly becoming a residue (Sayilgan et al., 2009).

Alkaline batteries have in its constitution materials with economic value like steel, zinc and manganese; also, they might have heavy metals, which represent a threat both to human and environmental health (Nogueira and Margarido, 2015, Sayilgan et al., 2009, Almeida et al., 2006).

Some of the existing industrial recycling practices are based essentially on pyrometallurgical processes that do not require several physical pre-treatment processes, but which present as a drawback high energy consumption and gaseous emissions (Sobianowska-Turek et al., 2016).

Hydrometallurgical routes are considered more economically feasible and environmentally friendly, since they require low energy and do not contribute for air pollution (Buzatu et al., 2013, Leite et al., 2019); however, the generation and treatment of wastewaters must be addressed (Sayilgan et al., 2009).

The hydrometallurgical processing of spent batteries focus essentially on the recovery of metals, disregarding other constituents (Sayilgan et al., 2009). For this reason, and considering a circular economy approach, it seems extremely important to develop a process that combines the recovery of metals as well as of other components present.

According to Sayilgan et al. (2009), the electrolyte solution (KOH) present in the alkaline battery powders could be easily separated from the other insoluble elements since it is highly soluble in water. Indeed, KOH is highly soluble in water contrarily to metallic zinc, zinc oxide and the manganese oxides, which present low solubility in water. Therefore, the washing of the black mass (term used from now on for means of simplicity) seems to be essential for the recovery of KOH both to allow its further use and to facilitate the recovery of other elements (note that an alkali material leads to a higher chemical consumption in the hydrometallurgical processes of recovery that follow, which more commonly employ mineral acids). Taking into account correct water utilization, the washing should be carried out with the minimum of water and thus producing a solution as most concentrated as possible. This last aspect is also important under the perspective of using KOH which could require solution evaporation or using the solution as it is obtained, in which case, by reasons of transportation economy, it should present the highest density possible. The so-called methodic washing corresponds to the concept of using few water to take the solute from the solids and it is carried out frequently by a counter-current process. In case of a multi-stage discontinuous process, the fresh solids contact with the solution at its final stage of the washing cycle (in a condition close to its saturation), immediately before it goes out; the first stage is that where the solids are washed the last time, in this case by fresh water, and following the cleaned material gets out the system. This first solution goes to the next step of washing where it is concentrated by contacting a solid from which it takes some solute. Then, the resulting solid goes to the first step where it is washed by a fresh solution. By washing black mass using a methodic washing, the solution will increase KOH concentration on the successive washings, consequently increasing its density. Ideally, methodic washing should originate a saturated KOH solution. However, black mass has very fine particles, namely carbonaceous particles which means that increasing density makes more difficult to separate solids from liquids, particularly the less dense and finer particles. Therefore, washing the black mass might not be an efficient process in the point of view of water use, since it is limited by the density of the solution. So, the concentration necessary for efficient separation should be much lower than that corresponding to KOH saturation. In short, black mass washing must guarantee: (i) a residual mass with as less as possible KOH, close to nil; (ii) a solution that easily separates from the solids, if possible by a simple method, as it is sedimentation; and, (iii) a solution with the maximum density that corresponds to an efficient solid/liquid separation, thus with the maximum possible KOH concentration.

The present study aims to evaluate a methodic pre-treatment process to recover, not only the metallic components present in the alkaline batteries, but also other inorganic compounds such as the potassium hydroxide electrolyte, by using methodic washing, to contribute for the development of an integrated process for full recovery of such materials.

2 MATERIALS AND METHODS

2.1 *Pre-treatment*

Spent alkaline batteries processing usually begins with physical methods, consisting in steps such as sorting (e.g. manual/magnetic separation), dismantling, size reduction (namely grinding) and size screening. Although the grinding step is used to improve the efficiency of the leaching process in the hydrometallurgical route (by reducing the specific area of contact with the extractant), it is energy intensive, therefore leading to high costs (Sayilgan et al., 2009). Hence, it is important to study this step to make it more sustainable, by using simple processes that can easily separate major components and further enable the recovery of the less accessible materials/elements.

Spent domestic alkaline batteries (AA type) used in this work were obtained from different collection points around the Porto city after manual sorting (separating alkaline from saline). Alkaline batteries were manually dismantled by cutting the top and bottom extremes with a saw and further separating the three different parts (top, bottom and body of the battery). The outer steel cylinder (outer case) and the negative and positive connectors of the body were manually separated. The inside black mass containing the anodic (rich in zinc) and cathodic (rich in manganese) fractions with the electrolyte (potassium hydroxide) were dried at 105°C until constant weight before characterization and further use.

2.2 *Characterization of the black mass*

For the black mass characterization, an acid digestion was first performed using an aqua regia solution (1:3 v/v of nitric acid and hydrochloric acid) according to the standard ISO 11466:1995.

Digestions were performed in triplicate with about 1 g of the black mass and 10 mL of solution, during 3h at 90°C. Then, the slurries were filtered, the solid residue was washed with distilled water and the final volume of solution was adjusted to 100 mL using such water.

Metals quantification was performed by atomic absorption spectrometry (AAS) using an UNICAM 969 AA spectrometer.

2.3 *Neutral leaching*

Samples of the black mass were subjected to washing cycles. The washing process was conducted at room temperature during 20 minutes in flasks with 30 g of the black mass and distilled water at defined S/L ratios (1:3 and 1:5) shacked in a Frequrol-U100 shaker regulated to 17 rpm and after separated by centrifugation (3500 rpm, 12 min).

The procedure to ensure washing of the mass which was used in the gravity concentration process used only the S/L ratio of 1:5, whereas in the procedure which aimed at concentrating the washing water by methodic washing both 1:3 and 1:5 ratios were used. In agreement, under the present study, results from 2.3.1 reveal the effectiveness of a washing process (which was not integrated with the countercurrent process) whereas results from 2.3.2 should provide answers to the development of methodic washing to concentrate the washing water. Since the idea was to evaluate the maximum KOH reached in the water, only the washing water was monitored (in the future, the objective is to integrate both steps).

2.3.1 *Washing of the mass for gravity concentration*

To wash the black mass, washing cycles with fresh water were conducted using a 1:5 S/L ratio, until the density of the washing water was close to that of the distilled water, since pH monitoring was not effective to control the process. In addition, the first and last washing water was subjected to volumetric titration with HCl (0.239 M) to validate the effectiveness of the washing.

2.3.2 *Concentration of washing water for further recovery*

A countercurrent washing process was established (batch) to concentrate the water. In such scheme, the process started by washing the fresh mass with washing water which has already passed a defined number of washing cycles, as explained in the introduction. To start the process, a first washing cycle was conducted, using the defined amount of water and fresh black mass (1:3 or 1:5 S/L ratio); then, the water from this cycle was again contacted with fresh black mass (so that the fresh black mass enters at the end of the washing cycles). This procedure was repeated 4 times for the highest S/L ratio and 7 times for the lowest, to allow proper phase separation, due to the restrictions in phase separation previously explained. This procedure was used to concentrate as much as possible the water aiming its further recovery and will be deeper studied in the future.

The residual washed black mass was dried at 105°C until constant weight in order to evaluate mass loss resulting from the pre-treatment. Water loss was also evaluated.

2.4 *Black mass concentration*

For the concentration of metallic zinc in order to facilitate its further hydrometallurgical recovery, the washed black mass was first screened using four sieves (from 90 μm to 1700 μm). The particles which presented adequate caliber to be submitted to gravity concentration in a Wilfley shaking table (intermediate fraction, particles size of 90–850 μm) were first washed and after submitted to the process at standard conditions (previous studies performed by Ventura et al. (2018) indicate the following conditions – water flow around 11 L/min and around 4° slope) aiming to obtain a fraction rich in metallic zinc which was then characterized by AAS. The tailing fraction was also characterized in order to evaluate the effectiveness of the concentration process. Further, additional concentration processes, adequate for higher caliber materials (ex. gigging) will be performed together with the determination of the chemical composition of the fractions (before and after the process).

3 RESULTS AND DISCUSSION

3.1 *Characterization of the black mass*

Table 1 shows the results obtained for the composition of the initial black mass that resulted from the dismantling of alkaline batteries previously described.

As expected, the main elements are manganese, zinc and potassium. Iron is in small quantity, and its presence should be related with the particles resulting from cutting the steel cylinder. Cadmium is a vestigial element in the samples, probably due to its well known paragenesis with zinc.

Martha de Souza et al. (2001) reported that in the alkaline battery powder the typical metals composition was in the ranges of 12–21% for Zn, 26–33% for Mn, 5.5–7.3% for K and iron appeared at a concentration of around 0.17%, values in the range of those obtained in the present study. Falco et al. (2014) and Furlani et al. (2009) also reported a residual amount of Cd in the alkaline battery powder, in agreement with the present study.

3.2 *Neutral Leaching*

3.2.1 *Washing of the mass for gravity concentration*

The variation of the density of the washing water from the first to the last cycle (fourth) was minimum, in the order of 1% (from 1009 kg m^{-3} to 998 g m^{-3}); although low, it stabilized in such value indicating that all KOH was leached. The titration of the water revealed that the first washing cycle led to a K content of 2%, relative to black mass weight, showing that around 30% of the K in the black mass was leached in the first cycle. The final washing cycle showed 0.05% of K, in terms of black mass weight, representing less than 1% of the K present in the black mass, thus revealing effective washing and providing relevant information for further integration with results from 3.2.2. It should also be highlighted that, during the process where difficulties in phase separation were found, some black mass was lost as well as some water.

3.2.2 *Concentration of washing water for further recovery*

For the concentration of the washing water in a countercurrent scheme aiming its further recovery, two different S/L ratios were tested (1:3 and 1:5) as previously mentioned (section 2.3.2). Figure 1 shows the evolution of the water density (kg m^{-3}) throughout the water washings for both S/L ratios.

Using a S/L ratio of 1:3, the maximum density reached was 1064 kg m^{-3} (at the fourth washing), whereas the maximum obtained using the 1:5 S/L was 1070 kg m^{-3}, thus representing a 7% increase of density compared to that of the distilled water. Some mass and water losses during the process were also observed, as in 3.2.1.

Using both S/L ratios, the first washing water dissolved 2% of K, relative to black mass weight, hence showing the leaching of around 30% of K from the mass, agreeing with was found in 3.2.1.

For the purpose of KOH recovery, the concentration of the solutions varied from an initial value of around 5 g KOH/L up to around 50 g KOH/L in the case of the highest ratio (1:3) and to around 34 g KOH/L in the case of the lowest ratio (1:5), thus a concentration of 10 times was performed in the first case, whereas in the second a concentration of around 7 times was performed during methodic washing.

In terms of K dissolution, using the highest ratio, in the end, the solution showed 10.2% of K, relative to the black mass used in the last cycle, whereas using the lowest ratio, 11.5% of K (also

Table 1. Mean composition and standard deviation of analyzed metallic elements from the batteries black mass, wt.%.

Elements	Zn	Mn	K	Fe	Cd
Content	12 ± 1	28 ± 1	6.9 ± 0.4	0.13 ± 0.07	$(18 \pm 5) \times 10^{-5}$

relative to the mass used in the last cycle) was obtained. From this perspective the differences are not high, although the use of the lowest ratio brings less problems in terms of phase separation thus making the process more appealing from a technical point of view.

Metals dissolution during neutral leaching was also verified and the concentrations of zinc, manganese and iron measured by AAS spectrometry in the initial and final washing waters gave all values extremely low, as expected.

3.3 *Black mass concentration*

From the black mass fed to the shaking table, two fractions were obtained: a concentrate, corresponding to the denser material, and a tailing. It is expected that the concentrate presents the higher concentration of metallic zinc than both the feed material and tailing fraction, due to its density.

Table 2 shows the composition obtained by analyzing samples of both fractions by AAS. The concentrate presented higher zinc content than the tailing; however, this last is not a so poor zinc material to be considered a true tailing. Anyway, it is expected that the concentrate has most of its zinc in the metallic form, contrarily to the tailing, where zinc may be predominantly as zinc oxide, which is lighter (those cannot be distinguished by the employed analytical method). This issue will be deeper studied in the future.

Both concentrate and tailing fractions showed much less manganese than that found in the black mass, probably a result of eliminating the richer manganese fraction on the sieving step referred before (particle size >850 μm, section 2.4). This aspect will be further evaluated.

As expected, potassium content is very low in both fractions, which means efficient removal of KOH by the washing carried out. The differences between the values here presented and those obtained by titration should relate to the different method associated as well as the low concentrations measured (which should have a higher error associated).

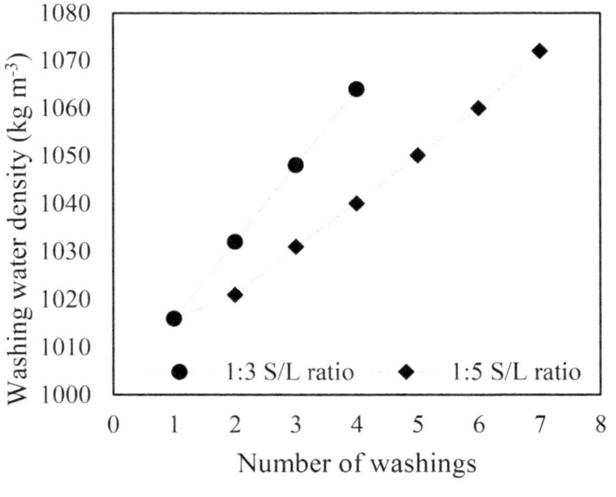

Figure 1. Evolution of water density during washing water concentration using two S/L ratios (30 g, 17 rpm, 20 min, room temperature).

Table 2. Composition of the concentrate and tailing fractions obtained from the black after gravity concentration in a Wilfley shaking table.

Sample \ Elements	Zn	Mn	K	Fe	Cd
Concentrated sample (wt. %)	16 ± 1	14 ± 2	0.12 ± 0.01	0.6 ± 0.1	$(25 \pm 6) \times 10^{-5}$
Tailing sample (wt. %)	12 ± 1	16 ± 2	0.10 ± 0.01	0.29 ± 0.07	$(24 \pm 10) \times 10^{-6}$

4 CONCLUSION

Alkaline batteries still lead the portable batteries market; in agreement, the recovery of wastes resulting from their use is of high relevance. Since they have in its constitution materials such as steel, zinc and manganese, they become a valuable resource aiming metals recovery.

The approach of this study was to develop a methodic process for the pre-treatment of spent alkaline batteries to obtain concentrated solutions of potassium hydroxide for further use and preparing the resulting mass to enable more effective zinc and manganese recycling.

The neutral leaching selectively removed the KOH, as expected and demonstrated by the very low concentrations of Zn, Mn and Fe in the washing waters.

Neutral leaching could be effectively conducted at a 1:5 S/L ratio (17 rpm, 20 min, room temperature), using four cycles with fresh water. The washing water density, although with small variation, could be effectively used to control the process. The first washing cycle leached 30% of the K on the black mass whereas the last washing water leached less than 1% of K.

For the purpose of concentrating the washing water for further recovery under a countercurrent washing process (methodic washing) using a S/L ratio of 1:3, only 4 cycles were performed whereas at 1:5, 7 cycles could be performed due to difficulties in phase separation and associated losses of mass and water. The concentration of the solutions varied from an initial value of around 5 g KOH/L up to around 50 g KOH/L in the case of the highest ratio (1:3) and to around 34 g KOH/L in the case of the lowest ratio (1:5), thus a concentration of 10 times was performed in the first case, whereas in the second a concentration of around 7 times was performed during methodic washing. From a practical point of view, the use of the 1:5 S/L ratio or lower seems more technically reasonable for future application. The monitoring of the water density should be further studied to be effectively used to control this process; the maximum density reached in the concentrated washing water at a 1:5 S/L ratio was 1070 kg m^{-3}.

ACKNOWLEDGMENTS

The authors thank Prof. Aurora Futuro for the support regarding the density concentration process. This work was financially supported by project UID/EQU/00511/2019 – Laboratory for Process Engineering, Environment, Biotechnology and Energy – LEPABE funded by national funds through FCT/MCTES (PIDDAC) and Project "LEPABE-2-ECO-INNOVATION" – NORTE-01-0145-FEDER-000005, funded by Norte Portugal Regional Operational Programme (NORTE 2020), under PORTUGAL 2020 Partnership Agreement, through the European Regional Development Fund (ERDF).

REFERENCES

Almeida, M. F., Xará, S. M., Delgado, J. & Costa, C. A. 2006. Characterization of spent AA household alkaline batteries. *Waste Management,* 26, 466–476.

Buzatu, M., Săceanu, S., Ghica, V., Iacob, G. & Buzatu, T. 2013. Simultaneous recovery of Zn and MnO2 from used batteries, as raw materials, by electrolysis. *Waste management (New York, N.Y.),* 33.

Falco, L., Quina, M., Ferreira, L., Thomas, H. & Curutchet, G. 2014. Solvent Extraction Studies for Separation of Zn(II) and Mn(II) from Spent Batteries Leach Solutions. *Separation Science and Technology,* 49, 398–409.

Furlani, G., Moscardini, E., Pagnanelli, F., Ferella, F., Vegliò, F. & Toro, L. 2009. Recovery of manganese from zinc alkaline batteries by reductive acid leaching using carbohydrates as reductant. *Hydrometallurgy,* 99, 115-118.

Leite, D. d. S., Carvalho, P. L. G., de Lemos, L. R., Mageste, A. B. & Rodrigues, G. D. 2019. Hydrometallurgical recovery of Zn(II) and Mn(II) from alkaline batteries waste employing aqueous two-phase system. *Separation and Purification Technology,* 210, 327–334.

Martha de Souza, C. C. B., Corrêa de Oliveira, D. & Tenório, J. A. S. 2001. Characterization of used alkaline batteries powder and analysis of zinc recovery by acid leaching. *Journal of Power Sources,* 103, 120–126.
Nogueira, C. A. & Margarido, F. 2015. Selective process of zinc extraction from spent Zn–MnO2 batteries by ammonium chloride leaching. *Hydrometallurgy,* 157, 13–21.
Sayilgan, E., Kukrer, T., Civelekoglu, G., Ferella, F., Akcil, A., Veglio, F. & Kitis, M. 2009. A review of technologies for the recovery of metals from spent alkaline and zinc–carbon batteries. *Hydrometallurgy,* 97, 158–166.
Sobianowska-Turek, A., Szczepaniak, W., Maciejewski, P. & Gawlik-Kobylińska, M. 2016. Recovery of zinc and manganese, and other metals (Fe, Cu, Ni, Co, Cd, Cr, Na, K) from Zn-MnO2 and Zn-C waste batteries: Hydroxyl and carbonate co-precipitation from solution after reducing acidic leaching with use of oxalic acid. *Journal of Power Sources,* 325, 220–228.
Tran, H. P., Schaubroeck, T., Swart, P., Six, L., Coonen, P. & Dewulf, J. 2018. Recycling portable alkaline/ZnC batteries for a circular economy: An assessment of natural resource consumption from a life cycle and criticality perspective. *Resources, Conservation and Recycling,* 135, 265–278.
Ventura, E., Futuro, A., Pinho, S. C., Almeida, M. F. & Dias, J.M. 2018. Physical and thermal processing of Waste Printed Circuit Boards aiming for the recovery of gold and copper. *Journal of Environmental Management*, 223, 297–305.

Wastes: Solutions, Treatments and Opportunities III – Vilarinho et al. (Eds)
 ISBN 978-0-367-25777-4

Eutrophication impact potential of solid waste management options in Harare

T. Nhubu & C. Mbohwa
University of Johannesburg, Johannesburg, Gauteng, South Africa

E. Muzenda
University of Johannesburg, Johannesburg, Gauteng, South Africa
Botswana International University of Science and Technology, Palapye, Botswana

ABSTRACT: Six municipal solid waste management options (A1–A6) in Harare were developed and analyzed for their eutrophication impact potentials under the Life Cycle Assessment (LCA) methodology. All the options started with waste collection and transportation to a centralized waste treatment centre where a combination of various municipal solid waste management and treatment methods were considered under the different options. Results show that landfilling and material recovery for reuse and recyle are the only MSW management processes that contributes to negative eutrophication potential giving options that had landfilling (A1, A4 and A6) an overall edge. The doubling of recycling rate under A5 and increasing it to atleast 25% under A6 result in below zero eutrophication impact potentials. Results reveal that anaerobic digestion and incineration contribute to increased eutrophication potential under all the options they were considered hence need for further assessments considering other impact categories to determine the most sustainable option.

1 INTRODUCTION

Harare, the capital city of Zimbabwe is experiencing enormous water challenges with regards to quality and quantity. The city is failing to consistently supply its residents with safe and clean potable water. Some residential areas are experiencing dry tapes for weeks or even months. Nhapi (2009) argued that these challenges are largely attributed to increased population, lack of necessary maintenance works on wastewater infrastructure, use of technologies that are expensive and institutional framework deficiencies. Residents have resorted to the drilling of boreholes and shallow groundwater wells at their households. However groundwater is not as safe as perceived due to nutrients, metals, acidity and coliform bacteria contamination emanating from diffuse pollution (Love et al., 2006, Eukay and Kharlamova, 2014, Kharlamova et al., 2016). The diffuse pollution largely results from poor Municipal Solid Waste (MSW) management practices and like any other urban environments in developing countries; Harare is facing MSW management challenges due to to rapid population growth and increased volumes of waste being generated by the public causing its failure to sustainably manage. Only 60 per cent of the MSW generated in Harare is collected and disposed of at dumpsites with the remaining 40% usually dumped illegally in open spaces, road verges, alleys and drainages for storm water (Tanyanyiwa, 2015, PASA, 2006, Jerie, 2006, Saungweme, 2012, Chirisa, 2013). This threatens the water situation in Harare and has been cited as the major cause of the annual outbreaks of Cholera and Typhoid in Harare (Manzungu and Chioreso, 2012, Tanyanyiwa and Mutungamiri, 2011, Chirisa, 2013)

The pollution of surface and groundwater in Zimbabwe's urban environments is a direct result of municipal solid waste management specifically MSW dumping in waterways and leachate from MSW dumpsites (Mangizvo, 2007, Mangizvo, 2010, Tsiko and Togarepi, 2012) together with the discharge of untreated or partially treated sewage into river systems (Nhapi et al., 2004). This has led to the high level eutrophic status of Lake Chivero, the source of potable water to Harare and Chitungwiza. The eutrophication of Lake Chivero is not a welcome development considering its threats to the availability of potable water for Harare and Chitungwiza. It has led to increased costs for potable water production partly contributing to the erratic potable water supplies in most parts of Harare currently being experienced. Bauman and Tillman (2004) described eutrophication as a phenomenon with the potential of affecting terrestrial and aquatic ecosystems due to nutrient enrichment namely Nitrogen (N) and phosphorus (P) in water systems. The MSW generated in Harare has an excess of biowaste constituting over 60% (UNEP, 2011) hence if improperly managed can lead to nutrient enrichment in water bodies leading to eutrophication. Magadza (2003) reported that the breakdown in hygiene has led to nutrient rich surface run-off from uncollected MSW and illegal MSW dumps significantly contributing to Lake Chivero eutrophication.

The design and development of sustainable MSW management option for Harare becomes a necessity to address the human health challenges and availability of freshwater that guarantee the long term consistent supply of potable water for Harare residents. Life Cycle Assessment (LCA) has proven to be an effective tool for designing and developing sustainable and integrated MSW management as it aids the assessment of environmental loads of different MSW options (Miliute and Kazimieras Staniškis, 2010; Rives et al., 2010; Koci and Trecakova, 2011; Stucki et al., 2011; Gunamantha and Sarto, 2012; Fernández-Nava et al., 2014). Therefore, this work is an LCA based comparative study to assess the eutrophication impact potential of different MSW management options for Harare. The objective being to determine the option with the least eutrophication impact potential in light of the reported eutrophic status of the potable water sources in Harare.

2 MATERIALS AND METHODS

2.1 *Description of study area*

The study area is Harare the capital city of Zimbabwe with an estimated population of 1,485,231 (Zimstat, 2013) and covering an area of 960.6 km^2 at an altitude of 1483 m. An estimate of 325,266 tons of MSW is generated per year (Mshandete and Parawira, 2009, Muchandiona et al., 2013, Pawandiwa, 2013, Mbiba, 2014, Hoornweg and Bhada-Tata, 2012, Emenike et al., 2013) in Harare with 60% indiscriminately collected and dumped at Pomona dumpsite the only official dumpsite whose capacity is exhausted by 2020 (Chijarira, 2013). The waste that is generated in Harare is estimated to have average composition of 42% biodegradable waste, 33% plastics, 8% metals, 14% paper and 3% glass (Nyanzou and Steven, 2014, Mudzengerere and Chigwenya, 2012). Harare sits upstream and on the catchment of its potable water source (Lake Chivero) making all the MSW management activities in Harare contributing towards the reported super eutrophic levels of the Lake. Underground water in Harare has also been reported to have been contaminated with nutrients, metals, acids and coliform bacteria (Muchandiona et al., 2013, Love et al., 2006, Eukay and Kharlamova, 2014, Kharlamova et al., 2016).

2.2 *MSW management option 1 – A1*

The entire 325,266 tons of MSW generated per year in Harare is indiscriminately collected before any treatment (both biodegradable and nonorganic MSW) and landfilled in a sanitary landfill with biogas recovery and landfill leachate treatment. The recovered biogas is fed into Combined Heat and Power (CHP) plant to produce electricity.

2.3 *MSW management option 2 – A2*

The entire 325,266 tons of MSW generated per year in Harare is indiscriminately collected before any treatment (both biodegradable and nonorganic MSW) and incinerated in an incinerator with energy recovery, flue gas treatment and treatment of leachate produced during the recovery of the incinerator bottom ash. The incinerator bottom and fly ash is used as material for road construction considering the road infrastructural needs of the country.

2.4 *MSW management option 3 – A3*

Biodegradable MSW generated amounting to 136,612 tons is digested in an anaerobic digester producing biogas. The biogas is fed into Combined Heat and Power (CHP) generation plant to produce heat and electricity. The remaining non-biodegradable fraction 188,654 tons mixed bag MSW (107,338 tons plastics, 26,021 tons metals, 45,537 tons paper and 9,758 tons glass) is incinerated as in A2 with energy recovery, flue gas treatment and treatment of leachate produced during the recovery of the incinerator bottom ash. The incinerator bottom and fly ash is used as material for road construction considering the road infrastructural needs of the country.

2.5 *MSW management option 4 – A4*

As in A3 difference being that the remaining non-biodegradable fraction 188,654 tons mixed bag MSW (107,338 tons plastics, 26,021 tons metals, 45,537 tons paper and 9,758 tons glass) is landfilled as in A1 with biogas recovery and landfill leachate treatment. The recovered biogas is fed into Combined Heat and Power (CHP) plant to produce electricity.

2.6 *MSW management option 5 – A5*

20% of the non-biodegradable MSW amounting to 37,731 tons (21,468 tons plastics, 5,204 tons metals, 9,107 tons paper and 1,952 tons glass) are recovered in the material recovery facility or sorting plant for reuse and recycling. The 80% non-biodegradable MSW remaining from the material recovery facility amounting to 150,923 tons (85,870 tons plastics, 20,817 tons metals, 36,430 tons paper and 7,806 tons glass) is incinerated as in A2 with energy recovery, flue gas treatment and treatment of leachate produced during the recovery of the incinerator bottom ash. The incinerator bottom and fly ash is used as material for road construction considering the road infrastructural needs of the country.

2.7 *MSW management option 6 – A6*

20% of the non-biodegradable MSW amounting to 37,731 tons (21,468 tons plastics, 5,204 tons metals, 9,107 tons paper and 1,952 tons glass) are recovered in the material recovery facility or sorting plant for reuse and recycling. The 80% non-biodegradable MSW remaining from the material recovery facility amounting to 150,923 tons (85,870 tons plastics, 20,817 tons metals, 36,430 tons paper and 7,806 tons glass) is landfilled as in A1 with biogas recovery and landfill leachate treatment. The recovered biogas is fed into Combined Heat and Power (CHP) plant to produce electricity.

2.8 *Life Cycle Assessment*

The eutrophication impact potential for the six MSW management options was estimated using the LCA methodology with the ISO 14040 standards applied as the basis of the LCA. Simapro version 8.5.2 analyst software and update 852 database were used for the LCA under the ReCiPe 2016 v1.02 endpoint method. The yearly MSW generation of 325,266 tons was used as the functional unit (Fernández-Nava et al., 2014, Beigl and Salhofer, 2004, Cherubini et al., 2009). Waste collection and transportation, landfilling, incineration, anaerobic digestion material recovery, CHP generation, landfill leachate treatment and incineration flue gas treatment were the considered life cycle stages.

Table 1. Process Contributions to eutrophication.

	MSW management options					
	A1	A2	A3	A4	A5	A6
Process	Species.yr					
Waste transportation	1.39E-04	1.39E-04	1.39E-04	1.39E-04	1.39E-04	1.39E-04
Landfilling	−1.53E-02	–	–	−8.87E-03	–	−2.09E-03
Anaerobic digestion of biowaste	–	–	5.43E-03	5.43E-03	5.43E-03	5.43E-03
Incineration	–	6.96E-06	4.04E-06	–	9.51E-07	–
Materials recovery	–	–	–	–	−2.78E-03	−2.78E-03
Total	−1.52E-02	1.46E-04	5.57E-03	−3.30E-03	2.79E-03	6.99E-04

3 RESULTS AND DISCUSSIONS

Figure 1 shows the LCIA results with regards to the eutrophication impact potential of the six MSW management options. MSW management options A1and A4 leads to reduced extinction rate of species thus reduced eutrophication impact potential with A2 to A6 bringing about an increased species extinction rates.

MSW option A4 is the most favorable with eutrophication impact potential reduction of −3.30E-02 species.yr. Figure 2 and Table 1 shows detailed contribution of the processes constituting the six MSW management options to the eutrophication potential. An assumption was made that all the waste is collected and transported to a central MSW management facility where the various MSW management options were considered giving a constant waste collection and transportation impact potential. The collection and transportation of 325,266 tons of MSW generated in Harare per annum contributes to increased eutrophication potential of 1.39E-04 species.yr. Need for alternative MSW collection and transportation rather than the household to household collection system currently being practiced arise. Such strategies include the use of centralized or decentralized waste transfer stations, higher volume trucks or trains and citizens waste disposal facilities which will result in increased MSW collection and transportation efficiency from sources and per day collections. In addition, the distance travelled by the MSW trucks is also reduced by implementing these alternative MSW collection and transportation strategies.

The anaerobic treatment of biodegradable MSW fraction (136,612 tons per annum) is part of MSW management options A3–A6 and results indicate it contributes to increased eutrophication of 5.43E-03 species.yr for all the MSW management options. This was also observed by Bernstad and la Cour Jansen (2011) and Mendes (2003). Bernstad and la Cour Jansen (2011) cited digestate use as biofertiliser substituting inorganic fertilisers as the driver for increased eutrophication potential. Improper handling of AD feedstock is also another source of the eutrophication potential. There is however need to assess the trade-off between the eutrophication increase and the benefits derived from the biogas derived renewable electricity that has been found to be environmentally sustainable compared to fossil fuels derived electricity and organic fertilizer from the anaerobic digestion of biodegradable waste. With regards to landfilling, results indicate that for the three MSW management options A1, A4 and A6, there is reduced eutrophication in the magnitudes of −1.52E-02 and −3.30E-03 for A1 and A4 respectively with A6 bringing about increasing eutrophication of 6.99E-04 species.yr. Landfilling associated with the recovery of energy (biogas) and the treatment of landfill leachate is environmentally favorable as observed by Zaman (2010) and Hong et al (2010) despite its threat to land availability. Incineration which is considered under MSW management options A2, A3 and A5 contributes towards the increase in the species extinction rate of 6.69E-06, 4.04E-06 and 9.51E-07 species.yr respectively due to ammonia emissions despite the associated energy recovery and reduction of waste volume. Material recovery brings about the

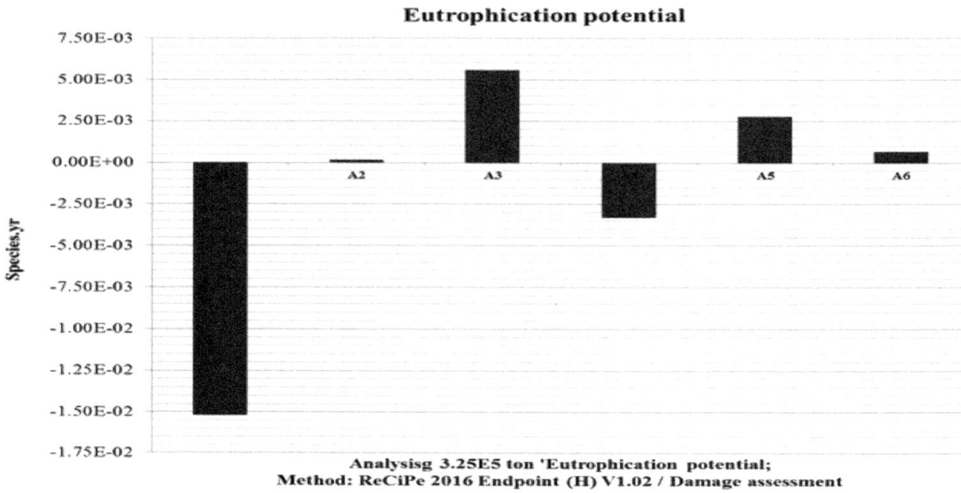

Figure 1. Eutrophication impact potential of MSW management options.

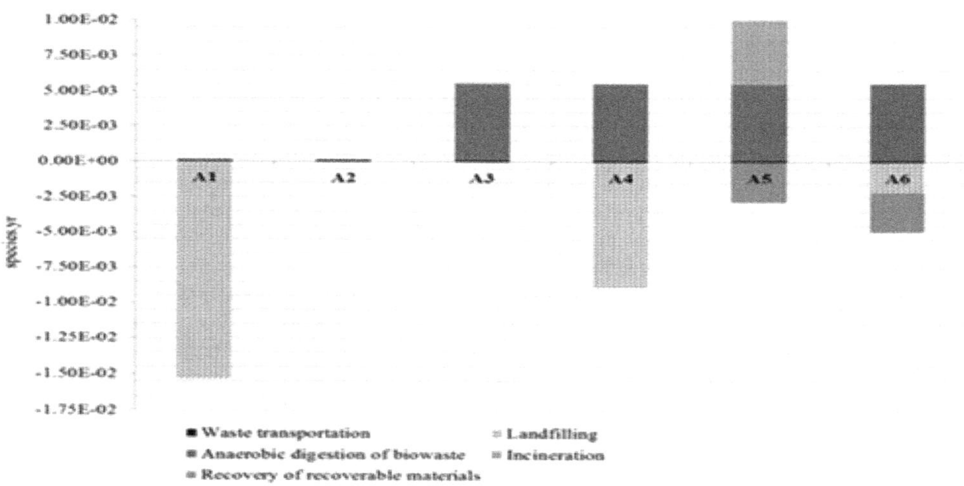

Figure 2. MSW management process contribution to eutrophication impact potential

reduction in species extinction rate of −2.78E-03 species per year for the two MSW Management options considered namely A5 and A6. Doubling recovery rate of recoverable materials under A5 and increasing it to atleast 25% under A6 results in below zero eutrophication impact potential.

4 CONCLUSION AND RECOMMENDATIONS

The study results show that landfilling and material recovery for reuse and recyle are the only MSW management processes that contributes to negative eutrophication potential i.e reduction in potential extinction rate of species due to its associated energy recovery and landfill leachate treatment. However, threats to land availability have led to increased global concern on waste landfilling. The anaerobic digestion, waste collection and transportation and incineration contribute to increased eutrophication potential. However in the case of anaerobic digestion of biodegradable MSW fraction other factors such as the benefits derived from the production of renewable electricity and the production of an organic fertiliser in the form of the digestate need also to be taken into consideration in opting for the best MSW management option as they have proved to be environmentally sustainable. Overally MSW management option 1 (A1) proved to be the best MSW management option when considering the eutrophication potential despite its threat to land availability and increasing global concerns on waste landfilling. However, there were limitations with regards to data availability on actual MSW generation and composition as well as transportation distance. Hence, this study utilized estimates from literature.

Detailed studies to quantify the MSW generation and composition in Harare as well as further life cycle assessments considering other impact categories to determine the most sustainable option are therefore recommended.

ACKNOWLEDGEMENTS

The authors acknowledge the Life Cycle Initiative for 2017 Life Cycle award that provided the Simapro software that was used to carry out the LCIA. The authors are also grateful to the National Geographic Society for the early career grant that enabled data collection for the study. The personnel and management at Harare municipality who assisted with the requisite information and data will always be appreciated without their support this work could not have been accomplished. The Universities of Johannesburg and South Africa are also acknowledged for funding the studies, and as well as conference registration and attendance.

REFERENCES

Bauman, H. & Tillman, A. 2004. The Hitch Hiker's Guide to LCA. Studentlitteratur AB, Sweden.

Beigl, P. & Salhofer, S. 2004. Comparison of ecological effects and costs of communal waste management systems. *Resources, Conservation and Recycling,* 41, 83–102.

Bernstad, A. & La Cour Jansen, J. 2011. A life cycle approach to the management of household food waste–a Swedish full-scale case study. *Waste Management,* 31, 1879–1896.

Cherubini, F., Bargigli, S. & Ulgiati, S. 2009. Life cycle assessment (LCA) of waste management strategies: Landfilling, sorting plant and incineration. *Energy,* 34, 2116–2123.

Chijarira, S. R. 2013. The impact of dumpsite leachate on ground and surface water: A case study of pomona waste dumpsite, Department of Geography. Bindura University of Science and Technology, Bindura, p. 63.

Chirisa, I. 2013. Solid Waste, the 'Throw-Away' culture and livelihoods: Problems and prospects in Harare. *Zimbabwe Journal of Environmental Science and Water Resources,* 2, 001–008.

Emenike, C. U., Iriruaga, E. T., Agamuthu, P. & Fauziah, S. H. 2013. Waste Management in Africa: An Invitation to Wealth Generation. *Proceedings of the International Conference on Waste Management and Environment, ICWME, University of Malaya, Kuala Lumpur, Malaysia, 26–27th August.*

Eukay, M. S. & Kharlamova, M. 2014. An assessment of the relationship between the spatial distribution of undesignated dumpsites and disease occurrence in Budiriro, Harare, Zimbabwe. In Science, Technology and Higher Education. Materials of the V International Conference, June 20th 2014, Westwood, pp. 236–242, ISBN 978-1-77192-087-22014.

Fernández-Nava, Y., J.D, R., Rodríguez-Iglesias, J., Castrillón, L. & Marañón, E. 2014. Life Cycle Assessment (LCA) of different municipal solid waste management options: A case study of Asturias (Spain). *Journal of Cleaner Production,* 2014.

Hong, J., Li, X. & Zhaojie, C. 2010. Life cycle assessment of four municipal solid waste management scenarios in China. *Waste Management,* 30, 2362–2369.

Hoornweg, D. & Bhada-Tata, P. 2012. What a Waste: A Global Review of Solid Waste Management. Urban development series;knowledge papers no. 15. World Bank, Washington, DC. © World Bank. https://openknowledge.worldbank.org/handle/10986/17388License:CCBY3.0IGO." IN BANK, W. (Ed.

Jerie, S. 2006. Analysis of institutional solid waste management in Gweru. EASSRR, Vol xxii No 1 January 2006.

Kharlamova, M. D., Mada, S. Y. & Grachev, V. A. 2016. Landfills: Problems, Solutions and Decision-making of Waste Disposal in Harare (Zimbabwe). *Biosciences Biotechnology Research Asia,* 13, 307–318.
Love, D., Zingoni, E., Ravengai, S., Owen, R., Moyce, W., Mangeya, P., Meck, M., Musiwa, K., Amos, A., Hoko, Z., Hranova, R., Gandidzanwa, P., Magadzire, F., Magadza, C., Tekere, M., Nyama, Z., Wuta, M. & Love, I. 2006. Characterization of diffuse pollution of shallow groundwater in the Harare urban area, Zimbabwe. IN XU, Y. & USHER, B. (Eds.) *Groundwater pollution in Africa.* Taylor & Francis.
Magadza, C. H. D. 2003. Lake Chivero: a management case study. *Lakes & Reservoirs: Research & Management,* 8, 69–81.
Mangizvo, R. V. 2007. Challenges of Solid Waste Management in the Central Business District of the City of Gweru in Zimbabwe. *Journal of Sustainable Development in Africa,* 9, 134–145.
Mangizvo, R. V. 2010. Illegal Dumping of Solid Waste in the Alleys in the Central Business District of Gweru, Zimbabwe *Journal of Sustainable Development in Africa,* 12, 110–123.
Manzungu, E. & Chioreso, R. 2012. Internalising a Crisis? Household Level Response to Water Scarcity in the City of Harare, Zimbabwe. *Journal of Social Development in Africa,* 27.
Mbiba, B. (2014) Urban solid waste characteristics and household appetite for separation at source in Eastern and Southern Africa. *Habitat International,* 43, 152–162.
Mendes, M. R., Aramakib, T. & Hanakic, K. 2003. Assessment of the environmental impact of management measures for the biodegradable fraction of municipal solid waste in São Paulo City. *Waste Management* 23, 403–409.
Mshandete, A. M. & Parawira, W. 2009. Biogas Technology Research in Selected Sub-Saharan African Countries—A Review. *African Journal of Biotechnology,* 8, 116–125.
Muchandiona, A., Nhapi, I., Misi, S. & Gumindoga, W. 2013. Challenges and opportunities in solid waste management in Zimbabwe's urban councils. Harare, University of Zimbabwe.
Mudzengerere, F. H. & Chigwenya, A. 2012. Waste management in Bulawayo City Council in Zimbabwe: In search of sustainable waste management in the city. *Journal of Sustainable Development in Africa* 14, 228–244.
Nhapi, I. 2009. The water situation in Harare, Zimbabwe: a policy and management problem. *Water Policy* 11, 221–235.
Nhapi, I., Siebel, M. A. & Gijzen, H. J. 2004. The impact of urbanisation on the water quality of Lake Chivero, Zimbabwe. *Water and Environment Journal,* 18, 44–49.
Nyanzou, P. & Steven, J. 2014. Solid waste management practices in high density suburbs of Zimbabwe: a focus on Budiriro 3, Harare. *The Dyke,* 8.3, 17–54.
PASA 2006. Proceedings of the Emerging Issues in Urban Waste Management Workshop. 10 February 2006 Harare. Harare: Practical Action Southern Africa.
Pawandiwa, C. C. 2013. Municipal Solid Waste disposal site selection the case of Harare. Bloemfontein, MSc Thesis, University of the Free State.
Saungweme, M. 2012. An Integrated Waste Management Approach as an Alternative Solid Waste Management Strategy for Mbare Township, Zimbabwe. http://wwwscribd.com.
Tanyanyiwa, V. I. 2015. Not In My Backyard (NIMBY)? : The Accumulation of Solid Waste in the Avenues Area, Harare, Zimbabwe. *International Journal of Innovative Research and Development,* 4, 122–128.
Tanyanyiwa, V. I. & Mutungamiri, I. 2011. Residents perception on water and sanitation problems in Dzivaresekwa 1 high density suburb, Harare. *Journal of Sustainable Development in Africa,* 13.
Tsiko, R. & Togarepi, S. 2012. A Situational Analysis of Waste Management in Harare, Zimbabwe. *American Journal of Science,* 8, 692–706.
UNEP 2011. Managing the environment in developing countries, United Nations Environmental Programme, Available online, www.unepwcmc.org.
Zaman, A. U. 2010. Comparative study of municipal solid waste treatment technologies using life cycle assessment method. *International Journal of Environmental Science & Technology,* 7, 225–234.
ZIMSTAT 2013. 2012 Zimbabbwe Census National Report, Zimbabwe National Statistics Agency. Harare.

Wastes: Solutions, Treatments and Opportunities III – Vilarinho et al. (Eds)

A review of municipal solid waste data for Harare, Zimbabwe

T. Nhubu & C. Mbohwa
University of Johannesburg, Johannesburg, Gauteng, South Africa

E. Muzenda
University of Johannesburg, Johannesburg, Gauteng, South Africa
Botswana International University of Science and Technology, Palapye, Botswana
University of South Africa, Gauteng, South Africa

B. Patel
University of South Africa, Gauteng, South Africa

ABSTRACT: Municipal solid waste (MSW) data sources in Harare metropolitan province show significantly varying data with regards to generation and composition. The sources of variations include data lumping; exclusion of MSW managed outside the formal system and remain uncollected, lack of a clear definition of what constitutes MSW within the Zimbabwean context as well as temporal variations. It is therefore important for waste generation and characterisation studies to be undertaken building upon the already existing datasets to ensure the accuracy and reliability needed for data credibility for use in MSW management planning.

1 INTRODUCTION

Reliable and accurate municipal solid waste (MSW) data both on generation rate and characteristics (composition, moisture content, density and calorific value) for a given temporal and spatial scale is critical in deciding and planning the most appropriate and sustainable MSW management strategies (Aleluia and Ferrão, 2016, Palanivel and Sulaiman, 2014, Suthar and Singh, 2015, Dangi et al., 2011, Zaman and Lehmann, 2013, Hanc et al., 2011, EMA, 2014). The lack of universally agreed definition of MSW and methods to estimate MSW per capita generation rates and composition bring challenges in comparing and or benchmarking the reliability and accuracy of MSW data from various sources and different geographical areas of varying lifestyles. Different MSW definitions exist with the Intergovernmental Panel on Climate Change (IPCC) (2006), United Nations Department of Economic and Social Affairs (UN DESA) (2008) and Hester and Harrison (2002) defined MSW as those waste streams generated in urban environments that are managed by or on behalf of municipalities or other urban local authorities. This MSW constitutes food, park and garden, cardboard and paper, wood, disposable diapers, textiles, leather and rubber, metals, plastics, glass, pottery and chinaware, ash, dust, soil, dirt and electronic waste usually excluding demolition and construction derived wastes. This definition is relatively universal though variations exist amongst jurisdictions.

There are significant environmental impacts and additional costs consequences that arise from under and or overestimation of MSW generation (Beigl et al., 2003). Accurate MSW data thus allows for the prioritization of materials and energy recovery opportunities, attraction of investors in MSW management, baseline development for continuous long-term monitoring and evaluation and the formulation of informed MSW management policies. Worryingly reliable accurate MSW data is lacking in developing countries (Buenrostro et al., 2001, Kawai and Tasaki, 2016). Available MSW data is inconsistent as it comes from different sources difficult to validate and not scientifically measured but assumption based (IPCC, 2006, Couth and Trois, 2011, Miezah et al., 2015). This is despite the existing enormous MSW management challenges and the economic opportunities that are possible from materials and energy recovery in the MSW management sector.

Zimbabwe is no exception in regard to the deficiency of reliable and accurate MSW data associated challenges. National and sub national level statistics on MSW generation and characteristics is generally lacking as there has not been any holistically and systematically conducted studies on waste generation and composition. MSW data for Harare metropolitan province of Zimbabwe covering the Capital City, Chitungwiza and Epworth comes from various database sources and studies whose reliability and accuracy has not been ascertained. Therefore, this study seeks to establish whether the MSW data for Harare was systematically obtained or not to ensure reliability and accuracy for its use as baseline data for sustainable MSW management planning.

2 MATERIALS AND METHODS

2.1 *Description of the study area*

Harare metropolitan province comprises of Harare, the Capital City of Zimbabwe and its 2 dormitory towns of Chitungwiza and Epworth with a total population of just over 2 million (Zimstat, 2013). The uniqueness of Harare metropolitan province is its location upstream in the catchment of its potable water sources. The mismanagement of MSW generated in Harare metropolitan province is contributing to the eutrophic status of Lake Chivero. At present, slightly over 400 thousand tons of municipal solid waste is generated in Harare metropolitan province (Makarichi et al., 2019) with reported collection falling from 52% in 2011 to 48.7% in 2016 (EMA, 2016) indicating that almost half of the MSW generated remaining uncollected. Solid waste generated in Harare metropolitan province is being indiscriminately collected and dumped at the three official poorly managed dumpsites which are unprotected without leachate infiltration into groundwater prevention mechanisms namely Pomona for Harare, Chitungwiza for Chitungwiza and Golden Quarry for Epworth. Pomona covers an area of 100 hectares and has been operational since 1985 (Chijarira, 2013). The City of Harare Management records of 2010 indicate that the disposal capacity of Pomona dumpsite is expected to be exhausted by 2020. This calls for the need to redesign and define future integrated and sustainable municipal solid waste management strategies. Such future management strategies can only be feasible if reliable and accurate MSW data on generation, composition, characteristics and properties is available. Hence need to assess the accuracy and reliability of the available data which is the purpose of this study.

2.2 *Review of few selected MSW generation and characterisation methodologies*

MSW constitutes household waste generally reported to constitute between 55 to 80% with markets and or commercials areas constituting between 10 to 30% and varying contributions from institutions, streets and industries (Nabegu, 2010, Okot-Okumu, 2012). Therefore, MSW data from these sources need to be accounted for in any MSW data to ensure its reliability and accuracy. Estimating MSW data should involve the collection of MSW from where it is generated (households, restaurants, streets, supermarkets, offices) according to the criteria established by Tchobanoglous and Kreith (2002) as well as ensuring that MSW managed outside the official management system is also incorporated as argued by Abel (2007).

Temporal variations on a seasonal, monthly and week day scale (Tchobanoglous et al., 1993, Vesilind et al., 2002, Hanc et al., 2011, Gómez et al., 2009, Denafas et al., 2014) and geospatial variations (Miezah et al., 2015) exist in the quantity and composition of MSW generated depending on the prevailing socio economic situation. Estimation of MSW generation and characterisation data therefore need to consider all the MSW streams, temporal and spatial variations and the socio economic or demographic profiling (low density or high income, high density or low income and medium density or medium income of households).

Palanivel and Sulaiman (2014) randomly collected three 20 kgs samples of MSW being disposed at a landfill per fortnight in winter and summer thereby considering seasonal variations and assumed

100% MSW collection efficiency which is rarely the case as there is also MSW that remains uncollected and managed outside the official systems. Suthar and Singh (2015) selected a sample of 144 households from 11 systematically identified blocks of varying socio economic status in Dehradun city of India. MSW generated from restaurants, supermarkets, hotels, schools, offices and streets was considered with no seasonal variations bringing some limitations regarding accuracy and reliability of the MSW data. Dali et al (2011) used three-stage stratified cluster sampling technique to analyse solid waste generated from 336 households that represented four socio-economic strata of Kathmandu Metropolitan City in Nepal considering MSW generated from restaurants, hotels, schools and streets as well and assuming the negligibility of temporal scale variations. Miezah et al (2015) considered three socio economic classes where households were determined using stratified, purposive and direct sampling technique in all the Capital Cities of the ten regions in Ghana without considering alternative MSW streams and temporal variations.

2.3 *Available MSW data for Harare metropolitan province*

Three sources of MSW data in Harare metropolitan province were obtained and analysed (Zimstat, 2016, EMA, 2014, Makarichi et al., 2019). The Ministry of Environment, Water and Climate (MEWC) in 2011 contracted the Institute of Environmental Studies (IES) of the University of Zimbabwe to undertake a baseline assessment of waste generation and management systems that characterised Zimbabwe in 2011 whose outcome facilitated the development of the national integrated solid waste management plan. The national biennial urban waste data collected by Zimstat (2016) is used by the United Nations Statistics Division (UNSD) and United Nations Environment Programme in the development of the UNSD International Environment Statistics Database. Makarichi (2019) estimated waste composition and generation to assess the suitability of MSW generated in Harare metropolitan province for thermochemical waste to energy conversion. The accuracy and reliability of these MSW data sources together with the appropriateness of the methodology used for data collection and estimation is vital in that the national integrated solid waste management plan was developed based on the EMA data, and also the UNSD International Environment Statistics Database is a source of data used by various stakeholders for decision making, research , and as well as thermochemical waste to energy conversion options in Harare.

3 RESULTS AND DISCUSSIONS

Tables 1–6 show the national, Harare metropolitan province and city specific MSW generation and composition for the three data sources.

Table 1 shows that the national MSW generation data possesses discrepancies possibly emanating from a number of factors. The Zimstat datasets only considers MSW collected and managed within the official systems of urban environments leading to underestimation. What constitutes MSW differs in both datasets with Zimstat datasets considering other sources apart from households waste namely waste generated from ISIC divisions 36, 37, 39 and 45 to 99 while excluding waste from ISIC 38 activities associated with waste collection, treatment and disposal and materials recovery. The EMA data includes all solid waste from households or residential areas including other solids that does not constitute MSW with annual solid waste figures from commercial, academic, medical institutions and industry also being lumped inclusive of MSW constituents as shown in Table 3. The lumping associated with the EMA dataset therefore brings along with challenges in extracting accurate and reliable MSW data. Both datasets in Table 1 are not for the same urban environments and do not cover all the national urban environments resulting in underestimation and distortions.

Makarichi et al (2019) reported a MSW generation for Harare metropolitan of 421,757 tons per annum with Tirivanhu and Feresu (2013) reporting a per capita daily generation rate of 0.361kg at household level translating to 279,751 tons per annum. This annual generation data from Tirivanhu and Feresu (2013) is expectedly lower as it projects a 33.6% increase in annual MSW generation in Harare from 2013 to 2017 considering an annual population growth of 2.2 reported by Zimstat

Table 1. MSW generation in Zimbabwean urban environments (Zimstat, 2016, EMA, 2014).

	Zimstat, 2016		EMA, 2014*****
	2014	2015	2011
Waste stream	1,000 tons		
Commercial activities	–	–	485,72
Academic activities	–	–	72,03
Medical activities	–	–	34,14
Industrial activities	–	–	442,84
Other economic activities	100.53*	126.16***	–
Residential areas or households	291.64**	293.18****	614.84
Total	392.16	419.34	1649.57

*Data refer to Bindura, Bulawayo, Chitungwiza, Epworth and Mvurwi only
**Data refer to Bindura, Bulawayo, Chitungwiza, Epworth, Kariba, Kwekwe, Masvingo, Mutare, Mvurwi, Norton, Nyanga and Plumtree only
***Data refer to Beitbridge, Bindura, Bulawayo, Chitungwiza, Epworth and Mvurwi only
****Data refer to Beitbridge, Bindura, Bulawayo, Chitungwiza, Epworth, Kariba, Kwekwe, Masvingo, Mutare, Mvurwi, Norton, Nyanga and Plumtree only
*****Data refer to Harare, Bulawayo, Chitungwiza, Mutare, Gweru, Masvingo, Chinhoyi, Chegutu, Ruwa, Epworth, Domboshava and Murehwa

Table 2. Harare metropolitan province MSW generation data (Zimstat, 2016).

Category	Unit	2014	2015
Total population of the Province	1,000 inhabitants	2,067.50	2,123.11
Average percentage population served by MW collection	%	61.40*	67.45*
Total amount of municipal waste generated		–	–
Municipal waste collected from households		239.12	181.98
Municipal waste collected from other origins		87.72	76.36
Total amount of municipal waste collected (=4+5)		326.84	258.34
Amounts going to recycling		17.27	23.39
Amounts going to Composting	1,000 tons	3.84	3.84
Amounts going to Incineration		0.26	14.65
Incineration with energy recovery		–	–
Amount going to landfilling		–	–
Landfilling with energy recovery and leachate treatment		–	–
Disposed at dumpsites		305.48	216.46

*Simple average for Harare and Chitungwiza only as Epworth population contribution to the metropolitan province is significantly low hence the 5% percentage served by municipal is negligible and would result in distortions

(2013). This low estimation is because other waste streams such as supermarkets, restaurants, offices, streets etc were not segregated at a city level. Instead they were lumped at a national level in the commercial, academic, industry and medical categories without extracting per capita data from these streams at a city regional or provincial level. The MSW generation data for Harare metropolitan province like other Zimbabwean urban environments reported to the United Nations Statistics Division and UNEP by Zimstat (2016) is silent on the total amount of municipal waste generated. This data is based on the municipal waste collected from households and other sources leaving out the municipal waste that remains uncollected and managed outside the official system. With almost 50% of municipal waste collection efficiency as reported by EMA (2016) it therefore

Table 3. Contribution of the waste streams to the national mean composition of solid waste generated in Zimbabwe from the EMA dataset (EMA, 2014, Tirivanhu and Feresu, 2013).

Waste stream	Composition by mass (tons/year)									
	Biowaste	Paper	Plastic	Textile	Metal	Glass	E-waste	Medical	Rubble	Other
Residential	345,809	62,197	80,985	31,997	39,406	23,542	2,851	26,713		1,341
Commercial	76,411	181,233	127,643	24,490	29,775	9,080	12,915	15,798		8,371
Academic	12,801	27,892	20,344	94	10,270	313				313
Medical	3,775	6,540	3,329		365	254		18,980		893
Industrial	91,951	129,346	70,739	44,284	28,808		7,842		39,794	30,076
Grant total	530,746	407,207	303,040	100,865	108,624	33,188	23,607	61,491	39,794	40,995
Mean % composition	32	25	18	6	7	2	1	4	2	2

Table 4. Composition of solid waste collected from Zimbabwean urban environments for years 2012 to 2016 reported to the United Nations Statistics Division and UNEP (Zimstat, 2016).

	2012*	2013**	2014***	2015****	2016*****
Waste type	% weight				
Paper, paperboard	12.88	10.26	14.96	14.44	12.50
Textiles	1.13	2.88	3.40	2.72	1.24
Plastics	20.75	16.00	15.50	13.25	22.50
Glass	4.33	3.81	5.28	5.62	6.00
Metals	6.10	4.69	5.06	7.08	3.50
Other inorganic material	9.70	12.48	10.66	8.03	5.52
Organic material	45.13	49.89	45.16	48.85	48.75
TOTAL	100	100	100	100	100

*Simple average of Harare, Epworth, Kwekwe and Kariba

**Simple average of Bulawayo, Chinhoyi, Chitungwiza, Epworth, Kariba, Kwekwe, Norton and Plumtree

***Simple average of Bindura, Bulawayo, Chitungwiza, Epworth, Gweru, Kariba, Kwekwe, Masvingo and Plumtree

****Simple average of Beitbridge, Bindura, Bulawayo, Chitungwiza, Epworth, Gweru, Kariba, Kwekwe, Masvingo, Mutare and Plumtree.

*****Simple average of Beitbridge and Gutu

means there is significant underestimation in the Zimstat datasets. Municipalities need monitoring to ensure accuracy of data they provide to Zimstat as they might be tempted to report false coverage of population served by MW collection to improve their image.

National average compositions of biowaste for the EMA datasets in relation to solid waste generated in 2011 shown in Table 3 and those reported to United Nations Statistics Division and UNEP for the years 2012 to 2016 in Table 4 with a minimum difference of 13.13% observed in 2012 indicating presence of inconsistences with regards to the datasets. Such inconsistences are a major cause for concern considering the importance of such data sets. The EMA dataset was used as baseline for the development of the national integrated solid waste management plan while the Zimstat dataset reported to UNSD and UNICEF are used for research and other planning purposes at national, regional and international levels in light of global concern on biodegradable waste landfilling. The same discrepancies are observed on the composition of paper with 25% recorded in the EMA dataset for 2011 in Table 1 and a maximum of 14.96% in the 2014 reported data by Zmstat to UNSD and UNEP in Table 3.

Table 5. Percentage composition of MSW in Harare metropolitan province wet and dry basis (Makarichi et al., 2019).

	Harare	Chitungwiza	Epworth	Harare	Chitungwiza	Epworth
Waste type	% weight (wet basis) original composition			% weight (dry basis)after reconstitution		
Food	28.00	40.00	46.40	14.50	27.30	40.60
Paper	13.00	4.00	3.30	18.20	7.60	7.70
Yard	12.00	11.00	2.30	12.20	17.60	5.00
Other fines	1.00	2.00	0.90	1.20	3.70	1.90
Plastics	23.00	10.00	12.40	38.70	23.00	33.20
Textiles	10.00	11.00	5.10	13.50	20.80	11.30
Rubber	1.00	–	0.10	1.70	–	0.30
Glass	4.00	3.00	4.00	–	–	–
Metals	4.00	1.00	1.90	–	–	–
Rubble	4.00	18.00	23.60	–	–	–

Table 6. The mean composition of household solid waste in Harare metropolitan province (Tirivanhu and Feresu, 2013, EMA, 2014).

	% weight (wet basis)								
Urban area	Biowaste	Paper	Plastic	Metal	Glass	Textile	E-waste	Sanitary	Other
Harare	62	10	11	5	4	4	2	4	0
Chitungwiza	71	7	9	3	1	4	0	5	1
Epworth	42	12	14	15	6	8	1	2	0
Mean	58	10	11	8	4	5	1	4	0

Discrepancies do exist as well with regards to Harare metropolitan province waste composition data obtained by Makarichi et al (2019) and that observed under the EMA dataset shown in Tables 5 and 6 respectively.

With regards to biowaste these discrepancies are pronounced with only data for Epworth exhibiting some near similarities. Even after considering yard waste under biowaste the differences remain pronounced.

4 CONCLUSION

The MSW data review revealed that the available MSW data sources in Harare metropolitan province show significantly varying data with regards to MSW generation and composition. All sources of significant variations in the datasets need to be eliminated to ensure reliability and accuracy of the data for use in the planning and designing of sustainable future MSW management options for Harare. Such sources of variations include data lumping; exclusion of MSW managed outside the formal system and remains uncollected, lack of a clear definition of what constitutes MSW within the Zimbabwean context as well as temporal variations. However it is interesting to note that Tirivanhu and Feresu (2013) considered temporal variations as they collected data from January to November 2011considering all the annual seasonal variations. It is therefore important for waste generation and characterisation studies to be undertaken building upon the already existing datasets to ensure the accuracy and reliability needed for data credibility for MSW management planning.

ACKNOWLEDGEMENTS

The authors acknowledge the Life Cycle Initiative for 2017 Life Cycle award that provided the Simapro software that was used to carry out the LCIA. The authors are also grateful to the National Geographic Society for the early career grant that enabled data collection for the study. The personnel and management at Harare municipality who assisted with the requisite information and data will always be appreciated without their support this work could not have been accomplished. The Universities of Johannesburg and South Africa are also acknowledged for funding the studies, and as well as conference registration and attendance.

REFERENCES

abel, A. 2007. An analysis of solid waste generation in a traditional African city: the example of Ogbomoso, Nigeria. *Environment and urbanization,* 19, 527–537.
Aleluia, J. & Ferrão, P. 2016. Characterization of urban waste management practices in developing Asian countries: A new analytical framework based on waste characteristics and urban dimension. *Waste Management,* 58, 415–429.
Beigl, P., Wassermann, G., Schneider, F., Salhofer, S., Mackow, I., Mrowinski, P. & Sebastian, M. 2003. LCA- IWM Report D2.1: Waste Generation Prognostic Model., The Use of Life Cycle Assessment Tool for the Development of Integrated Waste Management Strategies for Cities and Regions with Rapid Growing Economies LCA- IWM.
Buenrostro, O., Bocco, G. & Vence, J. 2001. Forecasting generation of urban solid waste in developing countries—a case study in Mexico. *Journal of the Air & Waste Management Association,* 51, 86–93.
Chijarira, S. R. 2013. The impact of dumpsite leachate on ground and surface water: a case study of Pomona waste dumpsite. *Department of Geography.* Bindura, Bindura University of Science and Technology.
Couth, R. & Trois, C. 2011. Waste management activities and carbon emissions in Africa. *Waste Management,* 31, 131–137.
Dangi, M. B., Pretz, C. R., Urynowicz, M. A., Gerow, K. G. & Reddy, J. M. 2011. Municipal solid waste generation in Kathmandu, Nepal. *Journal of Environmental Management,* 92, 240–249.
Denafas, G., Ruzgas, T., Martuzevičius, D., Shmarin, S., Hoffmann, M., Mykhaylenko, V., Ogorodnik, S., Romanov, M., Neguliaeva, E., Chusov, A. & Turkadze, T. 2014. Seasonal variation of municipal solid waste generation and composition in four East European cities. *Resources, conservation and recycling,* 89, 22–30.
EMA 2014. Zimbabwe's integrated solid waste management plan. Environmental Management Agency & Institute of Environmental Studies, University of Zimbabwe. Harare.
EMA 2016. Waste generation and management in Harare, Zimbabwe: Residential areas, commercial areas and schools. Unpublished internal report, Environmental Management Agency. Harare, Zimbabwe.
Gómez, G., Meneses, M., Ballinas, L. & Castells, F. 2009. Seasonal characterization of municipal solid waste (MSW) in the city of Chihuahua, Mexico. *Waste Management,* 29, 2018–2024.
Hanc, A., Novak, P., Dvorak, M., Habart, J. & Svehla, P. 2011. Composition and parameters of household bio-waste in four seasons. *Waste Management,* 31, 1450–1460.
Hester, R. E. & Harrison, R. M. 2002. *Environmental and health impact of solid waste management activities (Vol. 18),* Royal Society of Chemistry.
IPCC 2006. 2006 IPCC Guidelines for National Greenhouse Gas Inventories. Waste, vol. 5 <http://www.ipcc-nggip.iges.or.jp/public/2006gl/vol5.html> accessed 11.02.2019.
Kawai, K. & Tasaki, T. 2016. Revisiting estimates of municipal solid waste generation per capita and their reliability. *Journal of Material Cycles and Waste Management,* 18, 1–13.
Makarichi, L., Kan, R., Jutidamrongphan, W. & Techato, K. A. 2019. Suitability of municipal solid waste in African cities for thermochemical waste-to-energy conversion: The case of Harare Metropolitan City, Zimbabwe. *Waste Management & Research,* 37, 83–94.
Miezah, K., Obiri-Danso, K., Kádár, Z., Fei-Baffoe, B. & Mensah, M. Y. 2015. Municipal solid waste characterization and quantification as a measure towards effective waste management in Ghana. *Waste Management,* 46, 15–27.
Nabegu, A. B. 2010. An analysis of municipal solid waste in Kano metropolis, Nigeria. *Journal of Human Ecology,* 31, 111–119.
Okot-Okumu, J. 2012. Solid waste management in African cities–East Africa. *In Waste Management-An Integrated Vision,* IntechOpen.

Palanivel, T. M. & Sulaiman, H. 2014. Generation and composition of municipal solid waste (MSW) in Muscat, Sultanate of Oman. *APCBEE procedia,* 10, 96–102.
Suthar, S. & Singh, P. 2015. Household solid waste generation and composition in different family size and socio-economic groups. *Sustainable Cities and Society,* 14, 56–63.
Tchobanoglous, G. & Kreith, F. 2002. *Handbook of Solid Waste Management,* New York City, McGraw Hill.
Tchobanoglous, G., Theisen, H. & Vigil, S. 1993. *Integrated Solid Waste Management,* New York, USA, McGraw-Hill.
Tirivanhu, D. & Feresu, S. 2013. A situational analysis of solid waste management in Zimbabwe's urban centres. Institute of environmental studies.
UNDESA 2008. *United Nations Department of Economic and Social Affairs Statistics Division; International Standard Industrial Classification of All Economic Activities (ISIC) (No. 4)* New York, United Nations Publications.
Vesilind, P. A., Worrell, W. & Reinhart, D. 2002. *Solid Waste Engineering,* Pacific Grove, Califonia, USA, Books/Cole Thomson Learning.
Zaman, A. U. & Lehmann, S. 2013. The zero waste index: a performance measurement tool for waste management systems in a "zero waste city". *Journal of Cleaner Production,* 30, 123–132.
ZIMSTAT 2013. 2012 Zimbabwe Census National Report, Zimbabwe National Statistics Agency. Harare.
ZIMSTAT 2016. Zimbabwe biennial urban waste data collection for the United Nations Statistics Division (UNSD) International Environment Statistics Database. Harare.

Wastes: Solutions, Treatments and Opportunities III – Vilarinho et al. (Eds)

Testing scenarios for municipal waste management chasing carbon neutrality

A. Fernández-Braña
CERNAS, Instituto Politécnico de Coimbra, Coimbra, Portugal
Universidade de Santiago de Compostela, Galicia, Spain

G. Feijoo-Costa
Universidade de Santiago de Compostela, Galicia, Spain

C. Dias-Ferreira
CERNAS, Instituto Politécnico de Coimbra, Coimbra, Portugal
CICECO, Universidade de Aveiro, Aveiro, Portugal
Universidade Aberta, Lisboa, Portugal

ABSTRACT: A Life Cycle Assessment (LCA) of the municipal solid waste (MSW) management in a selected residential area of the Portuguese city of Aveiro was conducted. The results showed a poor environmental performance in terms of greenhouse effect gases (GHG) emissions, due to the high amounts of waste being landfilled and the low extent of separate collection. Alternative scenarios were tested, where separate collection is enhanced, in order to improve the environmental balance of GHG emissions until reaching a balanced situation between positive and negative effects. It was found that by using an adequate combination of several treatment options and increasing the separate collection of recyclable materials it is possible to turn MSW management neutral in terms of GHG emissions.

1 INTRODUCTION

In the Portuguese city of Aveiro, as well as generally in Portugal, it has been found in last years that the separate collection of recyclable materials has not been able to compensate the increasing generation of mixed municipal solid waste (MSW). At national level, the percentage of separate collection has only slightly increased in last years: +1.2% between 2015 and 2017, reaching 16.5%; meanwhile, overall MSW production is growing at a faster rate: +2.4% in the same period (APA 2017a, 2018a). In Aveiro, separate collection has decreased: 6.3% from total MSW generation in 2015, 6.1% in 2016, and 6.0% in 2017 (ERSUC 2016, 2017, 2018). This deficient source separation results in an excessive presence of heterogeneous materials within residual MSW, which are difficult to separate at the mechanical-biological treatment (MBT) facility where the mixed waste generated by the city is sent: 51% of the waste mass entering the MBT was disposed of in a sanitary landfill in 2015, 54% in 2016 and 64% in 2017 (APA 2016, 2017b, 2018b). Most of this landfilled waste corresponds to rejected materials, unsuitable for organic valorisation in form of biogas or compost, but also biowaste which is not recovered.

In view of this situation, the municipal government of Aveiro has shown interest in testing the adoption of a waste pricing scheme based on a Pay-as-you-throw (PAYT) strategy. PAYT systems have widely shown to be effective in reducing the amount of mixed MSW by transfer of recyclable materials from residual MSW to the separate collection fluxes, as shown by several authors (Morlok et al. 2017, Elia et al. 2015, Brown & Johnstone 2014). Under the scope of the EU funded LIFE PAYT project, a given residential neighbourhood was selected for pilot testing of the new policy (Dias-Ferreira et al. 2019).

Previously to the actual beginning of testing, a broad study of the current situation of waste management within the selected area was performed, making use of the Life Cycle Assessment (LCA) methodology. LCA has been useful to demonstrate that the environmental benefits derived from increasing recovery of waste materials are relevant for the environmental performance of waste treatment, in contrast to other treatment options (Cimpan et al. 2015, Montejo et al. 2013, Song et al. 2013, Rigamonti et al. 2010). However, a question remains open concerning to which extent is feasible to increase source separation. Several authors tested the performance of given MSW management systems comparing scenarios which assumed different levels of source separation, including separate collection of biowaste (Ripa et al. 2017, Giugliano et al. 2011, Massarutto et al. 2011). The results showed that an increased recovery of recyclable materials brings environmental benefits as a consequence, being preferable to incineration in waste-to-energy (WtE) facilities for electricity production. However, as pointed by Rigamonti et al. (2009), a limitation would be the likely loss of quality in recyclable materials due to higher contamination: in their study, this effect would render separate collection ineffective beyond a 50% level.

In the case of this work, the initial objective is to perform an LCA of the current baseline scenario, to show how a low source separation level, besides an MBT facility for mixed MSW working far from optimal conditions, can negatively influence the GHG emissions balance. But in turn, a great room of improvement would be opened with the introduction of incentives for the source separation of recyclables. Thence, as second objective, an alternative scenario is assessed, aiming at exploring the feasibility of turning the result into an overall neutral (or even positive) carbonic balance through an improvement of MSW management performance. This improvement consists mainly in an increased amount of recyclable materials being separately collected, whereas in this case the overall performance would not be compromised: even though the quality in source separated materials may decrease, the higher quality of the biowaste sent to the biologic treatment stage might on the contrary contribute to improve its own performance.

2 CASE OF STUDY AND METHODOLOGY

2.1 *Goal and scope*

The goal of the study is to analyse the environmental impacts derived of the collection and treatment of MSW (mixed and source separated) generated by the given neighbourhood in the Portuguese city of Aveiro. The functional unit selected for the LCA corresponds to the total annual production of MSW within the considered area. This was obtained after a field characterisation campaign, reported by Dias-Ferreira et al. (2019): 503 t of MSW, consisting of 449 t of mixed MSW and 54 t of separately collected materials (28.5 t of paper and cardboard, 19.9 t of plastic and metal packaging and 5.7 t of glass). Regarding the separate collection streams, the responsible company declared that 94% of the total weight is suitable for recovery. All these numbers are referred to the year 2017.

The environmental analysis was performed using the commercial software SimaPro (version 9.0.0.). Among the available impact categories, the contribution to climate change in form of GHG emissions (CO_2 eq.) was chosen for this study, since this is a category of current general concern, and always present in waste management LCA studies. The impact assessment method for evaluating this category was the 2013 IPCC characterisation methodology for a 100-year period (version 1.02).

2.2 *Life cycle inventory*

A Life Cycle Inventory (LCI) was built from the data previously gathered. For the LCI modelling process, an attributional framework was chosen. Even though the analysed system is considered to provide one main function – i.e. management of household waste – several products are obtained as a result of its activity: compost for land application, electricity production from biogas and secondary raw materials from recycling processes. Accordingly, a multifunctionality solving procedure through system expansion was needed to take into account the environmental credit derived from the substitution of primary resources by the secondary products obtained. When not from

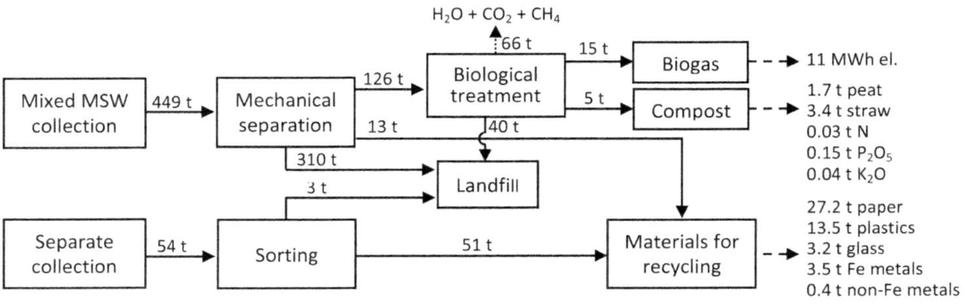

Figure 1. Scheme of the analysed system with main MSW mass flows.

direct sources, data were taken from life-cycle reference databases – namely ecoinvent 3.5®, and extrapolated for the area of study.

The whole MSW management system analysed is summarised in Figure 1, in form of a flow diagram. Solid arrows represent waste fluxes, dashed arrows represent primary resources being replaced by waste products and a dotted arrow represents waste outputs in form of gaseous emissions (mostly carbon dioxide, methane and water as a result of biologic processes).

2.2.1 *MSW collection*

MSW in the studied area is collected from twenty-six 800 L street containers with open access for mixed waste and five drop-off banks for separate collection of recyclables, organised in three fluxes: glass, paper/cardboard and plastic and metal packaging. The modelling of mixed MSW collection was mainly based on the data provided by the municipality for equipment and fuel consumption (Fernández-Braña et al., in prep.). For separate collection, a simpler approach using the literature data of Teixeira et al. (2014) for fuel consumption was deemed sufficient, given the minor contribution of this element to the overall environmental impact – as confirmed by results in Section 3.

2.2.2 *MBT facility*

The MBT facility responsible for treating MSW generated by the city of Aveiro was modelled combining the waste composition previously obtained (see Section 2.1) with the recovery efficiencies deduced from the mass balances annually published by the managing company (APA, 2018b), following the scheme shown in Figure 1. Composition of biogas and emissions from combustion were calculated following the approach of ecoinvent 3.5 database for biogas obtained from biowaste (Dinkel et al. 2012), as well as for biogas leaks and CH_4 diffuse emissions during composting, but this was completed with own composition measurements of the digested matter – published by Oliveira et al. (2016). Not all compost produced is used for agricultural application, a large part is internally applied on the sanitary landfill (see Figure 1). Consumption of utilities and consumable goods was obtained through personal communication or estimated with data from other facilities. Impacts related to the construction of buildings and equipment were taken from ecoinvent 3.5 database, adapted to the actual surface occupied by the facility.

2.2.3 *Assignment of environmental credits by system expansion*

A critical issue when performing LCA on waste management is the assignment of credits to the materials recovered for recycling in account for replacing primary raw materials. This may greatly influence results. In this work it was decided to follow the approach suggested by Bala Gala et al. (2015), which takes into account not only the lower quality of some recycled materials when compared to the primary ones, but also considers the fact that for materials where recycling is already well established – like glass or metals – it is more realistic to model recycling as a replacement for a mixture of primary and secondary (recycled) materials, rather than a 100% amount of primary material. These two correction factors are combined to obtain global substitution factors, presented in Table 1.

Table 1. Credits assigned to recovered materials.

Material	Quality loss factor	% recycled market share	Substitution factor
Glass	1	0.45	0.84
Fe-metals	1	0.50	0.59
Non Fe-metals	1	0.37	0.65
Paper/cardboard	0.80	0.47	0.47
Plastics	0.75	0.12	0.67

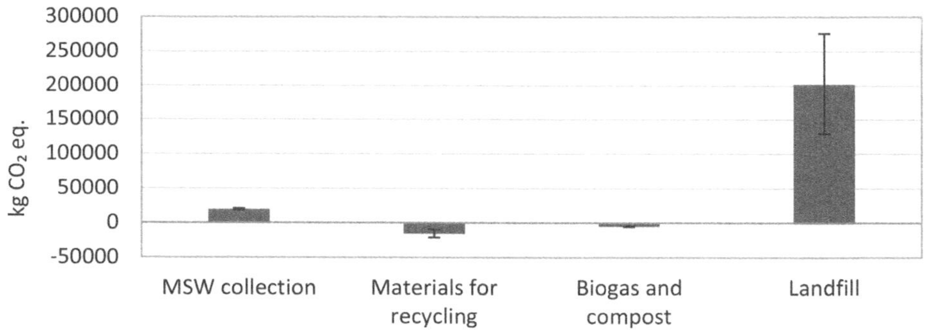

Figure 2. Baseline impact assessment: contribution to climate change (GHG emissions) in 2017.

For the credits derived from biogas utilisation in electric generators, a 1:1 substitution for an equivalent Portuguese electricity average production (as represented in ecoinvent 3.5) was applied. The electric generation efficiency was set at 40%. For compost use on land, it was chosen to follow the approach of Hermann et al. (2011): industrial compost substitutes as a soil conditioner in a 1:1 basis a mixture of 25% peat and 75% straw. Substitution capacity regarding fertiliser potential was set at 20% for N, 100% for P and 100% for K, as in Boldrin et al. (2010).

3 RESULTS AND DISCUSSION

The baseline scenario impact assessment, referred to GHG emitted in 2017, is shown in Figure 2.

Figure 2 represents a clear picture of an inefficient performance of MSW management, with GHG emissions derived from landfilling dominating the assessment (overall impact: $200{,}679 \pm 77{,}353$ kg CO_2 eq. emitted). Regarding the contribution of the biologic processes – anaerobic digestion and composting –, it is observed that the benefits of replacing conventional electricity production with generation based on biogas combustion hardly offset its own emissions. This could be explained by the already large influence of renewable energies in the Portuguese electric generation mix, and also due to the lack of any heat recovery application, thus limiting further environmental "credits" by replacing other energy sources. Taking this baseline as a starting point, several improvements on the system were introduced. Firstly, it was considered that the separate collection of MSW on the studied area increased from the previous 12% until 34%. If the waste composition initially considered is not altered, this would be the highest possible separation level without introducing a separate collection of biowaste, and is typical in cases of successful application of incentives for source separation. Consequently, as an undesired side effect, contamination of separate collection fluxes was increased until 30% for plastic and metal packaging, 10% for paper and cardboard and 5% for glass, typical reported values in Portugal (APA 2017b). Nonetheless, the recovery of non-recyclable materials as refuse-derived fuel (RDF) was introduced as a secondary valorisation option – although not currently usual in Portugal due to low price RDF concurrence

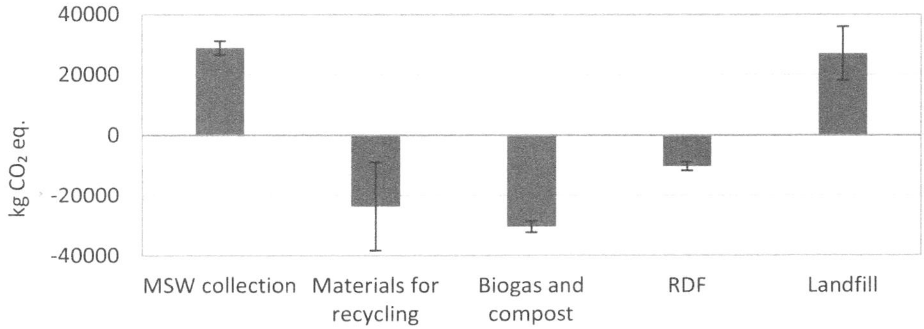

Figure 3. Impact assessment of alternative scenario.

from foreign countries, namely UK. The calorific power of RDF was set at 12 MJ/kg, and natural gas was chosen as replaced fuel.

Finally, production of biogas and application of compost were maximised – corresponding to a higher quality of the substrate waste used: 242 t of biowaste were assigned to biologic treatment (instead of 126 t in Figure 1), producing 30 t of biogas with a typical 65% of CH_4 in volume and 115 t of compost, of which 73 t are used on agricultural application and the rest is used in the sanitary landfill. However, great uncertainty exists on quantification of fugitive CH_4 emissions during composting, which, depending on their extent, might partially mitigate the positive effect of substituting fossil fuels for energy production.

The results are shown in Figure 3.

In this alternative scenario, the modifications introduced allowed to reach a negative value for the overall impact: -8494 ± 16839 kg CO_2 eq., meaning a positive environmental outcome, since GHG emissions are avoided. All major sources of impact in Figure 3 progressed in a more environmentally beneficial direction: less landfilling, more recycling and more energy recovery through biogas and RDF. The exception are the emissions of collection, now higher due to the increased separate collection effort. However, uncertainty also exists on evaluating how fuel consumption rates will evolve with variations in the collected amounts of MSW.

As a conclusion, this assessment demonstrates that it is technically viable to achieve environmental sustainability of MSW management even in complicated situations like the one analysed in this work. Nevertheless, a careful combination of all the available treatment options is required in order to reach this goal.

ACKNOWLEDGMENTS

The authors acknowledge the financial support of LIFE+ (financial instrument of the EU for the environment) for funding the LIFE PAYT project (LIFE 15/ENV/PT/000609). C. Dias-Ferreira acknowledges the support through FCT (Fundação para a Ciência e para a Tecnologia) of POCH (Programa Operacional Capital Humano) within ESF (European Social Fund) and of national funds (SFRH/BPD/100717/2014) from MCTES (Ministério da Ciência, Tecnologia e Ensino Superior).

The authors would also like to acknowledge the Aveiro Municipal Council and MSW management company ERSUC for the data they kindly provided for this work.

REFERENCES

APA 2016. *Relatório Anual Resíduos Urbanos 2015*. Amadora: Agência Portuguesa do Ambiente (APA).
APA 2017a. *Relatório do Estado do Ambiente 2017*. Amadora: Agência Portuguesa do Ambiente (APA).

APA 2017b. *Relatório Anual Resíduos Urbanos 2016*. Amadora: Agência Portuguesa do Ambiente (APA).
APA 2018a. *Relatório do Estado do Ambiente 2018*. Amadora: Agência Portuguesa do Ambiente (APA).
APA 2018b. *Relatório Anual Resíduos Urbanos 2017*. Amadora: Agência Portuguesa do Ambiente (APA).
Bala Gala, A., Raugei, M. & Fullana i Palmer, P. 2015. Introducing a new method for calculating the environmental credits of end-of-life material recovery in attributional LCA. *International Journal of Life Cycle Assessment* 20(5): 645–654. DOI: 10.1007/s11367-015-0861-3
Boldrin, A., Hartling, K.R., Kaugen, M. & Christensen, T.H. 2010. Environmental inventory modelling of the use of compost and peat in growth media preparation. *Resources, Conservation and Recycling* 54: 1250–1260. DOI: 10.1016/j.resconrec.2010.04.003
Brown, Z. S. & Johnstone, N. 2014. Better the devil you throw: Experience and support for pay-as-you-throw waste charges. *Environmental Science & Policy* 38: 132–142. DOI: 10.1016/j.envsci.2013.11.007
Cimpan, C., Rothmann, M., Hamelin, L. & Wenzel, H. 2015. Towards increased recycling of household waste: Documenting cascading effects and material efficiency of commingled recyclables and biowaste collection. *Journal of Environmental Management* 157: 69–83. DOI: 10.1016/j.jenvman.2015.04.008
Dias-Ferreira, C, Neves, A. & Braña, A. 2019. The setting up of a pilot scale Pay-as-you-throw waste tariff in Aveiro, Portugal. *WIT Transactions on Ecology and the Environment* 231: 149–157. DOI: 10.2495/WM180141
Dinkel, F., Zschokke, M. & Schleiss, K. 2012. *Ökobilanzen zur Biomasseverwertung*. Basel: Carbotech AG. Final report prepared for Bundesamt für Energie (BFE) – Forschungs- und Entwicklungsprogramm Biomasse und Holzenergie, Schwerpunkt Biomasse ohne Holzenergie.
Elia, V., Gnoni, M.G. & Tornese, F. 2015. Designing Pay-As-You-Throw schemes in municipal waste management services: A holistic approach. *Waste Management* 44: 188–195. DOI: 10.1016/j.wasman.2015.07.040
ERSUC 2016. *Relatório e Contas ERSUC 2015*. Coimbra: Resíduos Sólidos do Centro S.A. (ERSUC).
ERSUC 2017. *Relatório e Contas ERSUC 2016*. Coimbra: Resíduos Sólidos do Centro S.A. (ERSUC).
ERSUC 2018. *Relatório e Contas ERSUC 2017*. Coimbra: Resíduos Sólidos do Centro S.A. (ERSUC).
Fernández-Braña, A., Feijoo-Costa, G. & Dias-Ferreira, C. 2019. Looking beyond the banning of lightweight bags: analysing the role of plastic (and fuel) impacts in waste collection. Submitted to publisher.
Giugliano, M., Cernuschi, S., Grosso, M. & Rigamonti, L. 2011. Material and energy recovery in integrated waste management systems. An evaluation based on life cycle assessment. *Waste Management* 31(9–10): 2092–2101. DOI: 10.1016/j.wasman.2011.02.029
Hermann, B. G., Debeer, L., De Wilde, B., Blok, K. & Patel, M. K. 2011. To compost or not to compost: Carbon and energy footprints of biodegradable materials' waste treatment. *Polymer Degradation and Stability* 96(6): 1159–1171. DOI: 10.1016/j.polymdegradstab.2010.12.026
Massarutto, A., De Carli, A. & Graffi, M. 2011. Material and energy recovery in integrated waste management systems: A life-cycle costing approach. *Waste Management* 31(9–10): 2102–2111. DOI: 10.1016/j.wasman.2011.05.017
Morlok, J., Schoenberger, H. & Styles, D. 2017. The Impact of Pay-As-You-Throw Schemes in the Management of Municipal Solid Waste: The Case of the County of Aschaffenburg, Germany. *Resources* 6(1). DOI: 10.3390/resources6010008
Oliveira, V., Ottosen, L.M., Labrincha, J. & Dias-Ferreira, C. 2016. Valorisation of Phosphorus Extracted from Dairy Cattle Slurry and Municipal Solid Wastes Digestates as a Fertilizer. *Waste and Biomass Valorization* 7(4): 861–869. DOI: 10.1007/s12649-015-9466-0
Rigamonti, L., Grosso, M. & Giugliano, M. 2009. Life cycle assessment for optimising the level of separated collection in integrated MSW management systems. *Waste Management* 29(2): 934–944. DOI: 10.1016/j.wasman.2008.06.005
Rigamonti, L., Grosso, M. & Giugliano, M. 2010. Life cycle assessment of sub-units composing a MSW management system. *Journal of Cleaner Production* 18(16–17): 1652–1662. DOI: 10.1016/j.jclepro.2010.06.029
Ripa, M., Fiorentino, G., Vacca, V. & Ulgiati, S. 2017. The relevance of site-specific data in Life Cycle Assessment (LCA). The case of the municipal solid waste management in the metropolitan city of Naples (Italy). *Journal of Cleaner Production* 142: 445–460. DOI: 10.1016/j.jclepro.2016.09.149
Song, Q., Wang, Z. & Li, J. 2013. Environmental performance of municipal solid waste strategies based on LCA method: A case study of Macau. *Journal of Cleaner Production* 57: 92–100. DOI: 10.1016/j.jclepro.2013.04.042
Teixeira C.A., Russo M., Matos C. & Bentes I. 2014. Evaluation of operational, economic, and environmental performance of mixed and selective collection of municipal solid waste: Porto case study. *Waste Management & Research* 32(12): 1210–1218. DOI: 10.1177/0734242X14554642

Wastes: Solutions, Treatments and Opportunities III – Vilarinho et al. (Eds)
© 2020 Taylor & Francis Group, London, ISBN 978-0-367-25777-4

2035 MSW Targets: A scenario analysis for the Portuguese MSW management system

A. Lorena & S. Carvalho
3Drivers – Engineering, Innovation and Environment, Lisboa, Portugal

ABSTRACT: The 2018 revision of the Waste and Landfill Directives established that, by 2035, the preparing for re-use and the recycling of municipal waste must be increased to a minimum of 65% by weight and the landfilling of municipal waste to a maximum of 10%. Given the risks that Portugal faces of not meeting these new targets, three possible scenarios for MSW management were developed and analyzed with a MSW Allocation model to determine what would it take to meet EU waste management targets. Results show that compliance will require very high sorted collection rates, around 75–90% for all waste types, nearly zero residuals from sorting operations and waste to energy capacity to ensure that residuals are not landfilled. Falling behind in a single waste type will force unrealistic sorted collection rates in the remaining ones. These conclusions hold even with successful waste prevention actions.

1 INTRODUCTION

1.1 *Motivation*

In December 2015, the European Commission proposed a Circular Economy package that included a set of revised waste directives which, after a final ratification, became law in July 2018.

The revised Waste Directive proposed an increase on the MSW management targets established in Directive 2008/98/EC, thus reflecting the European Union's ambition to move to a circular economy. According to the new legislation, by 2035, the preparing for re-use and the recycling of municipal waste shall be increased to a minimum of 65% by weight. In addition, the revised Landfill Directive requires Member States to reduce the landfilling of municipal waste to a maximum of 10% by 2035, as well as to introduce a ban on landfilling of separately collected waste, including biodegradable waste. Portugal was one of the 14 Member States to be identified as at risk of missing the municipal waste preparing for reuse/recycling target of 50% by 2020 (European Commission 2018).

In 2017, the country produced around 5 million tonnes of municipal waste, 57% of which was landfilled, and had a recycling rate of 38%. However, it is important to point out that in 2025, a new calculation method will be in place where the recycling rate will be determined by the ratio between recycled waste and the total municipal waste as opposed to the recyclable waste, which is currently used. This results in much lower recycling rates for Portugal, resulting in an even harder challenge to ensure compliance to European targets.

Portugal's poor results in light of the new ambitious EU management targets has led the Portuguese government to propose a new Strategic Plan for Municipal Waste for 2020-2025, PERSU 2020+, which intended to revise PERSU 2020 in order to correct the current trajectory and plan out the work necessary to achieve the newly defined targets.

The draft of the plan which was available for public consultation until January 25th (APA 2018a), demonstrated a lack of a detailed plan on how to achieve the EU waste targets by 2025, merely presenting a general discourse that it is necessary to increase separate collection, decrease the amount of landfilled waste and start implementation of biowaste collection. It is imperative to address this knowledge gap and assess possible scenarios for the Portuguese MSW management from 2020 to 2035 and determine the amount of effort and investment required to do so.

1.2 *Objectives and scope*

Given the challenges that lay ahead, it is crucial to resort to strategic and robust tools that support decision-making at a policy level. One of such tools is strategic scenario analysis. The present work aims at assessing possible scenarios for the Portuguese MSW management system in 2035, focusing on the allocation of waste management activities to ensure the compliance of EU waste management targets. To determine these scenarios, the authors have resorted to previous work in this field, namely in Lorena & Carvalho (2017), which looked into critical factors and risks for compliance with 2020 targets and the possible pathways for the Portuguese MSW management from 2020 to 2030, and in 3drivers (2017), which determined three different scenarios for residual waste management in 2030, but without the knowledge of the final versions of the waste-related directive.

2 METHODOLOGY

2.1 *Introduction to input-output modelling*

Building from previous work, the authors use a decision-support system that considers technical, environmental, economic and social aspects related to waste management. A detailed description of the EPRIO model is outside the scope of this paper, but a short introduction to its Allocation module is needed to understand the modelling process.

Consider the Leontief demand-driven input-output model, or IO model for short (Leontief 1941, Oosterhaven 1996), that allows the quantification of the effect of final demand stimulus in output and primary production factors (de Haan & Keuning 1996, Leontief 1970, Wiedmann et al. 2009), and can be represented by the matricial equation:

$$\vec{x} = (\mathrm{I} - \mathrm{A})^{-1} \vec{f} \tag{1}$$

where x is the total output vector, I represents the identity matrix with the same dimensions as A, A is the matrix of technical coefficients and f is the final demand vector. To fully understand input-output models, the reader is recommended to read Miller & Blair (2009).

It is possible to extend the Leontief IO to include direct primary inputs (which may include, among others, environmental bads such as waste) as opposed to just gross outputs. In many cases, one is more interested in knowing the environmental (e.g., CO_2 emissions) or employment impacts as a function of final demand. The direct primary input of a given type, represented by r, can be used in equation (2) to obtain total variation of direct primary input:

$$\Delta\vec{r} = \hat{r} (\mathrm{I} - \mathrm{A})^{-1} \Delta\vec{f} \tag{2}$$

where $\hat{r}$ is the diagonal matrix of $\vec{r}$. The vector $\vec{r}$ can be used freely and creatively to account for physical units and flows (e.g., GHG emissions), but also accounting variables (e.g., contribution to waste management targets).

The matrix of technical coefficients A represents the uses and supplies of each modelled sector, i.e., what each sector uses in terms of products and supplies to the rest of the economy (hence the input-output designation). It is possible to consider sector to sector inputs and outputs, but a more practical approach to waste management systems is to consider sectors and goods separately. The later approach, succinctly identified as a rectangular matrix in opposition to square matrices, describes the use and supply of each good for each sector. Table 1 represents the technical coefficient matrix used in the proposed model.

Finally, vector $\vec{f}$ stands for the final demand vector. When modelling the whole economy, usually $\vec{f}$ stands for the consumption of households, public administration, fixed capital formation and exports, implying that these are considered as external to the model.

Table 1. Submatrix A_{tw} with allocation coefficients.

	OW	GW	PC	Pl	Me	Gl	Wd	Tx	ST	Cm	HW	BW	Fi	OW
OWC	0.8													
GWC		0.9												
PCC			0.75											
PMC				0.7	0.9					0.5				
GlC						0.9								
OSC							0.9	0.5			0.5	0.9		
UnC	0.2	0.1	0.25	0.3	0.1	0.1	0.1	0.5	1	0.5	0.5	0.1	1	1
ORe	0.8	0.9												
MRe			0.8	0.8	0.9	0.9	0.9	0.5		0.4	0.4	0.4	0.7	
WtE	0.2	0.1	0.2	0.2	0.1	0.1	0.1	0.5	0.9	0.5	0.4	0.4		1
Lf									0.1	0.1	0.2	0.2	0.3	

The model represented by equation (2) can be easily used as a basis for a Waste Allocation model, in which one considers $\vec{f}$ as waste streams, $\vec{r}$ as contribution to waste management targets, as this is the major restriction to the model, and A is the allocation matrix.

2.2 *Model definition*

The general structure of the model can be depicted by equation (3). The *w* subscript stands for the waste types (goods) and the *t*subscript stands for the waste treatment options (sectors).

$$\underbrace{\begin{pmatrix} x_w \\ x_t \end{pmatrix}}_{\vec{x}} = \underbrace{\begin{pmatrix} I & -A_{wt} \\ -A_{tw} & I - A_{tt} \end{pmatrix}}_{(I-A)}^{-1} \underbrace{\begin{pmatrix} f_w \\ f_t \end{pmatrix}}_{\vec{f}} \tag{3}$$

Looking first at A, the submatrix A_{tw} stands for the distribution of the different waste streams between waste management options, whereas submatrix A_{wt} stands for the generation of waste streams from the waste management options, including, for example, residuals from sorting or ashes from incineration. This aspect was partially circumvented by allocating residuals in submatrix A_{tw} and by defining ash production from WtE in submatrix A_{tt}.

The model considers the following waste types and waste management options:

- Waste types: organic waste (OW), garden waste (GW), paper and cardboard (PC), plastics (Pl), metal (Me), glass (Gl), wood (Wd), textiles (Tx), diapers, wipes and other sanitary textiles (ST), composite waste (Cm), hazardous waste (HW), bulk waste (Bk), fine fraction (Fi), other wastes (OW);
- Waste management options: organic waste collection (OWC), garden waste collection (GWC), paper and cardboard collection (PCC), plastics and metal collection (PMC), glass collection (GlC), other sorted collection (OSC), unsorted collection (UnC), Organic recovery (ORe), material recycling (MRe), waste to energy (WtE), landfill (Lf).

In this case, since the focus is on MSW, waste types were defined according to the reported global MSW composition by the Portuguese Environmental Agency (APA, 2018b). The universe of waste management options was defined according to the previous knowledge of the Portuguese MSW management system and essentially correspond to existing infrastructure.

Table 1 show the structure of submodule A_{tw}. The interpretation of submatrix A_{tw} is relatively straightforward. The first column, for example, represents how organic waste (OW) will be collected and treated. In this case, 80% will be collected separately (thus 20% will be collected in mixed waste), and 80% will be treated in dedicated organic recovery facilities, whereas the remaining 20% will end up in WtE facilities.

Table 2. Transpose of the final demand vector (f') according to MSW composition (%).

	OW	GW	PC	Pl	Me	Gl	Wd	Tx	ST	Cm	HW	BW	Fi	OW
MSW	36.5	1.9	13.6	10.7	1.9	6.7	0.9	3.5	5.7	3.6	0.3	0.8	11.2	2.6

As for the remaining submatrices, A_{wt} is a zero matrix and submatrix A_{tt} only has one non-zero value. The intersection between Landfill row and WtE column (A_{LfWtE}) has a value of 0.1 which can be interpreted as ash production per tonne of waste entering WtE.

Finally, final demand is modelled according to MSW composition, either in percentages or mass values. Taking into consideration the 2017 MSW composition, as reported by the Portuguese Environmental Agency (APA, 2018b), one obtains the final demand vector represented in Table 2.

3 RESULTS

The presented Allocation model was used to assess different waste management options and MSW production rates. These scenarios were designed in such a way that they all complied with the EU waste management targets, namely preparation for reuse and recycling and landfill diversion, thus answering the question 'What will it take'.

Several key indicators were then calculated to determine the more tangible consequences related to each scenario.

3.1 *Scenario 1*

The first scenario takes the reference MSW composition and production rate and considers a very high performance MSW management system, heavily based on sorted collection and recovery options. In most waste types, 75% to 90% will be collected separately, including organic waste. Waste types such as sanitary textiles, fine fraction and other wastes will continue to be mostly collected in mixed waste. Waste types such as hazardous wastes or textiles, which are expected to have specific waste collection systems, will have a significant share of sorted collection.

The scenario also considers high recycling rates. Nearly all sorted waste will be treated in dedicated facilities with virtually zero residual rates. Paper and plastics will still be recovered from unsorted waste in mechanical treatment facilities. Waste not recovered will be treated in WtE facilities, but some categories, such as bulk waste and the fine fraction, can be directed to landfill.

The considered allocation coefficients and MSW composition for Scenario 1 were already presented in Table 1 and Table 2.

In Scenario 1, Portugal meets the EU waste management targets with preparation for reuse and recycling reaching 70% and landfilling will be around 7%. Over 60% of all MSW will be collected through sorted collection systems, already considering the 31% of biowaste sorted collection. Waste to energy will represent around 25% (1.13 million tonnes) of MSW final management options, while recycling and organic recovery will represent 40% (1.81 million tonnes) and 31% (1.40 million tonnes), respectively.

3.2 *Scenario 2*

In Scenario 2, it is assumed that MSW prevention will be successful, with total MSW production dropping 28% in mass, particularly due to specific waste types that are today the focus of waste prevention programs (food waste, packaging, particularly plastic, and sanitary textiles). The corresponding MSW composition also changes, as shown in Table 3. Scenario 2 considers the same allocation as in Scenario 1 (Table 1).

In Scenario 2, despite the considerable reduction in MSW production, Portugal still needs to meet the very high-performance standards, with around 75%-90% of all waste types being collected

Table 3. Transpose of the final demand vector (f') for Scenario 2.

	OW	GW	PC	Pl	Me	Gl	Wd	Tx	ST	Cm	HW	BW	Fi	OW
MSW	30.7	2.7	17.2	9.0	2.3	5.7	1.2	4.4	7.2	2.5	0.3	0.9	12.6	3.3

Table 4. Submatrix A_{tw} with allocation coefficients for Scenario 3.

	OW	GW	PC	Pl	Me	Gl	Wd	Tx	ST	Cm	HW	BW	Fi	OW
OWC	0.55													
GWC		0.55												
PCC			0.95											
PMC				0.95	0.95					0.5				
GlC						0.95								
OSC							0.95	0.5			0.1	0.1		
UnC	0.45	0.45	0.05	0.05	0.05	0.05	0.05	0.5	1	0.5	0.9	0.9	1	1
ORe	0.55	0.55												
MRe			0.95	0.95	0.95	0.95	0.95	0.5		0.4	0.1	0.1	0.7	
WtE	0.45	0.45	0.05	0.05	0.05	0.05	0.05	0.5	0.9	0.5	0.7	0.7		1
Lf									0.1	0.1	0.2	0.2	0.3	

separately and nearly zero residual rates. Preparation for reuse and recycling reaches 70% whereas deposition in landfill will be around 8%. Nearly 60% of all MSW will be collected through sorted collection systems, already considering the 27% of biowaste sorted collection. Waste to energy will represent around 26% (1.18 million tonnes) of MSW final management options, while recycling and organic recovery will represent 42% (1.90 million tonnes) and 27% (1.22 million tonnes), respectively.

3.3 *Scenario 3*

In Scenario 3, there is also an improvement of the MSW management system, but performance in some waste types will lag. Organic and garden waste collection and recovery is set at 55%. Other waste types, such as textiles, hazardous waste and bulk waste all have their sorted collection and material recovery rates dropped in relation to Scenario 1 and 2. To compensate, the more traditional collection systems will need to perform almost faultlessly to meet the waste management targets. Collection and recycling of paper and cardboard, plastic and metal waste and glass is set at 95%, arguably unrealistic figures. Table 4 summarizes the allocation assumptions. MSW production was assumed to be as in Scenario 1, i.e. without waste prevention.

In Scenario 3, the difficulty to implement sorted collection and recycling systems for some specific waste types will need to be compensated by the increase in the performance of more typical sorted collection systems. Portugal will meet the targets with preparation for reuse and recycling reaching barely above 65%, whereas deposition in landfill will be around 8%. Around 56% of all MSW will be collected separately, already considering the 18% of biowaste sorted collection. Waste to energy will represent around 31% (1.38 million tonnes) of MSW final management options, while recycling and organic recovery will represent 46% (2 067 000 tonnes) and 18% (830 000 tonnes), respectively.

4 DISCUSSION AND CONCLUDING REMARKS

The presented model was used to determine what would it take to meet EU waste management targets but considering the whole system and as a function of MSW composition. All three scenarios

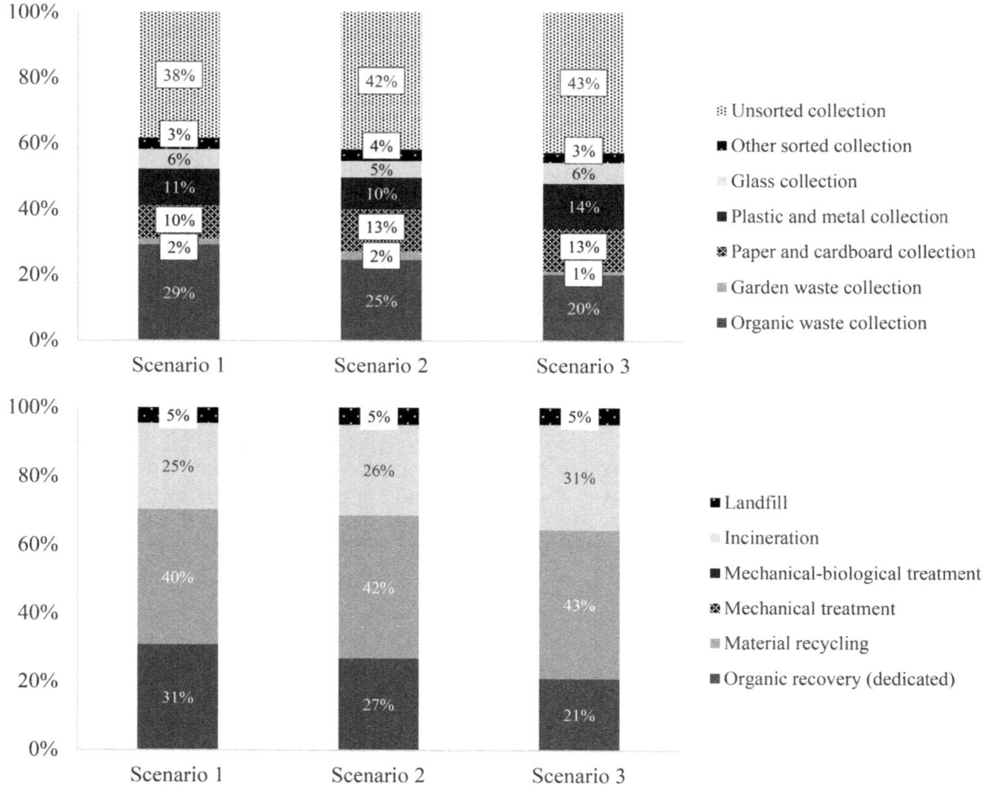

Figure 1. Results for each scenario.

were modelled as complying to the EU waste management targets but taking different assumptions into consideration.

The general (and predicted) conclusion is that it will be difficult to meet these targets, since performance, particularly of sorted collection systems, will need to be very high. Collection rates around 75%–90% in all major waste types and nearly zero residual rates are necessary conditions. It is also shown that MSW prevention will not ease these performance conditions but can still reduce the necessary investments in waste collection and treatment activities. Also, the 'new' waste types for which the Waste Directive calls for sorted collection systems, will need to be addressed with same high-performance standards. If these are unsuccessfully implemented, the pressure will fall on the more typical waste types (e.g., packaging waste), to an unrealistic level.

Figure 1 shows the allocation in each scenario, both in terms of collection and final waste treatment, demonstrating that few degrees of freedom that Portugal, or any other Member State for that matter, has in the pursuit of the 2035 waste management targets. It is also noteworthy that WtE capacity should be between 0.8 and 1.3 million tonnes per year, depending on the scenario considered.

REFERENCES

3drivers. 2017. Characterisation of the Production of Residual Waste in Portugal and Assessment of the Recovery Potential. Study developed for the Smart Waste Portugal Association (in Portuguese).

APA. 2018a. PERSU 2020+ – Document for public discussion. (in Portuguese). Retrieved March 15th, 2019 from https://www.apambiente.pt/_zdata/DESTAQUES/2019/PERSU2020/PERSU2020%20_Audicao_Publica_dez2018.pdf.

APA. 2018b. Municipal Solid Waste Annual Report 2017. Agência Portuguesa do Ambiente, I.P.

de Haan, M. & Keuning, S. 1996. Taking the Environment into Account: The Namea Approach. *The Review of Income and Wealth*, volume 42, issue 2, 13–148.

European Commission. 2018. Report from the Commission to the European Parliament, the Council, the European Economic and Social Committee and the Committee of the Regions on the implementation of EU waste legislation, including the early warning report for Member States at risk of missing the 2020 preparation for re-use/recycling target on municipal waste, COM(2018) 656 final. Brussels, 24 of September of 2018.

Leontief, W. 1941. *The structure of American economy, 1919–1929: an empirical application of equilibrium analysis.* Harvard University Press.

Leontief, W. 1970. Environmental Repercussions and the Economic Structure: An Input-Output Approach. *The Review of Economics and Statistics*, volume 52, issue 3, 262–271.

Lorena, A. & Carvalho, S. 2017. Going Beyond 2020: Development Pathways for the Portuguese MSW Management System Post-2020. *4th International Conference: WASTES: Solutions, Treatments and Opportunities*.

Miller, R. & Blair, P. 2009. *Input-Output Analysis Foundations and Extensions.* 2nd Edition, Cambridge University Press, Cambridge.

Oosterhaven, J. 1996. Leontief versus Ghoshian Price and Quantity Models. *Southern Economic Journal*, volume 62, issue 3, 750–759.

Wiedmann, T. et al. 2009. Companies on the Scale – Comparing and Benchmarking the Sustainability Performance of Businesses. *Journal of Industrial Ecology*, volume 13, issue 3, 361–383.

Wastes: Solutions, Treatments and Opportunities III – Vilarinho et al. (Eds)
 ISBN 978-0-367-25777-4

Physical separation applied to WPCB recycling process to optimize copper enrichment

J.O. Dias, A.G.P. da Silva & J.N.F. Holanda
LAMAV – Laboratory of Advanced Material, Universidade Estadual do Norte Fluminense Darcy Ribeiro, Rio de Janeiro, Brazil

A.M. Futuro
Department of Mining Engineering, Faculty of Engineering, University of Porto, Porto, Portugal

S.C. Pinho
LEPABE – Laboratory for Process Engineering, Environment, Biotechnology and Energy, Faculty of Engineering, University of Porto, Portugal

ABSTRACT: The recycling of Waste Printed Circuit Boards (WPCBs) constituent materials has been of increasing relevance, especially when associated to the also increasing volume of WPCBs, caused by factors such as the technological advancement and the reduction of estimated useful life of products. The WPCBs recycling plants use physical processing operations such as dismantling, crushing, sieving, magnetic, gravity and density separation for the recovery of metals. The main objective of this work was the evaluation of the efficiency of the copper enrichment of fractions obtained in the physical recycling process of WPCBs. According to the results obtained in the present study, it is indicated the use of the process of physical separation in fractions with particle size +1 mm, aiming at the enrichment of metallic fractions, especially the nonmagnetic fraction, once the concentrate NM + 1 mm fraction, presented a concentration of 78% of copper in its composition.

1 INTRODUCTION

Printed circuit boards (PCBs) are an integral part of electrical and electronic equipments (EEE), composed by metallic, ceramic and polymer materials (Ernst et al. 2003). According to Cobbing (2008), waste of electrical and electronic equipment (WEEE), especially waste printed circuit boards (WPCB), are chemically and physically distinct concerning to other industrial wastes. It contains a great diversity of metals, such as copper, iron, nickel, chromium, silver and gold, which means not only high economic value but also an imminent risk for the environment. In addition, the WPCB also contains large quantities of nonmetals such as epoxy resin, glass fiber and various plastics (Yamane 2011, Silvas 2015).

The recycling of WPCBs constituent materials has been of increasing relevance, mainly when associated to the also increasing volume of WPCB, caused by factors such as the technological advancement and the reduction of the estimated useful life of products (Wang, 2018). The recycling processes include mechanical, thermal and chemical processing, or a combination of these processes (Birloaga et al. 2014, Cui & Zhang 2008, Wang et al. 2017). The WPCB recycling plants use physical operations such as dismantling, crushing, sieving, magnetic and density separation, gravity separation, air classification, corona electrostatic separation and eddy current separation. This set of operations can be defined as a pretreatment for the recovery of metals using the hydrometallurgical or pyrometallurgical processes (Estrada 2016, Huang 2009, Zhang 2017). The separation of metal from the non-metal fraction of WPCBs is an essential step to recover noble metals.

Gravity concentration methods can be defined as a separation of different particles from one another based on differences in their specific gravities using the gravity force, and other dynamic forces, which can be augmented by centrifugal forces (Galvão 2017).

By using gravity concentration methods, a large amount of the materials can be pre-concentrated in a very efficient and economical way.

The main objective of this paper is to study the efficiency of the physical process applied to WPCBs, evaluating the enrichment of the fractions in terms of concentration in copper. For a more robust analysis, the gravity separation, by the oscillatory table, and the magnetic separation of the analyzed sample was also carried out, obtaining the differentiation by class, size fraction and magnetic property.

2 METHOD

2.1 *Dismantling and shredding process*

The WPCBs used in all experiments were taken from obsolete computers, from which the large components as the universal serial bus (USBs) and the high-definition multimedia interface (HDMIs) were removed. The WPCBs were cut in pieces of 10 × 10 cm using a guillotine. Following, the WPCBs were comminuted using a hammer mill (EWZ M400/I-200 single-shaft shredding machine) with 8 mm diameter grid to reduce sample size-primary-crushed and a RETSCH SM 200 knife mill with a grid of 4 mm in diameter to obtaining smaller sizes. The resultant material was classified according to three sizes, respectively, +1 mm, −1 mm +0.45 mm, −0.45 mm with a vibrating sieve shaker (RETSCH AS 200) and the corresponding sieves.

2.2 *Gravity separation*

After separation by size fraction, the sample was submitted to gravity separation, using an oscillatory table (model Wilfley). This process allowed the generation of three classes of products, denominated: Concentrate, Middling and Tailing. The Concentrate and Middling classes were analyzed due to the higher concentration potential in metals.

The shaking table consist of a slightly inclined deck with a small protrusions (riffles) positioned in parallel, along its length, and a wash water is distributed along the balance of the feed side from lauder. The table is vibrated longitudinally, using a slow forward stroke and a rapid return, which causes the particles to move along the deck parallel to direction of motion.

Besides the separation operation by gravity, the magnetic separation was performed using a manual magnetic device, in order to obtain fractions potentially enriched in copper.

Thus, the magnetic (M) and non-magnetic (NM) classifications of the analyzed samples were obtained.

2.3 *Chemical analysis*

For copper quantification, the samples were subjected to chemical attack with aqua regia, according to ISO 11466:1995 standard. The digestions were carried out in *Kjeldahl* tubes, with approximately 1 g of each sample and 15 mL of aqua regia during 3 h at 90°C. The tests were carried out in triplicate. After cooling, the solutions were filtered and diluted with deionized water until a volume of 100 mL. Copper was determined in the solutions by AAS analysis using a UNICAM 969 AA spectrometer. To characterize the concentrate obtained, the following metals were analyzed: Zinc, Gold, Silver, Manganese, Iron and Nickel.

The organic content was determined by the Total Carbon (TC) with a TC analyzer (TOC-V, SHIMADZU), according to EN 13137:2001 standard.

Table 1. Percentage of the weight of each product obtained by gravity separation.

Class	*+1 mm (%)*	*−1 mm +0.45 mm (%)*	*−0.45 mm (%)*
Concentrate	17	3,5	2
Middling	27	4,4	5
Tailing	22	9,1	10
Total	66	17	17

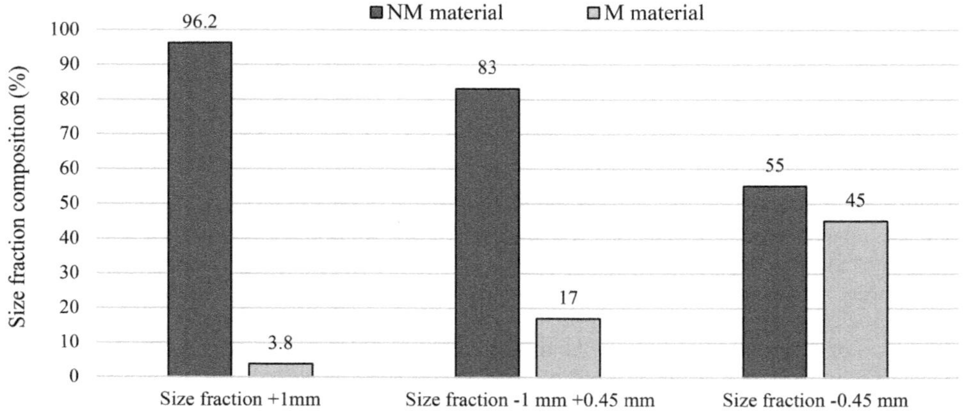

Figure 1. Composition of each fraction size according to M and NM material.

3 RESULTS AND DISCUSSION

After the crushing process, an amount of about 3.9 kg of WPCB was used for the gravity separation. The fractions obtained in this operation are shown in Table 1. According to a quantitative analysis, the concentrate fraction corresponds to 22.5%, the middling fraction represents 36.4% and tailing fraction 41.1%, (distributed over the three particle sizes) of the total weight of the sample processed. The higher sample quantity was in the class of larger particle size.

The granulometry of the sample adopted in this study was related to the gravity separation process, since it acts by means of the relative density between the fluid and the submitted material, thus influencing the obtaining of potentially enriched fractions.

The tailing class is characterized by low content of the sample, according to the dynamics of the oscillatory process of the table. Thus, the concentrate and middling classes are the main potential sources of metals, among them, copper.

Trough magnetic separation, applied to the concentrate and middling classes, six different classes were obtained, constituted for three particles sizes and magnetic and non-magnetic metals, as shown in Figure 1.

All three particle size fractions are constituted mainly of NM material, by weight. However, the finer the particle size the smaller amount of NM material. On the other hand, the M material is more expressive in the smaller particle size classes.

It was observed the significant participation of the middling class in the composition of NM products, contributing with values of 60.2%, 50.5% and 47.4%, while the concentrate class is responsible for 36%, 32.1% and 7.6%, respectively, Table 2.

The NM fraction with particle size +1 mm of the concentrate class presents 78% of copper in its composition, thus facilitating the recovery of copper by metallurgical processes. The NM fraction −1mm + 0.45 mm of the middling class also presents relevant values to the concentration of copper

Table 2. Composition of each size fraction according to M and NM, Concentrate and Middling.

Size fraction	*Concentrate NM (%)*	*Middling NM (%)*	*Concentrate M (%)*	*Middling M (%)*
+1 mm	36	60.2	3	0.8
−1 mm +0.45 mm	32.1	50.5	12.4	5
−0.45 mm	7.6	47.4	21.5	23.5

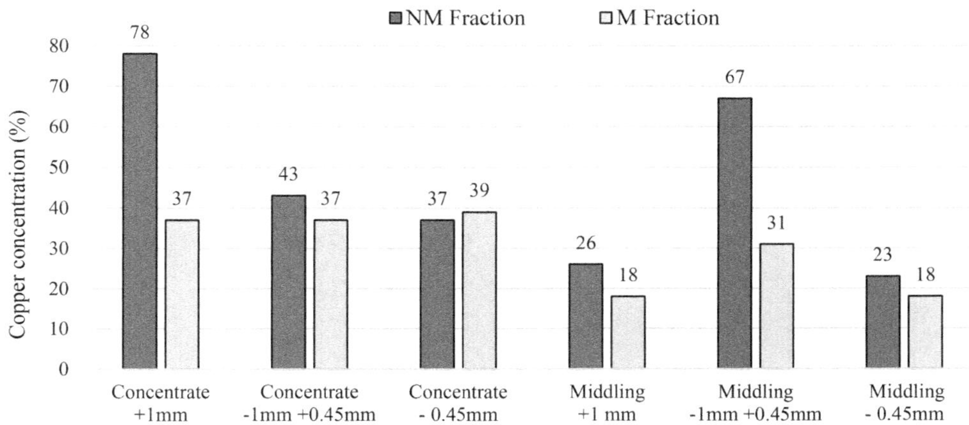

Figure 2. Concentration of copper for the concentrate and middling classes according to each size fraction.

in composition, reaching the value of 67%. As expected, the non-metallic fraction showed a higher amount of copper. Figure 2 shows the percent concentrations for each of the analyzed products.

In general, it was observed that the presence of copper in the NM fraction of the concentrate class decreases according to the reduction of the sample particle size. The same does not occur in the middling class, since the NM −1 mm fraction and NM + 0.45 mm fraction presents the highest value of copper concentration in this class. This can be attributed to the greater presence of nonmetallic materials in the composition of NM middling fractions +1 mm and −0.45 mm.

In order to verify the presence of non-metallic materials, total carbon analysis was assessed, allowing the identification of low percentages of non-metallic materials in the concentrate class, presenting a maximum value of 1.9%, as shown in Figure 3. The presence of non-metallic materials was more significative in the NM fraction compared with M fractions.

Comparing the concentrate and middling classes, it is possible to note a significant increase in the presence of non-metallic material in the middling class. The NM middling +1 mm fraction, stands out for presenting about 24% of a non-metallic material in its composition, followed by the middling NM −0.45 mm fraction, which shows 11.1%, and the middling NM −1 mm +0.45 mm with 5%.

It is still observed that the fractions −1 mm + 0.45 mm, in both the concentrate and middling classes present in general, a low presence of non-metallic materials.

Due to the low presence of non-metallic materials in the concentrate fraction, the presence of the metals zinc, gold, silver, iron, manganese and nickel, as well as copper, was also analyzed for this class (Figure 4).

In this analysis, it was found that iron is contained in large quantities in magnetic fractions, mainly in the concentrate classes of +1 mm M and −0.45 mm M fraction, representing 44.7% and 55.2%, respectively of the fraction composition.

The zinc also shows with significance in the general analysis, expressing its greater participation in the composition of the concentrate +1 mm NM fraction, with 15.5% of this fraction. The

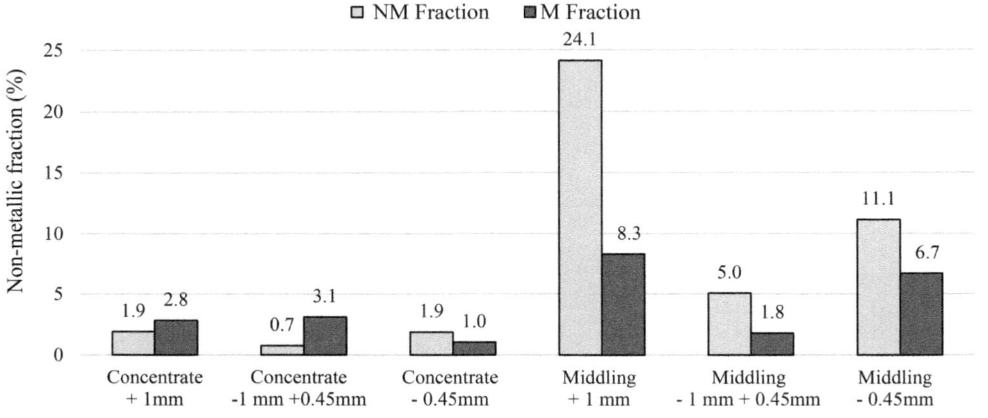

Figure 3. Non-metallic content in the concentrate and middling classes for each size fraction.

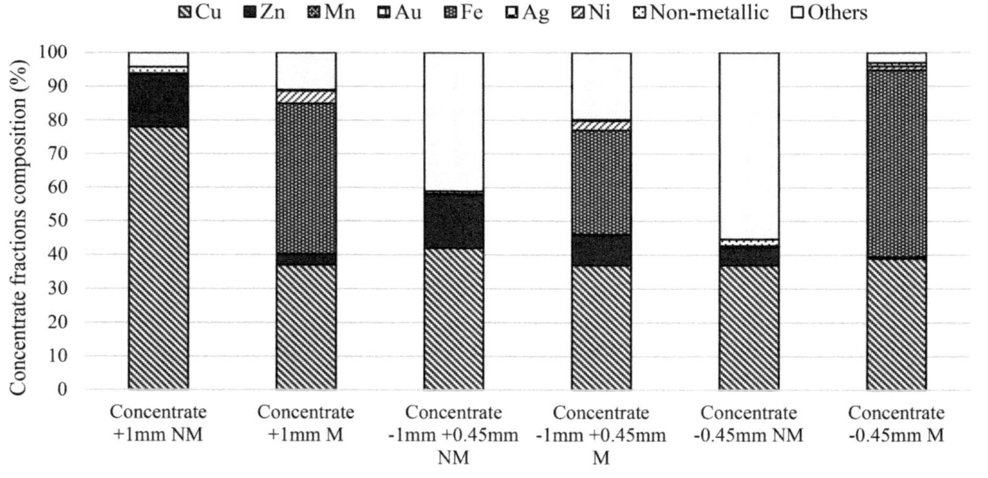

Figure 4. Chemical composition of the concentrate fractions.

nickel has expressive values in the M fractions, having its highest value in the composition of the concentrate +1 mm M fraction, corresponding to 3.7%. According to the reduction of the particle size of the sample of the concentrate M fraction, it observed the decline of the presence of nickel.

Specifically, to Au and Ag, which are characterized by their high economic value, it was identified a low content in the chemical composition of the concentrate class, since Au and Ag had its most important participation in the concentrate −0.45 mm fraction 0.08% and 0.04%, respectively.

It is still possible to infer that the concentrate +1 mm NM fraction and the concentrated −0.45 mm M fraction show the lowest values relative to other metals, representing 4.33% and 3.06%, respectively, obtaining high efficiency in the concentration of metals during the separation.

4 CONCLUSION

The physical separation operation can to contribute to the concentration of metals, especially copper, facilitating the subsequent operations of recovery and refining of these metals in metallurgical process. However, it is necessary to adjust the grain size to the gravitational separation process, considering the different behavior between the fractions analyzed.

According to the results obtained in the present study, it is indicated the use of the process of physical separation in fractions with particle size +1 mm, aiming at the enrichment of metallic fractions, especially the nonmagnetic fraction, once the concentrate NM + 1 mm fraction, presented a concentration of 78% of copper in its composition.

ACKNOWLEDGMENTS

This work was financially supported by: project UID/EQU/00511/2019 – Laboratory for Process Engineering, Environment, Biotechnology and Energy – LEPABE funded by national funds through FCT/MCTES (PIDDAC); Project "LEPABE-2-ECO-INNOVATION" – NORTE-01-0145-FEDER-000005, funded by Norte Portugal Regional Operational Programme (NORTE 2020), under PORTUGAL 2020 Partnership Agreement, through the European Regional Development Fund (ERDF). This work was conducted during a scholarship supported by the International Cooperation Program CAPES/COFECUB at the University of Porto. Financed by CAPES – Brazilian Federal Agency for Support and Evaluation of Graduate Education within the Ministry of Education of Brazil.

REFERENCES

Birloaga, I., Coman, V., Kopacek, B. & Vegliò, F. 2014. An advanced study on the hydrometallurgical processing of waste computer printed circuit boards to extract their valuable content of metals. Waste Manag. 34, 2581–2586. doi.org/10.1016/j.wasman.2014.08.0028

Cobbing M. 2008. *Toxic Tech: Not in Our Backyard.* Uncovering the Hidden Flows of e-waste. Report from Greenpeace Internatioal. http://www.greenpeace.org/raw/content/belgium/fr/press/reports/toxic-tech.pdf, Amsterdam.

Cui, J. & Zhang, L. 2008. Metallurgical recovery of metals from electronic waste: A review. J. Hazard. Mater. 158 (2–3), 228–256. doi.org/10.1016/j.jhazmat.2008.02.001

Ernst, T., Popp, R., Wolf, M. & Van Eldik, R. 2003. Analysis of eco-relevant elements and noble metals in printed wiring boards using AAS, ICP–AES and EDXRF. *Analytical and Bioanalytical Chemistry*, v. 375, n. 6, p. 805–814.

Estrada-Ruiz, R. H., Flores-Campos, R., Gámez-Altamirano, H. A., & Velarde-Sánchez, E. J. 2016. Separation of the metallic and non-metallic fraction from printed circuit boards employing green technology. *Journal of Hazardous Materials*, 311, 91–99. https://doi.org/10.1016/j.jhazmat.2016.02.061

Galvão, R. O. 2016. *Aplicação de um planejamento fatorial na recuperação de liga metálica (FeSiMn) de escória empregando-se mesa oscilatória do tipo wilfley.* Dissertação de mestrado, Universidade Federal de Pernambuco (UFPE), Recife, 2016.

Huang, K., Guo, J., & Xu, Z. (2009, May 30). Recycling of waste printed circuit boards: A review of current technologies and treatment status in China. *Journal of Hazardous Materials*. https://doi.org/10.1016/j.jhazmat.2008.08.051

Silvas, F. P. C., Jiménez Correa, M. M., Caldas, M. P. K., de Moraes, V. T., Espinosa, D. C. R., & Tenório, J. A. S. 2015. Printed circuit board recycling: Physical processing and copper extraction by selective leaching. *Waste Management*, 46, 503–510. https://doi.org/10.1016/j.wasman.2015.08.030

Wang, H., Zhang, G., Hao, J., He, Y., Zhang, T. & Yang, X. Morphology, mineralogy and separation characteristics of nonmetallic fractions from waste printed circuit boards (2018) *Journal of Cleaner Production*, 170, pp. 1501–1507. doi: 10.1016/j.jclepro.2017.09.280

Wang, H., Zhang, S., Li, B., Pan, D., Wu, Y. & Zuo T. 2017. Recovery of waste printed circuit boards through pyrometallurgical processing: A review. Resour. Conserv. Recycl. 126, 209–218.

Yamane, L. H., de Moraes, V. T., Espinosa, D. C. R. & Tenório, J. A. S. 2011. Recycling of WEEE: Characterization of spent printed circuit boards from mobile phones and computers. *Waste Management*, 31(12), 2553–2558. https://doi.org/10.1016/j.wasman.2011.07.006

Zhang, G., Wang, H., He, Y., Yang, X., Peng, Z., Zhang, T., & Wang, S. 2017. Triboelectric separation technology for removing inorganics from non-metallic fraction of waste printed circuit boards: Influence of size fraction and process optimization. *Waste Management*, 60, 42–49. https://doi.org/10.1016/j.wasman.2016.08.010

Wastes: Solutions, Treatments and Opportunities III – Vilarinho et al. (Eds)
© 2020 Taylor & Francis Group, London, ISBN 978-0-367-25777-4

Estimation of biogas production by a landfill in Luanda using first-order-decay model

C.J. Maria
Faculty of Engineering, University Agostinho Neto, Luanda, Angola

J.C. Góis
Association for the Development of Industrial Aerodynamics; Department of Mechanical Engineering, University of Coimbra, Coimbra, Portugal

A.A. Leitão
LESRA, University Agostinho Neto, Luanda, Angola

ABSTRACT: Landfills provide the most economical and generally accepted method in the World for the disposal of municipal solid wastes. However, the anaerobic degradation of organic matter based on the action of microorganisms in landfills leads to the emission of biogas, which has a great contribution to enhanced greenhouse gases. The estimation of biogas emissions requires comprehensive field measurements in landfill or the use of simulation models built under local climate conditions. One of the mathematical models used for evaluating the amount of biogas potential in landfills is the LandGEM model, which is applied in this research. The aim of this study is the estimation of biogas production in the Mulenvos landfill in order to assess the energy recovery potential and to reduce emissions into the atmosphere. The methane and carbon dioxide emissions were measured in situ. The results were compared with the emissions predicted using the first order decay model.

1 INTRODUCTION

Over the last past years Angola had a rapid increasing in urban population mainly due to rural exodus to get security and better life conditions. In 2014, Angola had a population of about 24 million and the estimated annual growth rate was 2.7%. Between 2000 and 2014, the population of Luanda (the capital and largest city of Angola) raised from 2,571,600 to 6,945,386 (INE 2014). Consequently, the production of solid urban waste (MSW) and landfill increased sharply.

The landfill is a large anaerobic biological reactor and is responsible for the production of biogas (landfill gas, LFG), a gas mixture combustible rich in methane (CH_4) and carbon dioxide (CO_2), and other gases in lower concentration, such as hydrogen sulphide (H_2S), nitrogen (N_2), water vapour and non-methane organic compounds (NMOC). Although, CH_4 has a high heat value, it has a high global warming potential and forms an explosive mixture with atmospheric air for concentrations between 7 to 17% v/v (Tchobanoglous et al. 1994).

According to the Intergovernmental Panel on Climate Change (IPCC) landfill accounts for about 5–20% of CH_4 emissions into the atmosphere (IPCC 1996). Estimations of biogas emissions at final disposal sites have been studied by several researchers (Machado et al. 2009; Aguilar-Virgen et al. 2014; Amini et al. 2013; Scarff & Jacobs 2006; De Bella et al. 2011).

Modelling the LFG generation allows predicting its production, but the biggest challenge is the determination of the factors that affect LFG generation within the mass of wastes disposed.

Over the years, various modelling approaches have been developed to estimate the methane and biogas generation rate of landfills, based on the first order decay method. Most of these models are based on two basic parameters; the methane generation potential (L_0) and the first-order decay

Table 1. Average composition of MSW in MSL (BAS 2014).

Waste category	Percentage of waste, % wt
Putrescible fraction (Food wastes and garden wastes)	25
Paper and paperboard	10
Plastics	15
Glass	7
Textiles	8
Metals (ferrous and nonferrous materials)	7
Sable	21
Others (fine fraction, inert materials, hazard waste)	7

Table 2. Annual disposed and accumulated wastes.

Year	Disposed wastes t	Accumulated wastes t
2007	37,287.40	372,187.40
2008	1,023,403.18	1,060,690.58
2009	1,545,596.00	2,606,286.58
2010	1,884,463.82	4,490,750.40
2011	1,931,405.83	6,422,156.23
2012	2,263,006.75	8,685,162.98
2013	2,632,375.31	11,317,538.29
2014	2,169,611.52	13,487,149.81
2015	1,809,240.00	15,296,389.81
2016	1,511,804.00	16,808,193,81
2017	2,277,259.61	19,085,453.42
2018	2,190,000,00	21,275,453.00

constant (k). Among them, the LandGEM model is the most flexible, and it has been recognized as the approach most widely used (Aguilar-Virgen et al. 2014).

The United State Environmental Protection Agency (USEPA) developed this first order equation model, which provides a very acceptable and accurate estimation of the methane amount produced rests on L_0 and k (USEPA 1997). The model has a database containing series of default information based on two sets of landfill criteria: Clean Air Act (CAA) is one of the landfills requirements, and the other criteria are based on Agency's Compilation of Air Pollutant Emissions Factors (AP-42), USEPA. It has been recommended to use AP-42 default values for standard landfills. The CAA default values have a high methane generation potential of 180 m^3 $CH_4.t^{-1}$ of waste (USEPA 2005)

Although, Mulenvos Sanitary Landfill (MSL) has not a facility provided for methane recovery, hence, this study estimates the capacity of methane generating by anaerobic digestion of organic fraction of MSW during landfill lifespan, using the latest LandGEM model version (3.02).

The aim of this study is the estimation of biogas production in MSL in Luanda, in order to assess the energy recovery potential and to reduce emissions of methane into the atmosphere.

1.1 *Landfill waste disposed, composition and production*

Luanda has been served only by MSL, which is a well-engineered facility operating under strict regulations to ensure protection to human health and environment. Lots of trucks with more than 6000 t of waste collected from all municipalities of Luanda are daily unloaded in MSL.

The composition of deposited waste expressed as a percentage by weight of total mass is shown in Table 1.

The compositions of the disposed waste were provided by the landfill operator and it is based on a field characterization analysis carried out on late 2014 (BAS 2014).

In Luanda the MSW generation rate increased from about 1,000,000 tonnes in 2008 to over 2 million of tonnes in 2018 as shown in Table 2.

The accumulated total mass of wastes disposed at MSL reaches about 21 million of tonnes. From 2019, the amount of MSW was estimated based on the population growth (2.7%) and the per capita generation. The net mass of MSW accumulated will reach over 47 million of tonnes in 2038, at the closure period of the landfill.

2 METODOLOGY

Methane and carbon dioxide emissions were measured at MSL at the end of 2018 using the mass-balance method. These measurements were taken weekly by a suction analyser meter. The measured emissions are compared with emissions predicted using the first order decay model.

2.1 *The landfill site*

The Mulenvos Landfill is located 12 km from the city centre of Luanda. It covers an overall area of about 270 ha. The MSW disposal started in December 2007, and the site has a lifespan until 2030. Currently around 34 ha have been used for waste landfilling. The landfill consists of seven layers of 7 m high each. The layer basis at the bottom remains 30 m deep below the soil surface. The top layer cover rises at about 40 m above the sea level, and the thickness of the total waste column reaches about 70 m at the end of disposal process.

The MSL is equipped by a basic waterproof system, leachate collection system, biogas drainage and treatment system, rainwater drainage network and environmental monitoring system for soil and groundwater. This confers the fundamental structures allowing the environmental sustainability and economic efficiency of landfill.

A reinforced composite liner system is used at the bottom and on the lateral slopes of the trench to prevent soil contamination. Temporary covers are made using a single layer of soil. Today MSL has over 70 deep biogas wells but biogas collection is inoperational.

2.2 *Biogas tests measurement*

All data of extraction biogas, referred in this study, in terms of composition, were collected along the investigating period.

The device used for the biogas extraction testes included a portable biogas analyser, GEM5000, an anemometer and a thermometer. The GEM5000 biogas analyser was designed, specifically, for use on landfills to monitor LFG extraction systems, flares, and migration control systems. The device was configured by supplier to measure high and low levels of biogas. This analyser allows simultaneous, measurements of four gases such as methane, carbon dioxide, oxygen and hydrogen sulphide.

During this research the gas analyser was calibrated according the gases to be measured. It was used to calibrate a gas mixture of CH_4 (60%), CO_2 (40%) and O_2 (0%). This calibration mixture raises the best gas level of CH_4 and CO_2 measurement in situ conditions (Geotech 2014).

At the landfill site, to perform readings, the analyser was prepared connecting two tubes for gas inlet or outlet, the temperature sensor and the anemometer for flow metering. For each sampling point the analyser measurement included: composition (CO_2, CH_4, O_2 and H_2S), biogas flow, biogas temperature and its heat content.

Currently MSL has more than 70 biogas wells, in the already closed cells, and only 17 holding the system prepared to take up readings. About 200 readings were obtained in 4 sampled months, namely, September, October, November and December 2018.

The wells were selected firstly based on the principle that each closed cell should be included by at least a well. Secondly according the shape of the liner layers and the easily of access to wells. All readings were recorded in real-time on analyser and daily were transferred into computer where all the results are presented in a simple excel format. The total biogas flow rate of the landfill was obtained by adding the biogas flow rates of each well.

2.3 *Estimation of biogas generation*

The biogas prediction was carried out following the guidelines proposed by USEPA's LandGEM model.

The USEPA model determines the amount of methane generated using the methane generation capacity and the amount of waste disposed.

The LandGEM model can be described mathematically by the equation presented in Eq. 1:

$$Q_{CH_4} = \sum_{i=1}^{n} \sum_{j=0,1}^{1} kL_0 \left(\frac{M_i}{10}\right) e^{-kt_{ij}} \quad (1)$$

Where: Q_{CH4} is the annual generation of CH_4 rate, in the year of calculation, in (m^3 $year^{-1}$), i is the time increment in one year, n is the total number of years being modelled, j is the time increment in 0,1 years, k is the methane generation rate in ($year^{-1}$), L_0 is the methane yield in (m^3 $CH_4.t_{waste}^{-1}$), Mi is the mass of waste, on a total weight basis, disposed in the ith year (t), and t_{ij} is the age of the jth portion of M_i of waste disposed in the ith year.

The L_0 represents the total volume of methane generated from specified quantity of disposed waste, which is dependent on the organic content of the waste, the age of waste, and can vary widely (6.2 to 270 m^3 t^{-1}) (USEPA 2005; Hamid et al. 2012).

Machado et al. (2009), specified, for tropical landfill, a L_0 of 70 m^3 t^{-1}, based on both, laboratory and on site measurement.

Otherwise using LandGEM model in Central American countries, USEPA estimated potential capacity of methane emission for different areas between 78–101 m^3 t^{-1} in 2009.

The value of k is dependent on moisture, pH, temperature, and other environmental factors, as well as landfill operating conditions.

In this paper values for the variables L_0 and k were adopted in basis of approach from USEPA (1997; 2005) for the CH_4 generation rate, k according to USEPA (2005), and its value can be calculated using equation indicated if the landfill meteorological data is available.

3 RESULTS AND DISCUSSION

3.1 *In situ measurements*

The composition of waste expressed as percentage by weight of total mass is shown in Table 1. The organic fraction of MSW is about 25% wt. This relatively high proportion of organic waste is considered to be a characteristic of MSW in Angola, as well as in several developing countries in Africa (Couth & Trois 2010).

The more organic waste presented in a disposed mass more LFG is produced by microorganism decomposition under certain conditions.

Over 6000 tonnes of unselected household, commercial and public residues collected in Luanda have been daily sent to Mulenvos Landfill since 2007.

This significant amount of wastes provides a strong reason to study the released gases generated at landfill. The analyses of the biogas composition and productivity parameters are shown in Table 3.

Regarding biogas composition, in-situ measurements confirms that CH_4 is the major component in biogas produced at MSL. The LFG composition varies slightly throughout the different wells selected for study.

The lower percent of oxygen recorded indicates that the anaerobic environment is suitable and it has real conditions to improve and stabilise biogas and methane generation. About the corrosive gas, H_2S in LFG the value recorded is typically between 25 and 100 ppm. A few wellheads have a kind of higher value.

In general the temperature within a landfill must be higher than ambient air temperatures. Landfills with temperature that exceeds 60°C are known as hot landfills.

Over the period of collection parameters in situ, the LFG temperature ranged between 37 and 64°C, which is agreement the literature's values.

Table 3. Biogas contents and parameters measured in 2018.

	Parameter								
Well ID	CH_4 %	CO_2 %	O_2 %	H_2S ppm	Bal. %	T_{biogas} °C	Q_0 m^3/h	Power kW	HHV_{biogas} kWh/m^3
ASMC2P30	50.42	34.38	0.40	4.83	14.80	37.64	8.18	43.98	5.37
ASMC3P22	57.70	38.08	0.03	61.17	4.18	45.44	14.38	86.50	6.01
ASMC4P25	58.30	37.98	0.02	77.83	3.70	53.88	30.85	187.98	6.09
ASMC4P11	56.10	39.89	0.04	79.71	3.97	40.72	27.23	159.67	5.86
ASMC4P12	56.19	39.80	0.03	89.25	3.99	49.71	14.45	85.03	5.88
ASMC4P13	56.13	39.88	0.05	86.38	3.95	50.57	31.38	184.28	5.87
ASMC4P14	55.17	40.04	0.06	80.00	4.73	47.05	11.09	63.91	5.77
ASMC5P03	57.06	39.02	0.04	79.20	3.88	48.88	24.02	143.62	5.98
ASMC5P05	55.26	40.93	0.04	93.14	3.77	50.42	17.86	103.30	5.78
ASMC5P07	56.27	39.80	0.03	104.14	3.39	50.92	31.43	185.01	5.89
ASMC5P09	55.10	40.86	0.01	75.29	4.03	47.32	17.73	102.23	5.77
ASMC6P00	57.50	39.30	0.00	129.20	3.20	59.08	23.38	140.60	6.01
ASMC6P02	56.71	38.93	0.03	68.00	4.33	46.07	21.96	130.54	5.95
ASMC6P49	54.93	39.37	0.03	101.67	5.67	55.02	20.13	115.73	5.75
ASMC7P33	55.91	39.93	0.03	100.71	4.13	52.73	20.04	1116.93	5.83
ASMC7P41	56.30	40.00	0.05	93.00	3.65	49.50	17.85	105.25	5.90
ASMC7P70	56.85	39.15	0.05	92.00	3.95	57.00	70.20	417.95	5.95
Average	55.99	39.25	0.06	83.27	4.70	49.53	23.66	139.56	5.86

The biogas flow rate measured in situ shown values in a range between 8.18 and 70.20 $m^3.h^{-1}$.

The large variation of biogas flow rate can be due to the different age of biogas wells used as sample of this study, and also due to the dynamic equilibrate between the production and flow of the biogas because of temporal variation of the porosity of deposited waste.

A typical normal cubic meter of LFG has a higher heat value (HHV) around 4.4 kWh/m^3, supposing that biogas composition has 50% v/v of CH_4.

The last column in Table 3 shows the HHV measured in each well.

According to Table 3 the landfill gas flow average value is about 23.66 ± 0.001 (m^3 $hour^{-1}$ $well^{-1}$) in 2018 and this means that for the 71 wells, at the already closed cells, the annual LFG generation of the landfill in 2018 is given using the average flow rate, total sampling wells and time, by:

$$Q_{LFG}=23.66x71x24x365 \cong 14{,}715{,}573.60 \ m^3year^{-1}$$

Then, in 2018, the biogas generation was 14,715,573.60 m^3 $year^{-1}$. Therefore, as the average content of methane is 55.99%, it annual generation is calculated to be 8,239,249. 65 m^3year^{-1}.

3.2 *Landfill gas estimation*

In the present work, the default values of the conventional CAA category were used because there do not exist any currently record of studies on the actual composition of the wastes deposited in the landfill and it is overwhelming accepted by several Nation United Agencies the values suggested for parameters L_0, k and NMVM.

For instance, using local data as the percentage of methane obtained in 2018, by the average of in-situ direct measurements made at the landfill wells (Table 3), the potential methane generation capacity suggested by (USEPA and IPCC) is $L_0 = 170\ Nm^3\ t^{-1}$ (calculated based on the physical composition of the waste on a dry-basis), the methane generation rate, $k = 0.05\ year^{-1}$, and the concentration of NMVOCs, 4000 ppmv.

The carbon dioxide and methane content was 44.01% v/v and 55.6% v/v of total landfill biogas respectively. The predicted LFG curve was obtained using values for tropical wet climate conditions, as recommended by the model guidelines.

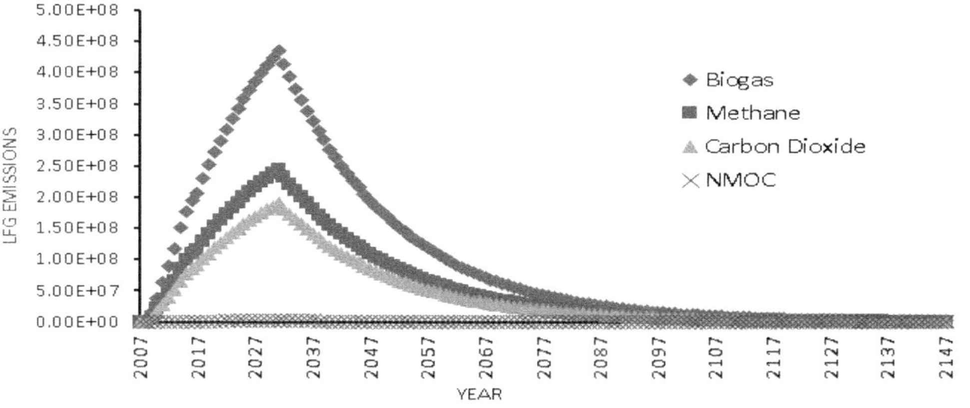

Figure 1. LFG emissions (m^3year^{-1}), obtained from LandGEM model using k and L_o referred.

Table 4. Readings and estimation of production and recovery biogas.

	Biogas				Biogas		
Year	Yield m^3year^{-1}	Collected (70%) m^3h^{-1}	Power MW	Year	Yield m^3year^{-1}	Collected (70%) m^3h^{-1}	Power MW
2007	0.00	0.00	0.00	2023	325,608,714.80	26,018.96	50.32
2008	552,054.28	44.11	0.09	2024	341,841,468.39	27,316.10	52.82
2009	15,677,004.72	1,252.73	2.42	2025	357,282,541.23	28,549,97	55.21
2010	37,795,570.53	3,020.19	5.84	2026	371,970,544.07	29,723.67	57.48
2011	63,852,457.24	5,102.37	9.87	2027	385,942,204.56	30,840.13	59.64
2012	89,333,528.92	7,138.52	13.80	2028	399,232,459.13	31,902.14	61.69
2013	118,481,354.23	9,467.69	18.31	2029	411,874,540.33	32,912.35	63.65
2014	151,676,273.22	12,120.25	23.44	2030	423,900,059.96	33,873.29	65.50
2015	176,400,858.17	14,095.96	27.26	2031	435,339,088.07	34,787.37	67.27
2016	194,584,179.82	15,548.96	30.07	2032	414,107,350.21	33,090.77	63.99
2017	207,477,036.21	16,579.22	32.06	2033	393,911,096.42	31,476.91	60.87
2018	231,073,955.82	18,464.81	35.71	2034	374,699,825.55	29,941.77	57.90
2019	251,917,224.00	20,130.37	38.93	2035	356,425,499.42	28,481.49	55.08
2020	271,743,954.00	21,714.70	41.99	2036	339,042,422.69	27,092.43	52.39
2021	290,603,722.96	23,221.76	44.91	2037	322,507,128.62	25,771.12	49.84
2022	308,543,690.15	24,655.32	47.68	2038	306,778,270.35	24,514.25	47.41

Fig 1 shows the results obtained using LandGEM model for the above values of k and L_0. The model assumes that there is no biogas generation in the first year of the waste disposal.

Anaerobic fermentation begins one year after grounding, nevertheless generation of biogas started in 2008 and is expected to continue for several years.

Based on the estimated quantities of biogas recovery from the landfill collection system and taking into account that the extraction efficiency suggested by studies on similar landfill is about 70% (Amini et al. 2013), the available power (MW) from the biogas over a period of 32 years (from 2007 to 2038) is calculated taking into consideration that the methane contented is about 55% v/v as shown in Table 4.

The recovery biogas rate average is about, 29,335 m^3h^{-1}. The power obtained for 2018 is about 35.71 MW. The results show that the available power has an increase trend associated to biogas

Table 5. Gas by in-situ measurement and modeling in 2018.

	Biogas m^3 $year^{-1}$	Methane m^3 $year^{-1}$	Carbon dioxide m^3 $year^{-1}$
LandGEM	231,073,956.00	129,724,919.00	101,349,037.00
Measurement "in situ"	14,715,573.60	8,239,249.65	5,775,862.64
Gas content by LandGEM	100%	55.99%	44.01
Rate LandGEM/In situ	15.70	15.74	17.55

production, which will increase until one year after the landfill closes, do not accepted any more wastes, and will decrease after that date.

Table 5 shows the annual generation of gases from the LandGEM model and direct measurements in situ in 2018. In 2018, the methane content represents 55.99% of the total landfill biogas while the carbon dioxide content corresponds to 39.25%.

Table 5 shows that the ratio between LandGEM calculations of annual gas generation and direct in-situ measurement is 15.74 and 17.55 for methane and carbon dioxide, respectively. Several factors may explain the high difference; such as the LandGEM does not take into account the characteristics and age of waste disposed. It considers all waste to be MSW. Otherwise, the direct measurements in the wells do not take into account the biogas released throughout the huge cover landfill surface.

4 CONCLUSIONS

This work provides a comparison between direct measured and modelled biogas and methane emissions from Mulenvos Sanitary Landfill in Luanda. The obtained results show that this model could be applied to estimate biogas and methane emission with slightly adaptation in order to consider specific local conditions. The results show a high deviation between LandGEM model prediction and in-situ measurement, but also show that direct measurements in landfill are the best way to calibrate estimation models to obtain a more reliable biogas and methane estimation and to prevent the direct release of a powerful pollutant and combustive gas as CH_4.

The amount of methane calculated can be considered in waste-to-energy programs and determine Angolan's contribution in global emission of greenhouse gases arising from the wastes.

REFERENCES

Aguilar-Virgen, Q., Gonzales, P. T., & Benitez, S. O. 2014. Power generation with biogas from municipal solid waste: Prediction of gas generation with in situ parameters. Renewable and Sustainable Energy Reviews, 30, 412–419.

Amini, H., Reinhart, D., & Mackie, K. 2012. Determination of first-order Landfill Gas modeling parameters and uncertainties. Waste Management, 32, 305–316.

Amini, H., Reinhart, D., & Niskanan, A. 2013. Comparison of first-order-decay modeled and actual field measured municipal solid waste landfill methane data. Renewable and Sustainable Energy Reviews, 33, 2720–2728.

BAS. 2014. Relatorio de Atividades referente ao mes de Fevereiro . Luanda.

Bella, G. D., Trapani, D. D., & Viviani, G. 2011. Evaluation of methane emissions from Palermo municipal landfill: Comparison berween field measurements and models. Waste Management, 31, pp. 1829–1826.

Couth, R., & Trois, C. 2010. Carbon emissions reduction strategies in Africa from improved waste management: A review. Waste Management, 30, pp. 2336–2346.

Geotech. 2014. Biogas 5000 Gas Analyser – Operating Manual. Geitechnical Instruments (UK) Ltd. England: Geotech.

IPCC, I. P. 1996. Guia para Inventarios Nacionais de gases de Estufa. USA: Livro de Trabalho.Vol 3.

INE 2014. Relatório Definitivo do Censo da População e Habitação em Angola. Luanda
Machado, S., Carvalho, M., Gourc, J., Vilar, O., & Nascimento, J. 2009. Methane generation potential in tropical landfills: simplified methods and field results. Waste Management, 29, pp. 153–161.
Scharff, H., & Jacobs, J. 2006. Applying guidance for methane emission estimation for landfills. Waste Management, 26, 417–429.
Tchobanoglous, G., Thenssen, G., & Vigil, S. 1994. Municipal Solid Waste Management. McGraw Hill.
USEPA. 2005. First Order Kinetic Gas Generation Model Parameters for wet Landfills. Washington DC: Report n° EPA -600/R-05/072, USEnvironmental Protection Agency.
USEPA. 1997. Compilation of air pollutant emissions factors, AP-42.

Wastes: Solutions, Treatments and Opportunities III – Vilarinho et al. (Eds)
 ISBN 978-0-367-25777-4

Tag & sensoring solutions overview for smart cities waste management

R.C. Madureira
Research Unit on Governance, Competitiveness and Public Policies (GOVCOPP), Department of Economics, Management, Industrial Engineering and Tourism (DEGEIT), Universidade de Aveiro (UA), Aveiro, Portugal

C. Dias-Ferreira
Universidade Aberta, Lisboa, Portugal
Research Center for Natural Resources, Environment and Society (CERNAS), Polytechnic Institute of Coimbra, Coimbra, Portugal
Aveiro Institute of Materials (CICECO), University of Aveiro, Aveiro, Portugal

ABSTRACT: This paper describes the technology behind the implementation of smart waste management systems at the starting point of collecting information about the user and the amount of produced waste. This analysis is based on new technological approaches framed by the smart city concept and often associated to the pay-as-you-throw (PAYT) conceptions. We begin with an overview of the current available technologies with examples of real applications, with a glance on the costs/benefit and system's limitations. We conclude with suggestions for further research on technology approaches to support best practices both for municipal policies and citizen's behavior. The results show how pertinent is the developing of reliable systems to help decision makers managing waste smartly towards a truly smart city concept.

1 INTRODUCTION

As time goes by, the theme of smart and sustainable cities is less a trend and more a mainstream issue. Urban centers have grown exponentially and somehow undisciplined, yet they have financial capacity and a propensity for learning and receptivity to innovation – they are therefore excellent living laboratories to experiment emerging technologies towards a higher quality of life of their citizens and contribute to the planet behalf. Management of urban waste is one such problems, especially in what concerns three main demands: the increasingly amounts of waste every year, the waste management cost's transparency and the environment care.

Unsorted waste is increasing due to excessive consumption and increasing population. Recycling appeared as a possible solution to diminish its impact in the environment, however mixed waste is difficult to separate and to recycle. Portugal has invested on Mechanical and Biological Treatment Units to ensure that part of the unsorted waste could be restored to recycling, as well as municipalities have invested on environmental education at schools so the waste separation at source starts to be a way of life among the younger generations. Those actions derive mainly from the demanding targets of the European Commission for 2020 to reduce significantly the landfill usage and to establish a circular economy where waste is seen as a resource. In Portugal, namely in Aveiro (Dias-Ferreira et al. 2019), as well as in other parts of the globe, the search of innovative sustainable solutions that benefit the environment is driven by the will to find reliable technologies and systems that help decision makers managing waste smartly, towards a truly smart city concept.

Waste collection is a complex and costly process that requires elaborate logistics. In order to achieve both optimization and transparency on the handling of municipal solid waste, the smart waste management requires interconnection among heterogeneous devices and teams: the

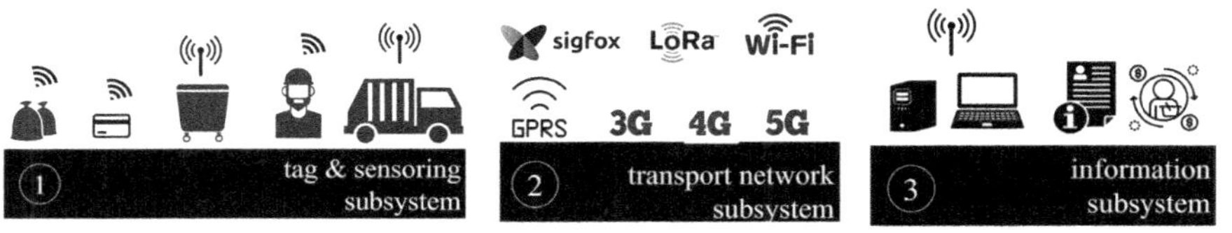

Figure 1. Smart waste management system: 1) the tag and sensoring subsystem, 2) the transport subsystem and 3) the information subsystem.

population, the local authorities, the service providers and the country governments. Technology is then a great ally to implement solutions that can help all of them to achieve a higher environmental awareness. One way to increase the individual participation in community projects is to combine technology with financial incentives, leading the consumers to feel the return of their actions. The PAYT concept is one way to create a more efficient and fairer waste invoice bill calculation (Hanf 2008). With this approach the waste is individually measured and that information can lead to build an individual invoice with the actual value of waste each family produces. This way the waste tariff ceases to be associated with other factors (e.g. the water consumption) and becomes much more accurate and transparent. The implementation of this type of system was noted at least in 10 of the 28 European capitals (ERSAR 2017), as well as in cities all over North America and Asia (D'Amato et al. 2013). A reliable PAYT system should acquire the exact amount of waste each family produces either in weight or volume units. A global measurement per groups of containers is possible but granularly is not enough to define each family's consumptions. In order to achieve this scenario several technological approaches have been done along the last decade but there are still some technological limitations. Moreover, collection routes are not widely optimized, most of the time some bins are filled while others are empty when collected, leading to excessive fuel usage, costs and greenhouse gas emissions.

The problems identified above are not new and for many of them there has already been new technological approaches, scientific research and some commercial products. However, the solutions presented are not exhausted and sometimes not even consolidated and that is why the response to the aimed quality of life for smart cities is far from being comprehensively analyzed. Several technical solutions have already been implemented to measure, for example, the container's fill-level per consumer/family and several types of data communication networks have been used with different levels of success.

The complete smart waste management system, from waste deposition to invoice, can be divided into three subsystem blocks, each one with its peculiarities and worries of implementation. Figure 1 describes these three fundamental technological modules. The first one – closer to the citizens (and waste producers) – is tag & sensoring, providing the identification and quantification of the user and/or respective produced waste; the second module is the transport network that will ensure that the collected information will be delivery from the pickup point to the final module that is the data analysis system at a central processor to process the monthly waste bills.

The purpose of this article is to make an overview of technological solutions of the first subsystem tag & sensoring (Fig. 1) concerning their technical reliability, constrains in maintenance and their limitations regarding the interaction with the public. The other two subsystems (transport network and information) are analyzed in similar articles (Madureira & Dias-Ferreira 2019; Aguiar et al. 2019).

2 TAG & SENSORING WASTE SUBSYSTEM

The tag & sensoring module is probably the most sensitive point of all waste management system. It is here, exactly at the very beginning of the process that most of the smart solutions fail and is here where the demand for reliable and robust technology is most needed. Important data to

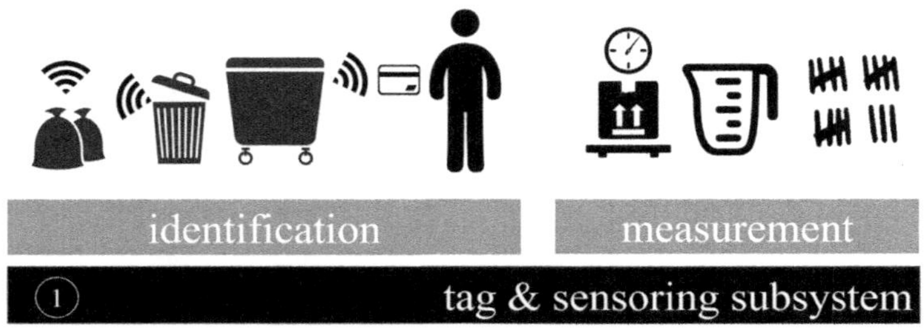

Figure 2. Smart waste management subsystem1, tag and sensoring, including identification (right) and measurement/quantification (left).

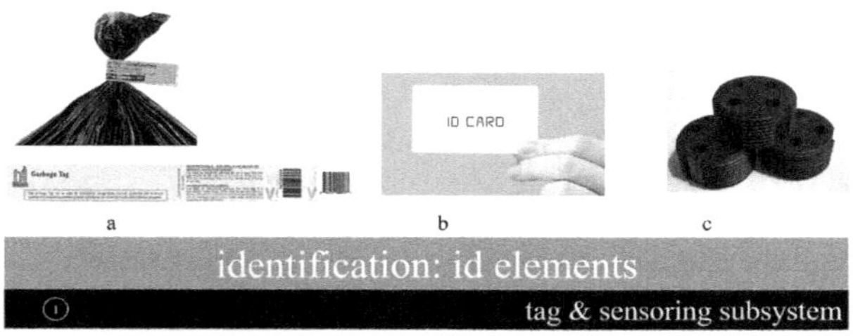

Figure 3. Examples of identification elements: a) Canada's garbage tags, on line: http://webaps.halton.ca/, access on January 2019; b) PAYT system in Aveiro with personnal RFID cards, www.microio.pt, c) bin tag transponders, www.hidglobal.com, access on January 2019.

gather is the user identification and/or the waste weight and/or the waste volume measurement. This module implies two distinct actions, the identification of the citizen and the unambiguous individual measurement/quantification of the deposited waste. If the identification and measurement process are not trustworthy the adjacent technological solutions can be weakened, not due to the technology itself but to the worthless information transported. Figure 2 depicts this model: the two main concepts of tag and sensoring subsystem with the identification and measurement modules.

2.1 *Identification sub module*

Nowadays identification in the sense *that establish or indicate who or what (someone or something)* is not the main technological challenge. Several types of identification are possible on each waste bag, individual bins or collective containers. The identification system includes the identification element, the tag and the reader. The identification elements can be passive optical bar codes printed in paper, and other more robust, the radio-frequency identification, RFID, tags. RFID can be passive or active tags depending if they have batteries or not. In this context passive RFID are most often used. A full analysis of several options for linking information technology to materials and products through the use of bar codes and radio-frequency identification (RFID) tags, and the implications for product life-cycle management can be read at Saar et al. (2002).

Real examples of identification systems can be observed in Figure 3. Figure 3a) shows a personal waste identification stickers (WIS) or paper code tags identification system that has already been used for some time in several regions in Canada. In this real example, each year 52 pre-paid paper tags are individually delivered to families and 86 for commerce to identify each waste bag. Only bags with this tag are collected. The middle image b) presents a card where each user is identified with a personal RFID card that is read by a sensor in the collective bin. Image c) shows a commercial bin tag transponders that can be installed into most waste bins to identify each container.

The paper tags are themselves very cheap (in the order of centimes of Euros) and it is relatively cheap to find a reading system to store that information. It is a cost-effective system to identify users, but one that shows serious limitations since paper tags are quite prone to damage. If used

Figure 4. Identification reading systems: a) example of tag reader handheld system for bins and bags, on line, www.waste360.com and b) RFID card reader in PAYT Aveiro Portugal pilot trial, photo authorized by Ubiwhere, www. ubiwhere.com and c) waste truck fully equipped with RFID reader.

in PAYT system, paper tags might easily get damaged, leading to lack of knowledge of how much waste (weight or volume) each family produces. RFID tags are a more robust solution, resistant to weather and adverse conditions. But once again, by themselves these tags cannot provide more information other than the user identification. However, these solutions can evolve to a PAYT solution if complementary technologic systems are used.

The next step is to have this information read and collected so that it can be sent to a cloud data platform. There is more than one way to perform this data acquisition and Figure 4 presents real examples for information acquisition: a) RFID reading gloves in workers garment to read the waste bag tag or individual containers, b) reading RFID card system in the collective bin used in Aveiro's PAYT pilot system and c) reading sensors in waste trucks that can read either the waste tags or the card information stored in the container (Wen et al. 2018).

When talking about RFID it is relevant to note that with the same name several solutions can be more or less consolidated leading to more or less efficiency of the system. New RFID's systems keep appearing in the industry and new references in the literature to developments in this area (Abdoli 2009). A drawback of this solution is not in the identification module but on the reading part because RFID tags readers might need to be often renewed due to the frequent handling of waste bins, thus increasing operating costs.

2.2 *Measurement sub-module*

Recalling Figure 2, the second part of the tag & sensoring system is the measurement module. There are several solutions already implemented to ensure the most accurate measurement of the collected waste per user. These systems can be based on either volume or weight measurement but often with a good deal of estimations.

The volume implemented solutions are relatively widespread for fullness monitoring for route optimization. The volume, the amount of space occupied by a three-dimensional object as measured in cubic units, can be obtained by using an ultrasonic sensor inside the waste container as depicted in Figure 5a). The sensor sends a high-frequency sound wave into the container and measures how long it takes for the echo of the sound to bounce back. This allows to determine the level of waste within the limited range of the sound wave. Ultrasonic volume measurement can be used to calculate waste volume, but its accuracy depends on the even waste distribution inside the container, which is rarely the case. Recently, studies have been made to increase the number of ultrasonic sensors and build a three-dimensional surface of the waste inside the container. As the number of sensors increase the cost also increases. Ultrasonic sensors might be useful to estimate the level of waste in a container and organize collection routes accordingly but are not helpful in the measurement of the volume deposited by a single user, as required for PAYT implementation.

Image based sensors can also provide volumetric information (Figure 5b), based on the analysis of image frames capture from time to time and is complemented with a transmission system that

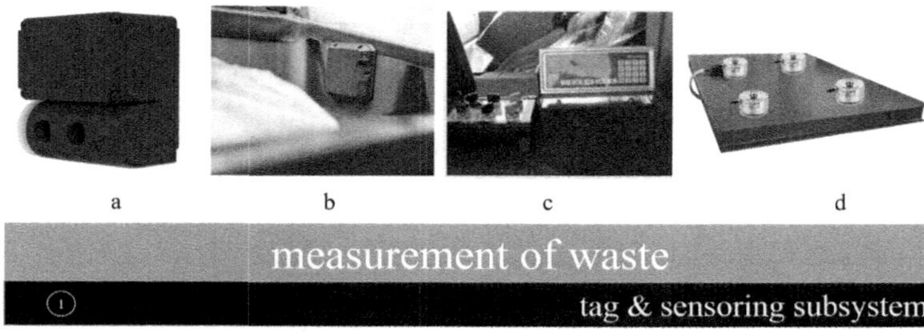

Figure 5. Examples of volumetric sensors to determine the waste volume, a) ultrasonic volumetric sensor by eculabs, www.ecubelabs.com, b) image sensor from Compology, compology.com, internet access February 2019, c) on-board weighing system, www.eilersen.com and d) .load cell, photo from the authors.

sends the information to a cloud system that performs the visual analysis (Nugroho 2013) This solution, even though more accurate, leads to a larger amount of transmitted data comparing to other systems, leading to high charges in communication and increasing OPEX expenses, and therefore is not suitable for large scale use within a PAYT system.

Weigh or volume estimation is also used in some PAYT systems. In this solution each family has one card and each time the container is opened, a "*discharged*" is registered. The system interprets this "*discharged*" as "one" usage with a predefined value of weight or volume. This measurement estimation was used in Aveiro's trial where each card usage and container opening was registered as a 30 L deposit. Weight system depend on the installation of scale's or load cells to measure the weight per bag or the weight per bin. The measurements/readings can be done both in the waste container (Fig. 5d) or alternatively inside the waste truck with an on-board weighing system (Fig. 5c). There are several factors affecting the selection and application of load cells used for accurate weight measurements, namely: relative size and cost, direction of force, range of measurement, vibration or signal noise, frequency of oscillation, temperature changes, fatigue, and the use in underwater, explosive, or other special environments (Dunn 2016). In consequence, measurements based on weight tend to be more expensive to implement and difficult to maintain, and therefore are not as common as volume-based systems.

Weight measurement systems can also operate on a volume-based approach using volume-to-weight conversion factors that are in constant discussion and evaluation (USEPA 2016). There are useful usages for both information: volume and weight, one can use the weight value for cost evaluation and volume for trucks routes optimization. However, they are not always correlated, so if both parameters need to be acquired with accuracy, two different sensors in each bin are required. Other empirical methods to estimate the same result can also be found in the literature, for example new methods to measuring household food waste by linking the food waste with its source (Elimelech et al. 2018).

3 INTERACTIONS BETWEEN THE TECHOLOGY AND THE CITIZENS

Taking the complete smart waste management system, tag & sensoring is the module closest to the citizens. This means that the technology should be robust, and any failures during usage will be a drawback and will undermine the initial will of the citizen to be a part of such system. The interface with the public should be carefully considered, in order to be simple to use and accessible for all, including non-technical users.

When systems are first installed undesirable effects of user fees for solid waste services might happen, including families forgetting to bring the RFID card when throwing the garbage out, leaving open the lids of the bin thus invalidating the next card reading, illegal dumping or vandalism. Part of these behaviours are consequence of a lack of trust of the citizen on the implemented technology. Intuitively the public feels weaknesses in the system as well the lack of trust in the governmental managers, not believing in the good intentions of fair tariff systems and in the environment protection. These aspects highlight that communication and transparency are essential in the introduction of smart waste systems for waste fees, but also that the technology must be reliable, in order to build confidence.

4 CONCLUSIONS

The intelligent cities paradigm arises with the investigation of spatial and urban dynamics associated with an emerging need to solve problems or even to predict them. It is consensual that waste management will be one of the challenges of the future cities. From time to time it is important to overview what has been achieved so far and understand and discuss the limitations and the unaccomplished, in order to be able to go further in the future.

There are technical solutions already on the market for both user identification and bin identification that can be used with PAYT systems. Paper (or plastic) tags are cost-effective but are prone to damage, resulting in the loss of information on how much waste each family produces. RFID tags are a more robust solution, resistant to weather and adverse conditions. However, in the case of the citizen, using an identification card which is specific for waste deposition is not practical and people sometimes forget to bring it when taking out the garbage so new, imaginative solutions have to be created. Sometimes good solutions have nothing to do with technical complexity, but with technical creativity and sensitivity in the use of technology by non-technical users.

The unambiguous individual measurement/quantification of the deposited waste is still a challenge. Commercial solutions are either volume-based or weight based. Measurements based on weight tend to be more expensive to implement and difficult to maintain, and therefore are not as common as volume-based systems. But research has been carried out to improve the mechanical robustness of the weight measurements load cells and load sensing devices, so new developments are expected in the future.

REFERENCES

Abdoli, S. 2009. RFID Application in Municipal Solid Waste Management System. *International Journal of Environmental Research 3(3)*: 447–454.

Aguiar et al 2019. This issue (submitted).

Nugroho, A., Tongthong T. & Shin, T. 2013. Measurement of the Construction Waste Volume Based on Digital Images. *International Journal of Civil & Environmental Engineering IJCEE-IJEN 13*(2):35–41.

D'Amato, A., Mazzanti, M. & Montini, A.(Ed.) 2013. *Waste Management in Spatial Environments*. New York: Routledge, Taylor and Francis,

Dias-Ferreira, C., Neves, A. & Braña, A. 2019. The setting up of a pilot scale Pay-as-you-throw waste tariff in Aveiro, Portugal. WIT Transactions on Ecology and the Environment, 31: 149–157.

Dunn, B. 2016. Load Cells – Introduction and Applications. BD Tech Concepts LLC. BD Tech Concepts LLC.

Elimelech, E., Ayalon, O. & Ert, E. 2018. What gets measured gets managed: A new method of measuring household food waste. *Waste Management 76*: 61–81.

ERSAR 2017. *GUIA TÉCNICO DE IMPLEMENTAÇÃO DE SISTEMAS PAY-AS-YOU-THROW (PAYT).*

Hanf, M.B. 2008. The fairness of PAYT systems: Some guidelines for decision-makers. *Waste Management, 28*(12): 2793-2800. doi: 10.1016/j.wasman.2008.02.031

Madureira, R.C. & Dias-Ferreira, C. 2019. Data communication solutions overview for smart cities waste management. This issue (submitted).

Saar, S. & Thomas, V. 2002. Toward Trash That Thinks: Product Tags for Environmental Management. *Journal of Industrial Ecology* 6(2): 133–146.

USEPA. 2016. *Volume-to-Weight Conversion Factors*. U.S. Environmental Protection Agency. Retrieved from

Wen, Z., Hu, S., De Clercq, D., Beck, M.B., Zhang, H., Zhang, H., Fei., F. & Liu, J. 2018. Design, implementation, and evaluation of an Internet of Things (IoT) network system for restaurant food waste management. *Waste Management 73*: 26–38. doi: 10.1016/j.wasman.2017.11.054

Wastes: Solutions, Treatments and Opportunities III – Vilarinho et al. (Eds)
© 2020 Taylor & Francis Group, London, ISBN 978-0-367-25777-4

Leaching of Cr from wood ash – discussion based on different extraction procedures

L.M. Ottosen, P.E. Jensen & G.M. Kirkelund
Department of Civil Engineering, Technical University of Denmark, Denmark

A.B. Ribeiro
CENSE, Departamento de Ciências e Engenharia do Ambiente, Faculdade de Ciências e Tecnologia, Universidade Nova de Lisboa, Caparica, Portugal

ABSTRACT: Experimental work with wood ash (WA) from incineration of virgin wood shows that about 45% of the initial 79 mg/kg Cr leaches out in water, and even after a three step washing process in water, the WA leaches Cr. The leaching is highest at highly alkaline conditions. Electrodialysis showed that the leached Cr was in anionic form, but at the present high pH, both Cr(III) and Cr(VI) can form anionic complexes, so the oxidation state cannot be determined from this investigation. Because the extraction of Cr is so dependent on pH under alkaline conditions, the use of acid in the first step of sequential extraction gave an error in assessing the Cr present in this step. More Cr is extracted from WA in water than in weak acid.

1 INTRODUCTION

The ongoing coal phase out in many EU countries result in conversion of the coal-fired power plants to biomass fired (mainly wood) in e.g. Denmark. Wood ash (WA) is inevitably generated. To obtain a resource efficient society, it is important to find use for residues like WA. In order to evaluate the possible use, the leaching properties of the heavy metals retained in the WA must be assessed. The leaching of heavy metals is pH dependent, and most heavy metals are extracted at low pH, however, Cr show amphoteric behavior and dissolve over the pH range 1–13 from WA [Pöykiö et al. 2009]. Cr thus stands out as a major problematic heavy metal not at least due to the possibility for leaching Cr(VI) which is carcinogenic and of high acute toxicity. This work is an experimental study, where results from different extraction methods are compiled and discussed to assess the leachability of Cr from WA under different conditions. The methods are washing in water, pH dependent extraction, sequential extraction and electrodialysis. The experimental work includes Cr mobility from both WA as received and WA washed in water (W-WA) for comparison.

2 MATERIALS AND METHODS

The wood fly ash was from Amagerværket, DK, which fires wood pellets from pine and spruce. The plant incinerates about 350.000 ton wood a year. The is about 1000–1100°C in the grate.

2.1 *Characterization of the investigated wood ash*

WA characterization was made with dried ash (105°C, 24 hours). Concentrations of Cr, Pb, Zn were and selected macro-elements were measured in accordance to DS259; 1.0 g ash and 20.0 ml (1:1) HNO_3 was heated at 200 kPa (120°C) for 30 minutes. Filtration through 0.45 μm filter and the concentrations were measured with ICP-OES. Ash pH was measured by suspending 10.0 g ash

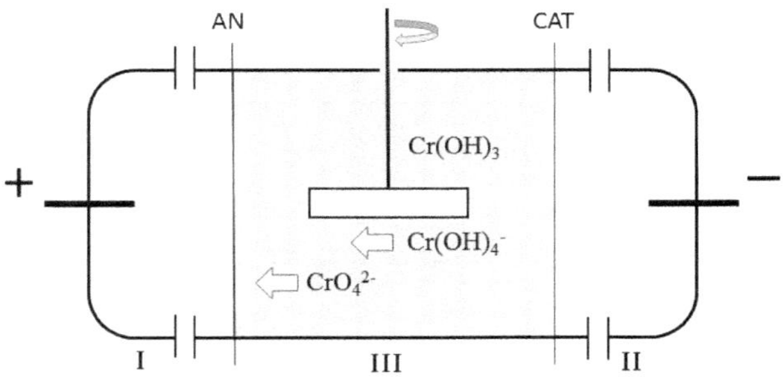

Figure 1. Electrodialytic cell. The ash is suspended in compartment III. AN = anion exchange membrane, CAT = cation exchange membrane.

in 25 ml distilled water. After 1 hour agitation pH was measured directly in the suspension. Loss on ignition (LoI) was found after 30 minutes at 550°C.

2.2 *Washing procedure, sequential extraction and pH dependent extraction*

Sequential extraction was made in four steps based on the three-step method (Rauret et al., 1999), with the forth step being analysis of the metal content after digestion as described above.

The pH dependent extraction from WA and W-WA was found from the following procedure: 5.0 g dry ash and 25 ml (0.5; 1.0; 1.5 M) HNO_3, distilled water or (0.5; 1.0 M) NaOH were agitated for 1 week in closed flasks (in duplicates). The suspensions were filtered (0.45 μm). The Cr concentration and pH was measured in the liquid phase with ICP–OES.

The washing procedure: 100.0 g WA suspended in 500 ml distilled water and agitated for 1 min. After settling, the water was decanted. New 500 ml distilled water added. This was repeated so the ash was washed 3 times. Finally, the suspension was filtered and solution and ash were saved for analysis. The W-WA was dried (105°C), weighed and characterized by the same methods as the raw ash. The pH and heavy metal concentrations were measured in the filtrate.

2.3 *Electrodialytic experiments*

The electrodialytic (ED) setup combines application of an electric DC field with the use of ion exchange membranes for separation of anions and cations from the Wa suspension. In the ED cell (fig. 1), the suspension of WA or W-WA in water was in compartment III. The setup was first developed for remediation of polluted soil [Ottosen et al., 1997], but in this work it is used to evaluate the charge of the leached Cr ions and complexes.

The electrodes had separate compartments. A cation exchange membrane and an anion exchange membrane, respectively, were barriers to compartment III. The internal diameter was 8 cm. The electrode compartments were 5 cm long, and the central compartment 10 cm. An overhead stirrer (RW11 basic from IKA) kept the WA suspended. In each electrode compartment, 500 ml 0.01 M NaNO3 adjusted to 2 with HNO_3 was circulated. Platinum coated electrodes were connected to the power supply (Hewlett Packard E3612A), which kept a constant current.

Four ED experiments were conducted (table 1), two with WA and two with W-WA. Experiment with liquid to solid ratio (L:S) of 10 and 7 were made with both. In all experiments, the ash was suspended in 350 ml distilled water, and the L:S was obtained by suspending 35 or 50 g, respectively. The experiments are named after the ash (WA or W-WA) and the L:S ratio.

The duration of the experiments with WA was 7 days. The experiments with W-WA lasted 14 days, as the conductivity in the suspension was too low, to apply 50 mA. Therefore, the current was halved to 25 mA and the duration doubled to 14 days to ensure the same charge transfer in all

Table 1. Overview of ED experiments.

Experiment	Current (mA)	L:S	Duration (d)	Ash
WA-10	50	10	7	WA
WA-7	50	7	7	WA
W-WA-10	25	10	14	W-WA
W-WA-7	25	7	14	W-WA

Table 2. Selected characteristics of the WA and W-WA. Standard deviations are given in brackets.

	pH	LoI (%)	Cr (mg/kg)	Cu (mg/kg)	Zn (mg/kg)	K (g/kg)	Ca (g/kg)	Cl (g/kg)	S (g/kg)
WA	13.0	3.5	79	84	860	192	127	58.6	26.2
	(0.01)	(0.06)	(1.9)	(3.3)	(24)	(7.6)	(4.9)	(5.8)	(0.7)
W-WA	11.8	6.6	49	110	1090	44	156	0.4	1.7
	(0.01)	(0.1)	(0.8)	(1.5)	(20)	(0.6)	(2.9)	(0.002)	(0.06)

experiments. During the experiments, the pH was adjusted to 1–2, and the voltage and the pH in the suspension were measured. At the end of the ED experiments, the contents of Cr, Pb and Zn were measured in the membranes, ash, solutions and at the cathode.

3 RESULTS AND DISCUSSION

3.1 *Wood ash characteristics before and after washing*

The concentrations of Cr, Cu and Zn and the 4 macro-elements with highest concentration (K, Ca, Cl and S) for WA and W-WA are shown in table 2, where also pH and LoI are given.

The WA solubility in water was 27%. The concentrations of K, Na, Cl and S decrease during washing with 231 g/kg in total (corresponding to 23% of the total weight). Thus, these four chemical elements constitute the major part of the soluble salts. The high pH shows that hydroxides were leached or formed. The pH decreased to 11.8 after washing due to wash out of hydroxide. The concentrations Al, Ca, Fe, Mg, Mn and P increased with a factor of 1.2–1.3 during washing (results not shown), showing that the minerals with these elements were little soluble as this factor corresponds to the solubility. The LoI increased as well after washing, representing the unburned fraction, which is not soluble in water. The heavy metal concentrations in the investigated WA was in the ranges previously compiled by [Kröppl et al. 2011].

3.2 *Cr extraction with the different methods*

The Cr chemistry differs from other heavy metals e.g. Cu and Zn, as the Cr concentration was decreased in the WA after washing, whereas the concentrations of other heavy metals increased as result of low leaching and the removal of soluble salts. The concentration in the washing waters were: 7.4 mg Cr/l, 2.2 mg Cr/l and 0.6 mg/l, i.e. decreasing with number of washing steps (the concentrations in the two last washing waters include Cr leached during the first washing, as the procedure involved decanting of excess water and thus some water from the first washing remained).

Fig. 2a shows the leached Cr (mg/kg) from WA and W-WA as function of pH from the chemical extraction experiments under slightly acidic and alkaline conditions. Alkaline conditions gives the highest Cr extraction. The Cr extraction in NaOH is higher than in water. The lowest concentrations

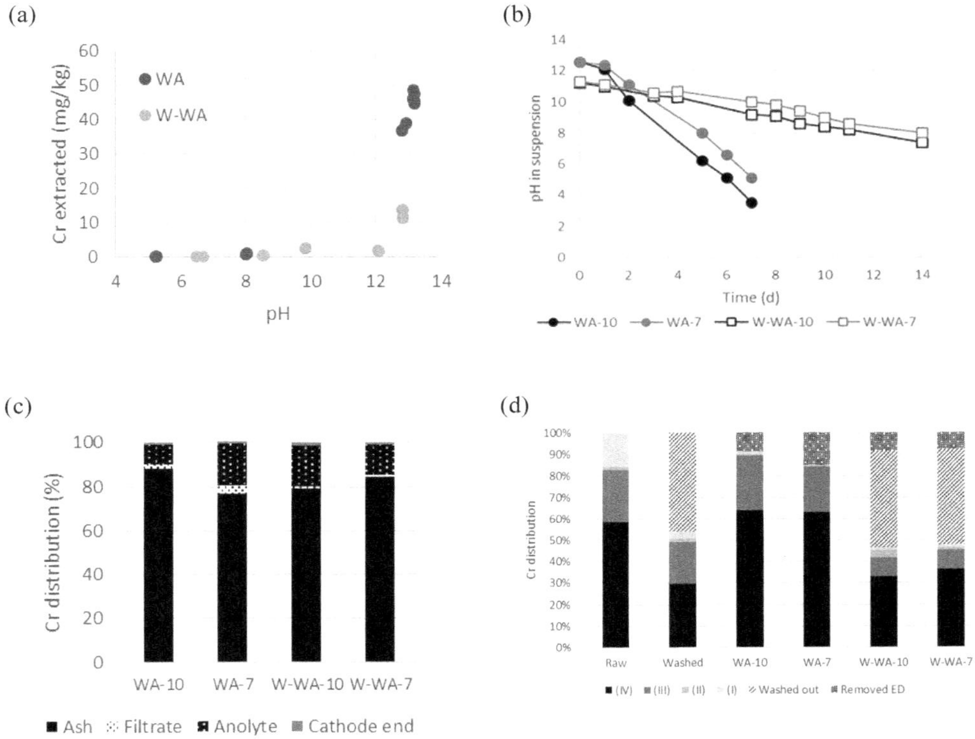

Figure 2. (a) pH in ash suspension during ED treatment, (b) distribution of Cr in the ED cell at the end of the experiments, and (c) Cr distribution from sequential extraction, wash and removed during ED.

measured in both WA and W-WA was 0.2 mg/kg, found at the lowest pH of the investigation (5.2 and 6.5, respectively). During ED, pH of the ash suspension decreased (fig. 2b). The decrease was due to water splitting at the anion exchange membrane and exchange over the cation exchange membrane of H^+ from the catholyte with ions in the ash suspension [Nystroem et al., 2005]. The pH in the suspension reached acidic conditions in the experiments with WA (3.5 and 4.5), but in the experiments with W-WA, the pH was neutral or slightly alkaline at the end of the experiments. That pH reaches the lowest pH for WA during ED is in consistency with the result from extraction in acid (fig. 2a), indicating a higher buffering capacity of W-WA. However, the differences in the acidification rates between ED experiments with WA and W-WA was among others also related to the extent of water splitting. In both ED experiments with WA and W-WA the pH was less in the experiment with L:S 10 than in L:S 7 (fig. 2b). The difference within each being due to the different ash amounts and thus buffering capacity of the ash suspension.

The Cr distribution in the ED cell at the end of the experiments is in figure 2c. The major part of Cr remained in the ash. Cr was extracted from both WA and W-WAs, showing that washing in water do not solve the issue of Cr leaching from the WA. The extracted Cr was electromigrating into the anolyte, i.e. in anionic form. In an aquatic environment in the absence of complexing agents other than H_2O or OH^-, the following trivalent and hexavalent Cr species may be found (Ščančar & Milačič, 2014): in solutions more acidic than pH 4.0, *Cr(III)* exists as hexa-aqua Cr^{3+}, in less acidic solution as $Cr(OH)^{2+}$) and $Cr(OH)_2^+$ species. In the neutral up to alkaline pH region, Cr is mainly precipitated as a sparingly soluble $Cr(OH)_3$(s), however, in alkaline solutions at pH higher than 11.5 the precipitate re-dissolves, resulting in formation of $Cr(OH)_4^-$ complex. *Cr(VI)* species present in aqueous solutions at pH more than 1 yielding $HCr_2O_7^-$ that occurs within the pH range from 1 to 6.5, while above pH 6.5 only CrO_4^{2-} ion exists. Thus, at the initial high pH of the ash suspensions in the ED experiments, both Cr(III) and Cr(VI) could be in anionic form. The initial pH of the suspension with W-WA was 11.2, so it is at the limit where Cr(III) forms $Cr(OH)_3(s)$, which could indicate that the removed anions are Cr(VI), but no conclusions can be drawn. However, at a lower pH, ED might be used to separate extracted Cr(III) and Cr(VI), as it is only at high pH (>11.5) that Cr(III) is in anionic form.

Figure 2c shows the summarized distribution of Cr from the four step of sequential extraction (I – Exchangeable and soluble in water/acid, II – Reducible, III – Oxidizable and IV – residual), washing (calculated on basis of concentration in the WA before and after washing and taking into account the WA solubility) and Cr removed during ED. More than twice as much Cr leached during water washing than found in step I in the sequential extraction. This underlines that sequential extraction schemes must be chosen with caution. Less Cr was found in the first step than actually present here, which can be explained from fig. 2a, showing that just lowering pH slightly at around 13 means significant decrease in extracted Cr, and thus this sequential extraction scheme was not optimal for Cr from WA. The L:S was 5 in both washing and pH desorption test, and the Cr concentration in the first washing water (7.4 mg Cr/l) corresponded to the 7.5 mg Cr/l extracted in distilled water in pH dependent desorption. The duration of the two tests was however very different; 1 minute and 1 week, respectively. This reveals that the chemical reactions determining the Cr solubility are almost instant and that reactions with slower kinetic does not play a major role in the overall Cr extraction in water.

Similar to the first step of sequential extraction less Cr was extracted during ED than during washing, most likely due to the decreasing pH all through the experiments. What was removed by ED from the WA correlates to the Cr found in step I of the sequential extraction. The percentage of the total Cr removed during ED was still in the range as before the washing. It seems as if it is Cr dissolved from step III – the oxidizable phase, which is removed from W-WA. WAs has a self-hardening ability in reaction with water caused by formation of new phases [Steenari et al 1997], and the overall self-hardening processes may influence the heavy metal including speciation of Cr, as less Cr was in the residual phase (VI) in W-WA than in WA.

4 CONCLUSIONS

When suspended in water, 29 wt% of WA dissolve, and about 55% Cr of the initial 79 mg/kg was washed out in a three step washing process. The WA leachate has a pH of 12.8, and most Cr leaches under strong alkaline conditions. When evaluating the chemical forms by sequential extraction for Cr in WA caution must be taken in relation to the first step, which corresponds to the least hard bound fraction. More Cr was extracted in water than in weak acid, and the latter is often used in sequential extraction, causing the result to be misleading by showing less Cr in this step than actually being the case. The extraction of Cr in water was at the same level after 1 minute and 1 week indicating that the kinetic of the Cr dissolution was fast. Electrodialytic experiments showed that the extracted Cr was in anionic form. At such high pH, both dissolved Cr(III) and Cr(VI) are in anionic form, so the method cannot be used to distinguish between the two oxidation steps here. During electrodialysis, Cr was removed from the washed WA revealing that washing in water does not solve the problem with Cr leaching even though the concentration is decreased in the washed ash compared to initially.

ACKNOWLEDGEMENTS

This work has received funding from the European Union's Horizon 2020 research and innovation programme under the Marie Skłodowska-Curie grant agreement No. 778045

REFERENCES

Kröppl, M.; Muñoz, I.L. & Zeiner, M. 2011. Trace elemental characterization of fly ash, *Toxicological & Environmental Chemistry*, 93:5, 886.

Nystroem G.M.; Ottosen L.M. & Villumsen A. 2005. Acidification of Harbor Sediment and Removal of Heavy Metals Induced by Water Splitting in Electrodialytic Remediation. *Separation Science and Technology* 49(11): 2245

Ottosen, L.M.; Hansen, H.K.; Laursen, S. & Villumsen, A. 1997. Electrodialytic Remediation of Soil Polluted with Copper from Wood Preservation Industry. *Environ. Sci. Technol.* 31: 1711

Pöykiö, R.; Rönkkömäki, H.; Nurmesniemi, H.; Perämäki P.; Popov, K. & Välimäki, I. 2009. Release of metals from grate-fired boiler cyclone ash at different pH values. *J. Chemical Speciation & Bioavailability* 21(1): 23

Rauret, G.; Loópez-Saánchez, J.F.; Sahuquillo, A.; Rubio, R.; Davidson, C. & Ureb, A.; Quevauvillerc, Ph. 1999. Improvement of the BCR three step sequential extraction procedure prior to the certification of new sediment and soil reference materials. *J. Environ. Monit.* 1: 57

Steenari, B.-M.; Lindqvist, O. 1997. Stabilisation of biofuel ashes for recycling to forest soil. *Biomass and Bioenergy* 13(1): 39.

Wastes: Solutions, Treatments and Opportunities III – Vilarinho et al. (Eds)
© 2020 Taylor & Francis Group, London, ISBN 978-0-367-25777-4

Innovation policy: Introduction and consolidation of PAYT waste tariffs in Portugal

B. Bringsken
CERNAS, Polytechnic Institute of Coimbra, Coimbra, Portugal

R. Madureira
Research Unit on Governance, Competitiveness and Public Policies (GOVCOPP), DEGEIT, Aveiro University, Aveiro, Portugal

C. Vilarinho
Metrics of University of Minho, Guimarães, Portugal

C. Dias-Ferreira
Universidade Aberta, Lisboa, Portugal
CERNAS, Polytechnic Institute of Coimbra, Coimbra, Portugal

ABSTRACT: Pay-as-you-throw (PAYT) waste tariffs are at an early stage of development in Portugal, and policy has a role in shaping this potential innovation management tool towards a national consolidation. To understand the dynamic policy process on PAYT tariffs, this study first gained insight into the current dominant state, insofar as governance mechanisms and public-private municipal solid waste (MSW) sector. Subsequently, it analyzed how the policies promoting PAYT are aligned with four fundamental attributes for policy design under concepts of Innovation Policy. These attributes are: resource-intensiveness, political risk, state and private market interventions, and targeting. For the elected officials (decision-makers), PAYT as an economic instrument represents a long-term investment with political uncertainty. Therefore, policymakers need to develop robust economic regulations to improve transparency and complementary actions for a more integrated MSW sector, which creates a transitional space for change, destroying pre-existing conditions that are not aligned to the future state.

1 INTRODUCTION

Pay-as-you-throw (PAYT) waste tariffs are based on the polluter-pays principle in which citizens pay flexible tariffs according to the amount of mixed waste produced, which may encourage source separation of wastes. In opposition to the current water-indexed waste tariff, PAYT scheme is considered a tool to improve separation at source, an important innovation considering that Portugal is 12% behind the targets for 2020 of the EU Waste Framework Directives for reuse and recycling of waste (APA 2017). A national strategic plan (PERSU 2020), influenced by the EU Directives, reinforces the use of economic instruments, such as PAYT tariffs. However, a national policy needs to create proper conditions for innovation to be implemented at the local level.

The overall objective of this study is to re-examine the national strategies related to the PAYT system in Portugal through the lens of four attributes of policy design to determine if the national strategies consider the conditions for change required for a transitional space and diffusion of the system. First, this article presents a snapshot to outline the current dominant state of the municipal solid waste (MSW) system; Then it uses four valuation criteria based on concepts of innovation policies to understand how the strategies and policies for the MSW system effectively create a transitional space by destroying or reinforcing the current state for PAYT schemes in Portugal.

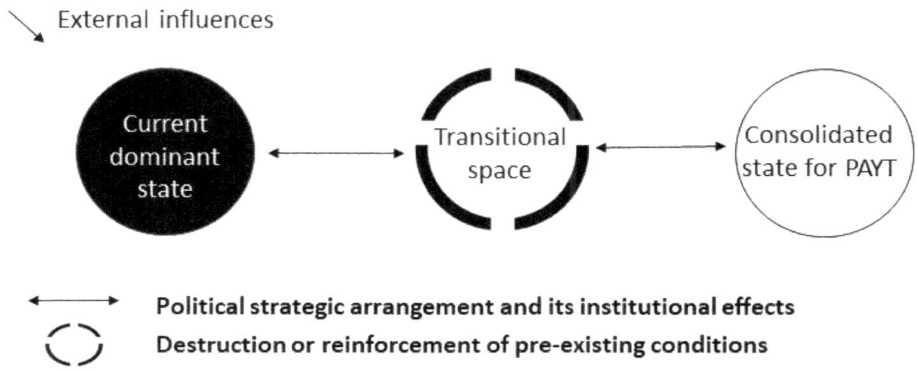

Figure 1. Dynamic structure of the policy process on PAYT tariffs.

One of the regulatory external influence of waste management practices in Portugal was the Waste Framework Directive developed by the European Commission, towards which Portugal harmonized its laws and targets. The current dominant state of the Portuguese MSW system establishes mechanisms of governance to open a space for change. This dynamic political structure (Fig.1) is continuously balancing stable cohesion with space for flexibility in a learning process, defined as transitional space. Therefore, it is important to understand the framework conditions of the regulations and governance mechanisms paving the way to an efficient introduction in the short-term and future consolidation of PAYT schemes in Portugal in the long-term.

This section also presents a brief theoretical understanding of Innovation Policy concepts to be explained as reference for the further policy process analysis. A policy is how a government can intervene in a society (Schneider & Ingram 1990) to enforce a change that could solve what politicians perceive as problematic situation. In addition, innovation brings 'problem-solving' ideas into society. Apart, the words policy and innovation are old phenomena, but when placed together 'innovation policy' constitute a term that became frequent at the end of the eighties as a new policy area and approach to analyze how innovation and policies relate (Weber & Truffer, 2017). Roy Rothwell (1982) a pioneer academic that influenced the terminology, characterized innovation policy as a 'fusion' of previous policy instruments under specific labels. The state interventions are important to enforce or promote solutions, preventing the lack of innovation. According to Van der Doelen (1989), instruments for the design and implementation of policy encompasses three classifications of tools: 1) Legal as regulations, licenses, permits, etc; 2) Economic as grants, subsidies, taxes and user charges; and 3) Communicative as public information, propaganda, cluster resolutions, etc. An innovation policy process is not only a technical change, but it transforms political arrangements and social structures, with a dynamic impact to create effective forms of governance. The aim of an innovation process should be learning to build the path towards the most promising effective solution, but it is not easy to evaluate its effects, because the instruments interact dynamically with multiple effects (Edler *et al.* 2012).

2 A SNAPSHOT OF THE CURRENT DOMINANT STATE OF THE MSW SYSTEM

The main actor and policymaker in this setting is the Portuguese Environmental Agency, known as APA. Besides this authority, the Portuguese government created an entity for water and waste regulation, known as ERSAR, to regulate and develop several reforms in the waste sector. Moreover, the municipal government has the responsibility to manage the waste from users producing less than 1100 litters of waste per day, but the operation is divided among the public-private MSW management system. There are several operators and service classifications, but the sector is composed by 23 Urban Waste Management Systems (SGRU) segregated within these two groups:

1. Municipal (3 associations) or Intermunicipal systems (8 indirect public management);
2. Multi-municipal systems (12 concessionaires: 1 BRAVAL + 11 multi-municipals with public-private partners joining the company *Empresa Geral do Fomento* EGF).

The waste collection and transport are usually not integrated with treatment and disposal, apart from a few municipalities and SGRU's operators as exceptions. Those non-corresponding activities are in most cases provided by different operators. The activities carried out by the operators or entities can be distinguished as: 1) "Upstream" system – covers final waste treatment and disposal, managed by intermunicipal or multi-municipal groups; 2) "Downstream" system – covers refuse collection and transportation. However, the mixed and the source-separated collection are in most cases part of different systems, as the source-separated collection is commonly under the responsibility of the 'waste treatment operator' responsible for the upstream system. As the role of an innovation policy is to promote efficient solutions, the Portuguese MSW sector still could improve the articulation between source-separated and mixed collections (horizontal integration), or upstream with downstream systems (vertical integration).

Regarding the introduction of PAYT schemes, the national strategic plans (PERSU I, II and 2020) for the MSW sector were influenced by the European policies, as chronologically presented here:

1997 – PERSU I: established a plan with initial support for source-separated waste collection and closed dump sites.
2006 – Waste Framework Directive 2006/12/EC from the European Parliament and Council reinforces the 'polluter-pays principle'.
2007 – PERSU II: promotes the 'polluter-pays principle', PAYT, separation at source, new financial model for tariffs and determines public information about waste tariffs and service.
2008 – EU Directive 2008/98/CE reinforces the concept of 'extended producer responsibility' and establishes an obligation for all the member-states to develop a National Waste Management Plan, in Portugal known as the PNGR 2014-2020.
2014 – PERSU 2020: confirms PAYT as an economic instrument and innovation tool; transposed the EU targets but presented few practical measures and actions.
2018 – EU action-plan '2018 Circular Economy Package' is an influence on the next Strategic Plan PERSU2020+.

The PAYT system needs a "*pertinent legislative foundation embedded into a conducive policy framework in which polluters responsibility, waste avoidance and material recycling take the priority*" (Reichenbach *et al.* 2004). Although, PERSU II and PERSU 2020 reinforces the relevance of the polluter-pays principle, it is the regulatory entity ERSAR that has been promoting economic instruments for waste tariffs by developing methods and technical measures for PAYT tariffs. The strategic plans elaborated by APA transcribed more principles than technically strong solutions, actions and specific components for PAYT implementation. Given that policies on PAYT schemes are complex, clear guidance towards public participation, robust legal resources and technical specific methodologies are required. During the public consultation process for establishing PERSU2020+, the policymakers received a critique published on a public letter from two important associations of the waste management sector: ESGRA and AVALER. These associations declared that the SGRU's operators are investing and fulfilling the regulations regarding the low tariffs reflecting on low investments for waste valorization. As they declared: '*if the sector is going through a financial crisis, the ones to be blamed are the policymakers*' (ESGRA & AVALER 2019). According to innovation policy concepts, policymakers need to promote a dialogue and construct relations that are dynamic (Dubina 2013), but it requires the development of appropriate coordination tools with all actors of influence (Kuhlmann & Rip 2014).

3 APPLYING THE FOUR VALUATION CRITERIA TO PAYT IMPLEMENTATION

This section presents tools attributed to innovation policy concepts in order to conduct an analysis of the regulatory framework of PAYT schemes in Portugal. In the book "Designing Government: From Instruments to Governance" (Eliadis *et al.* 2005), the authors considered four valuation criteria developed by Linder and Peters (1989) as fundamental for policy design. These four valuation criteria are related to legal, economic and communication tools, but the new approach is considered by the beforementioned scholars as innovative for the multidimensional comparisons, offering a diversity of reasons and decisions regarding public policies. These four criteria are instruments determining the fundamental attributes of the policy in order to meet objectives: Resource-intensiveness, Political risk, Financial and ideological constraints and Targeting. Therefore, these criteria will be contextualized in the next four sections and at the end of this section the results will be presented in a table according to the dynamic structure of the policy process.

3.1 *Resource-intensiveness: operating costs*

This criterion outlines regulations on the MSW operating costs, which are traditional challenges faced by local authorities. It is hard to determine the role of policies when often "*financing of MSW necessary infrastructure is a sensitive issue, where local conditions and preferences vary*" (Reichenbach *et al.* 2004). In general lines, the strategic plan PERSU II established that municipalities should guarantee the services and the financial sustainability of the waste systems. However, the tariffs in Portugal are still not fully reflecting the 'principles of full cost recovery', but the PAYT schemes could play a corrective role with a fair tariff model, as ERSAR recommends. Economic regulation is required for the waste sector, for it has been a challenge for the sector to cover costs when there is such a lack of integration between costs and tariffs. ERSAR as a regulatory agency elaborated recommendations in 2010 and 2014 for a new tariff regulatory model, but without strong enforcing measures. A study about the water-indexed waste tariff pointed that using 10 m^3 of water per month for one year represents in Portugal a cost ranging from 7,32 € (in Castanheira de Pera) to 122,04 € (in Póvoa de Varzim) (DECO 2017). Most municipalities charge waste in proportion to the water bills without great complaints, since the values practiced may not be reflecting the real costs. In 2005/6 only about 23% of the costs incurred by the municipality with the provision of the MSW management service were charged to the citizens (CESUR 2008). On the other hand, other municipalities may be charging tariffs above the recommendation to increase its revenues. During the public consultation process on the strategic plan PERSU 2020+, the two associations representing the SGRUs declared that the applied tariff systems were unsustainable (ESGRA and AVALER, 2019). They suggested to apply the polluter-pays principle and to establish a national calculation method that reflects the actual production of waste.

3.2 *Political risk: public visibility and impact on voters*

This criterion addresses how policymakers understand the perspective of municipal elected officials in order to reduce the risk of public opposition. In this case, decision-makers and citizens need to gain knowledge about PAYT schemes, but the lack of information can be a constraining factor to innovation. Communicating fairness is a demand, as citizens collaborate when they associate PAYT with fair charging schemes (Reichenbach *et al.* 2004). The waste sector has been failing to communicate transparently the MSW costs and the water-indexed waste tariffs. There is insufficient information available on the websites of management entities regarding the conditions of service provided and tariffs (ERSAR 2013). The political risks are even greater if the MSW tariffs used to be perceived as "free" or "already paid by taxes". In this case PAYT tariffs could have negative effects. "Hidden" water-indexed waste tariff prevents voters to complain against local taxes or any financial burden, but it is not fair for those sorting waste, since they may use more water. However, elected officials need to communicate the possibility to reduce costs with PAYT. Researchers from the European Commission examined public support for environmental taxes, and the results showed that perceived fairness influences the outcome. The same study found out that public support for PAYT was about 10% higher and with less resistance among respondents that experienced the system. This data gives more confidence for elected official to face initial opposition, knowing that residents tend to accept it when given the proper conditions (Brown 2014). According to Slavik et al. (2013) in the Czech Republic, the results of PAYT implementation showed effective changes in cities where technicians and politicians were involved with the population, as in cities where environmental awareness was a concern.

3.3 *Financial and ideological constraints: the role of state and private market interventions*

This criterion considers the political culture as an influence to determine the consistency of policies and interventions. This is necessary to discern the course of action between the state intervention or the private market unseen forces. However, a lack of leadership on interventions can have as consequence a lack of knowledge and transparency. A study supporting PAYT implementation had a point of concern about the private sector, due to often "lack of information availability" and a concern that profits might not be re-directed to the fee system (BiPRO & CRI 2015). A. *Ideological constraints on transparency and trust*: this analysis refers to a contested state intervention because of the EGF's monopoly, since this public-private company can use public infrastructures to reduce costs; but the private waste sector has a short history in public services (Cruz et al. 2013). The company used to have a public majority of shares, but 51% of the shares were sold in 2014 to a multinational company; the municipalities composing the minority participation were against this negotiation (Silva 2014). Although, the state could intervene to enforce more responsibilities on EGF for monopolizing the market, this context should not be isolated, as obstacles to innovation are produced through inter-related processes (Kuruppu & Willie 2015).

B. *Financial constraints – integration of collections:* a lack of synergy between mixed and source-separated collection (apart from few exceptions) has been pointed out as a potential area for efficiency gains by sharing infrastructures and services. An evaluation carried out by ERSAR identified potential savings resulting from the integration of both collections (ERSAR 2017). The strategic plan PERSU2020 (measure 8.1 action 7.6) also proposes the integration, but this national policy was not effective to promote it. Policymakers decide if the state is competent to enforce this integration or leave it to the private market. The ongoing discussion on the PERSU2020+ intends to promote an economy of scale, but this topic becomes a concern with the EU Council decision on bio-waste collection after 2024, a target that requires business integration and effective operational coordination. PERSU2020+ presents a limited policy intervention for optimizing collections, as it does not examine the adequate investments and responsibilities (Praça 2019).

3.4 *Targeting: how precisely and selectively policy instruments target beneficiaries and costs.*

This criterion gives a proper attention to equality of opportunities. Part of the political challenge is to assure that tariffs and services meet the capacity of all end users. The beneficiaries include citizens with low-income, in large families or within special medical needs group (elderly, infants and handicapped), as even with PAYT tariffs the access to waste services should remain universal. Due to the current hidden waste tariffs, a social tariff has not become a concern for politicians, as citizens might not be aware of tariffs or application process for benefits. Therefore, the number of users requesting a social tariff is low. Concerning the costs, policy instruments should target at social participation to improve separation at source. Instead of that, the national strategy for the MSW sector prioritized high investment for optimization of the recovery and recycling infrastructures (APA 2017), which resulted in low material recovery rates. As PAYT system reduces mixed waste collection, it represents a saving for the municipality; while this reduction has a negative impact on revenues for the upstream concessionaires, while the costs for separated collection increases. A situation that requires clear regulations and public contracts for separate collection.

To conclude, table 1 correlates the four policy criteria to the 3 concepts of dimensional states:

Table 1. Political strategic arrangements and its effects.

	Current state	Transitional space	Consolidated state
1. Resource-intensiveness	Most municipalities still lack fully covered operation costs.	Enforcement of full cost recovery principles in national calculations methods.	Strong economic regulation integrating costs and tariffs.
2. Political risk	Decision-makers fear negative impact on voters; lack of transparent information about waste tariffs.	Politicians and citizens receive more information and experiences with PAYT.	Broader public communication of fairness in charging waste tariffs schemes.
3. Financial and ideological constraints	Presents a constraint on EGF's monopoly and lacks integration of collections.	The state sets a robust reporting system to monitor the private market and take leadership.	Policies aids to increase trust and transparency.
4. Targeting	Hidden tariffs prevent interest for social benefits; Policies did not target social participation and recovery rate was low.	Policymakers need to be supported by experts' decisions, methods, calculations and prospective scenarios.	Social participation, clear regulations and public contracts for separate collection are the target.

4 CONCLUSION

To understand the dynamic structure of the policy process on PAYT tariffs, a snapshot of the dominant state of the MSW sector was briefly discussed, followed by a framework of innovation policy's concepts to guide the analysis on the national strategies related to PAYT schemes in Portugal. The analysis indicates that, even though, the regulations transpose values and principles to promote a transition path to a new tariff system, these are not powerful enough to tackle contextual rooted problems, namely: integration and correct reflection of the service costs within waste tariffs (1), lack of public communication concerning tariffs paid due to fear of public opposition (2), poor integration of waste collection (3), and social participation to improve separation at source (4). The innovation policies need to be aligned to opportunities and specific financial and technical solutions in order to promote investment. However, these constraints make it harder for a municipality to innovate and to risk investments for the PAYT scheme, because elements of the old system tend to reinforce the current institutional alignment, while transparency and trust are the foundations for the long-term investment on PAYT systems. The previous strategic plans (PERSU I, II and 2020) included concerns on these issues and pointed out general lines to address complex problems of the MSW management sector. Hopefully, the next strategic plan will be able to tackle the root and structural problems of the sector by implementing robust economic regulations to facilitate the implementation of PAYT schemes in Portugal.

ACKNOWLEDGEMENTS

The authors acknowledge LIFE+ for funding the LIFE PAYT project (LIFE 15/ENV/PT/000609).

REFERENCES

APA. 2018. *Relatorio Anual de Resíduos Urbanos 2017*. Amadora, Agência Portuguesa do Ambiente, IP.

BiPRO/CRI. 2015. Assessment of separate collection schemes in the 28 capitals of the EU, Final report.

Brown, Z. S. & Johnstone, N. 2014. Better the devil you throw: Experience and support for pay-as-you-throw waste charges. *Environmental Science & Policy*. 38: 132–142.

CESUR 2008. Sistemas Tarifários de RSU em Portugal RSU. Levy, J. de Q. & Pinela, A. *A Política Ambiental na Fiscalidade sobre os Resíduos. Porto, 21 de Julho de 2008.* Available at: http://www.geota.pt/xFiles/scContentDeployer_pt/docs/articleFile164.pdf Accessed at 25 February 2019.
Cruz N.F., Simões P & Marques R. 2013. The hurdles of local governments with PPP contracts in the waste sector. *Environment and Planning C: Government and Policy*; 31(2): 292–307.
DECO, 2017. *Associação Portuguesa para a Defesa do Consumidor & Proteste Magazine, 28 June 2017.*
Dubina, I. & Carayannis, E. 2015. Potentials of game theory for analysis and improvement of innovation policy and practice in a dynamic socio-economic environment. *Journal of Innovation Economics & Management*, 18, (3): 165–183.
Eliadis, P., Hill, M. & Howlett, M. 2005. Designing Government: From Instruments to Governance. *Choice of Policy Instruments: Confronting the Deductive and the Interactive Approaches.* McGill-Queen's University Press; 1st edition :106–131.
Edler J., Berger M., Dinges M. & Gök A. 2012. The practice of evaluation in innovation policy in Europe, *Research Evaluation*, 21(3): 167–182.
ERSAR. 2013. Implementação do princípio do poluidor-pagador no setor dos resíduos. ERSAR nº 1/2013
ERSAR. 2017. Estudo de avaliação de sinergias. Available at: http://www.ersar.pt/pt/site-comunicacao/site-noticias/documents/relatóriopreliminar_sinergias.pdf Accessed at 1 March 2019.
ESGRA & AVALER, 2019. Contributos, Consulta Pública, Desafios do Setor dos Resíduos Urbanos no Pós 2020 Proposta do PERSU 2020+. *Lisboa, 25 de janeiro de 2019.* Available at: http://www.esgra.pt/wp-content/uploads/2019/02/20190125_Comentarios-à-proposta-de-PERSU2020final-VF.pdf. Accessed at 25 February 2019.
Kuhlmann, S. & Rip, A. 2014. The Challenge of Addressing Grand Challenges. *Report to the European Research and Innovation Area Board.*
Kuruppu N. & Willie R. 2014. Barriers to reducing climate enhanced disaster risks in least developed country-small islands through anticipatory adaptation. *American Geophysical Union, Fall Meet. 2015.*
Linder, S. & Peters, B. 1989. Instruments of Government: Perceptions and Contexts. *Journal of Public Policy, 9*(1), 35–58. doi:10.1017/S0143814X00007960
Praça, P. 2019. Desafios dos Resíduos Urbanos no Pós 2020. O PERSU 2020+. *Ambiente Online.*
Reichenbach, J. et al. 2004. Handbook on the implementation of Pay-As-You-Throw as a tool for urban waste management.
Schneider, A. L. & Ingram, H. 1990. The behavioral assumptions of policy tools. Journal of Politics, 52.
Slavik J, Pavel J. 2013. Do the variable charges really increase the effectiveness and economy of waste management? *A case study of the Czech Republic. Resources, Conservation and Recycling* 70: 68–77.
Van der Doelen, F.J.C. 1989. Beleidsinstrumenten en energiebesparing: de toepassing en effectiviteit van voorlichting en subsidies, energiebesparing in de industrie 1977 tot 1987. Universiteit Twente.
Weber, K. M, Truffer B. 2017. Moving innovation systems research to the next level: towards an integrative agenda, *Oxford Review of Economic Policy*, 33 (1) :101–121.

Wastes: Solutions, Treatments and Opportunities III – Vilarinho et al. (Eds)
© 2020 Taylor & Francis Group, London, ISBN 978-0-367-25777-4

Benefits of introducing door-to-door separate collection in rural areas

J. Vaz
Ecogestus, Lda., Figueira da Foz, Portugal

V. Sousa
Instituto Superior Técnico, Lisboa, Portugal

C. Dias-Ferreira
Universidade Aberta, Lisboa, Portugal
CERNAS, Instituto Politécnico de Coimbra, Coimbra, Portugal

ABSTRACT: There is a pre-conception that in rural isolated areas, the production of recyclable waste is reduced and door-to-door collection schemes are not worth setting-up. This work describes the implementation of a pilot test for door-to-door collection of dry recyclable waste at a small rural site. It aims at improving the knowledge about the waste generated in rural areas and the impact on source segregation habits arising from the implementation of door-to-door collection. The strategy comprised the introduction of individual waste containers and bags at household level and setting up a waste generation monitoring and physical composition evaluation programs. The recyclable waste more than tripled after implementation of door-to-door collection, from 23 kg/inhab/year to 73 kg/inhab/year. Concomitantly, residual waste decreased from 230 kg/inhab/year to 180 kg/inhab/year. The strategy followed at the test site represented a positive incentive for householders and services to source segregate and produce less residual waste.

1 INTRODUCTION

Historically, there has been a higher concern with waste management in large cities. This can be explained by various factors, namely the complexity of the waste management (e.g., size of the utility, resources consumed) and the consequences of service failures (e.g., financial, social, environmental), driven by amount and concentration of waste produced. Amongst the factors underlying the priority given to the waste management of large urban centers, there is also a generalized idea that the waste production in smaller communities is distinct both in terms of quantity and quality. It is believed that most people living in rural areas have a small yard, or land where they cultivate vegetables and grow small farm animals, which are fed with organic waste (food leftovers). This leads to the general perception that rural areas do not discard so much residual waste into bins and containers. At the same time, income levels in rural areas are lower, population is older which leads to reason that "recyclables" (paper/cardboard, metal/plastic and glass packaging) are not so prevalent as in more urban areas.

However, despite the progressive migration of population to large urban areas, there is still a large proportion of the population living in smaller communities. In Portugal, according to the last census (2011), roughly 40% of the population resided in communities of less than 2000 inhabitants (Fig. 1). Additionally, there is no clear trend regarding the amount and quality of waste amongst the 23 bulk waste utilities operating in mainland Portugal depending on their degree of urbanization (Fig. 2). In fact, the only distinctive case is ALGAR, which operates in the most touristic region of Portugal and also has a significant proportion of foreign residents. Therefore, finding adequate solutions in small communities is also relevant for the overall performance of the waste management at national

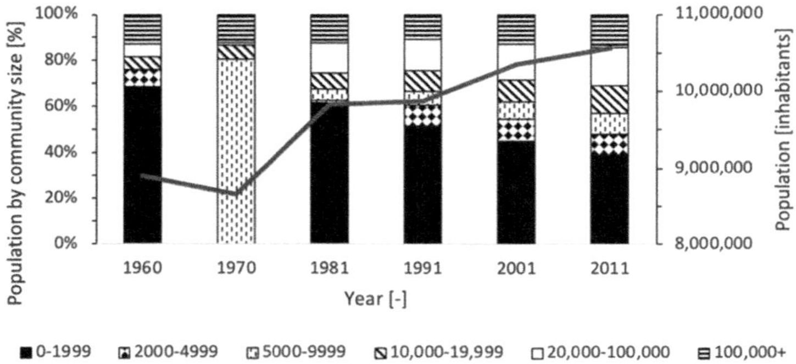

Figure 1. Evolution of the population and its distribution by community size in Portugal (data retrieved from PORDATA 2013). (Note: the information for the 1970 census is not available with the same level of detail of the remaining years).

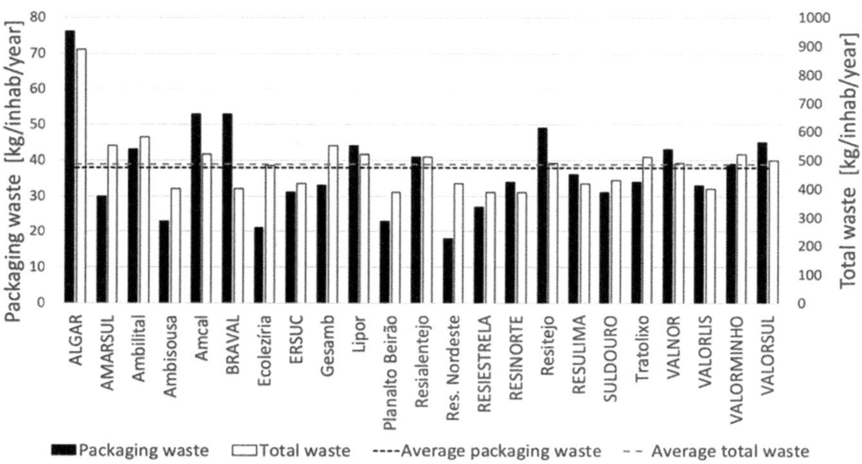

Figure 2. Packaging and total waste production in waste utilities operating in Portugal mainland in 2017 (data retrieved from APA 2018a).

scale. In this regard, this work aims at improving the knowledge about the waste generated in rural areas and the impact in the source segregation habits of implementing a door-to-door collection scheme for the dry recyclable waste fractions, comprising paper/cardboard, metal/plastic and glass packaging waste.

2 METHODOLOGY

2.1 *Study area*

ERSUC is one of the 23 bulk waste service utilities operating in mainland Portugal. It serves 36 municipalities in the center of Portugal through 7 transfer stations and 2 mechanical-biologic treatment units (TMB). Within the ERSUC area of operation, there are some differences regarding the amounts of waste produced, both mixed and separate (Table 1). The smaller municipalities produce less 16% mixed waste and less 10% separate waste, comparing to the large municipalities. The differences for the medium municipalities compared to the large ones are slightly inferior (less 12% and less 7% for mixed and separate waste, respectively).

Table 1. Mixed and separate waste capitation per municipality size (ERSUC 2018).

Municipality size	Average Population	Mixed waste (kg/inhab/year)	Separate waste (kg/inhab/year)
Small (inhab < 10,000) (n = 8)	5103	319	26
Medium (10,000 < inhab < 20,000) (n = 10)	14,058	336	27
Large (20,000 < inhab) (n = 18)	42,086	382	29

Figure 3. General view of the testing site, a rural small district located in the center of Portugal (left and center); bring-banks for dry recyclables located at the study site (right).

Figure 4. Waste bins to be distributed at the testing site (left); calendar and waste guide (center); plastic bags with the dry recyclable waste fractions put out for collection (right).

To evaluate the performance of a door-to-door collection scheme and evaluate the waste behaviour in small communities, two rural isolated small districts of the same characteristics were selected as study areas (Fig. 3). Site T is the test site and Site C serves as the control.

Both sites belong to the same municipality in the area of operation of ERSUC and have similar waste collection infrastructures: 9 collective curbside waste containers (900 L or 120 L) and 3 bring-banks for packaging waste (glass, paper, metal/plastic).

2.2 *Implementation of door-to-door collection for the dry recyclable waste fractions*

The test site has 128 inhabitants who live in 49 detached houses (44 with garden and 5 without garden) and 2 coffee/grocery shops. A set of containers for source segregation was distributed to every household and shop. The set comprises 3 bins of 30 L each: one for plastic/metal (yellow), one for paper/cardboard (blue) and one for glass (green). Bags of the same color were distributed, to be used together with the bins. The color code follows the one established in Portugal. In addition to the bins, a Waste Guide and a waste collection calendar were also distributed (fig. 4). The waste calendar clearly indicated the days and hours of the new door-to-door collection scheme.

Between 02.05.2017 and 27.06.2017, a door-to-door collection scheme was implemented in which the bags with the dry recyclable waste were placed outside by the householders at designated days and were collected by the municipal team. The amounts of residual waste and dry recyclables were monitored and the physical composition of the residual waste was determined.

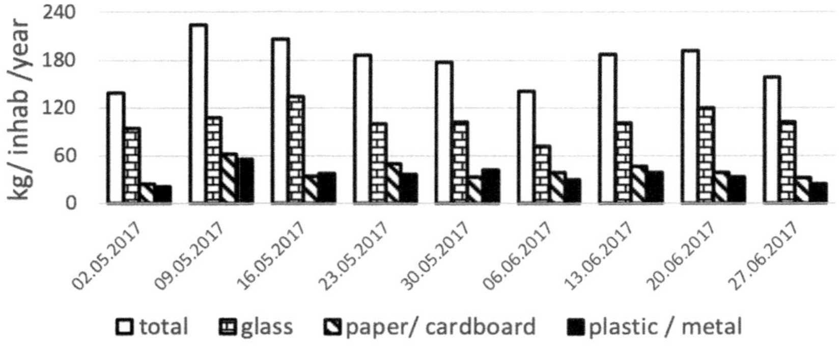

Figure 5. Amounts of recyclable waste (kg/inhab/year) collected door-to-door at the testing site.

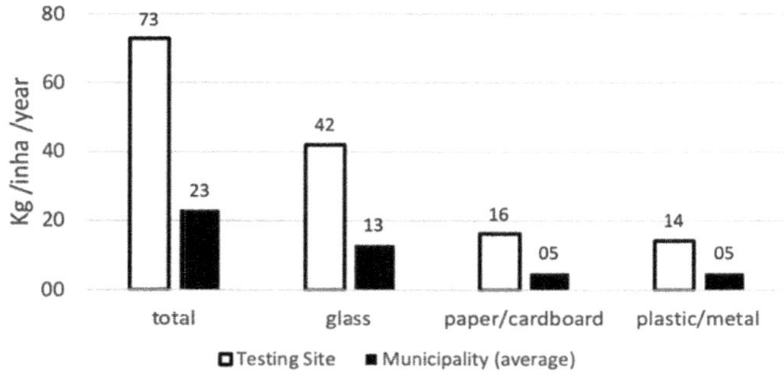

Figure 6. Annual amounts of recyclable waste (kg/inhab/year) estimated for door-to-door collection at the Site (128 inhabitants) and average values for the municipality (14,453 inhabitants).

3 RESULTS AND DISCUSSION

3.1 *Dry recyclable waste collection during the door-to-door trial*

The amounts of each waste fraction collected weekly are shown in figure 5. The recyclable waste collected amounted to 1610 kg, corresponding to 72.7 kg/inhab/year. Glass was the fraction collected in higher amounts, representing more than 50% of the total. The quantities of plastic and paper waste were similar.

Prior to the trial, the collection of dry recyclables was based on one bring-bank (Fig. 3, right) where the 3 fractions were accordingly sorted: paper/cardboard, plastic & metal packaging and glass. Even though 88% of the people had replied in the initial survey that they already source segregated their dry-recyclable waste using the bring-bank, the annual per capita values more than tripled after the implementation of door-to-door collection, from 23 kg/inhab/year to 73 kg/inhab/year (Fig. 6). Concomitantly, residual waste decreased from 230 kg/inhab/year (average for the municipality) down to 180 kg/inhab/year (at the test site). This means that the strategy followed at the test site, comprising the distribution of individual bins and the implementation of door-to-door collection services, represented a very positive stimulus for householders and services to source segregate recyclable waste, and resulted in less residual waste being disposed of.

3.2 *Composition of the unsorted waste*

Following the implementation of door-to-door collection of the dry-recyclable waste, the composition of residual waste changed, as expected (Fig. 7). The percentages of paper/cardboard in the

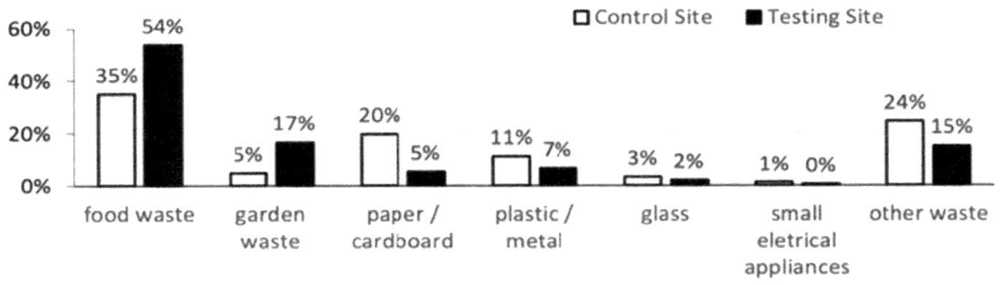

Figure 7. Physical composition of residual waste before (control site) and after (test site) the implementation of door-to-door.

Table 2. Presence of recyclable materials in residual waste.

Waste	Control site (rural site)	Urban Site*	Regional Average**	National Average***
Paper	19.7%	9%	13%	10.5%
Plastic/metal	11.2%	7%	12%	12.9%
Glass	3.3%	5%	5.4%	7.3%
Total	34.2%	21%	30.4%	30.7%

* retrieved from Dias-Ferreira (2019); ** from ERSUC (2014); *** from APA (2018b)

residual waste decreased 4 times, from 20% to 5%. The other fractions (glass and plastic/metal) also reduced, although not as significantly as paper did. Overall, the pilot test resulted in a reduction to 13% of dry recyclable materials in the residual waste, compared to 34.2% at the control site. The presence of plastic, paper, glass and metal in residual waste is unwanted because its recycling potential decreases due to difficulties in the recovery from residual waste with enough quality to allow its recycling.

The presence of paper, plastic, metal and glass in residual waste is presented in table 2 for the region, and country. Values found in the literature for an urban site located in the same region as the study site are also presented, highlighting that at the rural site the amount of packaging waste is similar to the regional and national averages.

As a result of the diversion of the dry recyclable waste, the percentages of biodegradable waste shifted from 40% to 70%. This effect increased the density of residual waste to 123 kg/m^3, meaning that the collection of the residual fraction is now more efficient (less volume, more mass transported by the collection vehicle).

3.3 *Potential of collection and capture rates of dry recyclable waste*

The estimated potential of recyclable waste at the testing site was 92.4 kg/inhab/year. This potential was calculated as the amount of recyclable waste present in the residual fraction (19.7 kg/inhab/year) plus the amount collected door-to-door (72.7 kg/inhab/year). The door-to-door collection strategy implemented at the study site was thus able to capture 79% of the estimated potential. Figure 8 represents the recyclable material in the residual waste in separately collected, according to the material. The capture rates achieved at the test site were 64% for paper, 66% for plastic/metal and 91% for glass.

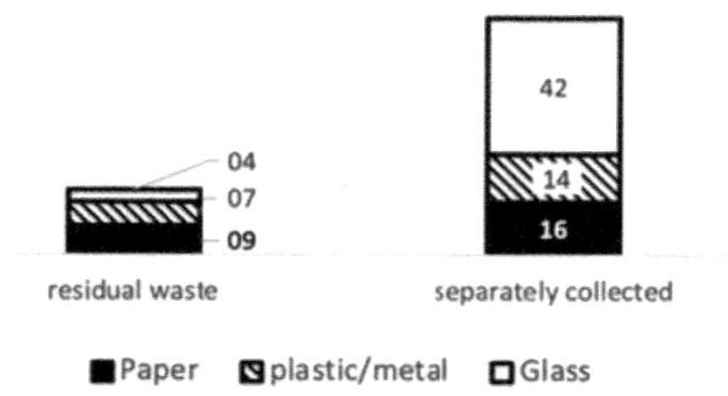

Figure 8. Annual amounts of recyclable waste (kg/inhab/year) in residual waste and in door-to-door collection at testing site.

3.4 *Obstacles to door-to-door collection of recyclable waste at rural locations: myth or reality?*

Door to door collection is perceived by municipalities as costly and very demanding. This arises from the day-to-day attitude to waste collection where the most important thing is to service the population and commerce, without regarding efficiency and adequate management indicators. At the same time, municipalities have no direct benefit in collecting recyclables, as no financial reward is available for improved recycling other than the indirect benefit arising from the reduced collection of residual waste. There is also no experience in Portugal, or sufficient studies, of door-to-door collection in rural areas, thus leaving decision-makers unease with a risky investment in such a waste collection method. Political risk has been highlighted by Bringsken et al. (2019) as one of the constraints in the setting of innovative solutions in the waste sector. The same authors refer that to overcome this problem more communication is required, and both decision-makers and voters need to gain knowledge about the advantages of the new system.

A very strong constraint to door-to-door collection is the present high collection frequency, twice per week, for residual waste, leading to high resource use (vehicles; fuel; human resources) and lack of capacity to invest those same resources in door to door collection. Reducing residual waste collection frequency is assumed as having a negative effect on population satisfaction, hence the political risk previously referred hindering the introduction of new collection systems. The door-to-door collection was discontinued after the study mainly because there was no financial incentive for the municipality to increase collection of recyclable materials, and decision makers are not supportive of changing something that is "working well enough" meeting inhabitants expectations i.e. "waste out of sight".

4 CONCLUSION

This work describes the implementation of a 2-month pilot test of door-to-door collection of recyclable waste (dry fraction) at a small, rural site. The strategy tested comprised the introduction of small, individual containers and bags at the household level, articulated with a once a week door-to-door collection of recyclable waste. The main conclusions are:

- The annual *per capita* collection more than tripled, from 23 to 73 kg/inhab/year after door-to-door collection was implemented.
- An overall 79% capture rate for dry recyclable waste was achieved, with material specific capture rates being: 64% for paper, 66% for plastic/metal and 91% for glass.
- The percentage of dry recyclable materials in waste materials at the rural site is similar to the regional average and to the national values.
- Biodegradable waste is 70% of the residual waste at the test site; this high value is an effect of the diversion of packaging waste materials to the door to door collection scheme.
- The specific weight of residual waste reaches 123 kg/m^3, allowing improved efficiencies in the collection.

ACKNOWLEDGEMENTS

The authors would like to thank J. Figueiredo and N. Oliveira, from the Penacova Municipality. This work was carried out in the framework of project, funded by POSEUR – European Union; Portuguese Government.

REFERENCES

APA, 2018a. Resíduos Urbanos. Relatório Anual 2017. Agência Portuguesa do Ambiente.

APA. 2018b. Desafios do setor dos resíduos urbanos no pós-2020. Proposta do PERSU2020+. Accessed 15.03.2019 at www.apambiente.pt/_zdata/DESTAQUES/2018/ApresentacaoPERSU2020/ PERSU2020_LNEC_A2.pdf

Bringsken, B., Madureira, R., Vilarinho, C. & Dias-Ferreira, C. 2019. Innovation policy: introduction and consolidation of PAYT waste tariffs in Portugal (this issue).

Dias-Ferreira, C., Neves, A., Braña, A. 2019. The setting up of a pilot scale Pay-as-you-throw waste tariff in Aveiro, Portugal. *WIT Transactions on Ecology and the Environment* 31:149-157 (DOI: 10.2495/WM180141).

ERSUC. 2014. Composição dos Resíduos, 2013.

ERSUC 2018. Relatório de Contas 2017.

PORDATA 2013. População residente segundo os Censos: total e por dimensão dos lugares. Acedido a 01.2019 em: https://www.pordata. pt/Municipios/População+residente+segundo+os+Cen sos+total+e+por+dimensão+dos+lugares-24

Wastes: Solutions, Treatments and Opportunities III – Vilarinho et al. (Eds)
© 2020 Taylor & Francis Group, London, ISBN 978-0-367-25777-4

Data communications solutions overview for smart cities waste management

R.C. Madureira
Research Unit on Governance, Competitiveness and Public Policies (GOVCOPP), Department of Economics, Management, Industrial Engineering and Tourism (DEGEIT), Universidade de Aveiro (UA), Aveiro, Portugal

C. Dias-Ferreira
Universidade Aberta, Lisboa, Portugal
Research Center for Natural Resources, Environment and Society (CERNAS), Polytechnic Institute of Coimbra, Coimbra, Portugal
Aveiro Institute of Materials (CICECO), University of Aveiro, Aveiro, Portugal

ABSTRACT: The current works describes the data transport networks within a smart waste management system. Existing technical solutions based on cellular networks and Internet-of-things (IoT) are described and the system's technological limitations are discussed from the smart city perspective. A comparison is made using as relevant parameters the rate of transmission (kilobits per second), the energy consumption, the coverage area and the costs. While cellular networks have higher rates of transmission and use pre-existing telecommunication infrastructures, the monthly cost can easily reach prohibitive values and prevent the use of cellular communications in smart waste systems, such as those used in pay-as you-throw (PAYT). Solutions based on IoT networks (LoraWAN and SigFox) might partially overcome communication costs, and seem especially promising, especially considering the emerging 5G technology that is expected to trigger the widespread development of IoT solutions for smart waste.

1 INTRODUCTION

Mega cities and overcrowded regions are increasing all around the world and therefore increasing the use of the natural resources and the outcome product of the consumer society. The theme of smart and sustainable cities is less a trend and more a mainstream issue mainly in order to deal with these new realities. The management of urban waste is one of the problems, especially in what concerns three main demands: the increasing amounts of waste, its environmental impact but also the waste management cost's transparency, accuracy and fairness. Smart cities tend to adopt smart ways to deal with difficult problems derived from overcrowded regions and inappropriate citizen's behavior. Therefore, there is a need to identify at the source the user and to measure the amount of waste individually discarded. In PAYT (Pay-as-you-throw) systems the main goal is to charge waste producers the amount of waste they discard, instead of constructing waste invoices using other indirect parameters, such as water consumption (widespread practice in Portugal), area of the building (widespread in Greece), number of elements in the family, etc). With PAYT users feel the fairness and the motivation to adopt behaviors that lead to the reducing of unsorted waste, most of which would end up in landfills and incinerators.

The technological approaches to implement smart waste management systems – from the waste deposition by the user to the invoice production – can be divided into three subsystems (Madureira & Dias-Ferreira, 2019): the first block, closer to the consumer is tag & sensoring providing the identification and quantification of the user and/or respective produced waste (Fig. 1.1); the second module is the transport network that will ensure that the collected information will be delivered to

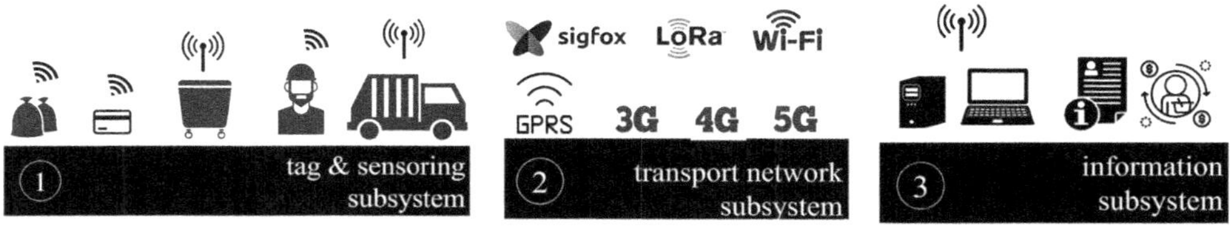

Figure 1. Smart waste management system: 1) tag & sensoring subsystem, 2) data transport subsystem and 3) information subsystem.

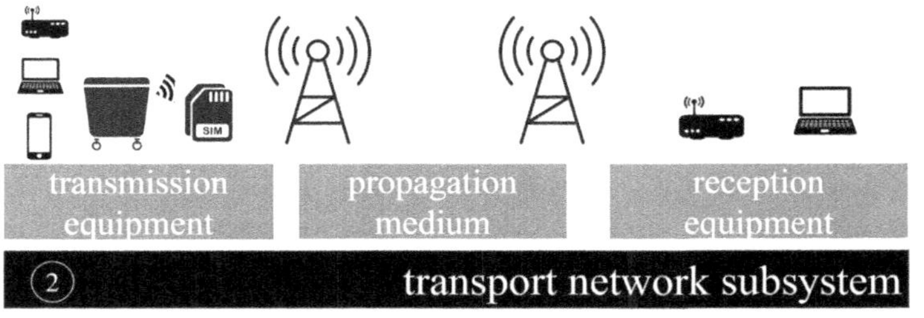

Figure 2. Transport newtwork subsystem for data transport from the waste deposition to the cloud.

a central platform (Fig. 1.2); the third block is the central waste information management platform (cloud-based) where data is analyzed and processed to calculate for instance, the individual monthly waste invoice (Fig. 1.3).

Previously, Madureira & Dias-Ferreira (2019) have described different technological solutions related to the first module, tag & sensoring, and detailed how user identification and measurement of individual waste amounts can be implemented, showing real examples from around the globe. The current work focuses on the second module, the data transport network. The main purpose is to give an overview of technological solutions available, discuss technological limitations and challenges and make comparisons in order to help technicians but also political decision-makers to define the best solution for each city.

2 DATA TRANSPORT NETWORK SUBSYSTEM

After data acquisition, no matter the adopted system, it is necessary to send the user identification and the waste measurement per user to the cloud platform for invoice production. The system in-between is the data transport subsystem and is based on a telecommunication network that is either already installed or can be installed specifically for this purpose.

A transport network includes a transmission system (with some electronic hardware and a transmission antenna), a propagation medium (the air or cables) and an equivalent reception equipment (Fig. 2). The main parameters that characterize communication networks are based on the quantity of data delivered per second (bit rate), the coverage range and the energy consumption of the transmission equipment.

In waste management systems, the transmission equipment can be located in the container or in the waste truck and they are a complement of the reading elements of the first module (tag & sensoring). This means that to the limit, each container needs to have a transmission equipment coupled to it. The reception equipment will be on the other side of the system, where an information system platform will receive the collected bits (Fig. 1.3). Because the data is communicated between two electronic devices without any human intervention it is called *machine-to-machine* (M2M).

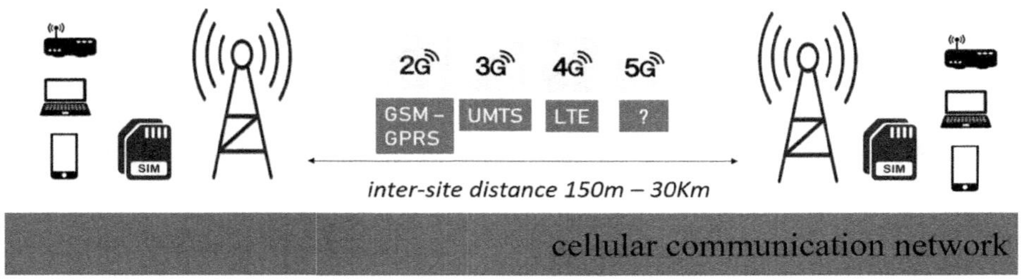

Figure 3. Cellular mobile communications network diagram.

Other M2M communication examples are traffic control, security systems, vending machines or healthcare.

Another specificity to take into account in the waste management scenario is the low amount of information to be transmitted as well the fact that there is no need to transmit the data in real-time (except for systems based in real time video image surveillance). These requirements are less demanding from the communications network point of view and open the opportunity to several solutions.

In addition, waste containers can be spread randomly in wide areas among a region and for that reason wireless communications – that means communications without cables – are also the obvious solution to the detriment of wireline solutions. Wireless communications uses electromagnetic (radio) signals to transmit data over the air. Even though the basis of wireless network are the same considering the conceptual functioning and infrastructure, there are several differences that justify grouping them into families. The two main relevant wireless family groups are described in the next subsections, and comprise the cellular (sometimes called mobile, associating the idea of mobile service) networks and the internet of things (IoT) communications.

2.1 *Cellular networks*

Historically, since the digital era in the late 1980's, the first machine to machine solutions for waste management were based on a GSM network (global system for mobile communication, also known as the second generation, 2G). GSM networks (and subsequent generations) are a cellular type network with several receiving antennas+equipment (base stations) organized in a cellular grid with pre-defined inter-site distance depending on the capacity and coverage operator's objectives (Fig. 3).

On early days, in Europe, 2G systems were used to transport data at a 9.6 kbps (kilobits per second). These 2G systems quickly evolved to a GSM-GPRS (general packet radio service) system that used the same infrastructure as 2G systems but which allowed data transmissions up to 21.4 kbps (3GPP 1997).

Because these systems were developed mainly for voice transmission, they impose a large amount of energy consumption in the transmission device (the mobile phone or similar). This happens even when using it for machine to machine communications, that need much less energy. As a consequence, the transmission device should be recharged each 4–5 days, which is a problem in the waste sector, especially if multiple transmission devices are in use (e.g. one in each waste container) and they are widespread across a municipality or a larger region.

In 2001 with the advent of the massive internet usage and the need of bigger data requests, new commercial systems emerged and a third generation (3G) of communication networks appeared, named UMTS – universal mobile telecommunications system allowing 384 kbps (3GPP 1996). Further improvements in UMTS allowed transmission rates up to 84 Mbps (3GPP 2015). The drawback of the increasing bit rate is the consequent increase in energy consumption and therefore the time between recharges of the reception devices is reduced to several hours.

Nowadays commercial mobile networks evolved to the 4rd generation, named LTE – long term evolution (3GPP 2004). This evolution was motivated by the reduced cost per bit, the increased service provisioning – more services at lower cost with better user experience, flexibility of use of existing and new frequency bands, simplified architecture, open interfaces, allow for reasonable terminal power consumption and reaches the theoretically data download of 300 Mbps in recent versions of LTE Advanced (3GPP 2009). Reaching these bit rates depends on the operator's choice when they buy the spectrum from regulators. The broadband is quite expensive and most commercial operators offer around 12-30 Mbps that is enough for current daily life applications, so far. Once again, high bit rate solutions allow bigger amount of data transmissions per second, but they also demand huge amount of energy, leading to high operational and maintenance expenses during the mobile equipment lifetime.

The adoption of cellular networks as an option for a communication network for PAYT implies the usage of a transmitter (transmission equipment) with a sim card per container. This means that the PAYT manager will not have to invest in a dedicated network but the existing commercial network will be used instead, reducing the capital expenditure (CAPEX) to the equipment (technically similar to a mobile) per container. However, this solution implies as operational expenses (OPEX) a monthly fee per container, plus the guarantee of battery maintenance during its lifetime. Despite the inherent implementation and maintenance costs, this solution is commonly used in PAYT systems from Malasya (Arebey et al. 2011) to Portugal (Dias-Ferreira et al. 2019).

2.2 *Internet of things (IoT) networks*

Waste management systems, as well as other telemetric systems, typically do not need large bit rates. This is because the data that will be transmitted is only the user identification code and the respective weight/volume measurement. This information can be condensed in some hundreds of bits which means that the required bit rate will be lower than 1 Kbps.

Based on this type of needs, innovative wireless solutions have emerged on the market. Those are called internet of things (IoT) networks. Some of the most well-known are low-power wide area networking, LoraWAN and Sigfox. Both of them use what is called narrow band communications and due to the low bit rate combined with a specific signal characteristic called spread spectrum these systems allow significant large coverage using a single base station. On top of this, the low bit rate approach also significantly reduces the energy consumption of the transmission system and as such the battery lifetime is extended to several years.

LoraWan is an open source technology originally developed by the French company Cycleo and later acquired by Semtech. Today more than 500 companies belong to the nonprofit LoraWan Alliance (set up in 2015) and operate this type of low power-low bit rate- wide range communication network to develop IoT applications.

When this solution is chosen for PAYT implementation, the network has usually to be built from scratch. This means that, depending on the area range, one or more LoraWan sites have to be built. This implies setting up one or more towers, or to reduce CAPEX, using high buildings. Then the equipment for each container has to be acquired, but at much smaller cost then for a cellular solution. The OPEX expenses involve battery in the containers and LoraWan site maintenance.

Another option is Sigfox, a proprietary French technology that was developed in 2009. Sigfox uses Ultra Narrow Band (UNB) radio technology and operates in the unlicensed bands (ISM). Radio messages handled by the Sigfox network are small (12-bytes payload in uplink, 8 bytes in downlink) thanks to lightweight protocol (Sigfox 2009). In order to become a Sigfox operator it is necessary to pay an annual subscription. Several countries in the world, mainly in Europe have already Sigfox operators. According to the company site, Portugal has already a 100% full coverage. The implementation of this solution will imply a low budget CAPEX since the network is already built. On the other hand, and as the cellular network solution, the operational cost will be higher than LoraWan solution because an annual subscription fee is required, the value of which may vary depending upon the negotiated contract.

Table 1. Main indicators of communications networks for smart waste management systems.

	Communications network	max bit rate	battery lifetime	coverage range	capital costs	operational costs
Cellular	2G-GPRS	21 bps	++	++	low	high
	3G-HSDPA	20 Mbps	+	+	(1 sim card	(monthly
	4G LTE Advanced	300 Mbps	–	+	per container)	fee)
IoT	LoraWAN	50 kbps	++++	++++	high (network infrastructure)	low (no monthly fee)
	SigFox	600 bps	++++	++++	low (only individual transmitters)	medium (annual fee)

3 COMPARISON OF ALTERNATIVES

A summary comparison of the solutions presented in the previous section is presented in Table 1. The main indicators considered relevant and used in the comparison are: the bit rate, the energy consumption, the coverage area and the costs.

4 CONCLUSIONS

In this work the data transport network within a smart waste management system was described and existing technical solutions were analyzed and compared.

The two main groups of technology for data telecommunication are both cellular networks and Internet-of-things communications. While cellular networks allow transmission of considerably more data per second and uses existing telecommunication infrastructures (thus reducing the investment costs of setting up a dedicated network) they require a SIM card in every container and involve the payment of a monthly fee per container, which, depending on the number of containers within a city, can easily reach prohibitive monthly values for the waste management entity.

On the other hand, technological solutions based on the internet-of-things comprise either LoraWan or Sigfox networks. These have advantages compared to the cellular networks related to the extended battery lifetime, but can represent higher investment costs, especially LoraWan, because a dedicated communication infrastructure (antennas) needs to be set up from scratch. In opposite, SigFox will imply a low investment cost (because the network is already built in most European countries), but the payment of an annual fee is required.

In the future it is expected that the data transport network will rapidly evolve to 5G systems, an emerging technology that covers three main developments: increasing the bit rate, reducing the latency and massive widespread of IoT. This massive development of IoT expected with 5G introduction will definitely contribute to the development of cheaper data communication solutions that can be used in the context of smart waste management systems, such as those required in pay-as-you-throw, allowing to overcome current obstacles in this domain.

REFERENCES

3GPP. 1996. *Universal Mobile Telecommunications System, 3GPP Technical Specifications 25.101 (FDD) and 25.102 (TDD)*

3GPP. 1997. *GPRS (Release 97) and EDGE (Release 98)*

3GPP. 2004. *Long Term Evolution of the 3GPP radio technology*

3GPP. 2009. *LTE towards LTE-Advanced – LTE Release10.*

3GPP. 2015. *High Speed Downlink Packet Access (HSDPA); Overall description.*

Arebey, M. & Hannan, M.A. 2011. Integrated technologies for solid waste bin monitoring system. *Environmental Monitoring and Assessment* 177(1–4): 399–408
Dias-Ferreira, C., Neves, A. & Braña, A. 2019. The setting up of a pilot scale Pay-as-you-throw waste tariff in Aveiro, Portugal. WIT Transactions on Ecology and the Environment 31: 149-157
Pires, J.S. 2013. *Implementação do princípio do poluidor-pagador no setor dos resíduos.* Entidade Reguladora dos Serviços de Águas e Resíduos, Departamento de Engenharia – Resíduos Departamento de Estudos e Projetos. ERSAR.
Miranda, M.L. & Ealdy, J. 1998. Unit pricing of residential municipal solid waste: lessons from nine case study communities. *Journal of Environmental Management* 52(1): 79–93.

Wastes: Solutions, Treatments and Opportunities III – Vilarinho et al. (Eds)
© 2020 Taylor & Francis Group, London, ISBN 978-0-367-25777-4

Phosphorus flows in the Portuguese agriculture and livestock sectors

J.L. Rocha
Polytechnic Institute of Coimbra & CERNAS, College of Agriculture, Coimbra, Portugal

V. Oliveira
Polytechnic Institute of Coimbra & CERNAS, College of Agriculture, Coimbra, Portugal
CICECO, University of Aveiro, Aveiro, Portugal

C. Dias-Ferreira
Polytechnic Institute of Coimbra & CERNAS, College of Agriculture, Coimbra, Portugal
Universidade Aberta, Lisboa, Portugal

ABSTRACT: Phosphorus plays a vital role as a limiting nutrient for plant growth, but the majority of minable phosphate rock reserves are located in just a small handful of countries: South Africa, Jordan and Morocco. Therefore, Portugal is totally dependent on imports, with phosphate rock being added to the European Union list of critical raw materials it is essential to increase its sustainable use. Thus, the main objective of this work was to compute substance flow analysis of phosphorus in the agriculture and livestock production sectors in Portugal. The data was gathered from several statistical sources and computed into phosphorus contents employing mass conservation law. Phosphorus flows were computed and quantified using STAN software. As result, we identified phosphorus flows, losses, and sinks. We can conclude that, both in agriculture and livestock production sectors, it is possible to decrease losses and increase phosphorus efficiency use by applying best management practices.

1 INTRODUCTION

Phosphorus (P) is an essential element for the support of life on Earth; it forms part of deoxyribonucleic (DNA), adenosine triphosphate (ATP), teeth, and bones. To maintain the high yields of modern agriculture it is necessary to supply this nutrient to the soil. This is done by applying high amounts of mineral fertilizers, which contain P derived from phosphate rock. As a result of phosphate rock being included in the critical raw materials list of the European Union, there is growing interest in P scarcity and food security.

Previous studies show that phosphate rock reserves could be depleted within 100 years (Smit et al. 2009, Cordell 2010) and that the peak of P could take place as early as 2033 (Cordell et al. 2009). Regardless of timelines, it is indisputable that most of phosphate rock reserves are held by only a handful of countries, mainly Morocco, which controls approximately 75% of the world's reserves according to the latest United States Geological Survey estimates (Jasinki 2017). The reserves of most producing countries, such as China and USA will be depleted in the coming decades (Cooper et al. 2011). Additionally, increasing demand, lower quality of phosphate rock and the rising of production costs will drive up the future P price, consequently having a strong repercussion for food costs (Childers et al. 2011, Metson et al. 2016).

On the other hand, P is also a water pollutant. While fertilizers increase agricultural production, they simultaneous increase the load of nutrients in water bodies causing eutrophication: a phenomenon known for causing algae bloom, which depletes the dissolved oxygen in water and can lead to the death of a whole ecosystem. P can enter water bodies through point sources (effluents from wastewater treatment plants) and through non-point sources (soil erosion from agriculture or surface runoff due to the application of mineral fertilizers and manure) (Linderholm et al. 2012). Based in the Water Framework Directive, the European Union demands both the surface and groundwater bodies to reach a 'good chemical and ecological status', which can be achieved by reducing point and diffuse emissions of P.

A crucial starting point for developing a sustainable P governance system is to know where phosphorus is and how much P is transferred across different environmental compartments. Substance Flow Analysis (SFA) is a methodological approach that allows to determine the flows and stocks of a material-based system. This methodology can be used to report the key processes of the material's life cycle and to reveal the main accumulation stocks and losses to the environment (Brunner & Rechberger, 2004). Numerous SFAs on P have been computed at different scales: global scale (Smil 2000, Cordell et al. 2009), continental scale focusing on Europe (Ott & Rechberger 2012, van Dijk et al. 2016) and Africa (Cordell et al. 2009) and country scale like Australia (Cordell et al. 2009), Spain (Alvarez et al. 2018), Austria (Lederer et al. 2014) and Singapore (Pearce & Chertow 2017). In addition, some works have narrowed the scale focusing on the P flows in the agricultural sectors (Li et al. 2016, Wironen et al. 2018).

Portugal has no phosphate rock reserves and relies entirely on imports of phosphate fertilizers to maintain its required agricultural yields. Consequently, Portuguese agriculture depends on the affordable supply of mineral fertilizers containing P. For the sake of food security, Portugal should move towards a circular system that would reduce consumption, prevent losses and increase P-recycling rates (Abdulai et al. 2015, Schroder et al. 2011, Cordell & White 2013). However, the lack of information about P stock and flows in the primary sector in Portugal, namely in the agriculture and livestock sectors does not allow such a progress.

The aim of this study is to gain knowledge on how P flows between agriculture and livestock production sectors at Portuguese national level and to identify P stocks and sinks using the SFA approach. This work aims to contribute to the establishment of a sustainable agricultural P management. The outcomes are suggestions of practices to increase resource efficiency in the use of P and decrease losses to the environment, addressing both issues of P scarcity and its pollution impacts.

2 MATERIALS AND METHODS

2.1 *Conceptual model of the Agro-Livestock system*

The Agro-Livestock system modeled in this work is represented in Figure 1. This model is a simplified representation of the Portuguese agricultural and livestock production sectors.

The input flows considered to the agricultural sub-system were: fertilizers, manure, bio-solids, compost and P atmospheric deposition. The considered output flows of this sub-system were: agricultural production (as fruits and vegetables) and agricultural losses (as leaching and runoff). Concerning the livestock sub-system, the considered input flow was animal feed while the output flows were meat, eggs, milk, and livestock losses (as runoff). Within these sub-systems, there are two flows considered: manure, which flows from the livestock to agriculture sub-system, and forage, which flows from the agriculture to livestock sub-system.

2.2 *Data collection*

Table 1 and Table 2 present the categories and subcategories, which make up each flow in the livestock and agriculture sub-systems, respectively. The data collected from each source concerns the year of 2010 and the P content in each subcategory of foodstuffs was retrieved from United States Department of Agriculture (USDA) food composition database (USDA 2015).

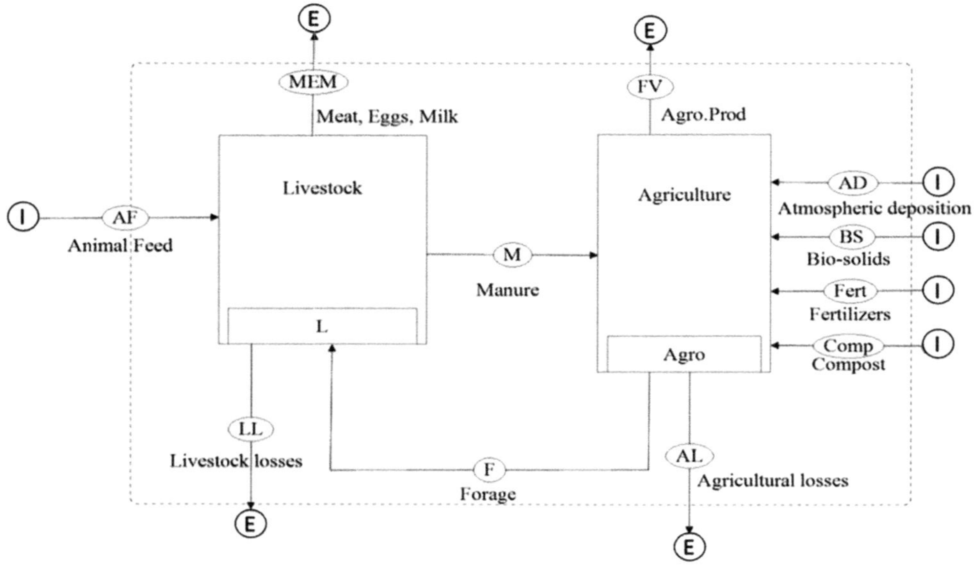

Figure 1. Representation of the Agro-Livestock system under study and main P flows and stocks (I – input flow; E – output flow; L – livestock P stock; Agro – agriculture P stock).

Table 1. Inventory for the livestock sub-system.

Category	Subcategory	Source
Animal Feed	Commercial feed	INE 2011, OECD 2017
Forage	Forage corn, wheat and oats	INE 2011
Meet	Beef, pork, lamb, chicken, duck, rabbit	INE 2011
Milk	Dairy and goat milk	INE 2011
Eggs	Hens' eggs	INE 2011
Manure	Cattle, swine, goats, sheep	OECD 2017, INE 2011
Livestock losses	Runoff	INE 2011, Cooper & Carliell-Marquet 2013

Table 2. Inventory for the agriculture sub-system.

Category	Subcategory	Source
Atmospheric deposition	Total annual transfer of P from the atmosphere to arable land	van Dijk et al. 2016, INE 2011
Biosolids	Treated sludge used as soil conditioner	Ambiente 2013
Fertilizers	Phosphate based synthetic fertilizers	INE 2011
Compost	Composted organic fraction of MSW	OECD 2017
Manure	Animal dung	OECD 2017
Fruits and vegetables	Legumes, apples, pears, peaches, cabbage, tomatoes, rice, among others	INE 2011
Forage	Forage corn, wheat and oats	INE 2011
Agricultural losses	Runoff and leaching	Ott & Rechberger 2012, INE 2011

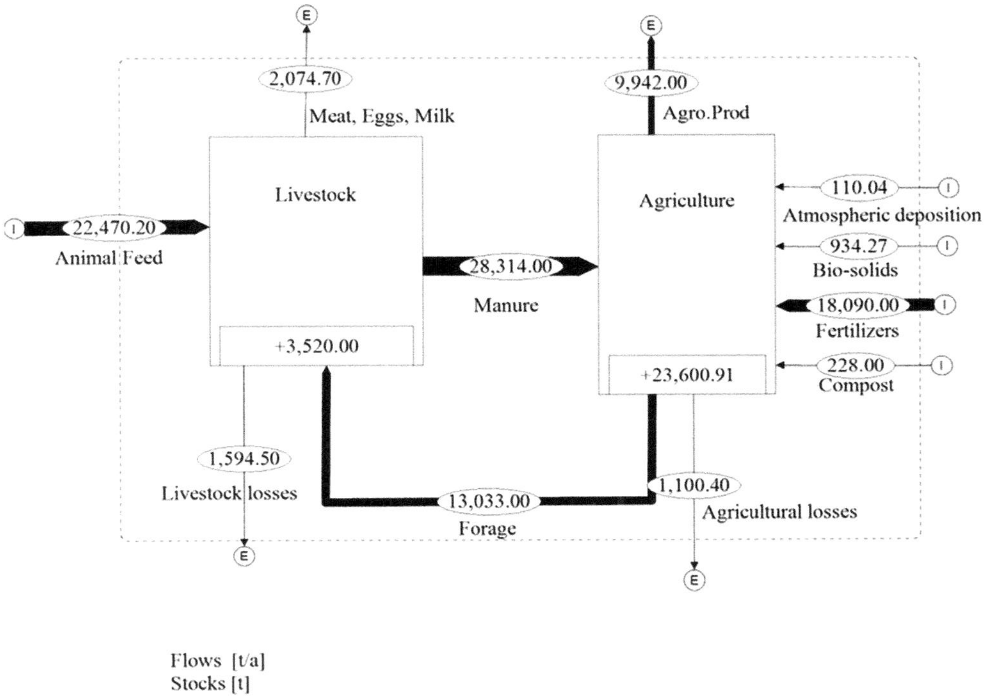

Figure 2. P flows (t/y) and stocks (t) within the agriculture and livestock sub-systems for the year of 2010.

2.3 *Agro-Livestock model for Portugal*

The software STAN was used to model the P flows. Total quantity and P concentration in each flow were used as inputs to the model. STAN then calculates the flows in terms of masses of P using Equation 1.

$$mP\ (t) = m_{flow}\ (t) \times [P]\ (tP.t^{-1}) \tag{1}$$

in which:

- mP is the mass of P (t).
- m_{flow} is the total mass of the flow (t).
- [P] is the P concentration of the flow (t P t^{-1}).

The software, using the principle of mass conservation, calculates any missing values and stocks.

3 RESULTS AND DISCUSSION

The Agro-Livestock system modeled is depicted on Figure 2. By analyzing the data for the year of 2010, it was observed that the utilized agricultural area (3 668 145 ha) has been maintained at a stable value (INE 2011).

In the agriculture sector, the inputs were much larger than the outputs. Therefore, the agriculture sub-system doesn't make an efficient use of P. If the use of fertilizers was removed all the agriculture needs would be met and surpassed by 5.5 t. The extensive use in the Portuguese agriculture of inorganic fertilizers means that in 2010 a P-stock of 23.6 kt was observed in the soils. In other words, for the year of 2010 the Portuguese soils witnessed the stocking of 6.4 kg P ha^{-1} in the agricultural utilized area. The P use efficiency, calculated as the ratio between the sum of outputs and the sum of inputs multiplied by 100, is 50% for the business as usual scenario. When the use of mineral fertilizers is removed this efficiency increases to 81%. Increasing the efficiency use of P does not only improve the profit of the agricultural unit because the farmers can save money on fertilizers purchase, but also increases its sustainability. When P is overused, it is very common to form stocks in the soils. Although, it is very common that this stocked P is not in plant available form. As a consequence, the farmers are using P that will not be used by crops and often it ends up in water bodies by runoff processes causing devastating eutrophication impacts in the ecosystems.

In the livestock sector, there was a stock of 3.5 kt of P, which is attributed to the number of live animals and also to the supply of commercial feed. To decrease the losses in this sector, manure storage facilities should be improved. There are still a considerable number of livestock units that have out-of-date and inefficient storage facilities, which increases leaching and runoff phenomena. In addition, livestock feeding can be planned according to the animal needs; the commercial feed is rich in P additives, the fodder also has a considerable P content. But on the other hand, the livestock has specific needs of P depending on the species and farming types. So if these animals are over fed in terms of P, this element will be present in their excreta and is not stored in their bodies.

4 CONCLUSIONS

The substance flow analysis of phosphorus in the agriculture and livestock production sectors highlights that Portugal still has a long path to pave to achieve a sustainable use of P in these sectors. We are presented with a common linear use of inputs that are P-rich. To increase the profitability and sustainability it is fundamental to move towards a circular use of P.

In the agriculture sector, the P use efficiency can be increased by smart fertilizing practices that take in consideration the crop needs. This will consequently decrease the dependence of mineral P based fertilizers. At the Portuguese national level, the composting material potential hasn't been fully achieved. Therefore, still leaving room for the further use of organic fertilizers to replace the P-based synthetic fertilizers. The losses by leaching to surface water bodies can be reduced by the use of buffer strips, so those plants can uptake the P before it reaches the water bodies and therefore avoiding eutrophication processes.

In the livestock sector, the P-use efficiency can be increased by upgrading manure-storing facilities, and by smart feeding systems, tailored to the needs to each specie and growth-development stage. Using these measures, the needs of the animals are covered but not exceeded.

Developing more best management practices according to the specific situation and environment will increase efficiency and decrease losses because there is no one size fits all.

ACKNOWLEDGMENTS

This work has been funded by project 0340-SYMBIOSIS-3-E co-funded by FEDER "Fundo Europeu de Desenvolvimento Regional" through Interreg V-A España-Portugal (POCTEP) 2014-2020. C. Dias-Ferreira and V. Oliveira have been funded through FCT "Fundação para a Ciência e para a Tecnologia" by POCH – Programa Operacional Capital Humano within ESF – European Social Fund and by national funds from MCTES (SFRH/BPD/100717/2014; SFRH/BD/115312/2016).

REFERENCES

Abdulai, M., Kuokkanen, A., Plank, B., Virtanen, E. & Zha, G. 2015. *Circular economy of Phosphorus flows.* HENVI Workshop 2015: Circular Economy and Sustainable Food Systems. Helsinki.

Alvarez, J., Roca, M., Valderrama, C. & Cortina, J. L. 2018. A Phosphorous Flow Analysis in Spain. *Sci Total Environ* 612: 995–1006.

APA 2013. *Gestão de Lamas de Estações de Tratamento de Águas Residuais Urbanas (2010–2013)*.
Brunner, P. H. & Rechberger, H. 2004. *Practical Handbook of Material Flow Analysis*, Lewis Publisher.
Childers, D. L., Corman, J., Edwards, M. & Elser, J. J. 2011. Sustainability Challenges of Phosphorus and Food: Solutions from Closing the Human Phosphorus Cycle. *BioScience* 61: 117–124.
Cooper, J. & Carliell-Marquet, C. 2013. A substance flow analysis of phosphorus in the UK food production and consumption system. *Resources, Conservation and Recycling* 74: 82–100.
Cordell, D., Drangert, J.-O. & White, S. 2009. The story of phosphorus: Global food security and food for thought. *Global Environmental Change* 19: 292–305.
Cordell, D. & White, S. 2013. Sustainable Phosphorus Measures: Strategies and Technologies for Achieving Phosphorus Security. *Agronomy* 3: 86–116.
INE, I. P. 2011. *Estatísticas Agrícolas 2010*. Lisboa, Portugal.
Jasinski, S. M. 2017. *Phosphate Rock*. USA: U.S. Geological Survey
Lederer, J., Laner, D. & Fellner, J. 2014. A framework for the evaluation of anthropogenic resources: the case study of phosphorus stocks in Austria. *Journal of Cleaner Production* 84: 368–381.
Li, G., Van ittersum, M. K., Leffelaar, P. A., Sattari, S. Z., Li, H., Huang, G. & Zhang, F. 2016. A multi-level analysis of China's phosphorus flows to identify options for improved management in agriculture. *Agricultural Systems* 144: 87–100.
Linderholm, K., Tillman, A.-M. & Mattsson, J. E. 2012. Life cycle assessment of phosphorus alternatives for Swedish agriculture. *Resources, Conservation and Recycling* 66: 27–39.
Metson, G. S., Cordell, D. & Ridoutt, B. 2016. Potential Impact of Dietary Choices on Phosphorus Recycling and Global Phosphorus Footprints: The Case of the Average Australian City. *Front Nutr* 3: 35.
OECD 2017. *Agri-Environmental indicators: Nutrients: Phosphorus balance* [Online]. Available: https://stats.oecd.org [Accessed 10 May 2017].
Ott, C. & Rechberger, H. 2012. The European phosphorus balance. *Resources, Conservation and Recycling* 60: 159–172.
Pearce, B. J. & Chertow, M. 2017. Scenarios for achieving absolute reductions in phosphorus consumption in Singapore. *Journal of Cleaner Production* 140: 1587–1601.
Schroder, J. J., Smit, A. L., Cordell, D. & Rosemarin, A. 2011. Improved phosphorus use efficiency in agriculture: a key requirement for its sustainable use. *Chemosphere* 84: 822–31.
Smil, V. 2000. Phosphorus in the environment: Natural flows and human interferences. *Energy and Environment* 25: 53–88.
Smit, A. L., Bindraban, P. S., Schroder, J. J., Conijn, J. G. & Van Der Horst, D. 2009. *Phosphorus in agriculture: global resources, trends and developments*. Wageningen: Plant Research International B.V.
USDA 2015. *USDA Food Composition Databases* [Online]. USA: USDA. Available: https://ndb.nal.usda.gov/ndb/search/list [Accessed 23 April 2017].
Van Dijk, K. C., Lesschen, J. P. & Oenema, O. 2016. Phosphorus flows and balances of the European Union Member States. *Sci Total Environ* 542: 1078–93.
Wironen, M. B., Bennett, E. M. & Erickson, J. D. 2018. Phosphorus flows and legacy accumulation in an animal-dominated agricultural region from 1925 to 2012. *Global Environmental Change* 50: 88–99.

Wastes: Solutions, Treatments and Opportunities III – Vilarinho et al. (Eds)
© 2020 Taylor & Francis Group, London, ISBN 978-0-367-25777-4

Winery by-products: Pomace as source of high value phenols

A. Tassoni
Department of Biological, Geological and Environmental Sciences, University of Bologna, Bologna, Italy

M. Ferri
Department of Civil, Chemical, Environmental and Materials Engineering, University of Bologna, Bologna, Italy

ABSTRACT: Grape pomace, the main by-product from winemaking industry, accounts for 20–30% of processed grape weight and still contains valuable compounds. Valorization of pomace by extracting phenols can be a further step towards a "near zero-waste society" direction. The present work aimed at the selection of the best extraction conditions for phenol recovery from Sangiovese and Montepulciano (*Vitis vinifera* L.) mixed winery pomace. Different solvents were screened in 2 h incubation at 30°C and 50°C, with 1:10 and 1:5 solid/liquid ratios, with or without ultrasound pre-treatment. Total phenol yields were measured in all extracts and best extracts were characterized for flavonoid, flavanol and anthocyanin contents and for total antioxidant activity. In conclusion, 75% acetone led to the highest compound extractions, followed by 50% acetonitrile and 50% ethanol. Higher yields were obtained with 1:5 solid/liquid ratio. The ultrasound pre-treatment did not significantly affect compound recovery, while better extractions were achieved with temperature increase.

1 INTRODUCTION

Agro-waste represents a huge amount of biomass resources that could be converted into sustainable bio-products according to the basic principles of the circular economy that aims to move towards a "near zero-waste society". However, energy consumption, degradation processes, complexity and variability in chemical composition of the waste feedstock, presence of contaminants, low quality and poor performance often characterize waste recovery and conversion products and could impair the effective exploitation of the waste. In particular, approximately 20–30% in weight of processed grape (*Vitis* sp.), the world's largest fruit crop mostly used in winemaking, ends up as pomace (Djilas et al., 2009). This large amount of by-products constitute a serious environmental and disposal problem. Grape pomace is characterized by a high content of polyphenol compounds that are only partially extracted during the winemaking process and whose range and extractability mainly depends on the technological parameters applied during vinification. Therefore it constitutes a rich but yet underutilized source of valuable compounds, which can be valorized following a biorefinery concept (Djilas et al., 2009; Ferri et al., 2016; Fontana et al., 2013). Grape pomace extracts can be exploited in the nutraceutical, pharmaceutical, cosmetic or biomaterial fields due to their biological activities, while solid residues can be valorized in nanofibers or biocomposite applications (Djilas et al., 2009; Ferreira et al., 2014; Fontana et al., 2013).

The present work aimed at the recovery of phenolic compounds from red (*Vitis vinifera* L., cv. Sangiovese and Montepulciano) grape pomace. Different solvents were screened and those best performing, in terms of released compound amount, were tested at increased incubation temperature and in combination with ultrasound pre-treatment.

2 MATERIALS & METHODS

Red wine pomace derived from a mix of *Vitis vinifera* L. Sangiovese and Montepulciano cultivars, was supplied by the Cantine Moncaro wineries (Jesi, Ancona, Italy). Pomace contained berry skins, seeds, petioles and stalks, was ground with a kitchen blender and stored at −20°C until used for following analyses.

A first set of experiments was performed by extracting the pomace with the following solvents: 100% water; 25%, 50%, 75% methanol; 25%, 50%, 75%, 95% ethanol; 25%, 50%, 75%, 80%, 90%, 100% acetone; 25%, 50%, 75% acetonitrile. Solvent was directly added to the ground pomace in two different solid/liquid ratios (1:10, 3 gFW pomace:30 mL solvent; 1:5, 5 gFW:25 mL) and the mixture was incubated at 30°C for 2 h in a shaking water bath at 150 rpm. In a second set of experiments, ground pomace was extracted with 50% ethanol, 50% acetonitrile or 75% acetone at two different incubation temperatures (30°C or 50°C), with or without ultrasound pre-treatment (US; 10 or 20 min; TranssonicTP690, Elma, Germany, HF 35 kHz). Liquid extracts were recovered after centrifugation (5000 rpm, 10 min, room temperature) and total phenols were quantified by the Folin–Ciocalteu spectrophotometric assay (Ferri et al., 2013; Singleton et al., 1999). Selected extracts were further characterized by spectrophotometric analyses by assessing flavonoid (Zhishen et al., 1999), flavanol (McMurrough et al., 1978) and anthocyanin (Ferri et al., 2009) contents and by quantifying the total antioxidant activity (ABTS assay with minor modifications; Ferri et al., 2013). The results are expressed as g of gallic acid (GA, for total phenols) or catechin (CAT, flavonoids and flavanols) or malvidin (MALV, anthocyanins) or ascorbic acid (AA, antioxidant activity) equivalent per L of extract.

All the extractions were performed twice, while the spectrophotometrical assays were performed at least twice with two technical replicates each. The results were expressed as the mean ($n = 2$) ±SD.

3 RESULTS & DISCUSSION

The recovery of phenolic compounds from Sangiovese and Montepulciano (*Vitis vinifera* L.) mixed red grape pomace was carried out by following a biorefinery concept aimed at the by-product valorization.

Different solvents at several concentrations were screened in 2h incubations at 30°C and total extracted phenols were quantified (Fig. 1). Two different solid/liquid ratios (1:10 and 1:5, gFW pomace:mL solvent) were tested. Generally, extracts coming from 1:5 ratio had double phenol concentration with respect to 1:10 samples treated with the same solvent, as a result of the dilution effect due to the use of a double volume of solvent in 1:10 processes. Therefore, 1:5 solid/liquid ratio was selected as it allowed to recover the same amount of phenols per pomace weight by using half solvent volume.

Among the tested solvents, higher yields were in general obtained with acetone (Fig. 1) which was tested at different concentrations, with 75% being the most efficient yielding a total phenol extracted content of 3.51 gGAeq/L with 1:5 solid/liquid ratio (corresponding to 17.55 gGAeq/kgFW of pomace). Other good yields were obtained with 50% ethanol (2.71 gGAeq/L) and 50% acetonitrile (2.40 gGAeq/L). The same grape pomace was previously used for phenol recovery via enzymatic digestions with maximum yield of 0.36 gGAeq/L (Ferri et al., 2016), similarly to the present water extraction (Fig. 1). Data reported in Fig. 1, also confirmed the previously obtained yield of 95% ethanol extraction (0.84 gGAeq/L both in Ferri et al. (2016) and Fig. 1). However, the present study was able to increase phenol recovery from Sangiovese and Montepulciano mixed pomace up to 4.2-times by using 75% acetone treatment (Fig. 1).

The three selected solvents (50% ethanol, 50% acetonitrile, 75% acetone) were used for extractions at increasing incubation temperature and in combination with ultrasound pre-treatment (Fig. 2). Higher temperature (50°C with respect to 30°C) allowed respectively 55.7%, 21.8% and 17.6% higher phenol yields with 50% ethanol, 75% acetone and 50% acetonitrile. The effect of temperature

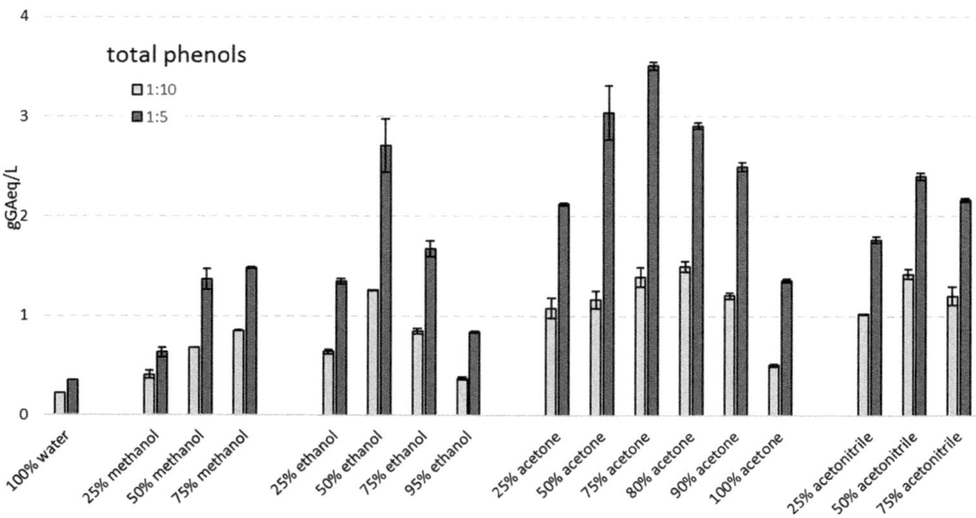

Figure 1. Total phenol levels in grape pomace extracts obtained with different solvents and with two solid/liquid ratios (1:10 or 1:5), in 2 h incubation at 30°C. The results are expressed as g of gallic acid (GA) equivalent per L of extract and they are the mean of at least two data ±SD.

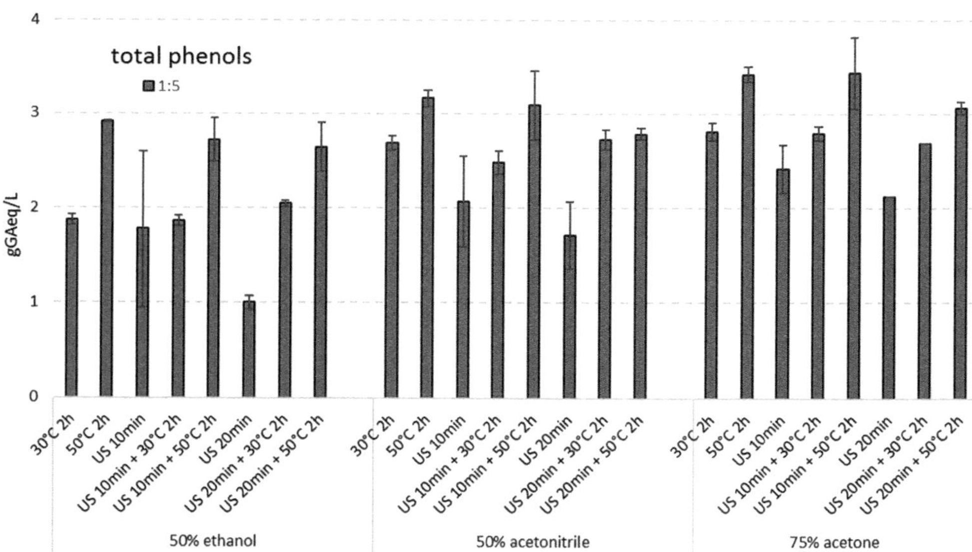

Figure 2. Total phenol levels in grape pomace extracts obtained with the three best solvents at two different incubation temperature (30°C or 50°C), with or without ultrasound pre-treatment (US; 10 or 20 min). The results are expressed as g of gallic acid (GA) equivalent per L of extract and they are the mean of at least two data ±SD.

was evident also when ultrasound pre-treatment was applied. Conversely, ultrasound pre-treatment itself seemed not to be effective in phenol release, with lower yield obtained at longer times (20 min with respect to 10 min). When ultrasound was followed by 2 h solvent incubation, samples at both pre-treatment times, showed similar phenol levels.

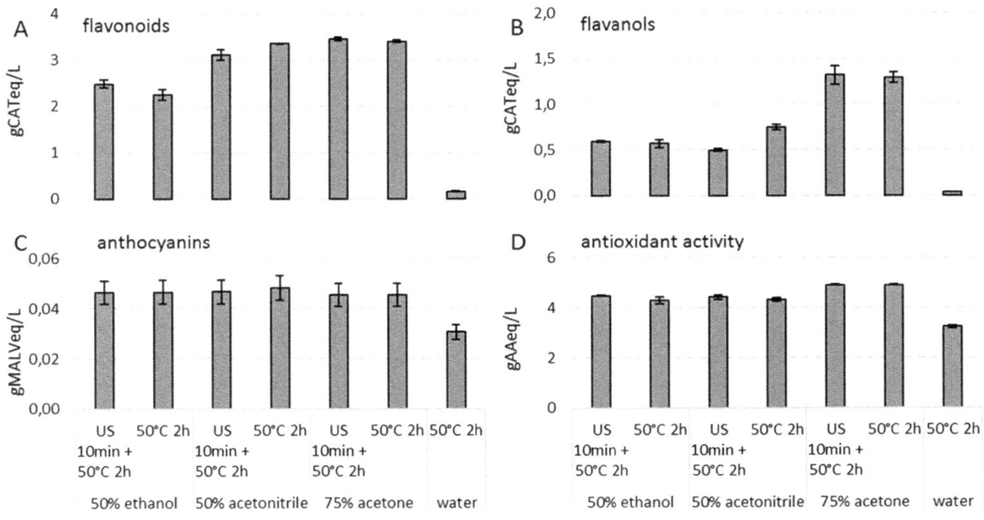

Figure 3. Total levels of flavanols (A), flavanols (B), anthocyanins (C), antioxidant activity (D) in selected extracts obtained with the three best solvents from 2 h incubation at 50°C, with or without ultrasound (US, 10 min), and in water extract as control. The results (mean of at least two data ±SD) are expressed as g of catechin (CAT) (A and B), malvidin (MALV) (C) or ascorbic acid (AA) (D) equivalents per L of extract.

Extracts obtained with the selected solvents from 2 h incubation at 50°C, with or without 10 min ultrasound pre-treatment, and water extract as control, were further characterized by spectrophotometric analyses. Total levels of flavonoids, flavanols and anthocyanins were quantified as well as antioxidant activity (Fig. 3). Consistently with total phenol results (Fig. 2), higher amounts of flavonoids and their sub-family flavanols, were found in 75% acetone extracts, followed by 50% acetonitrile samples (Fig. 3A-B). Maximum yields were 3.45 gCATeq/L of flavonoids and 1.32 gCATeq/L of flavanols, corresponding respectively to 17.25 and 9.59 gCATeq/kgFW of pomace. Water control showed 23-times and 34-times lower amounts of flavonoids and flavanols respectively than the highest detected levels (Fig. 3A-B). Anthocyanins seemed to be similarly solubilized at the same level by all selected solvents (0.046 gMALVeq/L on average), while the content in water extracts was 34% lower (Fig. 3C), as confirmed by visual observation where water samples were pink and solvent extracts were dark red.

The present results (Fig. 3) confirmed that the proposed chemical extractions from Sangiovese and Montepulciano mixed pomace were able to increase phenolic compound release with respect to enzymatic digestions and previous literature data (Ferri et al., 2016). The present best flavonoid yield obtained with ethanol extraction (Fig. 3A) was 3.6-times higher respect to previous data with the same solvent (Ferri et al., 2016); also the reported overall highest flavonoid yield (Fig. 3A, 75% acetone) was 3.4-times higher than previous data on the same type of pomace extracted by a two-step enzymatic-ethanol process (Ferri et al., 2016).

The levels of polyphenols (Fig. 2), flavonoids, flavanols and anthocyanins (Fig. 3A-C) were similar between the extracts obtained with the same solvent with and without ultrasound pre-treatment, confirming that, with respect to Sangiovese and Montepulciano mixed pomace, ultrasound extraction was not able to increase the release of phenolic compounds.

Total antioxidant activity of the extracts was assessed in view of their possible industrial exploitation. No significant differences were observed between samples obtained by the same solvent with and without ultrasound pre-treatment (Fig. 3D). The highest antioxidant activity was measured in 75% acetone extracts (4.90 gAAeq/L, corresponding to 24.50 gAAeq/kgFW), that were on average respectively 11.9%, 12.4% and 52.2% higher than 50% ethanol, 50% acetonitrile and water samples

(Fig. 3D). All solvent extracts (with the exclusion of water control) resulted to be more antioxidant than previously reported pomace extracts obtained from the same pomace, that possessed an activity lower than 3.0 gAAeq/L (Ferri et al., 2016).

4 CONCLUSIONS

The present research work considered several parameters for phenol recovery from Sangiovese and Montepulciano (*Vitis vinifera* L.) mixed winery pomace. Overall, the highest compound extractions were obtained by using 75% acetone after 2 h incubation at 50°C, with 1:5 solid/liquid ratio. Different acetone concentrations, other solvents, double solid/liquid ratio (1:10) and lower temperature (30°C) led to lower yields. Ultrasound pre-treatment did not significantly affect compound recovery.

The present study endorse the idea that agro-waste represents a valuable biomass resource that could be converted into sustainable bio-products, according to the basic principles of the circular economy that aims to move towards a "near zero-waste society". The results support the possibility of recovering phenols from grape pomace and, possibly, exploiting the extracts as ingredients for functional and innovative products in the nutraceutical, pharmaceutical, cosmetic or bioplastic fields.

REFERENCES

Djilas, S., Canadanovic-Brunet, J. & Cetkovic, G. 2009. By-products of fruits processing as a source of phytochemicals. *Chem. Ind. Chem. Eng. Q.* 15: 191–202.

Ferreira, A.S., Nunes, C., Castro, A., Ferreira, P. & Coimbra, M.A. 2014. Influence of grape pomace extract incorporation on chitosan films properties. *Carbohydr. Polym.* 113: 490–499.

Ferri, M., Bin, S., Vallini, V., Fava, F., Michelini, E., Roda, A., Minnucci, G., Bucchi, G. & Tassoni, A. 2016. Recovery of polyphenols from red grape pomace and assessment of their antioxidant and anti-cholesterol activities. *New Biotech.* 33: 338–344.

Ferri, M., Gianotti, A. & Tassoni, A. 2013. Optimisation of assay conditions for the determination of antioxidant capacity and polyphenols in cereal food components. *J. Food Comp. Anal.* 30: 94–101.

Ferri, M., Tassoni, A., Franceschetti, M., Righetti, L., Naldrett, M.J. & Bagni, N. 2009. Chitosan treatment induces changes of protein expression profile and stilbene distribution in *Vitis vinifera* cell suspensions. *Proteomics* 9: 610–24.

Fontana, A.R., Antoniolli, A. & Bottini, R. 2013. Grape pomace as a sustainable source of bioactive compounds: extraction, characterization and biotechnological applications of phenolics. *J. Agric. Food Chem.* 61: 8987–9003.

McMurrough, I. & McDowell, J. Chromatographic separation and automated analysis of flavanols. 1978. *Anal. Biochem.* 91 :92–100.

Singleton, V.L., Orthofer, R. & Lamuela-Raventos, R.M. 1999. Analysis of total phenols and other oxidation substrates and antioxidants by means of Folin-Ciocalteu reagent. *Method. Enzymol.* 299: 152-178.

Zhishen, J., Mengcheng, T. & Janming, W. 1999. The determination of flavonoid in mulberry and their scavenging effects on superoxide radicals. *Food Chem.* 64: 555–9.

Wastes: Solutions, Treatments and Opportunities III – Vilarinho et al. (Eds)
© 2020 Taylor & Francis Group, London, ISBN 978-0-367-25777-4

Stabilizing municipal solid waste incineration residues with alternative binders

J. Coelho
University of Minho, Guimarães, Portugal

B. Chaves
LIPOR – Intermunicipal Waste Management of Greater Porto, Portugal

M.L. Lopes
University of Porto, Porto, Portugal

L. Segadães & N. Cristelo
University of Trás-os-Montes e Alto Douro, Vila Real, Portugal

ABSTRACT: The rapid increase in world population in the last century has generated an increasing demand for raw materials, as well as an ever-growing production of municipal solid waste (MSW). Previous studies used cement to stabilise by-products resulting from the incineration of MSW, adding to the overall environmental toll, due to the high CO_2-eq associated with cement manufacturing. The present study targets the stabilisation/activation of the fly (FA) and bottom ash (BA) generated during incineration of MSW, using alkaline activation. Both precursors were milled and activated independently, using sodium hydroxide or sodium silicate solution. To establish a reference, the BA were also stabilised with Portland cement, and the analysis was based on uniaxial compression strength test, ultrasonic pulse velocity measurements and X-ray diffraction. The results showed that milled FA and BA resulting from incinerated MSW constitute an effective precursor for alkaline activation reactions, generating a competitive and more sustainable binder.

1 INTRODUCTION

A common treatment of solid waste produced in urban areas is incineration, which is a very effective procedure but, at the same time, produces very significant volumes of bottom ash (BA) and fly ash (FA). These by-products are mostly landfilled, with inevitable financial and environmental costs. Even though the BA/FA volume is much lower than the original waste volume, representing a weight reduction of about 80% (Wei et al., 2011), it still must be landfilled, which means their possible reintroduction in the production chain is an attractive possibility.

The application of BA in alkaline activation reactions is becoming an interesting solution, with most studies focusing on blends of BA and other well-known precursors, like coal incineration fly ash, blast furnace slag or glass powders (Monich et al., 2018; Gao et al., 2017); and/or activators simultaneously composed by sodium hydroxide and sodium silicate (Zhu et al., 2019; Shiota et al., 2017). The use of BA is a more interesting approach than FA, not only due to its intrinsic properties, more suited for the demands associated with the production of aluminosilicate-based pastes, but also because BA represents approximately 85% of the total incineration residue (Zhu et al., 2018), and it is significantly less polluting than FA (Li et al., 2012), although still possessing a considerable heavy metal content.

The present study addressed the viability of using the BA and FA, separately or jointly (although only the former results are here presented), resulting from the incineration of the MSW collected

Table 1. Chemical composition of the FA and BA.

Element	BA (wt%)	FA (wt%)
Al_2O_3	6.81	2.31
CaO	22.77	38.7
Cl	2.42	27.06
K_2O	3.12	8.13
Mg	5.11	1.67
Na_2O	7.00	11.57
P_2O_5	3.91	1.06
SO_3	2.43	4.59
SiO_2	43.7	3.23
Others	2.68	1.44

in the Greater Porto area. The initial stage of the project focused on the optimization of the milling process (not presented in the present paper) and continued with the direct activation of BA or FA with sodium hydroxide (SH, in different concentrations) or sodium silicate (SS). For comparison purposes, BA was also stabilized with cement, in different weight percentages of BA and cement. Each paste was assessed with uniaxial compression strength tests, X-ray diffraction and ultrasonic pulse P-wave velocity. Results show that the stabilization of the BA with SH and SS produced uniaxial compression strengths (UCS) up to 2 and 9 Mpa, while a cement content of 30% produced a UCS of approximately 3 Mpa.

2 MATERIALS AND METHODS

2.1 *Materials*

Both the fly ash and bottom ash were supplied by the company *LIPOR – Intermunicipal Waste Management of Greater Porto*, responsible for the management, recovery and treatment of the municipal solid waste produced in the 8 associated municipalities of the greater Porto area. Portland cement (OPC) type II/B-L, class 32.5 N was used. The activator was a sodium hydroxide (SH) or a sodium silicate (SS) solution. The sodium hydroxide was supplied in pellets, with a specific gravity of 2.13 at 20°C (99 wt.%), which was then dissolved in deionised water to previously defined concentrations of 4, 6, 8 and 12 molal. The SS presented a unit weight of 1.464 g/cm^3 (at 20°C), a SiO_2/Na_2O weight ratio of 2.0 (molar oxide ratio of 2.063) and a Na_2O concentration in the solution of 13.0%.

The chemical composition of the two residues, obtained by X-ray fluorescence (XRF), on a PHILIPS PW-1004 X-ray spectrometer, is shown in Table 1. Silica is the main element of the BA, although a significant calcium oxide content is also present (22.8%). The main element of the FA was calcium oxide (38.7%), but it also presented a very significant chlorine content (27.1%, against 2.4% in the BA).

The original materials (FA and BA) and the resulting pastes were characterised using X-Ray Diffraction (XRD), on a PANalytical X'Pert Pro diffractometer, using CuKα radiation. The scans covered a 2θ range of 10 to 60°, with a nominal step size of 0.017° and 100 s/step. The FA diffractogram showed the main presence of sylvite, halite and calcite, with traces of portlandite, anhydrite (calcium sulphate) and lime (I. Garcia-Lodeiro et al., 2016); while the diffractogram of the BA revealed mostly quartz, calcite, clinozoisite and akermanite (Yunmei Wei et al., 2011; I. Garcia-Lodeiro et al., 2016). The amorphous content of both precursors is evidenced by a halo between angles 25 to 33 °2θ (FA) and 18 to 35 °2θ (BA), approximately. The OPC diffractogram, as expected, shows traces of calcite, alite, belite quartz and gypsum.

Table 2. Identification and composition of the pastes tested (Solids = BA+FA+OPC).

ID	BA (wt%)	FA (wt%)	OPC (wt%)	Liquid/Solids (wt. ratio)	Liquid
PaB1	100	–	–	0.35	SH (4 molal)
PaB2	100	–	–	0.40	SH (12 molal)
PaB3	100	–	–	0.50	SS
PaF1	–	100	–	0.45	SH (6 molal)
PaF2	–	100	–	0.45	SH (8 molal)
PaF3	–	100	–	1.50	SS
PaC1	90	–	10	0.3	Water
PaC2	80	–	20	0.3	Water
PaC3	70	–	30	0.3	Water

2.2 *Preparation and testing*

The precursors were previously milled, on a ball mill with a 20 kg capacity, until 80% of the particles reached a size below 63 μm. The activator was then slowly added to the already homogenized solid components (FA, BA or BA+OPC), 4 to 6 h after preparation (required for the temperature to decrease, after the exothermic reaction between the pellets and the water), and a mixing period of 3 min followed, to guarantee an adequate homogenisation. After mixing, the pastes were left to rest for 10 minutes, and only then casted into 40 × 40 × 40 mm moulds. Each result presented is the average of 10 tested specimens.

Three different sets of paste were fabricated and tested (Table 2). The PaB pastes used only BA, and 2 different activators, one of them in two different concentrations. Pastes PaF were composed only by FA, activated also with 2 distinct solutions (again, one of them with two different concentrations). Finally, pastes PaC were exclusively prepared with Portland cement and water.

For the initial 72 hours, specimens were cured while still inside the moulds, under constant temperature and humidity of 30°C and 25% RH, respectively. After demoulding, they were kept under the same conditions (30°C and 25% RH) until the test, which was scheduled to occur 7 days after fabrication.

A servo-hydraulic testing machine, fitted with a 25 kN capacity load cell, was used for the uniaxial compression strength tests. These tests were carried out under monotonic displacement control, at a rate of 0.2 mm/min.

Non-destructive ultrasonic pulse velocity (P-wave) measurements were taken on purposely-built cubic specimens (100 × 100 × 100 mm), throughout their respective curing period. A commercially available equipment was used, which included a pair of axially aligned piezoelectric broadband transducers, with a nominal frequency of 54 kHz. The measurements started as soon as the specimens were demoulded (i.e. after 72 h).

3 RESULTS AND DISCUSSION

3.1 *Uniaxial compression strength*

Uniaxial compression strength (UCS) after 7 days is shown in Figure 1. It is notorious that the most effective combination was the activation of BA with sodium silicate. Other interesting nuances include the higher UCS obtained by the BA when activated with a lower concentration of SH; the higher UCS obtained by the FA when activated with a higher concentration of SH; the effectiveness of the SS that was confirmed when activating the FA (UCS of PaF3 was double the UCS obtained with SH); the increasing UCS with increasing cement content; and, finally, the fact that a significant cement content of 30% (wt.) was only able to produce a UCS value which is just slightly higher than the UCS obtained with a 4 molal SH solution.

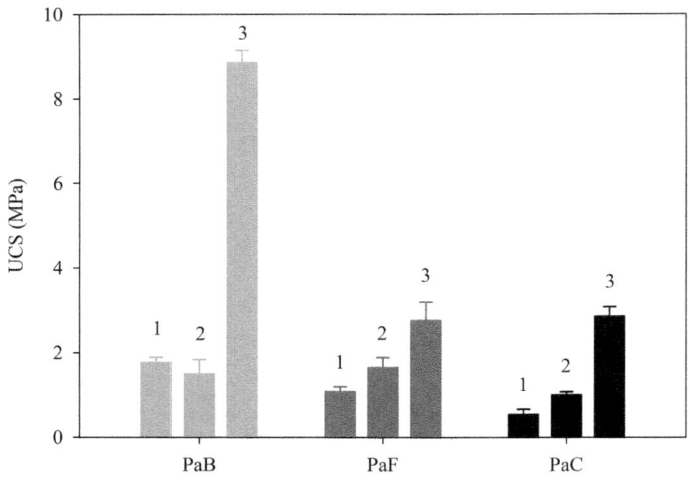

Figure 1. Uniaxial compression strength results after 7 days curing (average and standard deviation of 9 specimens).

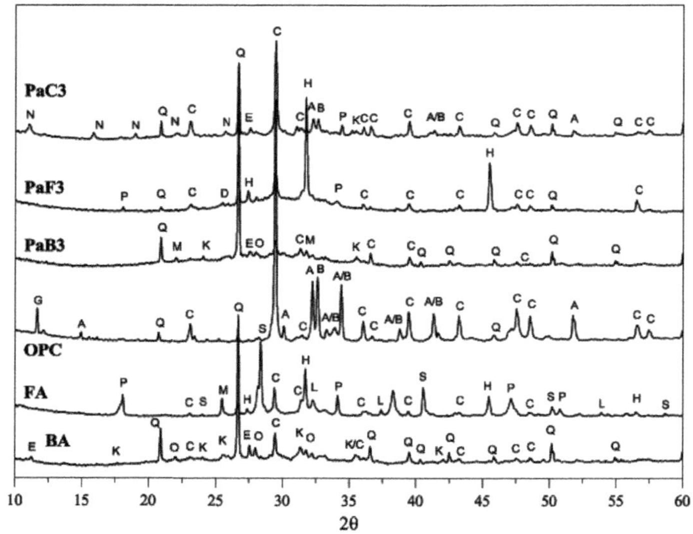

Figure 2. XRD diffratogram of the anhydrous materials (FA, BA and OPC) and pastes PaF3, PaB3 and PaC3 (A-alite; B-belite; C-calcite; E-clinozoisite; G-gypsum; H-halite; K-akermanite; L-lime; M-anhydrite; N-partheite; O-oligoclase; P-portlandite; Q-quartz).

This last comparison is very significant, and flattering for the BA-based paste, considering the low concentration of its SH solution (4 molal, which corresponds to a mere 160 g of SH per litre of water), which implies that the financial and environmental cost of this activator is very competitive; and also the higher residue (BA) content (100%) that this paste was able to include, compared with the mere 70% absorbed by the OPC-based paste.

3.2 *Minerology*

Figure 2 shows the diffractograms obtained with the anhydrous materials (FA, BA and OPC) and some selected pastes (PaB3, PaF3 and PaC3), also after 7 days curing.

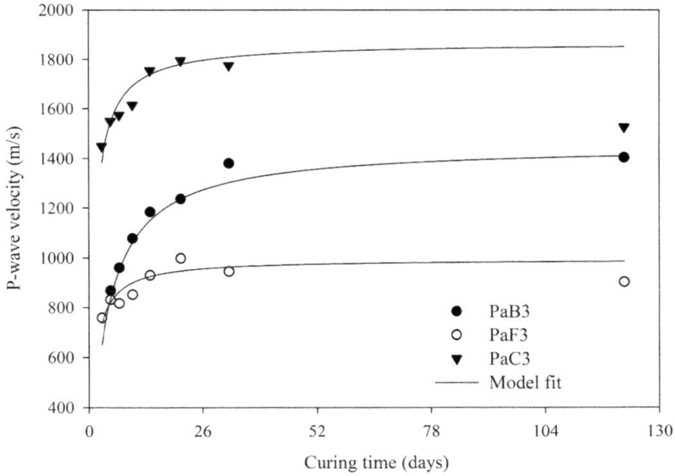

Figure 3. P-wave velocity along the vertical axis throughout curing.

The PaB3 diffractogram presented some differences regarding the anhydrous BA, in the sense that, although several phases remained mostly unchanged (like calcite, quartz, oligoclase and akermanite), the clinozoisite mineral disappeared. At the same time, the halo indicating the presence of amorphous matter slightly increased, indicating the formation of an essentially amorphous aluminosilicate gel (Ana Fernández-Jiménez et al., 2017).

A similar situation was registered for the PaF3, which kept the calcite, portlandite and anhydrite phases, while sylvite and lime disappeared. However, in the FA-based paste, the halite phase showed an intensity increase. Similar to what was observed for the BA-based paste, the amorphous content on the PaF3 appears to have increased, relatively to the anhydrous FA, considering the corresponding halo, which is now significantly more developed.

The PaC3 diffractogram shows phases resulting from both anhydrous materials (OPC and BA). The phases resulting from the BA appear mostly unchanged, with the exception of the clinozoisite, which disappeared. At the same time, a new calcium aluminium silicate phase was formed – partheite, from the hydration of the cement. This suggests that, contrary to what was observed for the alkaline activated pastes, the cement didn't react with the residue, instead it only surrounded its particles. This can have a significant influence on the leachate behaviour of these pastes, to be assessed in future studies.

3.3 *Ultrasonic pulse velocity measurements*

Figure 3 presents the data obtained from the velocity measurement of P-waves travelling through specimens fabricated with pastes PaB3, PaF3 and PaC3. Readings were taken after 3, 7, 14, 21, 32 e 122 days curing, and the results presented are the average velocities obtained with frequencies 24, 54, 150, 250 and 500kHz. Although the measurements were taken in the 3 axes of the cubes, only the vertical axis values are presented, for simplicity.

Paste PaF3 showed the lowest P-wave propagation velocities, for all curing periods, while the cement-based paste showed the highest initial value, although its increase rate, with curing time, was very similar to that shown by PaF3. Paste PaB3 showed a similar initial value to the PaF3, but its increase rate was the highest of the 3 pastes analysed. After 122 days, pastes PaB3 and PaC3 showed very similar velocities, which mostly results from the decrease, with time, of the cement-based material.

The model proposed by the ACI Committee 209 (1998) was applied to the 3 sets of data, to predict the strength development in cementitious binders. This model fit allowed to better illustrate that, although the initial consistency development of the pastes prepared with the MSW residues is significantly lower than that presented by the cement-based paste, the latter presented a progressive degradation, probably resulting from secondary chemical reactions (Cristelo et al., 2015). On the contrary, the BA and FA pastes appear to be more stable and, thus, more adequate to stabilise the complex and broad spectrum (in terms of chemical composition) residues generated by the incineration of municipal waste. Since, after 7 days curing, the UCS of the PaB3 is clearly higher than the UCS shown by the PaC3, these ultrasonic results possibly indicate that such difference in mechanical strength will increase with time.

4 CONCLUSIONS

The present paper describes the results obtained with the alkali activation of BA and FA generated by the incineration of the municipal solid waste collected in the Greater Porto area. Both residues were previously grinded, and activated with sodium hydroxide or sodium silicate. The effect of Portland cement on the bottom ash was also tested, to establish a reference.

Results showed that the sodium silicate is particularly effective in the stabilisation of the bottom ash, reaching a compression strength value approximately 3x higher than that achieved with 30 wt% of cement. Such results contribute to a progressive application of this residue in several cement-substitution applications, although more tests are required, especially regarding durability and environmental performance.

REFERENCES

ACI Committee 209 1998. Prediction of Creep, Shrinkage, and Temperature Effects in Concrete Structures. Committee Report ACI 209R-92. ACI Man. Concr. Pract. Part I.

Cristelo, N., Miranda, T., Oliveira, D., Rosa, I., Soares, E., Coelho, P. & Fernandes, L. 2015. Assessing the production of jet mix columns using alkali activated waste based on mechanical and financial performance and CO_2 (eq) emissions. *Journal of Cleaner Production* 102: 447–460.

Fernández-Jiménez, A., Cristelo, N., Miranda, T. & Palomo, A. 2017. Sustainable alkali activated materials: Precursor and activator derived from industrial wastes. *Journal of Cleaner Production* 162: 1200–1209.

Gao, X., Yuan, B., Yu, Q.L. & Brouwers, H.J. 2017. Characterization and application of municipal solid waste incineration (MSWI) bottom ash and waste granite powder in alkali activated slag. *Journal of Cleaner Production* 164: 410–419.

Garcia-Lodeiro, I., Carcelen-Taboada, V., Fernández-Jiménez, A. & Palomo, A. 2016. Manufacture of hybrid cements with fly ash and bottom ash from a municipal solid waste incinerator. *Construction and Building Materials* 105: 218–226.

Li, X.G., Lv, Y., Ma, B.G., Chen, Q.B., Yin, X.B. & Jian, S.W. 2012. Utilization of municipal solid waste incineration bottom ash in blended cement. *Journal of Cleaner Production* 32: 96–100.

Monich, P.R., Romero, A.R., Hollen, D. & Bernardo, E. 2018. Porous glass-ceramics from alkali activation and sinter-crystallization of mixtures of waste glass and residues from plasma processing of municipal solid waste. *Journal of Cleaner Production* 188: 871–878.

Shiota, K., Nakamura, T., Takaoka, M., Nitta, K., Oshita, K., Fujimori, T. & Ina, T. 2017. Chemical kinetics of Cs species in an alkali-activated municipal solid waste incineration fly ash and pyrophyllite-based system using Cs K-edge in situ X-ray absorption fine structure analysis. *Spectrochimica Acta Part B* 131: 32–39.

Wei, Y., Shimaoka, T., Saffarzadeh, A. & Takahashi, F. 2011. Mineralogical characterization of municipal solid waste incineration bottom ash with an emphasis on heavy metal-bearing phases. *Journal of Hazardous Materials* 187: 534–543.

Zhu, W., Chen, X., Struble, L.J. & Yang, E.H. 2018. Characterization of calcium-containing phases in alkali-activated municipal solid waste incineration bottom ash binder through chemical extraction and deconvoluted Fourier transform infrared spectra. *Journal of Cleaner Production* 192: 782–789.

Zhu, W., Chen, X., Zhao, A., Struble, L.J. & Yang, E.H. 2019. Synthesis of high strength binders from alkali activation of glass materials from municipal solid waste incineration bottom ash. *Journal of Cleaner Production* 212: 261–269.

Wastes: Solutions, Treatments and Opportunities III – Vilarinho et al. (Eds)
 ISBN 978-0-367-25777-4

Soil stabilized with alkali activated slag at various concentrations of activator

C. Pinheiro, S. Rios & A. Viana da Fonseca
CONSTRUCT-GEO, Department of Civil Engineering, University of Porto, Porto, Portugal

J. Coelho
Department of Civil Engineering, University of Minho, Guimarães, Portugal

A. Fernández-Jiménez
Instituto Eduardo Torroja (IETcc), CSIC, Madrid, Spain

N. Cristelo
CQVR, Department of Engineering, University of Trás-os-Montes e Alto Douro, Vila Real, Portugal

ABSTRACT: The study presented in this work comprises the alkaline activation of a slag, to be used as a binder for soil stabilization with the Deep Mixing Method. Activation of these materials is commonly made with Sodium Silicate (SS) and Sodium Hydroxide (SH). The SS provides immediate soluble silica, essential for the formation of the binding gel, which is responsible for the mechanical strength of these materials, but it is a very expensive material. Activation with SH alone is relevant, even if it does not generate the mechanical strength that a combination of SS and SH can achieve, it can still satisfy the current soil improvement requirements. For this purpose, a statistical analysis was performed, using the Design of Experiments (DoE) and 11 mixtures were tested to determine the optimum mixture in terms of mechanical behavior, which was evaluated the dates of compression strength tests with a response surface design.

1 INTRODUCTION

The increase in tourist activities along riverbanks and in coastal areas and the need for leisure areas have attracted more and more investment in the European Union (EU). These areas are generally characterized by soils with poor geomechanical characteristics such as fine sands, silts and sludges of low compactness and high-water content that do not guarantee the necessary support conditions for the works that are intended to be installed in them. The techniques currently used consist of foundation with piles up to a competent stratum, or in the treatment of these soils by jet grouting or Deep Soil Mixing, using large quantities of Portland cement that requires good quality raw material, and whose production releases about 1 ton of carbon dioxide (CO_2) to the atmosphere per ton of CO_2 produced.

The growth of concrete production has led to the need to increase cement production to quantities never attained. Cement production increases exponentially and is expected to reach the top of the emissions produced by human activity along with the energy and transport sector (Rattanasak & Chindaprasirt, 2009). New cementitious materials can be produced from calcined waste, but the hazardous elements present in their constitution need to be blocked in the three-dimensional framework of the geopolymer matrix.

Studies of the potentials of the alkaline activation (AA) technique in soil improvement show promising results (Cristelo *et al.*, 2012a; 2012b; 2013). Depending on the application, the quantities of water, binder and alkaline solutions can be adapted, considering various parameters such as Liquids/Solids ratio, silica/alumina ratio, among others. Therefore, it is essential to evaluate the best range of values of these parameters for application in soil improvement.

In this work, a mixture consisting of alkaline activated slag was used to stabilize a soil gathered from an experimental field located in Vila do Conde – Portugal. The soil in Experimental Field, being very close to a river mouth and therefore with very high-water levels. This comprises a very unfavorable situation for the soil stabilization process, due to the significant water content of the soil. To overcome this problem, the dry version of the Deep Mixing Method was simulated in the present study.

2 MATERIALS

2.1 *Slag*

The ladle slag used in this study was collected at the Megasa steel industry, in Maia, Portugal it is a white fine powder. Kambole et al. (2019) studied a geotechnical application for this slag, specifically in road bases, and concluded that it is a non-hazardous solid waste. According to the same author, the use of this slag produces a controlled environmental impact with limited leaching problems. It is very rich in calcium and, thus, ideal for alkali activation reactions (Table 1). This material was milled in cycles of 15 h, in a micro-deval apparatus, to increase the specific surface of the material, and enhance the quality of the reactions with the activator. After milling, more than 50% of the slag had a size of less than 10 μm, as shown in the Table 1.

2.2 *Soil*

The soil used in this study was collected in an area close to the center of Vila do Conde city, in the north of Portugal, located on the left bank of the Ave river, very close to its mouth, which means that it influenced by the tide. The tides in this region of the Atlantic coast are very wide, often reaching 4 m difference in height between low tide and high tide. The site presents coastal or estuarine situation, subject to sedimentation predominantly due to the sea agitation, but also by effect of tidal currents and fluvial inflows. In Cone Penetration Test (CPTu) tests performed by Rios *et al.*, (2018a) it was noticed that the soil had the groundwater level at surface for this reason the soil should be used in its saturated condition to prepare the mixtures. To reproduce this field behavior equation 1 was used, where the degree of saturation (S) was 1, and the void ratio (e) was assumed to be 0.8, based on the CPTu data. From this, a water content value of 30.55% was obtained which was kept constant throughout this experimental plan. The other geotechnical properties of the soil are resumed in Table 1.

$$w = \frac{S \cdot e}{G_s} \tag{1}$$

2.3 *Alkaline activator*

The alkaline activator solution was prepared in agreement with previous studies (Rios *et al.*, 2018b; Pinheiro *et al.*, 2018), consisting in a combination of sodium hydroxide and sodium silicate

Table 1. Geotechnical properties of Slag and Soil.

Property	Slag	Soil
Plastic Limit	NP	NP
Liquid Limit	NP	NP
D_{50}	0.01 mm	0.07 mm
Specific gravity	3.34	2.61
Fines fraction (sieve N° 200)	98.05%	50.52%
Uniformity Coefficient	1.81	10.02
Curvature Coefficient	1.03	2.72

solutions. The sodium hydroxide was originally in flake form, with a specific gravity of 2.13, at 20°C, and a 95–99% purity. The sodium silicate was already in solution form, with 26.50% SiO_2, 13.50% Na_2O and 60.00% H_20, a specific gravity of 1.5 and a SiO_2/Na_2O ratio of 2 by mass. Since the dry version of the Deep Mixing Method is to be simulated, the final water content in the mixtures results only from the above-mentioned water, present in the activator. In the laboratory, this was simulated by using the soil in its dry state, which resulted in a water content (in the activator) corresponding to 30.55% of the weight of dry soil. The final molal concentration of sodium hydroxide after mixing with sodium silicate is calculated taking into account the number of moles of NaOH (X_{NaOH}), derived from the Na_2O present in both the sodium silicate solution and in the sodium hydroxide flakes, with the total quantity of water present in the mixture in grams (w_t) – equation 2:

$$NaOH_{molal} = \frac{X_{NaOH}x1000}{w_t} \tag{2}$$

3 EXPERIMENTAL PLAN

3.1 *Response surface method for design of experiments*

The surface response method was followed as described in Rios *et al.*, (2018b) and Pinheiro *et al.*, (2018). This method is very useful to study pastes with several components, such as those resulting from alkaline activation of slag, where the best performing blend is the target. This is usually achieved by changing the amount of the components, one at a time, following the traditional method of one factor at a time. In fact, as the optimal response level approaches, a two-level factorial design no longer provides enough information to adequately model the true response surface. However, if central and axial points are added to the two-level factorial design, a composite design suitable for response surface method is obtained (Whitcomb & Anderson, 2004). In this work we have used a composite design centered on the face, which means that the axial points are on the face of the cube.

3.2 *Tested mixtures*

The parameters selected as the most relevant in the behavior of the mixtures were: sodium silicate percentage in relation to the activator solution (X) and final sodium hydroxide molal concentration (Y). The variation of only two components for this specific purpose was based in results obtained by Cristelo (2009); Rios *et al.*, (2016); Rios *et al.*, (2018b) and Pinheiro *et al.*, (2018). It was also observed that the material performed well at molar concentrations of the sodium hydroxide (Y) in the range of 6.5 to 8.0 molal. The combination of these two parameters will lead to different molar ratios for the four main elements responsible for the chemical reaction process: silica, alumina, sodium and water. The method defined the composition of the mixtures to be tested by changing the 2 parameters indicated above, in a range of values as expressed in Table 2. In the mixtures that were made, the Slag/Soil proportion was defined at a ratio of 1: 2 in weight. The higher ratios like 1:3 or 1:4 did not harden enough to be demolded after one day curing. This means that in a field application this material would be carried by the surrounding water before it could form a hardened binding gel. The compositions of the 11 mixtures generated are shown in Table 3.

Table 2. Range of variation entered in the program.

Variables	Minimum	Maximum
X	0.0	0.5
Y	6.5	8.0

Table 3. Design of Experiments with amounts of each constituent.

N°	X	Y	Soil (g)	Slag (g)	NaOH (g)	Silicate (g)	Water (g)
1	0.25	6.50	397.06	198.53	28.08	19.85	121.30
2	0.25	7.25	397.06	198.53	31.72	19.85	121.30
3	0.25	7.25	397.06	198.53	31.72	19.85	121.30
4	0.50	8.00	397.06	198.53	31.89	39.71	121.30
5	0.50	7.25	397.06	198.53	28.26	39.71	121.30
6	0.50	6.50	397.06	198.53	24.62	39.71	121.30
7	0.25	7.25	397.06	198.53	31.72	19.85	121.30
8	0.00	7.25	397.06	198.53	35.17	0.00	121.30
9	0.25	8.00	397.06	198.53	35.35	19.85	121.30
10	0.00	6.50	397.06	198.53	31.54	0.00	121.30
11	0.00	8.00	397.06	198.53	38.81	0.00	121.30

3.3 *Performed tests*

The activator solutions were prepared 24 h before the mixing of the pastes, to cool down, thus avoiding the heat of the reactions to affect the curing process, either by exothermic heat or by incomplete reactions in the dissolution of the activators. When the constituents were mixed, the slag was mixed with activation solution and after then the soil was added. The samples were prepared in metal molds of $4 \times 4 \times 16\,\text{cm}^3$ and demolded 24 h later (EN 2016). Bruce (2000), Bruce & Bruce (2001), BSI (2005), Kitazume *et al.* (2015), Yan *et al.* (2019) and Abdullah *et al.* (2019) used unconfined compression strength tests to evaluate the mechanical performance of DSM improved soils. For this reason, only compression strength tests were executed in the present work at 7 days of curing time.

4 RESULTS AND DISCUSSION

The results of compressive strength tests presented in Figure 1 were analyzed by the 'Design of Experiments' (DoE). Figure 2 illustrates the behavior of the test samples, by presenting the response surface for the proposed values of the model (left) and a contour plot representing a three-dimensional surface by plotting constant z slices, called contours, in a two-dimensional format (right). That is, given a value for z, the lines are drawn to connect the coordinates (x, y) where this z value occurs.

From Figure 2 (right) it seems that the hydroxide concentration does not have much influence on the strength since the contour lines are almost vertical. Of course, it is well known that the hydroxide concentration is very important as concluded by several researchers (Cristelo *et al.*, (2012a); (2012b); (2013); Rios *et al.*, (2016); (2018) and Pinheiro *et al.*, (2018)). However, for this specific and narrow range of sodium hydroxide concentrations [6.5–8.0] molal, increasing the concentration does not lead to higher strength. This means that the best composition is one with higher values of X and low value of Y, that is, NaOH with 6.5 molal and sodium silicate/sodium hydroxide of 0.5 (with a 50% of each solution), which gave a unconfined compressive strength value of 0.72 MPa at 7 days. It is expected that at 28 days the strength would be higher than 1 MPa, the value suggested by Bruce (2000) and Bruce & Bruce (2001). The surface plot indicated in Figure 2 (left) can be expressed by expression 3. However, it should be highlighted that this equation is just valid for this specific range of X and Y.

$$Cs_{7\,days} = 4{,}32 + 1{,}82\,X - 1{,}10\,Y - 0{,}654\,X^2 + 0{,}0728\,Y^2 - 0{,}072\,XY \qquad (3)$$

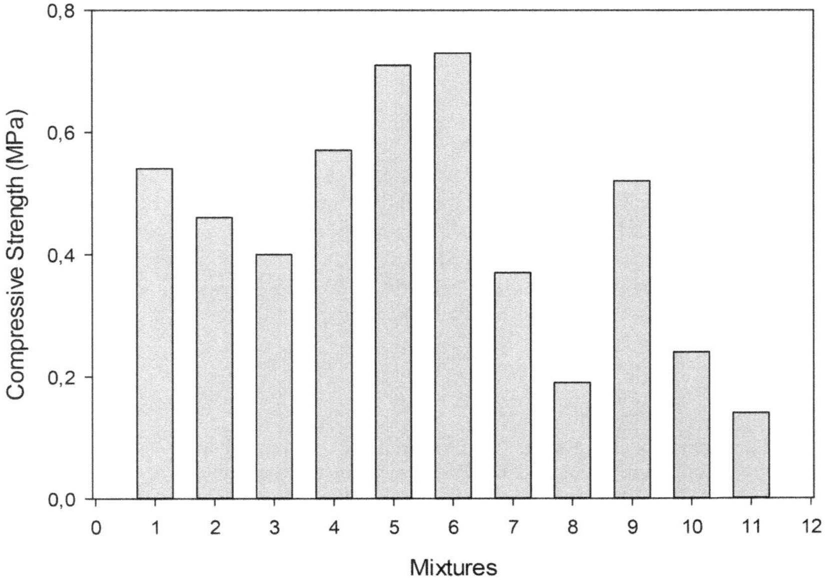

Figure 1. Results for compressive strength at 7 days curing.

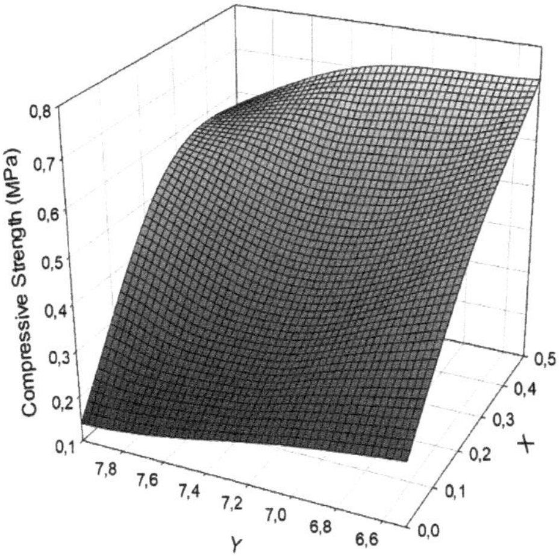

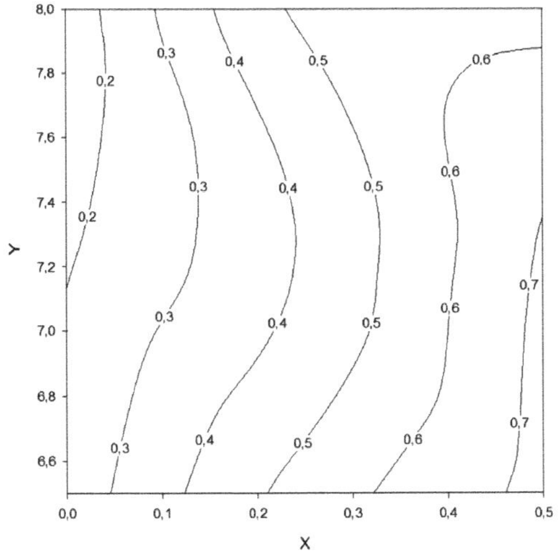

Figure 2. Surface plot on the evolution of the resistance against the variation of the inserted variables (left) and compressive strength at 7 days, showing the evolution of resistance with variation of input data (right).

5 CONCLUSIONS

The paper presents the use of the response surface method to optimize the composition of a soil-binder combination, fabricated with alkaline activated slag (including sodium silicate and sodium hydroxide) and soil collected in the coastal area of Vila do Conde (Portugal). The aim is to use this binder to stabilize the mentioned soil, on site. Results show that the optimum molar concentration of sodium hydroxide is 6.5, and that the mechanical strength increases with the silicate content.

Further studies are currently being developed to study mixtures with lower concentrations of sodium hydroxide.

ACKNOWLEDGMENTS

The authors would like to acknowledge the company Megasa, which runs the steel factory of Maia, for the slag supply. This work was financially supported by: UID/ECI/04708/2019- CONSTRUCT – Instituto de I&D em Estruturas e Construções funded by national funds through the FCT/MCTES (PIDDAC). It was also funded by CNPq (the Brazilian council for scientific and technological development) for its financial support in 201465/2015-9 scholarship of the "Science without borders" program.

REFERENCES

Abdullah, H. H., Shahin, M. A., & Sarker, P. 2019. Use of Fly-Ash Geopolymer Incorporating Ground Granulated Slag for Stabilisation of Kaolin Clay Cured at Ambient Temperature. *Geotechnical and Geological Engineering*, *37*(2), 721–740. https://doi.org/10.1007/s10706-018-0644-2

Bruce, D. A. 2000. *An introduction to the deep soil mixing methods as used in geotechnical applications.*

Bruce, D. A., & Bruce, M. E. 2001. Practitioner's guide to the deep mixing method. *Ground Improvement*, *5*(3), 95–100.

BSI. 2005. BS EN 14679: 2005: Execution of special geotechnical works–deep mixing. BSI London, UK.

Cristelo, N., Glendinning, S., Fernandes, L., Teixeira Pinto, A. 2012a. Effect of calcium content on soil stabilisation with alkaline activation. *Construction and Building Materials*, *29*, 167–174. https://doi.org/10.1016/J.CONBUILDMAT.2011.10.049

Cristelo, N. 2009. *Deep soft soil improvement by alkaline activation*. University of Newcastle Upon Tyne.

Cristelo, N., Glendinning, S., Miranda, T., Oliveira, D., & Silva, R. 2012b. Soil stabilisation using alkaline activation of fly ash for self compacting rammed earth construction. *Construction and Building Materials*, *36*, 727–735.

Cristelo, N., Soares, E., Rosa, I., Miranda, T., Oliveira, D. V, Silva, R. A., & Chaves, A. 2013. Rheological properties of alkaline activated fly ash used in jet grouting applications. *Construction and Building Materials*, *48*, 925–933.

EN, T. S. 2016. 196-1. Methods of testing cement–Part 1: Determination of strength. *European Committee for Standardization*, 36.

Kambole, C., Paige-Green, P., Kupolati, W. K., Ndambuki, J. M., & Adeboje, A. 2019. Comparison of technical and short-term environmental characteristics of weathered and fresh blast furnace slag aggregates for road base applications in South Africa. *Case Studies in Construction Materials*, *11*, e00239.

Kitazume, M., Grisolia, M., Leder, E., Marzano, I. P., Correia, A. A. S., Oliveira, P. J. V., Andersson, M. 2015. Applicability of molding procedures in laboratory mix tests for quality control and assurance of the deep mixing method. *Soils and Foundations*, *55*(4), 761–777.

Pinheiro, C., Molina-Gómez, F. A., Rios, S., Miranda, T., & Viana da Fonseca, A. 2018. Proportional Estatistics Analysis of the Constituents of an Alternative Binder Used In Soil Reinforcement. (p. 11). Punta Delgada.

Rattanasak, U., & Chindaprasirt, P. 2009. Influence of NaOH solution on the synthesis of fly ash geopolymer. *Minerals Engineering*, *22*(12), 1073–1078. https://doi.org/10.1016/J.MINENG.2009.03.022

Rios, S., Cristelo, N., Viana da Fonseca, A., & Ferreira, C. 2016. Stiffness Behavior of Soil Stabilized with Alkali-Activated Fly Ash from Small to Large Strains. *International Journal of Geomechanics*, *17*(3), 4016087.

Rios, S., da Fonseca, A. V., Cristelo, N., & Pinheiro, C. 2018a. Geotechnical Properties of Sediments by In Situ Tests (pp. 59–68). Springer, Cham. https://doi.org/10.1007/978-3-319-61902-6_6

Rios, S., Nunes, S., Viana da Fonseca, A., & Pinheiro, C. 2018b. Alkali-activated cement using slags and fly ash. *4th International Conference WASTES – Solutions, Treatments and Opportunities*, 161–166.

Whitcomb, P. J., & Anderson, M. J. 2004. *RSM simplified: optimizing processes using response surface methods for design of experiments*. CRC press.

Yan, B., Kouame, K.-J.-A., Tannant, D., Lv, W., & Cai, M. 2019. Effect of Fly Ash on the Mechanical Properties of Cemented Backfill Made with Brine. *Geotechnical and Geological Engineering*, *37*(2), 691–705. https://doi.org/10.1007/s10706-018-0640-6

Wastes: Solutions, Treatments and Opportunities III – Vilarinho et al. (Eds)

Incorporation of steel slag and reclaimed asphalt into pavement surface layers

I. Rocha Segundo & E. Freitas
Department of Civil Engineering, University of Minho, Guimarães, Portugal

V. Castelo Branco
Department of Transportation Engineering, Federal University of Ceará, Fortaleza, Brazil

S. Landi Jr.
Centre of Physics, University of Minho, Guimarães, Portugal
Federal Institute Goiano, Rio Verde – GO, Brazil

M.F. Costa & J. Carneiro
Centre of Physics, University of Minho, Guimarães, Portugal

ABSTRACT: There is an increasing concern about the recycling and reuse of wastes in different areas. This research aims to analyze the technical viability of the use of Reclaimed Asphalt Pavement (RAP) and Steel Slags (SS) for the composition of asphalt mixtures for surface layers. Therefore, three asphalt mixtures AC 10 were designed: without recycled materials (R), with 30% of RAP (F) and with 30% of SS (A). They were mechanically and superficially assessed. Their water sensitivity and the permanent deformation were similar. The mixture A presented higher stiffness modulus and lower fatigue resistance when compared to the other ones. The best sound absorption and mechanical impedance found were for A and the worst ones were for F. Mixture A had smoother macrotexture if compared to the other mixtures. The asphalt mixtures had similar friction. It can be concluded that these wastes can be incorporated in asphalt mixtures for surface layers.

1 INTRODUCTION

Currently, it is undeniable to say that there is an increasing concern about natural resources depletion and environmental damages. In Paving Industry, most used raw materials, generally, come from non-renewable sources, i.e. aggregates and oil products, which are used as binders. On the other hand, there has been growing motivation for recycling wastes by incorporating them into asphalt mixtures. With this it is possible to obtain more ecological pavements, guaranteeing greater sustainability. In this context, the potential of some wastes has been studied: Steel Slags (SS), Construction and Demolition (C&D) waste, tire rubber and some types of recycled polymers.

The main recycled waste incorporated in asphalt mixtures refers to the milled material from deteriorated asphalt pavement, which is called Reclaimed Asphalt Pavement (RAP). This material has already been studied in several replacement ratios in asphalt mixtures, and under different techniques of production, such as hot, warm and cold asphalt mixtures with or without additives. Its incorporation presents results which are even better than the ones found for conventional asphalt mixtures, while offering some benefits, for example, economical due to the use of aggregates and binder from RAP (Rocha Segundo *et al.*, 2016). In the United States, the most recycled material is RAP, with more than 80 million tons per year, representing of about twice the other fourth most recycled wastes: paper, glass, plastic, and aluminum (Brosseaud, 2011). When discarded improperly, it can cause damages to the environment. With its use as a recycled material, it is possible to save money with the transport of virgin materials to the worksite, since RAP can be found on the roads to be paved around the urban perimeter, where quarries are often not located.

For the aged asphalt binder presented in RAP, its incorporation can provide more rigidity to the asphalt mixtures. On one hand, this can improve permanent deformation resistance, on the other hand, its use can decrease fatigue cracking resistance (Hussain and Yanjun, 2013).

Other wastes from different sources have already been incorporated in the Paving Industry. It is worth mentioning the use of slags, the by-product of the Steel Industry. This residue has already been incorporated into asphalt mixtures, from base courses to surface layers. Generally, there are two types of slags produced by the industry: i) blast furnace: obtained directly from blast furnace; (ii) steel: resulting from the production of steel, obtained in electric furnaces and in oxygen converters. Some of its limitations are associated with its high density, causing higher transport costs, its high water absorption, which can increase the binder content, and its high expansion, requiring a curing period before the incorporation (Fakhri and Ahmadi, 2017).

Due to its metallic material content, mainly iron oxide (Fe_2O3), the SS present high electrical and thermal conductivity. Thus, asphalt pavements composed of this material, when irradiated by microwaves, can melt easier snow on its surface during the winter when compared to conventional asphalt pavements. Therefore, it can be portrayed as a new function of asphalt mixture, since this property is not considered as an essential one (Gao *et al.*, 2017).

The literature points out that most of the researches are about the mechanical characterization of recycled asphalt mixtures, but not the functional (superficial) characterization (mainly taking into account the use of SS). Thus, this research aims to analyze the technical viability of the use of RAP and SS for the composition of asphalt mixtures for the surface layer of road pavements.

2 MATERIALS

2.1 *Aggregates and Bitumen*

In order to design the asphalt mixtures, two granite aggregates (fine aggregate 0/4 mm and course aggregate 4/10 mm), filler, RAP 0/6 mm and SS 0/10 were used. Their main properties are presented in Table 1.

The Form 2D, Sphericity, Angularity, and Surface Texture were characterized by Digital Image Processing (DIP) technique using Aggregate Image Measurement System (AIMS) (Table 2) (Al Rousan, 2004; Araujo, Bessa and Castelo Branco, 2015). By Their Form 2D, the aggregates 4/10, 0/4 and RAP were classified as elongated, while SS and filler were characterized as semicircular and semi-elongated respectively. All the coarse aggregates were characterized as moderate sphericity by the parameter sphericity. Considering the angularity, all the aggregates were sub-rounded. The surface texture of the materials was smooth for 4/10 and RAP and low rough for SS. Regarding all these parameters, it is possible to mark that the best aggregate characteristics is for SS.

The binder used was a commercial one named Elaster 13/60 provided by Cepsa®. The modified bitumen was classified as 35/50 by penetration grade. The bitumen was characterized by (i) penetration of 50×10^{-1} mm (EN 1426/2015); (ii) Brookfield viscosity of 349 cP at 180°C (EN 13302/2010); (iii) Softening Point of 64°C (EN 1427/2015).

Table 1. Main properties of the natural and recycled aggregates.

Aggregate	Particles density (g/cm^3)	Water absorption (%)	Micro-deval abrasion loss (%)	Los Angeles abrasion (%)	Binder (%)
4/10	2.63	0.9	15	30	–
0/4	2.67	0.5	–	–	–
Filler	2.71	–	–	–	–
RAP 0/6	2.55	1.0	–	–	6.7
SS 0/10	3.38	2.4	11	25	–

Table 2. Form, sphericity, angularity and texture characterization.

Property	Aggregate	Average	Classification
Form 2D (fine)	4/10	8.0	Elongated
	0/4	8.4	Elongated
	RAP	8.1	Elongated
	SS	7.9	Semi-circular
	Filler	8.3	Semi-elongated
Sphericity (coarse)	4/10	0.764	Moderate sphericity
	RAP	0.737	Moderate sphericity
	SS	0.771	Moderate sphericity
Angularity (fine and coarse)	4/10	3619.0	Sub-rounded
	0/4	3911.4	Sub-rounded
	RAP	3825.5	Sub-rounded
	SS	3554.1	Sub-rounded
	Filler	2466.3	Sub-rounded
Surface Texture (coarse)	4/10	175.3	Smooth
	RAP	202.7	Smooth
	SS	345.4	Low roughness

Table 3. Macrotexture and friction results.

Asphalt mix	% 4/10	% 0/4	% Filler	% RAP 0/6	% SS 0/10	% binder	% virgin binder	MBD (g/cm^3)	BD (g/cm^3)	VC (%)
R	68	28	4	–	–	5.5	5.5	2.428	2.305	5.1
F	67	–	3	30	–	5.4	3.5	2.446	2.334	4.6
A	42	25	3	–	30	5.4	5.5	2.676	2.569	4.0

2.2 *Asphalt mixtures*

Three asphalt mixtures were designed by Marshall design method: i) R: reference mixture, ii) F: recycled mixture with 30% of RAP, iii) A: recycled mixture with 30% of SS. The asphalt mixtures have the same gradation and almost the same content of binder, of about 5.5%. Table 3 shows the composition of the asphalt mixtures, binder contents, Bulk Density (BD), Maximum Bulk Density (MBD) and Void Content (VC).

3 METHODS

3.1 *Mechanical characterization*

In order to characterize the asphalt mixture from a mechanical point of view, water sensitivity, permanent deformation, stiffness modulus, and fatigue resistance were assessed. The water sensitivity was evaluated by Indirect Tensile Strength Ratio (ITSR) test (EN 12697-12). The ratio between the wet samples (ITSw) and the dry samples (ITS) was calculated and defined as ITSR. The permanent deformation resistance of the asphalt mixtures was assessed by the Wheel Tracking Test (WTT) (EN 12697-22). The deformation curve versus cycle and the maximum rutting of the asphalt mixtures will be compared. For the stiffness modulus, the asphalt mixtures were assessed by the four-point bending test configuration (EN 12697-26) for different frequencies and temperatures. The master curve stiffness versus frequency (Hz) will be compared at 20°C. Considering the fatigue resistance (EN 12697-24), the samples at 20°C are submitted to sinusoidal loading at 10 Hz, in strain control mode of loading. The relationship between the strain level versus the number of cycles will be presented.

3.2 *Superficial characterization*

In order to characterize superficial properties of the asphalt mixtures, Mean Profile Depth (MPD), British Pendulum, sound absorption, and mechanical impedance tests were carried out. The same samples of WTT were used to characterize superficial properties. The macrotexture was assessed by MPD. Profiles of 25 × 25 cm^2 surface were acquired by a device equipped with a laser, and the MPD was calculated (ISO 13473-1:2019). The device can acquire profiles every 0.2 mm with a vertical resolution up to 0.01 mm. The baseline of the profiles is divided by half of the length, and the average of their two peak heights is used to calculate the MPD. To access the friction, British Pendulum test was carried out (EN 13036-4).

Related to tire-road noise, for sound absorption, a self-made impedance tube with 80 mm of diameter utilizing the two-microphone arrangement was used (Freitas *et al.*, 2014). A graph of absorption versus frequency (between 500 and 2000 Hz) will be presented. For mechanical impedance, damping is a measure for determining the capacity of the structure to dissipate energy. An accelerometer was bonded to the specimen and it was submitted to a hammer impact in the opposite side. The Graph damping versus frequency (Hz) will be presented. The configuration for the last two tests of noise indicators can be seen on Freitas et al. (2014).

4 RESULTS

The asphalt mixtures were mechanically characterized by water sensitivity, permanent deformation and fatigue resistance tests. The results of the water sensitivity test are shown in Table 4. On the one hand, the ITS of the recycled asphalt mixtures increased by 44% and 34% for those ones composed of RAP (F) and SS (A), respectively. The ITSw also increased, 37% and 35% for the same mixtures, respectively. On the other hand, the ITSR (water sensitivity) maintained for A but decreased 5% for F. All the asphalt mixtures respected the minimum value of ITSR (80%) required for Portuguese asphalt mixtures for the top layer of road pavements.

For the permanent deformation (Figure 1a), at the end of the test, R, F, and A mixtures respected the maximum value considering the Portuguese requirements (20 mm) and deformed 1.86, 1.93 and 2.00 mm. The recycled asphalt mixtures had similar behavior. F and R mixtures, respectively, deformed 4% and 8% more than the conventional AC 10. It can be concluded that the asphalt mixtures had a similar behavior considering the permanent deformation.

Regarding the results of the master curve of stiffness modulus at 20°C (Figure 1b), R (reference mixture), F (composed of 30% of RAP) and A (composed of 30% of SS) had values between 201 and 10,192 MPa, 858 and 14,504 MPa, 251 and 12,143 MPa, respectively. Thus, considering the low frequencies, on the one hand, A mixture had similar moduli when compared to R. On the other hand, F mixture had stiffness modulus of about 4 times higher than R for the lowest frequency. At high frequencies, the stiffness moduli of R and A were similar and the difference between the results of F and R was lower than that one found at low frequencies. F had stiffness modulus about 1.4 times higher than R for the highest frequency. In general, A and R had similar results, and F had

Table 4. Water sensitivity results.

Asphalt Mix	% RAP 0/6	% SS 0/10	VC (%)	ITS (MPa)	ITSw (MPa)	ITSR (%)
R	–	–	5.1	1.47	1.35	92
F	30	–	4.6	2.12	1.84	87
A	–	30	4.0	1.97	1.82	92

higher values. Probably this resulted from the aged binder of the RAP that composed F. Its content was 1.9%. For the fatigue resistance (Figure 1c), the strain for $N = 10^6$ cycles was 339×10^{-6}, 262×10^{-6} and 317×10^{-6} for R, F, and A, respectively. A and R had similar curves and F presented lower results. It can be concluded by the mechanical results that the asphalt mixture that used SS as aggregate (A) presented superior mechanical results. Probably, this fact is related to the use of the aged binder from RAP provided higher stiffness for the mixture, contributing to lower fatigue resistance and higher stiffness modulus.

Also, the asphalt mixtures were superficially characterized by sound absorption and mechanical impedance, macrotexture and friction. Considering the assessments related to tire-road noise, mechanical impedance, and sound absorption, the results are presented in Figure 1d. For sound absorption, the average value was 0.18, 0.16 and 0.20 for R, F, and A, respectively. Although the asphalt mixtures did not have a peak of maximum sound absorption within the interest frequencies, the best absorption was for A and the worst was for F. For damping, the investigated asphalt mixtures had 2 peaks. The first one was between 418 and 550 Hz and the second one was from 1,770 to 2,125 Hz. The same trend for sound absorption was reported. The best damping was found for A and the worst one for F.

For the macrotexture characterization, MPD was carried out and for friction characterization, the British Pendulum Test was used (Table 5). Analyzing the macrotexture, the mixture A had a macrotexture of about 17% smoother than the other mixtures. Regarding friction, A had 7% lower

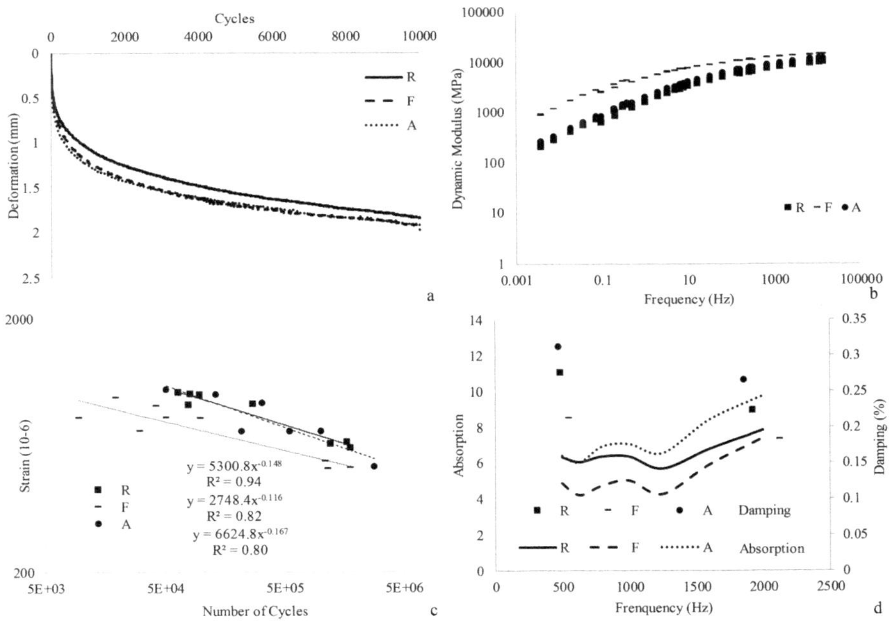

Figure 1. a) Permanent Deformation, b) Dynamic Modulus, c) Fatigue and d) Absorption and Damping.

Table 5. Macrotexture and friction results.

Asphalt mix	MPD (mm)	British pendulum value (PTV)
R	1.3	61
F	1.3	60
A	1.1	57

PTV and F had a similar PTV when compared to that of the R. Considering the superficial analysis, the best results for macrotexture and friction of the recycled asphalt mixtures is for F (30% RAP), but, considering the noise indicators, sound absorption and mechanical impedance, mixture A (30% SS) presented the best results.

5 CONCLUSIONS

This research aimed to analyze the viability of the use of RAP and SS in the composition of asphalt mixtures for surface layers. Three asphalt mixtures were designed, and mechanically and superficially characterized: without wastes (R), with 30% of RAP (F) and with 30% of SS (A). Based on the results of this research, the following conclusions can be drawn:

- ITS and ITSw of the recycled asphalt mixtures increased when compared to the conventional mixture, R. The water sensitivity of the asphalt mixtures was similar. All the asphalt mixtures had similar permanent deformation as well. In general, A and R had similar results of stiffness modulus, and F was stiffer. A and R had a similar fatigue curve and F had lower fatigue resistance. Probably, these results were due to the aged binder of RAP that composes F. Stiffer binder conducted to a stiffer asphalt mixture and to lower fatigue resistance.
- Regarding the surface characterization, the best absorption and mechanical impedance found was for A and the worst was for F. F and R had the same value of MPD (macrotexture). The mixture A had smoother macrotexture if compared to R and F. Regarding the friction results, the asphalt mixtures had similar PTV values.

Considering mechanical results, the best asphalt mixture was A (30% SS). Regarding the superficial analysis, the best macrotexture and friction results of the recycled asphalt mixtures was for F (30% RAP), but, considering the noise indicators, sound absorption and mechanical impedance, the best one was A. Thus, probably, mix A is more resistant and quieter than F, but F is safer than A. The aggregate characteristics of SS contributed to better mechanical results comparing to those ones observed for F. The use of SS did not impact the asphalt mixture properties significantly, differently of the use of RAP that conducted to lower fatigue resistance. Some superficial benefits were observed with the use of SS considering the noise indicators. By the fact that the recycled asphalt mixtures respected the characterization requirements, it can be concluded that SS and RAP can be used in asphalt mixtures for surface layers.

REFERENCES

Al Rousan, T.M. 2004. Characterization of aggregate shape properties using a computer automated system. Ph.D. Thesis, Texas A&M University.

Araujo, V.M.C., Bessa, I.S. & Castelo Branco, V.T.F. 2015. Measuring skid resistance of hot mix asphalt using the aggregate image measurement system (AIMS). *Construction and Building Materials*, 98: 476–481.

Brosseaud, Y. 2011. Recycling of asphalt mixtures: evolution after 20 years and the current situation of France (in Portuguese language). In: 3° Salão de Inovação ABCR – 7° *Congresso Brasileiro de Rodovias e Concessões*. Foz do Iguaçu.

Fakhri, M. & Ahmadi, A. 2017. Recycling of RAP and steel slag aggregates into the warm mix asphalt: A performance evaluation. *Construction and Building Materials*, 147: 630–638.

Freitas, E.F. Rodrigues, J.D., Rocha, J.A. & Silva, H.M.R.D. 2014. Innovative low noise surfaces: comparison of damping and absorption. In: *International Congress on Noise Control Engineering, Melbourne: Australian Acoustical Society (AAS)*.

Gao, J. Sha, A., Wang, Z., Tong, Z. & Liu, Z. 2017. Utilization of steel slag as aggregate in asphalt mixtures for microwave deicing. *Journal of Cleaner Production*, 152: 429–442.

Hussain, A. & Yanjun, Q. 2013. Effect of reclaimed asphalt pavement on the properties of asphalt binders. *Procedia Engineering*, 54: 840–850.

Rocha Segundo, I.G. Castelo Branco, V.T.F., Vasconcelos, K.L., Holanda, A.S. 2016. Hot mix recycled asphalt with the incorporation of a high percentage of reclaimed asphalt pavement as an alternative to high modulus layer (in Portuguese language). *Transportes*, 24(4): 85–94.

Wastes: Solutions, Treatments and Opportunities III – Vilarinho et al. (Eds)

A brief review on sustainable packaging materials

G. Silva & M. Oliveira
MDIP, FEUP, Porto, Portugal

B. Rangel
Design Studio FEUP, FEUP, Porto, Portugal

J. Lino
INEGI, Faculty of Engineering, University of Porto, Porto, Portugal

ABSTRACT: The demand to research sustainable packaging materials with increasingly better mechanical properties is growing because of the impact of packaging waste on the environment. Twenty five million tonnes of plastic waste are generated annually in Europe, and of these, only 30% are collected for recycling (Penca, 2018). The focus of this review is the biomaterials, with a highlight on the biodegradable materials. This choice is justified by the fact that they are a possible solution to the today's scenario. This research presents several biofoams that prove to be potential substitutes of EPS, a non-biodegradable fossil-based foam. Among them, it is proposed the Green cell foam that dissolves in water or Khel biofoam that is easily compostable in a household composter or in the backyard.

1 INTRODUCTION

An excessive production of plastic packaging materials with a purpose of only a single usage should mean an increasing use of biomaterials in the near future. The intention is to understand the field of biomaterials and to conduct an analysis to understand their possible application in packaging. Also, to analyse if the sustainable materials that are already in the market, or soon to be launched, are ready for replacing EPS (expanded polystyrene). The main goal is to find a material with mechanical properties that are similar to EPS. The solution of the EPS problem is encouraged by the European Union Plastics Strategy, which aims to protect the environment from pollution resulting from plastic production and disposal, while fostering growth and innovation. According to a new strategy for plastics, until 2030, all plastic packaging on the European Union market will be recyclable, the consumption of disposable plastic objects will be reduced and the intentional use of micro plastics will be restricted (ECHA, 2018). As the focus is directed towards the end-of-life, it is intended that, rather than finding a recyclable material, search for a biodegradable and compostable one in industrial facilities or at home. It is crucial to fight against the current environmental reality, in which more than 58 million tons of plastic are generated annually, being 40% used in packaging (European Commission, 2018). The plastic usage is so deep-rooted in our lives that it is difficult to avoid them. Plastics have a significant role in protecting and enabling safety and quality of the products (Bradley et al., 2011). Therefore, biomaterials – materials of biological origin – arise as a way to solve the end-of-life problem. One of which consists of collecting, identifying, separating, transporting, cleaning and reprocessing, making recycling too expensive. Consequently, the materials often go to incineration or to landfills where they remain untouched for hundreds of years. This leads to the contamination of oceans, lands and food chain (Souza and Fernando, 2016).

This led to wonder: what is a sustainable material? Although sustainability is a complex term and a concept that requires standards and documentation to evaluate the design of the packaging, the materials, processes and the entire life cycle, we need to focus on packaging disposal and end-of-life. The intention is to identify different materials that could compete with EPS, particularly in mechanical properties and that are versatile to a wide range of packaging functions, requirements and shapes.

2 BIOMATERIALS

There is still discussion about the definition of what is a biomaterial. Many companies are investing in new materials and on research. Consequently, biomaterials are still under study. They can be classified according to their interaction with biological systems – biocompatibility, and if they are bio-inert or biodegradable. Bio-inert when in contact with biological tissues do not interact with them, therefore, are commonly used in medical prosthetics. Examples of this are titanium, stainless steel and cobalt alloys. Biodegradable materials disintegrate when in contact with specific environments. They degrade under the action of microorganisms, such as fungi and bacteria. The cells of the microorganisms feed on these materials and are a carbon source.

The biodegradable materials are divided into three categories, namely metals, ceramics and plastics (Godavitarne et al., 2017). The idea is to focus on the (bio) plastics or biopolymers. Like biomaterials, these are also classified according to their production method.

2.1 *Biopolymers and bio composites*

A biopolymer can be defined by the production method. According to European Bioplastic they are bio-based, biodegradable or have both properties (European Bioplastics, 2016). When they are biodegradable or have both properties, the biodegrading will occur by the action of microorganisms, biological activity or by both (biodegradation and/or composting). To sum up, the biopolymers can be divided in three different groups, as mentioned. It should be noted that a bio-based material is different from a biodegradable material (Reddy et al., 2013). A material such as Bio-PE (polyethylene) can be bio-based, for example Spur's biofoam (called "I'm Green"). This foam is bio-based because one of its components is ethanol – made from sugarcane, a renewable resource. This material is not biodegradable but emits less greenhouse gases than traditional PE, being classified as a biomaterial or biopolymer. It is an alternative solution produced from renewable sources with properties similar to PE (SPUR®, 2018). On the other hand, a fossil-based material can be utilized and at the end of its use prove to be biodegradable, like polycaprolactone, an aliphatic polyester commercialized, for example, by Plastimake as material to manufacture prototypes (Plastimake, 2018).

Within biodegradable biopolymers, the compostable can also be found. These are divided in industrially and home compostable. A compostable plastic is one that undergoes biodegradation in a composting environment, producing carbon dioxide, water, inorganic compounds and biomass and does not release any toxic materials (Ravenstijn, 2010). A biopolymer is recognized as compostable when it meets the criteria of at least one of the international standards: ASTM D6400-12 or EN 14995: 2006 (for compostable plastics); D6868-12 or EN 13432: 2000 (for compostable packaging); or ISO 17088: 2012 (Niaounakis, 2015).

Despite all the advantages, biopolymers have some technical limitations that make their processability and use as a final product difficult. Researchers have been studying the modification of biopolymers to enable processing and their usage. For this purpose, they came up with the bio composites with the aim of improving properties such as, mechanical properties, gas permeability and, fundamentally, the rate of degradation. Bio composites are the ones which comprise one or more phases derived from a biological origin. In terms of reinforcement, they could include plant fibres, such as cotton, lignin, hemp, recycled wood fibres, paper waste or even by-products of food production. The matrices may be polymeric, ideally derived from renewable sources such as vegetable oils or starches. What justifies the investment in these materials is their biodegradability (Srebrenkoska et al., 2014).

The difference between a biodegradable and compostable material is determined by the rate of biodegradation, disintegration and toxicity. All compostable materials are biodegradable, but not all biodegradable ones are compostable (Gutiérrez, 2018).

2.2 *Industrial and home compostable*

In industrial compost, bacteria and fungi degrade biomass under aerobic conditions and in a controlled temperature environment. Compared to domestic composting, not only does it need higher temperatures, but also the biomass is more frequently involved in order to ensure a higher homogeneity and consequently a faster biomass degradation (Hermann et al., 2011). Industrial composting is used in the efficient processing of waste into fertilizers through optimized control of process conditions – moisture, oxygen, temperature – and the presence of contaminants in the feedstock (heavy or inert substances) or pathogenic microorganisms to agriculture. Carbon dioxide, methane and nitrous oxide emissions lead this process (PRS, 2013).

In domestic compost, bacteria and fungi degrade biomass under aerobic conditions and at room temperature (5–30°C). Compared to industrial composting, biomass remains at low temperatures and its mixing is less frequent, which leads to slower degradation (Hermann et al., 2011). This type of procedure is carried out in houses with gardens or terraces or indoors through a homemade composter. The local composting site directly influences the degradation time of the biomass. The compound generated in this process is mainly used as fertilizer for gardens. Domestic composting brings a great deal of environmental awareness and responsibility and helps in the significant reduction of the weight and volume of solid wastes that are not always transported and disposed correctly (PRS, 2013). The use of the domestic composting process has other advantages, such as reducing the amount of waste sent to composting plants and landfills. This reduces the costs of collecting and transporting waste, reducing the environmental impact caused by transportation and the pollutants produced during the disposal of organic materials in landfills.

3 COMPOSTABLE FOAMS REVIEW

EPS is a rigid cellular plastic material made by expanded polystyrene (2%) and air (98%), produced with thermoforming, allowing different shapes with low weight. It is a very durable material with excellent properties, particularly good resistance to cushioning, compression, flexural strength and traction (Plastimar, 2019). However, being a non-biodegradable and non-renewable material (fossil-based), its utilization in packaging has raised concerns about environmental pollution. Thus, there is a demand for safer management of such waste. Large amounts of EPS are produced every year and there are some apprehensions related to traditional methods for handling postconsumer, like recycling, incineration and landfilling. These methods generate greenhouse gases which pose a threat to our health and environment (Abdul Khalil et al., 2016). With the research criteria the following biomaterials were studied regarding the mechanical properties, composition, end-of-life, standards, certifications and the feasibility of packaging applications (Figure 1). It is intended to minimize the creation of packaging waste and consequently solve the waste disposal problem.

3.1 *Ecoflex® polymer + PLA*

Under the commercial name Ecovio® EA, this biofoam was developed by the prestigious German company BASF. This expandable closed-cell foam consists of the biodegradable BASF polymer ecoflex® (butanediol, adiptic acid and terephthalic acid) and a blend of polylactic acid (PLA), which is derived from corn or other sugar-generating plants like manioc. According to DIN EN 13432, its composition allows it to be industrially compostable in 5 months. In addition to its sustainable end-use, the biofoam has benefits such as easy production through thermoforming,

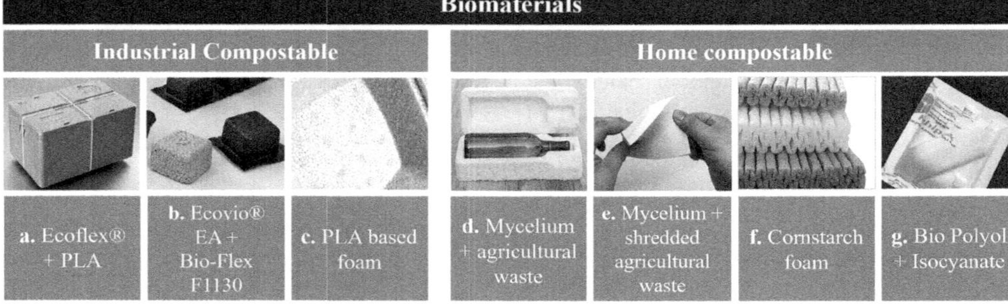

Figure 1. Biomaterials end-of-life (a. BASF®, 2016, b. Fibi-buffer, 2019, c. Synbra, 2019, d./e. Ecovative, 2019, g. Kehl, 2019, f. Green Cell Foam, 2018).

good thermal resistance, thermal conductivity (34 mW/mK) and great durability in warehouses and it can also be used as renewable energy – biogas (BASF®, 2016, Niaounakis, 2015). In short, this material has properties that are equivalent to EPS, it is industrial compostable, certified and suitable for packaging applications.

3.2 *Ecovio® EA (Ecoflex® polymer + PLA) + Bio-Flex F1130 (PLA based film)*

Developed by Fibi buffer in Netherlands, this solution is an industrial composting foam composed of two main elements: foam cubes in Ecovio® EA whose particles are purchased from BASF; and a biodegradable batch based on PLA that contains copolyester and additives, developed by Fkur, under the name Bio-Flex F1130. The *Ecovio®* particle cubes are produced by thermoforming and are subsequently sealed in a vacuum with the Bio-Flex F1130. This solution, in addition to being more environmentally friendly than conventional petroleum-based foams, also has great mechanical properties, functionality and adaptability, which gives it ample viability in the packaging sector. The geometry and design already incorporated in this sustainable foam allows it to be adapted to a wide variety of products and packaging of different dimensions, thus avoiding the need for a new design for each project (Fibi-buffer, 2019). To sum up, this material is made from a BASF foam and a Fkur film, its properties are equivalent to EPS, it is industrial compostable, but only the film is certified as compostable according the EN 13432/ASTM D 6400. It is also a suitable solution for a wide range of packaging applications.

3.3 *PLA based foam*

BioFoam is the commercial name of the PLA based foam developed by the Dutch company Synbra. This biofoam is produced by thermoforming, and it is bio-based which makes it an industrial compostable material that after 4 weeks at 70°C disintegrates completely. Cradle-to-Cradle environmental certificates and European standard EN 13501 govern its end-of-life This PLA based foam made from vegetable and biological materials is like conventional EPS, but it stands out as being biologically based because it contains 100% neutral carbon and it is free of carcinogenic, mutagenic and reprotoxic substances. The only drawback is its short shelf life. After a few months, the material will lose some properties (Ashter, 2016, Synbra, 2019). The properties of this foam are very similar to EPS, it is industrial compostable, certified and suitable for packaging in several applications.

3.4 *Mycelium + agricultural waste*

Made from mycelium and agricultural waste, Myco foam is the commercial name of this 100% natural and composite material. It is one of the most sustainable materials at all stages of its life cycle, because its production process is free of toxic emissions. In the processing, the mixture is placed in a mould for 5 days in a place devoid of light. Afterwards, it is possible to obtain a moulded material that, in spite of being light-weighted, it has good mechanical properties and good thermal and acoustic insulation. It is an easily disposable material that can be deposited in a garden, in a composter or even in the trash. This feature and its rapid degradation – in just 30 days – yielded Cradle-to-Cradle, the USDA-certified product and "ASTM D6400" certifications (Ecovative, 2019). This network of filiform fungal cells makes possible to link agricultural waste by using mycelium. Its properties are similar to EPS, it is home compostable, certified and suitable for packaging applications.

3.5 *Pure mycelium + shredded agricultural waste*

Myco flex is another novelty patented by Ecovative, developed mostly from corn bark, ground rice and pure mycelium. Myco flex presents itself as a 100% natural biofoam with great flexibility and mechanical properties. It is similar to Myco foam in terms of life cycle, manufacturing process and end-of-life domestic composting, but it takes few months to completely degrade. Myco flex has advantages over Myco foam, since it does not release particles during use and is indicated for a wide variety of applications – textiles, footwear, packaging and cosmetics (Ecovative, 2019). This material is made from pure mycelium, its properties are similar to EPS, it is home compostable, certified and suitable for some packaging applications.

3.6 *Cornstarch foam*

Under the name Green cell foam, this cornstarch material is the domestic composting foam developed by a company with the same name from USA. This biofoam is compostable in 4 weeks in a composter and has the particularity of being recycled with paper or dissolved in water. Its sustainable end-of-life has earned the ASTM D6400 certification. Green cell foam is a renewable 100% biologically based resource that requires 70% less energy and produces 80% less greenhouse gases than conventional petroleum-based foams. Its long shelf life and mechanical strength give it good properties for the packaging industry (Green Cell Foam, 2018) This material, naturally antistatic, made from cornstarch, has properties that are comparable with EPS, is home compostable, certified and feasible for a wide range of packaging applications.

3.7 *Bio polyol (castor oil, starch and soybean oil) + isocyanate*

Kehl bio foam, the commercial name for the foam developed by the Brazilian company Kehl, is a polyurethane household composting foam consisting of two main elements – bio polyol (castor oil, starch and soybean oil), patented by Kehl, and isocyanate. The polyol is the element responsible for making the foam biodegradable between 6 to 24 months. Although it does not yet have any environmental certificates, it had various positive tests according to DIN EN 13432 and ASTM D6400. In addition to being sustainable, this foam has the advantage that it can be produced through innovative foam-in-place technology and can therefore be adapted to any type of packaging and product. This biofoam comes to solve the issue of lack of sustainability in polyurethane foams. Although it resembles conventional polyurethane, regarding mechanical properties, it has a simple and low-cost production. It is notable for being bio-based as it does not interfere with the food chain and for having a wide variety of end-of-life possibilities, all without toxicities for the environment. This foam has a long shelf life after it is expanded and although it is bio-based, it is free of parasites. This material is made from Khel polyol and isocyanate, its properties are comparable to EPS, it is home compostable, patented but currently without certifications, and feasible for complex packaging applications.

4 CONCLUSION

The use of non-biodegradable materials has raised environmental concerns. To fight them, biopolymers and bio-composites have gained prominence and its usage in packaging is becoming increasingly common. The exposed solutions are a few of the different ones currently on the market. The objective is to motivate the use of these compostable materials, achieving a price reduction and encouraging development in this area, moving towards the sustainable disposal of packaging for the good of the ozone layer and the environment.

Nevertheless, there are still gaps to be addressed in terms of how to define and quantify the sustainability of a material. This is currently happening through European, American, Australian and German standards, but they do not include, for example, questions such as the bio percentage or the difference between industrial and home composting. This information is only presented in labels that unfortunately, in Europe, are attributed by private companies.

Another problem is the end of life of the compostable materials because it is still difficult to find containers or places to deposit these products and the collection system of these materials is still very primitive. In this way, domestic composting is the best solution for the conscience and daily habits of the current society. Home compostable materials present a greater diversity of end-of-life options, being able to go to inappropriate places in the ecosystem where the same biodegradation happens without harm to the environment.

It is necessary to be more technical and objective when it comes to the comparison of the mechanical properties between these materials. To achieve this, tests such as cushioning curves, impact weight tests or drop tests must be performed.

REFERENCES

Abdul Khalil, H. P. S., Davoudpour, Y., Saurabh, C. K., Hossain, M. S., Adnan, A. S., Dungani, R., Paridah, M. T., Islam Sarker, M. Z., Fazita, M. R. N., Syakir, M. I. & Haafiz, M. K. M. 2016. A review on nanocellulosic fibres as new material for sustainable packaging: Process and applications. *Renewable and Sustainable Energy Reviews*, 64, 823–836.

Ashter, S. A. 2016. 10 – New Developments. *In:* ASHTER, S. A. (ed.) *Introduction to Bioplastics Engineering.* Oxford: William Andrew Publishing.

BASF®. 2016. *ecovio® EA – certified compostable expandable particle foam* [Online]. Available: https://www.basf.com/global/en/products/plastics-rubber/fairs/BASFatK2016/must_sees/ecovio-EA.html [Accessed 2018 2018].

Bradley, E. L., Castle, L. & Chaudhry, Q. 2011. Applications of nanomaterials in food packaging with a consideration of opportunities for developing countries. *Trends in* Food Science & Technology, 22, 604–610.

ECHA. 2018. *ECHA to consider restrictions on the use of oxo-plastics and microplastics* [Online]. Available: https://echa.europa.eu/-/echa-to-consider-restrictions-on-the-use-of-oxo-plastics-and-microplasti-1 [Accessed 29-1-2019 2019].

Ecovative. 2019. *Ecovative* [Online]. https://ecovativedesign.com/: Ecovative. Available: https://ecovative design.com/ [Accessed 21-11-2018 2019].

European Bioplastics. 2016. *What are bioplastics? – Fact Sheet* [Online]. Available: https://docs.european-bioplastics.org/2016/publications/fs/EUBP_fs_what_are_bioplastics.pdf [Accessed 18-12-2018 2018].

European Commission. 2018. *European Commission* [Online]. European Commission. Available: http://europa.eu/rapid/press-release_IP-18-5_pt.htm [Accessed 29-01-2019 2019].

Fibi-buffer. 2019. *Fibi-buffer* [Online]. http://fibi-buffer.com/. Available: http://fibi-buffer.com/ [Accessed 23-01-2019 2019].

Godavitarne, C., Robertson, A., Peters, J. & Rogers, B. 2017. Biodegradable materials. *Orthopaedics and Trauma*, 31, 316–320.

Green Cell Foam. 2018. *Green Cell Foam* [Online]. Available: https://www.greencellfoam.com/ [Accessed 2-3-2019 2019].

Gutiérrez, T. J. 2018. Are modified pumpkin flour/plum flour nanocomposite films biodegradable and compostable? *Food Hydrocolloids*, 83, 397–410.

Hermann, B. G., Debeer, L., De Wilde, B., Blok, K. & Patel, M. K. 2011. To compost or not to compost: Carbon and energy footprints of biodegradable materials' waste treatment. *Polymer Degradation and Stability*, 96, 1159–1171.

Kehl. 2019. *Kehl* [Online]. http://www.kehl.ind.br/. Available: http://www.kehl.ind.br/ [Accessed 15-02-2019 2019].

Niaounakis, M. 2015. 1 – Definitions of Terms and Types of Biopolymers. *In:* NIAOUNAKIS, M. (ed.) *Biopolymers: Applications and Trends.* Oxford: William Andrew Publishing.
Penca, J. 2018. European Plastics Strategy: What promise for global marine litter? *Marine Policy,* 97f, 197–201.
Plastimake. 2018. Available: https://www.plastimake.com/ [Accessed 21-11-2018 2018].
Plastimar. 2019. *EPS* [Online]. Available: http://www.plastimar.pt [Accessed 10-02-2019 2019].
PRS. 2013. *Compostagem* [Online]. PRS – Portal Resíduos Sólidos. Available: https://portalresiduossolidos.com/compostagem/ [Accessed 6-03-2019].
Ravenstijn, J. 2010. Bioplastics in consumer electronics. *Industrial Biotechnology*, 6, 252.
Reddy, M. M., Vivekanandhan, S., Misra, M., Bhatia, S. K. & Mohanty, A. K. 2013. Biobased plastics and bionanocomposites: Current status and future opportunities. *Progress in Polymer Science*, 38, 1653–1689.
Souza, V. G. L. & Fernando, A. L. 2016. Nanoparticles in food packaging: Biodegradability and potential migration to food—A review. *Food Packaging and Shelf Life*, 8, 63–70.
SPUR®. 2018. Available: https://www.spur.cz/cs/ [Accessed 21-11-2018 2018].
Srebrenkoska, V., Bogoeva-Gaceva, G., Dimeski, D., Srebrenkoska, V. & Dimeski, D. 2014. *Biocomposites based on polylactic acid and their thermal behavior after recycing.*
Synbra. 2019. *BioFoam* [Online]. https://www.synbratechnology.com/. Available: https://www.synbratechnology.com/ [Accessed 23-1-2019 2019].

Wastes: Solutions, Treatments and Opportunities III – Vilarinho et al. (Eds)
© 2020 Taylor & Francis Group, London, ISBN 978-0-367-25777-4

Developing solutions for pay-as-you-throw information systems

J.P. Barraca, L.P. Almeida, A. Santos, H. Moreira, Ó. Pereira & R. Luís Aguiar
University of Aveiro, Telecommunications Institute, Aveiro, Portugal

C. Dias-Ferreira
Universidade Aberta, Lisboa, Portugal
Research Center for Natural Resources, Environment and Society (CERNAS), Polytechnic Institute of Coimbra, Coimbra, Portugal

ABSTRACT: The development of pay-as-you-throw (PAYT) systems – one of the strategies behind smart waste concepts – has a large set of challenges from the information technology (IT) point of view. The diversity of existing charging models in different towns already poses a complexity problem for a single universal IT solution. The situation is even more complex as the diversity of pay-as-you-throw systems is very large, with different tariffs and different objectives. This paper describes the development of an information system for supporting multiple approaches for PAYT systems and describes its implementation in the context of a European project. The design strategy and the use of best practices lead to a scalable and effective PAYT- specific Information System that has proved itself able to support a diversity of requirements across south Europe.

1 INTRODUCTION

The challenges of a sustainable society cover a wealth of aspects, from transportation to energy production, from materials reuse to proper disposal of unavoidable waste. As the world becomes more city-centric, urban centers are becoming megapolis, and population densities reach extreme values, increasing pressures.

Waste management becomes a critical activity in this context. The amount of waste produced in these megapolis is a major hindrance to society sustainability, and strategies for improving the eco-footprint of this waste are essential for our common future. These strategies resort both to less waste – through change of habits of the populations – and to improved recycling ratios – which are achieved by improved technologies and (again) by changing population habits. Overall, the targets of the European Commission for 2020 impose a significant reduction of the landfill usage and require the establishment of a consensual circular economy where waste is seen as a resource.

In this context, the strategies associated to pay-as-you-throw (PAYT) waste tariffs are deemed an essential economic instrument in waste management systems, promoting a change of perception on the generation and source-segregation of waste. The PAYT concept is one way to create a more efficient and fairer waste invoice bill calculation. Thus, waste tariffs cease to be collected in the traditional form of a fixed tariff (and/or associated with other factors) and are charged as a variable tariff, which depends on the quantity of waste produced. The advanced separation of waste that is inherently associated to a PAYT system acts as a positive reinforcement loop over the waste producers.

Recently, in Portugal, Greece and Cyprus a novel project (LIFE-PAYT) was launched to assess the deployment (and associated reaction to) of pay-as-you-throw concepts in a diversity of scenarios (Dias-Ferreira et al. 2019). A total of five municipalities were engaged in the process, from large municipalities to small towns: Aveiro, Condeixa-a-Nova, Lisbon, Larnaca and Vrilissia. Several waste producers were addressed in this project: big producers (namely industry producing more than 1100 L/d), small businesses (e.g. restaurants), and residential blocks. Different towns were

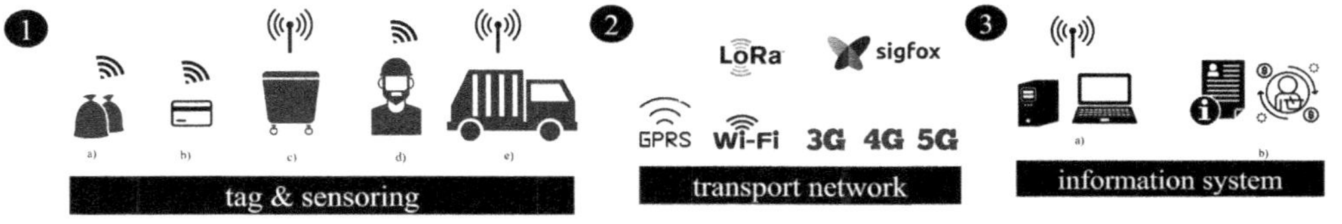

Figure 1. Technology flow for PAYT systems.

addressing different waste producers, in order to assess the reaction of different populations, and extract knowledge that may facilitate wider adoption in further locations. Given the differences among the target municipalities and waste producers, the technical solution adopted at the five locations were necessarily different. Regardless of these differences, all PAYT systems have the same main three components (Fig. 1): (1) the sensor and devices sub-system (identifying the user or the container); (2) a data network; (3) and the information system (collecting, storing, processing, and making available all information).

The tag & sensoring (module 1) and the transport network (module 2) have been previously addressed (Madureira & Dias-Ferreira 2019a, b). The current work describes the development of module 3, the information system, to be used within the LIFE PAYT project. In this particular case the Information System (IS), named *PAYT-IS*, is where data from all project sites is collected, processed and made available.

2 THE DESIGN OF PAYT-IS

2.1 *Users*

From the point of view of usage, the information system has to consider two major types of users: the end-user, who disposes the waste (citizen, industrial unit or commercial facility) and the waste system manager, which collects and manages the waste (either a town hall official, or a manager in a private waste collection company). Besides these, there is also the system administrator, which overall manages the system (manage users, grants access, manages system alerts, monitors platform usage and external interactions, etc). The administrator functions are focused on IT-specific management aspects, and as such will not be detailed in this paper.

It is obvious, in any waste management system, that the type of producer is an essential design parameter – addressing a professional company (e.g. a restaurant) or a residential neighborhood will lead to a different set of parameters required. Also, the same professional company might have several locations registered under a single contract and the applicable waste tariffs might be different than those of domestic waste producers.

The data that the information system needs to provide for each type of end-user are identified in table 1.

From the waste system manager, it is important to differentiate between door-to-door vs. collective waste collection. These vary substantially in the technology used to link the waste producer to the amount of waste produced. In door-to-door collection approaches each location has its own container. This means that if the container is properly identified (e.g. by an RFID tag) then the waste within this container can be linked to a specific waste producer. This is called *container identification*. Oppositely, if the container is collective (used simultaneously by several users) identifying the container will not allow to know who discarded the waste. In this case it is necessary to identify the individual user, generally through by only allowing access to the waste container to identified users (e.g. through a RFID card), and limiting the volume that can disposed per opening. This is called *user identification*. There is also the possibility of using pre-paid bags, but since the information system is not essential this situation was not considered.

Each collection strategy has its own relevant set of data that the waste system manager needs to retrieve from the information system, as shown in Table 2.

Table 1. Functions associated to the waste producers.

Functions	Non-domestic	Domestic
Access street map with waste disposal points	x	x
Amount of waste disposed	x	x
Expenses of the last month (tariff simulation)	x	x
Historic data on expenses and waste production	x	x
Information on waste containers under collection contract	x	
Add and manage other installations of the same entity	x	
Compare individual waste production with local averages	x	x

Table 2. Functions associated to the waste management entity.

Functions	Door-to-door collection	Collective waste containers
Access map with associated containers	x	x
Generate file with customer list	x	
Generate invoice data	x	x
Manage user access cards		x
Analyse:		
– producer with most/less waste production	x	
– producer with most/less separation ratio	x	
– location of installation	x	
– location of containers		x
Access information on:		
– total expenses per producer	x	x
– waste production per type	x	
– evolution of waste production		x
– installation details	x	
– container usage		x
– producer preferences		x

2.2 *Technical requisites*

Given the diversity of scenarios that are included in the project, and considering the intended use of the platform in other regions, the system needs to be:

- Scalable: the system may need to scale-up as the number of users, and as the PAYT concept, is gradually deployed in additional towns.
- Easy to deploy and operate: in the future, the system will be deployed by potentially different IT teams, associated to different municipalities, and has to be easy to operate by those (very different) professionals.
- Easy to configure and change: by the different requirements and potentially different evolution paths inside each municipality, *PAYT-IS* should be easy to change according to the specific requirements of each municipality.
- Easy to interact: *PAYT-IS* will have to cope with a diversity of users, with different abilities in interacting with IT technologies.
- Privacy-aware: the system should respect all relevant privacy and security requirements, at least matching those related to the European General Protection Data Regulation.
- Future proof: The system should not provide a lock-in solution, binding municipalities in their future IT strategies.
- Usable in different languages: given the different countries of the municipalities involved (Portugal, Greece, Cyprus), the European context (English) and the replication goals, *PAYT-IS* should be able cover any present as well as future usage, in different countries.

3 DEVELOPMENT OF PAYT-IS

The development of the information system was based on several best practices in the area. The architecture of the system developed (Fig. 2) was based in open-source software, relied in RESTfull services (Booth et al. 2004) and REST (Castillo et al. 2011) communications, in a micro-services environment, and using containers (Jaramillo et al. 2016). Some of the open-source components and information systems' best practices used in the architecture are clearly referenced (RabbitMQ, PostgreSQL, CKAN, ELK) in the picture. The system contains the following independent modules deployed according to a highly decoupled micro-services approach.

- Log Module – all interactions with *PAYT-IS* are recorded by this model, that may then provide a diverse set of reports.
- Authentication Module – This is the module that handles all security aspects of the (different) user credentials, accessing the security levels of the user. In particular, it implements a strict, label based, access control system to all data and resources.
- Data Module – This module contains the database(s) where all information is stored. We use scalable contextual approaches (Antunes 2016) to improve future data analysis over the system. The data models for the databases are essential for the scalability and flexibility of *PAYT-IS*. Also, data is strongly encrypted with municipality specific keys.
- Communication Module – all communication between *PAYT-IS* modules goes through this module, that manages and brokers the access. The only exception is the Open Data Module, that has a separate external interface, since it may have a separate access control. This module is agnostic to the networking technology, and can be supported even in novel networks for the Internet of Things (e.g. Suarez et al. 2016, Antunes et al. 2015).

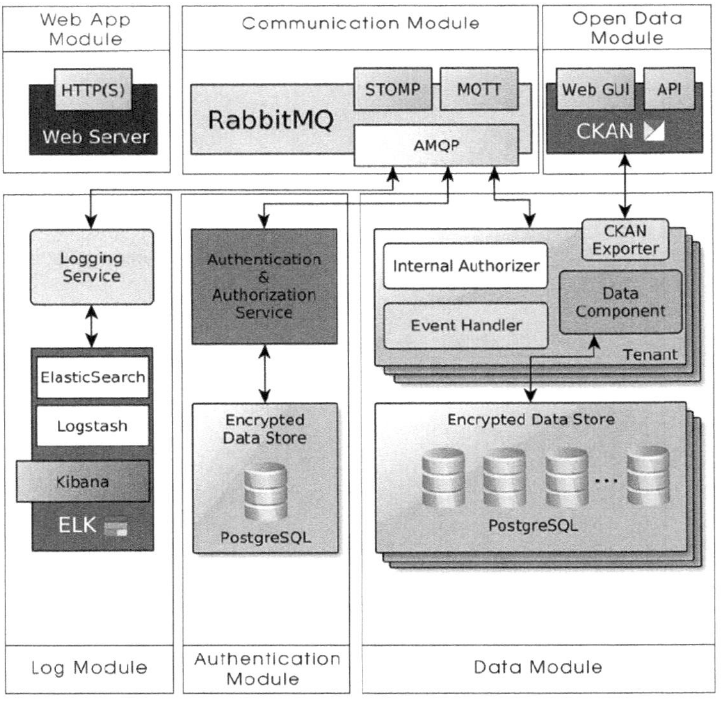

Figure 2. System architecture.

- Open Data Module – due to open data regulations, *PAYT-IS* contains a module that provides anonymized and summarized reports, that can be used by different entities. These reports may have separated distribution lists.
- Web Server Module – main interaction interface for user(s), both the waste producers, the PAYT administrators, and the waste managers.

One final word is needed for localization aspects. The requirement that the system is to be used in different languages (Portuguese, English and Greek) – some of which not even supporting western alphabet, posed a real requirement in terms of the design of the *PAYT-IS*. The solution found was resorting to a software pattern, Flux (Facebook 2017), which is based on Stores containing the application data and logic. They also contain the language selected by each active user (session). Changing the language in a view dispatches a change to all stores. Views that are observing those Stores will be promptly updated with the new state. This is also used to disseminate real time information, arriving from sensors, resulting in a responsive and reactive interface.

Overall, this design strategy and these Information System best practices lead to a scalable and effective information system for waste management.

4 SYSTEM ASSESSMENT

PAYT-IS is being used inside the LIFE-PAYT project, and fulfilling the requirements for the project. As a matter of illustration of its behaviour, we can show some of the interface screenshots for small and big producers (Fig. 3). Figure 4 illustrates screenshots for the waste managers, in particular

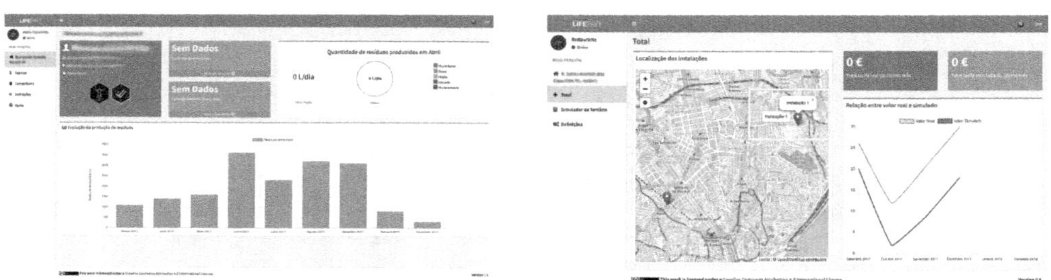

Figure 3. Producer views (left to right): small producer history page; big producer detailed view of "total production".

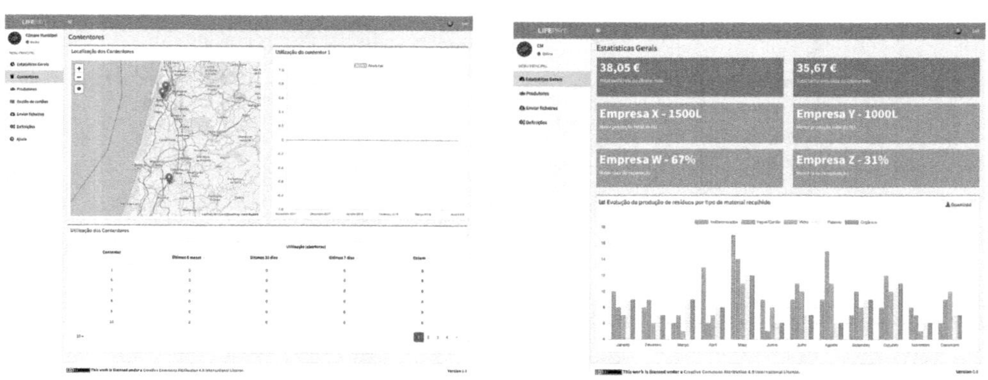

Figure 4. Waste Manager views (left to right): small producers summary; statistics per container.

the interfaces detailing comparative data for the different types of waste, and the indication of the top ranking waste producers, as well as their variation.

The system has been tested with real data (input from the different municipalities) and real users (small producers and waste managers). Overall performance has been measured to be within usability limits, with average web page loads inferior to 2 seconds (first page load taking almost 9 sec occasionally, depending on the network bandwidth of the client). After all caches in the browser are updated, the interaction flows with times in the order of the tens of milisseconds for each isolated interaction. Furthermore, a scalability assessment was performed with a simulated municipality with 10,000 users, and a peak 1000 accesses at the same time.

5 CONCLUSIONS

PAYT systems are an essential component for transforming the way people look at waste, and develop a more sustainable society. Given the diversity of technical solutions existing, any PAYT system will necessarily need to deal with technology complexity, which will impose the information system to be the clear system integrator for all underlying technology.

We developed an information system (*PAYT-IS*) that is able to cope with the different implementations of PAYT concept. This system was developed following a set of best practices and resorting to open source software for a future-proof implementation. The resulting system has been deployed to support the real-life scenarios of the LIFE-PAYT project, and satisfactory handled the diversity of PAYT approaches pursued in this project.

ACKNOWLEDGEMENTS

The authors acknowledge the financial support of LIFE+ (financial instrument of the EU for the environment) for funding the LIFE PAYT project (LIFE 15/ENV/PT/000609).

REFERENCES

Antunes, M., Gomes, D. & Aguiar, R.L. 2016. Scalable semantic aware context storage. *Future Generation Computer Systems* 56: 675–683.

Antunes, M., Barraca, J.P., Gomes, D., Oliveira, P. & Aguiar, R.L. 2015. Smart Cloud of Things: An Evolved IoT Platform for Telco Providers. *Journal of Ambient Wireless Communications and Smart Environments* (AMBIENTCOM) 1(1): 1–24.

Booth, D., Haas H., McCabe F., Newcomer E., Champion, M., Ferris, C. & Orchard, D. 2004. Web Services Architecture. W3C [Online] available: http://www.w3.org/TR/ws-arch/.

Castillo, P., Bernier, J., Arenas, M., Merelo, M. & Garcia-Sanchez, P. 2011. SOAP vs REST: Comparing a master-slave GA implementation. arXiv preprint arXiv:1105.4978.

Dias-Ferreira, C., Neves, A. & Braña, A. 2019. The setting up of a pilot scale Pay-as-you-throw waste tariff in Aveiro, Portugal. *WIT Transactions on Ecology and the Environment* 31: 149–157 (DOI: 10.2495/WM180141)

Facebook Inc. 2017. Flux – In Depth Overview [Online] available: https://facebook.github.io/flux/docs/in-depth-overview.html (accessed on 28/02/2019).

Jaramillo, D., Nguyen, D.V. & Smart, R. 2016. Leveraging microservices architecture by using Docker technology. *SoutheastCon*, IEEE: 1–5.

Madureira, R.C. & Dias-Ferreira, C. 2019a. Tag & sensoring solutions overview for smart cities waste management. *This issue* (accepted).

Madureira, R.C. & Dias-Ferreira, C. 2019b. Data communication solutions overview for smart cities waste management. *This issue* (accepted).

Reichenbach, J. 2008. Status and prospects of pay-as-you-throw in Europe – A review of pilot research and implementation studies. *Waste Management* 28(12): 2809–2814

Suarez, J., Quevedo, J, Vidal, I., Corujo, D., Garcia-Reinoso, J. & Aguiar, R.L. 2016. A secure IoT management architecture based on Information-Centric Networking. *Journal of Network and Computer Applications* 63: 190–204.

Wastes: Solutions, Treatments and Opportunities III – Vilarinho et al. (Eds)
 ISBN 978-0-367-25777-4

Application of ultrasounds in the extraction process for food waste valorisation

D. Lopes, A. Mota, J. Araújo & J. Carvalho
CVR – Centro para a Valorização de Resíduos, Guimarães, Portugal

C. Vilarinho
MEtRICs, Mechanical Engineering and Resource Sustainability Center, Department of Mechanical Engineering, University of Minho, Guimarães, Portugal

H. Puga
CMEMS-UMinho, Centre for Micro-Electro Mechanical Systems, Department of Mechanical Engineering, University of Minho, Guimarães, Portugal

ABSTRACT: Waste management is one of the major issues of developed world, given the growth of population and consumerism verified in the last years. The process of extraction can be used in the valorisation of food by-products and residues, namely in the recovery of bioactive components. However, traditional methods include some disadvantages, which can be diminished with the use of ultrasounds. This manuscript represents a review of the application of ultrasounds on the enhancement of the extraction process of components from food waste.

1 INTRODUCTION

As a consequence of the large population growth worldwide, which is estimated to have doubled over the last four decades (Goujon, 2019), the consumerism and the industrialization, waste management has become one of the major problems of the 21st century. In fact, waste generation is expected to triple by 2100 (Minelgaitė and Liobikienė, 2019), therefore, it is crucial to develop waste recovery techniques, for waste valorisation.

Since the primordial history, the extraction of substances from plant material has been used for food, aromatic and medicinal purposes, with the aim of obtaining a more specific, pure a standardized product. This process is interesting since it can be used to obtain polyphenols, antioxidants, proteins, lipids and other bioactive components from food waste. However, traditional extraction processes, such as maceration, infusion, percolation and decoction (Azwanida, 2015), entail some inconveniences such as high solvent and energy consumption, extended operating time, which makes them unattractive for this application. Novel techniques, such as supercritical fluid extraction (SFE), microwave-assisted extraction (MAE) and accelerated solvent extraction (ASE), Ionic-Liquid-Mediated Extraction, have emerged as alternatives to the conventional process, however these are also associated with high energy consumption, the release of carbon dioxide and harmful chemicals (Tiwari, 2015). All of these problems have provided an incentive to use new and sustainable techniques in the extraction process. The use of ultrasounds is one interesting alternative or complement to the conventional methods, given that it is an environmentally friendly technology, and generally, induces minimal damage to the properties of the extracts (Chemat et al., 2017).

In this manuscript, the ultrasound technology and its operating principles are presented, as well as a bibliographic review to the applications of this technology in the extraction process of compounds, particularly, from food waste by-products.

2 ULTRASOUND TECHNOLOGY

An ultrasound is a sound whose wave frequency exceeds 20 kHz, which corresponds to the maximum frequency that a human can detect (Sanderson, 2004). Figure 1 shows the frequency spectrum of the human ear and the range of ultrasound applications.

The majority of the first applications of the ultrasounds in food processing included non-invasive analysis to evaluate the quality of the products, where techniques similar to those developed in diagnostic medicine were used, within a range of 1 to 10 MHz Nowadays, most applications are in the conventional power ultrasound region, with the range of 20 to 25 kHz being the most usual (Paniwnyk, 2017).

The main mechanical effects of this technology are obtained when the power of the wave is high enough to induce cavitation in the propagation medium. When an ultrasound system is applied to a liquid medium, the propagation the sound waves causes cycles of high and low pressure (compression and rarefaction). If the wave is of sufficient intensity, small bubbles are formed in the medium, known as cavities. Generally, these bubbles grow in size and tend to collapse, during the cycles of high pressure, thus causing shock waves, which can increase local pressure and temperature values up to 101.3 MPa and 4726°C (Patist and Bates, 2008). The size of the bubbles of cavitation is inversely proportional to the frequency of the ultrasound wave, which justifies the fact that most food processing techniques use lower work frequencies.

Ultrasounds are typically generated by piezoelectric transducers, which consist of arrays of crystals that due to the piezoelectric effect, are able to convert electrical energy in mechanical and vice-versa, thus, in response to an electrical signal, these vibrate and generate high-frequency sound waves. The first ultrasound transducers were made of natural piezoelectric crystals, such as quartz, Rochelle salts and tourmaline. Nowadays, most transducers use synthetic crystals, such as lead zirconate titanate, also called PZT (Gray, 2019). The most used equipment's used for the processing of liquids are an ultrasonic cleaning bath, which sonicates samples that immersed in a bath, as shown in Fig. 2a, and a probe system, which is in direct contact with the samples, Fig. 2b.

Typically, the ultrasound power distributed by most commercial ultrasonic baths is in the 1–5 W/cm^2, while the probe system has applications with an intensity of up to 100 W/cm^2.

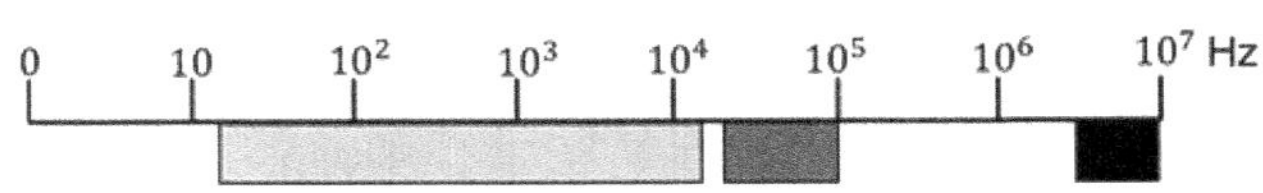

Figure 1. Frequency ranges of sound. □ Human hearing; ■ Conventional power ultrasound; ■ Diagnostic ultrasound. Adapted from (Patist and Bates, 2008).

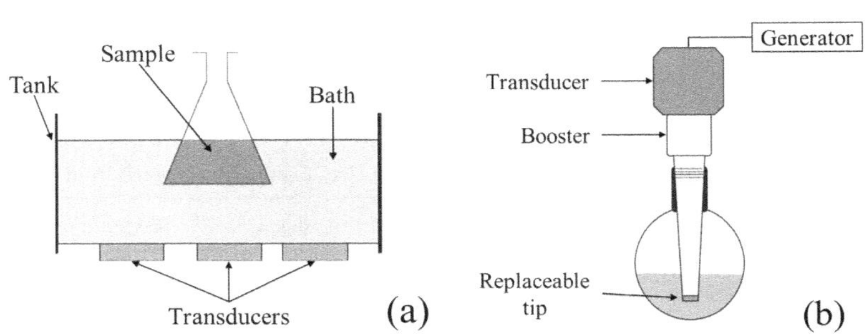

Figure 2. Ultrasound equipment. (a) Bath; (b) Probe system. Adapted from (Sun, 2005).

3 APPLICATIONS OF ULTRASOUNDS IN THE EXTRACTION PROCESS

The use of ultrasounds in extraction seeks to improve the efficiency and yield of the process. In this, it is expected that due to the cavitation phenomenon, the implosion of the bubbles cause the thinning of cell membranes and cell disruption, resulting in enhanced solvent penetration into cells and amplification of mass transfer of target compounds into the solvent (Tiwari, 2015). This technology enables the extraction of components at lower temperatures, compared to the conventional methods, thus maintaining the properties of extracts sensitive to thermal amplitudes.

(Chemat et al., 2011) carried out a study of the application of ultrasounds in the food industry, particularly with respect to the preservation and extraction process. From this, it is presented in Table 1, a comparison between the novel methods of extraction.

Of the four methods, although Ultrasound-Assisted Extraction (UAE) is the one that uses the largest quantity of solvent, it has another advantages over the others, namely with respect to initial investment. Any type of solvent can be used in the process, unlike in MAE, where the solvent must be capable of absorbing microwave energy. It is difficult to make comparisons between the methods, with regard to the extraction time variable, since it has a significant amplitude and depends on the type of application.

Vinatoru (2001) studied the influence of milling degree on the extraction of Eugenol from clove flowers, by comparing samples that were not milled, with samples that were grounded to a size of 0.1–0.5 mm. Furthermore, the author made the comparison between conventional extraction (CE) method with UAE, the results of which are show in Table 2.

From the analysis of Table 2, there is an increase of 22.8% of the efficiency of the process, when the ultrasound is added to a milled sample. The cavitation induced by the UAE technique causes a

Table 1. Advantages and disadvantages of the novel extraction techniques. Adapted from (Chemat et al., 2011).

	Ultrasound-assisted extraction	Microwave assisted extraction	Supercritical fluid extraction	Accelerated solvent extraction
Extraction time	10–60 min	3–30 min	10–60 min	10–20 min
Sample size	1–30 g	1–10 g	1–5 g	1–30 g
Solvent use	50–200 ml	10–40 ml	2–5 ml (solid trap) 30–60 ml (liquid trap)	15–60 ml
Investment	Low	Moderate	High	High
Advantages	Easy to use	– Rapid – Easy to handle – Moderate solvent use	– Rapid – Low solvent use – No filtration needed – Possible high selectivity	– Rapid – No filtration needed – Low solvent use
Disadvantages	– Large solvent use – Filtration required	– Extraction solvent must absorb microwave energy – Filtration required	– Many parameters to optimize	– Possible degradation of thermolabile analytes

Table 2. Influence of the milling degree in conventional and UAE extraction. Adapted from (Vinatoru, 2001).

Extraction time (min)	Extraction technique	Milling degree	Eugenol extracted (g/100 g)
30	Conventional	Not milled	4.10
30	Conventional	0.1–0.5 mm	25.20
30	UAE	Not milled	4.22
30	UAE	0.1–0.5 mm	32.66

Table 3. Yields obtained after the 1st and 2nd extraction steps for CE and UAE processes. Adapted from (Grassino et al., 2016).

	Conventional		Ultrasound assisted Extraction									
Temperature (°C)	60	80	60					80				
Extraction time (min)	1440	1440	15	30	45	60	90	15	30	45	60	90
Extraction yield after 1st step (%)	21.1	15.1	15.1	15.5	15.5	16.6	15.4	18.0	19.7	18.0	18.5	16.1
Extraction time (min)	720	720	15	30	45	60	90	15	30	45	60	90
Extraction yield after 2nd step (%)	31.0	31.2	33.7	34.6	34.6	34.8	35.7	35.8	35.9	34.0	34.1	33.9

greater breakdown of the cellular material, thus facilitating the diffusion of the intracellular material to the solvent.

Hossain et al. (2014) studied the UAE application in the extraction of steroidal alkaloids, which could serve as precursors to agents with apoptotic, chemo preventive and anti-inflammatory properties, from potato peel waste. His research focused on the investigation of the efficiency of UAE and solid liquid extraction (SLE) and the results showed that with UAE, 1102 μg of steroidal alkaloids/g of dried material were recovered relatively to 710.51 μg/g from the SLE process, which represents a higher yield of 35.53 % for the former. Recoveries of individual glycoalkaloids using UAE yielded 273, 542.7, 231 and 55.3 μg/g DPP for α-solanine, α-chaconine, solanidine and demissidine respectively, whereas for SLE yields were 180.3, 337.6, 160.2 and 32.4 μg/g DPP for the same components, respectively. The UAE presented superior yields across the board, and the optimal values of this extraction process were obtained with an intensity of 22.79 W/cm^2 and extraction time of 17 minutes, in intermittent operating periods of 5 s on and 5 s off.

Grassino et al. (2016) compared a conventional extraction (CE) process with UAE of pectin, a cell wall polysaccharide, from tomato waste. CE was performed at 60°C and 80°C during 24 h in the first step, and more 12 h after solvent renovation, in the second step. UAE was also carried at a frequency of 37 kHz and at the same respective temperatures, with a sonication time of 15, 30, 45, 60 and 90 minutes. Table 3 represents the yield results of both extraction processes.

After analysing Table 3, the extraction yield is found to be higher for CE, at a temperature of 60°C, however, the latter process takes significantly more time than the UAE. Expectably, after the second extraction step, yields increase for all situations and has higher values in the UAE, compared to CE processes.

He et al. (2016) compared the efficiency of CE with UAE for the extraction of phenolic compounds and anthocyanins from blueberry wine pomace (BWP). Their results evidenced that the highest yields of UAE were reached at 30 minutes, resulting in 4.27 mg/gBWP for total anthocyanins and 16.41 mg/gBWP for total phenolics, while the highest yields using CE methods were obtained after 35 minutes of extraction, with 1.72 mg/gBWP for total anthocyanins and 5.08 mg/gBWP for total phenolics. Furthermore, it was reported that the UAE extraction yield of total anthocyanins and phenolics was near 80 % of the total at 4 minutes, therefore, therefore, 80% of the extraction occurred in only about 13% of the total time of the process.

Puga et al. (2017) studied compared CE and UAE for the extraction of antioxidants and phenolics from coffee chaff, a by-product of the coffee roast industry. Hydroethanolic solvent was used in CE, whereas only water was used in the UAE. The results of their study showed a similar phenolics extraction yield for both methods, with 198.3 μg GAE/ml and 203.8 μg GAE/ml, for CE, with the duration of 60 min, and for the UAE, with 19.8 kHz, 500 W and 600 s, as process parameters, respectively. Regarding the concentration of ferric reducing antioxidants, the results were 2266 μg FSE/ml, and 2446 μg FSE/ml for CE and UAE, respectively. The authors concluded that the major advantages of ultrasound assisted extraction are the less energy and time requirements, the no need of organic solvents.

Table 4. Investment cost, energy consumption and carbon emissions of CE at industrial scale and of UAE at both laboratory and industrial scale (Plazzotta and Manzocco, 2018).

Extraction technique	Scale	Operative conditions	Maximum capacity (L)	Cost (€)	Energy consumption (kWh/kg of extracted polyphenols)	Carbon emission (kg CO_2/kg of extracted polyphenols)
CE	Industrial (medium)	15 min, 50°C	6.0	600	2286.0	964.7
UAE	Laboratory	2 min, 50°C	0.1	3000	2067.3	872.4
	Industrial (Small)	2 min, 50°C	2.0	10,000	248.1	104.7
	Industrial (medium)	2 min, 50°C	6.0	25,000	206.7	87.2

Plazzotta and Manzocco (2018) studied the effect of ultrasounds on the extraction of antioxidant polyphenols, in the valorisation of lettuce waste. These concluded that the application of US (400 W, 24 kHz) for 120 s led to polyphenol extraction yield (81 μg/mL) and antioxidant activity (101 μg TE/mL) significantly higher than those obtained by traditional solid-liquid extraction at 50°C for 15 min. The authors also estimated for the economic and environmental impact of both CE and UAE processes, which is represented in Table 4.

The results shown in Table 4 suggest that although the UAE technique entails higher initial investment, the saving in terms of energy expenditure and carbon emissions is significant, especially in a medium industrial scale, which makes it a more appropriate solution in the long term for the valorisation of lettuce waste,

Görgüç et al. (2019) studied the possibility of extracting proteins and antioxidants from sesame bran, a by-product of sesame seed production. It is estimated that are produced 360,000 ton of this product every year, from which 54,000 ton correspond to protein. A comparison was made between enzymatic extraction (Viscozyme L and alcalase), ultrasound assisted and combined ultrasound-assisted enzymatic extraction techniques. Among the extraction procedures, combined ultrasound-assisted enzymatic treatment revealed highest protein yield (87.9%), followed with enzymatic extraction by alcalase (79.3%), ultrasound-assisted extraction (59.8%) and enzymatic extraction by viscozyme L (41.7%). The highest protein yield, total phenolic compound and antioxidant capacities were found in ultrasound-assisted enzymatic extraction at 836 W ultrasound power, 43°C, 98 min, 9.8 pH value and 1.248 AU/100 g enzyme concentration.

4 CONCLUSION

This manuscript represents a bibliographic review of the application of the ultrasound technology in the extraction process. Several works that made the comparison between traditional extraction and ultrasound-assisted extraction methods, with the results being unanimous in the fact that the efficiency of the process was better in the latter. Perhaps the most important aspect of the UAE is the considerable reduction in the extraction time relatively to CE, as was clearly evidenced in the works of (Grassino et al., 2016), where it was reported that CE for a duration of 1440 minutes gave similar pectin extraction yields, compared to 15 minutes of UAE. This factor, coupled with higher extraction yield, lower solvent use, lower energy consumption and emission CO2, make this technology attractive to be used in the extraction of compounds from food residues and thus promote its valorisation.

ACKNOWLEDGEMENTS

This work has been co-financed by COMPETE 2020, Portugal 2020 and the European Union through the European Regional Development Fund – FEDER within the scope of the project,

POCI-01-0247-FEDER-033351, U2SCOFFEE: Cápsulas de Cafeína. Additionally, this work was supported by MEtRICs, Mechanical Engineering and Resource Sustainability Center (Reference 2019: UID/EMS/04077/2019),

REFERENCES

Azwanida, N., 2015. A Review on the Extraction Methods Use in Medicinal Plants, Principle, Strength and Limitation. Medicinal & Aromatic Plants 04. https://doi.org/10.4172/2167-0412.1000196

Chemat, F., Rombaut, N., Sicaire, A.-G., Meullemiestre, A., Fabiano-Tixier, A.-S., Abert-Vian, M., 2017. Ultrasound assisted extraction of food and natural products. Mechanisms, techniques, combinations, protocols and applications. A review. Ultrasonics Sonochemistry 34, 540–560. https://doi.org/10.1016/j.ultsonch.2016.06.035

Chemat, F., Zill-e-Huma, Khan, M.K., 2011. Applications of ultrasound in food technology: Processing, preservation and extraction. Ultrasonics Sonochemistry 18, 813–835. https://doi.org/10.1016/j.ultsonch.2010.11.023

Görgüç, A., Bircan, C., Yılmaz, F.M., 2019. Sesame bran as an unexploited by-product: Effect of enzyme and ultrasound-assisted extraction on the recovery of protein and antioxidant compounds. Food Chemistry 283, 637–645. https://doi.org/10.1016/j.foodchem.2019.01.077

Goujon, A., 2019. Human Population Growth, in: Encyclopedia of Ecology. Elsevier, pp. 344–351. https://doi.org/10.1016/B978-0-12-409548-9.10755-9

Grassino, A.N., Brnčić, M., Vikić-Topić, D., Roca, S., Dent, M., Brnčić, S.R., 2016. Ultrasound assisted extraction and characterization of pectin from tomato waste. Food Chemistry, Total Food 2014: Exploitation of agri-food chain wastes and co-products 198, 93–100. https://doi.org/10.1016/j.foodchem.2015.11.095

Gray, A.T., 2019. Atlas of Ultrasound-Guided Regional Anesthesia (Third Edition). Elsevier.

He, B., Zhang, L.-L., Yue, X.-Y., Liang, J., Jiang, J., Gao, X.-L., Yue, P.-X., 2016. Optimization of Ultrasound-Assisted Extraction of phenolic compounds and anthocyanins from blueberry (Vaccinium ashei) wine pomace. Food Chemistry 204, 70–76. https://doi.org/10.1016/j.foodchem.2016.02.094

Hossain, M.B., Tiwari, B.K., Gangopadhyay, N., O'Donnell, C.P., Brunton, N.P., Rai, D.K., 2014. Ultrasonic extraction of steroidal alkaloids from potato peel waste. Ultrason Sonochem 21, 1470–1476. https://doi.org/10.1016/j.ultsonch.2014.01.023

Minelgaitė, A., Liobikienė, G., 2019. Waste problem in European Union and its influence on waste management behaviours. Science of The Total Environment 667, 86–93. https://doi.org/10.1016/j.scitotenv.2019.02.313

Paniwnyk, L., 2017. Applications of ultrasound in processing of liquid foods: A review. Ultrasonics Sonochemistry 38, 794–806. https://doi.org/10.1016/j.ultsonch.2016.12.025

Patist, A., Bates, D., 2008. Ultrasonic innovations in the food industry: From the laboratory to commercial production. Innovative Food Science & Emerging Technologies 9, 147–154. https://doi.org/10.1016/j.ifset.2007.07.004

Plazzotta, S., Manzocco, L., 2018. Effect of ultrasounds and high pressure homogenization on the extraction of antioxidant polyphenols from lettuce waste. Innovative Food Science & Emerging Technologies 50, 11–19. https://doi.org/10.1016/j.ifset.2018.10.004

Puga, H., Alves, R.C., Costa, A.S., Vinha, A.F., Oliveira, M.B.P.P., 2017. Multi-frequency multimode modulated technology as a clean, fast, and sustainable process to recover antioxidants from a coffee by-product. Journal of Cleaner Production 168, 14–21. https://doi.org/10.1016/j.jclepro.2017.08.231

Sanderson, B., 2004. Applied sonochemistry – the uses of power ultrasound in chemistry and processing. By Timothy J Mason and John P Lorimer, Wiley-VCH Verlag, Weinheim, 2002, 303 pp, ISBN 3-527-30205-0. Journal of Chemical Technology & Biotechnology 79, 207–208. https://doi.org/10.1002/jctb.957

Sun, D.W., 2005. Emerging Technologies for Food Processing, 1st ed, Food Science and Technology, International Series. Elsevier Academic Press.

Tiwari, B.K., 2015. Ultrasound: A clean, green extraction technology. TrAC Trends in Analytical Chemistry, Green Extraction Techniques 71, 100–109. https://doi.org/10.1016/j.trac.2015.04.013

Vinatoru, M., 2001. An overview of the ultrasonically assisted extraction of bioactive principles from herbs. Ultrasonics Sonochemistry 8, 303–313. https://doi.org/10.1016/S1350-4177(01)00071-2

Wastes: Solutions, Treatments and Opportunities III – Vilarinho et al. (Eds)
© 2020 Taylor & Francis Group, London, ISBN 978-0-367-25777-4

May recycled concrete be used as an alternative material for asphalt mixtures?

H.M.R.D. Silva & J.R.M. Oliveira
CTAC, Centre for Territory, Environment and Construction, University of Minho, Guimarães, Portugal

ABSTRACT: Recycling of materials is a growing need nowadays, to ensure the sustainable development of the Society, reducing the use of natural resources and preventing wastes. In this context, some wastes obtained from construction and demolition works are likely to be utilized in new products. Therefore, this work aims at investigating the properties of recycled concrete, as one of the primary construction waste materials, to study the feasibility of its use (added value) as part of the aggregates applied in asphalt mixtures for road pavements. Laboratory tests were performed to characterize and compare the recycled concrete to natural granite. Then, the influence of using 30% waste material in asphalt mixtures was assessed through water sensitivity tests. As a conclusion, the use of recycled concrete as an alternative aggregate for asphalt mixtures is viable, but its application should be prudent due to its high water absorption and low wear resistance.

1 INTRODUCTION

In Portugal, some natural resources used in the construction sector, such as aggregates, are relatively abundant and cheap, leaving behind schedule the management of construction and demolition waste (CDW). However, preventive measures are urgent to be taken to increase sustainability in this activity because sooner or later these natural resources will become scarce.

In this context, the conceptual basis of this work is trying to change the current linear economy paradigm into the new EU Circular Economy Strategy (European Commission, 2015), also transposed to the Portuguese Circular Economy Action Plan (Presidência do Conselho de Ministros, 2017), and in particular, with its fifth action focused on developing a new life for wastes. The main objectives of that action are increasing the introduction of secondary raw materials into the economy, reducing the production of waste and promoting the reduction of the extraction of natural resources.

In particular, this work aims at investigating the properties of recycled concrete, which is one of the most significant CDW materials in Portugal (Coelho and Brito, 2010), to study the feasibility of its use in new asphalt mixtures for road pavements. According to Hendriks et al. (2004), concrete is generally the largest share of CDW materials in several countries (usually around 40%). The properties of the recycled concrete were evaluated through a series of tests in the lab to analyze if this material complies with the specifications related to aggregates for road paving works (Estradas de Portugal, 2014). Then, an asphalt mixture with 30% recycled concrete was produced to evaluate its durability (assessed by indirect tensile strength and water sensitivity tests) against a conventional asphalt mixture. The application of this waste on urban roads has a remarkable interest in the scope of cities circularity concept, by reducing the burdens associated with the transportation stage.

In 2008, construction was the economic sector responsible for the production of the most substantial amount of waste produced in the EU (Schrör, 2011), which amounted to 33% of the total. In the European Community, there has been a growing concern with waste generated by the construction industry. In 2010, only about 30% of the building materials were recycled as raw materials, a number that could rise to 90%, helping to support sustainable construction and generating a series of environmental and economic benefits (Sonigo et al., 2010).

More recent data presented by Eurostat (2019) showed that five tons of waste were generated per EU inhabitant in 2016, from which 46% were landfilled and only 38% were recycled. In 2016, the total waste generated in the EU-28 by all economic activities and households amounted to 2 533 million tons, which was the highest amount recorded during the period 2004–2016. The share of different economic activities and households in total waste generation in 2016, also presented by Eurostat (2019), assigned 36% of the generated waste to the construction area, which remains the economic activity with higher waste generation in EU. These numbers demonstrate the relevance of the work presented in this paper.

However, one of the common hurdles to recycling and reusing CDW in the EU is the lack of confidence in the quality of CDW recycled materials. There is also uncertainty about the potential health risk for workers using those materials. This lack of confidence reduces and restricts the demand for CDW recycled materials, which inhibits the development of CDW waste management and recycling infrastructures in the EU. Thus, the European Commission (2016) published a protocol aiming to increase confidence in the CDW management process and trust in the quality of CDW recycled materials, namely by defining methods to improve waste identification, source separation, and collection, as well as waste logistics and processing, quality management and appropriate policy and framework conditions.

From an economic point of view, the construction and demolition waste recycling worthwhile when the recycled product is competitive in comparison with natural resources in terms of cost and quality. Thus, recycled materials will be more competitive in regions with scarce raw materials, as they can reduce the transportation cost (Tam and Tam, 2006).

One example of previous research on the use of recycled aggregates in road construction was carried out in Nottingham in terms of mechanical and environmental performance (Hill et al., 2001). The authors concluded that recycled aggregates presented a high resistance and that its use could reduce the thickness of the pavement layers. Moreover, they did not give rise to environmental risks when used in asphalt mixtures. Other examples of research in this topic include studies performed in Coruña (Pasandín and Pérez, 2015, Pasandín et al., 2015) to evaluate the incorporation of recycled concrete in asphalt mixtures. In general, these studies concluded that recycled concrete could be an excellent alternative to natural aggregates.

2 MATERIALS AND METHODS

2.1 *Materials*

The CDW material used as aggregate in the present study was recycled concrete. This material was selected after analyzing the most common CDW materials in the waste management and treatment companies in northern Portugal. In this case, the CDW concrete was provided by *Braval*, a company for the recovery and treatment of solid waste, without any treatment after separation. The concrete pieces to be recycled provided were presented in asymmetric blocks with an average length of 15 cm (Figure 1a). The size reduction of the concrete was made in the crushing machine shown in Figure 1b. Figure 1c shows the final result of the material after crushing and ready for the subsequent characterization tests. At this stage, CDW concrete can already be considered as a final recycled product available to be used in asphalt mixtures.

In this study granite was used as natural material or standard aggregate, to make the comparison with the recycled concrete, and for the production of asphalt mixtures. This natural aggregate was selected because it is the most common in northern Portugal, being used for the production of asphalt mixtures. The granite was supplied by *Bezerras* quarry in the commercial fractions of 6/14, 4/6 gravel and 0/4 dust. Besides, limestone filer was used to improve the compatibility with bitumen and the workability of the asphalt mixture. The grading curves of the natural aggregates is shown in Table 1.

A commercial 50/70 bitumen from *Cepsa* was used in this study for the production of asphalt mixtures and determination of the aggregate-binder affinity, which must comply with the characteristics specified in EN 12591.

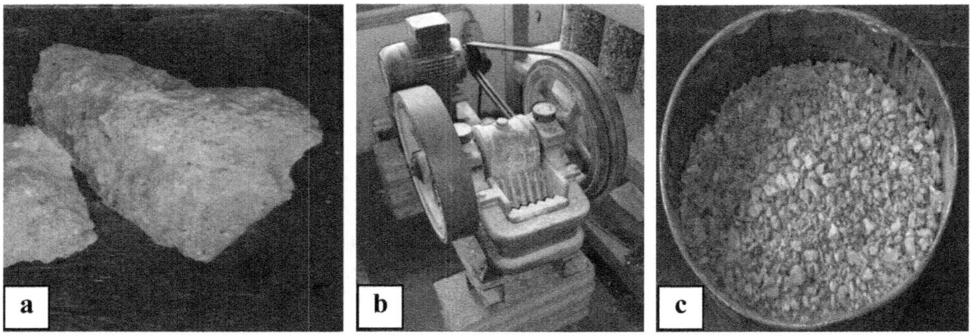

Figure 1. Concrete waste obtained from CDW separation (a), crushing equipment used for CDW treatment (b), and recycled concrete after treatment used in this work (c).

Table 1. Size distribution of the fractions of natural aggregates used to produce the asphalt mixtures.

Sieve dimension (mm)	20.0	14.0	10.0	4.0	2.0	0.5	0.125	0.063
Gravel 6/14	100.0	92.0	40.0	2.0	2.0	1.0	1.0	0.0
Gravel 4/6	100.0	100.0	100.0	10.5	3.9	2.7	1.9	1.3
Dust 0/4	100.0	100.0	100.0	94.0	73.0	39.0	17.0	11.0
Commercial filler	100.0	100.0	100.0	100.0	100.0	100.0	100.0	99.0

2.2 *Methods*

Although this work is mainly focused on characterizing the recycled concrete in comparison with the natural granite aggregate, the work began by evaluating the 50/70 pen bitumen used in some parts of the study. The penetration test was performed at 25°C through EN 1426 standard, and the ring and ball test was performed according to EN 1427 to determine the softening point. The viscosity test was also carried out at various temperatures with the Brookfield rotational viscometer according to EN 13302.

Then, several tests usually applied to characterize aggregates for paving works were carried out on the recycled concrete and the reference granite aggregate. The particle size distribution (grading) was initially evaluated by sieving according to EN 933-1. The shape index (BS 812 Part 105) was also measured since the excess of flaky or elongated aggregates decreases the resistance of the asphalt mixture. The absorption of water (EN 1097-5) is another quite important result because of its direct relation with the bitumen absorption of the aggregates. Then, the affinity between each aggregate and bitumen was assessed under more adverse conditions, according to EN 12697-11. The micro-Deval resistance to wear (EN 1097-1) is another fundamental property measured on coarse aggregates to ensure that a change in the performance of the mixture due to aggregate wear will not occur.

The next phase of the work was the design of the asphalt mixture to be produced with 30% recycled concrete as well as a reference mixture with only natural aggregates. The type of mix selected was an AC 14 surf mixture for road surface layers specified in Portugal by Estradas de Portugal (2014). The bitumen content of the mixture with natural aggregates was obtained by using the Marshall Mix design method, but it was essential to increase the binder content of the mix with recycled concrete to take into account its higher absorption of bitumen. After knowing the difference in water absorption of recycled concrete and natural granite, this value should be multiplied by 30% (CDW content in the mix) and then reduced by 50% to harmonize water and bitumen absorption, based on Lee et al. (1990) work. Finally, the abovementioned asphalt mixtures were produced at 150°C following EN 12697-35 and then several test samples were compacted in the impact compactor (EN 12697-30) with 75 strokes on each side. After determining their volumetric characteristics

Table 2. Size distribution of the crushed recycled concrete used in this work.

Sieve dimension (mm)	20.0	14.0	10.0	4.0	2.0	0.5	0.125	0.063
Recycled concrete	100.0	96.0	77.0	32.0	18.0	7.0	2.0	1.0

(densities and air voids volumes according to EN 12697-5 and EN 12697-6), the specimens were tested by evaluating the indirect tensile strength (EN 12697-23) and the sensitivity to water (EN 12697-12) in order to analyse the influence that the use of recycled concrete has on the durability of the asphalt mixtures.

3 RESULTS AND ANALYSIS

3.1 *Bitumen properties*

The 50/70 pen bitumen used in this work had an average penetration of 52.4 tenths of a millimetre, and a softening point of 52.2°C. These values are within limits designated in the current specifications for this type of binder (EN 12591).

The viscosity results of the 50/70 pen bitumen at elevated temperatures (production and compaction stage) varied from 6.21 Pa.s at 100°C to 0.16 Pa.s at 170°C. The production of asphalt mixtures should occur at the equiviscous (0.3 Pa.s) temperature of around 150°C.

3.2 *Recycled concrete vs. natural aggregate characteristics*

The recycled concrete has a well-graded size distribution (Table 2). The larger particles of the crushed recycled concrete have a 14 mm dimension, with 96% of material passing on that sieve. The minimum size (less than 10% of passing material) is minimal (0.5 mm), and thus this material corresponds to an aggregate fraction 0/14. The continuous curve of this material is relevant for an excellent fit to the grading curve of the asphalt mixture in the mix design phase.

The shape of the aggregate particles, which should be approximately cubic, was then evaluated, because the use of flaky or elongated particles, which are more brittle, is not desirable. The elongation and flakiness indexes of the recycled concrete (20% and 8%) were nearly equal to the corresponding values of the natural granite aggregate (19% and 7%). The elongation values were higher than the flakiness values, and both materials complied with the 25% limit specified for their use in an AC 14 surf mixture.

The water absorption of recycled concrete (9.8%) was much higher than that of natural granite aggregate (1.2%), either due to the cement used on its composition or its increased porosity. This result can be especially burdensome if asphalt plants are forced to use a more significant amount of energy for drying this aggregate, and also because it is a sign of greater absorption of bitumen, which implies a higher cost of the mixtures (as will be seen in the mix design phase).

The affinity of the aggregates to the bitumen is obtained through visual analysis, so it is essential to present the two aggregates studied before and after the test (Figure 2), namely after being in contact with water to remove part of the bitumen film that covered them. The result of this test corresponds to the surface area covered by bitumen at the end of the experiment, which was only 35% for natural granite and 76% for recycled concrete. As a porous aggregate, recycled concrete needs more bitumen for total wrapping, but has a better affinity with the binder.

The micro-Deval coefficient measures the wear loss of aggregates to be used in road pavements. At the end of the test, it was found that the recycled concrete has a rounded shape, due to its increased wear on the sharp edges. Thus, the wear of recycled concrete (33.8%) was much higher than that of natural granite (3.8%), not meeting the specified limit of 15% for an AC 14 surf mixture. Thus, the use of large quantities of this material in asphalt mixtures should be prudent, preferably in the substitution of fine fractions of aggregate.

Figure 2. Natural granite (left) and recycled concrete (right) before and after binder affinity test.

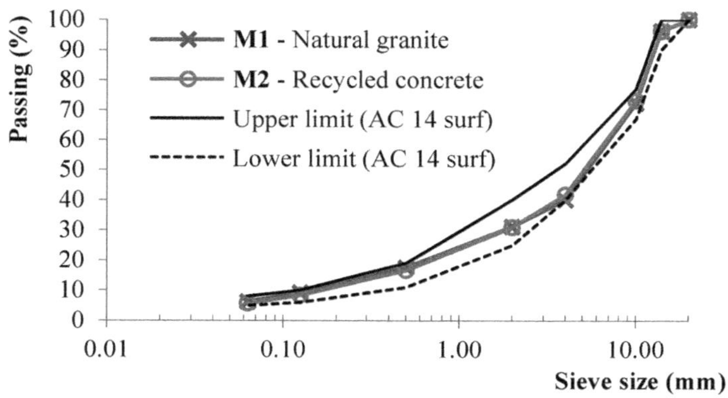

Figure 3. Grading curves of mixtures M1 (natural granite) and M2 (recycled concrete) within the specified limits.

3.3 *Mix design and evaluation of asphalt mixtures' performance*

After knowing the gradation of all the materials to be used, they were combined to obtain the final grading curves of the asphalt mixtures with natural granite and with 30% recycled concrete. The M1 mixture (only with natural granite) was produced with 46% of 6/14 gravel, 14% of 4/6 gravel, 38% of 0/4 dust and 2% of filler, while the M2 mixture (with 30% recycled concrete) was produced with 30% recycled concrete, 33% of 6/14 gravel, 4% of 4/6 gravel, 31% of 0/4 dust and 2% of filler. Figure 3 shows that the final grading curves of both mixtures (M1 and M2) meet the specified size distribution limits for the AC 14 surf mixture.

Subsequently, the M1 mixture with natural aggregates was designed with the Marshall Mix design method, and the final binder content obtained in that study was 5.0%. As explained in Section 2, this value was adjusted to the M2 mixture, with 30% recycled concrete, as a function of the increment of bitumen absorption estimated by the variation of water absorption (8.6%) between the two types of aggregates. Thus, it was concluded that the M2 mixture should be produced with a binder content of 6.3%.

After the design of both mixtures, samples were produced to initially determine their volumetric properties (bulk and maximum density, and air voids content). Then, tests were carried out to evaluate the indirect tensile strength (ITS) of the conditioned (wet) and unconditioned (dry) test samples (Figure 4). With those values, it was possible to estimate the water sensitivity (ITSR) of the mixes. These results are presented in Table 3 to assess the relative durability of the control mixture with natural granite (M1) against that with 30% recycled concrete (M2).

The results show that the two mixtures evaluated in this study have a good mechanical performance and have very similar properties concerning their expected durability (indirect tensile strength values between 2135 kPa and 2260 kPa and water sensitivity ITSR values between 73% and 75%). However, it is essential to take into account that the M2 mixture with recycled concrete

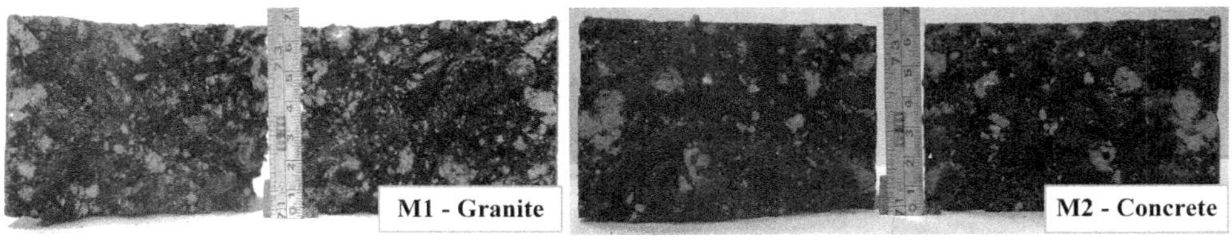

Figure 4. Samples of mixtures M1 (natural granite) and M2 (recycled concrete) after ITS tests.

Table 3. Results of water sensitivity and indirect tensile strength test.

Mixture	Natural granite (M1)		Recycled concrete (M2)	
Conditioning	Wet	Dry	Wet	Dry
Air voids content (%)	4.9	5.3	2.9	2.9
Bulk density (Kg/m^3)	2293	2273	2257	2261
Maximum density (Kg/m^3)	2406	2406	2330	2330
Indirect tensile strength, ITS (kPa)	1604	2135	1657	2259
Water sensitivity, ITSR (%)	75	–	73	–

used a higher binder content (an increase of 26%), which would have supported these good results for M2 mix and its lower porosity (2.9% vs. 5.1% of the M1 mixture). Although the typical reference goal of 80% for ITSR has not been reached, it can be concluded that M2 can be applied because it behaves as well as the control mixture only with natural granite (M1).

4 CONCLUSIONS

This work investigated the properties of recycled concrete, as one of the main CDW materials, to study the viability of its use as part of the aggregates (up to 30%) in asphalt mixtures.

It was concluded that recycled concrete could be easily crushed to obtained a well-graded aggregate with a 0/14 size distribution, which can easily be incorporated up to 30% in the design of asphalt mixtures. The recycled concrete shape complies with the specified limits, and its affinity with bitumen is better than that of natural granite. However, recycled concrete is a porous cement composite material, with high water absorption and low wear resistance, and does not meet the specified limits for these properties. Thus, higher amounts of binder should be used in mixtures with recycled concrete, thus assuring good durability (ITS and ITSR values) in the pavement. Therefore, although recycled concrete should be applied prudently, up to 30% of this CDW can be successfully used as an alternative material to produce new asphalt mixtures.

REFERENCES

Coelho, A. & Brito, J. 2010. Distribution of materials in construction and demolition waste in Portugal. *Waste Management & Research*, 29, 843–853.

Estradas de Portugal 2014. Caderno de Encargos Tipo Obra. Pavimentação – 14.03. Características dos materiais. Almada: Estradas de Portugal, S.A.

European Commission 2015. Communication from the Commission. Closing the loop – An EU action plan for the Circular Economy. *COM 614*. Brussels: European Commission.

European Commission 2016. EU Construction & Demolition Waste Management Protocol. Brussels: European Commission.

Eurostat 2019. Statistics Explained – Waste statistics. Brussels: European Commission.

Hendriks, C. F., Janssen, G., van den Dobbelsteen, A. & Xing, W. 2004. *A new vision on the building cycle*, Delft, Aeneas.
Hill, A. R., Dawson, A. R. & Mundy, M. 2001. Utilisation of aggregate materials in road construction and bulk fill. *Resources, Conservation and Recycling*, 32, 305–320.
Lee, D.-Y., Guinn, J. A., Khandhal, P. S. & Dunning, R. L. 1990. Absorption of asphalt into porous aggregates. SHRP-A/UIR-90-009. Washington, D.C.: National Research Council.
Pasandín, A. R. & Pérez, I. 2015. Characterisation of recycled concrete aggregates when used in asphalt concrete: A technical literature review. *European Journal of Environmental and Civil Engineering*, 19, 917–930.
Pasandín, A. R., Pérez, I., Oliveira, J. R. M., Silva, H. M. R. D. & Pereira, P. A. A. 2015. Influence of ageing on the properties of bitumen from asphalt mixtures with recycled concrete aggregates. *Journal of Cleaner Production*, 101, 165–173.
Presidência do Conselho de Ministros 2017. Resolução n.° 190-A/2017. *Plano de Ação para a Economia Circular em Portugal.* Lisboa: Diário da Républica.
Schrör, H. 2011. *Generation and treatment of waste in Europe 2008. Statistics in Focus 44*, Luxembourg, Environment and Energy Eurostat – European Commission.
Sonigo, P., Hestin, M. & Mimid, S. 2010. Management of construction and demolition waste in Europe. *Proceedings of the Stakeholders Workshop.* Brussels.
Tam, V. W. Y. & Tam, C. M. 2006. A review on the viable technology for construction waste recycling. *Resources, Conservation and Recycling*, 47, 209–221.

Wastes: Solutions, Treatments and Opportunities III – Vilarinho et al. (Eds)
© 2020 Taylor & Francis Group, London, ISBN 978-0-367-25777-4

Valorisation of steel slag as aggregates for asphalt mixtures

L.P. Nascimento, H.M.R.D. Silva & J.R.M. Oliveira
CTAC – Centre for Territory, Environment and Construction, University of Minho, Guimarães, Portugal

C. Vilarinho
MEtRICs – Mechanical Engineering and Resources Sustainability Centre, Department of Mechanical Engineering, University of Minho, Guimarães, Portugal

ABSTRACT: The construction industry is responsible for consuming significant amounts of raw material and causing severe environmental impacts. Natural aggregates are essential to build road pavements, and their extraction in nature causes several negative effects. Therefore, one way to reduce such impacts is to use recycled aggregates. In this work, two hot mix asphalts were studied, one incorporating 75% of steel slag, as aggregate substitutes, and an-other with 100% natural aggregates. Their mechanical performance was evaluated with permanent deformation and water sensitivity tests. The mixture incorporating steel slag showed higher resistance in both tests, proving that it is viable to use this waste material from the metallurgical industry as an alternative aggregate for producing asphalt mixtures, reducing the extraction of new materials and saving landfill space.

1 INTRODUCTION

Taking into account the environmental concerns that the Society is facing nowadays, changes are being made to the importance given to preserving the natural environment and minimising the impacts. Thus, the use of recycled material for civil engineering works has increased in recent years.

International environmental regulations are becoming more rigid about preserving natural resources and reducing waste production. In Portugal, the Decree-Law no. 73/2011 (2011) provides the basis for the approval of prevention programs and sets targets for re-use, recycling, and other forms of waste recovery, to be achieved by 2020. Given the importance of a strong incentive for recycling to accomplish these goals, but also with to pre-serve natural resources, this regulation states that at least 5% of recycled materials are to be used in public works contracts.

According to Araújo et al. (2013), a conventional asphalt concrete mixture for surface courses (AC14 surf 35/50) comprises about 95% aggregates and 5% bitumen. Thus, given the dominance of the aggregates in the asphalt mixtures, it can be concluded that recycling of paving materials, namely by substituting their aggregates, can reduce the environmental impact from mining/extraction of that natural resource, which would also lead to a reduction of cost.

One of the wastes that are being studied is the steel slag aggregate (SSA), which is de-rived from the steel production industry and are the most significant waste generated (more the 60%) in this type of industrial process (Chen et al., 2018, Nascimento et al., 2018). It is estimated that Portugal produces about 400 × 103 tons of steel slag from Electric Arc Furnaces (EAF) per year (Ferreira, 2010). Thus, with the amount of waste generated, its valorisation to produce new materials should be seriously considered.

Kim et al. (2018) investigated the mechanical behaviour of Hot Mix Asphalt (HMA) using SSA. In comparison to a conventional mixture, the asphalt mixture produced with SSA showed improved performance on the rutting resistance and a higher dynamic modulus, although, the air void content was higher than that of the conventional mix.

Kara et al. (2004) studied the incorporation of steel slag in HMAs to be applied in base courses (100% SSA), binder courses (70% SSA and 30% limestone aggregate) and surface courses

(50% SSA and 50% limestone) and concluded that steel slag can be used as an aggregate in asphaltic mixtures. A similar conclusion was obtained by Nguyen et al. (2018).

Martinho et al. (2018) studied the incorporation of 30% of SSA in Warm Mixture Asphalt (WMA) to decrease the energy consumed in the production process. In that study, it was observed that the Marshall Stability increased for the mixture with SSA incorporation, while the stiffness modulus reduced and fatigue resistance did not change. Thus, the authors concluded that this mixture showed satisfactory results when compared to a conventional mixture. Other authors (Masoudi et al., 2017) also concluded that the incorporation of SSA in WMA mixtures is generally recommended.

The steel slag may contain free CaO and MgO. Therefore it can expand and cause problems in the presence of water, decreasing the useful life of the pavement, which may start to show early distresses (Washington State DOT, 2015).

When working with SSA, some authors recommend using a curing process in open storage for a specific time and under humid climatic conditions to remove all free CaO and MgO (Horii et al., 2015, Choi et al., 2007, Chen et al., 2018). Fakhri and Ahmadi (2017) also recommend exposing the SSA to weather conditions for six months before using it in pavement applications.

Those studies show that SSA can replace a substantial amount of new materials, which means, a reduction on the landfill of these type of industrial wastes and the extractions of raw materials. Furthermore, their use can improve some pavement mechanical proprieties and mitigate the environmental impacts caused by the construction works.

In this study, experimental work was carried out to assess the possibility of incorporating high contents of SSA in HMAs, based on a series of mechanical tests, e.g., water sensitivity and rutting resistance, in comparison to a conventional HMA produced with natural aggregates.

2 MATERIALS AND METHODS

2.1 *Materials*

As previously mentioned, two types of aggregate were used in this work. A natural granitic aggregate separated in four granular fractions, namely, 0/4 mm, 4/10 mm, 10/14 mm, and 14/20 mm and a steel slag aggregate (SSA) divided in 0/20 mm and 10/14 mm fractions. The SSA is classified as "waste from steel slag processing" in chapter 10 of the European list of waste (OJEU, 2014), under code 10 02 01. The other materials used were a 35/50 pen bitumen, which is commonly used in Portugal and limestone filler.

The density and water absorption of each aggregate (EN 1097-6, 2013), and the penetration (EN 1426, 2015) and softening point (EN 1427, 2015) are presented in Tables 1 and 2.

Table 1. Physical properties of aggregates used.

	Density Mg/m^3	Water absorption %
Granitic aggregate	2.55	0.91
Steel slag aggregate	3.13	1.40

Table 2. Basic properties of bitumen.

	Penetration (EN 1426) 0.1 mm	Softening point (EN 1427) °C
Bitumen 35/50	36.4	47.3

2.2 *Methods*

2.2.1 *Mix design and production of the HMA with granitic and SSA aggregates*

Two asphalt mixtures were produced using the mentioned aggregates to be applied in a pavement base course. One of the mixtures was produced using 100% natural aggregates and the other using 75% SSA and 25% natural aggregates. According to the national specifications (Estradas de Portugal, 2014), the typical mixture used in such applications is an asphalt concrete (AC 20 base 35/50), which is produced with a maximum aggregate size of 20 mm and with a particle size distribution that fulfils a specified grading envelope. In order to determine the optimum binder content, a mix design study was carried out, based on the Marshall method, according to the EN 12697-12 (2018) standard.

After producing the asphalt mixtures, cylindrical test specimens were compacted using the impact compaction method, specified by EN 12697-30 (2018) standard. Slabs were al-so obtained for further testing, according to EN 12697-33 (2003), using the roller compactor.

2.2.2 *Water sensitivity*

One of the parameters usually used to assess the durability of asphalt mixtures is their water sensitivity. This parameter evaluates the influence of water in the loss of affinity be-tween the aggregate and bitumen, assessing its mechanical performance through indirect tensile strength (ITS) tests.

Two groups of specimens are conditioned in different scenarios to perform water sensitivity tests. One group (wet) is placed in a water bath for a period of 72 h at 40°C, and the other group of specimens is kept dry, at room temperature during the same period. Then, the specimens are tested using a uniaxial testing apparatus, according to EN 12697-12 (2018) standard, to obtain their indirect tensile strength. The ratio between the average results of the wet (ITSw) and dry (ITSd) groups is known as the indirect tensile strength ratio (ITSR) and is the primary indicator of the mixture's water sensitivity.

2.2.3 *Water sensitivity*

Wheel tracking tests (WTT) were used to assess the resistance to permanent deformation (also known as rutting resistance) of the asphalt mixtures, following the procedure de-scribed in the EN 12697-22 (2003) standard (procedure B, in air). This test was carried out in the laboratory using two slabs with the dimensions of $30 \times 30 \times 4\ cm^3$, and the test temperature is 60°C.

The test consists essentially of repetitively passing a wheel on these slabs until 10 000 load cycles have been applied, where the wheel load is 700N, and the frequency is 0.44 Hz. The results considered are the wheel tracking slope (WTSair) between the 5000th and the 10000th cycles, the proportional rut depth (PRDair), and maximum rut depth (RDair) at the end of the test. This test provides information on the stability of the mixture at high service temperatures, which also influences the durability of the material.

3 RESULTS AND DISCUSSION

3.1 *Mix design and production of the mixtures*

The mix design started with the analysis of the aggregates particle size distribution (according to the EN 933-1 (2012) standard) to combine the different aggregate fractions to fulfil the specified envelope and estimate the maximum amount of SSA that could be incorporated in the mixture. Thus, mixture 1 (M1) is a conventional mixture, comprising only natural aggregates and bitumen (5%), while mixture 2 (M2) was prepared with 75% SSA, 25% natural aggregates and bitumen (5%). The fulfilment of the grading envelope specified by Estradas de Portugal (2014) by both M1 and M2 can be observed in Figure 1. Regarding the determination of the optimum binder content, it was obtained by performing the Marshall mix design method, and the results indicated that both mixtures should be produced with a binder content of 5% (by mass of mixture).

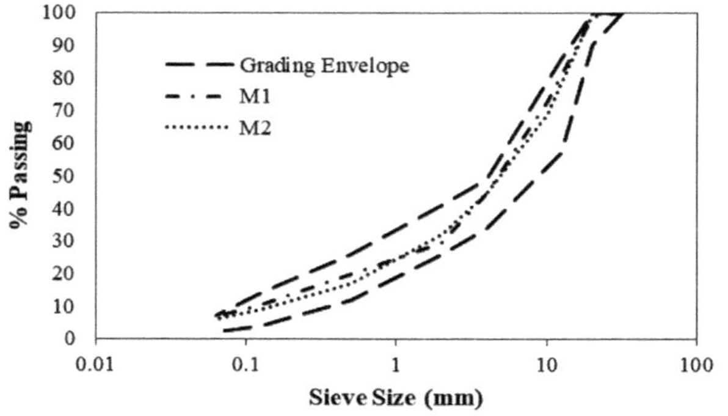

Figure 1. Particle size distribution of HMAs with 100% natural aggregate (M1) and 75% SSA (M2).

Table 3. ITSR results in the water sensitivity test.

Mixture	ITS dry kPa	ITS Wet kPa	ITSR (%)
M1	2936	1993	65
M2	3093	2326	75

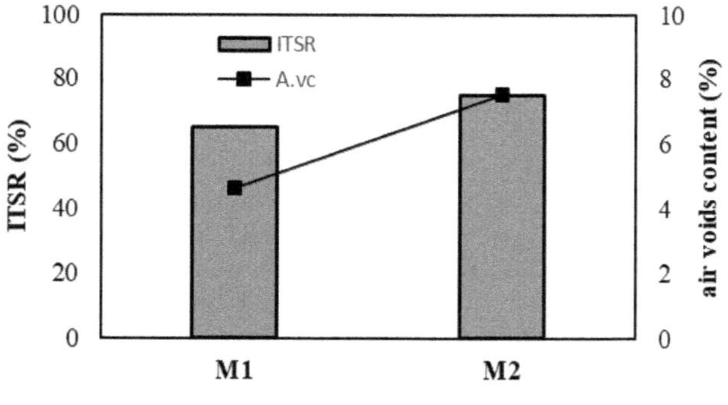

Figure 2. Results of the water sensitivity test: ITSR vs Air voids content.

3.2 *Water sensitivity*

The water sensitivity test method is the most relevant to study and understand the mechanical behaviour of asphalt mixtures in the presence of water, which influences the performance and durability of asphalt materials in service. The results of ITS and ITSR obtained in this test for both mixtures are presented in Table 3. The relation between the average air voids content of the tested specimens and the ITSR is shown in Figure 2 for both mixtures.

When analysing the results, it is possible to observe that mixture M2 presents a water sensitivity performance slightly better than M1, as both ITS and ITSR values are higher in mixture M2.

Another important factor that usually compromises the performance of the asphalt mixtures in this test is the air voids content. This parameter is directly influenced by the compaction of the mixture, the particle size distribution and also the porosity of the aggregates. Since mixture M1 was produced with a natural aggregate that is less porous than the SSA, it presents a lower air voids content (4.6%), in comparison to mixture M2 (7.6%). In conventional asphalt mixtures, a higher air voids content is generally associated with a less efficient performance in terms of water sensitivity. However, this was not the case in the present study. Mixture M2 showed a higher air void content and also a higher ITSR value, which may be explained by the better adhesion of the bitumen to the

Table 4. Wheel tracking test results.

Mixture	WTSair ($Mm/10^3$ cycles)	PRDair (%)	RDair (Mm)
M1	0.11	6.34	3.48
M2	0.80	4.95	2.73

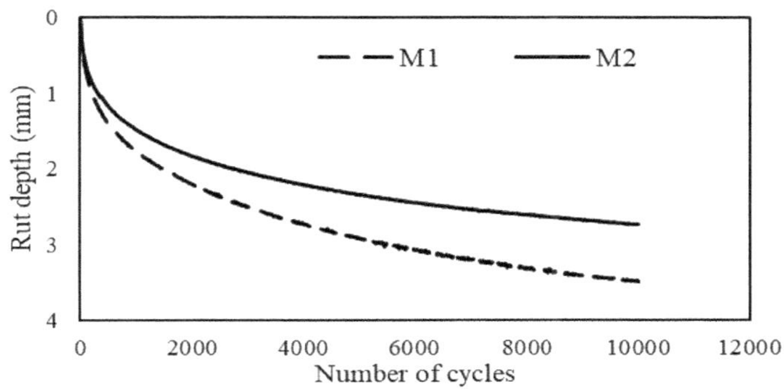

Figure 3. Rut depth evolution during wheel tracking tests.

SSA aggregates, due to the higher porosity of those aggregates, which makes it more difficult to the water to disturb the bond between the binder and the aggregates.

3.3 *Resistance to permanent deformation*

The resistance to permanent deformation or rut resistance is a significant property of asphalt mixtures, namely when they are applied in locations with high service temperatures. In this study, the rut resistance was evaluated by performing WTT tests on both mixtures. The results of that evaluation are presented in Table 4 and Figure 3.

When analysing the results presented in Table 4 and Figure 3, it can be concluded that M2 is the mixture with the highest resistance to permanent deformation. The lower rate of permanent deformation obtained for mixture M2 (with SSA) is visible in Figure 3, where the evolution of the rut with the number of cycles is plotted.

It is worth noting that this test was carried out at 60°C, which is a high service temperature since this parameter has a significant influence on the deformation. Thus, it can be confirmed that mixture M2 showed an adequate performance when regarding the resistance to permanent deformation at high service temperatures.

4 CONCLUSIONS

The valorisation of industrial by-products or wastes results in attractive solutions to environmental problems generated by industrial activities, such as overfilling of landfills and depletion of natural resources. Thus, research studies in these areas are essential to pro-mote more sustainable practices and reduce the environmental impacts caused by anthropogenic activities.

With the results obtained in this study, it can be concluded that the use of SSA in asphalt mixtures is a viable solution. Regarding the mechanical performance of the studied mixtures, namely concerning water sensitivity and permanent deformation, it was observed that the mixture with 75% SSA outperformed that with 100% natural aggregates. Thus, the use of this type of industrial waste as an aggregate substitute in asphalt mixtures has exceptional potential. Regarding the air voids content, it was observed that mixture M2 (with SSA) showed values slightly above the limits specified for the type of asphalt mixture produced in this work. Thus, although the mechanical performance was not affected by that parameter, its influence in other properties (e.g., fatigue resistance) should be further evaluated in future works.

REFERENCES

Araújo, J. P., Oliveira, J. & Silva, H. 2013. Avaliação da influência da camada de desgaste na sustentabilidade dos pavimentos rodoviários. *7° Congresso Rodoviário Português*. Centro Rodoviário Português.

Chen, Z., Jiao, Y., Wu, S. & Tu, F. 2018. Moisture-induced damage resistance of asphalt mixture entirely composed of gneiss and steel slag. *Con Buil Mat*, 177, 332–341.

Choi, S.-W., Kim, V., Chang, W.-S. & Kim, E.-Y. 2007. The present situation of production and utilization of steel slag in Korea and other countries. *Mag Korea Conc Inst*, 19.

Decree-Law no. 73/2011. 2011. In: Ministério do Ambiente e do Ordenamento do Território (ed.). Lisboa: Diário da República n° 116/2011, Série I.

EN 933-1. 2012. Tests for geometrical properties of aggregates – Part 1: Determination of particle size distribution – Sieving method. Brussels, Belgium: CEN.

EN 1097-6. 2013. Tests for mechanical and physical properties of aggregates – Part 6: Determination of particle density and water absorption. Brussels, Belgium: CEN.

EN 1426. 2015. Bitumen and bituminous binders – Determination of needle penetration. Brussels, Belgium: CEN.

EN 1427. 2015. Bitumen and bituminous binders – Determination of the softening point – Ring and Ball method. Brussels, Belgium: CEN.

EN 12697-12. 2018. Bituminous mixtures – Test methods – Part 12: Determination of the water sensitivity of bituminous specimens. Brussels, Belgium: CEN.

EN 12697-22. 2003. Bituminous mixtures – Test methods for hot mix asphalt – Part 22: Wheel tracking. Brussels, Belgium: CEN.

EN 12697-30. 2018. Bituminous mixtures – Test methods – Part 30: Specimen preparation by impact compactor. Brussels, Belgium: CEN.

EN 12697-33. 2003. Bituminous mixtures – Test methods for hot mix asphalt – Part 33: Specimen prepared by roller compactor. Brussels, Belgium: CEN.

Estradas de Portugal. 2014. Caderno de encargos tipo obra. 14.03-Pavimentos. Características dos materiais. Estradas de Portugal, S.A., Lisboa.

Fakhri, M. & Ahmadi, A. 2017. Recycling of RAP and steel slag aggregates into the warm mix asphalt: A performance evaluation. Con Buil Mat, 147, 630–638.

Ferreira, S. R. 2010. *Comportamento mecânico e ambiental de materiais granulares: aplicação às escórias de aciaria nacionais*. Doutoramento, Universidade do Minho.

Horii, K., Kato, T., Sugahara, K., Tsutsumi, N. & Kitano, Y. 2015. Overview of iron/steel slag application and development of new utilization technologies. *Nippon Steel & Sumitomo Technical Report*, 109.

Kara, M., Günay, E., Kavakli, B., Tayfur, S., Eren, K. & Karadag, G. 2004. The Use of Steel Slag in Asphaltic Mixture. *Key Engineering Materials*, 264–268, 2493–2496.

Kim, K., Jo, S. H., Kim, N. & Kim, H. 2018. Characteristics of hot mix asphalt containing steel slag aggregate according to temperature and void percentage. *Con Buil Mat*, 188, 1128–1136.

Martinho, F., Picado-Santos, L. & Capitão, S. 2018. Influence of recycled concrete and steel slag aggregates on warm-mix asphalt properties. *Con Buil Mat*, 185, 684–696.

Masoudi, S., Abtahi, S. M. & Goli, A. 2017. Evaluation of electric arc furnace steel slag coarse aggregate in warm mix asphalt subjected to long-term aging. *Con Buil Mat*, 135, 260–266.

Nascimento, L. P., Oliveira, J. R. & Vilarinho, C. 2018. Use of Industrial Waste as a Substitute for Conventional Aggregates in Asphalt Pavements: A Review. *International Conference on Innovation, Engineering and Entrepreneurship*. Springer, 690–696.

Nguyen, H. Q., Lu, D. X. & Le, S. D. 2018. Investigation of using steel slag in hot mix asphalt for the surface course of flexible pavements. *IOP Conference Series: Earth and Environmental Science*. IOP Publishing, 012022.

OJEU. 2014. Commission Decision 2014/955/EU – List of waste referred to in article 7 of Directive 2008/98/EC Brussels: Official Journal of the European Union.

Washington State DOT. 2015. WSDOT Strategies Regarding Use of Steel Slag Aggregate in Pavements. *A Report to the State Legislature in Response to 2ESHB 1299*. Washington: Washington State DOT, Construction Division Pavements Office.

Wastes: Solutions, Treatments and Opportunities III – Vilarinho et al. (Eds)
© 2020 Taylor & Francis Group, London, ISBN 978-0-367-25777-4

Construction and Demolition Waste (CDW) valorization in alkali activated bricks

I. Lancellotti, V. Vezzali, L. Barbieri & C. Leonelli
Department of Engineering "Enzo Ferrari", University of Modena and Reggio Emilia, Modena, Italy

A. Grillenzoni
GARC S.p.A., Carpi, Italy

ABSTRACT: This study investigates the synthesis of room temperature alkali activated (AA) bricks based on alkaline activation of construction and demolition waste coming from 2016 central Italy earthquake rubble. This waste consists in inert debris, which can be exploited as precursors or as fine and coarse aggregates in the production of AA bricks and cements. The material was ground and mixed to work with a representative portion. The forming process is uniaxial pressing. Typology, correct molarity of alkaline activators and curing conditions were also investigated. The consolidated bricks were subjected to integrity and water absorption tests to verify the effectiveness of the alkali activation. Apparent density tests were performed. Compressive strength and adhesion with concrete by means of three-point flexural tests were measured to choose the best samples. Furthermore, bottom ash from secondary raw material was added to CDW powders to evaluate the contribution of aluminosilicate source richer in amorphous phase.

1 INTRODUCTION

The present study investigates the synthesis of alkali activated (AA) bricks starting from powders of construction and demolition waste (CDW) recovered from 2016 central Italy earthquake rubble. These wastes are provided by GARC S.p.A., in the context of Emilia Romagna "PO FSE 2014/2020" regional project, and consists specifically in inert debris, as local rocks, red clay bricks and concrete. Such type of CDW is a very heterogeneous inert waste mix which is here fully exploited to limit as much as possible the complexity and costs of selection and separation processes. Currently, two of the main destinations of recycled CDWs are sub-bases/road embankments and aggregates in structural concrete and in filling mixtures (non-structural). Approximately 240,000 tons of processed rubble were generated during demolition operations of the area of the earthquake epicentre — the town of Amatrice, Italy, and the nearby- operated by GARC S.p.A. (2018) and main destinations were road backgrounds for highway and primary urbanization works for emergency housing solutions.

With this work, we want to contribute to an innovative use of earthquake debris as a precursor or as fine and coarse aggregates in the production of bricks and cements consolidated at room temperature after an alkali activation reaction.

In recent years, this innovative use has been investigated with different approaches by many authors. Use of cement and red clay brick fractions to produce alkali activated cements has been studied in literature (Allahverdi and Kani, 2009), (Komnitsas, 2015) (Vàsquez, 2016). The best results were generally obtained with the red clay brick fraction. Production of prefabricated insulating and radiating panels to be used in energy-efficient buildings starting from CDW fractions was the objective of European project (INNOWEE, 2016) with the focus of reaching energy efficiency standards and lowering environmental impact. Another European project (Re4 project) also deals with prefabricated energy-efficient building concept, with up to 65% in weight of recycled materials from CDW.

These practices can generally lead to several advantages, including: (i) savings in the use of non-renewable raw materials, (ii) given the cold forming process, a reduction in CO_2 emissions and energy consumption with consequent positive climate action; (iii) improving of waste management and increasing recycling in the context of a circular economy, in accordance with the European and Italian legislative framework aimed at reducing waste production.

In terms of originality, our research activity focused on the use of CDW withtout any selection or separation processes, without struiong milling and without the addition of metakaolin, main aluminosilicate precursor used in formulation of alkali activated materials.

2 MATERIALS AND METHODS

2.1 *CDW and AA bricks characterization*

Several parameters are currently being investigated during the different phases of the work. First of all the composition, mineralogy and particle size distribution of the CDW, aiming to a correct percentage of precursors in the final product. The material was collected at the ancient quarry of Carpelone, Posta municipality (Rieti, Italy) and comes from the demolition of the Amatrice city centre operated by GARC S.p.A.. The first step consists in grinding (hammer mill model C.I., HM/530) and homogeneously mixing in order to work with a representative portion of the amount of CDW. Several samples were obtained with two particle size distributions, one below 1 mm and one below 2 mm.

As alkaline activators, K-based (K_2SiO_3/NaOH) and Na-based (Na_2SiO_3/NaOH) solutions were used. Their activity was evaluated from 25 to 40% with respect to the solid powdery matrix. Water was added in order to allow a good workability of the mixture to be poured in a mould as the forming procedure chosen for this work was uniaxial pressing (0.1–0.5 MPa). Curing conditions were in air without any thermal treatment, except first 24 hours when samples were kept in sealed bag (100% humidity) in order to retain moisture and favour alkali activation.

In this work were also added bottom ash derived secondary raw material, in order to evaluate the contribution of another aluminosilicate source richer in amorphous phase with respect to CDW powders (Taurino, 2017, Lancellotti, 2015, Lancellotti, 2013). The bottom ash was added as fine (<250 μm) and course (<1 mm) aggregate.

Figure 1. Amatrice, Italy, CDWs representative mix before grinding.

The obtained bricks were firstly subjected to integrity tests (24 h in distilled water, at room temperature in static condition with solid/liquid ratio 1:100) and water absorption tests to verify the effectiveness of the alkali activation reaction. Apparent density was evaluated using a He pycnometer (Micromeritics, model GeoPyc 1360). Compressive strength (after 28 days of aging) and adhesion with concrete by means of three-point flexural tests were then measured to choose the best samples. To characterize the structure and the morphology of all the AA products, several techniques such as Scanning Electron Microscopy (SEM) and X-ray Diffraction (XRD) are used.

3 RESULTS AND DISCUSSION

No material release during the integrity test in water were observed, only slight cracks formation in the worst specimens. The best samples are: i) activated with K-based solution without aggregate and ii) Na-based solution activated samples with aggregate and iii) without aggregate (with 25% activating solution). On the other hands, the worst samples are K-based solution activated samples with fine aggregate.

From the mineralogical point of view, the starting CDW is mainly formed from calcite $CaCO_3$, quartz SiO_2 and anorthite $CaAl_2Si_2O_8$, while the activated bricks maintain the presence of quartz and calcite but show the formation of lisetite $Na_2CaAl_4Si_4O_{16}$ and orthoclase $KAlSi_3O_8$ when Na and K-based activation solutionsare used, respectively.

Na-based series has higher values of apparent density with respect to K-based series. Density of sample containing 30% solution and particle size distribution below 1 mm reaches values of 2.2 g/cm^3 if the solution is Na-based, while only 1.7 g/cm^3 if solution is K-based. Water absorption results show values below 3% for samples activated with Na and K-based solutions, particle size distribution below 1 mm and without introduction of the bottom ash aggregate, but many samples of the K-based series shows value higher than 5%.

As a proof of these observations and in order to investigate the microstructure of the samples, SEM analysis was performed. From the micrographs reported in Figure 3 appears evident the lower densification degree of the sample activated with K-based solution with respect to Na-based material.

Compressive test results show that K-based solution did not allow optimal activation of samples since the strength values are very low, around 1–2 MPa, for all the compositions.

The samples with acceptable compressive strength (comparable to traditional red clay brick) are the Na-based solution ones. For the sample containing 30% solution and particle size distribution below 1 mm activated with Na-based solution a value of 18 MPa was observed. Such a value can be considered satisfactory given the proximity to traditional bricks values (on average 20/25 MPa) and the fact that no metakaolin was used in the production of these samples.

1) 2)

Figure 2. 1) K-based solution activated samples without aggregate, 2) K-based solution activated samples with fine aggregate.

1) 2)

Figure 3. SEM micrograph of 1) Na-based solution and 2) K-based solution alkali activated bricks.

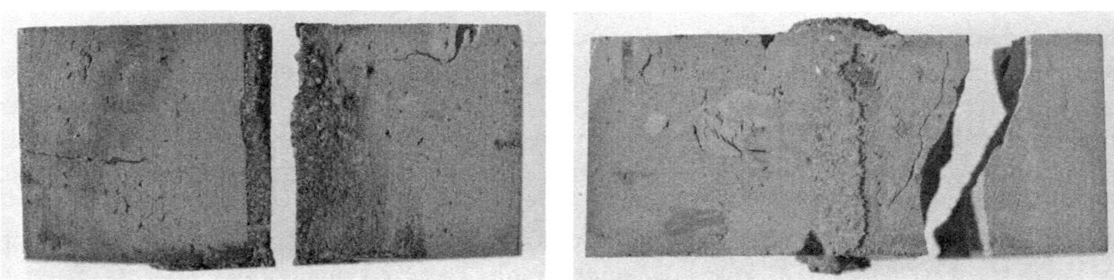

1) 2)

Figure 4. Samples of tradition bricks bound with 1) traditional OPC and 2) alkali activated Na-based mortar after three-point flexural tests.

Preliminary evaluation of the binding properties of the activated mixtures and comparison with CEM I 52.5 R Portland cement were performed through three-point flexural tests. Best results in terms of displacement-load were obtained with Na-based activating solution which showed a substantially higher adhesion strength in respect to Portland cement.

4 CONCLUSIONS

The experimental show that the K-based activating solution does not provide a good behaviour in terms of compressive/flexural strength. On the other hand, Na-based activating solution allows several possibilities of application: the replacing of traditional red clay bricks or binders, such as Portland cement, seems a promising solution in terms of economical and technical feasibility.

Sample with 25% of Na-based activating solution can be considered an example of circular economy and shows several advantages:

- no metakaolin and no aggregates were used (100% Amatrice mix),
- no separation of CDW were performed, a representative mix was used,
- the only treatment consisted in simple hammer mill grinding and a wide particle size distribution (<1 mm) were obtained,
- low % of Na-based activating solution (commercially available and low-cost) was used,

- no thermal treatment was applied,
- curing step in air and at room temperature was sufficient to reach good consolidation after 28 days.

Tests with small amounts of metakaolin are currently in progress on samples activated with Na-based solution in order to enhance mechanical properties.

ACKNOWLEDGMENTS

Authors are particularly grateful to Regione Emilia Romagna, Italy (Project: "Materiali Funzionali per uno sviluppo sostenibile, approvato con Deliberazione di Giunta Regionale n. 886/2016 PO FSE 2014/2020 Obiettivo Tematico 10") and the company GARC S.p.A.a, for financial support.

REFERENCES

Allahverdi, A. and N. Kani, E. 2009. Construction wastes as raw materials for geopolymer binders, Int. J. Civil Eng. 7 (3): 154–160.

Komnitsas K., Zaharaki, D., Vlachou, A., Bartzas, G., Galetakis, M. 2015. Effect of synthesis parameters on the quality of construction and demolition wastes (CDW) geopolymers, Advanced Powder Technology, 26(2): 368–376.

Vásquez, A., Cárdenas, V., A. Robayo, R., De Gutiérrez, R. M. 2016. Geopolymer based on concrete demolition waste, Advanced Powder Technology, 27(4): 1173–1179.

INNOWEE project, http://innowee.eu/it/

RE4 project, http://www.re4.eu/

Taurino, R., Karamanov, A., Rosa, R., Karamanova E. Barbieri L, Atanasova-Vladimirova S., Avdeev, G., Leonelli, C. 2017. New ceramic materials from MSWI bottom ash obtained by an innovative microwave-assisted sintering process; Journal of the European Ceramic Society 37(1): 323–331.

Lancellotti, I., Cannio, M., Bollino, F., (...), Barbieri, L., Leonelli, C. 2015. Geopolymers: An option for the valorization of incinerator bottom ash derived "end of waste"; Ceramics International 41(2): 2116–2123.

Lancellotti, I., Ponzoni, C., Barbieri, L., Leonelli, C. 2013. Alkali activation processes for incinerator residues management; Waste Management 33(8): 1740–1749.

Wastes: Solutions, Treatments and Opportunities III – Vilarinho et al. (Eds)
© 2020 Taylor & Francis Group, London, ISBN 978-0-367-25777-4

"2sDR" An innovative mini mill concept for EAF-dust recycling

M. Auer, J. Antrekowitsch & G. Hanke
Montanuniversität Leoben, Austria

ABSTRACT: Electric arc furnace dust (EAFD) from steel recycling is often highly loaded with zinc. Steel producers operating an electric arc furnace are using up to 100% scrap as raw material. Zinc is mainly used for corrosion protection of steel products, therefore EAFD from steel recycling shows high amounts of zinc. Melting one ton of scrap roughly produces 20 kg of zinc containing dust. Therefore, EAFD nowadays represents a well-known secondary resource in zinc metallurgy and the operating steel mills are keeping an eye on technologies to treat their own dust in the most effective way. This paper shows an innovative way to establish a regional recycling solution allowing "zero waste" multi metal recycling. The two-step-dust recycling-process ("2sDR") consists of a clinkering step and a subsequent carbo-thermal reduction to extract zinc oxide and an iron alloy. Furthermore, the clean slag should be used as construction material.

1 INTRODUCTION

Steel mill dust out of the electric arc furnace (EAF) route nowadays represents a well-known secondary resource in zinc metallurgy. Although a lot of research and development has been carried out within the last decades, no relevant breakthrough regarding an innovative recycling technology could be established. Roughly 90% of the produced dust is treated via the waelz kiln technology. However, globally less than 50% of this high zinc containing dust is recycled. The reasons for this still unsatisfying rate are disadvantages like the recovery of only one metal, the contamination of the produced zinc oxide with halides and the huge amount of newly generated residues (Schneeberger et al. 2012).

In times of limited dump capacities, increasing landfilling fees and strategies for a circular economy, the described way of treatment of potential by-products is hardly acceptable, even though the waelz process is still considered to be the best available technology.

The consequence of this situation is an increased research effort focusing on a multi metal recovery to allow a utilization of the present iron and, if possible, the contained lead. Furthermore, the use of the remaining slag in various sectors like cement industry up to building and construction industry is nowadays a crucial factor in process development (Antrekowitsch et al. 2015, Rütten 2011).

2 STATE OF THE ART

Within the last decades, researchers followed different developments to realize a more efficient EAF-dust treatment, especially to recover the iron and, in parallel, to minimize the newly generated residue.

Besides many theoretical approaches and some others that never made it beyond lab scale, three technologies could be named which were realized in industrial scale:

- the PIZO-process (Heritage)
- the Primus process (Paul Wurth)
- the rotary hearth furnace (ZincOx)

The PIZO-concept follows the principle of direct reductive melting in an inductively heated furnace, recovering zinc oxide via the off-gas and producing molten iron as well as slag. As for all of the described processes in this chapter, carbon is used as reducing agent. Difficulties regarding process technology and refractory lining, together with an incident in the only established unit, years ago, led to the end of this technology in industrial environment (Antrekowitsch et al. 2015, Rütten 2011).

The Primus process is a combination of a multiple hearth furnace together with an electric arc furnace. The multiple hearth furnace should do a part of the reduction, especially the one of zinc oxide and shall also allow a certain separation of halides. The remaining material is molten in the subsequent EAF to reduce the iron and the remaining zinc oxide. Therefore, zinc is brought to the off-gas, re-oxidized and collected in the filter system. The molten iron as well as the slag is tapped from the EAF. One pilot scale plant in Luxembourg and one industrial plant in Taiwan, that follows more or less the Primus concept, were erected. Due to many problems with the multiple hearth furnace and some other difficulties, the pilot plant was closed. So far, no further installation has been reported (Antrekowitsch et al. 2015, Rütten 2011).

The rotary hearth furnace follows a different strategy. EAF-dust is briquetted together with carbon and fed in a rotary hearth. At temperatures of about 1300°C the same principle for zinc separation as for all above mentioned concepts, including the waelz process, is applied. Zinc oxide is reduced, zinc evaporates and is re-oxidized in the off-gas and collected in the filter house. In case of the rotary hearth furnace process, the remaining material is not molten. Iron compounds are reduced too but form a kind of hot briquetted iron. Here the weak point of this technology becomes obvious: Because of the relation of iron to slag components, these produced briquettes can only be of rather poor quality and not be used easily in steel mills as proposed by the operators. Currently, one facility is in operation in South Korea, but no successful utilization of the iron briquettes in bigger scale is known to the author (Antrekowitsch et al. 2015, Rütten 2011).

In addition, it has to be stated that none of the described technologies offer a satisfying solution regarding the halogen impurities in the produced zinc oxide, which makes a utilization of this product in primary zinc industry difficult and lowers the possible revenues. Due to that, a new strategy is under development, which is described in the following chapters (Steinlechner 2013).

3 2SDR

3.1 *Process technology*

The technology, jointly developed by the Chair of Nonferrous Metallurgy at Montanuniversität Leoben and ARP, is a two-step process.

The first step – clinkering in a short rotary kiln – aims to get rid of various halogen compounds including critical fluorides like CaF_2 by pyrometallurgical treatment under oxidizing conditions. Because of the complexity of this process step, the following chapters are specialized on the clinkering of EAFD.

The second step is a reducing process that enables simultaneous recovery of different valuable metals based on a reducing iron bath using an electric arc furnace. An important novelty is the utilization of the produced slag for construction purposes. The presented values would fit to a dust with a zinc content of 25 to 28% and a higher load of halides, especially, chlorine (e.g. 5%). Not shown in Figure 1, is a pretreatment step for agglomeration to avoid too much carry over in the calcination step and to allow an easy transport and charging into the reduction furnace. This could be done by using a compulsory mixer (Rösler et al. 2014).

The technology clearly addresses small to mid-sized steel mills operating an electric arc furnace. The "2sDR" process offers an alternative to long transport routes and insufficient recycling by

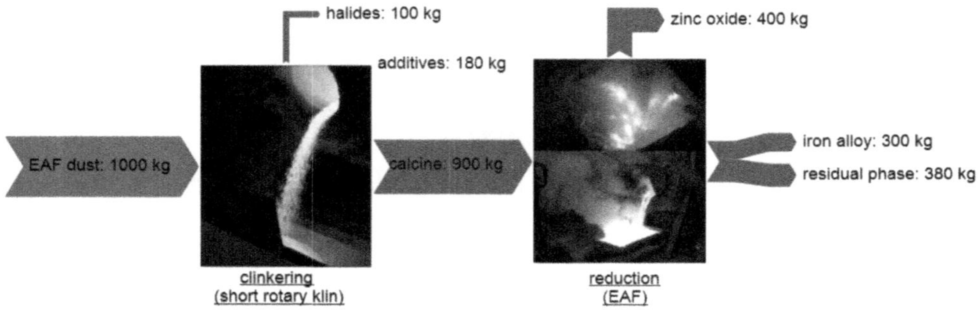

Figure 1. 2sDR process – clinkering and reduction.

providing a tailored mini-mill strategy, consequently following a zero-waste approach. The USPs of the technology can be clearly described as follows:

- Simultaneous recovery of zinc, iron and alloying elements
- Lowest number of remaining by-products (<30%) compared to alternative processes. In best case "zero-waste" is possible by transforming the residual slag into a product for construction industry
- Low/No transport costs
- High purity of zinc oxide and easy marketable products
- High flexibility due to the utilization of only one or two types of dust (no blending, mixing etc.)
- Economically viable even in small units

3.2 *Clinkering step*

Fluorine as well as chlorine are common elements in residues of metallurgical processes. In most cases they cause problematic reactions in the recycling process and decrease the product quality or require expensive off-gas treatment. A good option to get rid of the halogen problem is to install a metallurgical pre-processing at temperatures between 900°C and 1100°C, called clinkering. To operate the clinkering step in the most effective way, an in-depth metallurgical knowledge is necessary. The main goal is to remove chlorine, fluorine and lead by vaporization under oxidizing conditions.

The benefit of a preliminary clinkering process is to increase the product quality of zinc oxide, produced in the second step. If chlorine, fluorine or lead are not removed before, they are vaporized in the reduction process simultaneously with zinc and contaminate the final product. While lead is only decreasing the zinc content in the product, chlorine and fluorine are harmful elements for the further hydrometallurgical processing of the generated zinc oxide. Both elements are causing problems in the zinc electrolysis. Chlorine is responsible for an increasing corrosion of the electrodes and fluorine leads to a strong linkage between zinc and the cathode, which is counterproductive for the following stripping (Antrekowitsch 2004).

The high variation of reactive elements in the EAFD leads to a complex system of many chemical reactions running simultaneously. The behavior of the different compounds in the EAFD can be estimated in a simple thermodynamic calculation using HSC. Therefore, the vapor pressure of the individual halogens is shown, depending on temperature. The first step is to define the chemical reaction: Evaporation of a metal-chloride shown in Equation 1. The associated equilibrium constant (K) contains the partial pressure of the gaseous as well as the activity of the solid metal-chloride and is defined in Equation 2. To show the effect of increasing temperature on the equilibrium constant, the calculation was replied in the interval of 10°C between 0°C and 1500°C. Assuming the activity of a solid metal-chloride is 1, the partial pressure of the metal-chloride corresponds

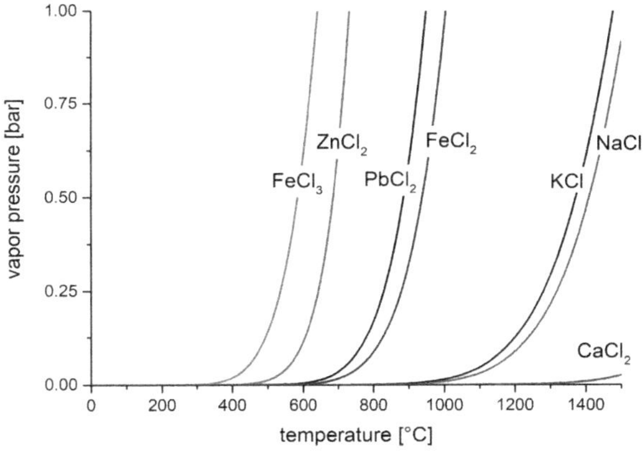

Figure 2. Vapor pressure of different chlorides.

with the equilibrium constant, described by Equation 3.

$$MeCl_X \rightarrow MeCl_{X_{(g)}} \tag{1}$$

$$K = \frac{p_{MeCl_X}}{a_{MeCl_X}} \tag{2}$$

$$K = p_{MeCl_X} \tag{3}$$

where $MeCl_X$ = metal-chloride; K = equilibrium constant; p = partial pressure; a = activity

Using the results of the HSC calculation, it is possible to picture the vapor pressure depending on the temperature. The graph is shown in Figure 2. $FeCl_x$ as well as $ZnCl_2$ and $PbCl_2$ are the first compounds starting to evaporate at temperatures between 350 and 700°C. But iron- and zinc-chlorides are not useful to remove chlorine of the EAFD. Iron as well as zinc are products of the "2sDR" process and therefore, a vaporization of $FeCl_3$, $ZnCl_2$ or $FeCl_2$ is attached to a loss of efficiency. The favorized compound to evaporate in the clinkering step is $PbCl_2$, because lead as well as chlorine have to be removed and the temperature can be set fairly low. KCl and NaCl are also suitable combinations to get rid of the chlorine content, but the temperature at which these compounds start to evaporate is much higher (900–1000°C). In case of temperature, Figure 2, also demonstrates, that the formation of $CaCl_2$ has to be avoided. The temperature limit of the clinkering process is set around 1100°C because reaching the sintering temperature of the material followed by an agglomeration reduces the reaction surface.

The same calculation (Equation 4–6) was done for fluorides and is pictured in Figure 3. Compared to the previous results it can be determined that the volatilization of fluorides takes place at higher temperatures and therefore it is more difficult to reach low fluorine contents by clinkering. The combination with lead (PbF_2) is again the best variation to remove fluorine from the EAFD. At temperatures above 1000°C ZnF_2 and KF start to evaporate. However, ZnF_2 should be avoided by the reason of maximum zinc output as mentioned before, in case of chlorides and KF is required. The compounds NaF, FeF_3 and CaF should be avoided because they are not volatile at set temperatures and cause additional product loss in the case of FeF_2.

$$MeF_X \rightarrow MeF_{X_{(g)}} \tag{4}$$

$$K = \frac{p_{MeF_X}}{a_{MeF_X}} \tag{5}$$

$$K = p_{MeF_X} \tag{6}$$

where MeF_X = metal-fluoride; K = equilibrium constant; p = partial pressure; a = activity

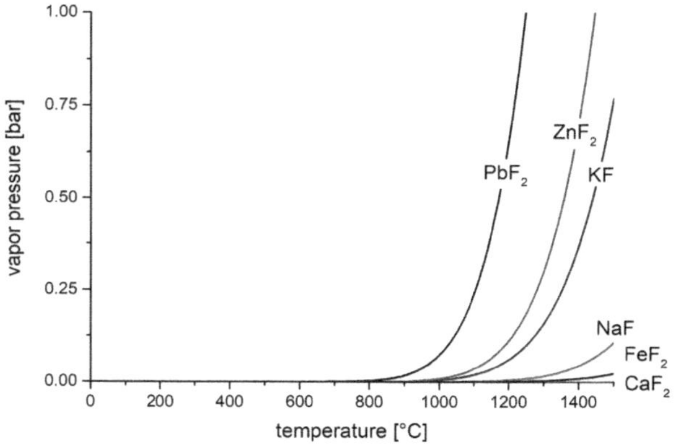

Figure 3. Vapor pressure of different fluorides.

Table 1. Average analysis out of the reduction step.

iron		zinc oxide		slag	
composition	wt%	composition	wt%	composition	wt%
Zn	0.01	ZnO	96.20	ZnO	0.41
Mn	0.02	MnO	0.21	MnO	5.2
Si	0.03	SiO_2	0.62	SiO_2	35.9
Fe	96.45	Fe_2O_3	0.82	Fe_2O_3	5.9
C	2.81	CaO	0.45	CaO	31.4
S	0.06	PbO	0.14	PbO	0.05
P	0.03	Cl	0.46	MgO	6.2
Cu	0.38	F	0.08	Al_2O_3	12.5
Cr	0.1				

3.3 *Reduction step*

Test trials at the Montanuniversität Leoben and ARP GmbH were made in the scale of 50 to 100 kg. The results of several campaigns are summarized and the average compositions of iron, zinc oxide and slag are shown in Table 1.

Halogens like fluorine and chlorine as well as lead show rather low contents in the zinc oxide product. Due to the fact, that the mentioned elements are the main impurities in other concepts based on carbo-thermal reduction, the zinc oxide shows a quite high purity. These results show the positive influence of the preliminary clinkering step.

Remaining iron content in the slag as well as the basicity (CaO/SiO_2) close to 1, cause no problem considering the further usage in the construction sector. The small trend to the acidic side leads to the advantage of a reduced melting point and a good interaction with the reducing iron bath.

3.4 *Economic considerations*

As described above, the 2sDR-process has been developed to allow the economic treatment of comparable low amounts of dust. Depending on the location and with this the specific costs for energy, labor and infrastructure, a minimum of 8000 to 12,000 tons of treated dust per annum should allow a feasible process.

First calculations were done on CAPEX (capital expenditures) considering the main process steps, off-gas cleaning, construction and buildings, storage and internal transport. A CAPEX-value of about 14 million USD was the result, based on according of 10,000 t/a.

OPEX (operational expenditures) calculations lead to a result of 205 USD/t. Zinc, of course, is the cash-cow, even though no special high value was considered for its comparable high purity. The calculation is based on the usual concentrate formula for primary zinc smelters based on the LME-price. Iron was calculated by comparison with low quality scrap and therefore a ton was assumed to have a value of 150 USD. Various trials and investigations are currently done, to make the slag usable in construction industry. However, the slag was calculated with zero value but also with no landfilling fee. The fee that has to be paid by steel mills when they give their dust to a recycler was calculated with 60 USD per ton including transport.

Nevertheless, based on an EAF-dust with 28% zinc and 30% iron a pay-back period of roughly 4 years could be realized. Even when the zinc price drops to 2000 USD/t, pay-back could be realized within 5.5 years. In case of rising prices (more than 3200 USD/t, in 2018) a value of 3 years is achievable.

4 PROSPECT

The bad product quality and the big amount of newly produced residues during the state-of-the-art dust recycling process hardly represents a future oriented solution. Because of the currently high zinc price, the recycling of electric arc furnace dust becomes more and more interesting. To reduce the amount of residues, iron recovery is necessary as well, but many processes failed because of high production costs and the low created value by the iron alloy. On the other hand, to prevent increasing disposal costs in the future, the recovery of iron is inevitable. For higher cost efficiency, it is also conceivable to create value by using the remaining slag in the construction sector.

The 2sDR-technology, developed by the Montanuniversität Leoben and ARP GmbH, gets rid of most of the difficulties in EAF-dust recycling:

- high quality zinc oxide by removing halogens in an upstream process (clinkering)
- iron recovery
- producing a residual slag which can be used in the construction sector
- mini mill solution

Future development of this process is dedicated to the clinkering step. Experiments in lab size as well as in technical scale are ongoing to prove the theoretical calculations presented in this paper. Based on these results, process parameters like temperature, treatment time, heating rate, etc. have to be optimized in industrial environment. First results of this research are expected shortly and will be published. In addition, it is considered to upscale the process to pilot scale together with an engineering company and a steel mill.

Therefore, establishing the 2sDR-technology, could help many steel mills to operate their own EAFD recycling at a most effective way. High quality zinc oxide as well as an iron alloy is created. In best case "zero waste production" is realized.

REFERENCES

Antrekowitsch J. 2004. Aufarbeitung zinkhältiger Stahlwerksstäube unter besonderer Berücksichtigung der Halogenproblematik. *Dissertation*, Leoben.

Antrekowitsch J., Rösler G., Steinacker S. 2015. State of the Art in Steel Mill Dust Recycling, *Chemie-Ingenieur-Technik*, 87 (11):1498–1503.

Rösler G., Pichler C., Antrekowitsch J., Wegscheider S. 2014. "2sDR": Process Development of a Sutain- able Way to Recycle Steel Mill Dusts in the 21st Century. *JOM* 66 (9):1721–1729.

Rütten, J. 2011. Various Concepts for the Recycling of EAFD and Dust from Integrated Steel Mills, *3rd Seminar, Networking between Steel and Zinc*, GDMB, Leoben.

Schneeberger G., Antrekowitsch J., Pichler C. 2012. Development of a New Recycling Process for High Zinc Containing Steel Mill Dusts including a Detailed Characterization of an Electric Arc Furnace Dust. *BHM* 157 (1): 1–6.

Steinlechner St. 2013. Amelioration and market strategies for zinc oxide with focus on secondary sources, *Dissertation*, Leoben.

Wastes: Solutions, Treatments and Opportunities III – Vilarinho et al. (Eds)
© 2020 Taylor & Francis Group, London, ISBN 978-0-367-25777-4

Assessment of by-products – from waste to values

W. Schatzmann & J. Antrekowitsch
Montanuniversität Leoben, Austria

ABSTRACT: The development of raw materials supply is characterized by decreasing metal grades and still quite low recycling rates. Metal bearing by-products such as slags, sludge or dust become more and more important for European industry as these residues carry a significant amount of valuables. Due to the lack of missing technologies and an accepted guideline for the assessment of landfilled material, this potential still lies idle. A new approach for the assessment of metal bearing by-products is currently followed by Montanuniversität Leoben and shortly presented. A case study for zero-waste processing of a slag from primary lead and zinc industry gives an insight on the ongoing research work and the upcoming challenges.

1 RAW MATERIALS SUPPLY AND IST LIMITS

Investigating the development of primary metal resources in general, two trends become obvious: The decrease of metal grades and the increase of undesired impurities. Figure 1 shows the development of ore grades for the base metals zinc and lead (on the left for the production in Australia) and for the minor metal gold (on the right for global production).

The development of zinc and lead ore within the last 100 years (Figure 1) clearly shows a decrease from roughly 12% in the sixties and seventies to about 4 to 6% in 2016. This must be seen as relevant change, especially facing only the moderate development of smelting technologies within this time span. In parallel, increasing iron values in zinc ores become apparent, leading to significantly higher amounts of residues. Due to missing reprocessing methods, these wastes currently have to be landfilled, representing a potential source of environmental pollution.

The same applies for gold deposits dropping from 10 g/t to 3–4 g/t within the same period. Additionally, some gold ores show higher contents of arsenic, leading to a reduced gold recovery due to missing proven concepts of liberation technologies.

Alternatively, metal recycling might offer a certain compensation for declining ore qualities and ensure the supply of raw materials for European industry. Recycling rates for the majority of

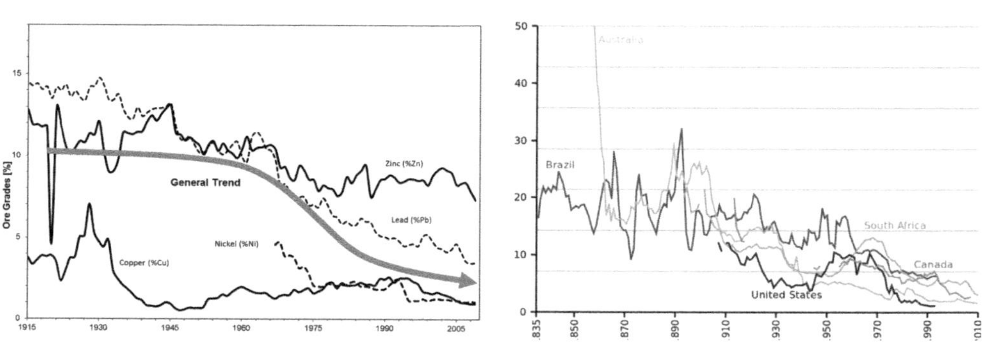

Figure 1. Decrease of metal grades in selected base metal and gold ores (Mudd 2009).

metals are below 50%, only Pb, Ru and Nb are exceeding this number. Although the recycling of metallic scrap is highly developed and applied in most regions of the world, iron and aluminum show recycling rates of <50%, copper and zinc are even below 25%. (Reuter 2013)

Having a closer look on metal processing circuits, it becomes obvious that certain instances are limiting the recycling rates and impede the dream of a circular economy:

- All processes applied produce significant amounts of residues
- Within pyrometallurgical processing, the targeted metal is partly lost within the slag.
- Regarding different alloys and complex end-of-use products (e.g. WEEE) significant amounts of secondary raw material leave Europe via legal (and partly illegal) channels.

Summing up, Europe faces a dilemma: On the one hand side, we are actively searching for own raw materials. On the other hand side, potentially secondary resources cannot be kept and processed in Europe. The reasons for this situation can be found in high costs for energy and labor, strict environmental legislations and safety regulations and the lack of innovative concepts for processing complex low grade materials.

2 A NEW APPROACH TO ASSESS BY-PRODUCTS

Searching for new raw material sources, industrial by-products got more and more in the focus of both, research and industry. For decades, residues from metal production such as slag, dust and sludge have often been qualified as waste and were landfilled. This fact was accepted by industry and involved authorities. In the eighties and nineties of the last century environmental concerns triggered more attention on residues as they were seen as a source of pollution.

With the development of multi-metal recovery processes for primary raw materials, these concepts are partially applied for secondaries. By-products of various metallurgical processes started to move into the focus of interest of both industry and research. Figure 2 shows different by-products and the overall metal contents of available in currently produced residues and elements that might be of economic interest.

These residues can be found in huge quantities either on heaps dating from the end of the last century or being currently produced and are therefore seen as an interesting secondary resource of metals. Due to the long European history in mining and metallurgy, existing landfills have to be considered as 'part of the market'. (Antrekowitsch & Steinlechner 2011, USGS 2018)

Some of the residues described in Figure 2 are already recycled to a certain extent. Although including these materials, the recycling rates are still below 50% due to the lack of optimized processes and missing information about the materials. Furthermore, the lack of a clear guideline describing how to evaluate residual materials is another limiting factor. For primary raw materials,

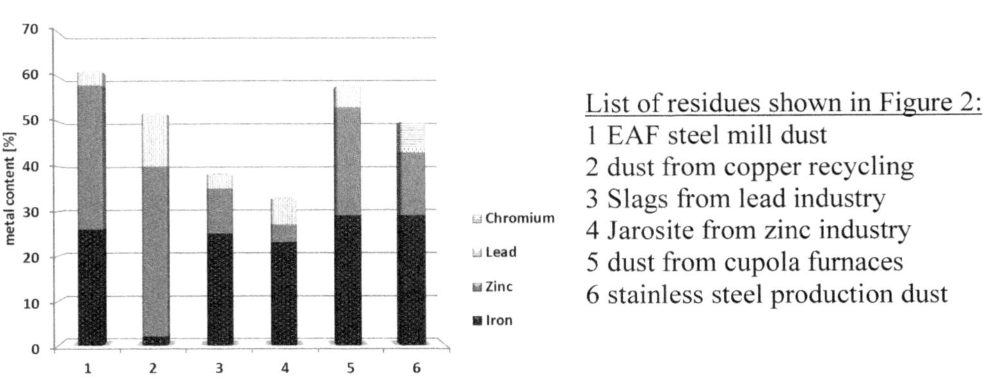

Figure 2. Examples for different residues and their metal content (except PGM).

codes do exist that serve as accepted and applied guidelines for the qualified evaluation of a resource (e.g. JORC Code, NI-43101), but cannot be applied to secondaries to a full extent. This leads to the fact that even if suitable concepts for the metal extraction are available, the monetary value of a slag heapcannot be defined. A consequent stepwise approach for both, the process development and the assessment of by-products, must be applied simultaneously.

Montanuniversität Leoben has started to develop principle concepts that should result in a guideline to allow a clear perspective on how to determine the valuable of the materials mentioned before. In October 2018 the project "COMMBY – competence network for the assessment of metal bearing by-products" has started. Financed by the Austrian Research Promotion Agency (FFG, COMET programme) and seven industrial partners, Montanuniversität Leoben as lead partner combines research competence in the area of geology/mineralogy, mineral processing and metallurgy. Within this project, two different approaches are followed:

2.1 *Characterization of by-products*

Detailed geochemical and mineralogical characterization of residues is of highest importance in order to develop appropriate mineral processing recipes and subsequently appropriate metallurgical methods to recover all metals of interest. Depending on the feed material and the treatment technology applied, various residues with different properties emerge that can be discriminated by their origin or properties (e.g. shape and texture, concentration of deleterious elements) by making use of their optical, electrochemical, thermal, chemical or mechanical characteristics. Similar to primary mineral deposits, processing results may widely vary, if these characteristics and potentially alteration processes are not taken into account.

Based on that, different possibilities for a treatment need to be determined. These results are mandatory for evaluating possible process steps to transform residues into products.

2.2 *Process development and product quality*

The two major concerns are the process development for a treatment of secondary resources and the optimization of existing processes. The challenge of the next years is to minimize newly generated residues and to achieve maximum product quality within the recycling processes. As a matter of fact, the quality of the products generated influence the targeted market and the revenues. Therefore, state-of-the-art processes have to be adapted regarding energy consumption, mass-balancing and product quality.

The characterization builds the basis for improving existing metallurgical processes. As demonstrated by several investigations carried out at Montanuniversität Leoben, all residues are highly complex and inhomogeneous, a previous separation by physical/mechanical techniques is required. Combining all information available and analyzing all results in detail, an initial outline for a standardized code for the assessment of by-products will be created. Within the ongoing research activities, a model for economic verification was already established and different heaps have already been investigated. To ensure the maximum benefit for industry and academia, Montanuniversität is closely working together with competent persons.

3 CASE STUDY – PROCESSING OF A LEAD AND ZINC CONTAINING SLAG

Within this chapter, only one of the projects within the context of reprocessing of metal bearing by-products is described exemplarily. Consequently following a zero-waste approach, the aim of this research is not only to extract contained metals from the slag, but also to find a proven solution for the residual oxidic phase.

3.1 *Slag heaps as environmental problem*

Slags are an inevitable waste produced in the recovery stages of pyro-metallurgical processing of non-ferrous metals from sulfide raw materials. Although most slags still contain significant amounts of valuables, they are accumulated in heaps because of the absence of accepted efficient technologies for further processing. Considerable attention has been given to the characterization and stability of lead slags, e.g. in France (Sobanska et al. 2016), the Czech Republic (Ettler et al. 2000) or Bulgaria (Atanassova & Kerestedjian 2009). The closed smelter and the associated slag dumpsite in Veles/Macedonia have been subject of investigations and are considered as a potential source of pollution. (EAR-MOEPP 2007, Stafilov et al. 2008)

3.2 *Short description of Veles smelter and available data*

From 1973 to 2003, MHK Zletovo operated a primary lead and zinc smelter in the municipality of Veles, Macedonia. The main production line comprised the sinter and sulphuric acid plants and the lead and zinc shaft furnace, applying the Imperial Smelting Furnace Technology (ISF). The slag was granulated and transported to the slag dump site. According to the Feasibility Study on MHK Zletovo 1,8 million tons of lead and zinc containing slag were landfilled,the surface of the dumpsite is about 3.3 ha. It was established that the content of Zn in the slag is about 7%, 1% of Pb and from 2 to 4000 mg/kg of the other heavy metals as Mn, Cu, Ni and As. (EAR-MOEPP 2007)

3.3 *Results of characterization*

As the available numbers in the literature cited do not allow an assessment of the slag at all, a sample of the Veles slag was characterized to serve as the basis for the development of a metallurgical route to convert the waste deposit into a source of raw materials.

The slag is typically dark grey to black colored fine grained with a bulk density of 2.1 t/m^3. Although 95% of the material is below 3 mm, coarser grains up to a maximum size of 20 mm can be found within the samples, predominately consisting of agglomerated slag. Several analytical methods were used to evaluate the slag to produce representative results including X-ray fluorescence analyses (XRF), inductively coupled plasma optical emission spectrometry (ICP-OES), scanning electron microscopy equipped with an X-ray energy dispersive spectrometry system (SEM-EDS) and X-ray diffraction analysis (XRD).

Table 1 shows the major constituents of the lead slag (analyzed by XRF and SEM-EDS). The primary constituents of the slag are Fe, Si and Ca, of which the oxides account for 70% of the weight. Table 2 presents the trace element content of the lead slag as a result from ICP-OES (performed by an accredited laboratory). The most common trace elements in the slag are Ba and As. Cd, Sb and Ag occur are in a range of <40 mg/kg.

As expected, the results of the XRD are very vague due to the high percentage of amorphous phase in the slag sample. Only a few compounds could be confirmed: Besides the expected phases due to the chemical composition of the primary ore processed and typical process parameter (e.g. zinc oxide, iron zinc oxide, ferrous sphalerite, iron silicon) calcium arsenate and pyrite could be identified.

Table 1. Major constituents of the slag.

Component	Na_2O	MgO	Al_2O_3	SiO_2	P_2O_5	SO_3	K_2O
[wt%]	1.469	1.786	5.662	18.587	0.305	6.622	0.663
Component	CaO	TiO_2	MnO	FeO	Cu	Zn	Pb
[wt%]	12.823	0.375	1.50	38.745	2.354	9.001	2.818

Table 2. Selected trace elements contents of the slag.

Component	Cr	Ni	As	Ag	Cd	Sn	Sb	Ba
[wt%]	0.04	0.02	0.2	0.004	0.003	0.03	0.04	0.2

Images taken with the scanning electron microscope (SEM) show that, although the matrix typically consists of a glassy and quite homogeneous phase formed by CaO-SiO_2-FeO, iron rich precipitations can be found within the range of a few microns. Zinc can be found homogeneously dispersed within the matrix, accumulated within regions that are significantly higher in aluminum. This indicates the presence of gahnite ($ZnAl_2O_4$) even though not identified by XRD. Lead can be found in sulfidic as well as in oxidic form. Arsenic occurs together with iron, forming small droplets within the matrix. To gain information on the thermal expansion and melting behavior, thermo-optical analyses were performed with a hot stage microscope. Reaching a temperature of 1099°C, the samples start to melt, the melting temperature under reducing conditions was determined at 1110°C.

3.4 *Proposed process*

Applying multi-metal recovery via a metal bath using EAF technology, lead and zinc is recovered as oxide in the flue gas. Liquid pig iron acts as reducing agent as well as a collector for molten iron or copper, whereas the clean residual phase is systematically modified during the smelting and cooling process. Figure 3 shows the principle idea of reprocessing slag.

Based on the previous described characterization, a thermo-chemical model of the smelting and reduction process was set up. To verify the assumptions and get additional information about temperature, limiting factors and generated products, first lab scale trials were performed in an induction furnace.

3.5 *Preliminary results of process development*

Using a graphite crucible, pig iron was molten, slag and CaO (to control the basicity) were added. The temperature was constantly kept at 1400°C, samples of the slag were taken within a fixed interval. Within the first 20 minutes of treatment, zinc fuming could be observed and the flue dust was caught within a condenser. After a treatment time of maximum 60 minutes the generated iron alloy and the cleaned residual phase were tapped and cooled.

Although detailed analyses of the generated products are pending, it can be stated that the quality of the results turned out to be better than expected: The flue dust mainly consists of ZnO and PbO (accounting for >85% of the weight) and some Fe and Si as "carry over". Concerning the quality, the flue dust can be further processed within primary zinc industry. The generated iron alloy shows typical 95% of Fe and up to 0.8% Cu and 3.5% C. Limiting factors for the subsequent use in construction steel production could be arsenic and sulfur. Using XRF and SEM-EDX, the residual

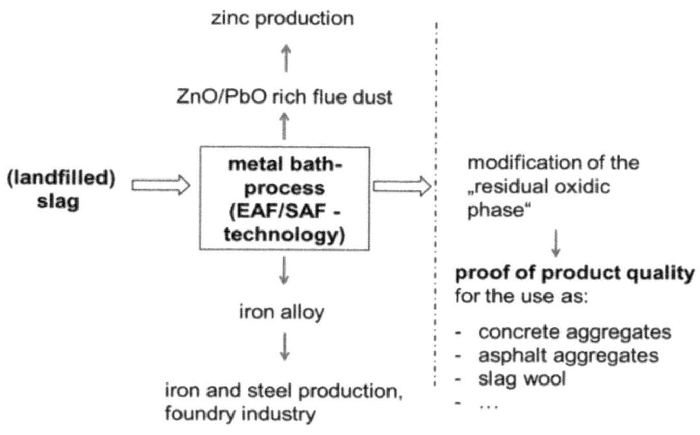

Figure 3. Principle scheme for effectively reprocessing slag.

phase shows CaO, SiO_2 and Al_2O_3 as the major constituents and is free of zinc, lead and arsenic. The iron content can be adjusted via the basicity and the treatment time. Summing up, the achieved yields for lead and zinc are >99.5% and between 60 and 90% for iron, respectively.

3.6 *Upcoming challenges*

As one of the aims of the project is to find a proven purpose of the residual phase within the construction industry, the research work will focus on this within the next months: Due to the lack of pan-European law concerning alternative construction materials, Austrian legislation is the basis for the assessment of the residual phase regarding the use as concrete or asphalt aggregate. As the bulk chemistry meets all requirements, leaching tests are performed (detailed results are pending). Different possibilities to modify physical properties are currently studied and evaluated. The ideas range from hot slag modification with additives to a controlled cooling and crystallization process. In order to test standardized parameters such as strength, abrasion resistance or room stability, a larger amount of residual phase has to be generated using a technical scale EAF.

Finally, the process will be validated in semi-industrial scale, meeting all technical and legislative requirements for the conversion of unused residues into products.

ACKNOWLEDGEMENTS

The authors would like to express the gratitude to the Austrian Research Promotion Agency (FFG) and the EIT RawMaterials for the financial support of the research. Special thank is given to the involved key researchers at Montanuniversität and the involved companies for their scientific input.

REFERENCES

Antrekowitsch J. & S. Steinlechner 2011. The Recycling of Heavy-metal Containing Wastes: Mass Balances and Economical Estimations. *JOM 63(1), 68–72.*

Atanassova R. & Kerestedjian T. 2009. Efflorescent minerals from the metallurgical waste heaps of the KCM non-ferrous metal smelter, Plovdiv. *Geochemistry, Mineralogy and Petrology 47, 51–63.*

EAR-MOEPP-Eptisa 2007. Feasibility Study – Volume II – MHK Zletovo – Veles, Development of Remediation Plans With Financial Requirements for Elimination of Industrial Hotspots (EURO-PEAID/123674/D/SER/MK), *European Agency for Reconstruction, Ministry of Environment and Physical Planning of the Republic of Macedonia, EPTISA, Skopje.*

Ettler V. & Johan Z. & Touray J.C. & Jelinek E. 2000. Zinc partitioning between glass and silicate phases in historical and modern lead–zinc metallurgical slags from the Pribram district, Czech Republic. *Earth and Planetary Sciences 331, 245–250.*

Mudd, G.M. 2009. The Sustainability of Mining in Australia – Key Production Trends and Their Environmental Implications for the Future. Figure based on the original data set: *http://users.monash.edu.au/~gmudd/sustymining.html*, retrieved 03/2019 development of gold grades: *http://www.wikiwand.com/en/Gold_mining*, retrieved 02/2019.

Reuter M.A. 2013. Metal recycling. *United Nations Environment Programme.*

Sobanska S. & DeneeleD. & Barbillat J. & Ledésert B. 2016. Natural weathering of lags from primary Pb-Zn smelting as evidenced by Raman microspectroscopy. *Applied Geochemistry 64, 107–117.*

Stafilov, T. & Sajn, R. & Panèevski, Z. & Boev, B. & Frontasyeva, M. V. & Strelkova, L. P. 2008. Geochemical Atlas of Veles and the Environs. *Faculty of Natural Sciences and Mathematics, Skopje.*

USGS 2018. Mineral Yearbooks: *http://minerals.usgs.gov/minerals/pubs/country/index.html#pubs*, retrieved 06/2018.

Wastes: Solutions, Treatments and Opportunities III – Vilarinho et al. (Eds)
© 2020 Taylor & Francis Group, London, ISBN 978-0-367-25777-4

Is the production of Kenaf in heavy metal contaminated soils a sustainable option?

B. Cumbane
MEtRICs, Faculdade de Ciências e Tecnologia, Universidade Nova de Lisboa, Caparica, Portugal
FCS, Universidade Zambeze, Mozambique

J. Costa
MEtRICs, Faculdade de Ciências e Tecnologia, Universidade Nova de Lisboa, Caparica, Portugal
ISEC, Lisboa, Portugal

M. Gussule
FEAF, Universidade Zambeze, Mozambique

L. Gomes, C. Rodrigues, V.G.L. Souza & A.L. Fernando
MEtRICs, Faculdade de Ciências e Tecnologia, Universidade Nova de Lisboa, Caparica, Portugal

ABSTRACT: Kenaf is a fiber crop producing high quality cellulose, suitable for the production of biomaterials. In order to avoid the conflict food *versus* fuel/biomaterials, the use of marginal land represents an alternative. Therefore, the aim of this work was to evaluate the environmental-socio-economic impact of the production of kenaf in heavy metal contaminated soils. To determinate the environmental-socio-economic sustainability, different categories were studied: energy savings/losses, emission of gases, cost savings/losses, employment potential creation and consumers/producers acceptance. Results suggest that the production of kenaf in heavy metal contaminated soils have positive and less positive aspects. The productivity loss in contaminated soils diminishes the energy, the carbon sequestered, the greenhouse savings and the economic balance but it may contribute to improve the quality of soil and waters and the biological and landscape diversity. But the production of kenaf in contaminated soils still involves much controversy, and not always have social acceptance.

1 INTRODUCTION

Soil contaminated by heavy metals is a major problem causing vast areas of agricultural land to become derelict and non-arable and hazardous for both wildlife and human populations (Barbosa and Fernando, 2018). Phytoremediation, the use of plants and their associated microbes for soil, water, and air decontamination, is a cost-effective, solar-driven, and alternative/complementary technology for physicochemical approaches. Plants can be used for extraction, or stabilization, of metals reducing their associate risks to humans, animals, and the environment (Barbosa et al, 2016). The application of a phytoremediation technology using industrial crops can represent an opportunity. In fact, bridging phytoremediation with the production of a multipurpose biomass could provide environmental, social and economic benefits, by improving the overall sustainability of the biosystem.

Kenaf (*Hibiscus cannabinus* L.) is an annual non-food fiber crop, suitable for the production of high quality cellulose for numerous uses (paper pulp, fabrics, textiles, building materials, biocomposites, bedding material, oil absorbents, etc.)(Pascoal et al, 2015) and that can be grown on soils of poor quality (Fernando, 2013). The crop is able for combating desertification, thanks to its high efficiencies in the use of resources and also to its strong and deep root system that promotes the control of runoff and erosion, and also aeration and organic matter increment in the soil (Fernando et al, 2015). Energy can be produced from the short wood fibers of the inner core, while the long fibers of the bark can be used for several high value fibre applications (Alexopoulou et al., 2013) offering the possibility to associate soil decontamination and restoration with the production of biomass for bioenergy and biomaterials with additional revenue. Additionally, when kenaf is cultivated in marginal/contaminated soils, land use conflicts with food crops are reduced (Dauber et al. 2012), minimizing direct and indirect negative effects due to Land Use Change (LUC). Considering kenaf morphological and anatomical features, and also its ability in the phytoremediation the aim of this work was to evaluate the environmental and socio-economic impact of the production of kenaf in heavy metal contaminated soils.

2 MATERIALS AND METHODS

In this study, it was assumed that kenaf bark fiber was used to produce thermal insulation boards and kenaf core fiber as solid fuel. To estimate the environmental and socio-economic impact on the production of kenaf in heavy metal contaminated soils, kenaf results obtained in the work of Cumbane et al. (2019a and b) were the basis for the assessment. In this work, kenaf (cultivar H328, developed by IBFC in China), in pots, was produced in heavy metal contaminated soils. The soils were artificially contaminated and the concentrations chosen were based on the limits established by the Decree Law 276 of 2009 (Portuguese regulation that establishes the regime for the use of sewage sludge in agricultural soils) – Zn: 450 mg/kg; Cr: 300 mg/kg; Pb: 450 mg/kg and Cu: 200 mg/kg. For the environmental impact assessment, the study focused on several categories, according to the methodology of Barbosa et al. (2013). Carbon sequestration (Mg C/ha) was calculated assuming a carbon content of 48, 4% for the whole plant (Vassilev et al., 2010). The emission of gases was calculated considering the relations described in the studies of Fernando et al. (2007) and IPCC (2006). Economic analysis was performed assuming the relations established between the productivities obtained in studies in field for Portugal (Fernando et al., 2007) and the extrapolations to our work, and also in the current prices found for fertilizers in fertilizer selling companies in Portugal.

3 RESULTS AND DISCUSSION

3.1 *Biomass productivity*

Results for productivity of kenaf are presented in figure 1.

According to these results, it was observed a productivity loss, significant in Zn and Cu contaminated soils (50–60% loss), and less significant in Cr and Pb contaminated soils (20–30% loss). According to Kabata-Pendias (2011), Zn, Cu, Cr, and Pb in excess in soils contribute to damages on the photosynthesis apparatus and belowground organs, leading to retarded growth of the entire plant. In the case of kenaf, damages were more significant due to Zn and Cu contamination than with Cr and Pb contamination. Yet, information on the interactions between the below ground organs of this crop and its growing medium is still lacking, which might provide more knowledge on the tolerance mechanisms associated with this plant on contaminated soils, explained differences among Zn/Cu and Cr/Pb.

3.2 *Energy balance*

The energy balance (GJ/ha) of kenaf (bark + inner core) produced in control and contaminated land is presented in Figure 2 and is based on several scenarios: bark is intended for the production of insulation boards or for fiber production; core fiber is used by the board producer in a small thermal power plant, in order to generate energy inside the industrial unity, or is channeled into the production of pellets (for home systems).

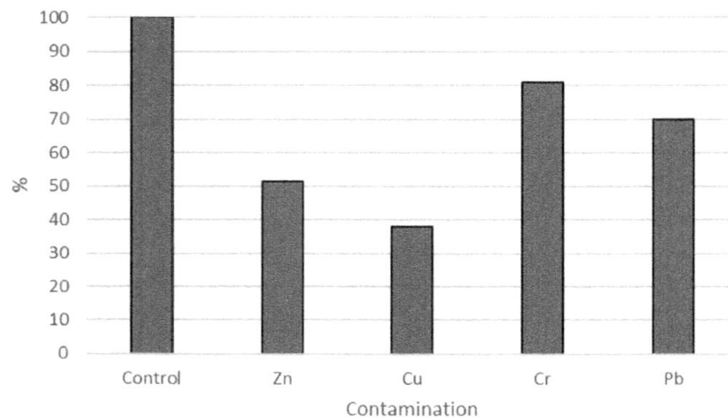

Figure 1. Yield of kenaf obtained in the different contaminated soils compared with the control (%).

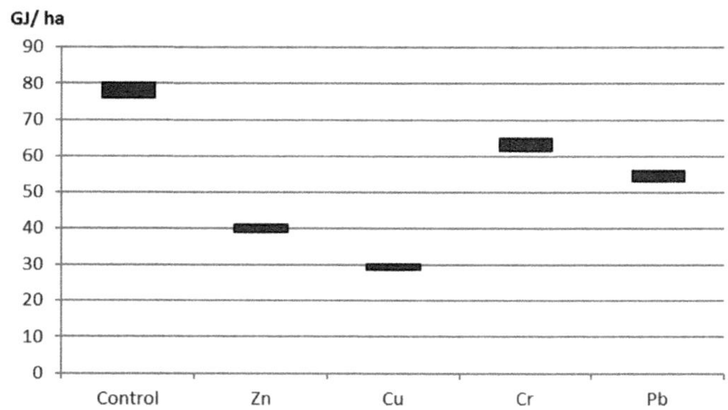

Figure 2. Energy balance (GJ/ha) for kenaf (inner core + bark material) obtained in the different contaminated soils compared with the control.

According to these scenarios, control sample show an energy production of 76–80 GJ/ha and samples obtained from Zn and Cu contaminated soils showed a significant lower positive energy balance than the control (30–40 GJ/ha). In soils contaminated with Cr and Pb, the energy production was lower than in the control (55–62 GJ/ha), but without significant differences. Nevertheless, even in soils that showed half of the yields obtained in non-contaminated soils, the energy balance was still positive, indicating that the production of kenaf in heavy metal contaminated soils has potential to reduce the dependence on fossil fuels. Indeed, the higher the yield of the crop, the higher is the energy efficiency and the amount of fossil energy saved.

3.3 *Carbon sequestration and phytoremediation*

Carbon sequestration (Mg C/ha) (Figure 3) was calculated assuming a carbon content of 48,4% for the whole plant (Vassilev et al., 2010) and the productivities in field in Portugal (Fernando et al., 2007). Then the relation obtained for the total carbon sequestration (Mg C/ha) are extrapolated for our results, obtained in pots. Control showed the highest total carbon sequestration (whole plant) and the highest carbon sequestration in the bark, leaves and roots. Samples from contaminated soils showed lower results for total carbon sequestration, especially in the case of Zn and Cu contaminated soils, following the pattern observed for the obtained yields. The carbon retained in the leaves contribute to the increase of the carbon content in soil when the leaves come into senescence and fall to the ground, increasing organic matter contents in the soil. Carbon sequestration in leaves and roots may contribute also for increasing soil structure and soil aeration, factors that contributes to

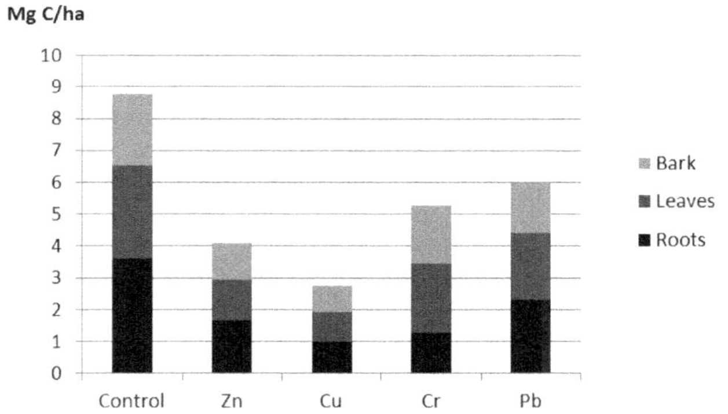

Figure 3. Carbon sequestration (Mg C/ha) by plant component obtained in the different contaminated soils compared with the control.

reduce soil erosion and to control desertification process (Barbosa et al., 2018). At this point we considered that bark material, through the production of insulation boards, with a life time of 30 years, represents a way to sequester CO_2. If insulation boards with bark fiber from kenaf, are being produced, there will be also a reduction in the use of polyurethanes, by substitution of that type of material, with environmental benefits. The vegetative cover in a derelict land presents also benefits towards the biological and landscape diversity, once there will be an increment in the soil organic matter and the crop being produce provide shelter for small animals (Fernando et al., 2018).

Kenaf also show good phytoremediation capacity for Zn. In fact, due to the high mobility of this element in the soil some zinc was absorbed by the roots and transported to the aerial fractions (stems and leaves). Yet, removal of Zn from the soil by the stems and leaves, only corresponded, in one growing season, to 5% of the total contamination. This means that to remove the entire contamination, *ca.* 20 years would be needed. Also, the amount of Zn in the stems and leaves (average 350 mg/kg), does not represent a threat to the microfauna and fauna once this element is not considered toxic at this level. Concerning the other metals, Cu, Pb and Cr, the amount extracted by the plants were below 0.1% of the total contamination. This means that those elements were not phytoextracted, and therefore the content in the stems and leaves is similar to what was observed by in the control, not threatening the fauna and microfauna.

3.4 *Greenhouse gas emissions reduction*

Reduction of greenhouse gases emissions (GHG) due to kenaf production and use is presented in Figure 4. According to the results estimated, the use of bark fiber as thermal insulation materials in building walls and roofs, and the use of inner core as solid fuel are relevant to reduce GHG emissions. Results also show that the production of kenaf in contaminated soils can be a relevant tool to improve emissions reduction, even in the case of Zn and Cu contaminated soils, where the yield reduction is more significant. Samples from Cr and Pb contamination showed GHG emissions reduction than control, but higher than in Zn and Cu contaminated soils, representing 70–80% of the emissions reduction estimated for the control.

3.5 *Economic and social considerations*

The economic balance per ha of kenaf production and use is presented in Figure 5. The costs of kenaf production in contaminated soils are higher than the profit in Zn, Cu and Pb contaminated soils, due to lower yields, and in Cr contaminated soils, the profit is equal to the costs. Therefore, the cultivation in kenaf contaminated soils is not profitable as in control soils. Yet, if we consider the cost of treating the contaminated soil, perhaps the economic balance will be improved in contaminated soils. Also, the production of kenaf in soils contaminated with heavy metals can provide to human

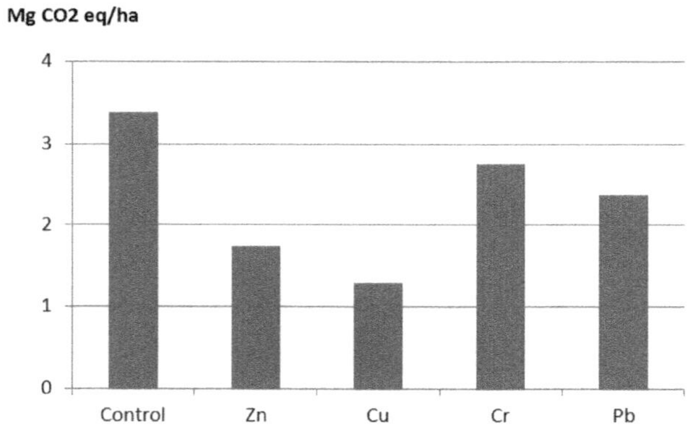

Figure 4. Estimated reduction of greenhouse gases emissions (Mg CO_2 eq/ha) due to kenaf production in the different contaminated soils and its use (core as solid fuel and bark as thermal insulation board).

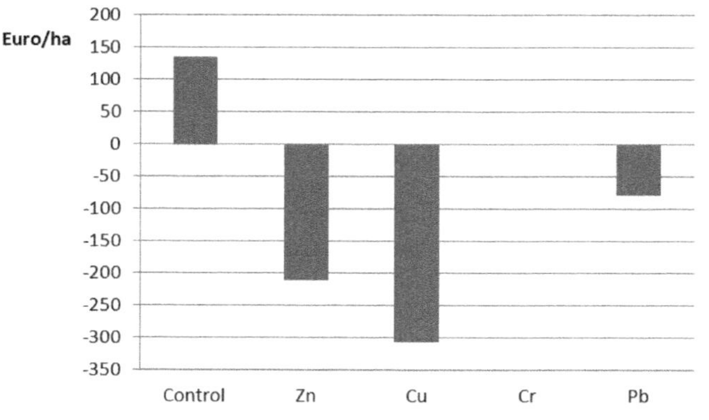

Figure 5. Economic balance of kenaf production, by type of soil contamination.

communities many social benefits once it reduces human and environment exposure to pollutants. Kenaf production and use present also a positive gain in terms of employment in small and medium-size enterprises and on the employment in rural areas. The positive influence on employment is also due to the contribution towards avoiding a rural exodus and to the contribution towards a more balanced rural development. Labour requirements per hectare for the production, in the farm, of kenaf is similar to the one showed for hemp (17 hours per hectare per year) (Correia and Fernando, 2013). Moreover, the cultivation in contaminated land represent an increase in agricultural activities, especially because extra value can be given to derelict land and so, it may generate extra employment.

4 CONCLUSIONS

Production of kenaf in soils contaminated is a relevant tool to improve energy performance by reducing the resources consumption (land use for food and feed) and the associated environmental burdens arising from the combustion of fossil fuels. Production of kenaf in contaminated soils also presented other environmental and socio-economic benefits. This approach involves less environmental impacts (lower GHG emissions, higher Carbon sequestration, reduced exposure to pollutants), and could provide social benefits in Mediterranean regions, namely employment opportunities in rural areas.

ACKNOWLEDGEMENTS

This work has the support of MEtRICs through FCT/MCTES [project number UID/EMS/04077/2019].

REFERENCES

Alexopoulou, E., Cosentino, S.L., Danalatos, N., Picco, D., Lips, S., van den Berg, D., Fernando, A.L., Monti, A., Tenorio, J.L., Kipriotis, E., Cadoux, S. & Cook, S. 2013. New Insights from the BIOKENAF Project. In: A. Monti & E. Alexopoulou (eds), *Kenaf: A Multi-Purpose Crop for Several Industrial Applications*: 177–203. Springer, London.

Barbosa, B., Costa, J., Boléo, S., Duarte, M.P. & Fernando, A.L. 2016. Phytoremediation of inorganic compounds. In: A.B. Ribeiro, E.P. Mateus & N. Couto (eds), *Electrokinetics Across Disciplines and Continents – New Strategies for Sustainable Development:* 373–400. Springer International Publishing, Switzerland.

Barbosa, B., Costa, J. & Fernando, A.L. 2018. Production of Energy Crops in Heavy Metals Contaminated Land: Opportunities and Risks. In: R. Li & A. Monti (eds.), *Land Allocation for Biomass*: 83–102, Springer International Publishing AG.

Barbosa, B. & Fernando, A.L. 2018. Aided Phytostabilization of Mine Waste. In: M. N. V. Prasad, P.J.C. Favas & S.K. Maiti (eds.), *Bio-Geotechnologies for Mine Site Rehabilitation*: 147–158, Elsevier Inc., UK.

Barbosa, B., Fernando, A.L. & Mendes, B. 2013. Environmental and socio-economic impact assessment of the production of kenaf (*Hibiscus cannabinus* L.) when irrigated with treated wastewaters. *Book of proceedings 2nd International Conference: Wastes: solutions, treatments and opportunities*: 643–648, Centro de Valorização de Resíduos.

Correia, F. & Fernando, A.L. 2013. Kenaf for a Sustainable Economy: Integrating Insulation Boards Production with Fossil Energy Savings. In: A. Eldrup, D. Baxter, A. Grassi & P. Helm, (eds.), Proceedings of the 21th European Biomass Conference and Exhibition, Setting the course for a Biobased Economy: 2022–2025, ETA-Renewable Energies and WIP-Renewable Energies.

Cumbane, B., Gomes, L., Costa, J., Cunha, J., Araújo, H., Pires, J., Rodrigues, C., Zanetti, F., Monti, A., Alexopoulou, E. & Fernando, A.L. 2019a. Understanding the potential of kenaf in heavy metals contaminated soils, 27th EUBCE, Lisbon, Portugal, 27th–30th May 2019.

Cumbane, B., Gomes, L., Costa, J., Cunha, J., Araújo, H., Pires, J., Rodrigues, C., Zanetti, F., Monti, A., Alexopoulou, E. & Fernando, A.L. 2019b. Phytoremediation potential of kenaf in heavy metals contaminated soils, 31st AAIC Annual Meeting, Tucson, Arizona, US, 8th–11th September 2019.

Dauber, J., Brown, C., Fernando, A.L., Finnan, J., Krasuska, E., Ponitka, J., Styles, D., Thrän, D., Groenigen, K.J.V., Weih, M. & Zah, R. 2012. Bioenergy from "surplus" land: environmental and social-economic implications. BioRisk 7: 5–50.

Decree Law 276. 2009. Anexo I, Valores limite de concentração relativos a metais pesados, compostos orgânicos e dioxinas e microrganismos, Diário da República, n° 192, I Série: 7154–7165.(in Portuguese)

Fernando, A., Duarte, M.P., Morais, J., Catroga, A., Serras, G., Lobato, N., Mendes, B. & Oliveira, J.F.S. 2007. Final report of FCT/UNL to the Biokenaf project, FCT/UNL, Lisbon, Portugal.

Fernando, A.L. 2013. Environmental Aspects of Kenaf Production and Use. In: A. Monti & E. Alexopoulou (eds) *Kenaf: A Multi-Purpose Crop for Several Industrial Applications*: 83–104. Springer-Verlag, London.

Fernando, A.L., Duarte, M.P., Vatsanidou, A. & Alexopoulou, E. 2015. Environmental aspects of fiber crops cultivation and use, Industrial Crops and Products 68: 105–115.

Fernando, A.L., Rettenmaier, N., Soldatos, P. & Panoutsou, C. 2018. Sustainability of Perennial Crops Production for Bioenergy and Bioproducts. In: E. Alexopoulou (ed) Perennial Grasses for Bioenergy and Bioproducts: 245–283, Academic Press, Elsevier Inc., UK.

IPCC – Intergovernmental Panel on Climate Change, 2006. 2006 IPCC Guidelines for National Greenhouse Gas Inventories. Prepared by the National Greenhouse Gas Inventories Programme, H.S. Eggleston, L. Buendia, K. Miwa, T. Ngara & K. Tanabe (eds), IGES, Japan.

Kabata-Pendias, A. 2011. Trace elements in soils and plants, 4th edn. CRC, Boca Raton.

Pascoal, A., Quirantes-Piné, R., Fernando, A.L., Alexopoulou, E. & Segura-Carretero, A. 2015. Phenolic composition and antioxidant activity of kenaf leaves, Industrial Crops and Products 78: 116–123.

Vassilev, S. V., Baxter, D., Andersen, L.K. & Vassileva, C.G. 2010. An overview of the chemical composition of biomass, Fuel 89: 913–933.

Wastes: Solutions, Treatments and Opportunities III – Vilarinho et al. (Eds)

Thermal and chemical characterization of lignocellulosic wastes for energy uses

G. Charis & G. Danha
Botswana International University of Science and Technology, Palapye, Botswana

E. Muzenda
Botswana International University of Science and Technology, Palapye, Botswana
University of Johannesburg, Johannesburg, Gauteng, South Africa
University of South Africa, Gauteng, South Africa

ABSTRACT: Two forms of waste lignocellulosic biomass- pine sawdust milling residues from Zimbabwe and Acacia tortilis, an encroacher species from rangelands and urban circles of Botswana- are characterized to evaluate pertinent thermal and chemical properties. Characterization is useful for Acacia tortilis, whose properties have scarcely been studied The ultimate analysis reveals that Pine has a CHNO composition of 45.76%, 5.54%, 0.039% and 48.66% respectively, while the Acacia elemental composition is C (41.47%), H (5.15%), N (1.23%) and O (52.15%). Thermogravimetry results showed that *A. tortilis*' ash, fixed carbon and volatiles matter composition was 3.90%, 15.59% and 76.51% respectively on a dry basis; while pine was 0.83%, 20% and 79.16% respectively. The high heating value for Acacia was found to be 17.27 MJ/kg compared to 17.57 MJ/kg for pine. The research establishes that A. tortilis' fuel properties are comparable to pin. *A. tortilis'* high ash content is below critical values.

1 GENERAL INTRODUCTION

1.1 *Global trends in the biomass to bioenergy industry*

The Biomass to Bioenergy (BtB) industry is growing fast driven by the global shift towards green/clean energy, rising environmental awareness, an emphasis on circular economies and government support (Sapp, 2017; World Energy Council, 2016). There have also been supportive policies from influential regions like the European Union (EU) and the United States (US), with ambitious goals of a 20% share of renewable energy sources in the overall energy mix by 2020 and 60bn litres of 2nd generation bio-fuel by 2022 respectively (Álvarez-Álvarez et al., 2018; IRENA, 2016; Walker, 2012). Given such policies, there is an increased consideration of forest biomass, forest residues and short-rotation biomass plantations as potentially important sources of renewable energy, in a drive to reduce dependence on fossil fuels and meet emission commitments (Álvarez-Álvarez et al., 2018). Lignocellulosic biomass, especially forest and agricultural residues, is regarded as a cheap and abundantly available feedstock that is carbon neutral and has a negligible effect on food security. The International Renewable Energy Agency (IRENA) stipulates that bioenergy is a strategic asset in the future of Africa, especially in the light of the fact that it comprises 50% of Africa's total primary energy supply (TPES) and more than 60% of sub Sahara's TPES (Stecher et al., 2013).

1.2 *Study context and rationale*

As Álvarez-Álvarez et al. (2018) contend, sustainable BtB development requires an adequate knowledge of both the supply capacity and quality of biomass (Álvarez-Álvarez et al., 2018).

Figure 1. Sawmill waste scenes from Manicaland, Zimbabwe 1. Receiving and sorting bay where bark is peeled off 2. Bark and small chips from receiving bay 3. Waste offcuts 4. Sawdust mound of around 5000t at a dysfunctional sawmill in Mutare. Many such mounds exist in Nyanga and Chimanimani.

Figure 2. Acacia encroacher bushes a) besides a railway line; b) at a river bank c) in commercial sites.

This study looks at 2 lignocellulosic wastes from Botswana and Zimbabwe, whose valorization for energy purposes could ease the environmental burden while offering socio-economic benefits.

Pine sawmill residues (Zimbabwe): Sawdust and shavings represent the most unutilized waste fractions (~70,000 tpa). Heaps are scattered all over the region, marring its aesthetic appeal and posing various ecological threats including fire, greenhouse gas (GHG) emissions and wood residue leachate which has high concentrations of Dissolved Organic Matter (Arimoro et al.,2006; Effah et al., 2015). Potential valorisation opportunities including combined heat and power (CHP), pyrolysis, gasification/ co-gasification and wood engineered products have not yet been explored. Figure 1 shows some of the mounds of waste from Manicaland region, Zimbabwe.

Encroacher bushes (Botswana): Together with other encroachers, *A. tortilis* has significantly reduced the size of quality rangeland available for both domestic and wild animals in Botswana. It costs local and national government millions of Pulas annually through various bush control programs like de-bushing to regenerate rangeland grasses or improve the aesthetic appeal in villages, towns, cities and highways (Kabajan et al., 2016). However, drought hardy *A. tortilis* shoots quickly spring up after de-bushing encouraged by overgrazing. Botswana can take a cue from Namibia which is successfully valorising its vast encroacher bush in charcoal, heat and power generation schemes (DECOSA, 2015). Figure 2 shows growing and debushed acacias in various places around a town in Botswana.

2 SIGNIFICANCE OF CHARACTERIZING THERMAL AND CHEMICAL PROPERTIES

Figure 3 shows various supply chain (SC) operations and the relevant physico-chemical properties at each point.

Table 1 then focuses on the importance of knowing the thermal and chemical properties of the biomass to be used in BtB projects.

The diversity of biomass properties from various tree species explains why property profiles for any 'new' feedstock possibility like *A. tortilis* have to be made. Although Pine-sawdust has been studied before, differences in tree cultivation, biomass pretreatment and storage schemes warrants characterization of feedstocks in different geographical locations.

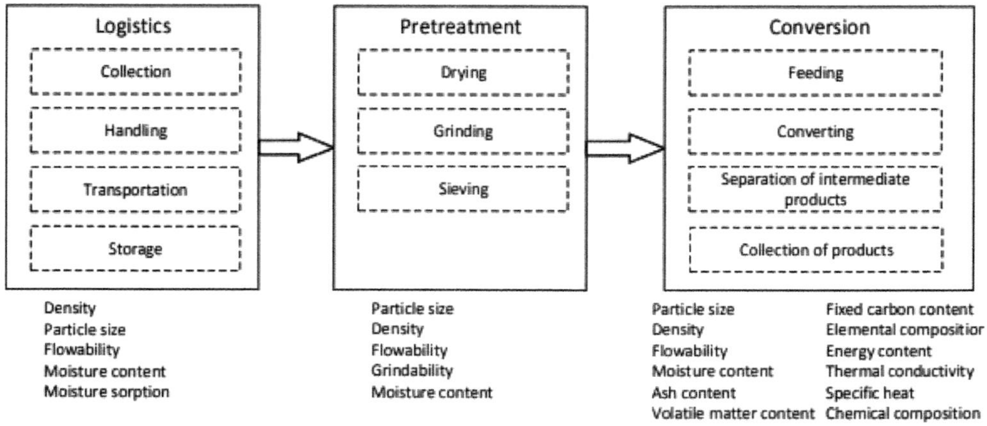

Figure 3. a) Relevant physico-chemical properties at various SC points (Cai et al., 2017).

Table 1. Relevance of Thermal and chemical properties studied in this research to various processes.

Thermal or chemical property	Relevance/Significance	Literature
Ash content	To assess potential risk of fouling or slagging during combustion or gasification	(Anukam et al., 2016; Kirsanovs et al., 2016)
Volatile matter	Conversion efficiency	(Cai et al., 2017)
Elemental (CHNO) composition	Conversion efficiency and product composition	(Anukam et al., 2016)
Fuel/Thermal properties	Thermochemical conversion efficiency; Heating capabilities of fuel (especially in combustion)	(Anukam et al., 2016; Kirsanovs et al., 2016)

3 METHODOLOGY

3.1 *Thermogravimetric analysis*

The moisture content (MC), Fixed Carbon (FC) and Volatile Matter (VM) of the pine and acacia samples was determined using LECO TGA 701 Thermogravimetric analyzer (TGA). The analyzer was set to comply with the ASTM E871 for moisture, D1102 & E830 for Ash, E872 and E897 for volatile matter (Acar & Ayanoglu, 2016). The thermogravimetric results show the behavior of the study material under heat and the proximate composition – moisture, volatiles, ash and fixed carbon. The TGA was programmed to perform all these tests at various temperatures. When a constant mass was achieved at the end of purging moisture and volatiles or ashing, the TGA would automatically go to the next stage, ramping up the temperature at pre-set rates. Otherwise, it would prompt for some manual adjustment (like opening of lids) prior to the start of the next stage. The LECO TGA 701 automatically weighs the samples giving the real time changes in mass as shown by the graph in Fig 4, 'Results and analysis' section. For the *A. tortilis*, the samples were placed in 3 categories – with bark, debarked and mixed. The objective was to see if there would be any significant difference in the TGA and proximate analyses. Pine sawdust comes in debarked state, so such classification was no necessary for it.

Moisture content – The TGA was set at 107°C and ramped from ambient temperature at 6°C/min according to ASTM E871, to remove all moisture. The drying process was carried out in an inert environment of nitrogen gas until a near constant 'moisture mass' was achieved.

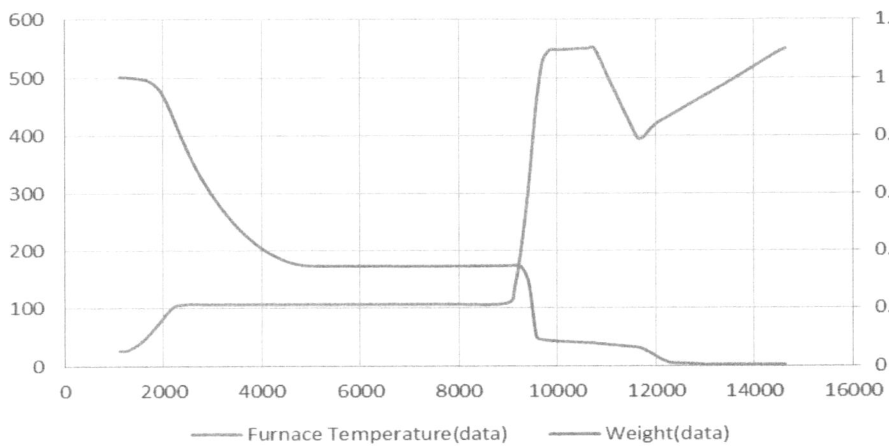

Figure 4. TGA for one of the pine dust samples.

Volatiles content – The TGA was set at 550°C for volatization, and ramped at 37°C/min from the 107°C. Volatization occurred in an inert environment of nitrogen gas until a near constant mass was achieved.

Ashing- After burning off the volatiles, the lids on the samples were removed, then the biomass samples were reheated to 550°C at a rate of 3°C, this time in an oxygen rich environment. The mass lost during *ashing* is the fixed carbon, which reacts with oxygen.

3.2 *Ultimate analysis*

Ultimate analysis was carried out using the Flash 2000 CHNS-O elemental analyzer (ThermoFisher Scientific, USA). The biomass samples were first weighed in containers, then the equipment's auto sampler system introduced them into the combustion reactor. After combustion in an oxygen rich environment, the gases given off are carried by a helium flow past a copper layer filled layer, through a GC column where the combustion gases are separated then detected by a thermal conductivity detector (Thermo Fischer Scientific, 2017).

3.3 *Calorimetry*

The High Heating Value (HHV) was determined using a bomb calorimeter, Bomb CAL2K-2, according to DIN 51900 T3 standards for testing of solid and liquid fuels, determination of gross calorific value. A crucible with a sample of about 1.000 g of biomass was placed in the calorimeter. The bomb was then closed and filled with oxygen pressurized to 30 bar. The sample was covered in an adiabatic jacket along with some known quantity of water. It was then ignited electrically, resulting in a rise of the water temperature, enabling the automatic evaluation of the HHV of the sample by the equipment. The calorimeter is calibrated using benzoic acid.

4 RESULTS AND ANALYSIS

4.1 *TGA, proximate and calorimetry*

Tests were done on samples as received- acacia was collected with bark while pine sawdust was mostly white wood. Figure 4 shows the TGA results for the pine samples. The proximate composition is determined from the plateau regions and shown in Table 2. Figure 5 on the other hand shows the actual proximate profile for the various categories of *A. tortilis* samples.

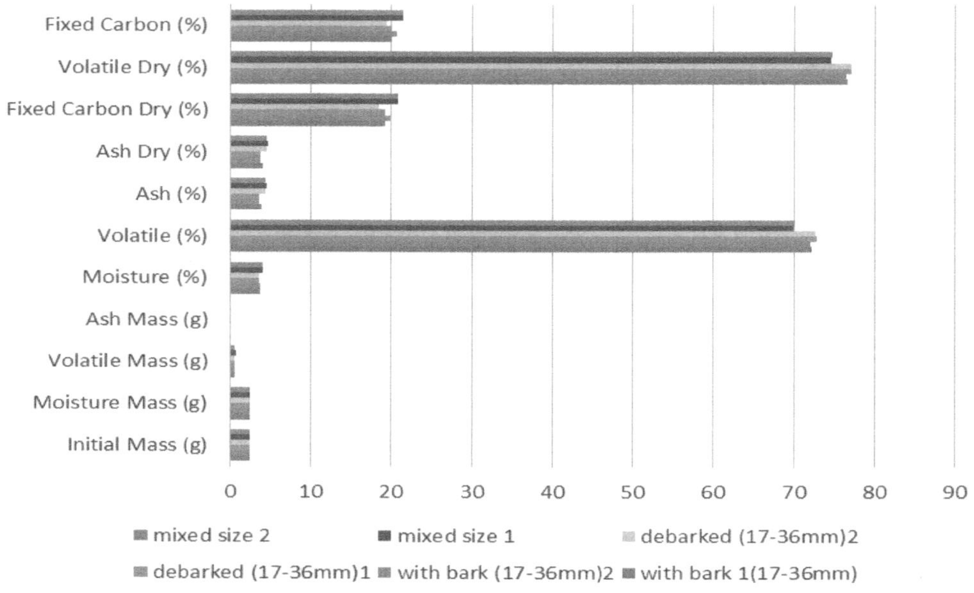

Figure 5. TGA for Acacia samples.

Table 2. Results of ultimate and proximate analyses (dry basis) for Acacia, compared to the common Pine. FC- Fixed Carbon; VM- Volatile Matter; MC- Moisture Content.

	Ultimate analysis- average of 2				Proximate (Dry basis)- averages				
%	C	H	N	O[a]	Ash	FC	VM	MC	HHV (MJ/kg)
ACACIA	41.47	5.15	1.23	52.15	3.90	19.59	76.51	3.72	17.267
PINEDUST	45.76	5.54	0.039	48.66	0.83	20.00	79.16	65.41	17.568

[a] *Calculated by difference*

Figure 5 on the other hand shows the actual proximate profile for the various categories of *A. tortilis* samples.

There was no fixed trend observed from the samples with bark, the mixed and debarked samples. This is probably because the Acacia bark is very thin compared to that of pine, therefore its contribution to the results could be insignificant. Table 2 shows a summary of the calorimetry, thermogravimetry proximate, and ultimate analyses results:

This comparative study of *A. tortilis* and *pine dust* shows that the latter generally has better fuel properties due to its higher HHV and fixed carbon with low ash and oxygen contents. However, since *A. tortilis'* thermochemical properties are not very far from those of pine and the ash content is still about half the critical slagging and fouling value of 6%, it can still be exploited effectively for combustion, gasification and pyrolysis (Anukam et al., 2016). The higher oxygen values in *A. tortilis* will further affect the quality of some products like pyrolysis oil, though marginally compared to the pine dust (Pratap & Chouhan, 2013).Notably, the HHV for *A. tortilis is* not very different from that of pine dust; though the Acacia has a considerably higher ash content and a slightly lower fixed carbon.The higher ash content in *A. tortilis* implies a higher mineral content, making it more prone to fouling and slagging compared to the pine; however the figure is below the critical 6% stipulated for gasification systems (Anukam et al., 2016).

5 CONCLUSION

Various literature sources have confirmed the importance of physicochemical characterization of biomass, a primary step in exploring the feasibility of exploiting it for bioenergy purposes. Since both biomass are wastes in their local contexts, and their thermochemical properties are within recommended ranges, they can be converted by any of the combustion, gasification and pyrolysis routes. They can also be torrefied or carbonized into charcoal to obtain solid fuels of higher heating value. Further experiments can be carried out for the various conversion routes to determine optimal conditions and the properties of products obtained. Intermediate size ranges seem more convenient for storage purposes given their higher compressibility indices. The knowledge of *A. tortilis'* properties relative to a more widely known pine sawdust helped ascertain that this biomass can be used in similar thermochemical purposes.

REFERENCES

Acar, S., & Ayanoglu, A. 2016. Determination of higher heating values (HHVs) of biomass fuels, (January 2012).

Álvarez-Álvarez, P., Pizarro, C., Barrio-Anta, M., Cámara-Obregón, A., María Bueno, J. L., Álvarez, A., . . . Burslem, D. F. R. P. 2018. Evaluation of tree species for biomass energy production in Northwest Spain. *Forests*, 9(4), 1–15. https://doi.org/10.3390/f9040160

Anukam, A. I., Mamphweli, S. N., Reddy, P., & Okoh, O. O. 2016. Characterization and the effect of lignocellulosic biomass value addition on gasification efficiency. Energy Exploration and *Exploitation*, 34(6), 865–880. https://doi.org/10.1177/0144598716665010

Arimoro, F. O., Ikomi, R. B., & Osalor, E. C. 2006. The impact of sawmill wood wastes on the water quality and fish communities of Benin River, Niger Delta area, Nigeria. *World Journal of Zoology*, 1(2), 94–102.

Cai, J., He, Y., Yu, X., Banks, S. W., Yang, Y., Zhang, X., . . .Bridgwater, A. V. 2017. Review of physicochemical properties and analytical characterization of lignocellulosic biomass. *Renewable and Sustainable Energy Reviews*, 76(March), 309–322. https://doi.org/10.1016/j.rser.2017.03.072

DECOSA. 2015. *Support to De-Bushing Project Value Added End-Use Opportunities for Namibian Encroacher Bush* (Vol. 264).

Effah, B., Antwi, K., Boampong, E., Asamoah, J. N., & Asibey, O. 2015. The management and disposal of small scale sawmills residues at the Sokoban and Ahwia wood markets in Kumasi – Ghana, 19(1), 15–23.

IRENA. 2016. Innovation Outlook Advanced Liquid Biofuels, 132. Retrieved from http://www.irena.org/DocumentDownloads/Publications/IRENA_Innovation_Outlook_Advanced_Liquid_Biofuels_2016.pdf

Kabajan, S., Kaunda, K., & Matlhaku, K. 2016. Shoot Production by Acacia tortilis under Different Browsing Regimes in South-East Botswana, (January 2011).

Kirsanovs, V., Blumberga, D., Dzikevics, M., & Kovals, A. 2016. Design of Experimental Investigations on the Effect of Equivalence Ratio, Fuel Moisture Content and Fuel Consumption on Gasification Process. *Energy Procedia*, 95, 189–194. https://doi.org/10.1016/j.egypro.2016.09.045

Pratap, A., & Chouhan, S. 2013. Critical Analysis of Process Parameters for Bio-oil Production via Pyrolysis of Biomass?: A Review, (July). https://doi.org/10.2174/18722121113079990005

Sapp, M. (2017, April). Technavio study says global advanced biofuel market will reach $ 44 . 6 billion by 2021. *BiofuelsDigest*, 2021. Retrieved from http://www.biofuelsdigest.com/bdigest/2017/04/17/technavio-study-says-global-advanced-biofuel-market-will-reach-44-6-billion-by-2021/

Stecher, K., Brosowski, A., & Thrän, D. 2013. Biomass potential in Africa. *Irena*, 44. Retrieved from www.dbfz.de

Thermo Fischer Scientific. 2017. Elemental Analysis: CHNS/O characterization of biomass and bio-fuels. Retrieved from thermofisher.com/OEA

Walker, G. M. 2012. *Bioethanol: Science and technology of fuel alcohol. Ebook.* Retrieved from http://bookboon.com/en/textbooks/chemistry-chemical-engineering/bioethanol-science-and-technology-of-fuel-alcohol

World Energy Council. 2016. World Energy Resources: Waste to Energy. Retrieved from https://www.worldenergy.org/wp-content/uploads/2013/10/WER_2013_7b_Waste_to_Energy.pdf

Wastes: Solutions, Treatments and Opportunities III – Vilarinho et al. (Eds)
 ISBN 978-0-367-25777-4

Technical and economical assessment of waste heat recovery on a ceramic industry

H. Monteiro, P.L. Cruz, M.C. Oliveira & M. Iten
Low Carbon and Resource Efficiency, R&Di, Instituto de Soldadura e Qualidade, Grijó, Portugal

ABSTRACT: Waste heat recovery is key for the improvement of several industrial sectors. It has the potential to be used as an energy source and, therefore, to reduce the current industrial energy consumption. The ceramic industry, in particular, encompasses significant thermal processes (the firing and drying) which generate substantial waste heat. This work studies the potential of waste heat recovery solutions for a ceramic floor tile plant. Direct re-use of hot air streams from kiln cooling zones as kiln combustion air is considered, as well as its use as drying agent and combustion air in dryers. The proposed strategies achieve natural gas savings up to 33% in the plant, reducing fuel consumption by 25–29% in kilns, by 44% in a dryer, and avoiding fuel consumption in the other dryer. Moreover, the payback period of all these strategies has been estimated as 0.12 years, which proves the attractiveness of waste heat recovery.

1 INTRODUCTION

The industry sector is accountable for 36% of final energy consumption (IEA, 2017), holding 33% of global GHG emissions. Nearly 70% of industrial energy demand is related to thermal processes, which result in a considerable amount of waste heat. The waste heat is the thermal energy lost in processes once the thermal carrier naturally cools down to the surrounding temperature. Despite its lower exergy, waste heat has the potential to be used as an energy source and thus to reduce current industrial energy consumption. The share of this heat that can be usefully recovered is called the waste heat potential. In EU, the industrial waste heat potential was estimated to be 300 TWh/year, 70 of which held by the non-metallic mineral industries, and with one third being considered as low-temperature (<200°C) (Papapetrou et al., 2018). Therefore, industrial waste heat is indeed an attractive topic.

Jouhara et al. 2018) present a comprehensive review of waste heat recovery methods and technologies for industrial processes addressing possible applications for three industrial sectors, namely: the steel and iron, food, and ceramic. The waste heat recovery can be defined as the transferring of heat from a waste heat stream to a process (direct use) or to an energy carrier to be used elsewhere (indirect use). The selected technology for the heat recovery depends on the existing waste heat temperatures and characteristics, and on the existing thermal needs (processes) that can potentially benefit from such recovery. Different heat recovery technologies have been identified and studied in the literature. For instance, conventional burners can be changed and upgraded to regenerative or recuperative burners, which preheat the combustion air European Comission, 2009; Jouhara et al., 2018). Also, economizers recover heat from exhaust gases (at low-medium temperature) for preheating liquids entering a system. Both sensible heat and latent heat can be recovered in a condensing economizer. Depending on the exhaust gas composition, the use of resistant materials may be required to withstand acid condensate and fouling. Rotary regenerators (as heat wheels) can be used for low-medium temperature applications (mostly in AVAC applications) and may offer a high heat transfer efficiency; however, these may not be fit for high temperature streams due to the risk of physical deformation (Jouhara et al., 2018; Panayiotou et al., 2017). For medium-high temperature streams, alternative heat exchangers may be used. The run around coil is used when two streams must be kept separate and/or are distant from each other. Plate heat exchangers can be applicable for liquids or vapor, working under high temperature and pressure operating limits. Moreover, heat pipes are heat transfer devices with an inner phase change fluid that works as an intermediate heat carrier and provides high thermal conductivity. This technology is currently being studied for its application in industrial waste heat recovery (Delpech et al., 2018; Jouhara et al., 2017). Heat pumps may be used to upgrade low temperature heat, for instance transfer heat from a lower temperature source to a higher temperature heat sink, requiring a small part of energy (electricity). Furthermore, some technologies can generate electricity from waste heat streams depending on the temperature level: i) thermionic generators for high temperature, ii) thermoelectric generators for medium-high temperature, and iii) piezoelectric materials for low temperature. However, these systems still present low efficiencies and high costs (Jouhara et al., 2018).

The ceramic industry is a non-metallic mineral industry that includes at least two highly energy-intensive thermal processes: the drying and the firing of ceramics. Moreover, some ceramic industries also include the spray-drying process of raw materials in atomizers. Such thermal processes generally are sourced by fossil fuels, such as natural gas, liquefied petroleum gas or fuel oil, whose consumption results in significant emissions to the environment. Additionally, in many ceramic plants, the firing process involves high temperatures (750–1200°C) and takes place continuously 24 h-a-day. This results in continuous waste heat streams that can represent almost 50% of energy loss (Delpech et al., 2018). Hence, the ceramic industry is a promising sector to consider when assessing opportunities to implement waste heat recovery.

The reference document on best available techniques (BREF) for ceramic manufacturing (European Comission, 2007) and literature on waste heat recovery opportunities (Delpech et al., 2018, 2017) refer that the heat from the cooling zones of tunnel kilns can be recovered to preheat the dryers or to preheat the combustion air. Additionally, depending on the temperatures and heat capacities of exhaust gases, their residual heat can be recovered for other thermal processes using heat exchangers. Furthermore, waste heat could also be used in a Combined Heat and Power generation installation or in an Organic Rankine Cycle to generate heat and electricity for different processes. Though, the economically feasibility of these installations requires careful study, since the capital investment is very high.

Despite the waste heat potential and the significant environmental benefit of waste heat recovery, its implementation depends mainly on the plant operational characteristics, physical constraints, and on attractive economic assessments. Generally, European industries consider an acceptable payback time of 3 years, under which a measure is considered economically viable (Weerdmeester & Tello, 2014).

Generally, industrials disregard the existing waste heat, showing resistance to heat recovery feasibility and its benefits. To allow a greater market uptake of heat recovery strategies, it is important to prove the economic potential of these measures, based on real case-studies' data.

This work presents a study of waste heat recovery strategies for a ceramic floor tile plant. To evaluate heat recovery feasibility, a thermal energy assessment is performed, identifying the existing waste heat streams and heat requirements of the plant. It includes the direct use of waste heat whenever possible, and proposes indirect use through a heat exchanger to produce domestic hot water. Jointly with the energy integration (recovery of waste heat), an economic evaluation to assess its viability is presented. The goal is to identify profitable measures to reduce fossil fuels consumption through waste heat re-use, reducing the associated environmental emissions.

2 MATERIALS AND METHODS

2.1 *Case study description*

The plant under study is a ceramic industry producing floor and wall tiles. The production process includes firstly the mixing of wet raw materials, its casting and pressing. Furthermore, the pressed ceramic tiles are dried in a tunnel dryer (working at 190–200°C). The tiles circulate in

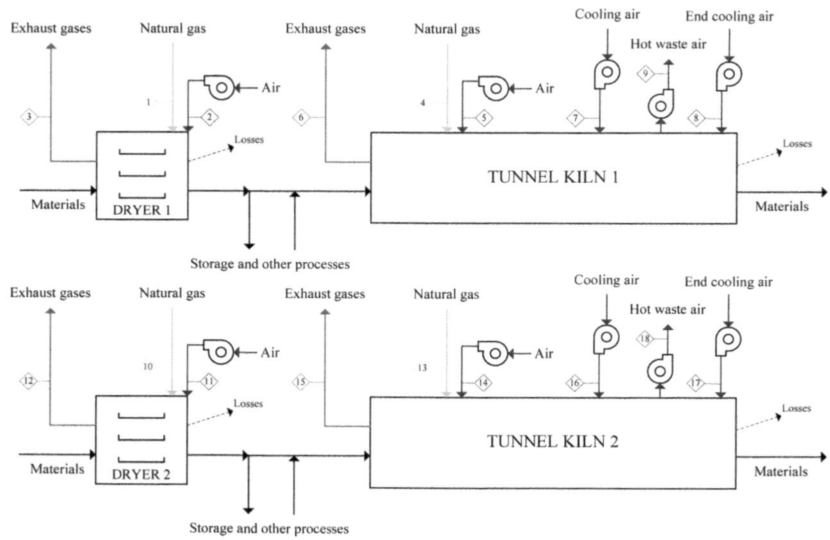

Figure 1. Process flowsheet of the base case process.

counter-current flow with the drier hot air, exiting the dryer with less than 2% of moisture content. After its glazing and decoration, the tiles are fired in a tunnel kiln following a specific gradient temperature, depending on the specific type. The end kiln counts with a cooling zone, which is equipped with cooling air injections that cools down the ceramic tiles, consequently producing a hot air waste stream that is rejected to the surrounding environment. The considered plant includes two production lines, one for special ceramic tiles and other for conventional tiles. Each production line has a tunnel dryer and a tunnel kiln, both running with natural gas. The kilns operate 8760 hours per year, while dryers 6000 hours per year. Figure 1 presents a schematic diagram of above explained process, also named in this work as base case process. Moreover, the base case process streams are further listed in Table 1.

2.2 *Waste heat recovery strategies*

Given the huge natural gas consumption of the plant, over 1,500 tons per year, opportunities for waste heat recovery have been identified. In particular, the significant hot waste air rejected (over 21 tons per hour at 123–160°C) from the cooling zone of the kilns to the surrounding environment. Therefore, its recovery to the process is expected to enhance the plant efficiency. In this sense, wasted hot air streams may be operationally used for two purposes: its direct utilisation as combustion air and its utilisation as drying agent. These strategies are analysed in this study, considering the mass and enthalpy balances, and taking into account the operational requirements of the process. Once the heat recovery and energy savings are determined, an economic assessment is performed to assess this potential improvement.

The 'available heat' to be supplied to the drying and firing processes before heat recovery strategies is considered as a constraint condition. The 'available heat' is determined by equation (1), in which $\dot{M}_{fuel}$ (kg/h) designates the fuel mass flow rate, LHV (kJ/kg) the fuel low heating value, $\dot{M}_{gas\ stream}$ (kg/h) the mass flow rate of the combustion air, $C_{P,gas\ stream}$ (kJ·°C^{-1}·kg^{-1}) the air heat capacity, T_{out} (°C) the outlet temperature of this gas stream and T_{in} the inlet temperature of this gas stream. The air-to-fuel ratio also constitutes a constraint to the implementation of these strategies, in order to ensure the thermal efficiency of burners remains the same.

$$q_{available} = \dot{M}_{fuel} \times LHV - \dot{M}_{gas\ stream} \times C_{P,gas\ stream} \times (T_{out} - T_{in}) \quad (1)$$

Considering the abovementioned constraints, the potential hot air recovery can be estimated. The global mass and enthalpy balances of dryers and kilns are determined considering the recycling of the air streams and a purging air stream. In the case of the tunnel kilns, the optimized fuel mass flow rate ($\dot{M}_{opt}$) is determined by equation (2), in which AF designates the air to fuel ratio and $T_{in,recovery}$ (°C) denotes the inlet temperature of the hot air recovered for combustion.

$$\dot{M}_{opt} = \frac{q_{available}}{LHV - C_{P,gas\ stream} \cdot \left(T_{out} - T_{in,recovery}\right) \cdot (1 + AF)} \tag{2}$$

$T_{in,recovery}$ is calculated by determining the heat loss on the air ducts that connect the kiln cooling zone and the heated air stream destination. According to equation (3), in which $\dot{M}_{air}$ (kg/h) designates the mass flow rate of the heated air streams and $C_{P,air}$ (kJ·°C^{-1}·kg^{-1}) the air heat capacity, the temperature difference is calculated according to equation (3).

$$T_{in,recovery} - T_{out,cooling\ zone} = \frac{q_{loss}}{\dot{M}_{air} \times C_{P,air}} \tag{3}$$

The heat losses are determined by estimating the convective heat transfer coefficient of the air through empirical correlations, considering the projected length, diameter and thickness of each duct and the estimated velocity of the air.

3 RESULTS

3.1 *Waste heat recovery*

Following the procedure detailed in section 2.2, hot air streams are integrated within kilns and dryers. Some assumptions are considered to ensure the operability of the process, namely:

i) heat flow supplied to the tiles and the air-to-fuel ratio, in the combustion zone of kilns and dryers, remain the same than for the base case process;
ii) temperature of exhaust gases in both kilns and dryers are assumed to remain constant, so as not to modify temperature profile and material requirements.

Considering these assumptions, mass flow and conditions of hot waste air streams are not affected by the integration, since temperature profiles and processed material rates remain constant. Hence, their potential is analysed and a new layout is proposed. Figure 2 shows this proposed layout (integrated process) and Table 1 details the mass flows and temperatures in comparison with the base case process.

As shown in the Figure 2, hot waste air from the kilns cooling zone is firstly used as combustion air. The thermal losses in the duct connecting the cooling zone to the combustion air-blower are determined (as presented before in equation 3) to determine the inlet temperature. This temperature is used to achieve the natural gas requirements, considering the abovementioned constraints. As a result, both fuel and combustion air flows are reduced in 29.1% and 24.9% for kilns 1 and 2, respectively. The remaining heat in the air streams is used in upstream dryers in different ways, depending on each one.

The waste heated air from kiln 2 is directly used as a drying agent in dryer 2. The required mass flow is calculated to ensure the heating supply in the dryer, considering the associated heat losses in the duct and the consequent temperature at the inlet of the dryer (157°C). It has been observed that it has the potential to provide all the required energy in the dryer and, hence, to avoid the use of natural gas in this equipment.

The remaining hot air from kiln 2 is mixed with excess hot air from kiln 1, and then fed as combustion air in dryer 2. However, as the heating potential of this mixed air does not achieve required energy in the dryer 1, it is assumed to be fed as the dryer's combustion air. Furthermore,

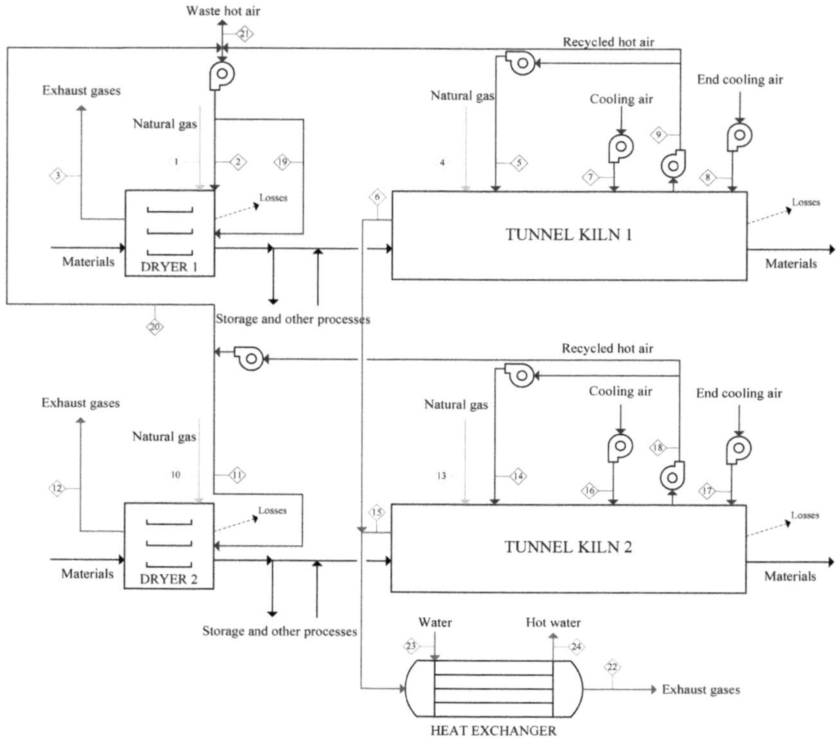

Figure 2. Flowsheet of the integrated process.

the higher air temperature at the inlet of the combustor implies a reduction of the natural gas and air flows, which may worsen the humidity removal in the material. Thus, in order to ensure drying capacity, the same exhaust gas mass flow than the base case process is assumed. This implies also the addition of a direct hot air stream in the dryer. As a result, the natural gas consumption is reduced in 44.2%.

Additionally, an energetic valorisation of the exhaust gases is considered. Although there is no real hot water needs in this plant, its potential production may be attractive in different industrial sectors. The exhaust gases exiting the kilns are mixed and fed into a heat exchanger with the aim of heating water up to 45°C. Thus, a potential for a hot water production of more than 6 tons per hour is calculated, considering the operational time of kilns. Exhaust gases containing acid compounds may produce corrosion problems if condensation occurs in the heat exchanger (Delpech et al., 2017), however the acids concentration in this process is relatively low due to a high quantity of excess air in the kilns (no vestigial quantity of SO_x; 26 mg/Nm^3 of NO_x; 4.62 mg/Nm^3 F^-) and these problems may be despised. Moreover, the temperature of the exiting exhaust gases (120°C) is considered enough to avoid condensation.

3.2 *Economic assessment*

The economic assessment of the proposed process integration is determined by estimating its investment costs, the potential economic savings and its associate payback period. The economic savings for each measure may be generically calculated by equation (4), in which t_{op} (h/year) designates the annual operational time of the concerned operation, C_{fuel} (€/kJ) the cost of the fuel per unit of energy and $\dot{M}_{initial}$ (kg/h) the initial fuel mass flow rate. The payback time may be calculated by equation (5), in which C_{Inv} designates the investment cost associated to each

Table 1. Mass flows and temperatures of process streams in the base case and in the integrated process.

Base case process			Integrated process		
Stream	Mass flow (kg/h)	T (°C)	Stream	Mass flow (kg/h)	T (°C)
1	15.7	25	1	8.8	25
2	1710.8	25	2	954.8	131
3	1816.8	80	3	1816.8	80
4	108.8	29	4	77.1	29
5	11 858.0	26	5	8405.7	119
6	11 966.0	198	6	8482.9	198
7	2336.3	26	7	2336.3	26
8	12 032.0	26	8	12 032.0	26
9	13 022.9	123	9	13 022.9	123
10	9.6	25	10	0.0	0
11	1041.5	25	11	4564.3	157
12	1084.3	80	12	4564.3	80
13	45.4	29	13	34.1	29
14	3711.0	26	14	2788.1	158
15	3756.0	152	15	2822.3	152
16	2336.3	26	16	2336.3	26
17	13 078.7	32	17	13 078.7	32
18	8934.0	160	18	8934.0	160
			19	853.3	131
			20	1581.6	157
			21	4390.7	131
			22	11 305.1	120
			23	6255.8	15
			24	6255.8	45

Table 2. Economic evaluation of the proposed integrated process.

Technology	Investment Costs (€)	Operation	Natural gas savings (GJ/year)	Savings (€/year)	Payback (year)
Heat exchanger	20 882.49	Kiln 1	13 069	121 934.77	0.12
		Kiln 2	4 658	43 461.38	
Ducts	4 915.21	Dryer 1	2 703	25 214.36	
		Dryer 2	1 962	18 304.85	
TOTAL	25 797.70	TOTAL	22 392	208 915.36	

measure (which includes installation costs). Maintenance costs associated with new equipment and process configuration are assumed not to be significantly different than existing ones, thus were not considered.

$$Sav = t_{op} \times C_{fuel} \times LHV \times \left(\dot{M}_{initial} - \dot{M}_{opt}\right) \tag{4}$$

$$PB = \frac{C_{Inv}}{Sav} \tag{5}$$

The implementation of the proposed solution involves an investment cost for the new and required ducting for the heat recovery and the heat exchanger itself and, as described by equation (4), a reduction on the operating costs related to the reduction of the fuel consumption. Hence, economic savings may compensate investment costs in a short period of time. Table 2 shows the economic evaluation performed, showing an overall pay-back period of 0.12 years. The achieved results show a profitable strategy; however, it is important to note that such payback strongly depends on natural gas price. For the current study the natural gas cost has been considered as 9.33 €/GJ (PORDATA 2018).

4 CONCLUSIONS

This work presents a techno-economic assessment of waste heat recovery for a ceramic tiles industry. Considering the base case process involved in the plant under study, potential heat recovery strategies are proposed and evaluated. Results shows that the direct recirculation of heated air from kilns cooling zones to kilns combustors can achieve natural gas reductions of 29% and 25% in kilns 1 and 2, respectively. Moreover, the use of the remaining hot air in dryers can avoid the use of fuel in dryer 2 and reduce the natural gas consumption by 44% in dryer 1. Additionally, an energy valorisation of exhaust gases is identified to potentially produce more than 6 tons per hour of domestic hot water. Overall, proposed recovery strategies reach 33% fuel savings in the studied plant corresponding to a payback period of 0.12 years. Thus, the economic feasibility is proven to be attractive for the current case study, namely a ceramic industry.

ACKNOWLEDGEMENTS

This publication has been funded by the European Union's Horizon 2020 research and innovation programme under grant agreement "No 810764".

REFERENCES

Delpech, B., Axcell, B. & Jouhara, H., 2017. A review on waste heat recovery from exhaust in the ceramics industry. E3S Web Conf. 22, 00034.

Delpech, B., Milani, M., Montorsi, L., Boscardin, D., Chauhan, A., Almahmoud, S., Axcell, B. & Jouhara, H., 2018. Energy efficiency enhancement and waste heat recovery in industrial processes by means of the heat pipe technology: Case of the ceramic industry. Energy 158, 656–665.

European Comission, 2007. Reference Document on Best Available Techniques in the Ceramic Manufacturing Industry. Ceram. Manuf. Ind. 210–211.

European Comission, 2009. Reference document on Best Available Techniques for Energy Efficiency. Eur. Com. 1–430.

IEA, 2017. Tracking Clean Energy Progress 2017.

INE-I.P./DGEG, 2018. PORDATA - Natural gas prices for households and industrial users (Euro). URL https://www.pordata.pt/en/Europe/Natural+gas+prices+for+households+and+industrial+users+(Euro)-1478 (accessed 4.1.19).

Jouhara, H., Almahmoud, S., Chauhan, A., Delpech, B., Bianchi, G., Tassou, S.A., Llera, R., Lago, F. & Arribas, J.J., 2017. Experimental and theoretical investigation of a flat heat pipe heat exchanger for waste heat recovery in the steel industry. Energy 141, 1928–1939.

Jouhara, H., Khordehgah, N., Almahmoud, S., Delpech, B., Chauhan, A. & Tassou, S.A., 2018. Waste heat recovery technologies and applications. Therm. Sci. Eng. Prog. 6, 268–289.

Panayiotou, G.P., Bianchi, G., Georgiou, G., Aresti, L., Argyrou, M., Agathokleous, R., Tsamos, K.M., Tassou, S.A., Florides, G., Kalogirou, S. & Christodoulides, P., 2017. Preliminary assessment of waste heat potential in major European industries. Energy Procedia 123, 335–345.

Papapetrou, M., Kosmadakis, G., Cipollina, A., La Commare, U. & Micale, G., 2018. Industrial waste heat: Estimation of the technically available resource in the EU per industrial sector, temperature level and country. Appl. Therm. Eng. 138, 207–216.

Weerdmeester, R. & Tello, P., 2014. SPIRE roadmap. Brussels.

Wastes: Solutions, Treatments and Opportunities III – Vilarinho et al. (Eds)
© 2020 Taylor & Francis Group, London, ISBN 978-0-367-25777-4

Impacts of waste management practices on water resources in Harare

T. Nhubu & C. Mbohwa
University of Johannesburg, Johannesburg, Gauteng, South Africa

E. Muzenda
University of Johannesburg, Johannesburg, Gauteng, South Africa
Botswana International University of Science and Technology, Palapye, Botswana
University of South Africa, Gauteng, South Africa

B. Patel
University of South Africa, Gauteng, South Africa

ABSTRACT: Poor municipal solid waste (MSW) management practices impact negatively on freshwater availability in terms of both quality and quantity. A review on the MSW practices currently being practiced and their impacts on water resources management in Harare was carried out to give recommendations towards sustainable MSW management and reduce the pollution of water bodies. There is urgent need for the development and implementation of a local level integrated MSW management plan tapping from the national plan that was pronounced in 2014. Such a plan should incorporate material and energy recovery with the best option combining anaerobic digestion of biodegradable MSW and incineration of the non-biodegradable MSW fraction. Legislative reforms to enforce source separation, the prohibition of dumbing waste at dumpsites together with the landfilling of biodegradable waste need to be urgently institutionalized.

1 INTRODUCTION

In developing countries, municipal solid waste (MSW) generation is increasing rapidly fuelled by the ever increasing population, economic growth, rapid urbanization and rising standards of living (Minghua et al., 2009). The increased municipal solid waste generation bring along challenges for municipal authorities to provide sustainable, effective and efficient MSW management systems due to increased burden of MSW management share on municipal budgets limited knowledge on different factors affecting different waste management stages and the enabling links for the functioning of the entire MSW handling system (Guerrero et al., 2013).

Tsiko and Togarepi (2012) noted MSW management amongst the greatest challenges faced by Harare municipality due to increased MSW generation rate making it difficult for municipality to raise adequate financial and technical resources to match the growth in MSW generation. Poor MSW management practices directly and indirectly contribute to both surface and groundwater quantity and quality impacts that drive freshwater scarcity within a given catchment. Engelman and LeRoy (1993) reported as early as 1993 that perennial water scarcity challenges characterised many nations with Nhapi and Hoko (2004) reporting the urgent water quality problem and imminent water scarcity in Lake Chivero and Manyame catchment, the potable water sources for Harare city, Chitungwiza, Epworth, Norton and Ruwa

Harare city, Chitungwiza, Epworth, Norton and Ruwa have been targeted for this study. They have a total population of 2,133.802 people, Harare at 1,485,231, Chitungwiza at 356,840, Epworth at 167,462, Norton at 67,591 and Ruwa at 56,678 people (ZIMSTAT, 2013). Population has increased

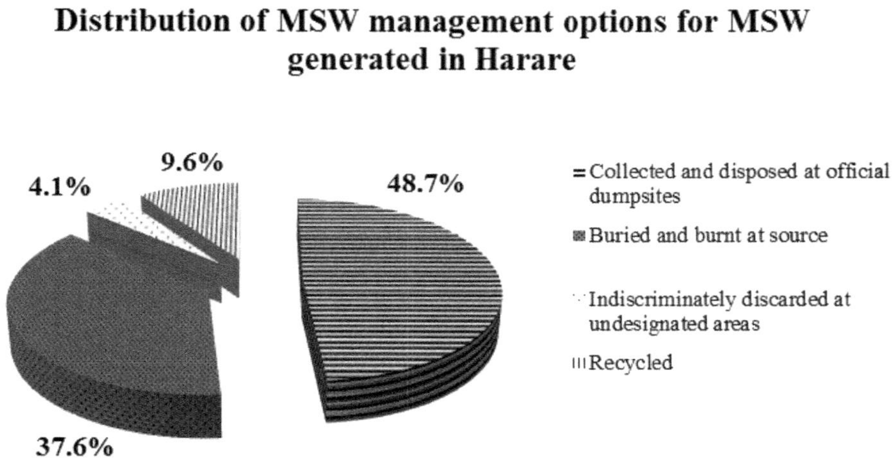

Figure 1. Management practices for MSW generated in Harare (EMA, 2016).

from an estimated 1.4 million in 2002 (CSO, 2003) to over 2.1 million in 2012 (ZIMSTAT, 2013) showing an estimated annual average increase of over 70,000 people over the 10 year period. The 1992 census estimated Ruwa population at 440 persons (Zinyama, 1994, CSO, 2003) with the 2012 census giving an estimate of 56,678 people (ZIMSTAT, 2013) indicating an average annual increase of over 2,800 persons over the 20 year span. Epworth had an estimated population of just over 110,000 in 2002 (CSO, 2003) increasing to just over 167,000 people in 2012 (ZIMSTAT, 2013) thus an annual average increase of over 5,700 people. These population increases have not been matched with corresponding MSW management infrastructure and systems. This study therefore reviewed the impacts of prevailing MSW management on water resources management in Harare, Zimbabwe.

2 CURRENT MSW MANAGEMENT PRACTICES

MSW is defined as waste that is managed by or on the behalf of municipalities (Hester and Harrison, 2002, Kawai and Tasaki, 2016). It includes waste generated from household constituting between 55 to 80%, waste generated from markets and other commercial areas contributing between 10 to 30% and waste generated from institutions, industries and streets of varying contributions (Nabegu, 2010, Okot-Okumu, 2012). Annual MSW generation in the study area is estimated at over 400,000 tons. Harare, Chitungwiza and Epworth have an estimated annual MSW throughput of 421,757tons (Makarichi et al., 2019). Tirivanhu and Feresu (2013) reported an annual MSW generation of 371,697tons for Harare city, 90% of which has potential for either reuse or recycle.

The Environmental Management Agency (EMA) (2016) observed that 48.7% of the MSW generated in the study area is formally collected and disposed at the three official dumpsites, 9.6% is recycled, 4.1% being indiscriminately discarded into undesignated areas with 37.6% being either buried or burnt at-source as shown in Figure 1. MSW collection was reported to have dropped from 52% in 2011 to 48.7% in 2016 (EMA, 2016) resulting in the accumulation of MSW awaiting collection.

Tevera and Masocha (2003) reported that at most 60 per cent of MSW generated in Harare is officially collected and disposed at dumpsites with the over 40% that remains uncollected usually discarded illegally in open spaces, alleys, storm water drains and road verges. The three formal MSW disposal dumpsites namely Chitungwiza, Pomona and Golden Quarry are unprotected and

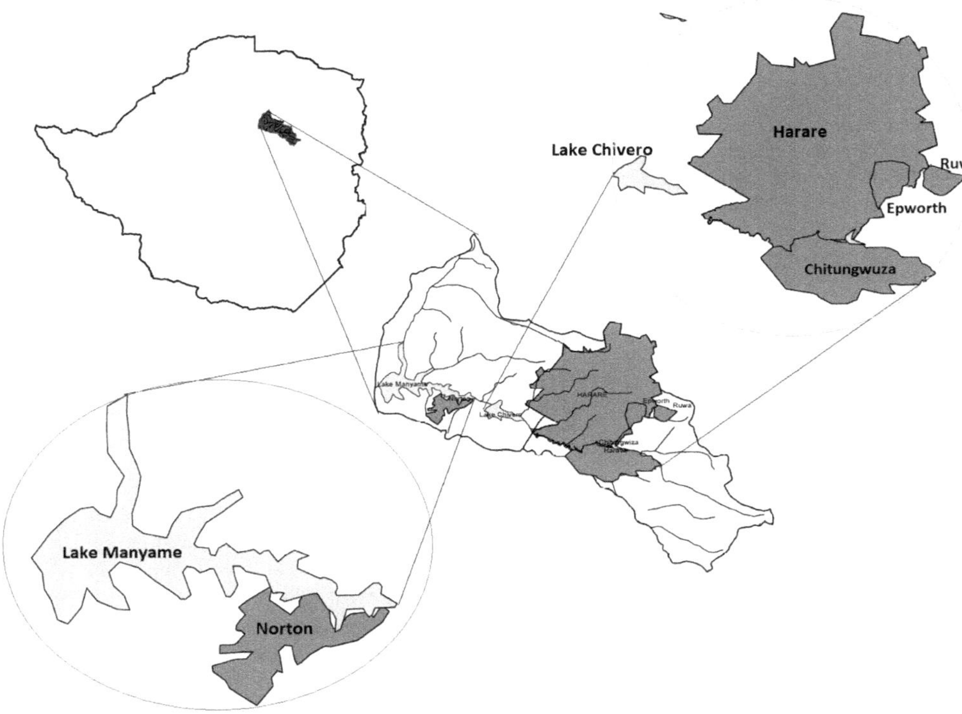

Figure 2. Urban environments located on lake chivero catchment.

poorly managed without impermeable lining to prevent dumpsite leachate infiltration to the groundwater. The Government of Zimbabwe (GoZ) (2014) planned the construction of a sanitary landfill for Harare anticipated to have become operational by end of the year 2018. However this has largely remained an idea with no actions on the grounds since the announcement of the National Integrated Solid Waste Management Plan in July of 2014 which provided for various intervention strategies from waste avoidance and minimisation, materials and energy recovery.

3 IMPACTS OF CURRENT MSW MANAGEMENT PRACTICES ON WATER RESOURCE MANAGEMENT

Four management practices for MSW generated in Harare have been identified namely, collec tion and disposal at official dumpsites, recycling, indiscriminate dumping in undesignated areas and the burning and burying at source. Amongst the four practices only material recycling is regarded not to have any harmful or negative impacts on water resources management. However material recycling must be expanded to other material constituents like glass, metals and others as it is currently limited to plastics and corrugated paper as noted by Makarichi et al (2019). The indiscriminate dumping of MSW in undesignated areas has negatively impacted the water situation in Harare leading to the annual and seasonal outbreaks of Typhoid and Cholera (Nhapi, 2009). MSW management problems in Harare are thus manifesting themselves through the pollution of both surface and groundwater emanating from the indiscriminate dumping of MSW in waterways and dumpsites leachate infiltration into groundwater. Although the discharge of raw or partially treated sewage into the river system that drains runoff from Harare has been cited as the major driver of the eutrophic status of Lake Chivero (Magadza, 2003, Rommens et al., 2003, Nhapi and

Table 1. Potable water treatment chemicals used by Harare city and their costs in 2017.

Name of chemical	Unit Metric tons	Unit cost US$/kg	Total cost US$
Powdered activated carbon	598.08	1.67	983,763.60
Granular Aluminium Sulphate	1317.00	0.65	856,050.00
Liquid Aluminium Sulphate	18,990.88	0.49	9,305,531.20
Poly Aluminium Chloride (granular)	580	0.76	440,800.00
Ammonia	3.43	2.65	9,089.50
Chlorine gas	443.15	1.27	562,800.50
White Hydrated Lime	244.85	0.56	137,116.00
Sodium Silicate	165.03	0.82	135,324.60
Sulphiric acid	3.42	0.51	1,744.20
Bulk Sulphiric acid	381.02	0.42	160,028.40
High-Test granular Calcium Hypochlorite (HTH)	96.4	3.05	294,020.00
Total			12,886,268.00

Tirivarombo, 2004, Nhapi et al., 2004, Nhapi, 2014), hygiene breakdown in the urban environments result in nutrient rich surface run-off from uncollected MSW and illegal MSW dumps significantly contributing to Lake Chivero eutrophication (Magadza, 2003). This is largely due to the location of the urban environments within and surrounding Harare City namely Chitungwiza, Norton, Ruwa and Epworth on the catchment that drains runoff into the Lake as shown in Figure 2. This therefore entails for a holistic solution that encompass the rehabilitation of existing wastewater treatment infrastructure, construction of new wastewater treatment infrastructure complimented with sustainable and integrated MSW management system to avert the threats from MSW laded runoff from residential areas and other undesignated areas.

The eutrophication of Lake Chivero has led to increased costs for potable water production due to increased chemicals and power used during potable water production. Filter clogging with colloidal algae calls for the need for filter backwashing using already filtered and partly treated water thus increased pumping and associated power for pumping during backwash and replenishing the water used during backwashing. In addition due to the eutrophic status of Lake Chivero, the City of Harare is forced to use nine chemicals namely powdered activated carbon, Aluminium Sulphate, High-Test granular Calcium Hypochlorite (HTH), Poly Aluminium Chloride, Poly Aluminium Chloride, Ammonia, Chlorine gas, White hydrated lime, Sodium Silicate and Sulphiric acid to treat raw water from Lake Chivero to meet potable standards. The 2017 City of Harare records shows that due to foreign currency shortages a third (US$12,886,268) was availed from the Reserve Bank of Zimbabwe against a planned budget of US$36 million to procure water treatment chemicals as shown in Table 1. Therefore, increased costs coupled with erratic foreign currency disbursements from the Reserve Bank of Zimbabwe have partly contributed to the erratic potable water supplies in most parts of Harare currently being experienced.

Dumpsites leachate infiltrates into groundwater polluting an alternative source of potable water supply which a significant population has resorted to due to the prevailing potable water shortages. Sood (2004) noted the potential of open and controlled MSW dumps to significantly contribute to the pollution of groundwater sources. Rainwater infiltrates refuse due to the absence of top soil cover with water percolating through decomposing MSW absorbing chemicals together with micro-organisms present thereby contaminating groundwater as well as surface waters posing public health and environmental risks to residents. Love et al (2006) reported that both Golden Quarry and Pomona dumpsites lack the impermeable engineering lining for water sources leachate pollution prevention. Further observations by Love et al. (2006) at Golden Quarry showed that the concentration of coliforms, nitrates, iron, cadmium, and lead were unexpectedly higher than the World Health Organisation acceptable potable water quality standards in the entire Westlea neighborhood area rendering the groundwater unsafe for potable and other domestic uses. High metal concentration was attributed to the indiscriminate nature of MSW disposal with industrial waste also being disposed at Golden Quarry dumpsite. Lower metal concentration was observed at Pomona dumpsite as compared to Golden Quarry as less industrial waste is disposed at Pomona than Golden Quarry. The metals, nitrate and coliforms concentration at Golden Quarry decreased westwards in the direction of groundwater flow. Leachate pollution of groundwater and water bodies in the vicinity of both dumpsites must have reached unexpectedly higher levels now as more MSW is being generated and continuously dumped at these dumpsites despite the fact that they have reached their design capacity limits and degraded into potential human health and environmental hazards.

4 CONCLUSION

MSW management practices currently being practiced in the urban environments within and surrounding Harare City are highly unsustainable as they contribute to the outbreak of water borne disease, threatening human health, polluting potable water sources and undermining the integrity of the environment. The development of integrated MSW management systems incorporating material and energy recovery is now urgent coupled with the construction of an engineered sanitary landfill. The developed national integrated solid waste management plan has largely remained a plan without implementation as it did not provide estimates of budgets and resources needed to implement action plans which could be used for fundraising purposes. It is also important to develop strategies that nurture a culture of responsible citizenship i regards to MSW generation and changing mindsets and behaviors to address the indiscriminate throwing away of litter and unfinished food in undesignated areas. The Presidential Cleanup Day Declaration that made every first Friday of a calendar month a National Environment Cleaning Day is one such gesture by the highest office in the country anticipated to nurture this culture on responsible MSW handling by citizens. However the monthly National Environment Cleaning Day need also to be complimented with tangible actions towards the prioritization of resource allocation towards the development of MSW management systems for municipalities since the waste that is being collected during the cleaning day is disposed at environmental unsafe dumpsites. Polluter pays legislation does exist and is subject to abuse as the polluting fines paid by polluters is not used for pollution remediation purposes as provided for in the legislation. There is need therefore for legislative adherence, enforcement and reform to facilitate waste minimization, source separation as well as the entry of private players into MSW management as current legislation gives MSW management mandate to municipalities only. However municipalities can hire other private players to manage MSW on their behalf. Opportunities for biogas production exist considering the biodegradable content of MSW generated which is in the excess of 40%. Makarichi et al (2019) reported that the composition of combustible fraction in MSW generated in Harare exceeds 75% by weight suitable for thermal treatment of MSW without adding supplementary fuel giving an annual energy potential of 3.8×10^6 GJ at a lower heating value of 10.1 MJ per kg. This can possibly bring about a 40% reduction of MSW sent to landfills and providing almost 112 GWh per year of electricity. This will ultimately increase the annual share of MSW and biofuels derived electrical energy from 1.3% to atleast 2.2%.

ACKNOWLEDGEMENTS

The authors acknowledge the Life Cycle Initiative for 2017 Life Cycle award that provided the Simapro software that was used to carry out the LCIA. The authors are also grateful to the National Geographic Society for the early career grant that enabled data collection for the study. The personnel and management at Harare municipality who assisted with the requisite information and data will always be appreciated without their support this work could not have been accomplished. The Universities of Johannesburg and South Africa are also acknowledged for funding the studies, and as well as conference registration and attendance.

REFERENCES

CSO 2003. Demographic and Health Survey. Harare, Central Statistics Office.

EMA 2016. Waste generation and management in Harare, Zimbabwe: Residential areas, commercial areas and schools. Harare, Unpublished internal report, Environmental Management Agency.

Engelman, R. & Leroy, P. 1993. Sustaining Water, Population and Future Renewable Water Supplies. Washington DC, Population Action International.

GOZ 2014. Zimbabwe's integrated solid waste management plan. Harare, Government of Zimbabwe, Environmental Management Agency, Institute of Environmental Studies, University of Zimbabwe.

Guerrero, L. A., Maas, G. & Hogland, W. 2013. Solid waste management challenges for cities in developing countries. *Waste Management,* 33, 220–232.

Hester, R. E. & Harrison, R. M. 2002. Environmental and Health Impact of Solid Waste Management Practices. *Issues in environmental science and technology.* Cambridge, Royal Society of Chemistry.

Kawai, K. & Tasaki, T. 2016. Revisiting estimates of municipal solid waste generation per capita and their reliability. *Journal of Material Cycles and Waste Management,* 18, 1–13.

Love, D., Zingoni, E., Ravengai, S., Owen, R., Moyce, W., Mangeya, P., Meck, M., Musiwa, K., Amos, A., Hoko, Z. & Hranova, R. 2006. Characterization of diffuse pollution of shallow groundwater in the Harare urban area, Zimbabwe. *Groundwater pollution in Africa.* CRC Press.

Magadza, C. H. D. 2003. Lake Chivero: a management case study. *Lakes & Reservoirs: Research & Management,* 8, 69–81.

Makarichi, L., Kan, R., Jutidamrongphan, W. & Techato, K. A. 2019. Suitability of *municipal* solid waste in African cities for thermochemical waste-to-energy conversion: The case of Harare Metropolitan City, Zimbabwe. *Waste Management & Research,* 37, 83–94.

Minghua, Z., Xiumin, F., Rovetta, A., Qichang, H., Vicentini, F., Bingkai, L., Giusti, A. & YI, L. 2009. Municipal solid waste management in Pudong new area, China. *Waste Management,* 29, 1227–1233.

Nabegu, A. B. 2010. An analysis of municipal solid waste in Kano metropolis, Nigeria. *Journal of Human Ecology,* 31, 111–119.

Nhapi, I. 2009. The water situation in Harare, Zimbabwe: a policy and management problem *Water Policy,* 11, 221–235.

Nhapi, I. 2014. *Options for wastewater management in Harare, Zimbabwe,* Delft, CRC Press.

Nhapi, I. & Hoko, Z. 2004. A cleaner production approach to urban water management: potential for application in Harare, Zimbabwe. *Physics and Chemistry of the Earth, Parts A/B/C,* 29, 1281–1289.

Nhapi, I., Siebel, M. A. & Gijzen, H. J. 2004. The impact of urbanisation on the water quality of Lake Chivero, Zimbabwe. *Water and Environment Journal,* 18, 44–49.

Nhapi, I. & Tirivarombo, S. 2004. Sewage discharges and nutrient levels in Marimba River, Zimbabwe. . *Water SA,* 30, 107–113.

Okot-Okumu, J. 2012. Solid waste management in African cities–East Africa. *In Waste* Management-*An Integrated Vision,* IntechOpen.

Rommens, W., Maes, J., Dekeza, N., Inghelbrecht, P., Nhiwatiwa, T., Holsters, E., Ollevier, F., Marshall, B. & Brendonck, L. 2003. The impact of water hyacinth (Eichhornia crassipes) in a eutrophic subtropical impoundment (Lake Chivero, Zimbabwe). *I. Water quality. Archiv für Hydrobiologie,* 158, 373–388.

Sood, D. 2004. Solid waste management study for Freetown, Sierra Leone. Component Design for the World Bank, Draft Report Project, (P078389).

Tevera, D. S. & Masocha, M. 2003. Urban waste pollutants and their threats to the environment and human health in Zimbabwe. Harare, World Health Organisation.

Tirivanhu, D. & Feresu, S. 2013. A Situational Analysis of Solid Waste Management in Zimbabwe's Urban Centres. Harare, Institute of Environment Studies, University of Zimbabwe.

Tsiko, R. G. & Togarepi, S. 2012. A situational analysis of waste management in Harare, Zimbabwe. *Journal of American Science,* 8, 692–706.

ZIMSTAT 2013. 2012 Zimbabwe Census National Report. Harare, Zimbabwe National Statistics Agency.

Zinyama, L. 1994. Urban growth in Zimbabwe—the 1992 census. *Geography,* 79, 176–180.

Wastes: Solutions, Treatments and Opportunities III – Vilarinho et al. (Eds)
© 2020 Taylor & Francis Group, London, ISBN 978-0-367-25777-4

Opportunities and limitations for source separation of waste generated in Harare

T. Nhubu & C. Mbohwa
University of Johannesburg, Johannesburg, Gauteng, South Africa

E. Muzenda
University of Johannesburg, Johannesburg, Gauteng, South Africa
Botswana International University of Science and Technology, Palapye, Botswana
University of South Africa, Gauteng, South Africa

B. Patel
University of South Africa, Gauteng, South Africa

ABSTRACT: Source separation of municipal solid waste (MSW) is an integral part of sustainable and integrated MSW management. In Zimbabwe, the national solid waste management plan of 2014 provide for source separation under goal 2 and public education for awareness raising on the importance of source separation under goal 7. This provides the necessary commitment and necessity for source separation at national level. This study reviewed the available opportunities and limitations for MSW source separation in Harare. Such opportunities for source separation that exists include the availability of a national plan, MSW composition which eases source separation of organics to either anaerobic digestion or composting, recyclables and non-recyclables of a low heating value amenable to incineration with energy recovery. However, limitations with regards to enabling legislation, municipal capacity to spearhead source separation in light of their failure to provide MSW receptacles to over 75% of households, equipment and technical expertise exist.

1 INTRODUCTION

United Nations Human Settlements Programme (UN-HABITAT) (2010) reported the emergency of Municipal solid waste (MSW) amongst the most urgent and serious challenge facing urban municipalities and local authorities in nations with transitional and developing economies. With global MSW generation projecting 2.2 billion tonnes per annum in six years' time from now (Hoornweg and Bhada-Tata, 2012), the enormity of the challenges is expected to rise. The provision of sustainable and integrated MSW management is hinged on MSW collection and transportation bringing in high costs consuming between 20 and 50% of municipal budgets (Un-Habitat, 2010) and operational problems that are highly complex and dynamic. The separation of waste at source is done subjectively by individuals sorting MSW where it is generated into defined types and placing them in different receptacles pending collection (Rousta et al., 2015) for reusing, recycling and environmental and economic burdens reduction to the MSW management system. Source separation positively contributes to the effectiveness of MSW management systems resulting from the significant changes it renders on quantity and quality of MSW at final management, treatment and disposal facilities (Hoornweg and Bhada-Tata, 2012). Source separation of MSW leads to increased recyclable and reusable waste collection (Boonrod et al., 2015) and consequently reduced volume and weight of MSW landfilled. In addition, it can lead to the recovery of substantial amount of MSW landfilled as energy and biofertiliser.

Source separation has been applied broadly in developed nations for sustainable MSW management purposes (Rousta et al., 2015, Tai et al., 2011). Tai et al (2011) noted that source separation in developing countries has been applied as a component of integrated MSW management system for pilot scale MSW source separation projects. Factors exist that leads to the low intake of source separation of MSW and volumes of sorted MSW in developing countries such as the unavailability of markets for reusables, recyclables and biodegradables. Hoornweg and Bhada-Tata (2012) and Boonrod et al (2015) cited the lack of public participation and their knowledge on the importance of source separation of MSW. Sukholthaman et al (Sukholthaman et al., 2017) cited the absence of updated laws, policies and regulations together with the unavailability of amenable facilities and infrastructures. Sustainable and successful MSW source separation implementation eases all the other post source separation processes namely collection and transportation, treatment and final disposal and requires sustainable public participation (Dhokhikah et al., 2015). Public participation in MSW separation is however dependent on human behavior and attitudes characterised by mixed feelings or contradictory ideas towards MSW management. The Government of Zimbabwe (GoZ) (2014) thus has developed a national integrated solid waste management pan whose seventh goal seeks to educate and raise awareness in all citizens of Zimbabwe to better understand and participate in source separation; resource recovery and conversion; and integrated and sustainable solid waste management. The overwhelming desire to increase correctly source separated MSW fractions (recyclable, reusable and biodegradable) requires the understanding of the factors that affect their quality and quantity at generation source together with the opportunities available for ease implementation of source separation. Goal 2 of the same plan seeks to separate solid waste at source through the establishment of appropriate systems for solid waste separation at source namely households, commerce, industries, schools, office blocks, hospital and service providers establishments.

Therefore, this study seeks to identify the available opportunities and limitations for MSW source separation in Harare to inform recommendations for possible interventions towards an enabling environment for source separation.

2 MUNICIPAL SOLID WASTE MANAGEMENT IN HARARE

The Capital City of Zimbabwe, Harare surrounded by urban and peri urban areas of Chitungwiza, Epworth, Norton and Ruwa with an estimated total population of 2,133.802 people (Zimstat, 2013). Historically Harare was largely designed to be the country's commercial and administrative city with traces of industrial activity which attracted people to migrate and stay in Harare in search of a living. Respective Waste Management Departments of the Harare Municipality, Chitungwiza town council, Ruwa, Epworth and Norton local boards are responsible for MSW management as provided under the Zimbabwean laws namely the Urban Councils Act [Chapter 29:15], Environmental Management Act [Chapter 20:27], Rural District Councils Act [Chapter 29:13], Hazardous Waste Management Regulations SI 10, 2007, Effluent and Solid Waste Disposal Regulations SI 6, 2007 and the local authorities' by-laws. MSW generation for Harare City, Chitungwiza and Epworth with a combined population of 2,009,533 is estimated at 421,757 tons per annum whose composition is shown in Figure 1 (Makarichi et al., 2019). In 2013 Harare City alone was reported to have an annual MSW throughput of 371,697 tons (Tirivanhu and Feresu, 2013).

Single stream indiscriminate collection of MSW is being practiced once per week for a given weekday for a particular residential area. However it is not always the case that MSW is collected once a week and on the given day of the week as variations occur and sometimes no collection is done for weeks evidenced by the Environmental Management Agency (EMA) (2016) report that MSW collection fell from 52% in 2011 to 48.7% in 2016. Therefor over 50% of the MSW generated remains uncollected and dumped indiscriminately in undesignated areas with 28%, 11%, 6% and 3% reportedly buried, burnt, illegally dumped and separated respectively (EMA, 2016). This has been pointed as a source of water pollution and cause of perennial outbreaks of Cholera and Typhoid (Nhapi, 2009). Uncollected MSW loads surface runoff with nutrients contributing to the eutrophication of the potable water source for Harare, Lake Chivero (Magadza, 2003).

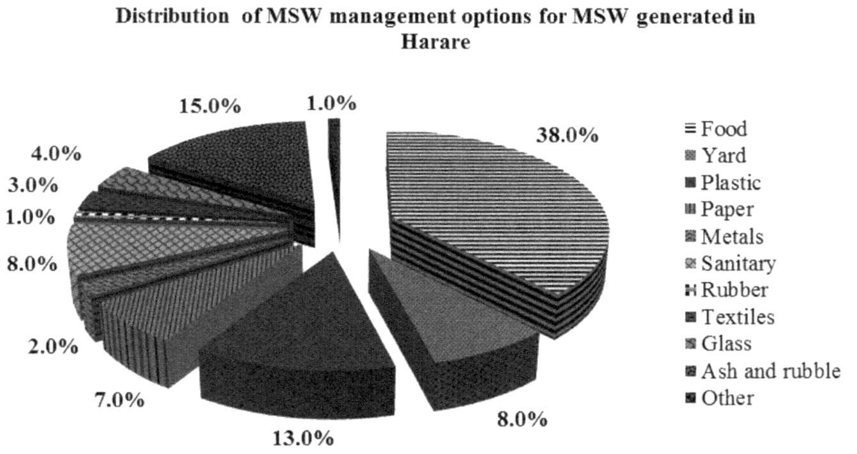

Figure 1. Composition of MSW in Harare (Makarichi et al., 2019).

The MSW officially being collected is indiscriminately dumped at poorly managed official dumpsites without impermeable lining to prevent leachate infiltration into groundwater. These dumpsites have reached their capacity limits with Pomona having been operational since 1985. Unexpectedly higher concentration of nitrates, coliforms, cadmium, iron and lead than the World Health Organisation acceptable potable water quality standards in the groundwater in surbubs surrounding two of the dumpsites, Pomona and Golden Quarry were observed rendering the groundwater unfit for potable and other domestic uses (Love et al., 2006). In light of the impacts the current MSW management system has on water resources and the environment, the exhaustion of the capacity of the official dumpsites and the pronouncement of the national integrated solid waste management plan, there is need to redesign and define future sustainable and integrated MSW management systems for Harare city and its surrounding urban and peri urban environments. Source separation of MSW is an integral and initial step after MSW generation for successful implementation of sustainable integrated MSW management (Reduce, Reuse and Recyle) system hence the need to identify its opportunities and limitations in Harare.

3 OPPORTUNITIES AND LIMITATIONS FOR SOURCE SEPARATION

Goal 2 of the national integrated solid waste management plan (GoZ, 2014) that was pronounced in July of 2014 provides for waste separation at source and goal 7 provides for public education for awareness raising on the importance of source separation of waste. This is a welcome development as it shows commitment at the national level on the need for source separation of waste and in that regard the plan is a springboard from which urban local authorities should develop their sustainable and integrated MSW management systems. However there hasn't been any movement five years after the pronouncement of the plan.

The composition of MSW generated in Harare and its surrounding urban environments shown in Figure 1 is ideal for source separation. Food waste and yard waste constitutes 46% generated at households, restaurants, food outlets and supermarkets followed by recyclables (paper, plastic, metals, rubber, sanitary etc) generated at institutions and households constituting between 30–35% which makes the source separation of the MSW management system relatively uncomplicated and easier to implement. There is need to take advantage of the literacy rate amongst the citizens which eases their better understanding and appreciation on the need for MSW source separation towards the development and implementation of sustainable integrated MSW management system which will address some of the human health, potable water supply and environmental challenges currently prevailing.

The high food and yard waste content is an indication for the potential of biodegradable waste anaerobic digestion (AD) and or composting. Composting and AD have largely been identified to be viable and sustainable alternative biodegradable waste management and treatment methods in the face of increasing global concern to biodegradable waste landfilling (Mendes et al., 2003). Both composting and AD of biodegradable waste lead to reduced GHG emissions and amount (weight and volume) of MSW landfilled thereby addressing the concerns associated with biodegradable waste landfilling. The compost from biodegradable MSW composting and digestate from the AD of biodegradable MSW are potential sources of biofertilisers which could possibly play a significant role in the Zimbabwean agricultural sector considering the agro based state of the Zimbabwean economy. However, AD is preferable over composting due to its renewable energy (biogas) production capacity. The biogas can produce electricity which could be used in water and wastewater treatment plants, residential areas and if excess sold to the national grid for other users. However the Zimbabwe integrated municipal solid waste management plan (GoZ, 2014) is silent with regards to the anaerobic digestion of biodegradable MSW only mentioning composting under goal 3 of the plan. In addition there are no biodegradable waste specific classifications, collection and treatment systems that have been established in Harare and Zimbabwe at large. The absence of AD options for the over 38% food waste (atleast 160,000 tons per annum) generated in Harare is a lost opportunities for renewable energy and biofertiliser production capable of stimulating local and national economic growth considering the AD associated benefits being derived in developed countries. The Linköping municipality in Sweden's AD system produces an upgraded biogas (reported by Svenselius (2018) in 2017 to have produced 120 GWh of energy equal to energy derived from 13 million litres of petrol) that fuels the entire transport system (International Energy Agency (IEA)) and a biofertiliser used in the agricultural sector. The Biffa's Poplars AD facility in Staffordshire in the United Kingdom treats 120,000 tons per annum of source separated food waste and other biodegradables to produce 6,000 kW of electricity adequate to power 15,000 homes. This calls for the need for biodegradable waste management directives and legislation to be enacted.

Findings from a life cycle impact assessment of various MSW management systems revealed that the best MSW management option with the least impacts on human health, acidification, eutrophication and global warming potential is a combination of anaerobic digestion of biodegradable MSW fraction with energy recovery, recovery of recyclables and incineration of the non-recyclables with energy recovery, treatment of flue gas and bottom ash (Nhubu et al., 2018). Such a MSW management system entails a robust MSW separation system with a source separation system being ideal. Makarichi et al (2019) observed that MSW generated in Harare is ideal for incineration with a reported low heating value (LHV) of 10100 kJ/kg from direct measurements and 9320 kJ/kg from the use of empirical formulae estimation giving a potential energy supply equivalent of 3.2% of energy currently being consumed in residential areas and 6.4% with regard to commercial and service areas. The LHV for MSW generated in Harare observed by Makarichi et al (2019) satisfies the suggestion by Zerbock (2003) that for incineration with energy recovery, the LHV must average a minimum of 7000 kJ/kg. Therefore considering that low LHV in MSW emanates from moisture and food waste constituents leading to incineration difficulties (unsteady unstable and incomplete combustion and increased air pollutant formation) source separation of food waste becomes paramount. For waste generated in Harare the source separation of biodegradable waste is expected to increase the LHV of MSW fraction sent to incineration possibly increasing the incineration efficiency and energy recovery per mass of MSW.

Sustainable integrated MSW management requires reliable and accurate data on per capita generation of recyclable, reusable and biodegradable MSW fractions. Source separation of MSW can be an avenue towards the establishment of a reliable inventory in this regard. The enactment of legislation that makes source separation mandatory need to be pursued with a view of enforcing residents to change their behavior and attitudes towards waste separation. This could possibly bringing about increased willingness to separate considering the national separation level of 3%

reported by EMA (2016). The legislation however needs to be complimented with awareness raising and capacity building on the importance of source separation. An opportunity in this regard is that at a national level there is commitment to educate the citizens as provided for under Goal 7 of the national integrated municipal solid waste management plan (GoZ, 2014)

Most urban local authorities by-laws provide for the provision of residents with council approved MSW receptacles. This is not being practiced as a mere 23% of the households surveyed by EMA (2016) had appropriate and approved metal and plastic bins. Source separation requires the provision of MSW receptacles for each sorted waste types. It is quite evident that the urban local authorities might lack the capacity to provide the bins for source separation considering their current failure to provide just one simple bin to households. This could be an opportunity for business and private sector to chip in with the production of durable bins that are affordable.

4 CONCLUSION

The development of sustainable integrated MSW management systems and its successful implementation requires a robust MSW source separation system to be in place. In Zimbabwe and Harare in particular, the development and implementation of a sustainable MSW management system has become an urgent issue considering the potable water, human health and environmental challenges currently prevailing. This therefore entails the need for MSW source separation. The national solid waste management plan of 2014 also provides for source separation under goal 2 and public education for awareness raising on the importance of source separation under goal 7 providing the necessary commitment and necessary for source separation at national level. The composition of the MSW generated in Harare eases source separation due to the high content of biodegradable waste and recyclables. In addition the LHV of the MSW generated in Harare makes it amenable to incineration with energy recovery hence separation of biodegradable fraction potentially increased the LHV thereby increasing incineration efficiency.

However the absence of legislation and directives specifically for biodegradable waste management is not a welcome development as its source separation will drive the interest on its composting and anaerobic digestion to produce a biofertiliser that could be used in the agricultural sector. Anaerobic digestion of the biodegradable MSW fraction will also produce biogas a form of renewable energy. The non-utilization of food waste constituting 38% of annual waste in Harare is a lost energy and material recovery opportunity. However the national solid waste management plan is silent on anaerobic digestion of MSW despite it being one of the strategies proposed under the national climate change response strategy and the Low Emissions Development strategy (LEDs) within Nationally Determined Contributions (NDCs) of the Paris Agreement. Reliable and accurate data on per capita generation of recyclable, reusable and biodegradable MSW fractions is lacking. Source separation of MSW will aid the development of reliable inventory data. The enactment of legislation for mandatory MSW source separation needs to be pursued. Local authorities are failing to provide households with MSW receptacles evidenced by the reported mere 23% of the households with receptacles surveyed by EMA (2016). In addition, municipal authorities do not have the equipment and technical expertise. It is therefore necessary to undertake initiatives towards tooling the local authorities with adequate resources to effectively plan and develop a sustainable integrated MSW management system in Harare.

ACKNOWLEDGEMENTS

The authors acknowledge the Life Cycle Initiative for 2017 Life Cycle award that provided the Simapro software that was used to carry out the LCIA. The authors are also grateful to the National Geographic Society for the early career grant that enabled data collection for the study. The personnel and management at Harare municipality who assisted with the requisite information and data will always be appreciated without their support this work could not have been accomplished.

The Universities of Johannesburg and South Africa are also acknowledged for funding the studies, and as well as conference registration and attendance.

REFERENCES

Boonrod, K., Towprayoona, S., Bonneta, S. & Tripetchkul, S. 2015. Enhancing organic waste separation at the source behavior: a case study of the application of motivation mechanisms in communities in Thailand. *Resources, Conservation and Recycling,* 95, 77–90.

Dhokhikah, Y., Trihadiningrum, Y. & Sunaryo, S. 2015. Community participation in household solid waste reduction in Surabaya, Indonesia. *Resources, Conservation and Recycling,* 102, 153–162.

EMA 2016. Waste generation and management in Harare, Zimbabwe: Residential areas, commercial areas and schools. Unpublished internal report, Environmental Management Agency. Harare, Zimbabwe.

GOZ 2014. Zimbabwe's integrated solid waste management plan. Harare, Government of Zimbabwe, Environmental Management Agency, Institute of Environmental Studies, University of Zimbabwe.

Hoornweg, D. & Bhada-Tata, P. 2012. *What a Waste: A Global Review of Solid Waste* Management. *Urban development series; knowledge papers,* Washington DC, World Bank.

IEA 100% biogas for urban transport in Linköping. Sweden biogas in buses, cars and trains. Biogas in the society Information from IEA bioenergy task 37 energy from biogas and landfill gas. International Energy Agency.

Love, D., Zingoni, E., Ravengai, S., Owen, R., Moyce, W., Mangeya, P., Meck, M., Musiwa, K., Amos, A., Hoko, Z. & Hranova, R. 2006. Characterization of diffuse pollution of shallow groundwater in the Harare urban area, Zimbabwe. *Groundwater pollution in Africa.* CRC Press.

Magadza, C. H. D. 2003. Lake Chivero: a management case study. *Lakes & Reservoirs: Research & Management,* 8, 69–81.

Makarichi, L., Kan, R., Jutidamrongphan, W. & Techato, K. A. 2019. Suitability of *municipal* solid waste in African cities for thermochemical waste-to-energy conversion: The case of Harare Metropolitan City, Zimbabwe. *Waste Management & Research,* 37, 83–94.

Mendes, M. R., Aramakib, T. & Hanakic, K. 2003. Assessment of the environmental impact of management measures for the biodegradable fraction of municipal solid waste in São Paulo City. *Waste Management,* 23, 403–409.

Nhapi, I. 2009. The water situation in Harare, Zimbabwe: a policy and management problem *Water Policy,* 11, 221–235.

Nhubu, T., Muzenda, E. & Mbohwa, C. 2018. Options for municipal solid waste management in Harare, Zimbabwe.

Rousta, K., Bolton, K., Lundin, M. & Dahlén, L. 2015. Quantitative assessment of distance to collection point and improved sorting information on source separation of household waste. *Waste Management,* 40, 22–30.

Sukholthaman, P., Chanvarasuth, P. & Sharp, A. 2017. Analysis of waste generation variables and people's attitudes towards waste management system: a case of Bangkok, Thailand. *Journal of Material Cycles and Waste Management,* 19, 645–656.

Svenselius, M. W. 2018. Linköping home to Sweden's largest biogas facility, https://liu.se/en/news-item/storsta-biogasanlaggningen-finns-i-linkoping. Linköping University.

Tai, J., Zhang, W., Che, Y. & Feng, D. 2011. Municipal solid waste source-separated collection in China: A comparative analysis. *Waste Management,* 31, 1673–1682.

Tirivanhu, D. & Feresu, S. 2013. A Situational Analysis of Solid Waste Management in Zimbabwe's Urban Centres. Harare, Institute of Environment Studies, University of Zimbabwe.

UN-HABITAT 2010. *Solid waste management in the world's cities,* Nairobi, Kenya, United Nations Human Settlements Programme (UN-HABITAT).

Zerbock, O. 2003. Urban solid waste management: Waste reduction in developing nations. Written for the Requirements of CE, 59935993 Field Engineering in the Developing World). Michigan Technological University, Houghton, MI.

ZIMSTAT 2013. 2012 Zimbabwe Census National Report, Zimbabwe National Statistics Agency. Harare.

Wastes: Solutions, Treatments and Opportunities III – Vilarinho et al. (Eds)
© 2020 Taylor & Francis Group, London, ISBN 978-0-367-25777-4

Carbonisation as a pre-treatment for RDF wastes prior to gasification

O. Alves, R. Panizio & M. Gonçalves
VALORIZA – Research Center for Endogenous Resource Valorisation, Polytechnic Institute of Portalegre, Portalegre, Portugal
MEtRICs -Mechanical Engineering and Resource Sustainability Center, Department of Science and Technology of Biomass, Faculty of Science and Technology, Universidade NOVA de Lisboa, Lisboa, Portugal

J. Passos, L. Calado, E. Monteiro & P. Brito
VALORIZA – Research Center for Endogenous Resource Valorisation, Polytechnic Institute of Portalegre, Portalegre, Portugal

ABSTRACT: The present work is intended to evaluate carbonisation as a pre-treatment for heterogeneous wastes containing polymeric fractions, before their gasification for energy production. Carbonisation had an upgrading effect in the raw wastes by yielding chars with higher apparent density and an improved behaviour during feeding operations. The performance of the gasification process was compared for the raw wastes and the corresponding biochars. The results obtained indicate that carbonisation helped to improve this performance by increasing the calorific value of syngas, and the process efficiency. Gasification of the chars produced less tars and gaseous HCl which is a positive effect concerning maintenance operations and overall costs. Therefore, the carbonisation pre-treatment can be seen as a valid option to be integrated in a gasification process to convert wastes into energy especially for low density wastes with high chlorine contents.

1 INTRODUCTION

Presently, the world is facing new challenges concerning the environmental protection and sustainability. New solutions for energy production must be defined in order to reduce the dependence on fossil fuels, due to the well-known global warming effect and the possible exhaustion of the available resources. On the other hand, huge amounts of wastes are produced every year, and the conventional treatment options like landfilling and incineration are being discouraged by legislation(e.g. EU directive 2008/98/EC) due to the associated environmental and social problems (Syed-Hassan et al. 2017).

Many typologies of wastes are composed by organic fractions that confer them interesting calorific properties when considering emerging waste-to-energy technologies such as gasification. In this process, solid wastes are converted into a syngas, rich in CO and H_2, that can be burned to produce heat. Gasification occurs at temperatures around 700–1000°C in the presence of limited concentrations of oxygen, lowerthan those required for stoichiometric oxidation. The syngas produced is more environment-friendly since it is produced from renewable sources and it contains less pollutants such as dioxins, NO_x and SO_x (You et al. 2016).

However, to use this technology it is necessary that the wastes to be processed present material and fuel properties compatible with an efficient gasification process. High moisture contents and the presence of polymers may lead to possible obstructions in the equipment, formation of gaseous HCl (that promotes corrosion), lower efficiencies during the conversion process, and high production of tars that may damage the unit and rise maintenance costs (Recari et al. 2016, 2017). Therefore, adoption of convenient pre-treatments for such wastes may be required before their conversion by gasification.

Carbonisation may be an eligible pre-treatment to fulfill these purposes that has been proposed by some authors as a strategy to reduce moisture and chlorine content and improve homogeneity and density of complex wastes (Bialowiec et al. 2017, Nobre et al. 2019). Carbonisation of these materials may be performed at mild temperatures (300–600°C) in the absence of oxygen, and usually improves their energy density, hydrophobicity and grindability (Lohri et al. 2016).Some works already demonstrated that the implementation of carbonisation before gasification generated a syngas with a better calorific value, lower HCl contents, and lower production of tars, which contributes for abetter performance of the overall system (Recari et al. 2017, Bach et al., in press).

The present work evaluated the effect of carbonisation of construction and demolition wastes containing lignocellulosic and polymeric fractions prior to their gasification. In particular, the properties of the produced syngas and overall efficiency were determined and compared with those obtained by direct gasification of raw wastes. Possible pathways for the valorisation of char by-products generated during gasification is also discussed in order to improve the overall sustainability of this process.

2 METHODOLOGY

An RDF sample were prepared by milling and mixing 80 wt% of lignocellulosic wastes (wood and paper/card) with 20 wt% of polymers (mainly plastics), all provided by a Portuguese waste management company and reproducing the proportions of those wastes processed monthly in this industry. A fraction of this sample was carbonised in an oven, at 400°C, 60 min, conditions that were selected according to prior trial experiments to ensure the production of homogeneously carbonised materials after some.

The raw RDF sample and the corresponding char (RDF1 and CRDF1, respectively) were gasified in a downdraft pilot unit (All Power Labs, PP20)according to the flowsheet schematically presented in Figure 1.

Pre-heated air entered the reactor and its flow was manually adjusted in the control board, in order to achieve a stable gasification temperature of around 800°C, in all tests. After stabilisation of the gasifier conditions, two syngas samples were collected in Tedlar bags for posterior analysis. HCl contained in the syngas was captured by scrubbing with a solution of NaOH 1M, using a trap system similar to that described inGai& Dong 2012. Chars and tars were collected after the experiments for further quantification and analysis.

Both RDF samples and the corresponding chars were characterised for apparent density (following EN 15103), higher heating value (HHV) in a bomb calorimeter (IKA C200), concentration of chlorine in a X-ray fluorescence analyser (Thermo Scientific Niton XL 3T Gold++), ultimate analysis (ThermoFisher Scientific Flash 2000 CHNS-O Analyser) and proximate analysis based on standards ASTM E949-88 (moisture), E897-88 (volatile matter) and E830-87 (ash); fixed carbon was calculated by difference. Before chlorine determination, the carbonised RDF(CRDF1) was suspended in water to remove chlorine adsorbed to the char surface. Composition of syngas was analysed in a gas chromatograph (Varian 450-GC) and the lower heating value (LHV) was calculated considering the volumetric proportions of identified compounds and the corresponding calorific values. Concentration of HCl in syngas was determined by measuring chloride concentration in the NaOH scrubbing solution according to standard USEPA 9212.The gasification char sub-products were also characterised for high heating value, chlorine content, ash content and ash composition using the methods described for the feed materials.

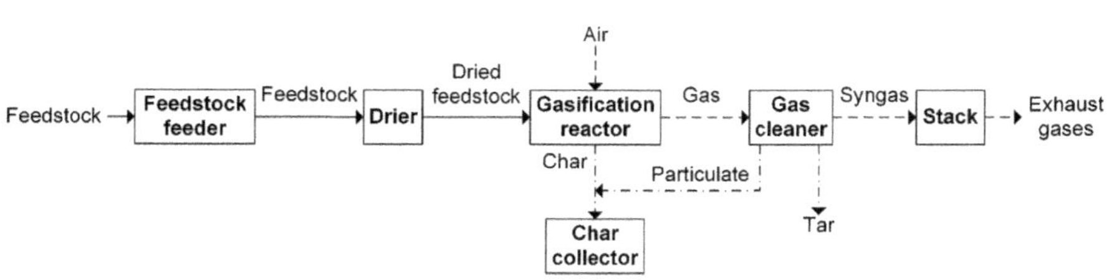

Figure 1. Schematic representation of the gasification process and product collection.

Gasification performance during the experiments was assessed by determination of equivalence ratio (ER), syngas yield (Y_{syngas}, m^3/kg) and cold-gas efficiency (CGE, %). ER was calculated by the ratio between the air flow entering the reactor and the air flow needed for the stoichiometric combustion of the feed. Syngas yield and CGE were calculated using equations 1 and 2, respectively (Allesina et al. 2018, Basu 2013):

$$Y_{syngas} = \frac{V_{air} \times 0.781}{m_{RDF} \times x_{N2}} \quad (1)$$

$$CGE = \frac{LHV_{syngas} \times m_{syngas}}{LHV_{RDF} \times m_{RDF}} \times 100 \quad (2)$$

with V_{air} representing the volumetric flow of air (m^3/s), m_{RDF} and m_{syngas} the mass flow es of RDF and syngas (kg/s), x_{N2} the volumetric fraction of N_2 present in the syngas, and LHV_{syngas} and LHV_{RDF} the lower heating values of syngas and RDF (MJ/m^3). For the determination of Y_{syngas} in equation 1, the concentration of N in RDF was not accounted because it was residual.

3 RESULTS AND DISCUSSION

From a visual analysis, the carbonised sample CRDF1 exhibited a more homogeneous aspect and by manipulation seemed more friable than the original RDF. These qualitative modifications induced by the carbonisation process were complemented by the analytical characterisation that indicated that the carbonised RDF (CRDF1) had higher carbon content, apparent density, fixed carbon content, chlorine and ash content than the raw RDF while moisture, volatile matter and high heating value were lower (Table 1). Moisture content reduction and improved hydrophobicity may contribute for a longer conservation during storage and a better decomposition during the gasification stage.

On the other hand, the raw RDF1 sample caused some obstructions in piping and vacuum pumps from the unit due to the high formation of tars during the tests, which increased cleaning and maintenance operations. The decrease of volatile matter and increase of ash contents observed for the CRDF1 sample justify its lower HHV when compared with RDF1. This negative effect may be moderated if lower carbonisation temperatures are to be used, preserving the carbon contents and reducing the concentration effect in the ash fraction (Haykiri-Acma et al. 2017, Buah et al. 2007).Reduction of temperature may be compensated with an increase of residence time and mixing of the materials during carbonisation, in order to ensure the formation of homogeneous chars. The large increase of the fixed carbon of the CRDF1 relatively to RDF1 is an evidence of the extensive re-organization of the materials structure that occurs during the carbonization pre-treatment (Haykiri-Acma et al. 2017). Both samples (RDF1 and CRDF1) had no sulphur and relatively low concentrations of nitrogen enabling to predict lower concentrations of gaseous pollutants (NO_x, SO_x, H_2S, etc.) after the gasification process. However, a slight rise of nitrogen was observed in CRDF1 which is in conformity with other works (Haykiri-Acma et al. 2017, Buah et al. 2007). Concentration of chlorine in CRDF1increased significantly as a result of the carbonisation process reaching a value (1.1 wt% db) but the washing step could efficiently remove chlorine absorbed at the char surface of grains, reaching a value of 0.6 wt.% db.

Table 1. Analytical properties of raw and carbonised RDF samples (RDF1 and CRDF1).

	Proximate analysis*				Ultimate analysis (wt% daf)							
Sample	M	VM	A	FC	N	C	H	S	O	HHV (MJ/kg db)	Chlorine (wt% db)	Density (kg/m^3)
RDF1	6.0	85.5	5.0	9.5	2.2	56.6	9.5	0.0	31.8	24.724	0.3	92.7
CRDF1	3.1	49.8	16.5	33.7	3.5	63.0	7.1	0.0	26.5	21.799	0.6	292.1

(*M – moisture in wt% ar; VM – volatile matter in wt% db; A – ash in wt% db; FC – fixed carbonin wt% db)

Carbonisation promoted the increase of the apparent density of CRDF1 to values around three times higher than those of raw RDF1. This result is relevant for transportation and storage operations since a higher density of feedstock is associated with lower costs for both operations.

The operating parameters and performance indexes of the gasification process and the characteristics of the obtained syngas samples are presented in tables 2 and 3, respectively.

Carbonisation improvedCGE (+15%) a fundamental index of gasification performance. The reduction of tars production (−32%), was also a positive effect of the carbonisation pre-treatment. Nevertheless, the CRDF1syngas yield diminished (−29%) and char production increased Significantly when compared to RDF1, indicating the higher resistance to thermal decomposition and oxidation of the carbonised RDF. This behaviour is expectable since chars usually have a higher degree of aromaticity than lignocellulosic materials, therefore require more aggressive conditions of temperature or oxidative potential to be decomposed (Alvarez et al. 2019). Increasing temperature or increasing the oxygen concentration in the gasifier are strategies that may improve the char decomposition and carbon recovery in the gas products.

Despite these disadvantages, performance of gasification was generally better when using carbonised RDF taking into account the increase of CGE, which was near the results found in the literature; in addition, values of ER's varied from 0.27 to 0.38approaching values previously reported for similar systems(Basu 2013, Khosasaeng & Suntivarakorn 2017, Recari et al. 2017). Carbonised RDF needed oxygen to be converted since ER was higher for this feedstock, as a consequence of a higher carbon content that affects stoichiometric ratio and the higher amount of fixed carbon that slows the reactivity of the feed (Basu 2013).

Regarding the composition of the syngas, contents of C_2H_4, CH_4 and H_2 increased slightly forCRDF1causing an improvement of the calorific properties, which rose from 3.1 to 5.3 MJ/m^3 (+71%);these results are in agreement with those reported by Recari and co-workers, for the gasification of SRF(Recari et al. 2017).BesidesN_2, that is the predominant syngas component when air is used as oxidant agent, the main compounds detected in the gas products were CO, CO_2 and H_2, with concentrations greater than 9 vol%. Contaminant H_2S was negligible in the syngas samples, which helps to reduce its toxicity. Also, gaseous HCl concentration in the syngas was halved whenCRDF1 was used as feed, a positive effect regarding corrosion of the equipment as stated by Recari et al. 2017.In short, adoption of carbonisation as a pre-treatment for RDF before gasification is advantageous since it allowed to obtain a syngas with better fuel properties and possibly may help to reduce maintenance costs when compared with the direct gasification of raw RDF. Since the production of chars as gasification sub-products is increased when the feed is a carbonised RDF, the properties of those gasification char products were investigated in order to discuss their sustainable management.

Tables 4 and 5 present some parameters for the characterisation of the gasification chars and the heavy metals content of the corresponding ash fraction, respectively.

Table 2. Operating and performance parameters of gasification tests.

Sample	Temperature (°C)	Equivalence ratio	Feedstock flow (kg/h)	Syngas yield (m^3/kg)	Tar yield (mL/kg)	Char yield (g/kg)	CGE (%)
RDF1	800	0.38	3.1	3.507	24.7	27.6	50.3
CRDF1	800	0.27	5.4	2.483	16.8	113.2	65.4

Table 3. Analysis of syngas samples from gasification tests.

Syngas sample	Composition (vol%)							LHV (MJ/m^3)	HCl (ppmv)
	CO_2	C_2H_4	H_2S	N_2	CH_4	CO	H_2		
RDF1	11.6	0.0	0.0	64.3	0.4	14.2	9.0	3.1	1034
CRDF1	11.4	2.0	0.0	59.1	2.2	13.1	10.0	5.3	446

Table 4. Characterisation of chars obtained from gasification tests.

Char	HHV (MJ/kg db)	Chlorine (wt% db)	Ash (wt% db)	Ash composition (wt% db)		
				CaO	Fe_2O_3	SiO_2
RDF1	4.618	2.3	81.7	26.8	9.6	4.7
CRDF1	5.943	2.4	78.0	27.2	8.4	4.4

Table 5. Concentration of heavy metals in the ash fraction of chars.

	Heavy metals in ash (ppmw db)						
Char	Cd	Cr	Cu	Hg	Ni	Pb	Zn
RDF1	0	880	1908	0	661	325	1598
CRDF1	0	1322	2244	0	769	519	1377

The gasification char products presented low HHV (<6 MJ/kg db) and high ash levels (>75 wt% db),regardless of the feed material used (RDF1 or CRDF1). These properties indicate that these chars are not suitable for energy conversion and material applications should be found to promote their valorisation.

In addition, chlorine content was similar for both chars, indicating that this element was mainly retained in the tars, since the syngas produced from CRDF1had lower amounts of HCl then the syngas obtained from RDF1 in spite of the slightly higher concentrations of chlorine in the carbonised RDF.

Implementation of a carbonisation step did not influence significantly the composition of ashes present in the gasification chars, that showed comparable concentrations of main oxides CaO, Fe_2O_3 and SiO_2. for both feed materials. Chars are especially rich in Ca and Fe, so they can be valorised as catalysts for gasification or pyrolysis instead of being directed to landfills (Yao et al. 2016, Shen 2015).

Analysis of heavy metal concentrations revealed that levels of Cr, Cu, Ni, Pb and Zn are too high (>100 ppmw db) and above the limits imposed by Portuguese legislation regarding the application of fertilisers for agriculture purposes (decree-law 103/2015, of June 15th). Therefore, ashes from both gasification chars are not appropriate for soil fertilisation, unless further decontamination treatments are performed.

4 CONCLUSIONS

The present work evaluated the influence of carbonisation as a pre-treatment for RDF samples containing polymeric fractions, before their gasification to obtain a syngas for energy production.

Generally, the results demonstrated that carbonisation allowed to improve gasification performance manifested by the increase of the calorific value of the syngas and of the cold-gas efficiency. Furthermore, is also contributed to decrease emissions of gaseous hydrochloric acid and tars, that may cause various problems of equipment maintenance and process sustainability.

Apparent density of the raw RDF increased after carbonisation, allowing to reduce transportation and storage costs of feedstocks. Chars from gasification may be used as catalysts for gasification or pyrolysis, a strategy that prevents their deposition in landfills.

Therefore, the inclusion of a pre-treatment of carbonisation of RDF's prior to gasification to enhance performance and syngas properties is an option to consider in future plants, however it may be pertinent in future works to analyse energy requirements for all the process in order to assess the economic viability and to establish the optimum conditions for operation.

ACKNOWLEDGEMENTS

Authors acknowledge financial support from Fundação para a Ciência e Tecnologia – Ministério da Ciência, Tecnologia e Ensino Superior (grant no. SFRH/BD/111956/2015), co-financed by Programa Operacional Potencial Humano and União Europeia-Fundo Social Europeu, from project POCI-01-0145-FEDER-024020 (RDFGAS – Aproveitamento energético dos combustíveis derivados de resíduos e lamas secas), co-financed by COMPETE 2020 - Programa Operacional Competitividade e Internacionalização, Portugal 2020 and União Europeia through FEDER, and also from project 0330_IDERCEXA_4_E - RENEWABLE INVESTMENT, DEVELOPMENT AND ENERGY FOR THE IMPROVEMENT OF THE ENTREPRENEURIAL FABRIC IN THE REGION CENTRO, ESTREMADURA AND ALENTEJO, co-financed by INTERREG – European Regional Development Fund through the FEDER.

REFERENCES

Allesina, G., Pedrazzi, S., Allegretti, F., Morselli, N., Puglia, M., Santunione, G. & Tartarini, P. 2018. Gasification of cotton crop residues for combined power and biochar production in Mozambique. *Applied Thermal Engineering* 139: 387–394.

Alvarez, J., Lopez, G., Amutio, M., Bilbao, J., Olazar,M. 2019. Evolution of biomass char features and their role in the reactivity during steam gasification in a conical spouted bed reactor. *Energy Conversion and Management* 181: 214–222.

Bach, Q.V., Gye, H.R., Song, D. & Lee, C.J. (in press). High quality product gas from biomass steam gasification combined with torrefaction and carbon dioxide capture processes. *International Journal of Hydrogen Energy*.

Basu, P. 2013. *Biomass gasification, pyrolysis and torrefaction – Practical design and theory*. Academic Press.

Bialowiec, A., Pulka, J., Stepien, P., Manczarski, P. & Golaszewski, J. 2017. The RDF/SRF torrefaction: An effect of temperature on characterization of the product – Carbonized Refuse Derived Fuel. *Waste Management* 71: 91–100.

Buah, W.K., Cunliffe, A.M. & Williams, P.T. 2007. Characterization of products from the pyrolysis of municipal solid waste. *Process Safety and Environmental Protection* 85(B5): 450–457.

Gai, C. & Dong, Y. 2012. Experimental study on non-woody biomass gasification in a downdraft gasifier. *International Journal of Hydrogen Energy* 37: 4935–4944.

Haykiri-Acma, H., Kurt, G. & Yaman, S. 2017. Properties of Biochars Obtained from RDF by Carbonization: Influences of Devolatilization Severity. *Waste and Biomass Valorization*8: 539–547.

Khosasaeng, T. & Suntivarakorn, R. 2017. Effect of Equivalence Ratio on an Efficiency of Single Throat Downdraft Gasifier Using RDF from Municipal solid waste. *Energy Procedia* 138: 784–788.

Lohri, C.R., Rajabu, H.M., Sweeney, D.J. & Zurbrügg, C. 2016. Char fuel production in developing countries – A review of urban biowaste carbonization. *Renewable and Sustainable Energy Reviews* 59: 1514–1530.

Nobre, C., Alves, O., Longo, A., Vilarinho, C., Gonçalves, M. 2019. Torrefaction and carbonization of refuse derived fuel: Char characterization and evaluation of gaseous and liquid emissions. *Bioresource Technology* 285:121325.

Recari, J., Berrueco, C., Abelló, S., Montané, D. & Farriol, X. 2016. Gasification of two solid recovered fuels (SRFs) in a lab-scale fluidized bed reactor: Influence of experimental conditions on process performance and release of HCl, H2S, HCN and NH3. *Fuel Processing Technology* 142: 107–114.

Recari, J., Berrueco, C., Puy, N., Alier, S., Bartolí, J. & Farriol, X. 2017. Torrefaction of a solid recovered fuel (SRF) to improve the fuel properties for gasification processes. *Applied Energy* 203: 177–188.

Shen, Y. 2015. Chars as carbonaceous adsorbents/catalysts for tar elimination during biomass pyrolysis or gasification. *Renewable and Sustainable Energy Reviews* 43: 281–295.

Syed-Hassan, S.S., Wang, Y., Hu, S., Su, S. & Xiang, J. 2017. Thermochemical processing of sewage sludge to energy and fuel: Fundamentals, challenges and considerations. *Renewable and Sustainable Energy Reviews* 80: 888–913.

Yao, D., Hu. Q., Wang, D., Yang, H., Wu, C., Wang, X. & Chen, H. 2016. Hydrogen production from biomass gasification using biochar as a catalyst/support. *Bioresource Technology* 216: 159–164.

You, S., Wang, W., Dai, Y., Tong, Y.W.& Wang, C.H. 2016. Comparison of the co-gasification of sewage sludge and food wastes and cost-benefit analysis of gasification- and incineration-based waste treatment schemes. *Bioresource Technology* 218: 595–605.

Wastes: Solutions, Treatments and Opportunities III – Vilarinho et al. (Eds)
 ISBN 978-0-367-25777-4

Refuse derived fuel char as a low-cost adsorbent for the cationic dye methylene blue

C. Nobre & M. Gonçalves
MEtRICs, Department of Science and Technology of Biomass, Faculty of Sciences and Technology, NOVA University of Lisbon, Caparica, Portugal

C. Vilarinho
MEtRICs, Department of Mechanical Engineering, School of Engineering, University of Minho, Guimarães, Portugal

ABSTRACT: This work addresses the application of Refuse Derived Fuel (RDF) char as a low-cost adsorbent for methylene blue (MB) in batch mode, as an approach to remediate pigment contaminated effluents and a model for the adsorption of cationic analytes. The RDF char was tested without pre-treatment and after a cold alkali treatment to enhance adsorption. The adsorbents were characterized for proximate and ultimate composition, pH_{pzc} and N_2 adsorption/desorption behavior. Operation parameters such as pH, adsorbent dose, contact time, dye concentration and temperature were varied. Kinetic studies revealed that adsorption of methylene blue in the RDF chars followed a pseudo-second-order kinetic model. Adsorption isotherms showed differences in the possible adsorption mechanisms as a result of char activation. This work highlighted that RDF chars that are not suitable for energy conversion may be valorized as a low-cost adsorbents.

1 INTRODUCTION

More than 280,000 ton of untreated synthetic dyes are discharged into water bodies globally, seriously increasing water pollution (Bharti *et al.*, 2019). Methylene blue (MB), is one of the most used dyes in the industry, being applied for coloring paper, cotton, wood or silk, and it is often used as a model in adsorption studies (Manna *et al.*, 2017).

Adsorption is a very efficient technique for wastewater treatment, that operates at ambient temperature and pressure, but generates relatively high amounts of solid residues. Activated carbon is the most widely employed adsorbent, in spite of its high cost and low regenerability. Low-cost materials can also be applied as adsorbents. They should be easily and locally available, preferably from waste sources, with minimum processing, granting economic and environmental sustainability to the adsorption process (Rafatullah *et al.*, 2010).

Refuse derived fuel (RDF) is usually produced from non-hazardous solid wastes through a series of mechanical or mechanical and biological processes (Çepeliogullar *et al.*, 2016). RDF is meant to be used in energy production through incineration in energy intensive industries such as cement production (Brás *et al.*, 2017). Not all of the produced RDF is used for its intended purpose, mostly because it does not present enough quality. As such, this fuel generally benefits from an upgrading treatment such as torrefaction in order to homogenize its physical and chemical characteristics (Nobre et al., 2019). Torrefaction has been thoroughly studied for different feedstocks, mostly biomass (Chen *et al.*, 2018), but also for wastes like RDF or MSW (Edo *et al.*, 2017). This process is known to increase feedstock energy density, enhance its hydrophobic nature and improve its commercial use for energy production due to the reduction in associated transportation costs (Schipfer *et al.*, 2017). Nevertheless, torrefaction can increase ash content, because solid product yield decreases, leading to a concentration of its mineral component. High ash contents can hinder energetic applications due to operational problems like slagging and fouling (Niu *et al.* 2016).

Chars with high ash contents could be regarded as potential low-cost adsorbents, as described by Correia et al. (Correia *et al.*, 2017).

In this work, the adsorption capacity of a char produced from RDF through torrefaction at 300°C for 30 minutes was studied. The RDF char was applied as an adsorbent with and without alkali activation treatment. The adsorption tests included batch equilibrium experiments using MB dye and varying adsorbent dose, MB solution pH, contact time, MB solution initial concentrations and temperature in order to define equilibrium and kinetic parameters that may elucidate the adsorption mechanism.

2 MATERIALS AND METHODS

2.1 *Raw material and torrefaction process*

Industrial RDF was supplied by CITRI, S.A., a waste management company from Setúbal, Portugal. RDF lab-scale torrefaction was carried out in covered porcelain crucibles under oxygen-limited conditions using a muffle furnace (Nabertherm®L3/1106), for a fixed temperature of 300°C and a fixed residence time of 30 minutes. After torrefaction, the RDF char was further milled (DeLonghi mill) and sieved to a particle diameter <500 μm (Retsch sieve). Alkali treatment of the RDF char was done according to the method described by Regmi *et al.* (2012), with concentrated KOH The non-treated sample was designated RDFchar and the treated sample was designed RDFchar-KOH.

2.2 *RDF char characterization*

Moisture, volatile matter and ash contents were determined gravimetrically according to the procedures described in ASTM 949-88, 897-88 and 830-87, respectively. Fixed carbon (FC) was determined by difference, on a dry basis (db). Ultimate analysis (CHNS) was performed using an elemental analyzer (Thermo Finnigan – CE Instruments Model Flash EA 112 CHNS series). Oxygen content was obtained by difference, on a dry ash free basis (daf). Point of zero charge pH (pH_{pzc}) was determined through pH drift tests as described by (Faria *et al.*, 2004). Porosity assessment was done by nitrogen gas porosimetry (Micromeritics ASAP 2010).

2.3 *Batch adsorption tests*

A given amount of char was added 5 mL of MB dye aqueous solution having a specific concentration and pH value. Samples were subjected to agitation with a constant speed of 15 rpm (Heidolph REAX top shaker). After the established contact time, samples were centrifuged (Hettich Zentrifugen EBA 20) at 3000 rpm for 5 minutes, and analyzed using a UV spectrophotometer (Pharmacia LKB-Novaspec II), determining MB concentrations through a calibration curve prepared from several dilutions of a MB stock solution (1 g/L).

The effect of operating parameters on the extent of dye removal, namely, pH of MB dye solution (2–12), adsorbent dosage (1–6 g/L), time (3 min–48 h), MB dye concentration (100–800 mg/L) and temperature (293.15–313.15 K) have been investigated. Removal efficiency (R) was calculated according to Equation 1.

$$R\ (\%) = [(C_0 x C_f)/C_0] x100 \quad (1)$$

where C_0 and C_f (mg/L) are the initial and final MB concentrations, respectively.

Kinetic studies were conducted applying the pseudo-first-order (PFO) and pseudo-second-order (PSO) models, in their linearized forms, as described by Lagergren (1898) and Ho and Mckay (1999), respectively. Adsorption isotherms, namely Langmuir, Freundlich and Temkin were applied to the experimental data as described by Dada *et al.*, 2012.

Table 1. Physical-chemical characterization of RDF and RDF chars.

Parameter	Unit	RDF	RDFchar	RDFchar-KOH
Moisture	wt.%	6.97 ± 1.00	1.69 ± 0.03	1.23 ± 0.05
Volatile matter		76.09 ± 2.44	61.09 ± 0.60	63.66 ± 0.09
Ash	wt.%, db	10.65 ± 0.44	21.94 ± 0.13	23.51 ± 0.23
Fixed carbon		13.26 ± 2.28	16.97 ± 0.53	12.82 ± 0.26
C		38.20 ± 1.54	59.72 ± 0.76	66.80 ± 0.22
H		5.20 ± 0.34	5.91 ± 0.05	6.45 ± 0.09
N	wt.%, daf	2.34 ± 0.03	1.05 ± 0.03	1.44 ± 0.05
S		0.40 ± 0.04	0.22 ± 0.01	0.0 ± 0.00
O		53.86 ± 1.21	32.41 ± 0.76	25.31 ± 0.27
O/C		1.06	0.41	0.28
H/C		1.63	1.16	1.19
pHpzc		—	6.5	6.8
Surface area$_{BET}$	m^2/g	—	4.049	4.265

3 RESULTS AND DISCUSSION

3.1 *RDF char characterization*

Results for the characterization of the original RDF, the RDFchar and the RDFchar-KOH are depicted in Table 1.

Moisture content of RDFchar decreased by 4.1 times, whereas RDFchar-KOH decreased by 5.7 times, relatively to that of the original RDF. The chars presented lower volatile matter content and higher ash content than the original RDF. The KOH activation treatment promoted an increase in volatile matter and ash contents of the RDFchar-KOH relatively to the non-treated RDF char, an indication that the activation process involves dissolution of a fraction of the organic components of the char and possibly contributed to further decomposition of the char (Regmi *et al.*, 2012). Both chars presented ash values above 20 wt.%. Although this is a negative feature as a solid fuel, high ash may be advantageous for adsorption applications since it may contribute to greater efficiencies in pollutant removal (Buah and Williams, 2010).

Carbon and hydrogen contents increased markedly, whereas nitrogen, sulphur and oxygen decreased with the torrefaction treatment. These changes were more significant for the sample RDFchar-KOH, to due the basic treatment, the exception was nitrogen content, which increased after the treatment. Both chars presented low O/C and H/C ratios when compared to the original RDF, and these reductions are mainly caused by water release and partial elimination of oxygen by decarboxylation, decarbonylation and dehydration reactions that occur during the torrefaction process (Bergman *et al.*, 2005).

At $pH > pH_{pzc}$ char surface becomes negatively charged favoring the adsorption of cationic species (Faria *et al.*, 2004). As such it is expected that MB adsorption will be favored for both chars at higher pH (above 6). The BET surface area of sample RDFchar was 4.049 m^2/g, whereas sample RDFchar-KOH presented a value of 4.265 m^2/g. These are very low surface areas that corroborate the absence of a porous structure, and suggest the possibility of a superficial adsorption process mediated for example by electrostatic interactions.

3.2 *Batch adsorption tests*

The effect of different operational parameters in the efficiency of MB removal is illustrated in Figure 1.

Regarding the effect of solution pH on MB removal (Figure 1a), a high adsorption capacity of MB by the RDFchar-KOH was observed from pH 6 to pH 12, with removal efficiencies between 95.3–99.0%. In the case of the RDFchar, a significant increase in removal efficiency (80.9–97.5%) was

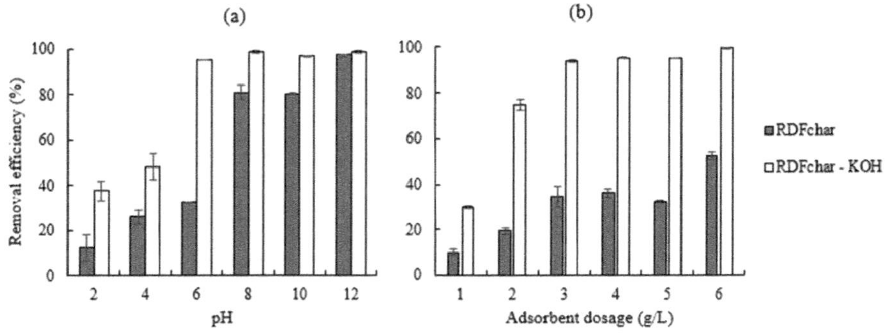

Figure 1. Effect of different parameters on MB removal efficiency (%) (a) pH (MB = 100 mg/L, Adsorbent dose = 5 g/L, t = 60 min); (b) Adsorbent dosage (MB = 100 mg/L, t = 60 min, no pH adjustment).

Table 2. Kinetic parameters for the adsorption of MB onto RDFchar and RDFchar-KOH (MB = 100 mg/L, adsorbent dose = 5 g/L, no pH adjustment, contact time between 3 min and 48 h).

	Sample	
Kinetic model	RDFchar	RDFchar-KOH
$q_{e(experimental)}$ (mg/g)	15.08	17.32
Pseudo-first order		
k^1 (min^{-1})	−0.0009	−0.0012
q_e (mg/g)	2.6988	1.2687
R^2	0.8309	0.6362
Pseudo-second order		
k^2(g/mg.min)	0.0006	0.0186
q_e (mg/g)	15.1745	17.2117
h_0 (mg/g.min)	0.1307	5.5157
R^2	0.9777	1.0000

also observed for solution pH in the range of 8–12. These results may be related to the deprotonation of functional groups from the char surface, when exposed to high pH aqueous solutions, therefore creating negatively charged active sites that favor adsorption of cationic analytes such as MB.

Increasing the adsorbent dose increased MB removal efficiency for both chars (Figure 1b). RDFchar-KOH showed increased removals between 1–3 g/L but remained fairly constant for higher adsorbent doses. Maximum removals for both chars were observed at 6 g/L (99.9% for RDFchar-KOH and 52.5% RDFchar). This effect can be ascribed to the increased surface area and availability of more adsorption sites (Kallel *et al.*, 2016).

The results for the kinetic studies are depicted in Table 2. For the pseudo-first order model (PFO), the calculated q_e values did not agree with experimental q_e values, and the determination coefficient (R^2) was considered low (0.8309 for RDFchar and 0.6362 RDFchar-KOH). For the pseudo-second order model (PSO) the R^2 values were above 0.90 for both tested chars. Furthermore, experimental q_e values were significantly closer to the theoretical q_e values determined with the PSO model, meaning that the obtained data is well-fitted by this kinetic model. The applicability of the PSO model indicates that the main adsorption mechanism involved in the adsorption of MB onto both chars is chemical sorption or chemisorption involving valence forces through sharing or exchange of electrons between adsorbent and adsorbate (Ho and MacKay, 1999).

The nature of the interaction between the MB molecules and the adsorbents may be investigated through adsorption isotherms. The obtained data were fitted into three isotherm models, namely, Langmuir, Freundlich and Temkin, and the results are shown in Table 3.

Table 3. Model parameters estimated for the Langmuir, Freundlich and Temkin adsorption isotherms of MB onto RDFchar and RDFchar-KOH ([MB] = 100–800 mg/L, T = 293.15 K, 303.15 K and 313.15 K).

	Sample					
	RDFchar			RDFchar.KOH		
Isotherm parameters	293.15 K	303.15 K	313.15 K	293.15 K	303.15 K	313.15 K
Langmuir						
K_L (L/mg)	0.004	0.010	0.016	0.006	0.007	0.006
q_{max} (mg/g)	100.000	84.746	76.923	208.333	256.410	204.082
R^2	0.8998	0.9817	0.9813	0.8512	0.5806	0.8841
Freundlich						
K_f	1.101	0.487	0.265	1.562	1.710	0.574
1/n	0.721	0.572	0.481	1.653	0.890	0.998
R^2	0.9467	0.9332	0.9041	0.4301	0.8504	0.7098
Temkin						
K_T (L/mg)	0.047	0.106	0.206	0.276	0.321	0.348
B_T (mg/g)	122.278	137.246	171.004	58.055	54.464	62.948
R^2	0.9602	0.9511	0.9067	0.9473	0.8970	0.9542

The Langmuir isotherm resulted in poor fits for RDFchar-KOH over the whole range of MB concentrations and adsorption temperatures. RDFchar presented a better correlation with the Langmuir isotherm, with R^2 values from 0.89 to 0.98 at the different temperatures tested.

Freundlich isotherms presented R^2 values >0.90 for RDFchar adsorption at the different temperatures but gave a bad fit for RDFchar-KOH with very low R^2 values demonstrating the weak applicability of this model to this adsorbent. The calculated 1/n was found to be in the range of 0–1 for both chars, which suggests that the adsorption process had a chemical nature (Zhang *et al.*, 2012).

The R^2 values obtained with the Temkin isotherm were higher than 0.9 for all temperatures tested and more homogeneous for both chars than those obtained with the previous isotherms. The fact that both chars present significant B_T values is also so an indication of the chemical nature of the adsorption process (Kallel *et al.*, 2016; Fan *et al.*, 2017). Furthermore, the Temkin isotherm model mainly describes chemical adsorption process as resulting from electrostatic interactions. This type of interaction may be particularly important for the adsorption of MB on the RDF char, since this char presented the highest B_T values.

4 CONCLUSIONS

In this work, the application of RDFchar as a low-cost adsorbent was studied, using the non-treated char and an RDF char subject to alkaline treatment. For both adsorbents the torrefaction process increased carbon content but it also increased ash content, decreasing the suitability of these chars for energy applications and justifying the study of their use as low-cost adsorbents.

MB adsorption onto RDFchar and RDFchar-KOH was investigated in batch mode and kinetic and equilibrium data was modelled and discussed. The kinetic process was better described by the pseudo-second-order kinetic model for both chars with R^2 values above 0.97. Adsorption isotherm was fairly described by the Langmuir isotherm at 303.15 K and 313.15 K for RDF char, while RDFchar-KOH was better described by the Temkin isotherm with R^2 values above 0.90 for all temperatures. Nevertheless, none of the tested isotherms presented excellent correlations for both chars ($R^2 > 0.99$), denoting differences between the adsorption process in both adsorbents, but indicating that chemical interactions play a decisive role in these adsorption systems. The obtained results indicate that RDF char is an adequate adsorbent for MB dye, and treating the char with a concentrated alkali solution improved adsorption.

REFERENCES

Bergman, P., Boersma, R., Zwart, R. and Kiel, J. 2005. Torrefaction for biomass co-firing in existing coal-fired power stations – 'Biocoal'.

Bharti, V., Vikrant, K., Goswami, M., Tiwari, H., Kumar, R., Lee, J., Tsang, D. C. W., Kim, K. and Saeed, M. 2019. Biodegradation of methylene blue dye in a batch and continuous mode using biochar as packing media. *Environmental Research* 171:356–364.

Brás, I., Silva, M., Lobo, G., Cordeiro, A., Faria, M. and Lemos, L. 2017. Refuse Derived Fuel from Municipal Solid Waste rejected fractions – a Case Study. *Energy Procedia* 120:49–356.

Buah, W. and Williams, P. 2010. Activated carbons prepared from refuse derived fuel and their gold adsorption characteristics characteristics. *Environmental Technology* 31(2):125–137.

Çepeliogullar, Ö., Haykırı-Açma, H. and Yaman Istanbul, S. 2016. Kinetic modelling of RDF pyrolysis: Model-fitting and model-free approaches. *Waste Management* 48:275–284.

Chen, D., Gao, A., Ma, Z., Fei, D., Chang, Y. and Shen, C. 2018. In-depth study of rice husk torrefaction: Characterization of solid, liquid and gaseous products, oxygen migration and energy yield. *Bioresource Technology* 253:148–153.

Correia, R., Gonçalves, M., Nobre, C. and Mendes, B. 2017. Impact of torrefaction and low-temperature carbonization on the properties of biomass wastes from *Arundo donax* L. and *Phoenix canariensis*. *Bioresource Technology* 223:210–218.

Dada, A., Olalekan, A., Olatunya, A. and Dada, O. 2012. Langmuir, Freundlich, Temkin and Dubinin–Radushkevich isotherms studies of equilibrium sorption of Zn^{2+} unto phosphoric acid modified rice husk. *IOSR Journal of Applied Chemistry* 3(1):38–45.

Edo, M., Skoglund, N., Gao, Q., Persson, P. and Jansson, S. 2017. Fate of metals and emissions of organic pollutants from torrefaction of waste wood, MSW, and RDF. *Waste Management* 68:646–652.

Fan, S., Wang, Y., Wang, Z., Tang, J., Tang, J. and Li, X. 2017. Removal of methylene blue from aqueous solution by sewage sludge-derived biochar: Adsorption kinetics, equilibrium, thermodynamics and mechanism. *Journal of Environmental Chemical Engineering*. 5(1):601–611.

Faria, P., Órfão, J. and Pereira, M. 2004. Adsorption of anionic and cationic dyes on activated carbons with different surface chemistries. *Water Research* 38:2043–2052.

Gouamid, M., Ouahrani, M. and Bensaci, M. 2013. Adsorption equilibrium, kinetics and thermodynamics of methylene blue from aqueous solutions using Date palm Leaves. *Energy Procedia* 36:898–907.

Ho, Y. and McKay, G. 1999. Pseudo-second order model for sorption processes. *Process Biochemistry*. 34:451–465.

Kallel, F., Chaari, F., Bouaziz, F., Bettaieb, F., Ghorbel, R. and Chaabouni, S. E. 2016. Sorption and desorption characteristics for the removal of a toxic dye, methylene blue from aqueous solution by a low-cost agricultural by-product. *Journal of Molecular Liquids* 219:279–288.

Lagergren, S. 1898. About the theory of so-called adsorption of soluble substances. *Kungliga Svenska Vetenskapsakademiens Handlingar* 24:1–39.

Manna, S., Roy, D., Saha, P., Gopakumar, D. and Thomas, S. 2017. Rapid methylene blue adsorption using modified lignocellulosic materials. *Process Safety and Environmental Protection*. 107:346–356.

Niu, Y., Tan, H. and Hui, S. 2016. Ash-related issues during biomass combustion: Alkali-induced slagging, silicate melt-induced slagging (ash fusion), agglomeration, corrosion, ash utilization, and related countermeasures. *Progress in Energy and Combustion Science*. 52:1–61.

Nobre, C., Alves, O., Longo, A., Vilarinho, C., Gonçalves, M. 2019. Torrefaction and carbonization of refuse derived fuel: Char characterization and evaluation of gaseous and liquid emissions. *Bioresource Technology* 285:121325.

Rafatullah, M., Sulaiman, O., Hashim, R. and Ahmad, A. 2010. Adsorption of methylene blue on low-cost adsorbents: A review. *Journal of Hazardous Materials*. 177(1–3):70–80.

Regmi, P., Garcia Moscoso, J. L., Kumar, S., Cao, X., Mao, J. and Schafran, G. 2012. Removal of copper and cadmium from aqueous solution using switchgrass biochar produced via hydrothermal carbonization process *Journal of Environmental Management* 109:61–69.

Schipfer, F., Vakkilainen, E., Proskurina, S. and Heinim, J. 2017. Biomass for industrial applications: The role of torrefaction. *Renewable Energy*, 111:265–274.

Zhang, Y., Liang, H. and Lu, R. 2012. Adsorption of chromium (VI) from aqueous solution by the iron (III)-impregnated sorbent prepared from sugarcane bagasse. *International Journal of Environmental Science and Technology* 9:463–472.

Wastes: Solutions, Treatments and Opportunities III – Vilarinho et al. (Eds)
© 2020 Taylor & Francis Group, London, ISBN 978-0-367-25777-4

A critical analysis on the gasification of lignocellulosic and polymeric wastes

R.M. Panizio, O. Alves & M. Gonçalves
VALORIZA – Research Center for Endogenous Resource Valorisation, Polytechnic Institute of Portalegre, Portalegre, Portugal
MEtRICs–Mechanical Engineering and Resource Sustainability Center, Department of Science and Technology of Biomass, Faculty of Science and Technology, Universidade NOVA de Lisboa, Lisboa, Portugal

L. Calado & P. Brito
VALORIZA – Research Center for Endogenous Resource Valorisation, Polytechnic Institute of Portalegre, Portalegre, Portugal

ABSTRACT: The present study compares the energetic potential of the gasification gasproduced from the gasification of lignocellulosic wastes (*Eucalyptus*) and its co-gasification withpolymeric wastes (Refuse-Derivate-Fuel pellets) at an incorporation rate of 30 wt%. The gasification tests were performed using a downdraft reactor, at 800°C,and the gasification gas was collected at equilibrium conditions. The gasification by-products (biochars and tars) were also collected at the end of each test. The results demonstrated that these two types of fuels, may be successfullyconverted producing a syngas with a heating value exceeding 5 MJ/Nm^3. The results also demonstrate the advantage of adding polymeric waste to the lignocellulosic biomass, since feeding operations are facilitated by mixing those fuels and the calorific value of the gasification gasincreased by 6% to a final value of 5.4 MJ/Nm^3.

1 INTRODUCTION

The increase in population, industrial development, services caused a steady grow of heat and electricity consumption. Electricity is important for increasing economic growth, but it is necessary to find cleaner and greener sources of electricity, in order to reduce the effects of climate change and boost sustainable development. There is a growing debate about the generation of electricity based on fossil fuels and their possible harmful effects on the environment. These effects may be detrimental not only to the environment but also negatively impact economic growth as a whole for societies (International Energy Agency, 2016). Thinking about the diversification of electricity sources, the European Union has proposed to member countries goals of reduction on the use of fossil fuels and of increase of renewable energy generation and improvement of energetic efficiency. Directive 2009/28 / EC of the European Parliament and of the Council, whose objectives for 2020 are: (i) 20% reduction of greenhouse gas emissions in the EU; (ii) 20% of EU energy from renewable energy sources; and (iii) a 20% improvement in EU energy efficiency. In fact, EU countries have designed and implemented public policies to develop and increase the deployment of wind, solar photovoltaic (PV), bioenergy and hydroelectric power in their electricity production systems (Marques, Fuinhas and Pereira, 2018).The Directive (EU) 2018/2001 of the European Parliament and of the Council of 11 December 2018 on the promotion of the use of energy from renewable sources (EU, 2018), reinforces these goals and establishes as priority policies the valorisation of wastes and the use of advanced conversion technologies, such as pyrolysis and gasification, for renewable energy production.

Thermal gasification is a very promising thermochemical conversion process that transforms solid biomass into combustible gases, i.e. a mixture of hydrogen, methane, carbon monoxide, carbon dioxide and light hydrocarbons. This mixture is called gasification gas andit can be used directly as a gaseous fuel to produce electricity and generate heat or upgraded for the synthesis of liquid fuels. In addition, gasification converts low-value and heterogeneous raw materials into high value-added gas products. Since it allows the production of energy from unconventional and diversified sources such as forest residues, agricultural residues, poultry residues and urban solid waste, this technology plays a decisive role in coupling waste valorisation and renewable energy production (Mutlu and Yucel, 2018).

The thermochemical processes pyrolysis and gasification are the most used, for the conversion and energy recovery from different wastes, includingbiomass residues, because they are versatile and efficient processes (Fernandez *et al.*, 2019). Residues from forest maintenance and timber processing are often sent to landfill because they are not adequate for the production of solid biofuels. Low density, ash composition and formation of harmful substances (PAHs, chlorine compounds, etc.) are characteristics that may hinder their direct valorisation in combustion boilers. The processing of Refuse Derived Fuels (RDF) is also a problem of great interest because they represent a large fraction of municipal solid waste and industrial wastes. However, its thermal behaviour often leads to the destabilization of the conversion processes (Donskoi, 2018). As an alternative to minimize these phenomena the co-processing of these fuels may present some advantages.

This work intended to compare the gasification of eucalyptus wastes and its co-gasification with RDF incorporated at a 30% concentration in the feed. The composition and calorific value of the gasification gas as well as the production of condensates and ashes was evaluated. The tests were performed at temperatures about 800°C in a downdraft fixed bed gasifier.The co-gasification of RDF and eucalyptus wastes, may enable their energetic valorisation with multiple positive effects from the environmental and economic perspectives.

2 METHODOLOGY

The *Eucalyptus* biomass chips were provided by a local forestry operator and underwent mechanical and manual sorting in order to select particles with dimensions between 1 cm and 4 cm, suitable for this type of reactor. The RDF was supplied by a waste management company and was also screened to separate thefraction with diameter of 1–4 cm.

2.1 *Characterisation of raw materials*

Proximate and ultimate analyses were carried out to characterize the waste samples;their calorific value and ash composition were also evaluated.

Proximate composition (moisture, volatile matter, fixed carbon and ash contents) was evaluated using thermogravimetric data. The thermogravimetric profile was determined under an oxidative atmosphere, using a PerkinElmer STA 6000 thermogravimetric analyzer. Elemental composition was determined using a Thermal Analyser ThermoFisher Scientific Flash 2000 CHNS-O Analyser. The composition of the inorganic fraction was determined by X-ray fluorescence emission using Niton™ XL3t XRF Analyzer. The high heating value (HHV) was determined by calorimetry (Calorimeter IKA C200), and the low heating value (LHV) was calculated taking into account the hydrogen and oxygen contents.

2.2 *Gasification and co-gasification tests*

For the gasification and co-gasification tests, the PP20 Power Pallets – 20 kW gasifier from AllPowerLabs was used.This equipment is a combination of a downdraft fixed bed reactor, an electric power generator and an electronic control unit supplemented with a fuel storage compartment, a particle separation cyclone and a biomass filter. The fuel is supplied from the top while the air moves downwards, being preheated through contact with the walls of the reactor. Ash collection is carried out in a separate tank in the lower zone of the reactor, while the synthesis gas produced passes through a cyclone to remove fine particles. The gasification gas is then conducted through the fuel storage compartment,passed through a filter composed by biomasses of various granulometries, and from here it can be collected for analysis or directly injected into the generator. The condensates are collected in the biomass filter. Biochars are collected in the cyclone and in the reactor bottom, in the end of each test, after cooling the system.

Gasification and co-gasification tests were performedin duplicate, at a temperature of approximately 800°C with a duration of 5 hours. In the case of the co-gasification a mixture of 70% of *Eucalyptus* and 30% of RDF was used. During the tests, the values of temperature and pressure in the upper and lower parts of the reactor (i.e. oxidation and reduction zones, respectively), pressure in the biomass particle filter, inlet air flow and finally the amount of fuel consumed during the test were monitored. The gasification gas samples were withdrawn from the biomass particle filter into suitable bags with the help of a vacuum pump when the gasification process was stabilized.

2.3 *Composition of the gasification gas*

Analysis of the gasification gas samples was performed by gas chromatography with TCD detection. The samples collected in Tedlar bags were injected in a Varian 450-GC gaseous chromatograph with two TCD detectors and the gas components were separated at two specific columns using helium and nitrogen as entrainment gases. Identification and quantification of the gas components (CO, CO_2, H_2, CH_4 and light hydrocarbons) was performed by calibration with appropriate standards.

2.4 *Composition and properties of the char by-products*

The biochar samplescollected in the end of each gasification teste were characterised for mass yield, composition of the ashes and thermogravimetric profile. The composition of the inorganic fraction was determined by X-ray fluorescence emission using XRF. Equipment and analytical methods were equivalent to the ones described above, for the analysis of raw materials.

3 RESULTS AND DISCUSSION

Characterization of the fuels.

The average values for proximate and ultimate composition, high heating value and chlorine content of the fuels used in the tests are presented in Table 1. The corresponding variation coefficients were lower than 10%.

A first examination of the results demonstrates that both fuels have similar elemental composition and low chlorine content. The HHV obtained for RDF(24.0 MJ/kg) was comparable to other studies (Vounatsos *et al.*, 2013), andsubstantially higher that ofeucalyptus biomass (17.8 MJ/kg). The heating value of the eucalyptus biomasswasanalogousto that of other forestry biomasses (Uzun *et al.*, 2017).

Proximate composition of both fuels presented significant differences.Moisture content varied between 6.8% for eucalyptus and 3.1% for RDF, values that are suitable for gasification applications.Nevertheless, this parameter influences LHV, and may affect the equilibrium temperature of the gasification process, because part of the heat released in exothermic reactions is used to evaporate the feed moisture. The relatively high volatile matter content of the RDF sample (82.7%) makes this material more readily devolatilized than fuels with lowervalues of this parameter.Raw materials with high volatile matter contents are generally associated with lower productionof biochar sub-product, what makes the RDF a suitable fuel for pyrolysis and gasification. As a general rule, fuelswith more than 10% VM are appropriate to feed downdraft gasifiers since with increasing volatile matter content, less heat is required for the thermochemical devolatilization reactions. As a

Table 1. Analytical properties of used fuels.

Parameters	Fuels	
	Eucalyptus	RDF
C	43.7	42.8
H	5.6	5.9
N	0.5	0
S	0	0
O	42.7 h	46.8
HHV (MJ/kg)	17.8	24
Moisture (%)	6.8	3.1
Volatiles (%)	41.7	82.7
Fixed Carbon (%)	44	4.5
Ashes (%)	7.5	12.8
Cl (%)	0.3	0.7

result, it was possible to gasify the RDF sample at relatively lower temperatures. Fixed carbon content for the RDF (4.5%) is much lower than that of eucalyptus biomass (44%), a characteristic that influences the energy balances in the gasifier, since the raw materials with high fixed carbon contentsalso have higher energy densities resulting in high energy throughputs of the gasifier. The ash contents of the two fuels were very different, 7.5% for eucalyptus and 12.8% for RDF. Biomass fuels with ash contents below 5–6% do not exhibit slagging tendencies. Biomasses with more than 6–7% ash are prone to induce slagging in gasification systems if the temperature is not kept below 1000°C. Slagging can lead to excessive tar formation and/or complete blockage of the reactor thus affecting smooth operation of the gasifier.

From the analysis of table 1, it can be also observed that the chlorine content of eucalyptus is lower than the content of RDF. The co-gasification of these two fuels may be beneficial because it would lower formation of HCl in the gasification gaswhen compared to the gasification of 100% RDF, thus reducing the tendency for equipment corrosion. The gasification performances of eucalyptus biomass and its mixture with 30% RDF are compared in Table 2.

The results showed that,for both fuels,the main components of the gasification gas were nitrogen, followed by carbon monoxide and hydrogen. Methane and other hydrocarbon gases presented higher concentrations in the gasification gas obtained with the mixture of eucalyptus biomass and RDF, a clear advantage of the co-gasification option. The high percentage of nitrogen is due to the fact that experiments were carried out with atmospheric air.

The elemental compositions of the RDF and the eucalyptus residues were similar a characteristic that contributed to the comparable values of LHV of the gasification gases obtained with those fuels.As mentioned before, the higher volatile matter content of the RDF may facilitate carbon recovery in the gas phase, what justifies the higher concentrations of hydrocarbon gases obtained for the mixture of RDF and eucalyptus chips, a tendency that contributes to the improvementofthe LHV of the gasification gas.

However, incorporation of RDF also increased ash and condensate production when compared with the gasification of 100% eucalyptus chips.

Gasification of eucalyptus chips yielded a carbon-rich gasification gaswith a calorific value of 5.1 MJ/Nm^3, a value that can be upgraded if air nitrogen is separated before or after gasification. An interesting aspect of the obtained results is that addition of 30% RDF to the eucalyptus chips, led to a significant increase in the calorific value of the gasification gas (near 6%), as a consequence of the significant capitalization in the production of C1-C2, as reported by other authors (Block

Table 2. Gasification performance and composition of gas products.

Operation Parameters	100% Eucalyptus	100% Eucalyptus + 30% RDF
Oxidation Zone Temperature (°C)	803	790
Reduction Zone Temperature (°C)	505	521
Oxidation Zone Pressure (mbar)	−13.6	−9
Reduction Zone Pressure (mbar)	−36.7	−25
Filter Pression (mbar)	−50.35	−37
Air Volumetric Flow (m^3/h)	11,1	10.3
Biomass Flow Rate (kg/h)	6	4.1
Gasification gas Volumetric Flow (m^3/h)	15.7	14.3
Biochar Produced (kg/h)	0.096	0.134
Tars Produced (L/h)	0.05	0.135
Gasification gas Composition (%)		
H_2 (%)	12.94	12.47
CO (%)	20.2	17.85
CH_4 (%)	1.61	3.13
C_2H_2 (%)	0	0.12
C_2H_4 (%)	0.37	1.16
CO_2 (%)	10.76	10.85
N_2 (%)	55.09	56.3
LHV Gasification gas (MJ/Nm^3)	5.1	5.4

et al., 2018). This fact may also be related with the higher devolatilization degreeobserved for RDF at low temperatures a condition that favors formation of hydrocarbon products. The feed materials that require higher temperatures and inlet air flows tend to undergo more extensive oxidation reactions namely C_nH_m oxidizing with O_2 (reforming reactions) therefore reducing hydrocarbon concentrations in the gas products.

Similar observations were made by other authors for the gasification of mixtures of lignocellulosic materials and polymeric ones, namely, the increase of the polymeric fraction was found to increase the C_nH_m concentrations in the gas products, as well as the yields of tar and biochar (Ephraim, 2016; Čepeliotullar and Pütün, 2014). Different authors also reported that the nature of the biomass feed has an important influence on the decomposition of co-gasified polymers (Brebu *et al.*, 2010; Ephraim, 2016; Widayatno *et al.*, 2016; Lathouwers and Bellan, 2001).

4 CONCLUSIONS

Gasification of lignocellulosic residues (Eucalyptus) and their co-gasification with refuse derived fuels, at a 30% incorporation levelwere evaluated.

Co-gasification of polymeric wastes and forest biomass residues has several advantages: it can help to overcome difficulties with seasonal availability of biomass, allowsenergy recovery from these residues, improves the position of polymer treatment within the waste treatment hierarchy and it can also potentially solve technical problems related to difficult feeding of polymer materials to gasifiers.

In both gasification tests, the gas product concentrations were similar with H_2 between 12.5 and 13%, CO varying between 17.5 and 20%, CH_4 between 1.5 and 3.5%, CO_2 around 11% and N_2 between 55 and 56.5%.The calorific value of the gas product varied from 5 to 5.5 MJ/Nm^3.

Incorporation of the RDF in the feed caused a very significant increase in the yields of condensates and biochars, however, there was also a 6% increase in the LHV of the gasification gas, that could be related to the increase of C_nH_m concentrations.

Furthermore, the synergies observed during co-gasification of biomass and polymeric waste need to be further studied in order to better understand the mechanisms involved in this process.

ACKNOWLEDGEMENTS

The authors are grateful for the financial support given to the project 0330_IDERCEXA_4_E – Renewable Investment, Development and Energy for the Improvement of the Entrepreneurial Fabric in the Region Centro, Estremadura and Alentejo, co-financed by ERDF – European Regional Development Fund through the INTERRREG and also from Fundação para a Ciência e Tecnologia – Ministério da Ciência, Tecnologia e Ensino Superior (grant no. SFRH/BD/111956/2015), co-financed by Programa Operacional Potencial Humano and União Europeia-Fundo Social Europeu.

REFERENCES

Block, C. *et al.* 2018. 'Co-pyrogasification of Plastics and Biomass, a Review', *Waste and Biomass Valorization*, 10, pp. 483–509.

Available at: https://link.springer.com/article/10.1007/s12649-018-0219-8.

Brebu, M. *et al.* 2010. 'Co-pyrolysis of pine cone with synthetic polymers', *Fuel*. Elsevier Ltd, 89(8), pp. 1911–1918. doi: 10.1016/j.fuel.2010.01.029.

Čepeliotullar, Ö. and Pütün, A. E. 2014. 'Products characterization study of a slow pyrolysis of biomass-plastic mixtures in a fixed-bed reactor', *Journal of Analytical and Applied Pyrolysis*, 110(1), pp. 363–374. doi: 10.1016/j.jaap.2014.10.002.

Directive (EU) 2018/2001 of the European Parliament and of the Council of 11 December 2018 on the promotion of the use of energy from renewable sources. (2018) Official Journal of the European Union, L328/82, 21.12.2018.

Donskoi, I. G. 2018. 'Process Simulation of the Co-Gasification of Wood and Polymeric Materials in a Fixed Bed', *Solid Fuel Chemistry*, 52(2), pp. 121–127. doi: 10.3103/s0361521918020027.

Ephraim, A. 2016. 'Valorization of wood and plastic waste by pyro-gasification and gasification gas cleaning', *Http://Www.Theses.Fr*. doi: 10.1111/bph.12440.

Fernandez, A. *et al.* 2019. 'Macro-TGA steam-assisted gasification of lignocellulosic wastes', *Journal of Environmental Management*, 233, pp. 626–635. doi: https://doi.org/10.1016/j.jenvman.2018.12.087.

International Energy Agency, I. 2016. 'Energy and Air Pollution', *World Energy Outlook – Special Report*, p. 266. doi: 10.1021/ac00256a010.

Lathouwers, D. and Bellan, J. 2001. 'Yield Optimization and Scaling of Fluidized Beds for Tar Production from Biomass', *Energy Fuels*, 15, pp. 1247–1262. doi: 10.1021/ef010053h.

Marques, A. C., Fuinhas, J. A. and Pereira, D. A. 2018. 'Have fossil fuels been substituted by renewables? An empirical assessment for 10 European countries', *Energy Policy*, 116, pp. 257–265. doi: https://doi.org/10.1016/j.enpol.2018.02.021.

Mutlu, A. Y. and Yucel, O. 2018. 'An artificial intelligence based approach to predicting gasification gas composition for downdraft biomass gasification', *Energy*, 165, pp. 895–901. doi: https://doi.org/10.1016/j.energy.2018.09.131.

Uzun, H. *et al.* 2017. 'Improved prediction of higher heating value of biomass using an artificial neural network model based on proximate analysis', *Bioresource Technology*. Elsevier Ltd, 234, pp. 122–130. doi: 10.1016/j.biortech.2017.03.015.

Vounatsos, P. *et al.* 2013. 'Report on RDF / SRF gasification properties', *Life*, p. 40 p. Available at: http://www.energywaste.gr/pdf/D4.1 – Report on RDF-SRF gasification properties .pdf.

Widayatno, W. B. *et al.*2016. 'Fast co-pyrolysis of low density polyethylene and biomass residue for oil production', *Energy Conversion and Management*. Elsevier Ltd, 120, pp. 422–429. doi: 10.1016/j.enconman.2016.05.008.

Wastes: Solutions, Treatments and Opportunities III – Vilarinho et al. (Eds)
 ISBN 978-0-367-25777-4

Giving cigarettes a second life: The E-Tijolo project

M. Soares, N. Valério, A. Ribeiro, A. Ferreira, P. Ribeiro, R. Campos & J. Araújo
CVR – Centre for Wastes Valorisation, University of Minho, Guimarães, Portugal

A. Mota, J. Carvalho & C. Vilarinho
CVR – Centre for Wastes Valorisation, University of Minho, Guimarães, Portugal
Departamento de Engenharia Mecânica, Universidade do Minho, Guimarães, Portugal

M. Iten, R. Dias, J. Henriques & D. Pinheiro
ISQ – Low Carbon and Resource Efficiency, R&Di, Instituto de Soldadura e Qualidade, Grijó, Portugal

ABSTRACT: Portugal presented 1,46 million smokers in 2014, resultingin about 2220 tonnes of cigarette butts per year. Such waste generates several harmful consequences to environment and public health. In this alignment, the current study purposes the incorporation of this waste in ceramic material to minimize its impacts. Cigarret buts were incorporated in clay fired bricks (0%, 2,5% and 5% (w/w)) and tested regarding theircompreensive strength, water absorption and density The results achieved with the waste incoporation indicatedas expected that the main properties are reduced by the incorporation of cigarrette butts. Moreover, thereduction indry density implies a decrease ofthe claycontent shoiwing also that this waste has an interisting potential in the production of clay fired bricks.

1 INTRODUCTION

Nowadays, the problematic of cigarette butts is increasingly evident.

Cigarette butts waste is a huge pollutant, the problem begins with tobacco manufacturing and transporting. Researchers estimate the annual global environmental costs of tobacco manufacturing results in 2 million metric tonnes of solid waste, 300 000 metric tonnes of nicotine-contaminated waste and 200 000 metric tonnes of chemical waste (Lee, Botero, & Novotny, 2016).The Carnegie Mellon University's Green Design Institute performed an Economic Input-Output lifecycle assessment (EIOLCA) concluding that USA's tobacco industry by its own is responsible for releasing 16 million metric tonnes of CO_2 (Proctor, 2012). Deforestation is the first negative consequence of the tobacco industry, being responsible for 4% of the worldwide deforestation (Lee et al., 2016).

In a worldwide perspective, in 2012, 6.25 trillion cigarettes have been consumed, releasing through the smoke denotative amounts of toxicants and pollutants directly into the environment.In a single year, the global tobacco smoke contributed with thousands of metric tonnes of carcinogens, other toxicants, and greenhouse gases. The toxic emissions include 3000–6000 metric tonnes of formaldehyde; 12000–47000 metric tonnes of nicotine; and the three major greenhouse gases found in tobacco smoke – carbon dioxide, methane, and nitrous oxides (DeBardeleben, 1981; Aeslina Abdul Kadir & Mohajerani, 2010).

The post-consumer waste cigarette butts also impact the environment.About two thirds of used cigarettes are being thrown away on the ground, resulting, each year, between 360 and 680 million kilograms of waste tobacco productsin the world.

Although the waste volume is frightful, the main harmful problem is the 7000 toxic chemicals contained in burnt cigarettes. Those toxic chemicalsaccumulate in the environment, ending up on the streets and contaminating fresh and marine water. Cigarette butts discharged contain dangerous

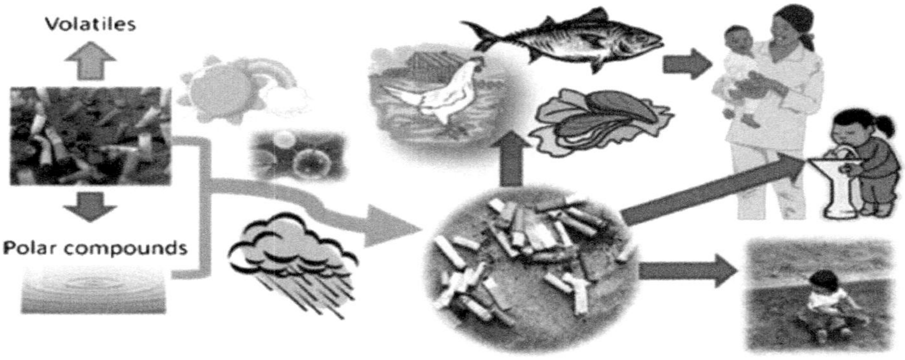

Figure 1. Possible pathways for human health risks due to the tobacco product wastes (Novtny& Slaughter 2014).

chemicals including nicotine, arsenic, heavy metals (such as lead, copper, chromium, and cadmium) and PAHs, which can induce public health problems, such as cancer (Novotny & Slaughter, 2014; Wright, Rowe, Reid, Thomas, & Galloway, 2015). The possible pathways for human health risks due to the tobacco product wastes are presented in Figure 1.

In terms of life cycle, nowadays,cigarette buts are the final waste of tobacco products. The cigarette butts are considered the major single type of litter by count (Novotny & Slaughter, 2014). They constitute about 30–40% of all items caught in urban and coastal clean-ups. Considering that in 20 cigarettes are present 3,4 g of filters, the estimated discarded from global cigarettes consumption was about 340–680 million kg in 2014, excluding the weight of tobacco scraps and other byproducts (Novotny & Slaughter, 2014; Witkowski, 2014).

The discarded cigarette butt consists of unsmoked remnant tobacco, the paper wrap remnants, and the filter. The filters may aggravate the potential environmental effect of chemicals leached from butts, being essentially a nonbiodegradable plastic compilation of cellulose acetate fibers. The plastic particles and their toxicants may never disappear from water or soil and may continue leaching chemicals for up to 10 years, even if exposed to UV rays.Individually, each one presents a private environmental concern (Novotny & Slaughter, 2014).

In Portugal, 1,46 million smokers have been accounted for in 2014 (INE, 2014). Figure 2 shows the average of cigarettes smoked in Portugal, daily. From the inquired smokers, 45% reported smoking up to 10 cigarretes per day, 46% smoke between 11 and 20 cigarretes, and 9% reported smoking a pack or more. Considering the best possible scenario (45% of Portuguese smokers consume 1 cigarette, 46% smoke 11 cigarettes and 9% smoke 21 cigarettes), it corresponds to aproximately 10,8 millions of cigarettes *per day*. Moreover, as the calculated weight of 10 cigarette butts is 3.5 g, it translates to3780 kg of cigarette butts being daily produced, meaning that in one year, approximately 1380 tonnes of cigarettes buttswould be produced in Portugal.

On the other hand, considering the worst case scenario (45% Portuguese smokers consume 10 cigarettes, 46% smoke 20 cigarettes and 9% smoke 30 cigarettes), thatwould correspond to approximately 23.9 millions of cigarettes consumed *per day*, 8 380 kg of cigarettes butts being wasted daily andtherefore, producing annually approximately 3059 tonnes of cigarettes butts in Portugal.

Accordingly to these premises, a middle ground can be considered. In this way, an average of 2220 tonnes of cigarette butts are produced, anually, in Portugal.

Nowadays, the main destination of cigarette butts correponds to the landfill, and its disposal together with muncipal solid wastes (World Health Organization, 2008, 2017). Considering the cigatte buts potential, this disposal is not environmentally and economically sustainable. The highest disadvantage is the fact that this waste remain as an environmental hazard, containing

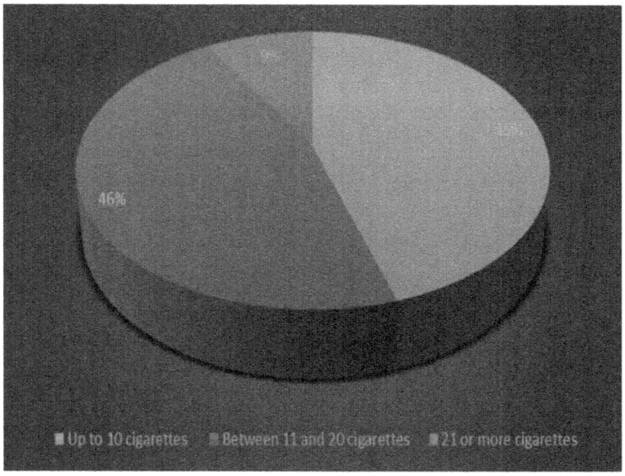

Figure 2. Quantification of cigarettes smoked daily, in Portugal.

toxic substances and organic content which are becoming increasingly expensive and problematic (Aeslina Abdul Kadir & Mohajerani, 2010).

In 2010, a study of Aeslina Abdul Kedir and Abbas Mohajerani from University of Melbourne suggested anatractive alternative incorporating cigarette butts into building material composites, such as a brick.In this studyit was observed that: (1) the density has been reducted up to 30%; (2) a reduction of compressive strength up to 90%; and (3) the absense of metals in leachate (Aeslina Abdul Kadir & Mohajerani, 2010). The main reason appointed for such a descresa in compressive strength is the cigarret buts sized that leads to clay fired bricks with many holes, and consequently with less resistence. Therefore and because it can be contaminated, it is importante to considered its pre treatment in order to stabilized and homogeneized the waste for incorporation in ceramic materials.

The main goal of this study is to perform an inicial evaluation of the influenced of incorporatin pre-treated cigarrette buts in clay fired bircks and how it performs regarding water absortion, density and compressive strength.

2 MATERIALS AND METHODS

The study presents an experimental testing of cigarette butts incorporation in bricks, namely water absortion, density and compressive strength.The production of bricks with this incorporation needs a pretreatment for the raw materials preparation. The process is described in below. The second part is the process itself, being the mix of waste with the clay and the brick production. In the end, several complementary tests were performed to analyze the bricks.The raw materials used to produce clay fired bricks were, clay provided by local producer, water and cigarette butts collected in local shops and by *Laboratorio da Paisagem* from Guimarães municipal city hall.

2.1 *Pretreatment*

The pretreatment of the cigarette buts is fundamental for the correct mixture of raw materials, in order to have an homogeneous and non-contaminated raw material.Thus, the first step is the sterilization of cigarette butts in the autoclave. Posteriorly, the cigarette butts are dried in an oven at 100°C during 2 h and crushed in a mill with a 2 mm sieve.

Regarding the clay, it must be dried in an oven during 24 h at 100°C and passed in a roller mill.

Table 1. Raw material used in bricks production.

Sample	Clay mass (g)	Cigarette butts mass (g)	Cigarette butts number
1	2500	0	0
2	2438	63	175
3	2375	125	350

2.2 *Incorporation in bricks*

For the incorporation in bricks, firstly clay and the waste should be mixed with water for a complete homogenization, during 15 min. Then the mold, whose dimensions are 22 × 11 × 7 cm (L × W × H), is filled with the mixture, which is subsequently pre-dried at 40°C overnight. A final drying at 110°C during 2 h is required. The cooking of the fresh brick is performed at 1000°C in a muffle. In order to perform the testing, three different percentages of cigarette butt incorporation have been produced. Namely a prototype with i) 0% of waste incorporation into the brick (1) ii) 2.5% of waste incorporation into the brick (2) and iii) 5% of waste incorporation into the brick (3). Table 1 details the brick production procedure with the approximate number of cigarette butts.

2.3 *Complementary assays*

The experimental testing of the three prototypes included the: the mechanical resistance, the dry density andthe water absorption. All tests have been carried out following the specific norms forclay masonry units and the results have been expressed using the mean of two values.

2.3.1 *Mechanical strength*

For the mechanical strength, the compressive strength has been used as mechanical resistance indicator, according to EN 771-1. This test was performed using the ImpactTest Equipment Building 21 Stevenston Ind. Est. Stevenston Ayrshire KA20 3LR C104 2005 version. The testing equipment is software-controlled with the testing specification and configuration data acquisition system. The specimen tested have been placed in the testing area, then increasing force in KiloNewton (kN) has been applied. The testinghas been ended when the software has detected a loss of strength due to the breakage of the specimen.In order to improve accuracy ofthe results, the specimens, evaluated in duplicate,have been carefully aligned and placed at the testing area to ensure uniform loading over the total area of the brick.

The compressive strength of each specimen has been calculated by the followed equation:

$$C = Ka \times (1000P/A)$$

Where, C = Compressive strength (MPa); P = Total failure load (N); A = Area (mm^2) and Ka = Determined

2.3.2 *Dry density*

The dry density of the standard clay brick and the clay-cigarettes butts bricks have been measured according to EN 772-13. The samples have been manufactured with the standard dimensions of 210 × 110 × 70 mm, evidencinga mean value near to 0.5 mm. In order to measure the dry density, the specimens have been dried at 110 ± 8°C with intervals at least of 4 h until constant weighthas been observed. Consecutive weight has been taken as weight of the sample after cooling to room temperature. These tests were performed in duplicate.

2.3.3 *Water absorption*

The water absorption test of the bricks has been done according to EN 772-11. In order to perform this test, the specimens have been prepared and dried in an oven at 105 ± 5°C until constant mass. Then, they the samples have been rested to cool to room temperature before their weighting.

Table 2. Complementary assays.

Sample	Incorporation rate (%)	Compression strength (MPa) ± STDV	Dry density (Kg/m^3)	Water absorption (%)
1	0	4.47 ± 0.704	1545.55 ± 12.11	19.91
2	2.5	2.31 ± 0.035	1412.04 ± 10.98	24.26
3	5	1.05 ± 0.35	1340.11 ± 5.70	27.85

Each unit was placed in the tank with water at room temperature. Firstly, the test has been done with 1/3 of the volume of the brick immersed in water. Then, the brick has been immersed until 2/3 of its volume. Finally, the specimens have been submerged for 24 h. Before weighing the bricks, all the excess water has been removed from all brick surfaces.To calculate the water absorption, the followed equation has been used:

$$w\_m = (m\_w - m\_d)/m\_d \ \ x\,100$$

Where, w_m is water absorption (%); m_w is wet mass (g) and m_d is dry mass (g).

3 RESULTS AND DISCUSSION

Table 2 lists the main results of the tests described before.According the compression strength, the incorporation of milled cigarette buts influences negatively this property being reduced in the range between 52% to 23%, respectively for incorporations of 2.5% and 5%. This decrease is in accordance with the literature (Aeslina Abdul Kadir & Mohajerani, 2010). However, the compression strength values are bellow the reported by Abdul Kadir study (2010) (Aeslina Abdul Kadir & Mohajerani, 2010).This is explained by the differences in the raw material used andthe procedure test, namely the compression velocity. Nevertheless, the results are comparable among samples and the main trend is similar to the minimum reported values for clay fired bricks in the standard NBR 7170 (ABNT-Associação Brasileira de Normas Técnicas, 1983).

Relatively to the dry density, all samples are considered as High Density units, according to EN 771-1 (Eurocode, 2003).For instance, the sample 1 presented a value of 1545.55 kg/m^3. This result was the highest of the tested bricks, demonstrating that, as expected, the cigarette butts incorporation reduce the dry density. The sample 3 exhibited less 13% regarding e sample 1. Also, sample 2 shows a reduction of the dry density less evident but still about 9%. Comparing to the Kadir's study (2010), the standard brick presents a dry density of 2118 Kg/m^3.The 2.5% and 5% of incorporation presents a density of 1941 kg/m^3 and 1611 kg/m^3, respectively (Aeslina Abdul Kadir & Mohajerani, 2010). These changescan be justified by the different raw material, cooking times and also by the incorporation of crushed cigarette butts, different to Kadir's study inwhich the cigarette butts have been incorporated in the whole way.

Furthermore, observing the water absorption test results it is possible to claim that the bricks are capable to absorb water. For instance, sample 1 presented 19.91% of water absorption, corresponding to the minimum absorption percentage. The water absorption increased with increasing waste incorporation. The most probable explanation lies in the porosity increase due to cigarette butts incorporation and volatilization during the cooking process.

4 CONCLUSIONS AND RECOMENDATIONS

The incorporation of cigarette butts in bricks is a potential solution for theenvironmental issue and health concerns.Thus, the consequences of this waste presence in the environment can be minimized. As shown by the results of the performed testing, the cigarette butts incorporation in bricks reduce the main properties (compressive strength, density and water absorption) of the clay fired brick. This may pose problems in the construction of higher structural walls but not in interior walls. Nevertheless, more studies are being made in order to obtain the ideal amount of milled cigarette butts that can be incorporated without causing a high decrease of mechanical properties.

ACKNOWLEDGEMENTS

This project has received funding from the European Union's Horizon 2020 research and innovation programme under grant agreement "No 810764".

REFERENCES

ABNT-Associação Brasileira de Normas Técnicas 1983. NBR 7170 Tijolo maciço cerâmico para alvenaria.

DeBardeleben, M. 1981. An overview of sidestream smoke: its components, its analysis, some influencing factors.

Eurocode2003. European Standard Norme 771-1.

INE2014. Inquérito Nacional de Saúde 2014 -Instituto Nacional de Estatística. https://doi.org/10.1017/CBO9781107415324.004

Kadir, Aeslina A, & Mohajerani, A. 2016. Recycling cigarette butts in lightweight fired clay bricks. *Proceedings of the Institution of Civil Engineers – Construction Materials*, 164(5), 219–229. https://doi.org/10.1680/coma.900013

Kadir, Aeslina Abdul, & Mohajerani, A. 2010. Possible Utilization of Cigarette Butts in Light-.

Lee, K., Botero, N. C., & Novotny, T. 2016. ' Manage and mitigate punitive regulatory measures , enhance the corporate image , influence public policy': industry efforts to shape understanding of tobacco- attributable deforestation. *Globalization and Health*, 1–12. https://doi.org/10.1186/s12992-016-0192-6

Novotny, T. E., & Slaughter, E. 2014. Tobacco Product Waste: An Environmental Approach to Reduce Tobacco Consumption. *Current Environmental Health Reports*, *1*(3), 208–216. https://doi.org/10.1007/s40572-014-0016-x

Proctor, R. N. 2012. Golden Holocaust: Origins of the Cigarette Catastrophe and the Case for Abolition: By Robert N. Proctor. *American Journal of Epidemiology*, *175*(9), 970–971. https://doi.org/10.1093/aje/kws175

Witkowski, J. 2014. Holding Cigarette Manufacturers and Smokers Liable for Toxic Butts: Potential Litigation-Related Causes of Action for Environmental Injuries/Harm and Waste Cleanup. *Tulane Environmental Law Journal*, *28*, 1–36.

World Health Organization 2008. Tobacco industry interference with tobacco control.

World Health Organization.(2017. *Tobacco and its environmental impact: an overview*. https://doi.org/10.1002/mrm.26081

Wright, S. L., Rowe, D., Reid, M. J., Thomas, K. V., & Galloway, T. S. 2015. Bioaccumulation and biological effects of cigarette litter in marine worms. *Scientific Reports*, *5*, 1–10. https://doi.org/10.1038/srep14119

Wastes: Solutions, Treatments and Opportunities III – Vilarinho et al. (Eds)

MANTO project – design for social inclusion in Póvoa de Lanhoso

P. Trigueiros & A. Broega
EAUM, University of Minho, Guimarães, Portugal

J. Carvalho
CVR – Centro para a Valorização de Resíduos, Guimarães, Portugal
MEtRICs, Mechanical Engineering and Resource Sustainability Center, Department of Mechanical Engineering, University of Minho, Guimarães, Portugal

ABSTRACT: MANTO (meaning mantle) is the name of an activity dedicated to social integration through design, carried out between May and July 2018, within the framework of the project Generation Lanhoso. This project takes part in an integrative program encompassing a set of activities which highlights solidarity upcycling initiatives to promote sustainable consumption of eco-friendly products, useful and accessible to all, as well as economic integration. It supports the creation of sustainable occupations for populations that are somehow disadvantaged (particularly women) and highlights handcraft knowledge. The foreseeable activities support their training to allow them to evolve in their work, gain competencies and develop their environmental awareness.

1 INTRODUCTION

1.1 *Circular economy working hand in hand with social inclusion*

This project takes part in an integrative program encompassing a set of activities which highlights solidarity upcycling initiatives to promote sustainable consumption of eco-friendly products, useful and accessible to all, as well as economic integration. This initiative launches partnerships among academy, from interdisciplinary sciences and craftsmen aiming to create innovative solutions for the environment and supports the emergence of green production sectors based on waste materials. It supports the creation of sustainable occupations for populations that are somehow disadvantaged (particularly women) and highlights handcraft knowledge. The foreseeable activities support their training to allow them to evolve in their work, gain competencies and develop their environmental awareness. The revenues generated are a source of empowerment and social inclusion, and the network helps them take responsibilities.

1.2 *Framework and objectives of the generation lanhoso project*

MANTO (meaning mantle) is the name of an activity dedicated to social integration through design, carried out between May and July 2018, within the framework of the project Generation Lanhoso. MANTO was carried out within the scope of a line of action called creative workshops of the referred project, aiming to promote the social inclusion of citizens of Póvoa de Lanhoso under the responsibility of Associação Sol do Ave, an institution of social solidarity (IPSS) based in Guimarães. The project aims to combat persistent poverty and social exclusion, often related to unemployment situations, in the context of a multi-sectoral and integrated approach.

The people covered by the project are, to a large extent, women over 50 years old, unemployed or inactive, in a socially vulnerable situation. Some of these people have worked in the textile and confectionery industry or, traditionally, have the skills to embroider, crochet and other textile products. Others, however, present depressed or low self-esteem, or other problems that are translated and manifested by some inertia, difficulties of attention and motivation, problems of fine motor skills, among others. In common, they also share a very fragile economic situation and are dependent on social benefits provided by the ISS (Social Security Institute).

Since the sponsoring association does not possess human recourses with design competences, this task was handed over to the Design Institute, under the coordination of Paula Trigueiros. It was an opportunity to develop a project of interaction between the University and the surrounding social fabric, demonstrating the power of design for social inclusion.

2 BACKGROUND

2.1 *Maria da Fonte – history and source of inspiration*

The implementation of the project began with an approach to the community of Póvoa de Lanhoso, involving the entire technical team in the areas of design and social service. In order to understand the work context local stakeholders, namely Associação Sol do Ave and Municipality, which demonstrated interest in the continuity of the activities and quality of the results of previous Creative Workshops were also involved in the initial diagnosis, namely through its culture and tourism sector, responsible for the CIMF- Centro Interpretativo Maria da Fonte (Interpretative Centre Maria da Fonte).

2.2 *Historical fair and recreation of Maria da Fonte*

In the first phase of this work, the community was studied and its values, references and a little of its history were carefully analyzed. From this analysis, Maria da Fonte excels: a local heroic figure, known for having spearheaded a popular revolt, mostly carried out by the women of Póvoa de Lanhoso in 1846 (Capela & Borralheiro, 1996).

The theme of Maria da Fonte was also the focus of the first set of activities with the group. The work proposal initially intended that people learn to make "sellable" products. Knowing about the first historical recreation of Maria da Fonte revolt (in 2016, on the day of the municipality - September 22) it was decided to emphasize its importance and take advantage of the local calendar date. Thus, it was decided to orientate the theme and typology of products to be realized in the Creative Workshops in relation to the props and costumes related to the theme. In this way they would be visible in the local festival – so that they could be proud of the people who created them and those who could wear them.

In addition to an annual party, one of the busiest fairs in the city is also taking place on this date. Thus, there was an additional opportunity for the products to be carried out by the trainees at this stage, or later on their own initiative, to be sold – directly or in a municipality-sponsored space, right in the center of the city.

On the other hand, the profusion of events and fairs in which the recreation and manufacture of appearance products and medieval and/or historical production techniques is valued, aroused as an opportunity market for this kind of actions and products – which can broaden the scope and scale of their results.

2.3 *Before doing: learning*

In order to deepen the historical and context study, the involved technicians were formed to obtained explanations, documents, books and a visit to the Interpretation Center of Maria da Fonte in Póvoa de Lanhoso. In view of the theme and structure of work, a very experienced trainer in the creation of textile products and costumes for events and performances of a historical and artistic nature was recruited.

In order to emphasize and value the heritage of local traditional vests and encourage people to share and progress in the use of their skills – namely to make traditional lace and embroidery –

Figure 1. Visit with the participants to the costume museum illustrating the genuine aspects of the materials, shapes and adornments of the time.

the group visited Museu do Traje (Costume Museum), in Braga. In this space there are numerous genuine examples of regional costumes, embroidery and some other textile products very familiar to the involved people (Figure 1). The explanation provided by the local technician of the museum added knowledge about details, functions, and other stories related to the customs associated with those artefacts – deserving the attention of the visitors.

2.4 *Forming – Other content implicit in creative activities*

During the sessions of the Creative Workshops the trainer transmitted some notions about the importance of the genuine references of history and the rigor of their representation for the creation of value in their own future creations.

In addition to the selection of embroidery techniques and overall handcraft, some materials were also chosen and worked for the preparation of props for the historical representation – avoiding knits and synthetic materials and trying to use natural dyeing fabrics and techniques that best recreate flax, cotton and other fibers that existed at the time – especially the most characteristic of characters such as the common women who played part in those historical events.

2.5 *Social contingencies and the closure of the activity – 1st phase*

For reasons unrelated to this whole structure and organization, many people who were attending this action were called by Social Security for other training actions – which are compulsory for beneficiaries to maintain entitlement to social benefits. In addition to these other training actions promoted by the Employment Institute, this Creative Workshop is free of charge and does not confer the right to monetary remuneration or payment of expenses (travel or meals) of the involved people. This led to the early closure of this activity, after only 30, of the 100 hours of the foreseen planned activities. This was only resumed almost a year later – in a second phase that will be further described.

2.6 *Widening horizons – Maria de Hoje logo*

In the context of a strategy of continuity and expansion of intervention scale, a brand was created, and different packaging designed for different types of products to be carried out in this and other social contexts similar to the case study in point.

The proportion of the feminine gender and particularly vulnerable women in these isolated communities from the interior of the country, led us to emphasize the reference to the figure of Maria da Fonte for the creation of this brand. Thus, the brand Maria de Hoje (meaning Mary of

Figure 2. Maria de Hoje logo brand.

Today) was born, inspired by history and also by the cultural values of the region, namely referred, the Maria da Fonte revolution.

The inverted triangle design, symbolically associated with the female uterus, (Chevallaier & Chevallier, 1997) is filled with a regular and functional scratched graphic in its reproduction – which intends to leave in each product the identity of an essentially artisanal work and simultaneously enhances the role and the affirmation of the Woman of today (Gamma, Trigueiros, & Broega, 2018).

3 MANTO PROJECT

In face of the further developments and in order to guarantee the maintenance of the cohesion of the group, it was also planned at this stage to establish a strategy stability and involvement of people in the activity. The idea of creating a "Mantle" is not alien to all these facts that preceded it, and although it departs from the historical theme, it maintains much of its purpose and must be understood as a continuity of it – constituting a second phase the Creative Workshops.

As in the previous phase, it was necessary to meet with a few (new) persons responsible for the institutions involved – at this stage the sessions were housed in a community center of a small village, the Associaton "Em Diálogo" displaced from the center of the Municipality of Póvoa de Lanhoso – Monsul. For this phase, Cláudia Ribeiro, a costume designer very qualified in staging and performing events in public spaces related to music and theater, was selected for direct work with those women. The MANTO was held during the months of May to July 2018, in weekly sessions. In parallel, a workshop was also held involving students from different creative areas and several cycles of university studies (product design, textile and architecture) that discussed the inclusion proposal by design and participated actively in the concretion of a fence piece of 7m extension - in an event that took place in a single day. Some of them participated later in the date of the assembly of the MANTO on the spot and also contributed with the realization of some interviews to the participants.

3.1 *Concept and description*

The MANTO concept was inspired by crochet or patchwork blankets, made mostly by women of some age who thus occupy free time. These are made up of modular pieces made separately, taking advantage of the waste remains of wool from textile industry and other used raw materials offered by several members of the community. Those are then sewn forming blankets, shawls or other utilitarian products. Based on a triangulated network made of PVC pipes, participants in the initiative contributed with their own creations to fill triangular modules of the structure. In each triangle, the participants carry out various works, some of them result of their own creative experiences and others with support during the sessions led by the trainer, to learn and practice some

Figure 3. Final assembly of the MANTO, in the churchyard of the local church, with the participation of the community.

techniques of weaving, embroidery, macramé, among others. With the support of "Em Diálogo" Association, at the beginning of July a session was held that brought together other people from the community of Monsul and surrounding parishes. In this session many simple triangular modules were produced with various colored fabrics donated by textile companies in the region.

3.2 *Notes*

In the days prior to the pilgrimage and with the support of the local Parish Council, a wire structure was created, which was incorporated, radially, centered on a large pole on the local. The contributions of all were grouped together creating a large mantle (MANTO), raised from the ground about 4 to 6 meters, forming a great sunshade. This was mounted on the churchyard of Saint Tiago and Saint Luzia, to shade the space during the annual pilgrimage of Monsul in Póvoa de Lanhoso. This gave visibility to the collective installation, promoting the pleasure of concretization and the pride among the participants.

4 PROCESS DESIGN

4.1 *Participated process*

The history of participation in design was also highlighted by the defense of civil rights in the United States and the United Kingdom in the 1960s (Sanoff, 2008). This political context inspired the creation of community design centers, where designers and urban planners assisted in the design and demand of their needs and planning aspirations(Carvalho, 2018) Design for democracy, social design, design for participation, design for/of/in a democracy, political design, design for politics, design activism are examples of the increasing social/civic/political relevance of research and practice of self-directed and societally oriented design (Binder et al., 2015). This decentralization refers to the concept of co-creation (Carvalho, 2018).

4.2 *Systematization of the method*

4.2.1 *Study and community approach*

This task consisted in a deep and integrated study of themes, of history and local references, logistics, material and policy requirements as well as the preparation and coordination of work in collaboration and in harmony with partners and all involved institutions, at social, industrial, and local authorities.

4.2.2 *Inclusive creation: tolerant in individual participation and valued by the collective*

In accordance with the proposed strategy, the methodology involved the development of a module – or isolated element, executed individually. Each module gives meaning to the content of the training/techniques developed during the work sessions. The object of the training component and the practice is the experience of various techniques (weaving/embroidery, among others), welcoming diversity – from the work of the most virtuous to the apprentice.

4.2.3 *Community visibility and the symbolism of the collective festive experience*

The meeting of the elements in a large facility helps to unite the collective common mission – technicians and trainees, as well as authorities and local references are proud to be represented in these works with local visibility on festive dates. A symbolic and relevant date in the community calendar and an objective implementation schedule, contributed to the dynamics both project and the people.

4.3 *3rd Phase: future work*

The third and final phase of this project is ongoing, resuming the intention of the creation of utilitarian and salable products by the people themselves - stimulating the creation of income and promotion of self-esteem. Product design at this stage incorporated the concept and spirit of collective or participatory construction, which emerged and was successful in the project MANTO. For this purpose, products have been designed benefit from the learning provided by all the previous process – the notion that the final product can result from the set of individual participations. We call these products "Outros Mantos", (meaning other mantles) by incorporating these values and principles, regardless of their actual function.

5 REPLICABILITY, SCALE UP AND FINAL REMARKS

Under the concept of circular economy, a new definition for the concept of social sustainability arose. This definition includes society, culture and politics, involving social justice, social cohesion, poverty reduction, equality, labor rights, education, social care, social responsibility, the right of self-determination, but also social networking, security, democracy, health and social development. The social sustainability mainly refers to ensuring equality within and between generations, combating all forms of discrimination, social inclusion and cohesion, tackling all forms of exclusion, political participation, social mobility and respect for cultural diversity. The project Manto was the basilar stone for the definition of a future strategy in which, the exchange of ideas and practices under the concept of 'circular knowledge', along with the concept of seizing waste as an opportunity, act as a core function that can be fundamental to a true metabolism of society and culture, and lately of economy. It was demonstrated that at social and organizational contingencies level, design can participate, unite and give meaning to the whole. It was a successful view of an open and tolerant project that welcomed different visions, from that of local technicians, participants and experience and skills of the trainers. Ethics and the value of art and training was highlighted, showing that their role should not be neglected nor misunderstood. Even in the cases when trainees do not understand it, they were able to recognize the value of the work/art and gather consensus around it, always underlying the determination to give discarded materials a second life.

REFERENCES

Capela, J. V., & Borralheiro, R. 1996. *A Maria da Fonte na Póvoa de Lanhoso-Novos Documentos para a sua História*. Póvoa de Lanhoso: Câmara Municipal de Póvoa de Lanhoso.

Carvalho, C. P. 2018. *Utopia nas Margens – o papel do Design na cocriação de alternativas num contexto de exclusão social*. (F. do Porto, Ed.) Porto.

Chevallaier, J., & Chevallier, A. G. 1997. *Dictionnaire des Symboles*. (R. Laffont, Ed.)

Gama, C., Trigueiros, P., and Broega, C. 2018. *Criação de uma marca no âmbito da intervenção do design em projetos de inclusão social*. Proceedings of CIMODE 2018. Madrid: UMinho.

Wastes: Solutions, Treatments and Opportunities III – Vilarinho et al. (Eds)
© 2020 Taylor & Francis Group, London, ISBN 978-0-367-25777-4

Electrokinetic remediation technology applied to municipal sludge decontamination

A. Ribeiro, J. Araújo & J. Carvalho
CVR – Centre for Waste Valorisation, Guimarães, Portugal

C. Vilarinho
Mechanical Engineering and Resources Sustainability Center, Department of Mechanical Engineering, UMinho, Guimarães, Portugal

ABSTRACT: In this study, the remediation of Municipal Sewage Sludge by electrokinetic process, coupled with activated carbon as permeable reactive barrier, was investigated. In each experiment it was observed the removal of cadmium, lead, copper, chromium, nickel and zinc from the sludge. An electric field of 3 V cm^{-1} was applied and was used an adsorbent/sludge ratio of 30 g kg^{-1} for the preparation of the permeable reactive barrier. Results proved that that this process is perfectly suited for the removal of chromium, nickel and zinc metals from the sludge. In the end of the operation time it was achieved a maximum removal rate of 56% for chromium, 73% for nickel and 99% for zinc, with initial concentrations of 2790 mg kg^{-1}, 2840 mg kg^{-1} and 94200 mg kg^{-1}, respectively. Based in these results, the viability of the new coupling technology developed, to treat sewage sludges was demonstrated.

1 INTRODUCTION

The production of excess sewage sludge has increased substantially with the development of wastewater treatment (WWT) plants in recent years (Ščančar et al., 2000). Therefore, the disposal of these wastes is a growing problem worldwide. The European Community has developed the draft of "Working document on sludge" with the aim of updating the regulatory system for the implementation of a sustainable development concept which includes the prevention, re-use, recycling, incineration (with energy recovery) and landfill of sewage sludge (EWA., 2000)

Municipal sewage sludge is in liquid or semisolid form depending on operations and processes treatment used, with a solids content varying from 0.25 to 12 percent by weight. Typically, the sludge composition consists of high organic matter, some nutrients, microorganisms and heavy metals, which can cause serious environmental and public health impacts. Nevertheless, in WWT there are different kinds of sludge depending on the WWT scheme. In WWT, usually three kinds of sludge are generated, primary sludge from physical and/or chemical treatment, secondary sludge from secondary treatment, mainly biological, and the tertiary sludge often by nutrient removal (Fytili & Zabaniotou, 2008).

Over the past decades, many techniques have been developed for the removal and recovery of heavy metals from sludge. They include chemistry extraction, chlorination method, electrokinetic remediation, ion exchange, membrane separation and bioleaching methods (Pathak et al., 2009). Electrokinetic remediation is an emerging technology that has attracted increased interest among scientists and government officials in the last decade, due to several promising laboratory and pilot-scale studies and experiment. This method aims to remove heavy metal contaminants from low permeability contaminated soils under the influence of an applied direct current (Virkutyte et al., 2002). In fact, the low operational costs and potential applicability of this technology to a wide range of contaminants promoted the execution of several studies of heavy metal remediation (Reddy & Ala, 2007). The electrokinetic approach generally requires low-level DC current densities in the order of a few mA/cm^2 between suitably located electrodes, which any way induce physicochemical changes in the applied media. Several transport phenomena are generated due to the application of an electric field. The main phenomena are electromigration (movement of ionic species towards the electrode of opposite charge) and electroosmosis (net flux of water, and the species in solution, relative to a stationary charged surface) (Yeung & Gu, 2011).

Nowadays, the use of permeable reactive barriers (PRB) has gained acceptance in the remediation field, due to its high efficiency, low cost, and simple operation procedures. These barriers are composed by reactive materials that once in contact with the contaminated water, can degrade, adsorb or precipitate the targeted contaminants. Recent researchers have shown that several solid waste materials (red mud, bagasse fly ash, carbon slurry, eggshell) generated in industries, are efficient adsorbents for the removal of heavy metals (Yeung & Gu, 2011). Although, in this study it will be used commercial activated carbon.

Considering previous studies (Ribeiro et al., 2018), the aim of the present work was to evaluate the application of an innovative combined system, which couples electrokinetic remediation with specific permeable reactive barriers, to treat municipal sewage sludge (MSS) contaminated with heavy metals. The permeable reactive barrier is composed by activated carbon (AC). Results after the experiments were compared with the Portuguese legislation, in order to validate the proposed technology.

2 MATERIAL AND METHODS

2.1 *Municipal Sewage Sludge (MSS) characterization*

The experimental samples are dewatered sludge from a local wastewater treatment plant localized in Portugal. The basic characteristics of sludge can be seen in Table 1. The analysis of heavy metal total content was carried by flame atomic absorption. The process of digestion and subsequent release of the heavy metals into an aqueous phase (analyzed phase) was performed according to ISO 11466, where soils samples contact with aqua regia for 24 hours. This standard refers that contact time with aqua regia should be 4 hours. However, experiments proved that the contact time with aqua regia for 24 hours promotes better results.

As can be seen in Table 1, this sludge presents a very high chromium, nickel and zinc concentrations. In fact, this characterization indicates that this sludge has concentrations of heavy metals well above the limits imposed by DL 276/2009. Therefore, it can be concluded that this sludge does not present conditions for its incorporation in agricultural soils and presents a serious threat to the environment due to the high concentration of heavy metals. The contents of these heavy metals in the sludge are relatively high because the wastewater treatment plant that produced the sludges in the study most probably receives industrial wastewater. Same results were observed by Dongdong et al (2016). According to these authors, the municipal sludges from a wastewater

Table 1. Municipal Sewage Sludge (MSS) characterization.

Parameters	Results	Limit imposed by Portuguese Decree-Law 276/2009
pH	7,2	–
Density (kg/m^3)	436	–
Cadmium (mg/kg)	0,4	20
Lead (mg/kg)	45	750
Copper (mg/kg)	456	1000
Chromium (mg/kg)	2790	1000
Nickel (mg/kg)	2840	300
Zinc (mg/kg)	94200	2500

treatment plant localized in China also presented high concentrations of zinc because that plant receives 30% industrial wastewater (Dongdong et al., 2016).

Thus, the study carried out focused in the removal of the elements presented in higher concentration, especially zinc, nickel, and chromium, but also in the behavior of the other heavy metals in study, such as cadmium, lead, and copper.

2.2 *Experimental apparatus*

Electrokinetic remediation experiments were performed in acrylic cells (Figure 1) with 25 cm of length and 10 cm of diameter filled with 1.5 kg of contaminated municipal sewage sludge. Thus, the initial concentrations of the metals under study were: [Pb] 45 mg kg^{-1}, [Cr] 2790 mg kg^{-1}, [Zn] 94200 mg kg^{-1}, [Cd] 0.4 mg kg^{-1}, [Pb] 45 mg kg^{-1} and [Cu] with 456 mg kg^{-1}.

Then, a cathode and anode electrode chambers were coupled at the end of the column, isolated from the matrix with nylon net. Graphite electrodes were used for both chambers and three auxiliary electrodes allowed the measurement of the electric field through the column. Graphite electrodes are used to measure the conductivity through digital voltmeters. The remaining two graphite electrodes were used for introducing the electric current into the system, which is supplied by a current rectifier. In these experiments, it was introduced 60 volts, which represents a voltage/length ratio of 3 V cm^{-1}.

The pH in both chambers was controlled to maintain the value around 5. The adjustment was made with sodium hydroxide [0.1 M] and citric acid monohydrate [0.5 M] for anode and cathode chambers, respectively. It was used an AC/soil ratio of 30 g kg^{-1} of contaminated soil for the preparation of the permeable reactive barrier (PRB). Experiments were performed at 168 hours of operation time. During the operation time of each test, 20 ml of aqueous solutions from the anode and cathode reservoirs and the electrode assemblies were collected and volume measurements were made. The determination of total heavy metals content in aqueous solution was made by flame atomic absorption. Soil samples were again analyzed by flame atomic absorption. The process of digestion and subsequent release of the heavy metals into an aqueous phase (analyzed phase) was performed according to ISO 11466, where soils samples contact with aqua regia for 24 hours. The percentage of pollutant removed (R/%) from the solutions was calculated using Eq. 1.

$$R = \frac{(C_0 - C_{e)} \times 100}{C_0} \tag{1}$$

where R (%) is the removal rate, C_0 and C_e (mg L^{-1}) are the initial an equilibrium concentration, respectively.

3 RESULTS

3.1 *Heavy metals removal from Municipal Sewage Sludge (MSS)*

The variation of removal rate for each heavy metal in study with the normalized distance from the anode during the electrokinetic treatment is showed in Figure 1. The normalized distance is defined as the distance to the specific location from the anode divided by the total distance from the anode to the cathode [27]. Figure 1 also shows the legal limit of each metal concentration for further utilization of sludges in agricultural fields, according to the Portuguese legislation (DL 276/2009).

As can be seen in figure 1, it was obtained high percentages of removal efficiency for each metal in study. Results demonstrate that higher removals rates were achieved in zinc (99%), followed by nickel (73%), cadmium (60%), chromium (56%), lead (50%) and finally copper with 25%.

Results of the evolution of zinc removal over the operating time clearly indicate that zinc removal is completely favored by electrokinetic remediation. At the end of the operation time, the concentration of zinc in the sludge was approximately 300 mg kg^{-1}. This result could meet the Portuguese legislation for agricultural application of the sewage sludge. These results clearly indicate that electrokinetic remediation system with reactive barriers presents high affinity in the removal of zinc,

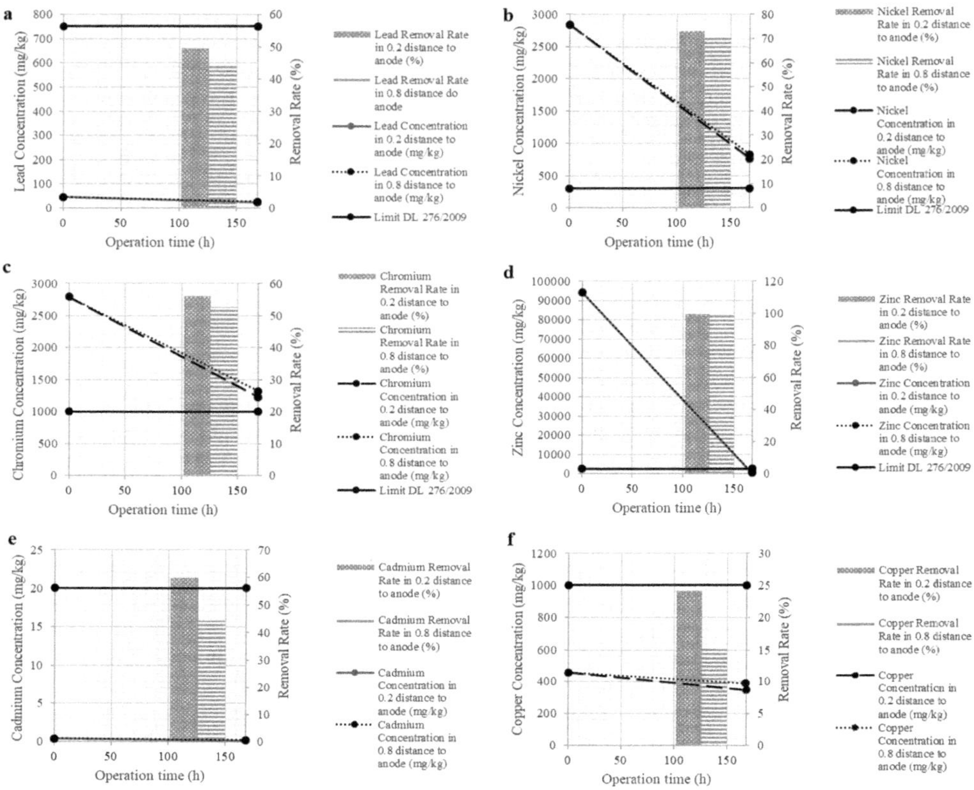

Figure 1. Variation of the removal rate of each heavy metal in study with different normalized distance from the anode and comparison with the Portuguese legal limit: a) lead removal rate; b) nickel removal rate; c) chromium removal rate; d) zinc removal rate; e) cadmium removal rate; f) copper removal rate.

and may even be studied, conditions with higher concentrations of this element. As already mentioned, these results present a potentially promising character, essentially due to the high toxicity of this metal. Peng et al (2011) also studied the removal of copper and zinc from sewage sludge with a combination of bioleaching and electrokinetic remediation technology. Authors stated that after six days of treatment, the contents of Cu and Zn in sewage sludge decreased from 296.4 mg kg^{-1} and 3756.2 mg kg^{-1} to 63.4 mg kg^{-1} and 33.3 mg kg^{-1}, respectively, which represents almost 80% of removal rate for copper and 99% for zinc. According to these authors, these results could meet the Chinese standard for land application of the heavy metals in sewage sludge (Peng et al., 2011).

Results of nickel removal efficiency also demonstrate that this process is quite effective in the migration and retention of nickel from the solid matrix. At the end of the operation time, it was achieved a reduction of this element to approximately 750 mg kg^{-1} of nickel. This result is slightly above the limit imposed by the Portuguese legislation. However, it should be noted that according to the results of figure 1, approximately 200 mg l^{-1} of nickel was not retained by the permeable reactive barrier composed by active carbon. Lee et al (2003) also studied the removal of copper and nickel from municipal sludge using the electrokinetic remediation process. These authors reported that under the optimal conditions (pH value of 3.8), the removal efficiencies for copper and nickel were 82.9% and 75.1%, respectively, after 132 hours of operation time (Liu et al., 2017).

As can be seen in Figure 1, results of chromium removal efficiency also demonstrated that electrokinetic remediation is quite effective in migration and retention of chromium from the

MSS. In the final of operation time, it was obtained a decreasing of this element to approximately 1200 mg kg^{-1} of chromium. These results are quite similar to other studies. According to Fonseca et al (2011), removal values of 60% and 79% were obtained when electrokinetic treatment was coupled with zeolite and activated carbon biobarriers, respectively, for a test period of 18 days. Authors also referred that the system with activated carbon showed higher uptake (0.074 mg $g^{-1)}$ (Fonseca et al., 2011). Fu et al (2017) studied the eletrokinetic remediation of Cr (IV) using two electrolytes (citric acid and polyaspartic acid). Authors reported that the removal efficiency of the total Cr using citric acid was significantly improved to (49.54%) when compared to the utilization of polyaspartic acid (29.24%) (Fu et al., 2017).

Regarding cadmium removal, as can be seen in Figure 1, results proved that electrokinetic remediation was suitable for this metal uptake. According to the results, at the end of the reaction period, it was obtained removal of 0,24 mg kg^{-1} of this element, which represents a removal efficiency of 60%. Gao et al (2013) also reported removal rates of 53% of cadmium from sludges through electrokinetic process. These researchers indicate that the main obstacles to Cd removal were the directional variations in electro-osmotic flow and the formation of anion complexes resulting from the high pH in cathode, causing them to move toward anode (Gao et al., 2013).

Results of copper removal efficiency proved that this element was quite difficult to remove with electrokinetic remediation, since it was only achieved 25% of removal efficiency. Kirkelund et al (2010) also studied the treatment of sediments contaminated with copper and zinc by electrokinetic remediation. Authors stated that the removal of Cu was strongly dependent on its dominant organic fraction, which can decrease its removal efficiency. In their report, 86% of Zn and 61% Cu were removed in soil sediments after 28 days of electrokinetic remediation. Zhua et al (2015) also reported removal rates of 80% for zinc and 68% for copper in eletrokinetic remediation of high organic content effluent (Zhua et al., 2015).

Results of Figure 1 also showed that higher removal rates were achieved in the nearest section of anode cell. It is important to refer that the dominant and most important electron transfer reactions that occur at electrodes during the electrokinetic treatment is the electrolysis of water. During this process, the electrolysis of water results in the formation of H^+ ions at the anode and OH^- ions at the cathode, and, primarily due to electromigration, these ions tend to migrate towards the oppositely charged electrode(s). Therefore, desorption of metal species from sludge particles occurred along with the migration of the acid front, during the treatment, and metal contaminants were gradually transported towards the cathode by electromigration and electroosmotic purging (Fonseca et al., 2011).

4 CONCLUSIONS

In this study, the removal of cadmium (Cd), lead (Pb), copper (Cu), chromium (Cr), nickel (Ni) and zinc (Zn) from a municipal sewage sludge were evaluated. From the results obtained in this work, it can be concluded that the utilization of citric acid as buffer solution and neutralizing agent was very effective in pH variation since pH values were kept close to the optimal values for metal adsorption (low pH). The negative effect of OH^- production in cathode was dissipated since the heavy metals flowed with the electroosmotic flow from anode to cathode and the utilization of sodium hydroxide prevented the formation of extreme acid sludge.

The combined technology of permeable reactive barrier and electrokinetic remediation removed the heavy metals presented in the municipal sewage sludge. At the end of the 168 hours of operation time, it was obtained high removal rates for each metal in study. Results demonstrate that higher removals rates were achieved in zinc (99%), followed by nickel (73%), cadmium (60%), chromium (56%), lead (50%) and finally copper with 25%.

Results of the evolution of zinc removal over the operating time clearly indicate that zinc removal is completely favored by electrokinetic remediation. At the end of the operation time, the concentration of zinc in the sludge was approximately 300 mg kg^{-1}. This result allows the fulfillment of the Portuguese legislation for the application of the municipal sewage sludge. agricultural fields.

It should be noted that at the end of the operating time, the metals are adsorbed into the reactive barrier, composed by the activated carbon. Therefore, more experiences must be performed, in order to valorize this material.

ACKNOWLEDGEMENT

This work has been co-financed by Compete 2020, Portugal 2020 and the European Union through the European Regional Development Fund – FEDER within the scope of the project UMinhoTech – technology for future (POCI-01-0246-FEDER-026795).

REFERENCES

Ferri, Violetta, Ferro, Sergio, Martínez-Huitle, Carlos and Battisti, Achille. 2009. Electrokinetic extraction of surfactants and heavy metals from sewage sludge. *Electrochimica.* 54 (7): 2108–2118.

EWA – European Water Association. 2000. *Working document on sludge.* 3rd draft (2000). URL: http://ec.europa.eu/environment/waste/sludge/pdf/sludge en.pdf.

Fonseca, Bruna, Pazos, Marta., Tavares, T, and Sanromán, Maria. 2011. Removal of hexavalent chromium of contaminated soil by coupling electrokinetic remediation and permeable reactive biobarriers. *Environ Sci Pollut Res* 19: 1800–1808.

Fu, R, Wen, D, Xia, X, Zhang, W and Gu, Y. 2017. Electrokinetic remediation of chromium (Cr)-contaminated soil with citric acid (CA) and polyaspartic acid (PASP) as electrolytes. *Chemical Engineering Journal* 316: 601–608.

Fytili, Despoina and Zabaniotou, Anastasia. 2008. Utilization of sewage sludge in EU application of old and new methods – A review. *Renewable and Sustainable Energy Reviews.* 12 (1): 116–140.

Gao, Jie, Luo, Qishi, Zhang, Changbo, Li, Bingzhi and Meng, Liang. 2013. Enhanced electrokinetic removal of cadmium from sludge using a coupled catholyte circulation system with multilayer of anion exchange resin." *Chemical Engineering Journal* 234: 1–8.

Kirkelund, Gunvor, Ottosen, Lisbeth and Villumsen, Arne. 2010. Investigation of Cu, Pb, and Zn partitioning be sequential extraction in harbor sediments after electrodialytic remediation. *Chemosphere* 79 (10): 997–1002.

Lee, S, Lee, K, Kim, S, and Ko, H. 2003. Effects of soil buffering capacity and citric acid in electrolyte on electrokinetic remediation of mine tailing soils. *J. Ind. Eng. Chem.* 9 (4): 360–365.

Pathak, Ashish, Dastidar, Manisha and Sreekrishnan, Trichur. 2009. Bioleaching of heavy metals from sewage sludge by indigenous iron-oxidizing microorganisms using ammonium ferrous sulfate and ferrous sulfate as energy sources: a comparative study. J. *Hazard. Mater.* 171 (1–3): 273–278.

Peng, Guiqun, Tian, Guangming, Liu, Junzhi, Bao, Qibei and Zang, Ling. 2011. Removal of heavy metals from sewage sludge with a combination of bioleaching and electrokinetic remediation technology. *Desalination* 271 (1–3): 100–104.

Reddy, Krishna and Ala, Prasanth. 2007. Electrokinetic Remediation of Metal-Contaminated Field Soil. *Separation Science and Technology.* 40 (8): 1701–1720.

Ribeiro, André, Mota, André, Soares, Margarida, Castro, Carlos, Araújo, Jorge and Carvalho, Joana. 2018. Lead (II) removal from contaminated soils by electrokinetic remediation coupled with modified eggshell waste. *Key Engineering Materials* 777: 256-261.

Ščančar, Jane, Milačič, Radmila, Stražar, Marjeta and Burica, Olga. 2000. Total metal concentrations and partitioning of Cd, Cr, Cu, Fe, Ni and Zn in sewage sludge. *Sci. Total Environ.* 25: 9–19.

Virkutyte, Jurate, Sillanpää, Mika and Latostenmaa, Petri. 2002. Electrokinetic soil remediation – critical overview. *The Science of the Total Environment* 289 (1–3): 97–121.

Yeung, Albert, Gu, Ying-Ying. 2011. A review on techniques to enhance electrochemical remediation of contaminated soils. *J Hazard Mater* 195: 11–29.

Zhua, Neng-Min, Chenc, Mengjun, Guo, Xu-Jing, Guo-Quan and Yu-Denga, Hua. 2015. Electrokinetic removal of Cu and Zn in anaerobic digestate: Interrelation between metal speciation and electrokinetic treatments. *Journal of Hazardous Materials* 286: 118–126.

Wastes: Solutions, Treatments and Opportunities III – Vilarinho et al. (Eds)
 ISBN 978-0-367-25777-4

Densification and combustion of biomass pruning residues

P. Ribeiro, J. Araújo & J. Carvalho
CVR – Centre for Waste Valorisation, Guimarães, Portugal

C. Vilarinho
Mechanical Engineering and Resources Sustainability Center, Department of Mechanical Engineering, UMinho, Guimarães, Portugal

ABSTRACT: This paper addresses the potential of biomass pruning residues for energetic valorization by combustion. For this, five representative types of woody biomass pruning residues were collected and its proximate and ultimate composition determined. After a period of natural drying, the biomass was densified by briquetting technique. The produced briquettes were characterized and burned in a downdraft wood gasification boiler. Flue gas was sampled and analyzed in terms of temperature, flow speed, moisture and O_2 and CO_2 composition. Pollutant emission was measured, namely CO, NO_x, TOC and PM. The results are discussed and compared to conventional briquettes made from pine sawdust and show that briquetting is an effective technique to densify the pruning biomass and its possible to burn on conventional boilers. Pollutant emission it's only a concern for NO_x, for which adequate NO_x emission reduction techniques are advised.

1 INTRODUCTION

Problems related to public health, environment and climatic change instigate the creation of guidelines for a low carbon based economy in Europe, (European Commission 2010). This politics lead to a significant increase in the share of renewable sources in Europe's energy mix on the last decade (Bioenergy Europe 2018), being the bioenergy the most significant renewable energy source, with more than 60% of that portion, in 2017 (Bioenergy Europe 2018). Heating and cooling represented near 75% of all the bioenergy consumed in Europe in 2016 (Bioenergy Europe 2018) in which solid biomass represented more than 80% (European Environment Agency 2018). Solid biomass represented in 2016 circa 13% of the gross final energy consumption on the heat and cooling energy sector in Europe (Bioenergy Europe 2018). Thus, this numbers suggest that biomass will play a key role in decarbonizing European energy.

The demand on biomass in present and in the future, especially from agriculture, may evidence sustainability problems (Faaij 2018). So, it's important to improve efficiency in the conversion systems and to develop new pathways for the enhancement of biomass valorization, reducing pressure on land use for food. Woody pruning residues from agriculture and urban park maintenance are usually not recovered, being chipped and returned to the soil or burned without any heat recovery, because of the difficulty to directly burn in conventional combustion systems.

In this paper, 5 types (vine, kiwi, olive, apple and urban parks) of woody pruning residues are proximate and ultimate analyzed, milled and fed to a briquetting machine to produce energetic briquettes. Briquetting willingness is evaluated, the products burned on a conventional log wood downdraft boiler and flue gas is investigated. The results are compared to pine wood briquettes and show that it's possible to produce good quality briquettes from woody biomass and they can be burned on conventional systems without major problems, except for apple pruning residues which show a significantly higher emission of NO_x, thus being recommended to be used in low NO_x emision combustion equipment.

2 MATERIALS AND METHODS

According to the biomass availability on the regions in study, it were selected 5 different sources of pruning residues: vineyard, kiwi plantation, olive grove, apple orchard and urban parks (Figure 1). The vineyard pruning residues were collected using a dedicated round baller (CAEB International, Quickpower 1230) and the bales were stored inside the facilities for 12 months. Olive tree, apple tree and kiwi pruning residues were handpicked and stored inside facilities, the first ones for 18 months and the other two for 12 months. The time that biomass rested on the ground before it was collected was not controlled, due to containments of the producers. Urban parks pruning residues were collected, chipped on site using a Vandaelle 12–20 chipper and stored inhouse for 8 months.

Every woody biomass source was proximal characterized both when received and before the densification process. It was determined the moisture content according to BS EN ISO 18134-3:2015 standard. Ash content and volatile mater were calculated following the principles of BS EN ISO 18122:2015 and BS EN ISO 18123:2015 standards, respectively. Fixed carbon (d.b.), C_f, were calculated by the differential method, using the expression (4):

$$C_f = 100 - A_d - V_d \quad (1)$$

Elementary constitution of the biomass in terms of C, H and N was determined using a LECO True Spec CHN analyzer. Heating values of the biomass were determined following the procedures described on the BS EN 14918:2009 standard and using a LECO AC500 calorimeter, both for higher (HHV) and lower (LHV) heating values.

After the stage of inhouse natural drying of all the biomass types, the densification was done by *briquetting* technology. Vine, kiwi, apple and olive pruning residues were crushed on a Bosch AXT2000 shredder and, after that, chipped using a Retsch SM2000 cutting mill and a 6 mm screen. Urban pruning residue were directly fed to the cutting mill because it was previously chipped on site. The resulting chipped wood was used in a RUF 4 briquetting machine to produce 60×40 mm^2 rectangular section briquettes, with length of 95 ± 10 mm. The produced briquettes were measured using a caliper rule, weighted on a precision scale, its density calculated and stored on a dry inhouse room.

The combustion experiment was performed on an ATMOS DC25SP 25 kW nominal output, downdraft boiler (Figure 2). A pitot tube with a probe is introduced on the stack for exhaust gas

Figure 1. Biomass as received. a) bale of vineyard pruning residue; b) loose sticks of kiwi plantation pruning residue; c) loose sticks of olive grove pruning residue; d) loose sticks of apple orchard pruning residue; e) chips of urban parks pruning residue.

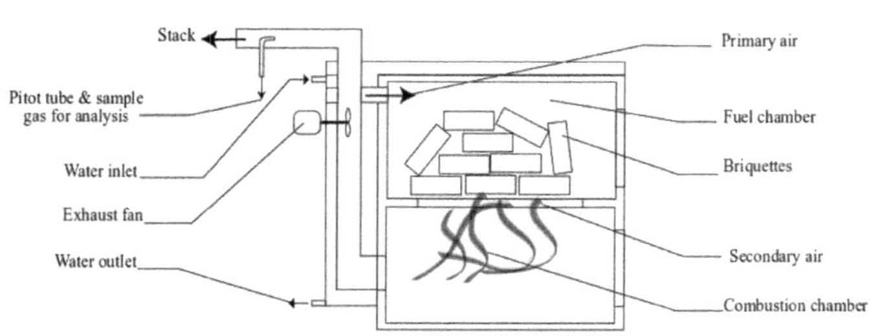

Figure 2. Diagram of the experimental setup for the combustion tests.

analysis. It was measured the moisture via a Dadolab ST5 analyzer, that also recorded the exhaust gas temperature. Flue gas velocity was calculated from the differential pressure on the pitot tube and the volume concentration on CO_2, O_2, NO_x and CO was determined using a Horiba PG 350 E (SRM) multi-gas analyzer. Total organic compounds (TOC) were determined by means of a Signal 3010 chromatographer and particulate matter (PM) was quantified using a Dadolab ST5 isokinetic sampler and an AE PW254 analytic scale.

The boiler was operated at nominal output during all the experiments. For each performed test, it was load with fuel and pre-heated until the exhaust temperature reached circa 230°C and water temperature was above 70°C. After this point, the boiler was loaded again with fresh fuel and the combustion tests started. In some cases, when was observed a higher-to-expected level of oxygen on flue gases (monitored in real time), the boiler was opened and quickly reloaded with more fuel. The combustion tests finished when 30 min passed from the beginning.

3 RESULTS AND DISCUSSION

The results of proximate and ultimate analysis of the biomass addressed in this study are summarized on Table 1. It shows a significant difference in the biomass moisture content as received, that can be justified by the time biomass spent on the ground before collection, the weather conditions when produced and the period stored inhouse before delivery to the experimental facilities. In respect to ash content, all the pruning residues present a superior level when compared to pine sawdust. This is attributed to the higher bark/trunk proportion that is natural on pruning residues. Volatile matter and fixed carbon evidenced to be very similar between all the types of biomass addressed in this study. In respect to the heating value, pruning residues proved inferior when compared to pine sawdust, a behavior that may be associated to the higher levels of ash content. Nitrogen content in pruning residues, a key element in respect do NO_x formation on solid fuels combustion (Perez-jimenez 2015), revealed significantly higher than pine sawdust. This may be attributed to the inherent type of biomass in study. Plant leaves are naturally rich in nitrogen, due to questions related to the photosynthesis capability (Ecarnot, Compan, and Roumet 2013) and also bark tissue and small branches (Dickson 1989), especially in trees that are fertilized with nitrogen to improve grow rate, thus justifying the findings.

Biomass was natural dried, and its moisture content determined prior to *briquetting*. The moisture content at the briquetting stage can be observed on Table 2. It shows that all the biomass sources had a moisture content equivalent to natural dried log wood, ranging from 9.5 to 12.7%, except

Table 1. Proximate and ultimate analysis of the biomass.

			Biomass Source (as received)					
			Vineyard	Kiwi	Olive	Apple	Urban	Pine
	Storing shape		Bale	Loose sticks	Loose sticks	Loose sticks	Chips	Sawdust
Prox. Analysis	Moisture content	*(%) wb*	47.1	58.1	12.3	16.0	47.5	11.5
	Ash content (550° C)	*(%) db*	3.29	3.19	3.72	5.11	4.84	0.8
	Volatile matter (900°)	*(%) db*	80.12	79.70	78.73	78.70	79.09	83.7
	Fixed Carbon	*(%) db*	16.59	17.11	17.55	16.19	16.07	15.5
	LHV	*MJ/kg*	15.77	16.90	17.03	14.65	15.56	18.6
	HVV	*MJ/kg*	17.03	18.63	18.57	15.8	16.89	–
Ult. Analysis	Carbon	*(%) db*	50.5	48.2	48.8	46.5	47.7	47.3
	Oxygen*	*(%) db*	39.4	41.4	38.5	40.9	39.8	45.2
	Hydrogen	*(%) db*	5.84	6.40	7.14	6.20	6.13	6.4
	Nitrogen	*(%) db*	0.95	0.70	1.80	1.20	1.45	0.13
	Sulfur	*(%) db*	0.043	0.150	0.077	0.080	0.070	0.99

*Estimated by the difference to the total, including ash

Table 2. Moisture content of the pruning residues biomass prior to briquetting.

Biomass Source	Vine	Kiwi	Olive	Apple	Urban
Moisture content (%)	12.2	19.5	10.6	12.7	9.5
Standard deviation	0.26	0.45	0.69	0.31	0.18

Figure 3. Briquettes produced. a) vineyard pruning residue; b) kiwi plantation pruning residue; c) olive grove pruning residue; d) apple orchard pruning residue; e) urban pruning residue.

Table 3. Density of the briquettes according to the biomass source.

Biomass Source	Vine	Kiwi	Olive	Apple	Urban Res.
Density ($\times 1000$) [kg/m^3]	1.13	0.84	1.17	1.09	0.95

kiwi pruning residue which registered a higher value (19.5%). This can be explained by the past time since the collection of the biomass, which was much shorter on the kiwi pruning residue and, thus, the natural drying process was not properly accomplished.

The briquetting process was realized with success for all the biomass types and the result is illustrated on Figure 3. No binder additives were used in the process in order to promote consistency of the products. Briquette toughness was good in vine, olive and apple samples. Urban and, especially, Kiwi pruning residues shown inferior mechanical properties to the other biomass types, although satisfactory. The calculated density of the briquettes is recorded on Table 3. From its analysis it can be concluded that the briquetting process improved the density of all the biomass sources, especially vine, olive and apple pruning residues. Kiwi pruning's briquettes shown the lowest density, a fact that can be explained by the higher moisture content of the biomass at the time of the process. On the other hand, the biomass that presented the lower moisture also shows a low density. These two situations suggest that moisture content is a key factor on the briquetting process and must be controlled in order to get the best quality briquettes, in respect to both density and mechanical properties.

The result of the combustion tests, namely the average flue gas composition, is summarized on Table 4. In the same figure are presented the results of a control test made with pine wood *briquettes* for comparison. Vineyard, olive and urban pruning residues combustion recorded an oxygen composition of around 7%, denoting the lowest excess air in the all the combustion tests and, thus, the higher exhaust gases temperatures (circa 550 K). Apple tree pruning residues briquettes combustion test presented some problems to stabilize, taking much time and fuel. Because of that, the produced mass of briquettes gone short and, at the end of the test, the level of excess air increased, resulting in a higher average oxygen content (8.3%) and a flue gas temperature circa 20 K lower. The reference test, made with pine wood briquettes, registered the higher level of excess air and the lowest flue gas temperature of all the tests. Because the boiler is a downdraft wood gasification system, this difference can be attributed to a very disparate initial fuel load on the test, a observation that is also true for kiwi pruning residues, which registered 8.0% of O_2.

Table 4. Flue gas composition on the combustion tests.

	O_2 (% vol)	CO_2 (% vol)	H_2O (% vol)	Temperature (K)	Flow speed (m/s)
Vineyard	7.1	15.2	10.2	547	2.08
Kiwi	8.0	14.0	13.7	503	2.85
Olive	6.8	15.3	7.0	553	2.25
Apple	8.3	7.6	7.7	534	2.33
Urban	6.9	15.4	8.3	547	3.23
Pine	13.4	6.2	12.9	440	1.3

Table 5. Pollutant emission on flue gas in the combustion tests.

	CO (mg/m^3 @ 6% O_2)	NO_x (mg/m^3 @ 6% O_2)	TOC (mg/m^3 @ 6% O_2)	PM (mg/m^3 @ 6% O_2)
Vineyard	275	615	126	137
Kiwi	265	1140	112	189
Olive	285	433	301	133
Apple	438	946	160	83
Urban	96	613	182	69
Pine	989	169	217	72

The average pollutant concentration on dry flue gas, namely CO, NO_x, COT and PM, corrected to an oxygen content reference of 6%, can be seen on Table 5. A very low level of carbon monoxide emission was registered on urban pruning residues (under 100 mg/m^3 @ 6% O_2), by contrast with pine wood briquettes, which registered an emission level 10 times superior (almost 1000 mg/m^3 @ 6% O_2). Vineyard, kiwi and olive grove pruning residues presented similar levels of CO emission (below 300 mg/m^3 @ 6% O_2) and apple pruning residues a little above the last two. CO emission is linked to incomplete combustion, that can be caused by several factors in solid fuel combustion, namely unsuitable level of excess air, combustion chamber temperature, reagent mixture, etc. The high emission rate of CO on pine wood briquettes may be linked to the high level of excess air in test that causes low combustion chamber temperature, proven by the lowest recorded exhaust gases temperature. The same conclusion can be drawn for apple pruning residues, which registered the higher excess air level and CO emission of all the pruning residues. The remaining biomass types presented CO emission in flue gas in accordance with the level of sophistication of the combustion equipment used in the experiment. These facts prove that the briquettes made from typical woody pruning residues can be burned without major concerns, in respect to the combustion behavior, when compared to briquettes made from log wood.

Nitrogen oxide emission on the combustion of the residual pruning biomasses ranged from 433 to 1140 mg/m^3 @ 6% O_2, values that contrast with the 169 mg/m^3 @ 6% O_2 of the reference pine wood briquettes. NO_x can be formed in the combustion of biomass by two main mechanisms, that are directly related to the fuel N content and to the combustion temperature (Perez-jimenez 2015). Therefore, this significant difference in NO_x emission between the pruning biomass and pine wood can be explained by the superior N content on the fuel, ranging from 5 to 14 times more in the pruning residues when compared to pine wood sawdust. Despite that, comparing the pruning residues NO_x emission it seems the emission level does not depend on the fuel N content but on the combustion temperature, which is inversely proportional to the excess air level (oxygen content on flue gas). These conclusions indicate that pruning residues biomass tend to have higher NOx emission levels than log wood but can be minimized with correct NO_x emission control techniques, like, for example, the air staging.

Organic compounds emission level is inferior in all the residual biomass sources when compared to the pine wood reference briquettes, except for olive pruning residues. During the production of the densified fuel it was observed that the raw material was composed by a significant amount of leaves and the produced briquettes presented a sticky surface to the touch. Given the biomass source, this could mean a presence of residual oils that may result in the emission of organic compounds in combustion, explaining the results.

In respect to particulate matter, apple and urban pruning residues registered an emission level like those observed on the combustion of pine wood briquettes. Vineyard and olive pruning residues briquettes presented a particulate matter emission level circa 80% superior to the others, and kiwi presented the highest value. Given that particulate matter emission in wood combustion is significantly affected by fly ash (Obaidullah, Bram, and Ruyck 2018), and pruning residues of vine and olive present lower ash content that apple and urban (contrary to the PM emission levels), this difference can be attributed to accumulated ash in the boiler chambers. This led to the conclusion that the cleanliness of combustion equipment is a key player on the control of PM emission, and not the ash content to the extent of the levels registered in this study. The higher PM emission on kiwi pruning residues can also suggest that the mechanical toughness of the briquettes may linked to the segregation of particles during combustion, resulting in PM emission.

4 CONCLUSIONS

Woody pruning residues from 5 different types of biomass are evaluated, in accordance to the regional availability: Vineyard, kiwi plantation, olive grove, apple orchard and trees and bushed from urban park maintenance. Its proximate analysis show a superior ash content when compared to pine wood and consequent inferior LHV, contrary to volatile matter that seems similar in all the biomasses. In respect to its ultimate composition, the biggest difference from the pruning residues to pine wood was found to be the higher composition on N, which may cause higher levels of NO_x emission.

Briquetting shown possible to be made for every biomass tested and so the combustion on typical wood biomass boiler. CO and PM pollutant emission is similar in all the samples, being affected mainly the equipment functioning. NO_x emission revealed high for apple and kiwi pruning residues, for which is only recommended the combustion on a low NO_x emission capable boiler. TOC were found to be lower than pine wood, except for olive tree pruning residues a phenomenon attributed to the presence of leaves on the pruning residues.

AKNOWLEDGMENT

This work has been co-financed by the European Union through the European Regional Development Fund (ERDF) under the INTERREG V-A Spain-Portugal Program (POCTEP) 2014–2020 (project n° 0390_MOVBIO_2_E).

REFERENCES

Bioenergy Europe. 2018. *Statistic Report 2018 – Key Findings*.

Dickson, R.E. 1989. Carbon and Allocation in Trees. *Ann. Sci. For Ensevier* 46: 631–47.

Ecarnot, Martin, Frédéric Compan, and Pierre Roumet. 2013. Assessing Leaf Nitrogen Content and Leaf Mass per Unit Area of Wheat in the Field throughout Plant Cycle with a Portable Spectrometer. *Field Crops Research* 140: 44–50. http://dx.doi.org/10.1016/j.fcr.2012.10.013.

European Commission. 2010. *Energy 2020: A Strategy for Competitive, Sustainable and Secure Energy – Communication from the Comission to the European Parliament, the Council, the European Economic and Social Committee and the Committee of the Regions.*

European Environment Agency. 2018. *Renewable Energy in Europe — 2018: Recent Growth and Knock-on Effects*.

Faaij, Prof André P C. 2018. Securing Sustainable Resource Availability of Biomass for Energy Applications in Europe: Review of Recent Literature. (November): 1–26.

Obaidullah, M, S Bram, and J De Ruyck. 2018. An Overview of PM Formation Mechanisms from Residential Biomass Combustion and Instruments Using in PM Measurements. 12.

Perez-jimenez, Jose Antonio. 2015. Gaseous Emissions from the Combustion of Biomass Pellets. 85: 85–99.

Wastes: Solutions, Treatments and Opportunities III – Vilarinho et al. (Eds)
© 2020 Taylor & Francis Group, London, ISBN 978-0-367-25777-4

Oyster shells' processing for industrial application

T.H. Silva & M.C. Fredel
Ceramic and Composite Materials Research Group (CERMAT), Department of Mechanical Engineering, Federal University of Santa Catarina (UFSC), Florianopolis, Brazil

J. Mesquita-Guimarães
TEMA–Centre for Mechanical Technology and Automation, Department of Mechanical Engineering, University of Aveiro, Aveiro, Portugal

ABSTRACT: Considering that oyster shells are mostly made of calcium carbonate, their reintroduction in the industry becomes easier, thus it can be used in several sectors such as building materials, food supplement, pharmaceutical industry, animal feed, ground correction among others. Besides that, in marine culture, especially oysterculture, in 2014, approximately 438 billion tons of these were produced and most of these shells were improperly discarded. This article proposes a method for oyster shells reusing, excluding the organic material from them and making them raw materials for the industry. The FTIR results from oyster shells calcium carbonate are the same as the ones found in literature. Other analyses, such as XRD and DTA corroborated this result.

1 INTRODUCTION

The seafood production is responsible for sustaining a global economy. At a worldwide level, the commercialization of mollusks occupies the second place of marine culture, with 16.1 million tons (USD 19 billion). In the year of 2014, 73.8 million tons were produced, estimating a sales value of 160 200 million dollars. In the previous year of 2013, approximately, 54% of the total production of mollusks was related to bivalve mollusks, such as oysters and shellfish (FAO, 2016). According to the Food and Agriculture Organization of the United Nations (FAO), the value of the global production of this species for the year of 2016 was the USD 3.69 billion (FAO, 2018; Paris et al., 2016).

Among the oyster's market, the most produced species is *Crassostrea giga* known widely as the Pacific oyster, due to their fast-growing potential and their wide tolerance to diverse environmental conditions.

However, what is worrying scientists and the communities around oyster production is a large number of shells generated in terms of volume of produced waste, because their shells correspond to more than 70% of its weight, besides that its disposal occurs improperly (Silva *et al.*, 2019).

The cleaned oysters can be reused in several applications. It is worth mentioning that the principal component of the oysters' shells, with a share of approximately 96%, is calcium carbonate ($CaCO_3$), which can be used in various sectors of the industry, from the construction industry as an aggregate of limestone for cement, to the pharmaceutical industry as a calcium enriched supplement (Alvarenga *et al.*, 2012; Yang *et al.*, 2005).

In this article, it will be shown the waste treatment procedure implemented for the oyster shells in the region of Florianopolis, Brazil and their transformation in raw material for industrial applications.

2 MATERIAL AND METHODS

2.1 *Starting materials*

For this study, it was used oyster shells of the species *Crassostrea giga*; collected from restaurants of the *Ribeirão da Ilha* community, Florianópolis, Santa Catarina, Brazil.

This region is responsible for about 60% of the production of the whole state, being the principal oyster producer state in Brazil (EPAGRI, 2015). During the cultivation, the oysters undergo a cleaning process, eliminating the marine organisms embedded in their shells, before they can be marketed for the restaurants. In the restaurants, most oysters pass through a preparation process in boiled water; this process facilitates the step of cleaning of the shells in further uses.

2.2 *Waste treatment and recycling*

As the shells were collected from restaurants, cross-contamination with other food waste occurs, so it was decided to do the cleaning combining two different methods: degradation of organic matter combined with bio decomposition. For this purpose, initially, it was realized the manual cleaning to separate the shells from other food traces. Posteriorly, the cleaning of the shells happened through the following process: shells were washed with water under pressure eliminating the impurity excess. Afterwards, the shells were placed in a 0.08 wt.% chlorine solution, for two hours. The chlorine was utilized as a chemical agent accelerator for degradation of organic matter, besides avoiding the generation of contaminating liquid effluents; granulated chlorine was used to treat water containing 65% active chlorine. Soon after, the shells were left to dry in the sun for 48 hours at ambient temperature (~30°C) during the day. The drying process can be done though oven, if necessary.

Depending on the application it is necessary to transform the shells in powder, for this purpose, after the cleaning, the shells undergo a process of fragmentation, with the assistance of hydraulic press of 10tons, where pressure was applied under the entire shells. After this, all particles pass for the second step of milling, in a hammer mill (Servitech, CT-058), for this work a range of granulometry was select (8 mm–4.76mm, 850 μm–500 μm, 300 μm–150 μm and below of 150 μm). The oyster powder passes undergo granulometric separation process by sieving. The sieving process presented an approximately 90% efficiency. The waste treatment process for the oyster shell powder obtention is illustrated in Figure 1.

2.3 *Technical analysis*

The samples were characterized using several technique analyses. For the powder morphology and microstructural analysis of the oyster shell, it was used as a low-vacuum scanning electronic microscope (TM3030, Hitachi, Japan). The true densities (ρt) of the powdered samples were determined by using a helium pycnometer (Ultrapycnometer1200 P/N, Quantachrome Instruments), and it was compared with the theoretical value for calcium carbonate. The chemical analysis was performed in powdered samples using X-ray diffractometry, XRD (Philips X'Pert). The radiation source was a CuKα line of copper ($\lambda = 0.15141$ nm), with 40 kV and 30 mA. The scan was done continuously in 2-theta setting from 0 to 80° with a step size of 0.02° and a step time of 2 s. For the identification of the crystalline phases, the software (X'pert high score plus, PANalytical, Netherlands) was used. The analysis of composition was realized with the FTIR instrument (Bruker, model TENSOR 27); this analysis allows identifying the bands of absorption characteristic of the calcium carbonate. The thermal analysis was done in order to determine the quantity of calcium oxide after calcination, for that it was used an SDT 2860 simultaneous DSC-TGA, TA Instruments, USA, using a heating rate of 10°C/min until 1000°C.

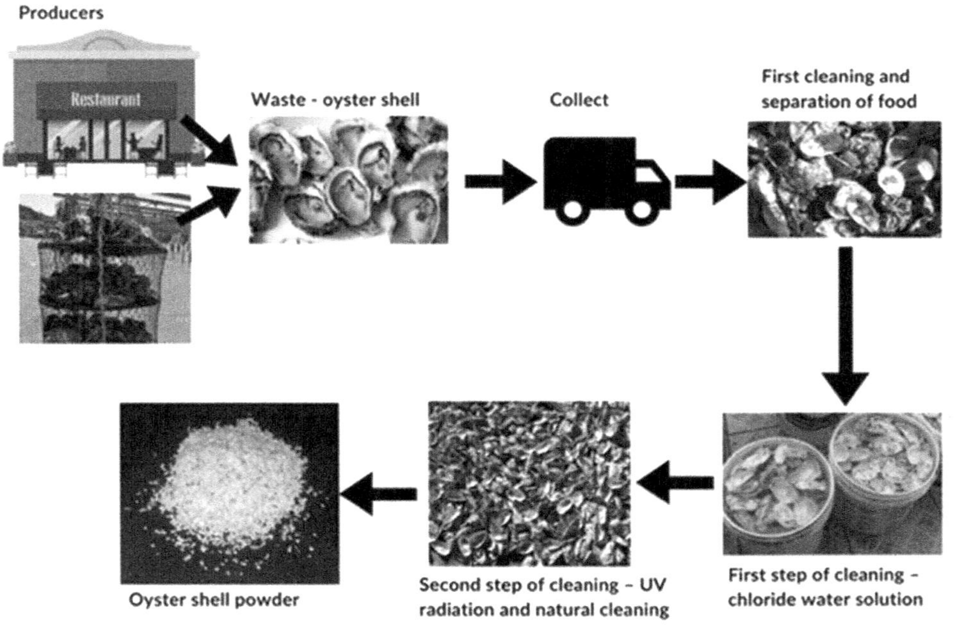

Figure 1. Scheme of the waste treatment process for oyster shell powder.

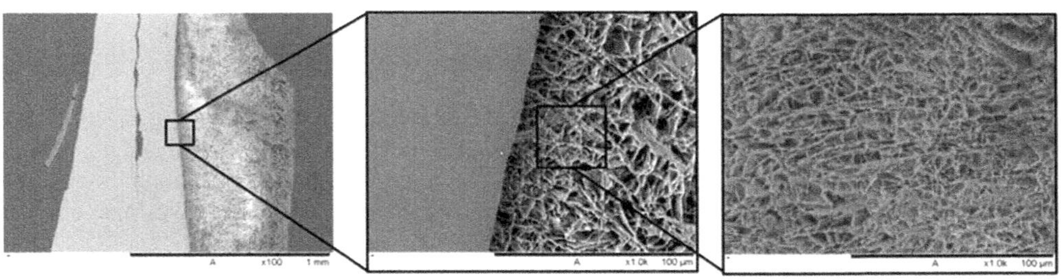

Figure 2. SEM micrographs of the oyster shell parts.

3 RESULTS AND DISCUSSION

3.1 *True densities*

For the true density analysis, the powder with size below 150 μm, was used. Five measures within an average volume of 4.98 ± 0.008 cm^3 were made, obtaining an average value of 2.49 ± 0.004 g/cm^3 for density. This value is relatively close to the theoretical value of industrial calcium carbonate, with 2.93 g/cm^3, high purity, commercially sold. The obtained result was the expected, as the calcium carbonate was obtained from oyster shells.

3.2 *MEV analysis*

As can be seen in Figure 2 the microstructure of oyster shells is composed of two phases: a sheet phase layer and a bulky porous layer. The sheet phase layer grows in lamellar form and orients

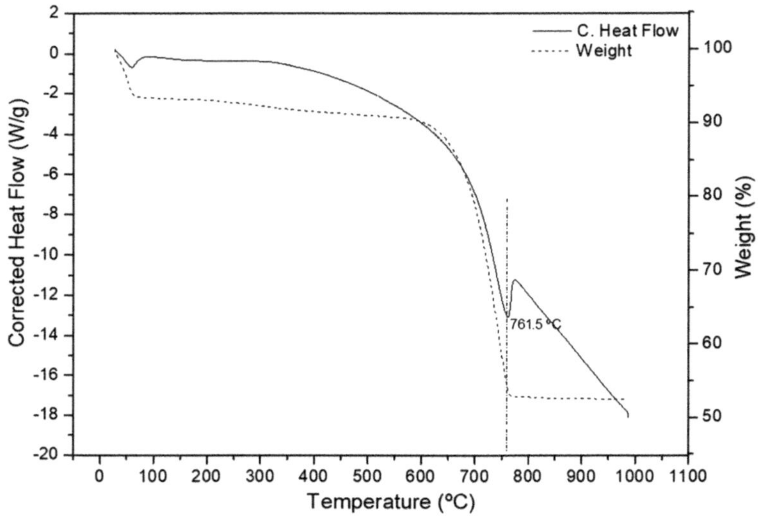

Figure 3. Calorimetry and thermogravimetry (DSC/TG) curves of the of oyster shells until 1000°C.

the growth of the shell. Porous bulky layer grows between the lamellas and does not show defined orientation. Both parts have calcium carbonate ($CaCO_3$) as chemical composition, with calcite as a crystalline phase.

3.3 *ADT*

The samples were cleaned to remove organic residues and afterwards, a thermal analysis was made to characterize the thermal events occurring on the samples. Normally, the transformation of the shells as a source of calcium carbonate ($CaCO_3$) is followed by a great loss of carbon dioxide that is related to the chemical decomposition of the calcium carbonate of the shells. This phenomenon is known as calcination and results in the formation of two products, calcium oxide (CaO) and carbon dioxide (CO_2). The calcination step is represented in the calorimetry and thermogravimetry (DSC/TG) curves of Figure 3.

In the thermogravimetric curve, it is possible to see two periods of weight losses. The first one between 100 and 200°C due to loss of humidity from the samples, the second one is around 300°C, possibly due to degradation of the organic matter remaining in the sample. The following DSC/TG results show that the cleaned oyster shells contained mainly calcium carbonate. Present a dissociation reaction peak of $CaCO_3$ into $CaO + CO_2$ at 761.5°C, which is a very typical temperature for this material. Also, this calcination step was followed by a lost weight of ~47.5%, which means that for 100 gr of $CaCO_3$ after calcination it will be obtained ~52.5% of pure CaO.

3.4 *FTIR*

In order to demonstrate the efficacy of the cleaning process FTIR analyses was performed, due to these analyses, organic matter can be detected. The samples were analyzed before and after the cleaning process, as illustrated in Figure 4. Before the cleaning process, the peaks referring to the calcium carbonate have less intensity. In the region around 3500 cm^{-1}, there are several other peaks in addition to a band, typically associated with organic compounds, so there is still organic matter in the shell, possibly decomposing since the material was collected from restaurant waste. The spectra of the cleaned oyster shell show peaks there are compatible with those found in the

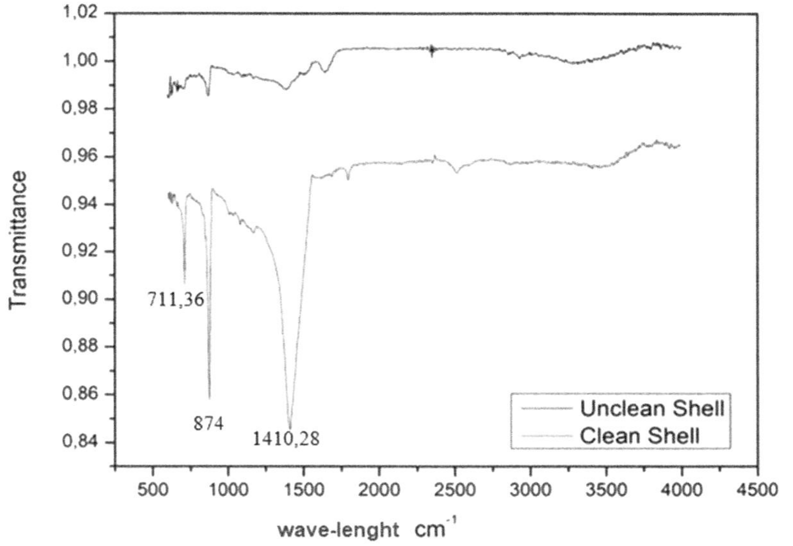

Figure 4. FTIR analyses, identify peaks and comparison between clean and unclean shell.

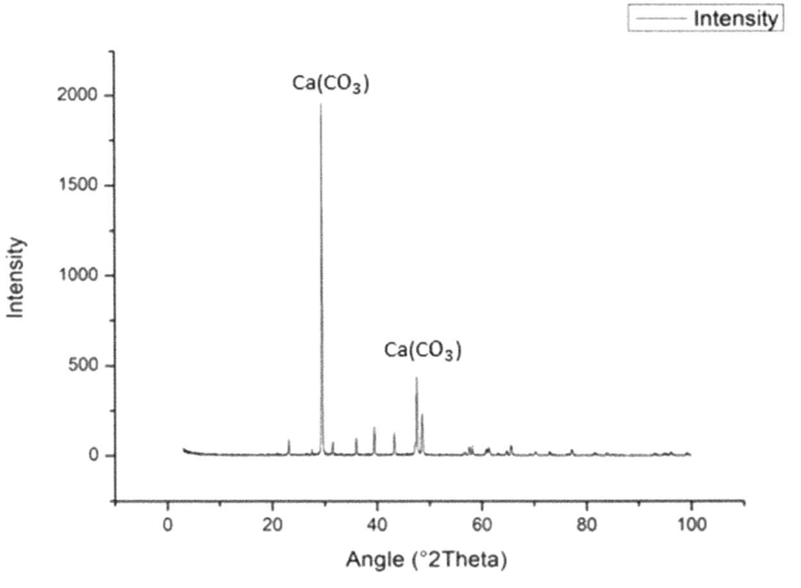

Figure 5. DRX diffractogram of the oyster shell powder.

literature for anhydrous calcium carbonate. The peak of 711 is related to calcite, and the peaks of 874 and 1410 refer to the ion (CO_3^{2-}).

3.5 *DRX analysis*

From the DRX analysis, it was possible to identify the presence of only one crystalline phase, related to calcium carbonate, as illustrated in Figure 5. This phase was identified by the pattern number

01-086-2339, corresponding to the calcium carbonate of a species of mollusk Guam (Oceania), with a rhombohedral crystalline structure, and with the following stoichiometry Ca.998 Mg.002.

From the obtained result, we can affirm that the waste treatment process reached its objective since it was only detected crystalline phase referring to the calcium carbonate of a mollusk in the material.

4 CONCLUSIONS

The results demonstrate the effectiveness of the waste treatment process as all the organic matter was eliminated. The properties of oyster shells can be used in several sectors such as civil construction, pharmaceutical industry and even in bio applications such as hydroxyapatite production. The utilization of oyster shells represents not only recycling material but also adds value to the waste and avoids the extraction of calcium carbonate from nature.

REFERENCES

Alvarenga, R.A.F. de, Galindro, B.M., Helpa, C. de F., Soares, S.R. 2012. The recycling of oyster shells: An environmental analysis using Life Cycle Assessment. *J. Environ. Manage* 106: 102–109. doi:10.1016/j.jenvman.2012.04.017

Barbosa, L.C., 2008. Espectroscopia no Infravermelho.

Barros, M.C., Bello, P.M., Bao, M., Torrado, J.J. 2009. From waste to commodity: transforming shells into high purity calcium carbonate. *J. Clean. Prod.* 17: 400–407. doi:10.1016/j.jclepro.2008.08.013

Calcium carbonate BioUltra, precipitated, ≥99.0% (KT) | Sigma-Aldrich [s.d.]. URL https://www.sigmaaldrich.com/catalog/product/sigma/21061?lang=pt&region=BR&cm_sp=Insite-_-prodRecCold_xviews-_-prodRecCold10-1 (acessado 3.20.18).

EPAGRI 2015. Síntese Informativa da Maricultura 2016. Empres. Pesqui. Agropecuária e Extensão Rural St. Catarina (Epagri). 2016, 1–7. doi:10.163

FAO 2018. Food and Agriculture Organization of the United Nations, for a world without hunger. Fish. Aquac. Dep.

FAO 2016. The State of World Fisheries and Aquaculture 2016. Contributing to food security and nutrition for all, Food and Agriculture Organization of the United. Rome.

FAO 2010. The State of World Fisheries and Aquaculture, Food and Agriculture Organization of the United Nations.

His, E., Seaman, M.. N.L. 2015. FAO – (Crassostrea gigas). *Mar. Biol.* 114: 277–279.

His, E., Seaman, M.N.L. 1992. Effects of temporary starvation on the survival, and on subsequent feeding and growth, of oyster (Crassostrea gigas) larvae. *Mar. Biol.* 114: 277–279. doi:10.1007/BF00349530

Silva, T., Mesquita-Guimaraes, J., Henriques, B., Silva, F., Fredel, M. 2019. The Potential Use of Oyster Shell Waste in New Value-Added By-Product. Resources (MDPI), 8(1), 13; https://doi.org/10.3390/resources80100131

Wastes: Solutions, Treatments and Opportunities III – Vilarinho et al. (Eds)

Development of alkali activated ceramic residue and fly ash blends

N. Gaibor, D. Leitão & T. Miranda
University of Minho, Guimarães, Portugal

N. Cristelo
University of Trás-os-Montes e Alto Douro, Vila Real, Portugal

ABSTRACT: Environmental concerns are becoming increasingly more significant worldwide, framing a scenario that points to the urgent development of new and sustainable alternatives for the industrial sector. The present study assesses the possible contribution of ceramic residue (CR) for alternative binders, for applications in the construction industry. The behavior of the CR alone in alkaline activated systems was initially tested, followed by a combination with a well-known precursor, fly ash (FA), considering different CR/FA weight ratios. Mechanical and microstructural analysis was conducted, including uniaxial compression strength (UCS) tests, Scanning Electron Microscopy (SEM), X-ray Energy Dispersive Analyzer (EDX), X-ray diffraction (XRD) and Fourier Transform Infrared Spectroscopy (FTIR). Results obtained showed that the most effective blend was a CR+FA combination, activated with sodium silicate, reaching UCS values higher than 20 MPa, after 90 days of ambient curing.

1 INTRODUCTION

Nowadays, industrial waste generation is a significant concern in terms of the environment, health, and its final disposal. Recycling and using such wastes in innovative construction materials appears to be a viable solution not only to the pollution problem but also an economical option in the construction sector by providing a potentially sustainable source. (Jindal et al., 2018). Still, the construction industry shows an important possibility to apply, in a direct or indirect strategy, significant percentages of several types of by-products and/or waste in the way of contributing to such sustainability. Some of the best common examples are mentioned: fly ash, construction, and demolition waste, and blast furnace slags, which have been used as a cement or aggregate replacement, embankment fill, road and railway pavement foundation, among others (Cristelo et al. 2017).

Ceramic materials are manufactured abundantly in different parts of the world. China and Europe are the world leaders in ceramic production. It is also exported to other regions due to its durability and variety in design. Except for the ceramic industry's vast economic benefits, it causes adverse environmental impacts through high large pollutant release and energy consumption (Wang et al., 2018). Ceramic residues are hard, durable, extremely resistant to chemical, physical and biological degradations and highly thermally stable (Khan et al., 2016). Ceramic producers are looking for valuable waste disposal ways due to an increase in a pile-up of them. There is a wide research work in reprocessing and reusing of the waste materials obtained from marble and ceramic industries in different areas such as building materials, cement additives (Aruntaş et al., 2010), infiltration (Davini, 2000), desulphurization techniques (De Bresser et al., 2005), clay-based materials (Acchar et at., 2006). Polymer-based waste ceramic composites were proposed for high voltage outdoor performance (Aman et al., 2013). For the synthesis of geo-polymers, waste ceramics as an efficient alternative have been employed (Sun et al., 2013).

To manage and protect the natural resources, the European Union has consolidated its management policy in the documents Directive 2006/12/EC and Directive 2008/98/EC, which provides the "framework for the handling of waste in the Community". (UE, 2008). In the legislation, specifically in Decree-Law No. 73/2011 (Article 7), the recycling and re-use of waste are identified in the scope of the 2020 targets. The aim is to achieve a recycling rate of at least 50% of the value of the waste produced. Thus, Decree-Law No. 18/2008 of January 29, established the obligation to use at least 5% of the recycled material in public works construction. In Portugal exists the Decree-Law (DL 73/2011) transcribed by the Ministry of Environment and Planning. The Directive 2008/98/EC is applied on the same basis as its own management.

In 2014, the total waste generated by all activities for the twenty-eight members of the European Union (EU-28), reached 2503 million tonnes (being the highest amount during the last 10 years reported). Regarding each waste component, the construction and demolition waste (CDW) increased by 57.2% during the 2004–2014 period, thus showing an opposite behavior compared with other sectors. The manufacturing waste generation, in the same period, decreased by 32.2%. However, ceramic production and its corresponding waste generation are strongly related to both sectors. The numbers confirm the relevance of waste valorization in order to propose new management and treatment alternatives (Eurostat, 2018).

The conjugation and transformation of the wastes into binder materials will be achieved through the alkaline activation technique, which has known as a strong development in the last 15 years, it is increasingly recognized as a strong alternative to Portland cement (Torres, 2015). Based on the activation (alkaline solution) of one or more precursors (moderately amorphous residues), it is possible to obtain a binder matrix of high mechanical quality and durability, with the incorporation of residues having significant economic and environmental advantages. This paper aims to understand how the CR fundamental properties can be better used in future incorporations. In this way, the behavior of the ceramic residue and fly ash in alkaline activated systems was analyzed, using compression tests and microstructural analysis.

2 MATERIALS AND METHODS

2.1 *Materials description*

The blends tested within the scope of this research were prepared with the following solid precursors: ceramic residue (CR), fly ash (FA); which were alkali activated with sodium silicate (Na_2SiO_3). The ceramic residue was provided by a Portuguese licensed waste management operator. Based on the technical sheet, it is part of a batch whose material originates from the demolition operations. It is considered a selected waste, specifically floor and wall tiles and sanitaryware waste. The ceramic residue was mixed with different percentages of fly ash type F (low CaO content), which was provided by a Portuguese thermo-electric power plant.

The particle size distribution of the original ceramic residue was determined based on the test for geometrical properties of aggregates, Sieving Method: NP EN 933-1: 2000. The maximum and minimum nominal sieve opening used was 8 mm and 0.063 mm, respectively. Most of the particles were retained until the 1 mm sieve size, which represents the 70.91% of the sample, meaning that there is no fine particle presence. Therefore, the original ceramic residue was dried, and it was mechanically milled for 32 hours until to get a powder form, more than 90% of particles passed through a nominal sieve opening of 63 μm.

The chemical composition of the starting materials was gained by X-ray fluorescence (XRF). The main components of the CR are silica (65.3%) and alumina (18.62%), with low calcium content (5.23%). Similar results are presented for FA, an important percentage correspond to silica (56.11%) and alumina (21.44%), with a calcium content of just 1.30%.

In Table 1 is presented the activator/precursor ratio. Sodium silicate (Na_2SiO_3) is used in its commercial presentation.

2.2 *Binders preparation*

The composition of the mixtures was defined based on results of a preliminary test, where the variation of the solid/liquid ratio allowed to determine the most effective ratio, regarding the

Table 1. Activator/precursor nominal ratios (wt%).

ID	Al_2O_3	CaO	Na_2O	SiO_2	CaO/SiO_2	Na_2O/Al_2O_3	SiO_2/Al_2O_3	SiO_2/Na_2O
100CR	0.19	0.49	0.07	0.71	0.70	0.38	3.79	10.00
75CR25FA	0.19	0.37	0.14	0.69	0.55	0.70	3.55	5.05
50CR50FA	0.20	0.25	0.15	0.67	0.38	0.73	3.33	4.59
25CR75FA	0.21	0.13	0.15	0.65	0.21	0.75	3.12	4.18
100FA	0.21	0.01	0.16	0.63	0.02	0.77	2.93	3.82

Table 2. Mixture composition.

ID	Precursor		Precursor/ Activator Ratio
	CR (%)	FA (%)	
100CR	100	0	0.40
75CR25FA	75	25	0.43
50CR50FA	50	50	0.46
25CR75FA	25	75	0.49
100FA	0	100	0.52

workability of the mixtures. For each of the series shown in Table 2, three cubic specimens of $40 \times 40 \times 40\,mm^3$ were made. All the specimens were subjected to the uniaxial compression test after 1, 14, 28 and 90 days curing.

Preparation of the ceramic residue (CR) and fly ash (FA) mixtures consisted of weighing the defined quantities of each one and add the activator. This was then followed by mechanical mixing, for a period of 3 minutes. The homogenized paste was poured inside a cubic stainless-steel mold, with nominal height, length and width are $40 \times 40 \times 40\,mm^3$, respectively. Mechanical vibration was then applied, for 2 minutes, and the initial curing stage was concluded with 24 h inside an oven, at 70°C.

2.3 *Uniaxial compressive strength test*

The uniaxial compression tests were conducted at the civil engineering laboratory of University of Trás-os-Montes and Alto Douro (UTAD) and the University of Minho, at 1–14, and 28–90 days curing time, respectively. Each specimen was measured and weighed prior to testing. A servo-hydraulic testing machine with two different actuators with load cell capacity equal to 50 kN and 100 kN were used. These tests were carried out under monotonic displacement control, at a rate of 0.3 mm/min, and both the peak load and displacement were obtained from each test result. Stress-strain curves were plotted for all tests performed.

2.4 *Mineralogical and microstructural characterization*

The present study is complemented by a microstructural analysis of the initial materials and the different blends. To this end, an additional specimen was fabricated for each case, compacted and cured under the same conditions. It should be noted that just one specimen was enough since the analyzes require a very small volume of material, which is "cut" from this purposely fabricated specimen, after each curing time desired. The microstructural analysis techniques used were Scanning Electron Microscopy (SEM), X-ray Energy Dispersive Spectroscopy (EDX), X-ray diffraction (XRD) and Fourier Transform Infrared Spectroscopy (FTIR). Only the highest curing time (90 days) data is presented in this paper.

3 RESULTS AND DISCUSSION

3.1 *Compression strength*

The UCS results significantly varied among the five different mixtures considered in this study. Each specimen was tested after 1, 14, 28 and 90 days curing. The properties of the five mixtures allowed its demolding, without difficulty, after 24 h. Figure 1 shows that the curing time increases the strength of the mixtures. It can be also observed that, in general, the higher strength values were obtained with the ceramic residue (CR) alone (i.e. with no fly ash). However, the higher UCS obtained in this study was developed by the 75C25F mixture (20.26 MPa), after 90 days curing. This only happened after 90 days, with curing times of 1, 14 and 28 days showing the 100% ceramic mixture as the most effective. The reason for this is most likely the well-known lower rate of strength development usually showed by low calcium precursors (like FA). The presence of 25% FA enables a higher UCS at the longest curing period, while the higher CR present in the 100CR mixtures was responsible for the higher UCS after 1, 14 and 28 days.

3.2 *Mineralogical and microstructural characterization*

Different mixtures, cured at room temperature for 90 days, were analyzed. Punctual chemical composition analysis in 10 different points was performed.

Figure 2 shows the SEM image of the selected sample, allowing an understanding of the general elemental chemical composition of the mixtures. A similar structure was identified over the ten points in the samples of CR-FA/SS. The presence of iron in point number 10 is highlighted, as an example. The gel developed in these mixtures is an aluminosilicate, with some calcium incorporate, which is basically determined by the starting materials.

The previous elemental chemical composition analysis, Table 3, helped to identify the chemical compounds formed in each mixture. Figure 3 shows the XRD diffractograms of the different

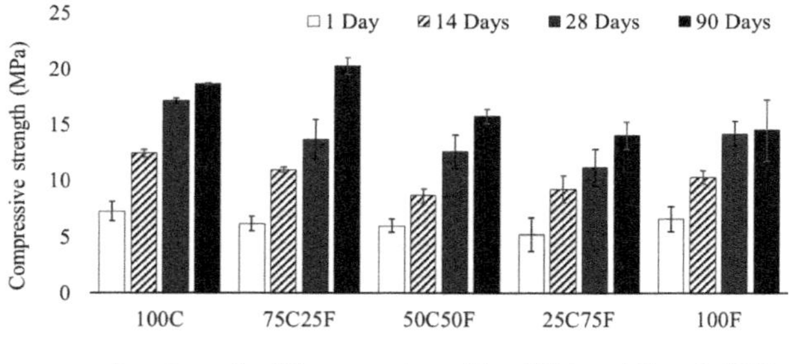

Figure 1. Uniaxial compression strength of the ceramic residue (CR) and fly ash (FA) mixtures, at 1, 14, 28, 90 days curing.

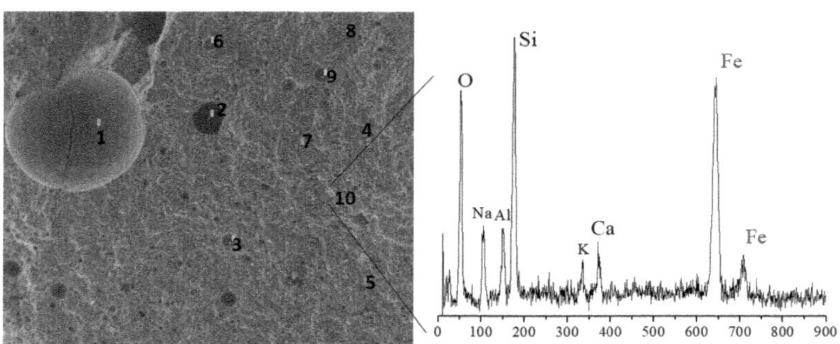

Figure 2. SEM images of CR-FA/SS mixtures, after 90 days curing time.

Table 3. Chemical composition of CR-FA/SS mixtures at 90 days curing time, (wt.%).

Al	Ca	Fe	K	Mg	Na	Si	Ti
15.66	3.28	1.28	3.34	1.78	16.64	57.42	0.60

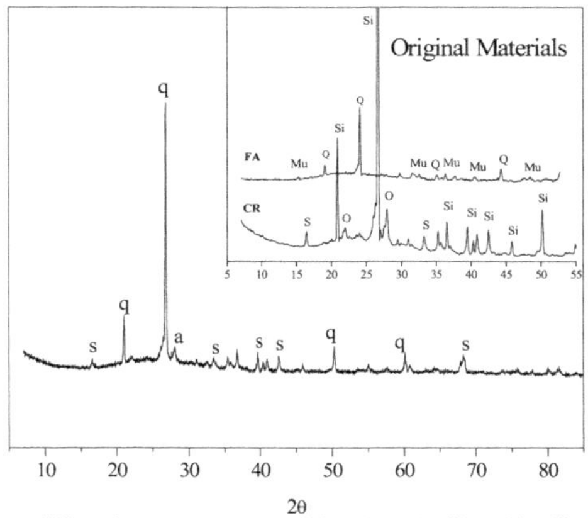

Figure 3. XRD patterns of the CR-FA/SS mixture, after 90 days curing time.

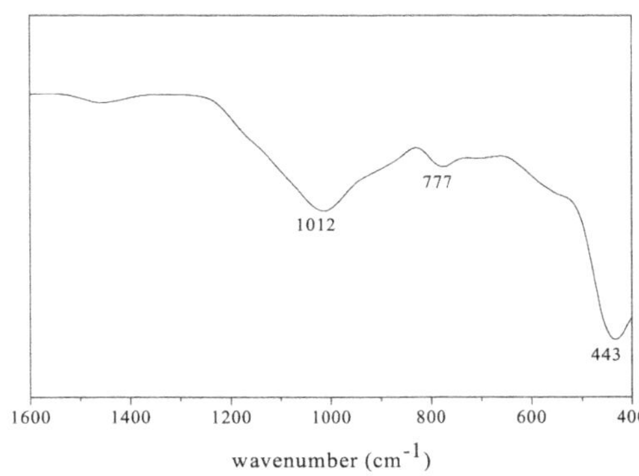

Figure 4. FTIR spectra of CR-FA mixtures, after 90 days curing time.

specimens after a curing time of 90 days. The products resulting from the activation reaction present similar peaks which were also found in the initial materials. However, some new crystalline phases were identified. The albite formation is observed in a hump in the 26° to 29° (2θ) angle range. The main registered halo between values of $2\theta \approx [25\text{–}30]°$ (quartz) and the presence of sillimanite along of the diffractograms, make them the chief crystalline phases.

The FTIR spectra is shown in Figure 4 where it is observed a strong and wide main band that appears around 1012 cm^{-1}. This is typical of asymmetric tension vibrations, T-O (Table 4). It can be related to the formation of aluminosilicate gel. Another band is also identified at 443 cm^{-1} which is associated with deformation vibrations of the O-Si-O bonds (internal vibrations of tetrahedra). Furthermore, the presence of the band at 777 cm^{-1} is characteristic of quartz (Si-O) identification, crystalline phases previously detected in these materials by XRD, Figure 3.

Table 4. Bands assignation of the FTIR spectrum of the CR-FA/SS mixture, after 90 days curing time.

Frequency (cm^{-1})	1012	777	443
Assignation	ν_{as}T-O (T: Si o Al)	ν s Si-O	δ O-Si-O

ν_{as}: asymmetric tension vibrations; νs: tension vibrations; δ: deformation vibrations

4 CONCLUSIONS

The fundamental properties of a ceramic residue (CR) from a Portuguese operator and its performance in combination with fly ash (FA), in a precursor role, activated with sodium silicate (SS), were studied. Based on uniaxial compression strength tests (UCS), after 1, 14, 28 and 90 days, it is possible to conclude that the strength decreases proportionally to the FA content. For 1, 14 and 28 days curing age, higher compressive strength values were observed with 100% of the ceramic residue in comparison with the other four compositions. However, the highest UCS (20.26 MPa) was reached by the combination of 75CR-25FA/SS after 90 days curing.

REFERENCES

Acchar, W., Vieira, F. A., & Hotza, D. 2006. Effect of marble and granite sludge in clay materials. *Materials Science and Engineering: A*, *419*(1–2), 306–309. https://doi.org/10.1016/J.MSEA.2006.01.021

Aman, A., Yaacob, M. M., Alsaedi, M. A., & Ibrahim, K. A. 2013. Polymeric composite based on waste material for high voltage outdoor application. *International Journal of Electrical Power & Energy Systems*, *45*(1), 346–352. https://doi.org/10.1016/J.IJEPES.2012.09.004

Aruntaş, H. Y., Gürü, M., Dayı, M., & Tekin, İ. 2010. Utilization of waste marble dust as an additive in cement production. *Materials & Design*, 31(8), 4039–4042. https://doi.org/10.1016/J.MATDES.2010.03.036

Cristelo, N., Fernández-Jiménez, A., Miranda, T., & Palomo, Á. 2017. Sustainable alkali activated materials: Precursor and activator derived from industrial wastes. *Journal of Cleaner Production*, *162*, 1200–1209. https://doi.org/10.1016/j.jclepro.2017.06.151

Davini, P. 2000. Investigation into the desulphurization properties of by-products of the manufacture of white marbles of Northern Tuscany. *Fuel*, *79*(11), 1363–1369. https://doi.org/10.1016/S0016-2361(99)00277-X

De Bresser, J. H. P., Urai, J. L., & Olgaard, D. L. 2005. Effect of water on the strength and microstructure of Carrara marble axially compressed at high temperature. *Journal of Structural Geology*, *27*(2), 265–281. https://doi.org/10.1016/J.JSG.2004.10.002

Eurostat. 2018. Waste statistics – Statistics Explained. Retrieved May 3, 2018, from http://ec.europa.eu/eurostat/statistics-explained/index.php/Waste_statistics

Jindal, A., & Ransinchung R.N., G. D. 2018. Behavioural study of pavement quality concrete containing construction, industrial and agricultural wastes. *International Journal of Pavement Research and Technology*. https://doi.org/10.1016/J.IJPRT.2018.03.007

Khan, M. S., Sohail, M., Khattak, N. S., & Sayed, M. 2016. Industrial ceramic waste in Pakistan, valuable material for possible applications. *Journal of Cleaner Production*, *139*, 1520–1528. https://doi.org/10.1016/J.JCLEPRO.2016.08.131

Sun, Z., Cui, H., An, H., Tao, D., Xu, Y., Zhai, J., & Li, Q. 2013. Synthesis and thermal behavior of geopolymer-type material from waste ceramic. *Construction and Building Materials*, *49*, 281–287. https://doi.org/10.1016/J.CONBUILDMAT.2013.08.063

Torres, C. M. 2015. *Reutilización de residuos vítreos urbanos industriales en la fabricación de cementos alcalinos: Activación, comportamiento y durabilidad*. Retrieved from https://repositorio.uam.es/handle/10486/670399

Wang, H., Chen, Z., Liu, L., Ji, R., & Wang, X. 2018. Synthesis of a foam ceramic based on ceramic tile polishing waste using SiC as foaming agent. *Ceramics International*, *44*(9), 10078–10086. https://doi.org/10.1016/J.CERAMINT.2018.02.211

Author index

Wastes: Solutions, Treatments and Opportunities

Conference Selected Papers

ISSN 2640-9623
eISSN 2640-964X

1. WASTES 2015 – Solutions, Treatments and Opportunities
Edited by Cândida Vilarinho, Fernando Castro & Mário Russo
ISBN: 978-1-138-02882-1 (Hbk)
ISBN: 978-0-429-22577-2 (eBook)

2. WASTES – Solutions, Treatments and Opportunities II
Edited by Cândida Vilarinho, Fernando Castro & Maria de Lurdes Lopes
ISBN: 978-1-138-19669-8 (Hbk)
ISBN: 978-1-315-20617-2 (eBook)

For Product Safety Concerns and Information please contact our EU representative GPSR@taylorandfrancis.com Taylor & Francis Verlag GmbH, Kaufingerstraße 24, 80331 München, Germany

Batch number: 10726408

Printed by Printforce, the Netherlands